Les Evans

# Chemical Modeling
# of Aqueous Systems II

ACS SYMPOSIUM SERIES **416**

# Chemical Modeling of Aqueous Systems II

**Daniel C. Melchior,** EDITOR
*EBASCO Services*

**R. L. Bassett,** EDITOR
*University of Arizona*

Developed from a symposium sponsored
by the Division of Geochemistry
at the 196th National Meeting
of the American Chemical Society,
Los Angeles, California,
September 25–30, 1988

American Chemical Society, Washington, DC 1990

**Library of Congress Cataloging-in-Publication Data**

Chemical modeling of aqueous systems II
  Daniel C. Melchior, R. L. Bassett.
    p.    cm.—(ACS Symposium Series, ISSN 0097–6156; 416).

  Previous symposium published under title: Chemical modeling in aqueous systems. 1979.

  "Developed from a symposium sponsored by the Division of Geochemistry at the 196th National Meeting of the American Chemical Society, Los Angeles, California, September 25–30, 1988."

  Includes bibliographical references.

  ISBN 0–8412–1729–7

  1. Chemistry—Mathematical models—Congresses.
2. Solution (Chemistry)—Congresses. 3. Water chemistry—Congresses.

  I. Melchior, Daniel C., 1958–  . II. Bassett, R. L., 1948–  . III. American Chemical Society. Division of Geochemistry. IV. American Chemical Society. Meeting (196th: 1988: Los Angeles, Calif.) V. Chemical modeling in aqueous systems. VI. Series.

QD39.3.M3M52  1989
541.3′422′015118—dc20                           89–28446
                                                   CIP


The paper used in this publication meets the minimum requirements of American National Standard for Information Sciences—Permanence of Paper for Printed Library Materials, ANSI Z39.48–1984.

∞

Copyright © 1990

American Chemical Society

All Rights Reserved. The appearance of the code at the bottom of the first page of each chapter in this volume indicates the copyright owner's consent that reprographic copies of the chapter may be made for personal or internal use or for the personal or internal use of specific clients. This consent is given on the condition, however, that the copier pay the stated per-copy fee through the Copyright Clearance Center, Inc., 27 Congress Street, Salem, MA 01970, for copying beyond that permitted by Sections 107 or 108 of the U.S. Copyright Law. This consent does not extend to copying or transmission by any means—graphic or electronic—for any other purpose, such as for general distribution, for advertising or promotional purposes, for creating a new collective work, for resale, or for information storage and retrieval systems. The copying fee for each chapter is indicated in the code at the bottom of the first page of the chapter.

The citation of trade names and/or names of manufacturers in this publication is not to be construed as an endorsement or as approval by ACS of the commercial products or services referenced herein; nor should the mere reference herein to any drawing, specification, chemical process, or other data be regarded as a license or as a conveyance of any right or permission to the holder, reader, or any other person or corporation, to manufacture, reproduce, use, or sell any patented invention or copyrighted work that may in any way be related thereto. Registered names, trademarks, etc., used in this publication, even without specific indication thereof, are not to be considered unprotected by law.

PRINTED IN THE UNITED STATES OF AMERICA

# ACS Symposium Series

## M. Joan Comstock, *Series Editor*

### *1989 ACS Books Advisory Board*

Paul S. Anderson
Merck Sharp & Dohme Research Laboratories

Alexis T. Bell
University of California—Berkeley

Harvey W. Blanch
University of California—Berkeley

Malcolm H. Chisholm
Indiana University

Alan Elzerman
Clemson University

John W. Finley
Nabisco Brands, Inc.

Natalie Foster
Lehigh University

Marye Anne Fox
The University of Texas—Austin

G. Wayne Ivie
U.S. Department of Agriculture, Agricultural Research Service

Mary A. Kaiser
E. I. du Pont de Nemours and Company

Michael R. Ladisch
Purdue University

John L. Massingill
Dow Chemical Company

Daniel M. Quinn
University of Iowa

James C. Randall
Exxon Chemical Company

Elsa Reichmanis
AT&T Bell Laboratories

C. M. Roland
U.S. Naval Research Laboratory

Stephen A. Szabo
Conoco Inc.

Wendy A. Warr
Imperial Chemical Industries

Robert A. Weiss
University of Connecticut

# Foreword

The ACS SYMPOSIUM SERIES was founded in 1974 to provide a medium for publishing symposia quickly in book form. The format of the Series parallels that of the continuing ADVANCES IN CHEMISTRY SERIES except that, in order to save time, the papers are not typeset but are reproduced as they are submitted by the authors in camera-ready form. Papers are reviewed under the supervision of the Editors with the assistance of the Series Advisory Board and are selected to maintain the integrity of the symposia; however, verbatim reproductions of previously published papers are not accepted. Both reviews and reports of research are acceptable, because symposia may embrace both types of presentation.

# Contents

**Preface** ............................................................................................................ xiii

**Memorial to Robert M. Garrels** ...................................................................... xv

1. **Chemical Modeling of Aqueous Systems: An Overview** ....................... 1
   R. L. Bassett and Daniel C. Melchior

   AQUEOUS THERMODYNAMICS AND THEORETICAL ADVANCEMENTS

2. **Activity Coefficients in Aqueous Salt Solutions: Hydration Theory Equations** ............................................................................... 16
   Thomas J. Wolery and Kenneth J. Jackson

3. **Ion-Association Models and Mean Activity Coefficients of Various Salts** ..................................................................................... 30
   David L. Parkhurst

4. **Models for Aqueous Electrolyte Mixtures for Systems Extending from Dilute Solutions to Fused Salts** ............................... 44
   Roberto T. Pabalan and Kenneth S. Pitzer

5. **Solubility of Volatile Electrolytes in Multicomponent Solutions with Atmospheric Applications** .............................................................. 58
   Simon L. Clegg and Peter Brimblecombe

6. **Modeling Solid–Solution Reactions in Low-Temperature Aqueous Systems** ..................................................................................... 74
   Pierre D. Glynn

7. **Effect of Pressure on Aqueous Equilibria** ............................................. 87
   Pradeep K. Aggarwal, William D. Gunter, and Yousif K. Kharaka

## CODE DEVELOPMENT AND DOCUMENTATION

8. Current Status of the EQ3/6 Software Package for Geochemical Modeling ...................104
   Thomas J. Wolery, Kenneth J. Jackson, William L. Bourcier, Carol J. Bruton, Brian E. Viani, Kevin G. Knauss, and Joan M. Delany

9. Geochemical Modeling of Water–Rock Interactions Using SOLMINEQ.88 ...................117
   Ernest H. Perkins, Yousif K. Kharaka, William D. Gunter, and Jeffrey D. DeBraal

10. Application of the Pitzer Equations to the PHREEQE Geochemical Model ...................128
    L. Niel Plummer and David L. Parkhurst

## APPLICATIONS TO MODELING: EQUILIBRIUM AND MASS TRANSFER

11. Reconstruction of Reaction Pathways in a Rock–Fluid System Using MINTEQ ...................140
    Hannah F. Pavlik and Donald D. Runnells

12. Hydrogeochemical Interactions and Evolution of Acidic Solutions in Soil ...................154
    Patrick Longmire, Douglas G. Brookins, and Bruce M. Thomson

13. Geochemistry of Organic Acids in Subsurface Waters: Field Data, Experimental Data, and Models ...................169
    Paul D. Lundegard and Yousif K. Kharaka

14. Carbon Isotope Mass Transfer as Evidence for Contaminant Dilution ...................190
    Laura Toran

15. Transport of $^{14}CO_2$ in Unsaturated Glacial and Eolian Sediments ...................202
    Robert G. Striegl and Richard W. Healy

## APPLICATIONS TO MODELING: TRANSPORT AND COUPLED CODES

16. Evolution of Dissolution Patterns: Permeability Change Due to Coupled Flow and Reaction .......... 212
    Carl I. Steefel and Antonio C. Lasaga

17. Modeling Dynamic Hydrothermal Processes by Coupling Sulfur Isotope Distributions with Chemical Mass Transfer: Approach .......... 226
    David R. Janecky

18. Coupling of Precipitation–Dissolution Reactions to Mass Diffusion via Porosity Changes .......... 234
    Chalon L. Carnahan

19. Simulation of Molybdate Transport with Different Rate-Controlled Mechanisms .......... 243
    Kenneth G. Stollenwerk and Kenneth L. Kipp

## APPLICATIONS TO MODELING: SURFACE CHEMISTRY

20. Numerical Simulation of Coadsorption of Ionic Surfactants with Inorganic Ions on Quartz .......... 260
    Rebecca L. Rea and George A. Parks

21. Constant-Capacitance Surface Complexation Model: Adsorption in Silica–Iron Binary Oxide Suspensions .......... 272
    Paul R. Anderson and Mark M. Benjamin

22. Influence of Temperature on Ion Adsorption by Hydrous Metal Oxides .......... 282
    Michael L. Machesky

23. Coagulation of Iron Oxide Particles in the Presence of Organic Materials: Application of Surface Chemical Model .......... 293
    Liyuan Liang and James J. Morgan

## ADVANCEMENTS IN MODELING: MODELING SENSITIVITIES

24. **Uncertainties in Ground Water Chemistry and Sampling Procedures** ..................310
    Michael J. Barcelona

25. **Using Chemical Analyses and Assessing Quality in Aqueous Environmental Monitoring Programs** ..................321
    Thomas R. Wildeman, Leslie S. Laudon, Roger L. Olsen, and Richard W. Chappell

26. **Expert Systems To Support Geochemical Modeling** ..................330
    F. J. Pearson, Jr., B. Skytte Jensen, and Andreas Haug

27. **Numerical Modeling of Platinum Eh Measurements by Using Heterogeneous Electron-Transfer Kinetics** ..................339
    J. Houston Kempton, Ralph D. Lindberg, and Donald D. Runnells

28. **Use of Model-Generated $Fe^{3+}$ Ion Activities To Compute Eh and Ferric Oxyhydroxide Solubilities in Anaerobic Systems** ..................350
    Donald L. Macalady, Donald Langmuir, Timothy Grundl, and Alan Elzerman

29. **Energetics and Conservative Properties of Redox Systems** ..................368
    Michael J. Scott and James J. Morgan

30. **Rates of Inorganic Oxidation Reactions Involving Dissolved Oxygen** ..................379
    L. Edmond Eary and Janet A. Schramke

## ADVANCEMENTS IN MODELING: THERMODYNAMIC AND KINETIC ADVANCES

31. **Revised Chemical Equilibrium Data for Major Water–Mineral Reactions and Their Limitations** ..................398
    Darrell Kirk Nordstrom, L. Niel Plummer, Donald Langmuir, Eurybiades Busenberg, Howard M. May, Blair F. Jones, and David L. Parkhurst

32. Solubilities of Aluminum Hydroxides and Oxyhydroxides in Alkaline Solutions: Correlation with Thermodynamic Properties of $Al(OH)_4^-$ ......................... 414
John A. Apps and John M. Neil

33. Aluminum Hydrolysis Reactions and Products in Mildly Acidic Aqueous Systems ......................... 429
John D. Hem and Charles E. Roberson

34. Effect of Ionic Interactions on the Oxidation Rates of Metals in Natural Waters ......................... 447
Frank J. Millero

35. Role of Reactive-Surface-Area Characterization in Geochemical Kinetic Models ......................... 461
Art F. White and Maria L. Peterson

### ADVANCEMENTS IN MODELING: ORGANIC COMPOUNDS

36. Quantitative Structure–Activity Relationship Models for Predicting Aqueous Solubility: Comparison of Three Major Approaches ......................... 478
Nagamany N. Nirmalakhandan and Richard E. Speece

37. Equilibrium Model for Organic Materials in Water ......................... 486
Frank R. Groves, Jr., and Majd El-Zoobi

38. Importance of Organic–Inorganic Reactions to Modeling Water–Rock Interactions During Progressive Clastic Diagenesis ......................... 494
Donald B. MacGowan and Ronald C. Surdam

39. Copper Complexation by Natural Organic Matter in Ground Water ......................... 508
Thomas R. Holm and Charles D. Curtiss III

40. Kinetics of Rare Earth Metal Binding to Aquatic Humic Acids ......................... 519
Sue B. Clark and Gregory R. Choppin

41. Microscale Processes in Porous Media: Transport of Chlorinated Benzenes in Porous Aggregates .................................. 526
   Roger C. Bales and James E. Szecsody

INDEXES

Author Index .................................................................................................... 540

Affiliation Index ............................................................................................... 541

Subject Index .................................................................................................... 541

# Preface

CHEMICAL MODELING OF AQUEOUS SYSTEMS is an area of active research that has been applied to environmental problems, evaluation of sedimentary diagenesis, mineral deposition and mineral recovery, geothermal energy, and radioactive waste processes, to name only a few applications. The science behind chemical modeling and the numerous applications of chemical modeling have evolved in the past 20 years, since Werner Stumm organized a symposium on equilibrium concepts in natural water systems in 1967. That symposium resulted in the publication of *Equilibrium Concepts in Natural Water Systems,* Advances in Chemistry 67. Many of the issues addressed in that symposium still confront modeling practitioners today.

In 1978, Everett A. Jenne organized a symposium on chemical modeling, which resulted in the publication of ACS Symposium Series 93. *Chemical Modeling in Aqueous Systems* contains 38 chapters spanning six major topics; it firmly established the usefulness of chemical models in addressing a wide range of aqueous chemical issues.

The solutions to problems facing modelers today have been facilitated by increased confidence in the methods of chemical analysis, a better understanding of sampling pitfalls, the evolution of aqueous solution theory, and an increased awareness of the geologic and hydrologic processes influencing solution chemistry.

The goal of the 1988 symposium, on which this book is based, was to bring together the scientists active in chemical modeling to exchange ideas and highlight the advances made since the 1978 conference. This symposium was organized to reassemble the original group of researchers and their students, many of whom are major contributors in the field of aqueous geochemistry and are recognized as initiators of early chemical model development, as well as current developers and users of chemical models. The purpose was to summarize and evaluate progress, discuss issues critical to modeling, and address new areas of fundamental research.

The volume that has resulted from this symposium has been divided into two broad areas: the current status of models and the areas in which significant progress has been made over the past 20 years; and new concepts and approaches to address future modeling issues. In addition, several chapters address theoretical approaches to evaluating the behavior of dissolved species in aqueous systems.

We dedicate this volume as a memorial to Robert M. Garrels, who is credited with initiating the mathematical descriptions of natural aqueous systems and whose efforts, along with those of his students and protégés helped to shape geochemical modeling as it is recognized today.

The timeliness of this symposium has increased in importance as the world has focused a great deal of attention on the causes of the degradation of the atmosphere and hydrosphere. We hope that works contained in this volume will lead to a better understanding of aqueous environmental chemistry and the importance of properly using geochemical principles to solve these and other geologic and hydrologic problems.

We would like to thank the authors for their efforts, patience, and diligence in completing each chapter and the many outside reviewers who donated their time to provide technical reviews. Each chapter received technical review by outside referees and the editors to ensure technical quality. We also thank Cheryl Shanks for her help and continuous patience throughout the review and acquisition effort and Paula M. Bérard for her help during production.

In addition, we would like to thank both EBASCO and the University of Arizona for support throughout. We would also like to acknowledge several people at EBASCO and the University of Arizona's Department of Hydrology and Water Resources. Melchior expresses his gratitude to Dorothy Farrah, who handled correspondence, filing, and continuous requests with professional dedication. Bassett expresses his appreciation to Anne Lauver and Rebecca Marable for their secretarial and word processing assistance.

DANIEL C. MELCHIOR
EBASCO Services
Santa Ana, CA 92704

R. L. BASSETT
University of Arizona
Tucson, AZ 85721

October 5, 1989

# Memorial to Robert M. Garrels

## August 24, 1916–March 8, 1988

WITH THE PASSING OF BOB GARRELS last year, we lost the foremost researcher of his generation on the chemistry of natural waters, not to mention his outstanding work on other aspects of earth science. He was talking about ion-pairing in solution long before anyone had any idea of its geologic importance. In fact, his Ph.D. thesis, completed at Northwestern University in 1941, was an experimental study of complex formation in aqueous solution between lead and chloride ions. Soon after writing his thesis, he presented his results to the Geological Society of America, generalized to emphasize the role of complex ions in the transport of metals in "underground waters", but the revolutionary ideas were incomprehensible to most geologists present.

Bob published a book *Mineral Equilibria* in 1960 (later expanded to *Solutions, Minerals, and Equilibria* in 1965 and coauthored with Charles Christ), which brought about a quantum jump in our understanding and approach to the physical chemistry of natural waters. At about the same time, he conducted a series of outstanding studies with Mary Thompson and Ray Siever at Harvard University; especially noteworthy is the work on a chemical model for seawater. In this study, Garrels and Thompson laid the foundation for the calculation of activities in seawater, based primarily on the concept of ion pairing. (This paper has been so influential that it has been designated a classic by *Science Citation Index*.)

During his long career, Bob and his co-workers also published pathfinding papers on the origin of the composition of springs and lakes, reverse weathering and the control of seawater composition, and the theoretical treatment of rock–water interaction. All in all, Bob showed us how to look at published analyses of natural waters and how to deduce all sorts of wondrous things from a little data. Also, most importantly, he set an unusual example of good humor, brilliance, humility, patience, optimism, and a deeply ingrained sense of fairness.

We all mourn his loss.

ROBERT A. BERNER
Department of Geology and Geophysics
Yale University
New Haven, CT 06511

August 31, 1989

# Chapter 1

# Chemical Modeling of Aqueous Systems

## An Overview

### R. L. Bassett[1] and Daniel C. Melchior[2]

[1]Department of Hydrology and Water Resources, University of Arizona, Tucson, AZ 85721
[2]EBASCO Services, 3000 West MacArthur Boulevard, Santa Ana, CA 92704

Chemical modeling in hydrologic systems has become an area of active research with immediate opportunities for application to environmental problems, interpretation of diagenetic processes, and in evaluating the deposition of minerals to name a few. Researchers working in these and other areas have identified the need to quantitatively evaluate the chemical changes in geologic systems as a result of both natural and man-induced processes. To expedite these evaluations, fundamental theories of aqueous chemistry have been incorporated into computer codes along with the required constants and thermochemical data, to perform various simulations.

Over the past several decades the equations and resultant codes have evolved greatly. Many of the codes that today serve as the foundations for geochemical calculations have increased in complexity both as a result of new theories in aqueous chemistry and the technical issues which needed to be resolved. The technical issues currently being addressed are often more complex in scope than originally imagined, and as a result the limits of application and the inadequacies are revealed. Researchers have evaluated the models under a wide variety of circumstances including the adequacy of the aqueous theory, usefulness for a broad range of applications, and the impact of solution-solid interactions such as is affected by mineral surface chemistry.

Advancements in modeling capabilities have resulted from the work of many researchers. These give rise to new concepts in modeling such as evaluating model sensitivities, improvements to the thermodynamic and kinetic data bases, and the introduction of organic compounds to existing codes. The resulting changes have greatly enhanced the flexibility and utility of most aqueous geochemical codes.

## HISTORICAL FRAMEWORK

The concept of chemical modeling of natural hydrologic systems was introduced by Garrels and Thompson (1) in a paper that described the distribution of chemical species in seawater. Their approach was to construct a rigorous thermodynamically based model that was (1) mathematically but not conceptually decoupled from flow, and (2) could provide quantifiable information about the chemical processes active in an aqueous system, such as seawater or groundwater. Their initial model considered 17 species, was restricted to 25 °C and remarkably enough, clearly quantified the predominant ion and ion pair speciation in seawater. This work set the framework for a number of the computer codes used today.

The historical evolution of chemical codes has not been documented nor will it be attempted here, yet several publications are available that describe the variety of codes from which to choose (2-4). One distinct trend of this past decade has been the cessation of the rapid increase in the number of models being developed; this has been replaced with a more focused effort to document and improve existing models. A small number of these codes have become the mainstay for geochemical applications. An abbreviated pedigree of computer codes is illustrated in Figure 1. The initial conference on this subject (2) was convened during the early phase of model development, and

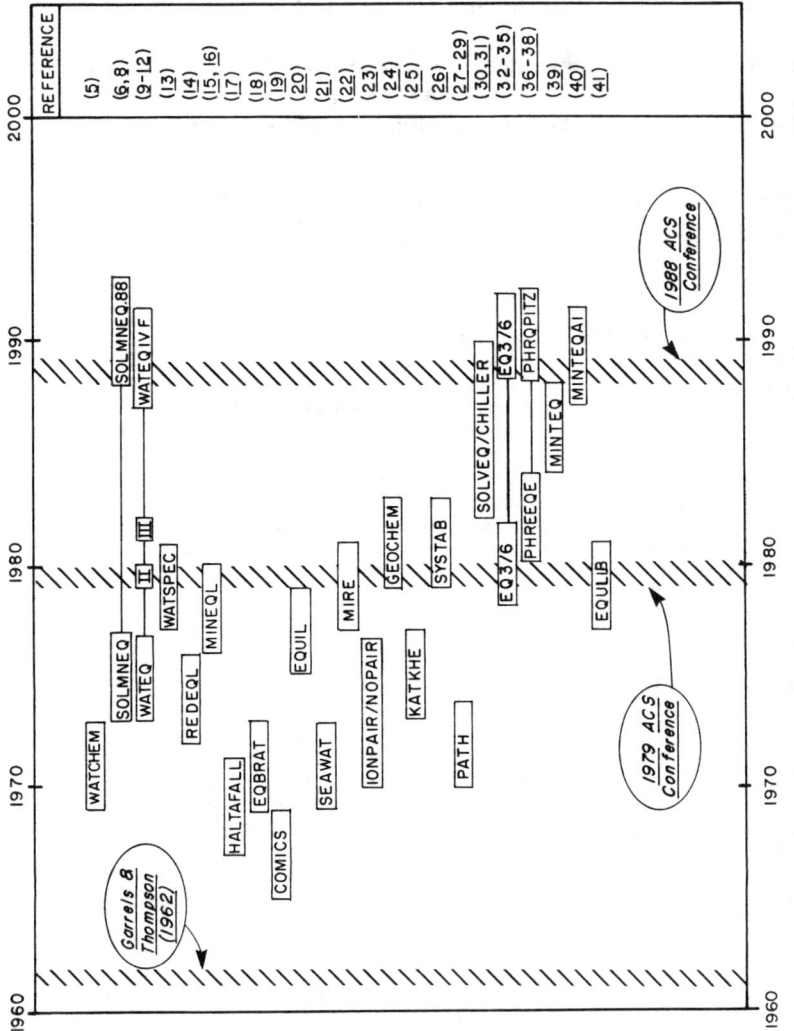

Figure 1. An abbreviated documentation of chemical model development. Numerous additional codes currently exist as modifications or extensions of these referenced codes; most are unpublished specific purpose codes and are consequently not given here.

up to that time the focus had been on equilibrium modeling and on the expansion of the number of species in the models. Model development was rapid, machine dependent, often simultaneous and generally poorly documented.

The recent emphasis in code improvement centers on issues such as more comprehensive and more complete thermodynamic data, better transportability between computer systems, the capability for evaluating systems at higher temperature, pressure and ionic strength, and more specific utility of the codes for environmental, industrial and research applications. Not surprisingly, in view of the large effort required to develop, document and calibrate these codes, only a few models have emerged which offer a reasonable possibility for surviving the scrutiny of QA/QC rigor, and legal qualification which will be required for future widespread use of these codes in the arenas of both research and the private sector. Issues of liability and legal defensibility are important considerations, especially in the more complex areas of geochemical and hydrochemical research such as hazardous waste, low and high level radioactive waste and groundwater contamination.

The technical issues confronting researchers and applications-oriented scientists have changed dramatically over the past several decades. The current models tend to be used most frequently for critical applications and problem solving investigations as opposed to investigations of general concepts of the science. As a result of these applications-oriented approaches new concerns have evolved which have set the direction for future research (Table I).

Table I. Broad categories examined at the symposium which embody the current status of chemical modeling and the potential advances

| A. CURRENT MODELS | B. NEW CONCERNS |
|---|---|
| Aqueous Theory | Modeling Sensitivity |
| - Ion Interaction | - Sampling/Analysis |
| - Model Documentation | - Computational |
| Applications | - Redox/Metastability |
| - Mass Transfer | Thermodynamic and Kinetic/ |
| - Isotope Fractionation | Data Advancements |
| - Redox | Organic Compounds |
| - Organic Compounds | - Macromolecules |
| - Coupled Models | - Cosolvents |
| Surface Chemistry | - Partitioning |
| | Coupled Hydrology/Chemistry |

It is the intent of this paper to discuss several relevant details of the models currently used and to indicate some of the most pressing research problems that will most likely be pursued in the near future.

## CURRENT MODELS

### Aqueous Theory

Ion-Association and Ion-Hydration. Aqueous solutions of electrolytes have been chemically described using a variety of theories. The original theoretical approach used by geochemists to model aqueous systems was based on the concept of ion-pairing or ion-association. The ion association approach as described by Garrels and Thompson (1) accurately depicted the speciation of seawater and later many other aqueous solutions. This approach was subsequently found to be inadequate for defining the chemistry of more complex and more concentrated aqueous solutions or those solutions near the critical point of water. This deficiency led to the use of other theoretical approaches to describe these systems, such as the ion-interaction, mean salt, and ion-hydration theories.

The ion-association concept relies on the use of Debye-Hückel based activity coefficients to calculate aqueous activities and is one that is employed most frequently in the models used today. A primary assumption of this approach is the use of the MacInnes convention such that for an aqueous solution containing equimolal concentrations of $K^+$ and $Cl^-$, their activities and hence activity coefficients are equivalent. This approach and convention were reexamined by Parkhurst in this volume, who explored the issue of mean salt based activity coefficients and the apparent discrepancies between values calculated by each method. This enlightened examination merits further evaluation in the future as new approaches to calculating electrolyte activity coefficients are addressed.

The ion-hydration approach for describing electrolyte chemistry has been described here by Wolery and Jackson. This alternative approach is based on some of the pioneering work done by Stokes and Robinson (43) and attempts to examine electrolyte speciation in terms of the degree of solvation. Since this approach and concept is relatively new, it remains a question whether these methods can be applied to modeling natural systems.

Modeling of natural systems is indeed more complex than those prepared in a laboratory setting. Geochemists recognize that many variables will influence the chemistry of natural aqueous systems and minerals present. The importance of temperature and pressure on aqueous chemistry is well recognized. Aggarwal discusses in this volume the significance of pressure effects on aqueous thermodynamics and presents an approach for addressing this issue. Glynn presents a thermodynamic framework for evaluating the effect of mineral solid-solutions and aqueous solution compositions in Chapter 6.

The ion-interaction theory in contrast to the above was developed by Pitzer (42) as an outgrowth of work done by Guggenheim (44). This phenomenological methodology was based on the concept that ions electrostaticly interact in solution and that these interactions were based on a statistical likelihood of collision, hence the ionic strength dependency. Several papers in this volume discuss aspects of the importance of this approach to modeling the chemistry of complex systems.

Ion Interaction. Ion-interaction theory has been the single most noteworthy modification to the computational scheme of chemical models over the past decade; this option uses a virial coefficient expansion of the Debye-Hückel equation to compute activities of species in high ionic strength solutions. This phenomenological approach was initially presented by Pitzer (42) followed by numerous papers with co-workers, and was developed primarily for laboratory systems; it was first applied to natural systems by Harvie, Weare and co-workers (45-47). Several contributors to the symposium discussed the ion interaction approach, which is available in at least three of the more commonly used codes; SOLMNEQ.88, PHRQPITZ, and EQ 3/6 (Figure 1).

The ion-interaction model is a theoretically based approach that uses empirical data to account for complexing and ion pair formation by describing this change in free ion activity with a series of experimentally defined virial coefficients. Several philosophical difficulties have resulted from the introduction of this approach: the lack of extensive experimental database for trace constituents or redox couples, incompatibility with the classical ion pairing model, the constant effort required to retrofit solubility data as the number of components in the model expand using the same historical fitting procedures, and the incompatibility of comparing thermodynamic solubility products obtained from model fits as opposed to solubility products obtained by other methods.

A controversy exists between the proponents of ion-association versus the ion-interaction approach. This controversy usually revolves around the issue of chemical realism since many known ion pairs such as $CaSO_4^o$ were not explicitly defined by the ion-interaction approach. However, the impact of the strong ionic interaction can be reflected in the magnitude of the virial coefficient terms. More recently, this deficiency has been addressed for the carbonate system, however, questions still remain whether mixed methodologies reflect the true solution chemistry or are simply forced fits of experimental data.

Clearly the ion-interaction methodology is superior to the ion-pairing approach at high ionic strength, and can also be used at low-ionic strengths as well. With this formulation, it is possible to compute the activities of many electrolytes up to 20 m. Pabalon and Pitzer illustrated the utility of this approach by examining salt solutions from moderate concentrations to fused salts (this volume). Clegg and Brimblecombe have employed the model in combination with Henry's Law constants to calculate gas solubilities for several volatile electrolytes (this volume). The ion interaction approach is limited by the available experimental data, both in terms of elements considered and the availability of data for fitting solubility of solid phases at elevated temperature. Another caution discussed here by Plummer and Parkhurst is the requirement to not only be internally consistent with the interaction parameters used in the model, but to be consistent with the choice of the activity coefficient scale, if measured pH values are used.

For solution compositions in which the virial coefficients are well defined, the mineral phase boundaries, ionic activity, and the activity of water, can be modeled remarkably well. Numerous applications are already benefiting from the existing database, such as the ability to predict the solubility of minerals in brine environments. Despite the advancement in the description of high ionic strength solutions, the incompatibility of the ion-pairing model and the semi-empirical phenomenological model is an unresolved issue and will need to be resolved for a unified theoretical description of the solution for many applications.

Model Documentation. Geochemical computer codes have increased in complexity over the past decade, as options were added and as theoretical developments were incorporated. Several of the more widely used models have now been updated and documented (EQ3/6, PHRQPITZ, SOLMNEQ.88, MINTEQ, WATEQ4F). Little change has occurred with respect to the more standard computations such as: speciation to determine single ion activities, distribution of redox species based of the "system Eh," or based on specified redox couples, and the determination of the degree of saturation of a water with respect to a predetermined table of minerals. The principal components present in most geochemical codes are illustrated in Figure 2. Additionally, some of the most common options which have been incorporated into the more generally used models during this past decade and the major evolutionary changes that have been observed are also given in Figure 2. Detailed descriptions of many of these options can be found elsewhere in the users manuals of specific codes.

Other options, such as water mixing subroutines, incorporation of adsorption submodels, organic ligand complexation, and transportability of codes to run on a microcomputer are now available. In the future, subroutines that incorporate algorithms for computing the solubility of anthropogenic organic compounds and partitioning of dissolved organic species between phases, will be of great interest to environmental scientists.

A Computer Workshop was convened during the 1988 ACS Symposium on this topic, during which time software related to the conference theme was demonstrated. Participants provided brief descriptions of these models which are given in the Appendix.

Applications

Mass transfer or reaction path modeling probably represents the most extensive evolutionary change in the past decade. The first code that had the capability to model mass transfer, by simulating the change in solution composition as minerals dissolve or precipitate was PATH. This complex model was cumbersome and was restricted to mainframe computers.

Reaction path modeling is now routinely performed using new codes such as PHREEQE, PHRQPITZ, EQ3/6, PATH, SOLMNEQ.88 and MINTEQ, which now are executable on microcomputers and workstations. The ability to simulate the evolution of a water composition for points along a flow path, provides information about reaction mechanisms and the thermodynamic reaction state of the solution, without being bound to the rigorous constraints of the flow equations. Papers by Steefel and Lasaga, Janeky, Toran, Longmire, Striegl and Healy, and Carnahan in this volume, describe this process which now also includes the computation of isotope fractionation, redox speciation and even coupling to flow and transport models.

Surface Chemistry

The modeling of mineral/solution-interfacial reactions represents an area of geochemistry which has been shown to be critical in understanding the mobility of elements in natural systems. A recent ACS Symposium (48) focused on this topic and provided an excellent summary of advances in the field. Several key aspects of interfacial chemistry are presented in this volume by Stollenwerk and Kipp; Rea and Parks; Machesky; and Anderson and Benjamin. Submodels to the large aqueous speciation codes now go far beyond early attempts at simulating surface chemical reactions (ion exchange, Freundlich, Langmuir, etc.), by incorporating sophisticated adsorption subroutines, such as the constant capacitance or the triple-layer model (15,39,40). The modeling of the partitioning of inorganic species between the aqueous solution and a single solid phase is well documented. Adsorptive additivity for multiple phases is, however, not experimentally observed; therefore, the predictive simulation of sorption on natural materials such as soils and sediments using surface complexation concepts is at present unworkable. Anderson and Benjamin do, however, propose an explanation for the non-additive behavior in the binary suspensions of Al or Fe hydroxides and amorphous silica, by attributing the non-additive aspects of sorption to surface coverage of silica. This stands along side the assumptions of Honeyman (49) which attribute the non-ideality to particle interaction. Numerous research efforts are currently underway to rigorously model sorption in aqueous suspensions with multiple adsorbents.

The adjustment of adsorption parameters with respect to changing temperature is generally not addressed in most modeling exercises, owing to the uncertainty of the adsorption modeling

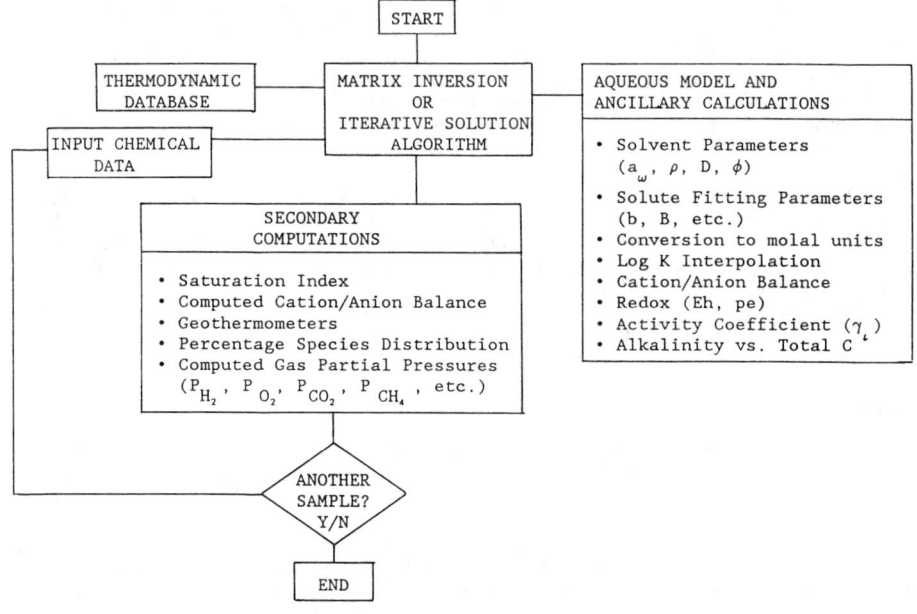

Figure 2. Fundamental components of most aqueous chemical models and the options which are available with many of the more commonly used codes.

parameters themselves. Machesky examined the magnitude of the effect of changing temperature on sorption and provides a generalized method for correcting the effects of temperature.

## NEW CONCEPTS IN MODELING

New concepts in modeling which provide innovative advancements in the modeling of geochemical systems are also presented in this volume. Ideas were also discussed which reveal the deficiencies that must be addressed in the future to make the models more realistic. In the future, advancements in modeling will most likely be realized in four broad areas: the application of error analysis to the modeling, speciation of organic ligands and macromolecules with the adjustment for cosolvents, options to the aqueous theory currently used, and the more efficient coupling of flow models with chemical codes.

### Model Sensitivities

A serious pitfall to many modeling efforts is the lack of a rigorous evaluation of model sensitivities to the errors in sampling, chemical analysis, and system condition; and the propagation of this error throughout the modeling process. In the future, geochemical codes used for environmental issues, industrial applications, low and high level nuclear waste disposal, etc., will require verification for accuracy, and the capability of estimating error. These key components to geochemical modeling are subject to error, which is usually ignored in model computations.

Sampling and Analytical Error. The methods used to collect samples has been shown to dramatically influence analytical results. In addition, chemical data that serve as input to modeling natural systems also have some degree of analytical error. The methods used in collecting groundwater samples have received greater scrutiny recently as geochemists take a closer look at how contaminants move in an aquifer. Barcelona, in this volume, points out that sample quality is a direct reflection of the drilling and sampling methods used, well materials, sampling equipment, and the well design. The results presented in this paper and the references to other related issues, set the framework for minimizing the variability of sample quality. Procedures which reduce error should be practiced by all geochemists and hydrologists before a field program begins and samples are obtained for laboratory analyses.

The usual gauge of the quality of analytical data is the ionic charge balance, which is only a gross indicator of error. To rigorously define the magnitude of inherent analytical error, the precision for the analysis of each analyzed species must be known from repeated measurements, spiked samples, historical evidence of laboratory precision and accuracy, etc., which then could be incorporated into the computational scheme. As discussed by Wildeman in this volume, these errors can be large, even in government regulated programs. The error attributable to sampling alone has been shown to be significant, and can also be evaluated with blanks and spikes, this is especially important for organic compounds as pointed out by Barcelona.

Most of the commonly used models compute a cation/anion equivalent balance, but even this information is seldom translated into error for any subsequent calculation. No codes currently incorporate known analytical precision into evaluation of the error and uncertainty of speciation, saturation index, or mass transfer calculations. Future codes such as INTERP (Appendix) or the expert system of Pearson and others described in this volume, should include an optimization routine which would calculate the propagation of these reported errors, and compute bounding values that clearly define the magnitude of uncertainty.

Computational Errors. Errors in computation can be attributed to two specific circumstances: convergence criteria and improper selection of parametric equations. Most geochemical codes require that mass balance and mass action equations be solved to obtain an acceptable mathematical model of the aqueous solution. Minimal error can be attributed to the mathematical formulation of the equation solvers, because such narrow convergence criteria are used in the matrix inversion or iterative solution algorithms. Numerical dispersion or out of bounds computations generally do not occur unless phase boundaries are exceeded.

The more significant error is associated with the specific mathematical model chosen to describe the system. Each computer code is based on assumptions concerning the equations, algorithms and constants used to compute basic quantities such as the activity coefficients, solvent parameters, (activity of water, dielectric constant, density, etc.), and how these quantities vary with temperature and pressure. None of the commonly used models report the likely range of error which is a consequence of these assumptions. A rigorous comparison of simulated results to actual data is needed, however, to evaluate the present magnitude of error.

Pearson and others (Chapter 26), have proposed a method for using an expert system approach for selecting an optimum method for solving a geochemical computational problem. The approach they have presented here is a first step toward completion of a system that can be used by both practicing geochemists and others for selecting the proper approach to solve a problem. The goal of their approach is to attempt to make all modeling efforts as rigorous and defensible as possible while maintaining a critical objectivity. The approach presented is a promising application of expert systems for addressing geochemical issues which will assist the user in identifying problems when modeling water/rock interactions in natural systems.

System Condition: Redox and Metastability. Two major sources of uncertainty in modeling aqueous systems are the redox potential and metastability; these are frequently acknowledged as conceptual problems, but discussion of the error which results from improper assumptions and calculations is generally avoided. The Eh or electrochemical potential which is computed from a potential measured with a platinum electrode, is used in almost all geochemical models as a system parameter.

The issue of whether or not redox conditions for a system can be accurately reflected by an Eh value is no longer a question. Most geochemists recognize that many aqueous systems are not in redox equilibrium for all sensitive species. This realization has led many researchers to question the use of the standard platinum electrode measurement techniques.

Unfortunately, most redox species do not actually distribute according to the potential indicated by the platinum electrode measurement. Kempton and Runnels, Macalady et al., and Scott and Morgan discuss this situation in this volume and indicate that only iron and possibly manganese are affecting this measured potential. Models which force a system Eh to be used in calculating the distribution of the redox species may generate large errors in the distribution of other redox couples which act independently. This uncertainty must be evaluated because of the likelihood of large error in critical speciation calculations especially involving transition metals.

The non-equilibrium condition of most groundwater systems with respect to many primary minerals, or similarly the metastability which exists with respect to many semi-crystalline or amorphous phases are common problems, especially for silicates. Some clear identification is needed for system reaction time, or the rate at which equilibrium is approached, and similarly identification is needed for metastable plateaus of pseudo-equilibrium, especially for compounds such as amorphous silica, cristobalite, quartz, clay minerals, etc. The likely magnitude of saturation indices which could apply to a given mineral could be specified for a variety of conditions. In this volume, Glynn, and elsewhere others, have recently shown that some error occurs in the calculated saturation values for trace elements when pure end member minerals are assumed to be present, when actually the phases are solid solutions. The consensus among modelers appears to be that error is present and significant; the challenge is to develop procedures that quantify the error, so models become tools that provide realistic and interpretable results.

Thermodynamic and Kinetic Data Advancements

Early codes contained externally referenced thermodynamic databases which were compilations of constants selected by the code authors. These values could be replaced, updated or deleted by the user which led to the proliferation of codes for special applications containing their own customized files.

Current model development is becoming more of a multidisciplinary team effort; recognizing that a large source of error may well be the thermodynamic database. The three most advanced efforts to standardize these data are SUPCRT (29), CODATA and the EQ3/6 data management system (50), and NEA TDB (51). Delaney (50) is building a database management system that will have the capability to reconfigure the EQ3/6 database into a format that can be read by other popular geochemical codes. INTERA (52) conducted a performance assessment comparison using the models PHREEQE and EQ3/6 by generating a single database that both models could access. This exercise underscored the value of a database management system which will not only aid in standardizing the databases but will assist in verifying code calculations with published algorithms.

Reference files containing the thermodynamic database used by most computer models are constructed from both experimental data and from data which have been estimated. Many rely heavily on extrapolation techniques for determining the equilibrium constants or Gibbs Free Energy values at higher temperatures. These estimation techniques are essential, but the error in the estimated constants is generally not reported and is large for extreme conditions of T and P. The computed results derived from a specific model run may not always be affected by these poorly known parameters, but no indication of such uncertainty is reported.

Numerous efforts are currently underway to provide internally consistent thermodynamic data, such as the one presented in this volume by Nordstrom and others. Once attained, even an internally consistent database is still heavily dependent upon extrapolation techniques to provide the needed constants where none exist. The problem remains that the user is not provided with an indication of the magnitude of the error expected from this collection of thermodynamic data. A fundamental question still exists, however; can a truly internally consistent data base ever be assembled that addresses a wide range of geochemical problems? It is more likely that a database will evolve that is less than consistent, but contains known and evaluated error.

A central issue in the attempt to establish a reliable database is the requirement of critically evaluated thermodynamic data for several key species. One such pivotal element is aluminum, which has an extensive literature of solubility and thermochemical data from which to choose, for each of the aqueous species or complexes. The aluminum species are fundamental to the calculation of solubility and reaction state with respect to many silicates and aluminum oxides and hydroxides and are principal components in numerous surface chemical reactions in the environment. Two key chapters in this volume address this fundamental problem: Apps and Neil give a critical evaluation of the data for the aluminum system and Hem and Roberson present the kinetic mechanisms for hydrolysis of aluminum species.

<u>Organic Compounds</u>

Organic compounds in aqueous systems have taken on increased importance over the past decade both as a research topic and as a topic public concern. The issue of anthropogenic organic contamination of aquifers and hydrophilic ligand complexation are the focus of much current work. Early models (REDEQL, MINEQL, GEOCHEM, SOLMNEQ) recognized the need to incorporate hydrophilic organic ligands into the thermodynamic database; and conveniently, the speciation could be treated in a numerically identical fashion to that of inorganic ligands. Anticipated advancements are the modeling of three new classes of organic interactions in aqueous systems: macromolecules and their multiple binding sites, organic cosolvents, and partitioning of organic and inorganic species in a two phase system.

<u>Macromolecules.</u> Humic substances are ubiquitous in natural systems and are generally recognized as a significant component of natural water compositions. The structure and chemistry of these macromolecules are complex such that it is almost impossible to characterize a complete molecule. As a result, thermochemical data on these molecules are unavailable for modeling.

The currently available geochemical models do not have the capability to easily incorporate these substances into speciation calculations. Humic substances, are large macromolecules for which the existing expression for activity coefficients such as Debye-Hückel and the attendant extensions, do not apply, however, significant binding with metals is known to occur. Additionally, as pointed out in this volulme by Choppin and Clark, and Holm and Curtis, metals bind to specific sites on the molecule with binding energy dependent on site location and specific functional group; kinetic effects which impact this association are at present not well understood. Because of the importance of these large molecules in facilitating transport of many trace metals, and in partitioning organics between themselves and the aqueous phase, these substances must be part of the computational scheme of future models.

<u>Cosolvents.</u> The effect of organic solvents on the physical properties of water is a historically useful phenomenon and well defined for countless industrial and scientific applications. This phenomenon has always been ignored in the chemical models available for general usage. Cosolvents affect the dielectric constant of the water, impacting activity coefficient calculations, hydration of species, and adsorption computations. Application of current chemical codes to environmental contamination investigations will require a capability for simulating the effect of a wide variety of miscible solvents on the aqueous system. As the currently available codes are refined, subroutines should be added that address this issue.

<u>Partitioning of Organic Molecules.</u> Organic molecules in natural aqueous systems have a tendency to partition themselves depending on the structure, size, and functionality. This area of concern has taken on greater importance recently from two divergent interest groups. One group involved with the extraction of petroleum has focused on the partitioning of hydrophilic and hydrophobic petroleum components and the impact hydrophilic components have on mineral surfaces.

A variant of the above problem is the presence of immiscible liquids such as oil, dissolved hydrophilic organics, or hydrophobic organic matter on mineral surfaces. As discussed by MacGowan and Surdam in this volume, aluminum and other inorganic species important to our understanding of mineral systems, partition between immiscible liquids such as oil and water.

Another group concerned with the partitioning of organic molecules is focusing on the environmental impact these compounds have on the biosphere. This area of research is in its infancy and requires great attention in the future. This issue of partitioning between immiscible water/organic solvent systems or between aqueous solutions and organic substrates as described in this volume by Bales and Szecsody, also is significant and cannot be easily simulated at present. Nirmalakhandan and Speece presented a model in this volume based on structural parameters for predicting aqueous solubility of organic compounds; whereas Groves and El-Zoobi used empirical relationships to predict solubility and subsequently used activity coefficient relationships to arrive at activity values. Much more activity will be seen in this area as aquifer contaminant problems are remediated and the focus of engineering efforts moves toward restoring groundwater quality. Clearly a theoretically based approach will have longer lasting use and help us understand the effects each hydrophobic compound has on the systems as a whole.

Coupled Hydrology/Chemistry

Coupling of geochemical and hydrologic codes to model the impact geochemical reactions have on aquifer composition is a prime research effort at present. Both geochemists and hydrologists have recognized that aquifer properties such as permeability and fracture geometry can change as a result of mineral dissolution or precipitation. Changes in the physical properties of the aquifer can consequently affect the fluid residence times and thus impact aquifer mineralogy and solution chemistry. This area of concern is particularly crucial grown in importance in petroleum extraction, nuclear waste disposal, and in the general understanding of mineral deposition.

Carnahan, and Steefel and Lasaga (in this volume), have demonstrated that coupling of geochemical and hydrologic models is both conceptually and mathematically complex for just simple hydrologic systems. As the system components and hydrologic details increase, the demand for faster computer systems will be critical. Tripathi has discussed a highly sophisticated coupled model, HYDROGEOCHEM that is yet to be published.

Numerous discussions (53) have been held on the convergence of the two modeling disciplines: flow/transport models and aqueous chemical codes. The literature is extensive on attempts to merge these two approaches, but major conceptual problems remain, such as the conflict between the enormous detail provided in chemical codes regarding species, temperature adjustments and partitioning between gas and solid phase, vs. the conflict of transporting each specie individually from point to point in the hydrologic system.

Current research is advancing in several directions simultaneously, from suggestions to optimize the number of species to reduce the iteration time, execute the hydrologic and chemical models separately, or to completely integrate the models accepting the enormous computational times as a transient problem to be solved by faster machines.

CONCLUSIONS

Few new model codes are emerging at present and greater effort is underway to improve and qualify existing models. Model development is now more driven by application needs and the quality assurance and quality control constraints than by pure research needs, as in the past.

The future should witness much progress in the evaluation of error, better understanding of the effect of organic material and solvents and perhaps, a strong parallel development in the area of coupling chemistry with transport. Certainly decoupled geochemical modeling alone will continue to have strong application and will without question become a tool more useful for many new scientific and engineering applications.

LITERATURE CITED

1. Garrels, R.M.; Thompson, M.E. Am. J. Sci., 1962, 260, 57-66.
2. Jenne, E.D. ed.; Chemical Modeling in Aqueous Systems, ACS Symposium Series No. 93; American Chemical Society: Washington, DC, 1979.
3. Jenne, E.A.; Geochemical Modeling: A Review, Pacific Northwest Laboratory, PNL-3574, 1981.

4. Nordstrom, D.K.; Plummer, L.N.; Wigley, T.M.L.; Wolery, T.J.; Ball, J.W.; Jenne, E.A.; Bassett, R.L.; Crerar, D.A.; Florence, T.M.; Fritz, B.; Hoffman, M.; Holdren, G.R., Jr.; Lafon, G.M.; Mattigod, S.V.; McDuff, R.E.; Morel, F.M.M.; Reddy, M.M.; Sposito, G.; and Thrailkill, J. In Chemical Modeling in Aqueous Systems, Jenne, E.A. ed. Amer. Chem. Soc. Symp. Series 1979, vol. 79; pp. 857-892.
5. Barnes, I.; Clarke, F.E.; U.S. Geol. Survey Professional Paper 498-D, 1969.
6. Kharaka, Y.K.; Barnes, I. SOLMNEQ: Solution-Mineral Equilibrium Computations, NTIS Technical Report PB214-899, 1973.
7. Kharaka, Y.K.; Gunter, W.D.; Aggarwal, P.K.; Perkins, E.H.; DeBraal, J.D. U.S. Geol. Survey Water-Resources Invest. Report 88-4227, 1988.
8. Perkins, E.H.; Kharaka, Y.K.; Gunter W.D.; Debraal, J.D. 1989-, this volume.
9. Truesdell, D.H.; Jones, B.F. WATEQ: A computer Program for Calculating Chemical Equilibria of Natural Waters., NTIS Technical Report PB2-20464, 1973.
10. Ball, J.W.; Jenne, E.A.; Nordstrom, D.K. In chemical Modeling in Aqueous Systems; Jenne, E.A., Ed.; ACS Symposium Series No. 93; American Chemical Society: Washington, DC, 1979; pp 815-835.
11. Ball, J.W.; Jenne, E.A.; Cantrell, M.W. U.S. Geol. Survey Open-File Report 81--1183, 1981.
12. Ball, J.W.; Nordstrom, D.K.; Zachman, D.W. U.S. Geol. Survey Open File Report 87-50, 1987.
13. Wigley, T.M.L. Brit. Geomorph. Res. Group Tech. Bull. 20, 1977.
14. Morel, F.; Morgan, J.J. Env. Sci. Tech. 1972, $\underline{6}$, 58-67.
15. Westall, J.C.; Zachary, J.L.; Morel, F.M.M. MINEQL, A Computer Program for the Calculation of Chemical Equilibrium Composition of Aqueous Systems, MIT Technical Note 18, 1976.
16. Westall, J.C. MICROQL, A Chemical Equilibrium Program in BASIC. Report 86-02, Oregon State University: Corvallis, Oregon, 1986.
17. Ingri, N.; Kakolowlez, W.; Sillen, L.G.; Warnquist, B. Talenta, 1967, $\underline{114}$, 1261-1286.
18. Detar, D.F. Computer Programs for Chemistry; W.A. Benjamin: New York, 1969.
19. Perrin, D.D. Nature 1986, $\underline{206}$, 170-171.
20. Fritz, B., Ph.D. Thesis, Univ. Louis Pasteur, Strasbourg, France, 1975.
21. Lafon, G.M. Ph.D. Thesis, Northwestern University, Evanston, Illinois, 1969.
22. Holdren, G.R., Jr. Ph.D. thesis, Johns Hopkins University, Baltimore, Maryland, 1977.
23. Thrailkill, J. University of Kentucky Water Res. Inst. Res. Rep. 19, 1970.
24. Mattigod, S.V.; Sposito, G. In Chemical Modeling in Aqueous Systems; Jenne, E.A., Ed.; ACS Symposium Series, No. 93; American Chemical Society: Washington, DC, 1979; 837-856.
25. Van Breeman, N. Geochim. Cosmochim. Acta 1973, $\underline{37}$, 101-107.
26. MacCarthy, P.; Smith, G.C. In Chemical Modeling in Aqueous Systems; Jenne, E.A., Ed.; ACS Symposium Series No. 93-American Chemical Society: Washington, DC, 1979; 201-222.
27. Helgeson, H.C. Geochim. Cosmochim Acta, 1968, $\underline{32}$, 853-877.
28. Helgeson, H.C.; Brown, T.H.; Nigrini, A.; Jones, T.A. Geochim. Cosmochim. Acta 1970,$\underline{34}$, 569-592.
29. Helgeson, H.C.; Delany, J.M.; Nesbitt, H.W.; Bird, D.K.Am. J. Sci. 1978, $\underline{278A}$, 1-229.
30. Reed, M.H. Geochim. Cosmochim. Acta 1982, $\underline{79}$, 422-425.
31. Spycher, N.F.; Reed, M.H. Geochim. Cosmochim. Acta 1988, $\underline{52}$, 739-749.
32. Wolery, T.J. Calculation of Chemical Equilibrium Between Aqueous Solutions and Minerals, Lawrence Livermore Laboratory, UCRL-52658, 1979.
33. Wolery, T.J. EQ3NR, A Computer Program for Geochemical Speciation-Solubility Calculations, User's Guide and Documentation, UCRL-5314, 1983.
34. Wolery, T.J. EQ6, A Computer Program for Reaction-Path Modeling of Aqueous Geochemical Systems, UCRL-51.
35. Wolery, T.J.; Jackson, K.J.; Bourcier, W.L.; Bruton, C.J.; Viani, B.E.; Knauss, K.G., Delaney, J.M. 1989; this volume
36. Parkhurst, D.L.; Thorstenson, D.C.; Plummer, L.N. U.S. Geol. Survey Water-Resources Invest. Report 80-60, 1980.
37. Plummer, L.N.; Parkhurst, D.L.; Fleming, G.W.; Dunkle, S.A. U.S. Geol. Survey Water-Resources Invest. Report 88-4153, 1988.
38. Plummer, L.N.; Parkhurst, D.L. 1989; this volume.
39. Felmy, A.R.; Girvin, D.; Jenne, E.A. MINTEQ: A Computer Program for Calculating Aqueous Geochemical Equilibria. U.S. Env. Protect. Agency, 1984.
40. Brown, D.S.; Allison, J.D. U.S. Env. Prot. Agency, EPA 60013-871012, 1987.
41. Shannon, D.W.; Morrey, J.R.; Smith, R.P. Proc. International Symp. Oilfield and Geothermal Chemistry. 1977, 21-36.
42. Pitzer, K.S. J. Phys. Chem. 1973, $\underline{77}$, 268-277.

43. Stokes, R.H.; Robinson, R.A. J. Amer. Chem. Soc. 1948, 70, 1870-1878.
44. Guggenhein, E.A. Phil. Mag. 1935, 19, 588.
45. Harvie, C.E.; Weare, J.H. Geochim. Cosmochim. Acta 1980, 44, 981-997.
46. Harvie, C.E.; Moller, N.; Weare, J.H. Geochim. Cosmochim. Acta. 1984, 48, 723-751.
47. Weare, J.H. Reviews in Mineralogy, 1987, 17, 143-174.
48. Davis, J.A.; Hayes, K.F., Eds. ACS Symposium Series No. 168; American Chemical Society: Washington, DC, 1981, p 683.
49. Honeyman, B.D. Ph.D. Thesis, Stanford Univ., Stanford, CA. 1984.
50. Delany, J.M. written communication, Appendix this paper.
51. Muller, A.B. Rad. Waste Mgmt. Nuclear Fuel Cycle 1985, 6, 131-141.
52. INTERA. PHREEQE: A Geochemical Speciation and Mass Transfer Code Suitable for Nuclear Waste Performance Assessment 1983; Office of Nuclear Waste Isolation, Tech. Report ONWI-435, Columbus, Ohio.
53. Tripathi, V.S.; Yeh, G.T. Abstracts of Papers 1988; 196th ACS National Meeting, Los Angeles, California, GEOC 109.

**APPENDIX.** Computer models and supporting software demonstrated at the Modeling Workshop convened with the Chemical Modeling in Aqueous Systems II Symposium, September 25-30, 1988[1]

| PROGRAM NAME | DESCRIPTION | COMPUTER[2] | SOURCE |
|---|---|---|---|
| GBSSAS | A code for the Guggenheim parameterization of equilibrium relations in binary solid-solution aqueous solution systems. | PC | P.D. Glynn USGS-WRD 432 National Center Reston, VA 22092 |
| HYDRAQL | A version of MINEQL, a chemical speciation code, which has been expanded to simulate adsorption with the triple layer model. | PC | J.O. Leckie Dept. Civil. Eng. Stanford Univ. Stanford, CA 94305 |
| INTERP | A menu-driven geochemical utility that provides an interpretive environment for geochemical codes. One selects from several models, views, edits and graphically displays results. Output is examined as elemental percentage distribution, saturation index, multi-window plots for Pitzer computations, and reaction path plots with overlays for error and uncertainty. | PC | R.L. Bassett Dept. Hydrology Univ. of Arizona Tucson, AZ 85721 |
| LLE | A model for computing activity and activity coefficients for organic-aqueous liquid liquid equilibrium (LLE) in multicomponent systems using the UNIQUAC equations. | PC | F.R. Groves Dept. Chem. Eng. LSU Baton Rouge, LA 70803 |
| PHREEQE (w/PHRQ-INPT and Balance) | A geochemical computer code based on the ion-pairing model which calculates pH, redox potential and mass transfer. | PC | L.N. Plummer USGS-WRD 432 National Center Reston, VA 22092 |
| PHRQPITZ | A program adapted from the computer code PHREEQE which makes geochemical calculations in brines and other electrolyte solutions at high concentrations, using the Pitzer virial coefficient approach (see paper, this volume). | $M^3$ | WATSTORE PROGRAM OFFICE; USGS 437 National Center Reston, VA 22092 |
| SOL-MNEQ.88 | Geochemical code which is the most recent version of SOLMNEQ which runs on a microcomputer via a PC-shell (see paper, this volume). | PC | Y.K. Kharaka USGS - WRD, MS 427 345 Middlefield Road Menlo Park, CA 94025 |

*Continued on next page*

**APPENDIX.** *Continued*

| PROGRAM NAME | DESCRIPTION | COMPUTER[2] | SOURCE |
|---|---|---|---|
| SOLVEQ (with CHILLER) | SOLVEQ is a program for computing multicomponent homogeneous chemical equilibria in aqueous systems. For a given temperature and total composition of a homogeneous aqueous solution, it computes the activities of all aqueous species and the saturation indices of solids and the fugacities of gases. CHILLER is a program to compute multicomponent heterogeneous chemical equilibria among solids, gases, and an aqueous phase; suitable for modeling processes such as cooling, heating, mixing of solutions (aqueous, solid, or gaseous), mass transfer (titration of rock or of other reactants), boiling, condensation, and evaporation. | PC | M. Reed Dept. Geol. Sci. Univ. of Oregon Eugene, OR 97403 |
| WATEQ4F | A FORTRAN 77 version of the PL/1 computer program for the geochemical model WATEQ2 written for the personal computer. Limited data base revisions are included. | PC | J.W. Ball USGS - WRD, MS 420 345 Middlefield Road Menlo Park, CA 94025 |
| PTX | A program that calculates complete pressure-temperature (P-T), temperature-composition (T-$X_{H_2O-CO_2}$), and pressure-composition (P-$X_{H_2O-CO_2}$) diagrams. | PC | E.H. Perkins Alberta Res. Council P.O. Box 8330 Postal Station F Edmonton, Alberta CANADA T6H 5X2 |
| LLNL DATA BASE | A thermochemical data base for use in geochemical modeling calculations to simulate interactions between the groundwater, rock and a waste form in a geologic repository environment. Data are included for radionuclides, canister materials, key rock forming minerals and associated aqueous species coordination with other data groups, such as CODATA and NBS is being maintained. | M | J.M. Delany LLNL L-219 P.O. Box 808 Livermore, CA 94550 |

Notes

[1]All other models considered at the Symposium are either referenced in Figure 1 (this paper) or in the specific manuscript in which discussed (this volume).
[2]PC = personal computer, M = mini-computer.
[3]PC version available from R.L. Bassett, Dept. Hydrology, Univ. of Arizona, Tucson, AZ 85721.

RECEIVED August 23, 1989

# AQUEOUS THERMODYNAMICS AND THEORETICAL ADVANCEMENTS

# Chapter 2

# Activity Coefficients in Aqueous Salt Solutions

## Hydration Theory Equations

### Thomas J. Wolery and Kenneth J. Jackson

### Lawrence Livermore National Laboratory, L–219, P.O. Box 808, Livermore, CA 94550

A new set of phenomenological equations for describing the activity coefficients of aqueous electrolytes has been derived, based on the hydration theory concept of Stokes and Robinson (1), but using the "differentiate down" approach in which an expression is first defined for the excess Gibbs energy. Separate equations are given for the activity of water and the activity coefficients of ionic solutes. The new equations incorporate an empirical but thermodynamically consistent scheme for using an average ion size parameter in the Debye-Hückel part of the model. This permits the new equations to be applied to mixtures of aqueous electrolytes. The new equations, applied to the case of a pure aqueous electrolyte, do not reduce to the familiar Stokes-Robinson equation owing to a minor difference in how the Debye-Hückel model is presumed to apply to formally hydrated solutes. As a first step in evaluating the usefulness of the equations, we have fit a "two-parameter" (ion size plus hydration number) model to data for a number of pure aqueous electrolytes. The quality of the fits is excellent in many cases, but there are indications that more satisfactory results would be obtained by fixing reasonable values for the ion sizes and compensating for the loss of a fitting parameter by including ion pairs in the models, by including virial coefficient terms in the equations, or both.

Thermodynamic modeling of aqueous geochemical systems depends on functions which approximate the activity of water and the activity coefficients of the solutes. Ten years ago, the state of the art for addressing this problem in geochemical modeling codes was the use of simple extensions of the Debye-Hückel equation, such as the Davies equation or the "B-dot" equation (see ref. 2). These equations were useful up to concentrations approaching sea water ($I \simeq 0.7$ m, I = ionic strength), although they can be shown to depart from precise physical measurements in much less concentrated solutions. These approximations were used in the modeling codes in conjunction with a large number of ion-pairing and complexation constants to compute models of speciation and degree of saturation with respect to various minerals and other solids.

Ideally, a model for approximating activity coefficients in geochemical or engineering modeling codes would have the following properties: consistency with the laws of thermodynamics; compact mathematical form; high accuracy over wide ranges of temperature, pressure, and concentration; applicability to systems including most of the elements in the periodic table; consistency with generally all types of physical measurements bearing on activity coefficients, phase equilibria, and speciation; and existence of useful predictive means to estimate the model parameters, as opposed to obtaining them exclusively by data fitting. These predictive means might include estimation via empirical or semi-empirical correlation methods, which have been shown to exist for some models (3, 4). The best predictive means would be obtained if the parameters could be expressed as functions of the intermolecular forces (5-7). Some kind of predictive means is probably

NOTE: This chapter is Part II in a series.

essential to making substantial progress in modeling the behavior of aqueous systems containing many chemical elements, such as transition metals and actinides, that tend to exist as aqueous complexes more so than as dissociated ions.

No model currently exists which satisfies all of the above requirements. For modeling geochemical systems, the requirement of mathematical compactness has focused attention on phenomenological models guided to some degree by theoretical considerations. The most successful such example is Pitzer's equations (8-10), which are thermodynamically consistent and generally produce highly accurate results. The virial coefficients employed in this phenomenology have been obtained by fitting experimental data, originally only of vapor pressure equilibrium and electromotive force (emf) data (see ref. 9), and more recently also of solubility data (see ref. 10). Model development has been largely restricted to systems in which the solutes tend to exist in mostly dissociated form. Ion pair species are formally included in the models only when the data can not be adequately fit otherwise. Application of the Pitzer phenomenology has ignored some kinds of data. An important example is electrical conductance, the interpretation of which is generally ascribed to ion-pairing (see ref. 11). The picture obtained from such data is one in which there are many more ion pair species than are included in the present models based on Pitzer's equations. No generally useful predictive relations governing the Pitzer's equations parameters have been shown to exist; consequently, the usefulness of the existing models is largely restricted to systems that are part of the larger "sea-salt" system.

Some other kinds of models have shown parameters that seem to follow useful correlation relationships. Among these are the virial coefficient model of Burns (3), the interaction coefficient model of Helgeson, Kirkham, and Flowers (4), and the hydration theory model of Stokes and Robinson (1). The problem shared by all three of these models is that they employ individual ion size parameters in the Debye-Hückel submodel. This led to restricted applicability to solutions of pure aqueous electrolytes, or thermodynamic inconsistencies in applications to electrolyte mixtures. Wolery and Jackson (in prep.) discuss empirical modification of the Debye-Hückel model to allow ion-size mixing without introducing thermodynamic inconsistencies. It appears worthwhile to examine what might be gained by modifying these other models. This paper looks at the hydration theory approach.

PREVIOUS DEVELOPMENT

Hydration theory was first applied to aqueous electrolytes by Stokes and Robinson (1) in 1948. The approach is to correct for the fact that the actual solute species do not exist as bare ions as they are normally written, but as hydrated aqueous complexes. In the case of a solution of an aqueous non-electrolyte, this reduces to the assumption that the solution is ideal, or nearly so, if one writes the formula of the solute to include the waters of hydration and computes its concentration and that of the solvent on this basis. In the case of an aqueous electrolyte, it reduces to the assumption that the Debye-Hückel model becomes adequate (or at least more adequate) to describe the activity coefficients if this treatment is applied.

Hydration theory deals with aqueous solutions in terms of two parallel definitions of the set of components, the usual one in which the solutes are considered formally unhydrated and the amount of solvent is the nominal amount, and a second in which the solutes are formally hydrated and the amount of water is reduced from the nominal amount. The objective is to correct for hydration effects and obtain a model giving activity coefficients in terms of the usual set of components. Quantities pertaining to the second set of components will be denoted by an asterisk. The number of moles of water in the first set is given by

$$n_w = \Omega w_w \qquad (1)$$

where $\Omega$ is a constant, the number of moles of water having a mass of one kilogram ($\simeq$55.51), and $w_w$ is the number of kilograms of nominal water. The number of moles of water in the second set is given by

$$n_w^* = \left(\Omega - \sum_i h_i m_i\right) w_w \quad (2)$$

where $h_i$ is the hydration number (number of bound water molecules per molecule of solute) and $m_i$ is the molality of the i-th solute. The number of moles of a solute is the same in both component sets ($n_i^* = n_i$). The molality of a solute in the usual component set is given by

$$m_i = \frac{\Omega n_i}{n_w} \quad (3)$$

The molality of a solute in the second component set is given by the same equation, but with asterisks attached to the m and n quantities. It can be related to the nominal molality by

$$m_i^* = \frac{m_i}{D} \quad (4)$$

where

$$D = 1 - \frac{\sum_i h_i m_i}{\Omega} \quad (5)$$

Thus, the amount of "free" water is less than the nominal amount, and the molality of a solute, corrected for hydration, is greater than the nominal molality. The model encounters a singularity when all the "free" water has been used up.

Using an argument based on the Gibbs-Duhem equation, Stokes and Robinson (1) derived the following equation for a solution of a single aqueous electrolyte (A), presumed to be fully dissociated:\

$$\ln \gamma_{\pm,A} = -\ln\left(D + \frac{v_A m_A}{\Omega}\right) - \frac{h_A}{v_A}\ln a_w + \frac{|z_{A+} z_{A-}|}{2} f'(I, a_A) \quad (6)$$

(Note: the original equation has been translated into our notation). Here $\gamma_{\pm,A}$ is the mean activity coefficient of the electrolyte, $v_A$ is the number of ions produced by its dissociation, $m_A$ is its molality, $h_A$ is its hydration number, $z_{A+}$ is the charge on the cation, and $z_{A-}$ is the charge on the anion. The last term contains $f' \equiv \partial f/\partial I$, where $f(I, a_A)$ is a Debye-Hückel function which depends on I, the ionic strength, and $a_A$, an ion size parameter characteristic of the electrolyte. The Debye-Hückel function used by Stokes and Robinson is what Pitzer (8, 9) refers to as the DHC (Debye-Hückel-charging) form. Pitzer in his equations uses the alternative DHO (Debye-Hückel-osmotic) form. Equations for these forms are summarized in refs. 8 and 9. The ionic strength is given by

$$I = \frac{1}{2}\sum_i m_i z_i^2 \quad (7)$$

In such solutions, $D = 1 - (h_A m_A/\Omega)$. The hydration number of the electrolyte is easily identified as the stoichiometric sum of the hydration numbers of the individual ions. The relationship of the characteristic ion size parameter to the sizes of the individual ions was not addressed.

Stokes and Robinson (1) fit their equation to data reported for 35 pure aqueous electrolytes of the 1:1 and 2:1 types and obtained good results to fairly high concentrations. For example, for NaCl, the reported average difference in $\ln \gamma_\pm$ was only 0.002 for over a concentration range of 0.1-5.0 m (I = 0.1-5.0 m); for CaCl$_2$, 0.001 from 0.01-1.4 m (I = 0.06 to 8.4 m). These statistics can not be directly compared with those reported for fitting with Pitzer's equations (e.g., ref. 9) as the types of statistics employed are not the same; nevertheless, it is clear that the accuracy of this two-parameter model is close to that of Pitzer's model, which employs three virial coefficient parameters to fit a single aqueous electrolyte. Values of the characteristic ion size ranged from 3.48 to 6.18 Angstroms, values which appear reasonable. The hydration numbers ranged from 0.6 to 20.0 and also seem reasonable, although hydration numbers in the higher end of this range would require water molecules to be bound in other than the first hydration shell. Stokes and Robinson further proposed a correlation between the ion size parameter and the hydration number, and examined a "one-parameter" model in which the ion size was taken as a function of the hydration number. This was surprisingly successful, though less accurate.

The Stokes-Robinson model was later modified by its creators for use in more concentrated solutions by accounting for dehydration equilibria (12). The original model was also modified by Nesbitt (13), who made the hydration number a decreasing function of the ionic strength. Other workers too numerous to mention here have also attempted to do something with hydration theory.

Given that the Stokes-Robinson model was proposed well before the development of modern computers, and that the fitting accuracy of the "two-parameter" form is quite good, one might ask why this model was not used in geochemical modeling codes as soon as such codes were developed. One reason is that the mean activity coefficient can only be obtained if one knows the activity of water in the solution. Stokes and Robinson did not give a separate equation for this, as it was not necessary for what they were trying to do. In the majority of cases, a solution of a pure aqueous electrolyte is studied by the isopiestic method, which yields the osmotic coefficient ($\phi$), which is related to the activity of water ($\phi = -(\Omega/\Sigma_i m_i) \ln a_w$). The mean activity coefficient can be obtained by numerically integrating the following form of the Gibbs-Duhem equation:

$$\upsilon_A m_A \frac{\partial \ln \gamma_{\pm,A}}{\partial m_A} = -\upsilon_A - \Omega \frac{\partial \ln a_w}{\partial m_A} \qquad (8)$$

(sometimes the route is reversed, as when emf methods are used). In essence, Stokes and Robinson used their equation as an alternative to this numerical integration. Thus, in addressing the problem of interest to them, they already knew the activity of water.

Other problems are associated with the ion size parameter. The model could not be extended to electrolyte mixtures without introducing some means of allowing mixing of characteristic ion size parameters or losing accuracy by using a single value for all electrolytes in a given mixture. Many empirical schemes that have been proposed for ion size mixing can be shown to run afoul of thermodynamic consistency relations (see Wolery and Jackson, in prep.). Furthermore, the use of an ion size parameter characteristic to an electrolyte makes it difficult to see how the Stokes-Robinson equation can be adapted to a form giving single ion activity coefficients, which is generally useful and even necessary for some purposes in geochemical modeling codes.

It appears worthwhile to attempt to overcome these deficiencies, as the hydration concept has a powerful physical appeal. This does not mean that one must hold that the Debye-Hückel model is really adequate if one simply corrects for hydration. It has a number of deficiencies, as has been pointed out for example by Pitzer (8, 9). Like Pitzer's equations, the kind of model that we are addressing here incorporates some degree of theory as a guide but in essence remains phenomenological. We only expect that the hydration correction will turn out to be a useful part of such a model.

THERMODYNAMIC FRAMEWORK

A thermodynamic model that satisfies all of the necessary consistency relations can be devised by writing an expression for the excess Gibbs energy and applying appropriate partial differentiation with respect to the number of moles of the components (see Wolery, in prep.). If the excess Gibbs energy is defined with reference to mole fraction ideality, the differential equations are

$$\ln a_w = -\ln\left(1 + \frac{\sum_j m_j}{\Omega}\right) + \frac{1}{RT}\frac{\partial G^{EXx}}{\partial n_w} \tag{9}$$

$$\ln \gamma_i = -\ln\left(1 + \frac{\sum_j m_j}{\Omega}\right) + \frac{1}{RT}\frac{\partial G^{EXx}}{\partial n_i} \tag{10}$$

(Note- the solute components here are taken to be individual species; i.e., not neutral electrolytes.) Here R is the gas constant, T is the absolute temperature, and $G^{EXx}$ is the mole-fraction based excess Gibbs energy. The first term on the right hand side of the equation for the activity of water is equivalent to the logarithm of the mole fraction of that substance; the second term, the logarithm of its mole fraction activity coefficient. Alternative differential equations apply if one uses the excess Gibbs energy defined with reference to ideality based on molalities ($G^{EXm}$; see Wolery, in prep.), and such were used by Pitzer (8, 9) to develop his equations. There is a tendency to use the same expressions for the excess Gibbs energy regardless of the specific reference ideality employed. Therefore, it makes a difference in the resulting equations which set of differential equations one uses. As far as the development of phenomenological models is concerned, however, the choice is probably best viewed as a matter of taste. In this paper, we will use the set of differential equations given above. The work of Stokes and Robinson (1) was also consistent with mole fraction ideality.

EXCESS GIBBS ENERGY AND THE DEBYE-HUCKEL FUNCTION

The excess Gibbs Energy for a pure Debye-Hückel model is given by (see Wolery and Jackson, in prep.)

$$G^{EXx} = w_w RT\, f(I) \tag{11}$$

(Here the ion size parameter is taken as a fixed constant.) Partial differentiation gives

$$\ln a_w = -\ln\left(1 + \frac{\sum_j m_j}{\Omega}\right) - \frac{1}{\Omega}[\,If'(I) - f(I)\,] \tag{12}$$

$$\ln \gamma_i = -\ln\left(1 + \frac{\sum_j m_j}{\Omega}\right) + \frac{z_i^2}{2} f'(I) \tag{13}$$

ION SIZE AVERAGING

A simple, empirical, but thermodynamically consistent approach to ion size averaging has been recently proposed by Wolery and Jackson (in prep.). The excess Gibbs energy is written as

$$G^{EXx} = w_w RT\, f(I,\bar{a}) \tag{14}$$

where $\bar{a}$ is the average ion size defined by

$$\bar{a} = \frac{\sum_i m_i W_i \mathring{a}_i}{\sum_i m_i W_i} \qquad (15)$$

Here $\mathring{a}_i$ denotes the individual size of the i-th solute species. $W_i$ is an empirical charge-based weighting cofactor. One might choose $W_i = 1$, $|z_i|$ or $z_i^2/2$. Partial differentiation then gives

$$\ln a_w = -\ln\left(1 + \frac{\sum_j m_j}{\Omega}\right) - \frac{1}{\Omega}\left[\text{If}'(I,\bar{a}) - f(I,\bar{a})\right] \qquad (16)$$

$$\ln \gamma_i = -\ln\left(1 + \frac{\sum_j m_j}{\Omega}\right) + \frac{z_i^2}{2} f'(I,\bar{a}) + \frac{W_i f_a(I,\bar{a})}{\sum_j W_j m_j}\left(\frac{\mathring{a}_i - \bar{a}}{\bar{a}}\right) \qquad (17)$$

The equation for the activity of water is essentially unchanged, but a new term, the "averaging term," has been introduced into the equation for the solute activity coefficient. Here $f_a$ is a derivative Debye-Hückel function, defined as

$$f_a \equiv \bar{a}\left(\frac{\partial f}{\partial \bar{a}}\right)_I \qquad (18)$$

For both the DHC and DHO models, it may be shown that (Wolery and Jackson, in prep.)

$$f_a = 3\text{If}'(I,\bar{a}) - 2f(I,\bar{a}) \qquad (19)$$

The mean activity coefficient of electrolyte A (assumed to be completely dissociated) is given by

$$\ln \gamma_{\pm,A} = -\ln\left(1 + \frac{\sum_j m_j}{\Omega}\right) + \frac{|z_{A+} z_{A-}|}{2} f'(I,\bar{a})$$

$$+ \frac{W_A f_a(I,\bar{a})}{\sum_j W_j m_j}\left(\frac{\bar{a}_A - \bar{a}}{\bar{a}}\right) \qquad (20)$$

where $W_A$ is a characteristic weighting factor for the electrolyte defined by

$$W_A = \frac{\upsilon_{A+,A} W_{A+} + \upsilon_{A-,A} W_{A-}}{\upsilon_A} \qquad (21)$$

and $\bar{a}_A$ is a characteristic average ion size for the electrolyte defined by

$$\bar{a}_A = \frac{\upsilon_{A+,A} W_{A+} \mathring{a}_{A+} + \upsilon_{A-,A} W_{A-} \mathring{a}_{A-}}{\upsilon_{A+,A} W_{A+} + \upsilon_{A-,A} W_{A-}} \qquad (22)$$

In a solution of a pure electrolyte, the averaging term vanishes.

PROPOSED NEW HYDRATION THEORY MODEL

If we assumed that the Debye-Hückel model worked for a set of components in which the solutes were considered formally hydrated, we could write

$$G^{EXx*} = w_w^* \, RT \, f(I^*, \bar{a}) \qquad (23)$$

Note that the excess Gibbs energy carries an asterisk to denote that it pertains to the set of components with formally hydrated solutes. Thus, $w_w^*$ is the number of kilograms of "free" water ($w_w^* = w_w D$), and $I^*$ is the molal ionic strength, calculated in the usual way but with the nominal molalities replaced by molalities corrected for hydration (thus $I^* = I/D$). If it were desired to place less reliance on the Debye-Hückel model, other terms could be added to the right hand side in the usual way. We will presently ignore this as we are immediately interested in the hydration correction itself.

The objective is to obtain an expression for $G^{EXx}$ from that assumed for $G^{EXx*}$ and obtain expressions for the activity coefficients of the components in the usual component set by partial differentiation. First, however, we are motivated to make a change in the expression for $G^{EXx*}$. In the early stages of developing Debye-Hückel theory, the ionic strength is defined on a volumetric basis, and the molal ionic strength is later substituted as a convenient approximation. In general, we should expect the nominal molal ionic strength I to be closer to the volumetric ionic strength than is the molal ionic strength corrected for hydration. As did Stokes and Robinson (1), we will take the ionic strength to be the nominal molal quantity. We therefore change our starting equation to

$$G^{EXx*} = w_w^* \, RT \, f(I, \bar{a}) \qquad (24)$$

The total Gibbs energy is independent of how one chooses to define the components; therefore,

$$G^{EXx} = G^{IDx*} - G^{IDx} + G^{EXx*} \qquad (25)$$

The ideal chemical potential of the a-th solution component in the nominal set of components is given by

$$\mu_a^{IDx} = \mu_a^\circ + RT \ln x_a \qquad (26)$$

(the subscript a may denote either w or i). Here $\mu_a^\circ$ is the standard state chemical potential (for a standard state mole fraction of unity) and $x_a$ is the mole fraction of the component. This can be substituted into the first-order sum rule (see Wolery, in prep.) to obtain

$$G^{IDx} = \sum_a n_a (\mu_a^\circ + RT \ln x_a) \qquad (27)$$

A similar expression may be constructed for $G^{IDx*}$. It can then be shown that

$$G^{IDx*} - G^{IDx} = RT \left[ n_w \ln \left( \frac{x_w^*}{x_w} \right) + \sum_i n_i \ln \left( \frac{x_i^*}{x_i x_w^{*h_i}} \right) \right] \qquad (28)$$

where the mole fractions can be related to the nominal molalities by

$$x_w = \frac{\Omega}{\Omega + \sum_j m_j} \tag{29}$$

$$x_i = \frac{m_i}{\Omega + \sum_j m_j} \tag{30}$$

$$x_w^* = \frac{\Omega - \sum_j h_j m_j}{\Omega - \sum_j h_j m_j + \sum_j m_j} \tag{31}$$

$$x_i^* = \frac{m_i}{\Omega - \sum_j h_j m_j + \sum_j m_j} \tag{32}$$

The expression for the excess Gibbs energy can therefore be written as

$$G^{EXx} = RT \left[ w_w Df(I,\bar{a}) + n_w \ln\left(\frac{x_w^*}{x_w}\right) + \sum_i n_i \ln\left(\frac{x_i^*}{x_i x_w^{*h_i}}\right) \right] \tag{33}$$

The actual process of partial differentiation of this equation is quite tedious and will not be presented in detail. We note the following key intermediate results:

$$\frac{\partial}{\partial n_w}\left(\frac{G^{IDx*} - G^{IDx}}{RT}\right) = \ln\left(\frac{x_w^*}{x_w}\right) \tag{34}$$

$$\frac{\partial}{\partial n_i}\left(\frac{G^{IDx*} - G^{IDx}}{RT}\right) = \ln\left(\frac{x_i^*}{x_i x_w^{*h_i}}\right) \tag{35}$$

The final results are:

$$\ln a_w = -\ln\left(1 + \frac{\sum_j m_j}{\Omega D}\right) - \frac{1}{\Omega}\left[ DIf'(I,\bar{a}) - f(I,\bar{a}) \right] \tag{36}$$

$$\ln \gamma_i = -\ln D + (h_i - 1)\ln\left(1 + \frac{\sum_j m_j}{\Omega D}\right) + D\frac{z_i^2}{2} f'(I,\bar{a})$$

$$\ln \gamma_{\pm,A} = -\ln D + \left(\frac{h_A}{\upsilon_A} - 1\right) \ln\left(1 + \frac{\sum_j m_j}{\Omega D}\right) + D \frac{|z_{A+}z_{A-}|}{2} f'(I,\bar{a})$$

$$-\left(\frac{h_i}{\Omega}\right) f(I,\bar{a}) + D \frac{W_i f_a(I,\bar{a})}{\sum_j W_j m_j} \left(\frac{\hat{a}_i - \bar{a}}{\bar{a}}\right) \quad (37)$$

$$-\left(\frac{h_A}{\Omega \upsilon_A}\right) f(I,\bar{a}) + D \frac{W_A f_a(I,\bar{a})}{\sum_j W_j m_j} \left(\frac{\bar{a}_A - \bar{a}}{\bar{a}}\right) \quad (38)$$

Note that, compared with the counterpart equations without the hydration correction, the factor D appears in various terms. In the equations for molal activity coefficients, there are two additional terms, one containing ln D, the other containing the Debye-Hückel function f.

Combination of the equations for the activity of water and the mean activity coefficient for the case of a single aqueous electrolyte gives

$$\ln \gamma_{\pm,A} = -\ln\left(1 + \frac{\upsilon_A m_A}{\Omega}\right) - \ln D - \left(\frac{h_A}{\upsilon_A}\right) \ln a_w$$

$$+ D\left(\frac{|z_{A+}z_{A-}|}{2} - \frac{h_A I}{\upsilon_A \Omega}\right) f'(I,\bar{a}) \quad (39)$$

This is not the Stokes-Robinson equation, though some common elements may be discerned. Our result is different because we substituted I for I* in the Debye-Hückel equation for the excess Gibbs energy, whereas Stokes and Robinson (1) made the substitution in the Debye-Hückel equation for the activity coefficient of a formally hydrated solute ($\ln \gamma^*_{\pm,A}$). Differentiation of our equation for the excess Gibbs energy would yield a somewhat different equation for $\gamma^*_{\pm,A}$.

## PRELIMINARY MODEL DEVELOPMENT

The equations derived above constitute a "phenomenology" upon which specific models might be constructed. Using the same phenomenological equations, one can develop models that are distinct with regard to speciation. It is also possible to develop models that are distinct according to the observational data (e.g., isopiestic, emf, solubility, electrical conductivity, spectroscopic) that are to be explained. Within the framework of the equations given above, one may choose from among different Debye-Hückel submodels, which will also lead to distinct models (see ref. 14 for a review of advances in Debye-Hückel models). In addition, one might modify the equations, for example by including higher-order electrostatic terms (15) or adding empirical virial coefficient terms. Such modifications can be accomplished easily and safely by adding the requisite terms to the equation for the excess Gibbs energy and applying the partial differential equations given earlier in this paper.

A complete model development is beyond the scope of the present paper. We present here only some results of fitting our hydration theory equations to data for several pure aqueous electrolytes, ignoring ion pairing. The data fit are values for the osmotic coefficient as tabulated by Robinson and Stokes (11). The Debye-Hückel model employed was the DHC, and the Debye-Hückel parameters used were taken from ref. 16. The fitting was accomplished by unweighted least squares minimization, using as fitting parameters the characteristic average ion size and total hydration number of each electrolyte. The mean molal activity coefficient was then computed from equation 38 and compared with the values tabulated by Robinson and Stokes (11); the latter are related to the corresponding tabulated values for the osmotic coefficient via numerical integration of the Gibbs-Duhem equation.

Figures 1 and 2 show the results of fitting the osmotic coefficient of the aqueous electrolytes sodium perchlorate and potassium chloride, respectively. Analysis of the variance in fitting the osmotic coefficient indicates that the fits are about as good as those obtained using Pitzer's equations, despite the fact that our equations have one less fitting parameter. For sodium perchlorate, the standard deviation in our fit is 0.0011, whereas Pitzer (9) reports 0.001 using his equation. For potassium chloride, the standard deviation in our fit is 0.00036, that in Pitzer's, 0.0005 (for a maximum molality of 4.8 opposed to our 4.5).

Figure 3 shows the predicted mean molal activity coefficients evaluated from these fits. The fit for aqueous potassium chloride is essentially perfect in terms of both deviation in fitting the osmotic coefficient and the deviation between the activity coefficient predicted from our model and the same quantity obtained by numerical integration. The fit for aqueous sodium perchlorate, however, is not quite as good and is actually much more typical of the results obtained fitting data for 1:1 electrolytes. Not only are the residuals for fitting the osmotic coefficient greater, but the activity coefficient computed from our model rises noticeably (though not greatly) above the values obtained by numerical integration for an ionic strength greater than about three or four molal.

The results of fitting a number of 1:1 aqueous electrolytes are given in Table 1. The fits are noticeably less good for data likely to have been obtained in part or all by emf as opposed to isopiestic means. The data themselves are probably not as good. These cases include the hydroxides, significantly volatile acids, and salts of lithium. Values for the fitting parameters are physically reasonable in most cases. Exceptions are the negative hydration numbers of aqueous potassium, rubidium, and cesium nitrate. Except for these electrolytes and those mentioned above for which the data are probably not of highest quality, fairly regular and expected patterns are apparent. In the chloride series, for example, the characteristic ion sizes and the hydration numbers decrease going to alkali cations farther down the periodic table.

If single ion values are desired, a thermodynamic convention can be applied. For example, following the MacInnes convention, one might take the sizes of the potassium and chloride ions to be equal to the characteristic ion size of aqueous potassium chloride, and the hydration number of each of these ions to be equal to one-half the hydration number of the same aqueous electrolyte. A thermodynamically valid convention need not require physical (i.e., positive values) for ion sizes and hydration numbers of single ions.

The meaning of obtaining negative hydration numbers requires some discussion. The role of the hydration number is to cause the activity coefficient to increase at high ionic strength (assuming of course a positive value). The Debye-Hückel term only causes the activity coefficient to tend to decrease. The data for the potassium, rubidium, and cesium nitrate salts are consistent with an activity coefficient that is below that predicted by the Debye-Hückel model, hence the fit produces a negative hydration number. Ion pairing, which is ignored in the model, also tends to decrease the activity coefficient, and may be the real factor involved.

Reported ion pairing constants suggest that this is possible. Smith and Martell (17) report a log K of -0.15 for association of potassium with nitrate, in contrast to one of -0.7 for association of the same cation with chloride. One should keep in mind that the hydration numbers obtained for all of the aqueous electrolytes given in the table would be larger if ion pairing were explicitly treated in the model.

In general, attempts to fit the simple hydration theory model to 2:1 and 3:1 electrolytes were markedly less successful. In many cases, the fitting residuals were extremely large. Reducing the upper end of the concentration range fit in some cases resulted in acceptable fits. In many of these cases, however, we obtained physically unacceptable parameter values, typically negative characteristic ion sizes. In the case of 1:1 electrolytes, we found that varying the upper limit of the concentration range in the fitting produced little change in the fitting parameters. In the case of 2:1 and 3:1 electrolytes, we expected to see convergence of the fitting parameters as the upper

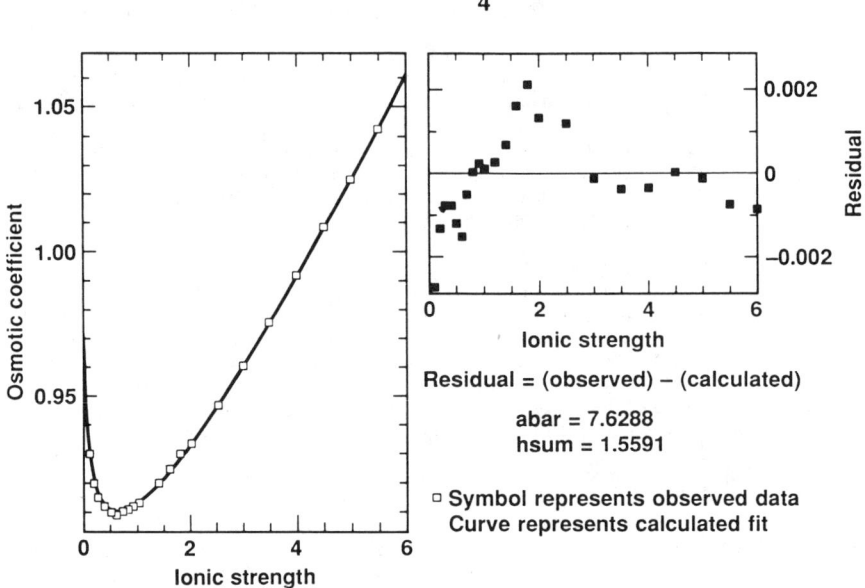

Figure 1. The fit of our hydration theory model to osmotic coefficient data for sodium perchlorate tabulated by Robinson and Stokes (11).

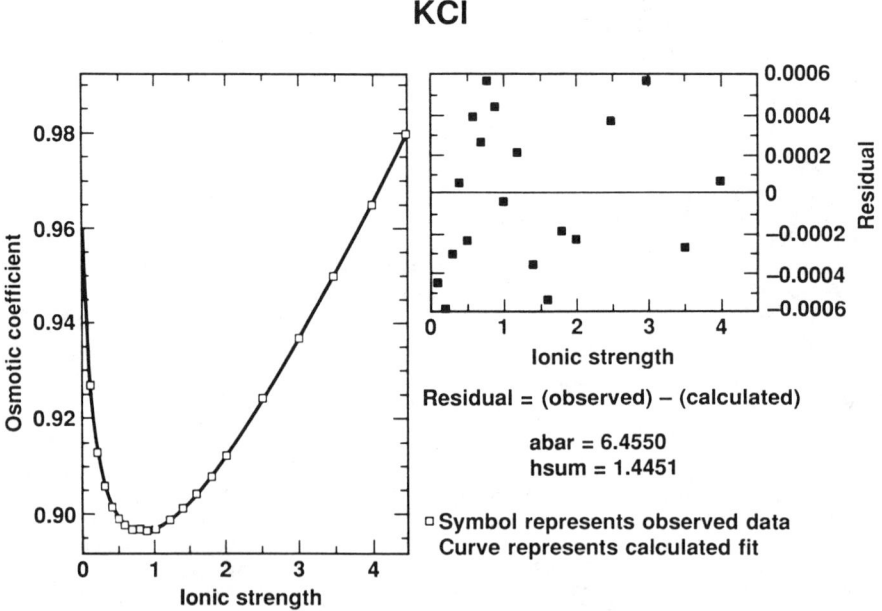

Figure 2. The fit of our hydration theory model to osmotic coefficient data for potassium chloride tabulated by Robinson and Stokes (11).

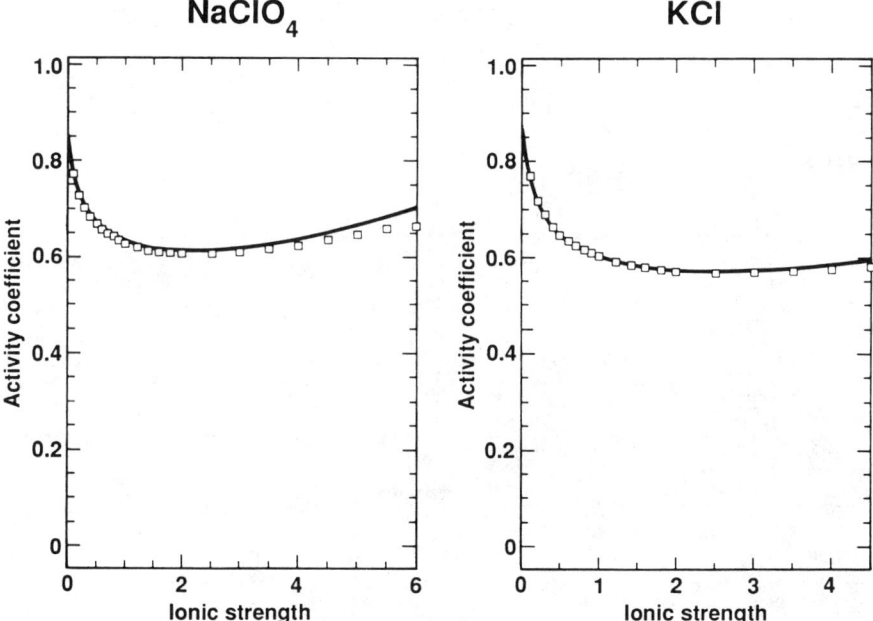

Figure 3. The mean molal activity coefficient of sodium perchlorate (left) and potassium chloride (right). The curves show the predictions of our model, fit to data for the osmotic coefficient. The squares represent the data tabulated by Robinson and Stokes (11).

Table 1. Results of fitting several 1:1 aqueous electrolytes

| Electrolyte | $a_A$ | $h_A$ | max.m | max $|\Delta\emptyset|$ |
|---|---|---|---|---|
| $HClO_4$ | 12.11 | 5.40 | 6.0 | 0.062 |
| $LiClO_4$ | 18.26 | 5.74 | 4.0 | 0.028 |
| $NaClO_4$ | 7.74 | 1.56 | 6.0 | 0.0028 |
| | | | | |
| HCl | 15.40 | 4.88 | 6.0 | 0.076 |
| LiCl | 11.25 | 4.53 | 6.0 | 0.0033 |
| NaCl | 7.15 | 2.51 | 6.0 | 0.0025 |
| KCl | 6.45 | 1.44 | 4.5 | 0.00059 |
| RbCl | 5.79 | 1.32 | 5.0 | 0.0017 |
| CsCl | 4.43 | 1.21 | 6.0 | 0.0058 |
| | | | | |
| HBr | 11.67 | 6.30 | 3.0 | 0.012 |
| LiBr | 12.03 | 5.15 | 6.0 | 0.068 |
| NaBr | 7.97 | 3.05 | 4.0 | 0.0033 |
| KBr | 7.14 | 1.53 | 5.5 | 0.00053 |
| RbBr | 5.89 | 1.13 | 5.0 | 0.0031 |
| CsBr | 4.20 | 1.14 | 5.0 | 0.0038 |
| | | | | |
| $HNO_3$ | 9.01 | 3.47 | 3.0 | 0.0025 |
| $LiNO_3$ | 14.51 | 3.38 | 6.0 | 0.041 |
| $NaNO_3$ | 5.45 | 0.29 | 6.0 | 0.0056 |
| $KNO_3$ | 2.98 | -0.94 | 3.5 | 0.0075 |
| $RbNO_3$ | 2.43 | -0.66 | 4.5 | 0.0083 |
| $CsNO_3$ | 2.72 | -1.27 | 1.4 | 0.0040 |
| | | | | |
| LiOH | 4.47 | 1.24 | 4.0 | 0.023 |
| NaOH | 7.20 | 3.11 | 6.0 | 0.018 |
| KOH | 10.72 | 4.14 | 6.0 | 0.038 |
| CsOH | 4.88 | 6.24 | 1.0 | 0.040 |

concentration limit was decreased. However, such behavior was generally not observed. We did note that for some 2:1 electrolytes, it was possible at selected upper limits of concentration to obtain reasonably good fits and physically realistic parameter values. However, at greater upper limits of concentration, either the deviation in the activity coefficient would increase and/or one or both of the parameter values would become unrealistic.

Lack of perfection is apparent even in the results for the 1:1 salts. Although the parameters show significant regularity, they do not satisfy additivity relations within the fitting error. Letting q stand for the characteristic ion size or hydration number of an electrolyte, relations such as q(KBr) = q(KCl) + q(NaBr) - q(NaCl) should be satisfied. For KBr, we predict values of 7.27 and 1.98 for the ion size and hydration number; the fitted values are 7.14 and 1.53. Although this is not too bad, if we attempt to predict the same parameters for CsBr from those of CsCl, NaBr, and NaCl, we obtain 5.25 and 1.75, compared to fitted values of 4.20 and 1.14. One could go further and test the ability of the models for pure aqueous electrolytes to be extrapolated to predict the properties of mixtures, such as the system $NaCl-KCl-H_2O$, as no additional fit parameters are required by the model. However, because the additivity relations are not precisely satisfied, there seems little point in doing so.

It seems likely that a major problem with the simple hydration theory model is the lack of explicit consideration of ion pairing, which is known to be stronger when the ions present have higher electrical charges. There may also be greater problems in the electrostatic part of the model. A more sophisticated electrostatic model than the DHC may required. Another factor may be kinetic repulsion, which is not addressed in electrostatic models such as Debye-Hückel. This, as well as other factors, might be addressed by including virial coefficient terms (hopefully to only second order) in the phenomenological equations.

## CONCLUSIONS

We have developed the basic equations for applying a thermodynamically consistent hydration correction for both pure aqueous electrolytes and mixtures of aqueous electrolytes. A thermodynamically consistent method of using a compositionally averaged ion size parameter is included in this treatment. A simple hydration theory model using the DHC version of Debye-Hückel theory and ignoring ion pairing works rather well for pure 1:1 aqueous electrolytes and offers encouragement to the development of more sophisticated models that might be useful in geochemical modeling calculations. The simple model does not work well for 2:1 and higher pure aqueous electrolytes.

Future efforts should be directed to develop more advanced hydration theory models than the simple form examined here. Ion pairing should be treated in an explicit manner, even for the 1:1 electrolytes. Other possible changes that might be explored include the use of a more sophisticated electrostatic model and the usage of virial coefficient terms in the phenomenological equations. These models must be tested against not only the ability to fit data, but also the abilities to satisfy additivity relationships and to predict the properties of mixtures of aqueous electrolytes.

## ACKNOWLEDGMENTS

Thanks are offered for encouragement at various stages in the development of this work to Judith Moody, Terry Steinborn, Paul Cloke, and Larry Brush. This work was begun with support from the Salt Repository Project Office and was continued with support from the Waste Isolation Pilot Plant. This work was performed under the auspices of the U.S. Department of Energy by Lawrence Livermore National Laboratory under contract No. W-7405-Eng48. We thank Paul Cloke, Richard Knapp, and three anonymous reviewers for their helpful comments.

Literature Cited

1. Stokes, R.H.; Robinson, R.A. J. Amer. Chem. Soc. 1948, 70, 1870-1878.
2. Nordstrom, D.K.; Plummer, L.N.; Wigley, T.M.L.; Wolery, T.J.; Ball, J.W.; Jenne, E.A.; Bassett, R.L.: Crerar, D.A.; Florence, T.M.; Fritz, B.; Hoffman, M.; Holdren, G.R., Jr.; Lafon, G.M.; Mattigod, S.V.; McDuff, R.E.; Morel, F.; Reddy, M.M.; Sposito, G.; and Thrailkill, J. In Chemical Modeling in Aqueous Systems, Jenne, E.A., ed. Amer. Chem. Soc. Symp. Series 1979, vol 79; pp. 857-892.
3. Burns, D.T. Electrochim. Acta 1964, 9, 1545-1547.
4. Helgeson, H.C.; Kirkham,D.H.,; Flowers, G.C. Amer. J. Sci. 1981, 281, 1249-1516.
5. Friedman, H. L. Ionic Solution Theory 1962; Intersicience Publishing, New York.
6. Friedman, H.L. Ann. Rev. Phys. Chem. 1981, 32, 179-204.
7. Dickson, A.G.; Friedman, H.L.; Millero, F.J. Appl. Geochem. 1988, 3, 27-35.
8. Pitzer, K.S. J. Phys. Chem. 1973, 77, 268-277.
9. Pitzer, K.S. In Activity Coefficients in Electrolyte Solutions 1979; Pytkowicz, R.M., ed., CRC Press, New York; pp. 157-208.
10. Harvie, C.E.; Moller, N.; Weare, J.H. Geochim. Cosmochim. Acta 1984, 48, 723-758.
11. Robinson, R.A.; Stokes, R.H. Electrolyte Solutions 1965; second edition, Butterworths, London.
12. Stokes, R.H.; Robinson, R.A. J. Soln. Chem. 1973, 2, 173-184.
13. Nesbitt, H.W. J. Soln. Chem. 1982, 11, 415-422.
14. Pitzer, K.S. Acc. Chem. Res. 1977, 10, 371-377.
15. Pitzer, K.S. J. Soln. Chem. 1975, 4, 249-265.
16. Helgeson, H.C.; Kirkham, D.H. Am. J. Sci. 1974, 274, 1199-1261.
17. Smith, R.M.; Martell, A.E. Critical Stability Constants, Volume 4, Inorganic Complexes 1976; Plenum Press, New York.

RECEIVED July 20, 1989

# Chapter 3

# Ion-Association Models and Mean Activity Coefficients of Various Salts

### David L. Parkhurst

**U.S. Geological Survey, Denver Federal Center, Mail Stop 418, Lakewood, CO 80225**

Calculations using the aqueous model from WATEQ and an aqueous model modified from WATEQ were compared to experimental mean activity coefficients for various salts to determine the range of applicability and the sources of errors in the models. An ion-association aqueous model was derived by least-squares fitting of ion-association stability constants and individual-ion, activity-coefficient parameters to experimental mean activity coefficients for various salts at 25°C. Salts of the following cations and anions were considered: $Al^{+3}$, $Ba^{+2}$, $Ca^{+2}$, $Cd^{+2}$, $Co^{+2}$, $Cs^+$, $Cu^{+2}$, $Fe^{+2}$, $H^+$, $K^+$, $Li^+$, $Mg^{+2}$, $Mn^{+2}$, $Na^+$, $Ni^{+2}$, $Pb^{+2}$, $Sr^{+2}$, $Zn^{+2}$, $Cl^-$, $ClO_4^-$, $F^-$, $OH^-$, and $SO_4^{-2}$. The stability constants of the the derived model and the WATEQ model were in agreement for most two-ion complexes but were not in agreement for most complexes containing three or more ions. The largest discrepancies in stability constants were for complexes of $Cu^{+2}$, $Mn^{+2}$, $Ni^{+2}$, and $Zn^{+2}$ with $Cl^-$. The derived-model calculations matched the experimental data for all salts to a concentration of about 2 molal, but the parameters of the model could not be defined uniquely by the fitting process. Alternative choices for the complexes included in the model and for the individual-ion, activity-coefficient parameters could fit the experimental data equally well.

Ion-association aqueous models have been used in geochemical modeling for many years, beginning with the seawater model of Garrels and Thompson ([1]). Several computer programs have been developed based on ion-association aqueous models to perform saturation-index calculations, including: SOLMNEQ ([2]), WATEQ ([3]), WATEQF ([4]), MINEQL ([5]), WATEQ2 ([6]), and EQ3NR ([7]). A comparison of these and other models was made by Nordstrom and others ([8]). The ion-association approach has been used for reaction-path calculations in several models, including; PATHI ([9]), PHREEQE ([10]), Reed ([11]), and EQ6 ([12]).

The components of an ion-association aqueous model are: (1) The set of aqueous species (free ions and complexes), (2) stability constants for all complexes, and (3) individual-ion activity coefficients for each aqueous species. The Debye-Huckel theory or one of its extensions is used to estimate individual-ion activity coefficients. For most general-purpose ion-association models, the set of aqueous complexes and their stability constants are selected from diverse sources, including studies of specific aqueous reactions, other literature sources, or from published tabulations (for example, Smith and Martell, ([13])). In most models, stability constants have been chosen independently from the individual-ion, activity-coefficient expressions and without consideration of other aqueous species in the model. Generally, no attempt has been made to insure that the choices of aqueous species, stability constants, and individual-ion activity coefficients are consistent with experimental data for mineral solubilities or mean-activity coefficients.

This chapter not subject to U.S. copyright
Published 1990 American Chemical Society

3. PARKHURST   *Ion-Association Models and Mean Activity Coefficients*

In this report, calculations made using ion-association aqueous models were compared to experimental mean activity coefficients for various salts to determine the range of applicability and the sources of errors in the models. An ion-association aqueous model must reproduce the mean activity coefficients for various salts accurately or it does not describe the thermodynamics of aqueous solutions correctly. Calculations were made using three aqueous models: (1) The aqueous model obtained from WATEQ (3), WATEQF (4), and WATEQ2 (6), referred to as the WATEQ model; (2) the WATEQ model with modifications to the individual-ion, activity-coefficient equations for the free ions, referred to as the amended WATEQ model; and (3) an aqueous model derived from least-squares fitting of mean activity-coefficient data, referred to as the fit model.

METHODS

The results of calculations made using ion-association aqueous models are the molality, activity, and individual-ion activity coefficient for each species in the model. The mean activity coefficient, $\gamma_\pm$, for a salt can be calculated as follows:

$$\gamma_{\pm\, M_{\nu_+}X_{\nu_-}} = \left( \frac{a_{M,F}^{\nu_+} a_{X,F}^{\nu_-}}{m_{M,T}^{\nu_+} m_{X,T}^{\nu_-}} \right)^{\frac{1}{\nu_+ + \nu_-}} \quad (1)$$

where $a$ is the ionic activity; $m$, molality; $M$, a cation; $X$, an anion; $\nu_+$, the number of cations (assuming complete dissociation); $\nu_-$, the number of anions; $F$ indicates free ion; and $T$ indicates total stoichiometric concentration.

Equations For Individual-Ion Activity Coefficients

The mean activity coefficient for a salt can be calculated from experimental data, but the individual-ion activity coefficients used in ion-association aqueous models cannot be determined experimentally. Several formulas were used for individual-ion activity coefficients in the calculations presented in this report. Three formulas were used in the WATEQ model: the extended Debye-Huckel formula including an ion-size parameter, $a_i$ (Equation 2); modified extended Debye-Huckel formula with two fitted parameters, $a_i$ and $b_i$ (Equation 3); and the Davies equation (Equation 4).

$$\log \gamma_i = - A z_i^2 \frac{\sqrt{I}}{1 + B a_i \sqrt{I}} \quad (2)$$

$$\log \gamma_i = - A z_i^2 \frac{\sqrt{I}}{1 + B a_i \sqrt{I}} + b_i I \quad (3)$$

$$\log \gamma_i = - A z_i^2 \left( \frac{\sqrt{I}}{1 + \sqrt{I}} - 0.3 I \right) \quad (4)$$

where $\gamma_i$ is the individual-ion activity coefficient of species $i$, $z_i$ is the charge of the species, $A$ and $B$ are constants equal to 0.5109 and 0.3287 at 25°C, $I$ is ionic strength, $a_i$ is the ion-size parameter, $b_i$ is an ion-specific parameter.

An individual-ion, activity-coefficient formula (Equation 5) derived by Millero and Schreiber (14) from the work of Pitzer (15) was used in the amended WATEQ and fit models:

$$2.303 \log \gamma_i = -0.392 z^2_i \left[ \frac{\sqrt{I}}{1+1.2\sqrt{I}} + \frac{2}{1.2}\ln(1+1.2\sqrt{I}) \right] + a_i I + b_i \left[ 1 - e^{-2\sqrt{I}}(1+2\sqrt{I}-2I) \right] + c_i I^2 \quad (5)$$

where $a_i$, $b_i$, and $c_i$ are ion-specific parameters. It may be convenient to use Equation 3 because WATEQ does not have the option to use Equation 5. The functions defined by Equation 5 for each ion were re-fitted with Equation 3. Parameters for Equation 3 that best reproduce individual-ion activity coefficients used in the amended WATEQ and fit models for each free ion are listed in Table I.

Equation 6, with one adjustable parameter, $b_i$, was used in the fit model. Equation 6 is equivalent to Equation 3 (used in WATEQ) with $a_i$ equal to 3.04:

$$\log \gamma_i = -A z_i^2 \frac{\sqrt{I}}{1+\sqrt{I}} + b_i I \quad (6)$$

Modifications to the WATEQ Models

The aqueous species; stability constants; and individual-ion, activity-coefficient parameters from WATEQ and WATEQ2 were used as published for one set of calculations. Perchlorate and $Co^{+2}$ were not included in the WATEQ aqueous models, but were added for these calculations. The individual-ion activity coefficient for $ClO_4^-$ and $Co^{+2}$ were calculated using Equation 2 and ion-size parameters ($a_i$) of 3.5 and 6.0, respectively (16). All calculations were made using a version of PHREEQE (10) modified to calculate mean activity coefficients for salts.

Technique For Fitting Ion-Association Parameters

Mean activity-coefficient data were obtained for 51 binary salts from the GAMPHI package developed by the National Bureau of Standards (17). All data were calculated at 25°C. The primary reference in GAMPHI was used for all of the salts. The cations considered in the study included the following species: $Al^{+3}$, $Ba^{+2}$, $Ca^{+2}$, $Cd^{+2}$, $Co^{+2}$, $Cs^+$, $Cu^{+2}$, $Fe^{+2}$, $H^+$, $K^+$, $Li^+$, $Mg^{+2}$, $Mn^{+2}$, $Na^+$, $Ni^{+2}$, $Pb^{+2}$, $Sr^{+2}$, and $Zn^{+2}$. The anions considered were: $Cl^-$, $ClO_4^-$, $F^-$, $OH^-$, and $SO_4^{-2}$. The GAMPHI package will estimate values of the mean activity coefficient for any concentration of a salt within the range of the data base. Generally, 17 values of the mean activity coefficient were used in the fitting process, corresponding to the following concentrations (in molality): 0.001, 0.003, 0.005, 0.007, 0.009, 0.02, 0.04, 0.06, 0.08, 0.1, 0.3, 0.5, 0.7, 0.9, 1.2, 1.6, and 2.0. All salts used in the fitting process are listed in Table II.

The individual-ion activity coefficients for the free ions were based on the MacInnis (18) convention, which defines the activity of $Cl^-$ to be equal to the mean activity coefficient of KCl in a KCl solution of equivalent ionic strength. From this starting point, individual-ion activity coefficients for the free ions of other elements were derived from single-salt solutions. The method of Millero and Schreiber (14) was used to calculate the individual-ion, activity-coefficient parameters (Equation 5) from the parameters given by Pitzer (19). However, several different sets of salts could be used to derive the individual-ion activity coefficient for a free ion. For example, the individual-ion activity coefficient for $OH^-$ could be calculated using mean activity-coefficient data for KOH and KCl, or from CsOH, CsCl, and KCl, and so forth.

All possible sets of salts that could be used to calculate the individual-ion, activity-coefficient parameters (Equation 5) were considered for each ion. The parameters that produced the largest individual-ion activity coefficients for an ion were used in the amended WATEQ and fit models. This choice of individual-ion activity coefficients insured that complexing could account, at least in part, for the differences between the calculated and experimental values of the mean activity coefficients, because the effect of adding a complex to the aqueous model is to decrease the calculated mean activity coefficient. The salts used to calculate the individual-ion, activity-coefficient parameters of the free ions are listed in Table I.

The salts listed in Table I were assumed to be completely dissociated in solution; all other salts considered in this report were assumed to be partially associated in solution. If the salts listed in

Table I. Salts used to fit ion-activity-coefficient parameters
and parameters for Equation 3 for each free ion

| Ion | Salts used to fit ion-activity-coefficient parameters | Parameters for equation 3 | |
|---|---|---|---|
| | | $a$ | $b$ |
| $Al^{+3}$ | $AlCl_3$, KCl | 6.65 | 0.19 |
| $Ba^{+2}$ | $BaCl_2$, KCl | 4.55 | 0.09 |
| $Ca^{+2}$ | $Ca(ClO_4)_2$, $LiClO_4$, LiCl, KCl | 4.86 | 0.15 |
| $Cd^{+2}$ | $Cd(ClO_4)_2$, $LiClO_4$, LiCl, KCl | 5.80 | 0.10 |
| $Co^{+2}$ | $Co(ClO_4)_2$, $LiClO_4$, LiCl, KCl | 6.17 | 0.22 |
| $Cs^+$ | CsCl, KCl | 1.81 | 0.01 |
| $Cu^{+2}$ | $Cu(ClO_4)_2$, $LiClO_4$, LiCl, KCl | 5.24 | 0.21 |
| $Fe^{+2}$ | $FeCl_2$, KCl | 5.08 | 0.16 |
| $H^{+2}$ | HCl, KCl | 4.78 | 0.24 |
| $K^{+2}$ | KCl | 3.71 | 0.01 |
| $Li^{+2}$ | LiCl, KCl | 4.76 | 0.20 |
| $Mg^{+2}$ | $Mg(ClO_4)_2$, $LiClO_4$, LiCl, KCl | 5.46 | 0.22 |
| $Mn^{+2}$ | $Mn(ClO_4)_2$, $LiClO_4$, LiCl, KCl | 7.04 | 0.22 |
| $Na^+$ | NaCl, KCl | 4.32 | 0.06 |
| $Ni^{+2}$ | $Ni(ClO_4)_2$, $LiClO_4$, LiCl, KCl | 5.51 | 0.22 |
| $Pb^{+2}$ | $Pb(ClO_4)_2$, $LiClO_4$, LiCl, KCl | 4.80 | 0.01 |
| $Sr^{+2}$ | $SrCl_2$, KCl | 5.48 | 0.11 |
| $Zn^{+2}$ | $Zn(ClO_4)_2$, $LiClO_4$, LiCl, KCl | 4.87 | 0.24 |
| $Cl^-$ | KCl | 3.71 | 0.01 |
| $ClO_4^-$ | $LiClO_4$, LiCl, KCl | 5.30 | 0.08 |
| $F^-$ | KF | 3.46 | 0.08 |
| $OH^-$ | CsOH, CsCl, KCl | 10.65 | 0.21 |
| $SO_4^{-2}$ | $Cs_2SO_4$, CsCl, KCl | 5.31 | −0.07 |

Table II. Association reaction, stability constant, ion-activity-coefficient equation, and ion-activity-coefficient parameters for each aqueous species for the WATEQ, amended WATEQ, and fit aqueous models

[Amended WATEQ model: WATEQ, indicates parameters from WATEQ model were used; Fit, indicates parameters from fit model were used in the amended model. Eqn: Number of ion-activity-coefficient equation in text. $a$, $b$, and $c$: Ion-activity-coefficient parameters for the specified ion-activity-coefficient equation. Source: Salt used to fit parameters for the species; WATEQ indicates parameters were adopted from the WATEQ model. Ionic strength: All salts used to fit the ion-activity-coefficient parameters for this species had data up to this ionic strength, in molality]

| Reaction | WATEQ Model | | | | Amended WATEQ model | Fit Model | | | | | Source | Ionic strength |
|---|---|---|---|---|---|---|---|---|---|---|---|---|
| | Log K | Eqn | $a$ | $b$ | | Log K | Eqn | $a$ | $b$ | $c$ | | |
| $Al^{+3} = Al^{+3}$ | 0. | 2 | 9.00 | - | Fit | 0. | 5 | 1.1086 | 2.6040 | 0.0050 | $AlCl_3$ | 4.8 |
| $Ba^{+2} = Ba^{+2}$ | 0. | 2 | 5.00 | - | Fit | 0. | 5 | 0.5713 | 0.6258 | −0.0246 | $BaCl_2$ | 4.8 |
| $Ba^{+2} + ClO_4^- = BaClO_4^+$ | - | - | - | - | - | −0.16 | 6 | - | 0.19 | - | $Ba(ClO_4)_2$ | 3.5 |
| $Ba^{+2} + H_2O = BaOH^+ + H^+$ | −13.36 | 4 | - | - | WATEQ | −13.14 | 6 | - | 0. | - | $Ba(OH)_2$ | 0.3 |
| $Ca^{+2} = Ca^{+2}$ | 0. | 3 | 5.00 | 0.165 | Fit | 0. | 5 | 0.6264 | 0.7744 | 0.0145 | $Ca(ClO_4)_2$ | 3.5 |
| $Ca^{+2} + Cl^- = CaCl^+$ | - | - | - | - | - | −2.21 | 6 | - | −0.02 | - | $CaCl_2$ | 3.5 |
| $Ca^{+2} + H_2O = CaOH^+ + H^+$ | −12.60 | 4 | - | - | WATEQ | −12.60 | 4 | - | - | - | WATEQ | - |
| $Ca^{+2} + SO_4^{-2} = CaSO_4$ | 2.31 | 4 | - | - | WATEQ | 2.25 | 6 | - | 0. | - | $CaSO_4$ [a/] | 2.0 [b/] |
| $Ca^{+2} + 2 SO_4^{-2} = Ca(SO_4)_2^{-2}$ | - | - | - | - | - | 2.62 | 6 | - | 0. | - | $CaSO_4$ [a/] | 2.0 [b/] |
| $Cd^{+2} = Cd^{+2}$ | 0. | 2 | 5.00 | - | Fit | 0. | 5 | 0.4186 | 1.0481 | 0.0453 | $Cd(ClO_4)_2$ | 3.5 |
| $2 Cd^{+2} + H_2O = Cd_2OH^{+3} + H^+$ | −9.39 | 4 | - | - | WATEQ | −9.39 | 4 | - | - | - | WATEQ | - |
| $Cd^{+2} + Cl^- = CdCl^+$ | 1.98 | 4 | - | - | WATEQ | 1.82 | 6 | - | 0. | - | $CdCl_2$ | 3.5 |
| $Cd^{+2} + 2 Cl^- = CdCl_2$ | 2.60 | 4 | - | - | WATEQ | 2.65 | 6 | - | 0. | - | $CdCl_2$ | 3.5 |
| $Cd^{+2} + 3 Cl^- = CdCl_3^-$ | 2.40 | 4 | - | - | WATEQ | - | - | - | - | - | - | - |
| $Cd^{+2} + H_2O = CdOH^+ + H^+$ | −10.08 | 4 | - | - | WATEQ | −10.08 | 4 | - | - | - | WATEQ | - |
| $Cd^{+2} + 2 H_2O = Cd(OH)_2 + 2 H^+$ | −20.35 | 4 | - | - | WATEQ | −20.35 | 4 | - | - | - | WATEQ | - |
| $Cd^{+2} + 3 H_2O = Cd(OH)_3^- + 3 H^+$ | −33.30 | 4 | - | - | WATEQ | −33.30 | 4 | - | - | - | WATEQ | - |
| $Cd^{+2} + 4 H_2O = Cd(OH)_4^{-2} + 4 H^+$ | −47.35 | 4 | - | - | WATEQ | −47.35 | 4 | - | - | - | WATEQ | - |
| $Cd^{+2} + H_2O + Cl^- = CdOHCl + H^+$ | −7.40 | 4 | - | - | WATEQ | - | - | - | - | - | - | - |
| $Cd^{+2} + SO_4^{-2} = CdSO_4$ | 2.46 | 4 | - | - | WATEQ | 2.19 | 6 | - | 0. | - | $CdSO_4$ | 3.5 |
| $Cd^{+2} + 2 SO_4^{-2} = Cd(SO_4)_2^{-2}$ | 3.50 | 4 | - | - | WATEQ | 2.94 | 6 | - | 0. | - | $CdSO_4$ | 3.5 |
| $Cl^- = Cl^-$ | 0. | 3 | 3.50 | 0.015 | Fit | 0. | 5 | 0.0967 | 0.1061 | −0.0013 | KCl | 4.8 |
| $ClO_4^- = ClO_4^-$ | 0. | 2 | 3.50 | 0 | Fit | 0. | 5 | 0.2883 | 0.1983 | −0.0096 | $LiClO_4$ | 3.5 |
| $Co^{+2} = Co^{+2}$ | - | - | - | - | - | 0. | 5 | 0.7110 | 1.1178 | 0.0390 | $Co(ClO_4)_2$ | 3.5 |
| $Co^{+2} + Cl^- = CoCl^+$ | - | - | - | - | - | 0.07 | 6 | - | 0.34 | - | $CoCl_2$ | 3.5 |
| $Co^{+2} + SO_4^{-2} = CoSO_4$ | - | - | - | - | - | 2.02 | 6 | - | 0. | - | $CoSO_4$ | 0.4 |
| $Co^{+2} + 2 SO_4^{-2} = Co(SO_4)_2^{-2}$ | - | - | - | - | - | 3.05 | 6 | - | 0. | - | $CoSO_4$ | 0.4 |
| $Cs^+ = Cs^+$ | - | - | - | - | - | 0. | 5 | 0.0233 | −0.0503 | 0.0024 | CsCl | 4.8 |
| $Cu^{+2} = Cu^{+2}$ | 0. | 2 | 6.00 | 0. | Fit | 0. | 5 | 0.7350 | 0.8887 | 0.0268 | $Cu(ClO_4)_2$ | 4.8 |
| $Cu^{+2} + Cl^- = CuCl^+$ | 0.43 | 4 | - | - | WATEQ | 0.28 | 6 | - | 0.22 | - | $CuCl_2$ | 3.5 |
| $Cu^{+2} + 2 Cl^- = CuCl_2$ | 0.16 | 4 | - | - | WATEQ | - | - | - | - | - | - | - |
| $Cu^{+2} + 3 Cl^- = CuCl_3^-$ | −2.29 | 4 | - | - | WATEQ | - | - | - | - | - | - | - |
| $Cu^{+2} + 4 Cl^- = CuCl_4^{-2}$ | −4.59 | 4 | - | - | WATEQ | - | - | - | - | - | - | - |
| $Cu^{+2} + H_2O = CuOH^+ + H^+$ | −8.00 | 4 | - | - | WATEQ | −8.00 | 4 | - | - | - | WATEQ | - |
| $Cu^{+2} + 2 H_2O = Cu(OH)_2 + 2 H^+$ | −13.68 | 4 | - | - | WATEQ | −13.68 | 4 | - | - | - | WATEQ | - |
| $Cu^{+2} + 3 H_2O = Cu(OH)_3^- + 3 H^+$ | −26.90 | 4 | - | - | WATEQ | −26.90 | 4 | - | - | - | WATEQ | - |
| $Cu^{+2} + 4 H_2O = Cu(OH)_4^{-2} + 4 H^+$ | −39.60 | 4 | - | - | WATEQ | −39.60 | 4 | - | - | - | WATEQ | - |
| $2 Cu^{+2} + 2 H_2O = Cu_2(OH)_2^{+2} + 2 H^+$ | −10.36 | 4 | - | - | WATEQ | −10.36 | 4 | - | - | - | WATEQ | - |
| $Cu^{+2} + SO_4^{-2} = CuSO_4$ | 2.31 | 4 | - | - | WATEQ | 2.18 | 6 | - | 0. | - | $CuSO_4$ | 3.5 |
| $Cu^{+2} + 2 SO_4^{-2} = Cu(SO_4)_2^{-2}$ | - | - | - | - | - | 3.13 | 6 | - | 0. | - | $CuSO_4$ | 3.5 |
| $F^- = F^-$ | 0. | 2 | 3.5 | - | Fit | 0. | 5 | 0.2269 | 0.0960 | 0.0041 | KF | 2.0 |
| $Fe^{+2} = Fe^{+2}$ | 0. | 2 | 6.00 | - | Fit | 0. | 5 | 0.7122 | 0.7923 | −0.0073 | $FeCl_2$ | 4.8 |
| $Fe^{+2} + H_2O = FeOH^+ + H^+$ | −9.50 | 4 | - | - | WATEQ | −9.50 | 4 | - | - | - | WATEQ | - |
| $Fe^{+2} + 2 H_2O = Fe(OH)_2 + 2 H^+$ | −20.57 | 4 | - | - | WATEQ | −20.57 | 4 | - | - | - | WATEQ | - |
| $Fe^{+2} + 3 H_2O = Fe(OH)_3^- + 3 H^+$ | −31.00 | 4 | - | - | WATEQ | −31.00 | 4 | - | - | - | WATEQ | - |
| $H^+ = H^+$ | 0. | 2 | 9.00 | - | Fit | 0. | 5 | 0.6133 | 0.1884 | 0.0037 | HCl | 4.8 |

Table II continued

| Reaction | WATEQ Model | | | | Amended WATEQ model | Fit Model | | | | | Source | Ionic strength |
|---|---|---|---|---|---|---|---|---|---|---|---|---|
| | Log K | Eqn | $a$ | $b$ | | Log K | Eqn | $a$ | $b$ | $c$ | | |
| $H_2O = OH^- + H^+$ | −14.00 | 2 | 3.50 | - | Fit | −14.00 | 5 | 0.4235 | 0.5239 | 0.0344 | CsOH | 1.2 |
| $H^+ + ClO_4^- = HClO_4$ | - | - | - | - | - | −0.50 | 6 | - | 0.46 | - | $HClO_4$ | 2.0 |
| $SO_4^{-2} + H^+ = HSO_4^-$ | 1.99 | 4 | - | - | WATEQ | 1.93 | 6 | - | 0.04 | - | $H_2SO_4$ | 4.8 |
| $K^+ = K^+$ | 0. | 3 | 3.50 | 0.015 | Fit | 0. | 5 | 0.0967 | 0.1061 | −0.0013 | KCl | 4.8 |
| $K^+ + H_2O = KOH + H^+$ | - | - | - | - | - | −14.09 | 6 | - | 0.56 | - | KOH | 2.0 |
| $K^+ + SO_4^{-2} = KSO_4^-$ | 0.85 | 4 | - | - | WATEQ | 0.75 | 6 | - | 0. | - | $K_2SO_4$ | 0.692 |
| $2 K^+ + SO_4^{-2} = K_2SO_4$ | - | - | - | - | - | 0.81 | 6 | - | 0. | - | $K_2SO_4$ | 0.692 |
| $Li^+ = Li^+$ | 0. | 2 | 6.00 | - | Fit | 0. | 5 | 0.5009 | 0.2013 | 0.0120 | LiCl | 4.8 |
| $Li^+ + H_2O = LiOH + H^+$ | - | - | - | - | - | −13.56 | 6 | - | 0.40 | - | LiOH | 2.0 |
| $Li^+ + SO_4^{-2} = LiSO_4^-$ | 0.64 | 4 | - | - | WATEQ | 0.78 | 6 | - | 0. | - | $Li_2SO_4$ | 4.8 |
| $2 Li^+ + SO_4^{-2} = Li_2SO_4$ | - | - | - | - | - | −0.97 | 6 | - | 0. | - | $Li_2SO_4$ | 4.8 |
| $Mg^{+2} = Mg^{+2}$ | 0. | 3 | 5.50 | 0.200 | Fit | 0. | 5 | 0.7464 | 0.9424 | 0.0283 | $Mg(ClO_4)_2$ | 3.5 |
| $Mg^{+2} + Cl^- = MgCl^-$ | - | - | - | - | - | −0.58 | 6 | - | 0.29 | - | $MgCl_2$ | 3.5 |
| $Mg^{+2} + H_2O = MgOH^+ + H^+$ | −11.79 | 4 | - | - | WATEQ | −11.79 | 4 | - | - | - | WATEQ | - |
| $Mg^{+2} + SO_4^{-2} = MgSO_4$ | 2.25 | 4 | - | - | WATEQ | 2.18 | 6 | - | 0. | - | $MgSO_4$ | 3.5 |
| $Mg^{+2} + 2 SO_4^{-2} = Mg(SO_4)_2^{-2}$ | - | - | - | - | - | 2.44 | 6 | - | 0. | - | $MgSO_4$ | 3.5 |
| $Mn^{+2} = Mn^{+2}$ | 0. | 2 | 6.00 | - | Fit | 0. | 5 | 0.7175 | 1.2908 | 0.0381 | $Mn(ClO_4)_2$ | 3.5 |
| $Mn^{+2} + Cl^- = MnCl^+$ | 0.61 | 4 | - | - | WATEQ | 0.40 | 6 | - | 0.36 | - | $MnCl_2$ | 3.5 |
| $Mn^{+2} + 2 Cl^- = MnCl_2$ | 0.04 | 4 | - | - | WATEQ | - | - | - | - | - | - | - |
| $Mn^{+2} + 3 Cl^- = MnCl_3^-$ | −0.31 | 4 | - | - | WATEQ | - | - | - | - | - | - | - |
| $Mn^{+2} + H_2O = MnOH^+ + H^+$ | −10.59 | 4 | - | - | WATEQ | −10.59 | 4 | - | - | - | WATEQ | - |
| $Mn^{+2} + 3 H_2O = Mn(OH)_3^- + 3 H^+$ | −34.80 | 4 | - | - | WATEQ | −34.80 | 4 | - | - | - | WATEQ | - |
| $Mn^{+2} + SO_4^{-2} = MnSO_4$ | 2.26 | 4 | - | - | WATEQ | 2.30 | 6 | - | 0. | - | $MnSO_4$ | 3.5 |
| $Mn^{+2} + 2 SO_4^{-2} = Mn(SO_4)_2^{-2}$ | - | - | - | - | - | 2.84 | 6 | - | 0. | - | $MnSO_4$ | 3.5 |
| $Na^+ = Na^+$ | 0. | 3 | 4.00 | 0.075 | Fit | 0. | 5 | 0.2093 | 0.1603 | 0.0051 | NaCl | 4.8 |
| $Na^+ + ClO_4^- = NaClO_4$ | - | - | - | - | - | −0.35 | 6 | - | 0.10 | - | $NaClO_4$ | 2.0 |
| $Na^+ + H_2O = NaOH + H^+$ | - | - | - | - | - | −13.94 | 6 | - | 0.42 | - | NaOH | 2.0 |
| $Na^+ + SO_4^{-2} = NaSO_4^-$ | 0.70 | 4 | - | - | WATEQ | 0.66 | 6 | - | 0. | - | $Na_2SO_4$ | 4.8 |
| $2 Na^+ + SO_4^{-2} = Na_2SO_4$ | - | - | - | - | - | 0.87 | 6 | - | 0. | - | $Na_2SO_4$ | 4.8 |
| $Ni^{+2} = Ni^{+2}$ | 0. | 2 | 6.00 | - | Fit | 0. | 5 | 0.7392 | 0.9646 | 0.0351 | $Ni(ClO_4)_2$ | 3.5 |
| $Ni^{+2} + Cl^- = NiCl^+$ | 0.40 | 4 | - | - | WATEQ | −0.07 | 6 | - | 0.33 | - | $NiCl_2$ | 3.5 |
| $Ni^{+2} + 2 Cl^- = NiCl_2$ | 0.96 | 4 | - | - | WATEQ | - | - | - | - | - | - | - |
| $Ni^{+2} + H_2O = NiOH^+ + H^+$ | −9.86 | 4 | - | - | WATEQ | −9.86 | 4 | - | - | - | WATEQ | - |
| $Ni^{+2} + 2 H_2O = Ni(OH)_2 + 2 H^+$ | −19.00 | 4 | - | - | WATEQ | −19.00 | 4 | - | - | - | WATEQ | - |
| $Ni^{+2} + 3 H_2O = Ni(OH)_3^- + 3 H^+$ | −30.00 | 4 | - | - | WATEQ | −30.00 | 4 | - | - | - | WATEQ | - |
| $Ni^{+2} + SO_4^{-2} = NiSO_4$ | 2.29 | 4 | - | - | WATEQ | 2.10 | 6 | - | 0. | - | $NiSO_4$ | 3.5 |
| $Ni^{+2} + 2 SO_4^{-2} = Ni(SO_4)_2^{-2}$ | 1.02 | 4 | - | - | WATEQ | 2.95 | 6 | - | 0. | - | $NiSO_4$ | 3.5 |
| $Pb^{+2} = Pb^{+2}$ | 0. | 2 | 4.50 | - | Fit | 0. | 5 | 0.3120 | 0.7514 | 0.0109 | $Pb(ClO_4)_2$ | 3.5 |
| $Pb^{+2} + Cl^- = PbCl^+$ | 1.60 | 4 | - | - | WATEQ | 1.54 | 6 | - | 0. | - | $PbCl_2$ | 0.12 |
| $Pb^{+2} + 2 Cl^- = PbCl_2$ | 1.80 | 4 | - | - | WATEQ | 1.91 | 6 | - | 0. | - | $PbCl_2$ | 0.12 |
| $Pb^{+2} + 3 Cl^- = PbCl_3^-$ | 1.70 | 4 | - | - | WATEQ | - | - | - | - | - | - | - |
| $Pb^{+2} + 4 Cl^- = PbCl_4^{-2}$ | 1.38 | 4 | - | - | WATEQ | - | - | - | - | - | - | - |
| $Pb^{+2} + H_2O = PbOH^+ + H^+$ | −7.71 | 4 | - | - | WATEQ | −7.71 | 4 | - | - | - | WATEQ | - |
| $Pb^{+2} + 2 H_2O = PbOH_2 + 2 H^+$ | −17.12 | 4 | - | - | WATEQ | −17.12 | 4 | - | - | - | WATEQ | - |
| $Pb^{+2} + 3 H_2O = Pb(OH)_3^- + 3 H^+$ | −28.06 | 4 | - | - | WATEQ | −28.06 | 4 | - | - | - | WATEQ | - |
| $Pb^{+2} + 4 H_2O = Pb(OH)_4^{-2} + 4 H^+$ | −39.70 | 4 | - | - | WATEQ | −39.70 | 4 | - | - | - | WATEQ | - |
| $2 Pb^{+2} + H_2O = Pb_2OH^{+3} + H^+$ | −6.36 | 4 | - | - | WATEQ | −6.36 | 4 | - | - | - | WATEQ | - |
| $3 Pb^{+2} + 4 H_2O = Pb_3(OH)_4^{+2} + 4 H^+$ | −23.88 | 4 | - | - | WATEQ | −23.88 | 4 | - | - | - | WATEQ | - |
| $SO_4^{-2} = SO_4^{-2}$ | 0. | 3 | 5.00 | −0.040 | Fit | 0. | 5 | 0.1902 | 0.8411 | −0.0105 | $Cs_2SO_4$ | 4.8 |
| $Sr^{+2} = Sr^{+2}$ | 0. | 3 | 5.26 | 0.121 | Fit | 0. | 5 | 0.5686 | 0.8993 | 0. | $SrCl_2$ | 3.5 |
| $Sr^{+2} + ClO_4^- = SrClO_4^+$ | - | - | - | - | - | 0.20 | 6 | - | 0.52 | - | $Sr(ClO_4)_2$ | 3.5 |
| $Zn^{+2} = Zn^{+2}$ | 0. | 2 | 6.00 | - | Fit | 0. | 5 | 0.7728 | 0.8014 | 0.0299 | $Zn(ClO_4)_2$ | 3.5 |
| $Zn^{+2} + Cl^- = ZnCl^+$ | 0.43 | 4 | - | - | WATEQ | 0.12 | 6 | - | 0. | - | $ZnCl_2$ | 3.5 |
| $Zn^{+2} + 2 Cl^- = ZnCl_2$ | 0.45 | 4 | - | - | WATEQ | −0.85 | 6 | - | 0. | - | $ZnCl_2$ | 3.5 |
| $Zn^{+2} + 3 Cl^- = ZnCl_3^-$ | 0.50 | 4 | - | - | WATEQ | - | - | - | - | - | - | - |
| $Zn^{+2} + 4 Cl^- = ZnCl_4^{-2}$ | 0.20 | 4 | - | - | WATEQ | - | - | - | - | - | - | - |
| $Zn^{+2} + H_2O = ZnOH^+ + H^+$ | −8.96 | 4 | - | - | WATEQ | −8.96 | 4 | - | - | - | WATEQ | - |

*Continued on next page*

Table II continued

| Reaction | WATEQ Model | | | | Amended WATEQ model | Fit Model | | | | | Source | Ionic strength |
|---|---|---|---|---|---|---|---|---|---|---|---|---|
| | Log K | Eqn | $a$ | $b$ | | Log K | Eqn | $a$ | $b$ | $c$ | | |
| $Zn^{+2} + 2\,H_2O = Zn(OH)_2 + 2\,H^+$ | −16.90 | 4 | - | - | WATEQ | −16.90 | 4 | - | - | - | WATEQ | - |
| $Zn^{+2} + 3\,H_2O = Zn(OH)_3^- + 3\,H^+$ | −28.40 | 4 | - | - | WATEQ | −28.40 | 4 | - | - | - | WATEQ | - |
| $Zn^{+2} + 4\,H_2O = Zn(OH)_4^{-2} + 4\,H^+$ | −41.20 | 4 | - | - | WATEQ | −41.20 | 4 | - | - | - | WATEQ | - |
| $Zn^{+2} + Cl^- + H_2O = ZnOHCl + H^+$ | −7.48 | 4 | - | - | WATEQ | - | - | - | - | - | - | - |
| $Zn^{+2} + SO_4^{-2} = ZnSO_4$ | 2.37 | 4 | - | - | WATEQ | 2.02 | 6 | - | 0. | - | $ZnSO_4$ | 3.5 |
| $Zn^{+2} + 2\,SO_4^{-2} = Zn(SO_4)_2^{-2}$ | 3.28 | 4 | - | - | WATEQ | 2.86 | 6 | - | 0. | - | $ZnSO_4$ | 3.5 |

$a/$ Data derived from gypsum solubility in sodium chloride, sodium sulfate, and calcium chloride solutions were used in addition to calcium sulfate solutions.

$b/$ All solutions except pure calcium sulfate solutions had data up to 2.0 molal ionic strength.

Table I are not completely dissociated in solution, then the individual-ion activity coefficients used in this report are too small. Further, the stability constants for the other salts also are too small. If the salts that are assumed to be associated in solution in this report are, in fact, completely dissociated in solution, then the ion-association aqueous model is inadequate. In this case, the ion-association aqueous model can not reproduce accurately the mean activity coefficients of all the salts and simultaneously maintain a correct physical description of the solutions.

Mean activity-coefficient data at concentrations sufficient to estimate stability constants through the fitting process are available only for a few $OH^-$ salts. Hence, most $OH^-$ complexes in the fit model were obtained from WATEQ, with the exception of the $OH^-$ complexes of $Ba^{+2}$, $K^+$, $Li^+$, and $Na^+$, which were fitted to the experimental data.

For the fit model, aqueous complexes were added as needed in order to resolve the discrepancies between the calculated and experimental mean activity coefficients. The stability constants for the complexes and the individual-ion, activity-coefficient parameters for the complexes were estimated through least-squares fitting. The program that estimated the parameters used the modified version of PHREEQE ([10]) as a function subroutine that calculated the mean activity coefficient. The stability constants and individual-ion, activity-coefficient parameters of a specified set of complexes were adjusted until a least-squares fit was obtained between the calculated and experimental mean activity coefficients for a series of salt-solution compositions.

In general, if a single complex was hypothesized to fit a series of data, the stability constant and the $b_i$ parameter (Equation 6) were adjusted to achieve a best fit. If two or more complexes were hypothesized, then only the stability constants were adjusted to achieve a best fit. The $b_i$ parameters for these complexes were assumed to be zero in order to limit the number of adjustable parameters and not to overfit the data.

The solubility of $CaSO_4$ is limited, therefore it was necessary to use data from sources other than pure salt solutions to determine parameters for $Ca^{+2}$ complexes with $SO_4^{-2}$. Mean activity coefficients for $CaSO_4$ derived from gypsum-solubility experiments in NaCl, $Na_2SO_4$, and $CaCl_2$ solutions (Rogers ([20]), using the data of Briggs ([21]) and Lilley and Briggs ([22])) were used to augment the data for pure $CaSO_4$ solutions.

RESULTS AND DISCUSSION

The set of complexes; the individual-ion, activity-coefficient parameters; and the stability constants are listed in Table II, for each of the three models: (1) The WATEQ model, (2) the amended WATEQ model, and (3) the fit model. The WATEQ and amended WATEQ models had the same set of complexes and stability constants. The two models differed only in the individual-ion activity coefficients of the free ions. The fit model contained different complexes or stability constants or both compared to the WATEQ and amended WATEQ models (except for the $OH^-$ complexes that were obtained from the WATEQ model). The activity coefficients for the free ions were the same in the fit and amended WATEQ models, but the activity coefficients of the complexes differed.

The results of the calculations will be discussed in terms of: (1) Goodness of fit for the three models between the calculated and experimental mean activity coefficients; and (2) differences between the set of complexes and the values of stability constants in the fit model and in the WATEQ and the amended WATEQ models.

Chloride Salts

The individual-ion activity coefficients for $Cl^-$, $Ca^{+2}$, $K^+$, $Mg^{+2}$, $Na^+$, and $Sr^{+2}$ were fitted from the mean activity coefficients of $Cl^-$ salts using Equation 3. Thus, the WATEQ model accurately reproduced the mean activity coefficients of $CaCl_2$, KCl, $MgCl_2$, NaCl, and $SrCl_2$. Equation 2 was used for the individual-ion activity coefficients for other free cations; the calculated mean activity coefficients diverged from the experimental data at concentrations greater than 0.1 molal. The individual-ion activity coefficient for $H^+$ is one of the ions for which Equation 2 was used, and large discrepancies were observed in WATEQ calculations of the mean activity coefficients in HCl solutions. In the amended WATEQ model, the individual-ion activity coefficients for $Ca^{+2}$, $Cd^{+2}$, $Co^{+2}$, $Cu^{+2}$, $Mg^{+2}$, $Mn^{+2}$, $Ni^{+2}$, $Pb^{+2}$, and $Zn^{+2}$ were derived from solutions of their respective

$ClO_4^-$ salt. For these ions, the amended WATEQ model fit the experimental data for $ClO_4^-$ solutions, but was not reliable in $Cl^-$ solutions. For $Al^{+3}$, $Ba^{+2}$, $K^+$, $Cs^+$, $Fe^{+2}$, $H^+$, $Li^+$, $Na^+$, and $Sr^{+2}$, the individual-ion activity coefficients were derived from the respective $Cl^-$ salt. The WATEQ and the amended WATEQ models failed to reproduce the mean activity coefficients for the $Cl^-$ salts of Cu, Mn, Ni, and Zn. The fit model reproduced the mean activity coefficients for all of the $Cl^-$ salts considered in this report because of the inclusion of additional ion-association complexes.

The WATEQ and amended WATEQ models contained no $Cl^-$ complexes for $Ca^{+2}$, $Mg^{+2}$, and $Co^{+2}$; two $Cl^-$ complexes for $Ni^{+2}$; three $Cl^-$ complexes for $Cd^{+2}$ and $Mn^{+2}$; and four $Cl^-$ complexes for $Cu^{+2}$, $Pb^{+2}$, and $Zn^{+2}$. The fit model contained a single $Cl^-$ complex for $Ca^{+2}$, $Co^{+2}$, $Cu^{+2}$, $Mg^{+2}$, $Mn^{+2}$, and $Ni^{+2}$. Two $Cl^-$ complexes were used to fit $Cd^{+2}$, $Pb^{+2}$, and $Zn^{+2}$.

The stability constants for $CdCl^+$, $CdCl_2^\circ$, $PbCl^+$, and $PbCl_2^\circ$ determined by the fitting process are similar to the constants in the WATEQ model.

The stability constants for $CuCl^+$, $MnCl^+$, $NiCl^+$, and $ZnCl^+$ are smaller in the fit model by 0.15 to 0.5 log unit. The stability constant for the $ZnCl_2$ complex is much smaller in the fit model and other multichloride complexes of $Cu^{+2}$, $Mn^{+2}$, $Ni^{+2}$, and $Zn^{+2}$ were not needed for the fit model. Spectroscopic studies have determined the existence of complexes of one or more of the types $MeCl_2$, $MeCl_3^-$, and (or) $MeCl_4^{-2}$ (where Me indicates a metal ion). The fit model was derived from data from stoichiometric solutions with a fixed ratio of metal to $Cl^-$ of 1:2, and it may be that the multichloride complexes are not important in these solutions. Further work is needed to insure that the set of complexes hypothesized for an ion-association aqueous model is consistent with the experimental evidence demonstrating the types and concentrations of complexes in solutions.

Calculations for $MnCl_2$ solutions are shown in Figure 1. The calculated mean activity coefficients were much smaller than the experimental values for the WATEQ and amended WATEQ models. These models used different individual-ion activity coefficients for the free ions yet the calculated results were similar, which indicates that the discrepancy was caused by the stability constants of the $Cl^-$ complexes. The calculations indicate that the WATEQ and amended WATEQ models do not reproduce the mean activity coefficients for $CuCl_2$, $MnCl_2$, $NiCl_2$, and $ZnCl_2$ at concentrations greater than 0.1 molal.

Fluoride Salts

Potassium fluoride was the only $F^-$ salt considered in this report. The WATEQ model uses Equation 2 for the individual-ion activity coefficient of $F^-$, which causes deviations from the experimental data at concentrations greater than 0.1 molal. The amended WATEQ and fit models used the parameters calculated from the parameters of Pitzer (19) for KF solutions, and both models reproduce the experimental data.

Perchlorate Salts

Perchlorate was added to the WATEQ model for this study. Because Equation 2 was used for the individual-ion activity coefficient for $ClO_4^-$ in the WATEQ model, the calculated mean activity coefficients for most $ClO_4^-$ salts diverged from the experimental values at concentrations greater than 0.1 molal.

In the amended WATEQ model, $ClO_4^-$ salts were used to derive the individual-ion activity coefficients for most of the divalent cations. Thus, the amended model reproduced the mean activity coefficients for these salts. The individual-ion activity coefficients for $Ba^{+2}$, $H^+$, $Na^+$, and $Sr^{+2}$, were derived from the $Cl^-$ salts, and the amended WATEQ model does not reproduce the mean activity coefficients for the $ClO_4^-$ solutions of these ions.

The fit model reproduces the mean activity coefficients for all of the $ClO_4^-$ salts considered. The fit model included $ClO_4^-$ complexes for the following ions: $Ba^{+2}$, $H^+$, $Na^+$, and $Sr^{+2}$. Stability constants for these complexes were -0.16, -0.5, -0.35, and 0.2, respectively. However, it is uncertain whether these $ClO_4^-$ complexes have any physical reality. No $ClO_4^-$ complexes were included in the WATEQ or amended WATEQ models.

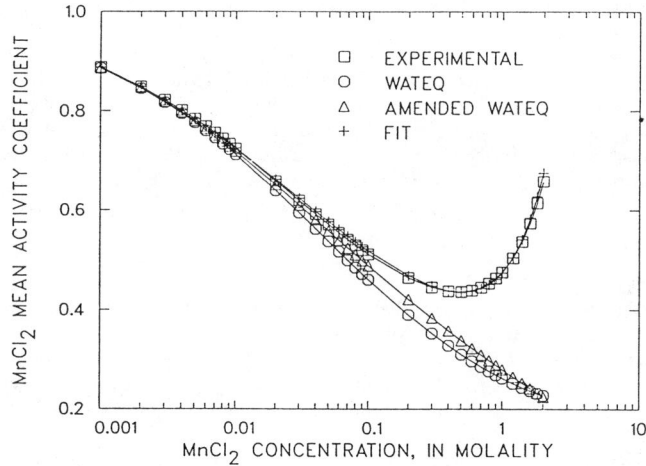

Figure 1. Comparison of experimental and calculated values of mean activity coefficients in MnCl$_2$ solutions.

## Hydroxide Salts

The individual-ion activity coefficient for $OH^-$ in the WATEQ model was not derived from mean activity-coefficient data. Instead, the coefficient was calculated by Equation 2, which is limited to solutions with low ionic strengths. The amended WATEQ model used mean activity-coefficient data from CsOH solutions to estimate individual-ion activity coefficients for $OH^-$. Neither the WATEQ model nor the amended WATEQ model were able to reproduce the mean activity coefficients of $OH^-$ solutions for the cations $Ba^{+2}$, $K^+$, $Li^+$, and $Na^+$. The amended WATEQ model, but not the WATEQ model, reproduced the experimental data for CsOH. Even with a different formulation for the individual-ion activity coefficient of $OH^-$, the amended WATEQ model calculated mean activity coefficients that were discrepant with the experimental data. The WATEQ model actually produced more accurate values than the amended WATEQ model for solutions of $OH^-$ salts except KOH, for which both models produced inaccurate values. The fit model, by including additional $OH^-$ complexes, reproduced the experimental data for all $OH^-$ salts considered.

It was possible to estimate parameters for $OH^-$ complexes of $Ba^{+2}$, $K^+$, $Li^+$, and $Na^+$ in the fitting process. Of these complexes, the WATEQ and amended WATEQ models included a $OH^-$ complex for $Ba^{+2}$, and the value for the stability constant was similar to that calculated by the fitting process. In the fit model, the $b_i$ (Equation 6) parameters for the $OH^-$ complexes of $K^+$, $Li^+$, and $Na^+$ were 0.56, 0.4, and 0.42 respectively. The physical significance of these large $b_i$ values is not clear. When the association reactions were written in terms of $OH^-$, the stability constants were -0.09, 0.44, and 0.06, respectively. Calculations indicate that, in a 1.0 molal KOH solution, the concentration of the $KOH°$ complex is 0.14 molal. Additional experimental evidence from titrations or spectroscopy is needed to determine whether this ion pair actually exists and whether this concentration is reasonable. If there is no evidence that the $OH^-$ salts are partially associated in solution, the large stability constants may indicate that specific-ion interactions are involved that can not be accounted for by the ion-association approach.

## Sulfate Salts

Most of the $SO_4^{-2}$ salts demonstrate substantial complexation in solution. The WATEQ model reproduced the experimental mean activity coefficients relatively well for all of the $SO_4^{-2}$ salts considered except for $Na_2SO_4$. Results for the amended WATEQ model were similar to those for the WATEQ model for the $SO_4^{-2}$ salts. Results for the amended WATEQ model were notably different only for $Li_2SO_4$. The fit model reproduced the mean activity coefficients for all of the salts considered.

The WATEQ and amended WATEQ models included single-sulfate complexes for $Ca^{+2}$, $Cu^{+2}$, $H^+$, $K^+$, $Li^+$, $Mg^{+2}$, $Mn^{+2}$, and $Na^+$. Two $SO_4^{-2}$ complexes were included for $Cd^{+2}$, $Ni^{+2}$, and $Zn^{+2}$. The fit model contained two complexes for each of the following cations: $Ca^{+2}$, $Cd^{+2}$, $Co^{+2}$, $Cu^{+2}$, $K^+$, $Li^+$, $Mg^{+2}$, $Mn^{+2}$, $Na^+$, $Ni^{+2}$, and $Zn^{+2}$. A single complex, $HSO_4^-$, was included in fitting $H_2SO_4$ data.

The stability constants for the single-sulfate complexes generally agreed within 0.3 log unit between the fit model and the WATEQ and amended WATEQ models. The agreement was less among the two-sulfate complexes. These complexes are important only at larger concentrations and may be more sensitive to differences in individual-ion activity coefficients and the range of concentrations used in the fitting process than are single-sulfate complexes.

Calculations for $Na_2SO_4$ solutions are shown in Figure 2. The WATEQ and amended WATEQ models included only the $NaSO_4^-$ ion pair and the calculated mean activity coefficients are too large. The fit model reproduces the experimental data with the addition of a second complex, $Na_2SO_4°$. The log of the stability constant for $NaSO_4^-$ estimated by the fitting process is 0.66, which is almost identical to the stability constant in the WATEQ model, 0.70.

The three models have been compared to calculations of the solubility of gypsum in 0.5 molal NaCl with varying concentrations of $Na_2SO_4$ (23) (Figure 3). Results of the WATEQ and amended WATEQ models were not similar to those of the Harvie and Weare model. However, results of the fit model were similar to those of the Harvie and Weare model. The fit model and the Harvie and Weare model used gypsum-solubility measurements to fit model parameters.

3. PARKHURST  *Ion-Association Models and Mean Activity Coefficients*  41

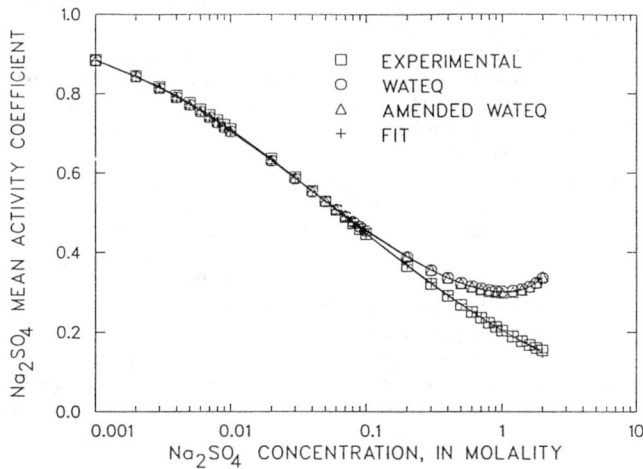

Figure 2. Comparison of experimental and calculated values of mean activity coefficients in $Na_2SO_4$ solutions.

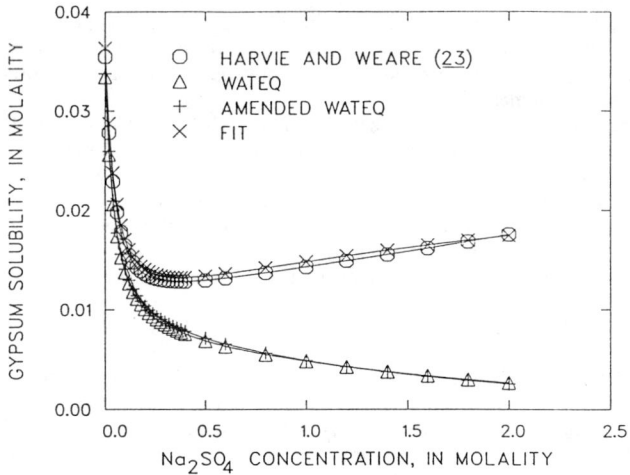

Figure 3. Comparision of gypsum solubilities calculated by the Harvie and Weare model (23) and ion-association aqueous models in 0.5-molal NaCl solutions with varying concentrations of $Na_2SO_4$.

## Limitations

Several limitations of the fitting approach used in this report need to be noted. First, the choice of complexes was based solely on modeling, not on any physical evidence that the complexes actually exist in solution. A more thorough approach needs to incorporate spectroscopic and titrimetric evidence for the hypothesized complexes.

Second, the fit model was not unique. Different sets of aqueous species and different individual-ion activity coefficients reproduced the experimental data equally well.

Third, the fit model was dependent on the set of salts used in its derivation. For example, if $ClO_4^-$ salts were not included, the individual-ion activity coefficient parameters for most of the divalent cations would have been different.

Fourth, in ion-association aqueous models, complexation only accounts for the attractive interactions among ions. All repulsive interactions among ions are included in the expressions for the individual-ion activity coefficients. Similarly, the individual-ion activity coefficients account for any hydration effects not specifically accounted for by complexation.

Fifth, the fit model has not been tested in mixed-salt solutions and against mineral solubility data (other than gypsum). It is not known whether an ion-association aqueous model, even with additional complexes, will be capable of accurately modeling these systems.

Last, the necessity for complexes in strong electrolyte solutions (in particular, the $OH^-$ and $ClO_4^-$ solutions) and large $b_i$ values with no apparent physical explanation (the $OH^-$ salts) may reflect fundamental problems with the ion-association aqueous models. More general geochemical models may require a more rigorous formulation and an approach that accounts for specific interactions between aqueous species.

## SUMMARY AND CONCLUSIONS

The calculations of mean activity coefficients for various salts using the WATEQ model indicate that if Equation 2 is used for individual-ion activity coefficients of free ions, the results are reliable only for concentrations of 0.1 molal or less. If equations 3 or 5 are used (the amended WATEQ model), the calculated mean activity coefficients are accurate for the salts used to derive the individual-ion activity coefficients of the free ions, but are not accurate for other salts unless additional complexes are included.

An ion-association aqueous model was derived by: (1) Selecting the set of salts used to calculate the individual-ion activity coefficients of the free ions; (2) hypothesizing an appropriate set of complexes; and (3) fitting the stability constants and individual-ion, activity-coefficient parameters for the complexes using the mean- activity-coefficient data.

Most stability constants from the fit model for two-ion complexes agreed with values used in the WATEQ model within 0.3 log unit. There was considerably less agreement between the fit model and the WATEQ model for the stability constants of complexes with three or more ions. The calculations indicated that stability constants are too large for the $Cl^-$ complexes of $Cu^{+2}$, $Mn^{+2}$, $Ni^{+2}$, and $Zn^{+2}$ in the WATEQ model, regardless of the choice of individual-ion activity coefficients of the free ions.

The derived model reproduced the experimental mean activity coefficients for a fixed set of salts to concentrations of about 2 molal. Thus, it is possible to construct an ion-association model that is consistent with the experimental data. However, the fitting process for the derived model could not determine uniquely all of the parameters of the model. Alternative choices for the complexes included in the model and for the individual-ion, activity-coefficient parameters could fit the experimental data equally well. Further work is needed to incorporate other physical evidence for the existence of aqueous complexes into the fitting process to insure that the ion-association model provides an accurate physical description of aqueous solutions in addition to reproducing experimental mean activity coefficients.

## LITERATURE CITED

1. Garrels, R.M.; Thompson, M.E. Amer. J. Sci. 1962 260, 57-66.
2. Kharaka, Y.K.; Barnes, Ivan NTIS Technical Report, PB214-899 1973, p 82.
3. Truesdell, A.H.; Jones, B.F. J. Res. U.S. Geol. Surv. 1974, 2, 233-248.
4. Plummer, L.N.; Jones, B.F.; Truesdell, A.H. U.S. Geol. Surv. Water-Resour. Invest. 76-13 1976, p 61.
5. Westall, J.C.; Zachary, J.L.; Morel F.M.M. Technical Note 18, Parsons Laboratory, Massachusetts Institute of Technology, 1976, p 91.
6. Ball, J.W.; Jenne, E.A.; Nordstrom, D.K. In Chemical Modeling in Aqueous Systems. Speciation, Sorption, Solubility, and Kinetics Jenne, E.A., Ed.; ACS Symposium Series No. 93: American Chemical Society: Washington, DC, 1979; pp 815-835.
7. Wolery, T.J. UCRL-53414, Lawrence Livermore Laboratory, 1983, p 191.
8. Nordstrom, D.K.; Plummer, L.N.; Wigley, T.M.L.; Wolery, T.J.; Ball, J.W.; Jenne, E.A.; Bassett, R.L.; Crerar, D.A.; Florence, T.M.; Fritz, B.; Hoffman, M.; Holdren, G.R., Jr.; Lafon, G.M.; Mattigod, S.V.; McDuff, R.E.; Morel, F.; Reddy, M.M.; Sposito, G.; Thrailkill, J. In Chemical Modeling in Aqueous Systems. Speciation, Sorption, Solubility, and Kinetics; Jenne, E.A., Ed.; ACS Symposium Series No. 93; American Chemical Society: Washington, DC, 1979; 857-892.
9. Helgeson, H.C.; Brown, T.H.; Nigrini, A.; Jones, T.A. Geochim. Cosmochim. Acta 1970 34, 569-592.
10. Parkhurst, D.L.; Thorstenson, D.C.; Plummer, L.N. U.S. Geol. Surv. Water-Resour. Invest. 80-96 1980, p 193.
11. Reed, M.H. Geochim. Cosmochim. Acta 1982, 46, 513-528.
12. Wolery, T.J. Lawrence Livermore Laboratory, 1984.
13. Smith, R.M.; Martell, A.E. Critical stability constants, v. 4, Inorganic complexes; Plenum Press: New York, 1976; p 257.
14. Millero, F.J.; Schreiber, D.R. Amer. J. Sci. 1982 282, 1508-1540.
15. Pitzer, K.S. J. Phys. Chem. 1973, 77 268-277.
16. Kielland, J. J. Am. Chem. Soc. 1937 59, 1675-1678.
17. Goldberg, R.N.; Manley, J.L.; Nuttall, R.L. Natl. Bur. Stand., Tech. Note 1206 1985, p 27.
18. MacInnes, D.A. J. Am. Chem. Soc. 1919 41, 1086-1092.
19. Pitzer, K.S. In Activity Coefficients in Electrolyte Solutions; Pytkowicz, R.M., Ed.; CRC Press Inc.: Boca Raton, Florida, 1979, 1, pp 157-208.
20. Rogers, P.S.Z. Ph.D. thesis, University of California, Berkeley, 1981.
21. Briggs, C.C. Ph.D. Thesis, University of Sheffield, Sheffield, England, 1978. (Cited in reference 20.)
22. Lilley, T.H.; Briggs, C.C. Proc. Roy. Soc. London 1976 A 349, 355. (Cited in reference 20.)
23. Harvie, C.E.; Weare, J.H. Geochim. Cosmochim. Acta 1980 44, 981-997.

RECEIVED August 31, 1989

# Chapter 4

# Models for Aqueous Electrolyte Mixtures for Systems Extending from Dilute Solutions to Fused Salts

## Roberto T. Pabalan[1] and Kenneth S. Pitzer

**Department of Chemistry and Lawrence Berkeley Laboratory, University of California, Berkeley, CA 94720**

Models based on general equations for the excess Gibbs energy of aqueous electrolyte mixtures provide a thermodynamically consistent basis for evaluating and predicting aqueous electrolyte properties. Upon appropriate differentiation, these equations yield expressions for osmotic and activity coefficients, excess enthalpies, heat capacities, and volumes. Thus, a wide array of experimental data are available from which model parameters and their temperature or pressure dependence can be evaluated. The most commonly used model for systems of moderate concentration is the ion-interaction approach developed by Pitzer (1) and coworkers. For more concentrated electrolyte solutions, including those extending to the fused salt, an alternate model based on a Margules expansion and commonly used for nonelectrolytes was proposed by Pitzer and Simonson (5). These two models are discussed and examples of parameter evaluations are given for some geologically relevant systems to high temperatures and pressures. Applications of the models to calculations of solubility equilibria are also shown for the systems $NaCl-Na_2SO_4-NaOH-H_2O$ and $NaCl-KCl-H_2O$ to temperatures up to 350°C.

Understanding various geochemical processes and industrial problems requires a thorough knowledge of the thermodynamic properties of aqueous electrolyte solutions. Chemical models that accurately describe the excess properties of aqueous electrolytes over wide ranges of temperature, pressure, and concentration, and which allow prediction of these properties for complex mixtures based on parameters evaluated from simple systems are essential to this understanding. Models based on general equations for the excess Gibbs energy of the aqueous solution provide a thermodynamically consistent basis for evaluating and predicting aqueous electrolyte properties. Upon appropriate differentiation, these equations yield other quantities, including osmotic and activity coefficients, excess enthalpies, entropies, heat capacities, and volumes. Thus, a wide array of experimental techniques provide data from which model parameters and their temperature or pressure dependence can be evaluated.

The purpose of this paper is to review two thermodynamic models for calculating aqueous electrolyte properties and give examples of parameter evaluations to high temperatures and pressures as well as applications to solubility calculations. The first model [the ion-interaction model of Pitzer (1) and coworkers] has been discussed extensively elsewhere (1-4) and will be reviewed only briefly here, while more detail will be given for an alternate model using a Margules expansion as proposed by Pitzer and Simonson (5).

[1]Current address: Southwest Research Institute, 6220 Culebra Road, San Antonio, TX 78228-0510

## ION-INTERACTION OR VIRIAL COEFFICIENT APPROACH

For systems ranging from dilute to moderate concentrations, the most commonly used model is based on a virial expansion developed by Pitzer (1). For a system with one solute, MX, having $v_M$ cations of charge $z_M$ and $v_X$ anions of charge $z_X$, the excess Gibbs energy can be written as

$$G^{ex}/n_w RT = f(I) + 2v_M v_X [m^2 B_{MX}(I) + m^3 v_M z_M C_{MX}] \quad (1)$$

where $n_w$ is the number of kilograms of solvent, $f(I)$ is a Debye-Huckel function, m is the molality and I is the ionic strength. $B_{MX}$ and $C_{MX}$ are, respectively, second and third virial coefficients representing short-range interactions between ions taken two and three at a time, and which can be determined from experimental data on single electrolyte solutions. Two additional virial coefficients ($\Phi$ and $\psi$) arise for ternary and more complex systems, but these can be evaluated from data on simple mixtures (2-4,6). The working equations for these solution properties are derived by differentiation of Equation 1, and are given by Pitzer (2).

In the case of systems for which a variety of experimental methods have been used, it is advantageous to develop general equations based on all critically evaluated data. Examples are the equations for the thermodynamic properties of NaCl(aq) to 300°C and 1 kb (7,8), of KCl(aq) to 325°C and 500 bars (9), of the alkali chlorides to 250°C at saturation pressure (10), and of the alkali sulfates to 225°C at saturation pressure (11). For many electrolytes, experimental activity/composition data at high temperatures are limited and parameter evaluations have to rely on other types of measurements. A good example is $Na_2SO_4$(aq) for which a chemical model was developed by Holmes and Mesmer (11) from their isopiestic measurements to 225°C and other literature data. Solubility calculations (12), however, indicated that their model can be improved by including additional experiments at temperatures above 180°C. These improvements were provided by heat capacity measurements (13) from 140 - 300°C at a pressure of 200 bars. Details of the parameter evaluations for $Na_2SO_4$(aq) at temperatures from 25° to 300°C and pressures to 200 bars are described in Pabalan and Pitzer (13), and applications of the derived model to solubility calculations are given below.

Prediction of mineral solubilities in electrolyte mixtures is an important use of chemical models for electrolyte solutions and is a stringent test of equations for activity and osmotic coefficients. Recent studies have shown that the ion-interaction model can be used successfully to predict solubility equilibria to high temperatures (12-14). In the examples given below, solubilities of the four stable sodium sulfate minerals are calculated and compared with experimental values. $Na_2SO_4$(aq) activity and osmotic coefficients, as well as standard state Gibbs energies, were calculated using the ion-interaction model (13). Gibbs energies of the solids were taken from the literature (15,16). In Figure 1, predicted solubilities of sodium sulfate solids in water to 350°C are compared to experimental values tabulated by Linke (17). Although the calculated values are higher than the experimental data below 241°C, the differences are not large, averaging 0.08 m $Na_2SO_4$.

For solubility calculations in ternary and more complex systems, the ion-interaction model requires two additional terms, $\Phi$ and $\psi$ (6). Whereas both these parameters undoubtedly vary with temperature, it was found adequate for solubility calculations to keep $\Phi$ at its 25°C value and to assign simple temperature functions to $\psi$ (12,13). For the ternary systems $Na_2SO_4$-NaCl-$H_2O$ and $Na_2SO_4$-NaOH-$H_2O$, comparisons of calculated and experimental solubilities up to 300°C are given by Pabalan and Pitzer (12,13). More stringent tests of the model are solubility calculations in the quaternary system $Na_2SO_4$-NaCl-NaOH-$H_2O$. Predicted and experimental solubilities at 200° and 300°C are compared in Figure 2. Other examples of solubility calculations using the ion-interaction approach are given in Refs. 4, 12, 14, and 18-26.

## MARGULES EXPANSION MODEL

An alternative to the ion-interaction approach is the Margules expansion model for systems which may range in concentration from dilute solutions to the fused salt. Aqueous solutions miscible over the whole concentration range at moderate temperatures are relatively uncommon, but there are electrolytes of geochemical and industrial interest which become extremely soluble in water at high temperatures and pressures. Although the ion-interaction model can represent electrolyte properties to very high ionic strengths by using additional virial terms (28), this treatment becomes more complex and unsatisfactory

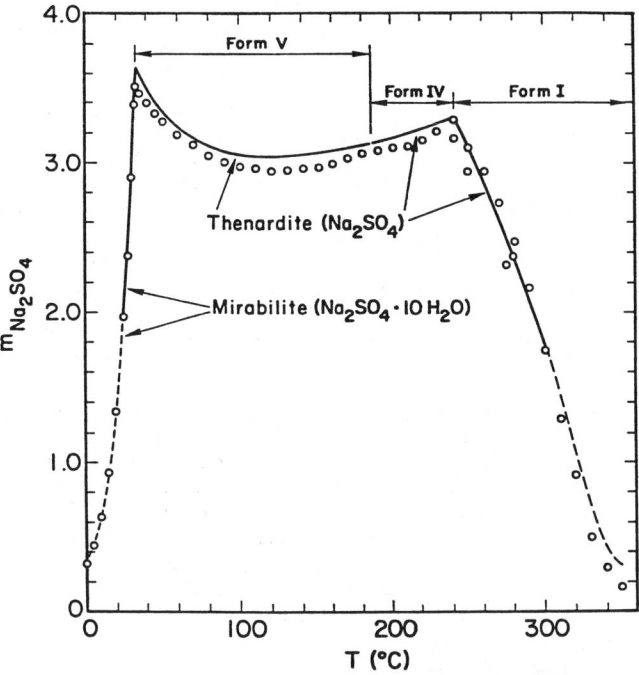

Figure 1. Solubilities of mirabilite ($Na_2SO_4 \cdot 10H_2O$) and thenardite ($Na_2SO_4$) in water as a function of temperature. The symbols are experimental data tabulated by Linke (17) and the curves are predicted values. (Reproduced with permission from Ref. 13. Copyright 1988 Pergamon Press.)

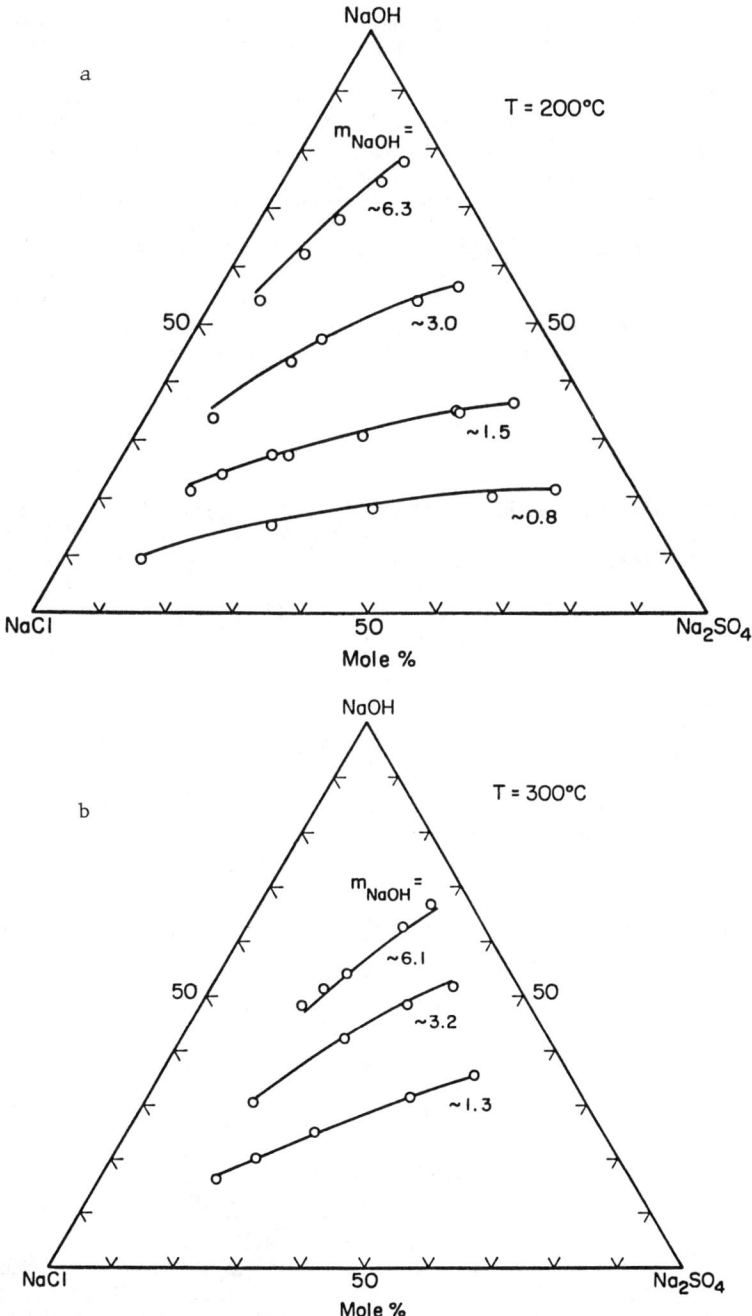

Figure 2. Calculated and experimental solubilities of thenardite ($Na_2SO_4$) at fixed molalities of NaOH in the quaternary system $NaCl-Na_2SO_4-NaOH-H_2O$ at: a) 200°C, and b) 300°C. The symbols are experimental data from Schroeder et al. (27). (Reproduced with permission from Ref. 13. Copyright 1988 Pergamon Press.)

at very high ionic strengths. In addition, the molality for a pure fused salt is infinite. An alternate model is required, therefore, for systems extending to the fused salt and for other systems of very high, but limited, solubility.

The model discussed below is analogous to models used for nonelectrolyte solutions. Various expressions have been used to describe the excess Gibbs energy of nonelectrolytes (29), and the Margules expansion has been used successfully by Adler et al. (30) for several nonelectrolyte systems. The theoretical basis for extending this approach to electrolyte solutions was discussed by Pitzer (31). Briefly, in electrolyte solutions where there is substantial ionic concentration, the long-range interionic forces are effectively screened to short-range by the pattern of alternating charges. Thus all of the interparticle forces are effectively short-range and the system properties can be calculated by methods similar to those for nonelectrolytes. In the dilute range, however, the alternating charge pattern and its accompanying screening effect is lost and the long-range nature of electrostatic forces must be considered. As with the ion-interaction model, this effect is described by an extended Debye-Huckel treatment.

The Gibbs energy per mole of solution can be written in terms of the sum of short-range and electrical effects:

$$G^{ex,x}/\sum_i n_i = g^{ex} = g^s + g^{DH}. \tag{2}$$

A choice must be made for the reference state for the solute: either the pure liquid (possibly supercooled), or the solute at infinite dilution in the solvent. The latter differs from the conventional solute standard state only in the use of the mole fraction scale rather than molality units. The activity coefficient of a symmetrical salt MX is either

$$\ln(\gamma_M^x \gamma_X^x) \to 0 \text{ as } x_1 \to 0 \tag{3}$$

or

$$\ln(\gamma_M^* \gamma_X^*) \to 0 \text{ as } x_1 \to 1, \tag{4}$$

where the asterisk denotes the infinitely dilute reference state on a mole fraction basis, and $x_1$ is the mole fraction of the solvent on an ionized basis. For example,

$$x_1 = n_1/(n_1 + 2n_2) \tag{5}$$

with $n_1$ and $n_2$ the numbers of moles of solvent and solute, respectively. In addition, the mole fractions are defined as:

$$x_M = x_X = n_2/(n_1 + 2n_2) = (1-x_1)/2, \tag{6}$$

$$x_2 = x_M + x_X = 1 - x_1 \tag{7}$$

and the ionic strength on a mole fraction basis is given by:

$$I_x = (1/2)\sum_i x_i z_i^2 \tag{8}$$

with $z_i$ the charge on the ith ion. For the present case with $z = 1$, $I_x = x_2/2$.

A system of equations for electrolytes based on the reference states expressed in Equations 3 and 4 was developed in detail for singly-charged ions by Pitzer and Simonson (5). Although they considered both types of reference states for the solute, most of their working equations are for the pure liquid reference state. This reference state was used by Pitzer and Li (32) for a study of the NaCl-H$_2$O system extending to 550°C. For the present research limited to 350°C, however, it seemed better to use the infinitely dilute reference state, and the equations below are derived on that basis. The short-range contribution to the excess Gibbs energy of an aqueous solution with a single solute MX can be written in terms of the Margules expression:

$$g^s/RT = -x_2^2(W_{1,MX} - x_1 U_{1,MX}) \qquad (9)$$

where $W_{1,MX}$ and $U_{1,MX}$ are specific to each solute MX and are functions of temperature and pressure. The Debye-Huckel term representing the long-range electrostatic contribution is given by:

$$g^{DH}/RT = -(4A_x I_x/\rho) \ln(1+\rho I_x^{1/2}). \qquad (10)$$

The Debye-Huckel parameter $A_x$ is related to the usual parameter $A_\phi$ (for the osmotic coefficient on a molality basis) by

$$A_x = \Omega^{1/2} A_\phi \qquad (11)$$

where $\Omega$ is the number of moles of solvent per kg.(~55.51 for water).

Appropriate differentiations of the Gibbs energy with respect to the numbers of moles of the solvent and MX (at constant T and P) yield the equations for the activity coefficients:

$$\ln \gamma_1 = 2A_x I_x^{3/2}/(1+\rho I_x^{1/2}) + x_2^2 [W_{1,MX} + (1-2x_1)U_{1,MX}], \qquad (12)$$

$$\ln(\gamma_M^* \gamma_X^*) = 2A_x\{(2/\rho) \ln (1+\rho I_x^{1/2}) + I_x^{1/2}(1-2I_x)/(1+\rho I_x^{1/2})\}$$

$$+ 2(x_1^2-1)W_{1,MX} + 4x_2 x_1^2 U_{1,MX}. \qquad (13)$$

The activities of the solvent and the solute are then given by:

$$\ln a_1 = \ln(x_1 \gamma_1), \qquad (14)$$

$$\ln a_{MX} = \ln(x_M x_X \gamma_M^* \gamma_X^*). \qquad (15)$$

The parameter $\rho$ in Equations 10, 12, and 13 is related to the distance of closest approach of ions. To keep the equations for the thermodynamic properties of electrolyte mixtures simple, it is desirable to have the same value of $\rho$ for a wide variety of salts and for a wide range of temperature and pressure. The functional forms of Equations 12 and 13 are relatively insensitive to variations in $\rho$ values. It has been found satisfactory (1,5) to take a standard value for $\rho$ and let the short-range force terms accommodate any composition dependency of $\rho$. In calculations for metal nitrates in water from 100-163°C (33), $\rho$ was given a fixed value of 14.9. For the systems considered in this study, a constant value of 15.0 was found to be satisfactory.

The Margules expansion model has been tested on some ionic systems over very wide ranges of composition, but over limited ranges of temperature and pressure (33,34). In this study, the model is applied over a wider range of temperature and pressure, from 25-350°C and from 1 bar or saturation pressure to 1 kb. NaCl and KCl are major solute components in natural fluids and there are abundant experimental data from which their fit parameters can be evaluated. Models based on the ion-interaction approach are available for NaCl(aq) and KCl(aq) (8,9), but these are accurate only to about 6 molal. Solubilities of NaCl and KCl in water, however, reach 12 and 20 m, respectively, at 350°C, and ionic strengths of NaCl-KCl-H$_2$O solutions reach more than 30 m at this temperature (35). The objective of this study is to describe the thermodynamic properties, particularly the osmotic and activity coefficients, of NaCl(aq) and KCl(aq) to their respective saturation concentrations in binary salt-H$_2$O mixtures and in ternary NaCl-KCl-H$_2$O systems, and to apply the Margules expansion model to solubility calculations to 350°C.

For purposes of developing general equations for the thermodynamic properties of electrolyte solutions, it is useful to recalculate experimental values to a single reference pressure. This allows experimental data for different solution properties (e.g., activities, enthalpies, and heat capacities) whose relationships with each other are defined on an isobaric basis, to be considered in the overall regression of the model equations. This procedure has been shown to be useful for NaCl(aq) (8), KCl(aq) (9), and Na$_2$SO$_4$(aq) (13). The parameters required to recalculate thermodynamic data from the experimental to the reference pressure can be determined from a regression of volumetric data. For the Margules expansion model, the pressure dependence of the excess Gibbs energy, osmotic coefficient and activity coefficient can be derived from volumetric data as shown below.

The apparent molal volume, $^\phi V$, of the solution is given by:

$$^\phi V = (A_{v,x}/\rho)\ln(1 + \rho I^{1/2}) + \overline{V}_2^\circ - 2RTx_2[W_{1,MX}^v - x_1 U_{1,MX}^v] \tag{16a}$$

where $\overline{V}_2^\circ$ is the partial molal volume of the salt at infinite dilution. Also,

$$A_{v,x} = -4RT\Omega^{1/2}(\partial A_\phi/\partial P)_T, \tag{16b}$$

$$W_{1,MX}^v = (\partial W_{1,MX}/\partial P)_T, \tag{16c}$$

and

$$U_{1,MX}^v = (\partial U_{1,MX}/\partial P)_T. \tag{16d}$$

Equations 16c and 16d relate the volumetric parameters $W_{1,MX}^v$ and $U_{1,MX}^v$ to the pressure dependence of the parameters $W_{1,MX}$ and $U_{1,MX}$ for the excess Gibbs energy and osmotic/activity coefficients. In this study values of $W_{1,MX}^v$ and $U_{1,MX}^v$ for NaCl(aq) were evaluated at temperatures from 25° to 350°C, and pressures from 1 bar or saturation pressure to 1 kb, based on apparent molal volumes calculated from the equations of Rogers and Pitzer (7), plus additional experimental data not considered in their study (36-38). Initial regressions of Equation 16 to isothermal and isobaric sets of experimental values indicated the P and T dependence of $\overline{V}_2^\circ$, $W_{1,MX}^v$ and $U_{1,MX}^v$. Excellent fits are normally obtained to isobaric-isothermal sets of data. Examples are shown in Figure 3 for apparent molal volumes at 350°C and pressures to 1 kb. The curves are calculated values at the indicated pressures using parameters listed in Table I.

The next step is to perform a simultaneous regression of NaCl(aq) apparent molal volumes from 25-350°C. Over this wide range of temperature, however, and particularly above 300°C, standard-state properties based on the infinitely dilute reference state exhibit a very complex behavior (7,8), which is related to various peculiarities of the solvent. Thus in their representation of NaCl(aq) volumetric properties, Rogers and Pitzer (7) adopted a reference composition of a "hydrated fused salt," NaCl • 10H$_2$O, to minimize the P and T dependence of the reference state volume and to adequately fit volumetric data to 300°C and 1 kb. In this study the (supercooled) fused salt is used as the reference state. The equation for the apparent molal volume on this basis can be easily derived from that for the excess Gibbs energy of Pitzer and Simonson (5), and is given by:

$$^\phi V = (A_{v,x}/\rho)\ln[(1 + \rho I_x^{1/2})/(1 + \rho(I_x^*)^{1/2})] + V_2^{*,fs} + 2RTx_1(W_{1,MX}^v + x_2 U_{1,MX}^v), \tag{17}$$

where $I_x^*$ represents $I_x$ for a pure fused salt and is 1/2 for salts of singly-charged ions, while $V_2^{*,fs}$ is the molal volume for a (supercooled) fused salt. Values for the infinitely dilute reference state can be calculated from $V_2^{*,fs}$ using the relation:

$$\overline{V}_2^\circ = V_2^{*,fs} + 2RTW_{1,MX}^v - (A_{v,x}/\rho)\ln[1+\rho(I_x^*)^{1/2}]. \tag{18}$$

Apparent molal volumes of NaCl solutions from 25-350°C were used to fit Equation 17 and the following P and T function for $\overline{V}_2^{*,fs}$, $W_{1,MX}^v$, and $U_{1,MX}^v$:

$$f(T,P) = A + BP + CP^2 \tag{19a}$$

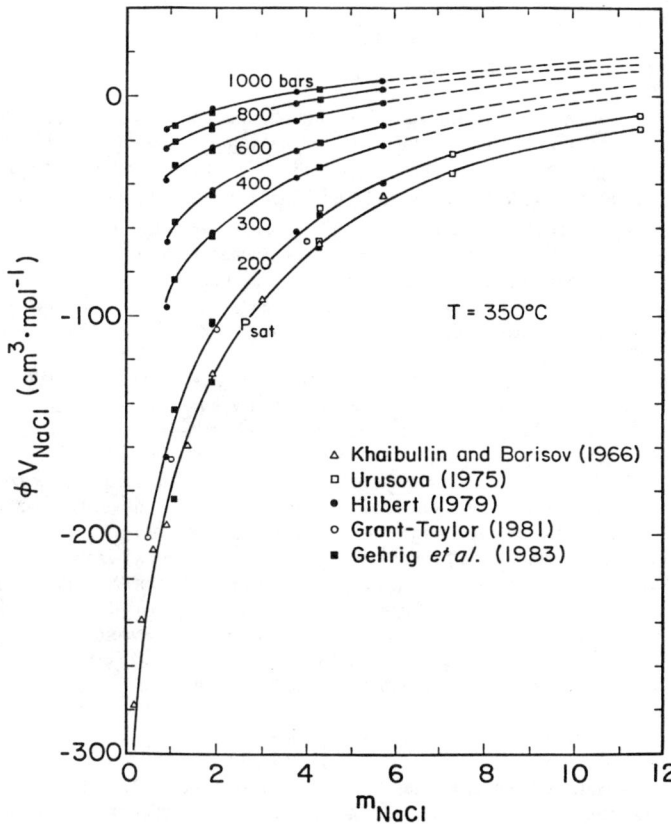

Figure 3. Apparent molal volumes of NaCl(aq) solutions at 350°C and pressures to 1 kb. The curves are calculated from Equation 16 and the parameters given in Table I.

Table I. Parameters for Equation 16 Evaluated from Isothermal-isobaric Apparent Molal Volumes of NaCl Solutions at 350°C and Various Pressures

| P(bars) | $\overline{V}_2^\circ$ | $W_{1,MX}^v$ | $U_{1,MX}^v$ |
|---|---|---|---|
| Psat | -4.46598E+02 | 3.14714E-03 | 2.62327E-03 |
| 200. | -3.86510E+02 | 2.15334E-03 | 1.29351E-03 |
| 300. | -2.29832E+02 | 1.57804E-04 | -9.31161E-04 |
| 400. | -1.57793E+02 | -1.14738E-04 | -5.35688E-04 |
| 500. | -1.15825E+02 | -4.49861E-04 | -5.68718E-04 |
| 600. | -8.92339E+01 | -5.79236E-04 | -5.00818E-04 |
| 800. | -5.90148E+01 | -2.77794E-04 | 1.79049E-04 |
| 1000. | -4.07166E+01 | -2.54388E-04 | 2.81882E-04 |

where

$$A = q_1 + q_2/T + q_3T + q_4T^2 + q_5/(647-T)^2 \quad (19b)$$

$$B = q_6 + q_7/T + q_8T + q_9T^2 + q_{10}/(647-T)^2 \quad (19c)$$

$$C = q_{11} + q_{12}/T + q_{13}T + q_{14}T^2 + q_{15}/(647-T)^2 \quad (19d)$$

Up to 300°C, the overall regression included values at pressures to 1 kb. At 350°C, however, only volumetric data at pressures less than or equal to 200 bars were included because the simple pressure function given by Equation 19a is inadequate to fit the 350°C data over the whole pressure range. The parameters for Equation 19 for $V_2^{*,fs}$, $W_1^v$, and $U_{1,MX}^v$ evaluated from the volumetric data are given in Table II. With this set of parameters it is possible to recalculate various thermodynamic properties of NaCl(aq) solutions to a reference pressure, chosen here to be 200 bars. Because the volumetric data for KCl solutions are more limited in concentration range and are less precise than those for NaCl solutions (9) a separate evaluation of KCl(aq) volumetric data was not done. Instead, the pressure dependence of KCl(aq) properties was approximated using values for NaCl(aq). This procedure has been shown to be successful for $Na_2SO_4$(aq) (13).

The principal interests in this study are osmotic and activity coefficients of NaCl(aq) and KCl(aq) solutions at temperatures to 350°C and up to saturation concentration. In the range 25-300°C and at 1 bar or saturation pressure, NaCl(aq) osmotic coefficients up to 4 m were taken from a comprehensive thermodynamic treatment of Pitzer et al. (9). Above 4 m, the values were taken from Liu and Lindsay (39). At temperatures above 300°C, osmotic coefficients were calculated from vapor pressure data of Wood et al. (40). Additional vapor pressure data are given in Refs. 41-47, but a critical evaluation of these data indicated that these are less precise measurements and were therefore given smaller weights in the regression. For KCl(aq), osmotic coefficients to 6 m at temperatures from 25-325°C at 1 bar or saturation pressure were taken from the ion interaction model of Pabalan and Pitzer (9). Additional values up to 350°C and saturation concentration were derived from Refs. 40, 41, 44, and 48.

Osmotic coefficients of NaCl and KCl solutions from these various sources were recalculated to the reference pressure of 200 bars. An overall regression to each isobaric set of data was done to determine values of $W_{1,MX}$ and $U_{1,MX}$ as functions of temperature using the equation:

$$f(T) = q_1 + q_2 \ln T + q_3/T + q_4T + q_5T^2 + q_6/(647-T). \quad (20)$$

The parameters for NaCl(aq) and KCl(aq) evaluated from the osmotic coefficients are given in Table III. These parameters, together with those of Table II, permit the calculation of osmotic and activity coefficients of NaCl and KCl solutions to saturation concentration at pressures to 1 kb from 25-300°C, and at pressures to 200 bars above 300°C.

For solubility calculations the Gibbs energy values for the solid (49) and for the aqueous species are needed. The latter can be derived from equations for standard state heat capacities given by Pitzer et al. (8) and Pabalan and Pitzer (9) for NaCl(aq) and KCl(aq), respectively, or from the equation of state for standard state properties of electrolytes given by Tanger and Helgeson (50). In the present calculations, the former equations were used to 300°C, and the latter above 300°C. Predicted solubilities for NaCl and KCl in water to 350°C are compared with various experimental values in Figures 4 and 5, respectively. These figures show good agreement between experimental and calculated solubilities, although there are substantial differences between various measurements for the solubility of KCl.

Of more interest are solubility calculations in the ternary system NaCl-KCl-$H_2O$. The equations for the excess Gibbs energy and activity coefficients in a mixture of a solvent and two salts with a common ion, MX and NX, and with cation fraction F of M are given by Pitzer and Simonson (5). Their equation for the activity coefficient of the solute MX in the ternary mixture MX-NX-$H_2O$ based on a pure fused salt standard state can be converted to one based on the infinitely dilute reference state. This is given by:

$$\ln(\gamma_M^* \gamma_X^*) = -2A_x\{(2/\rho)\ln(1+\rho I_x^{1/2}) + I_x^{1/2}(1-2I_x)/(1+\rho I_x^{1/2})\} + 2[x_1(1-x_IF)-1]W_{1,MX}$$

$$+ 2x_1x_I\{(1-F+2Fx_1)U_{1,MX} - (1-F)[W_{1,NX} + (x_I-x_1)U_{1,NX}]\}$$

$$+ x_I(1-F)\{(1-x_IF)W_{MX,NX} + X_I[3F-1+2x_IF(1-2F)U_{M,N}] + 2x_1(1-2x_IF)Q_{1,MX,NX}\}, \quad (21)$$

Table II. Parameters for Equations 17 and 19 Evaluated from Apparent Molal Volumes of NaCl Solutions up to 350°C and 1 kb

|  | $V_2^{\circ,fs}$ | $W_{1,MX}^v$ | $U_{1,MX}^v$ |
|---|---|---|---|
| $q_1$ | -6.434497E+02 | -1.724835E-02 | -2.9789818E-02 |
| $q_2$ | 0.0 | 3.347333E+00 | 5.7872110E+00 |
| $q_3$ | 3.554544E+00 | 1.996689E-05 | 3.5321836E-05 |
| $q_4$ | -4.612823E-03 | 3.861388E-09 | 6.7591836E-09 |
| $q_5$ | -3.254781E+05 | 4.273295E+00 | 4.4639601E+00 |
| $q_6$ | 4.496614E-01 | 3.318704E-06 | 1.2359971E-05 |
| $q_7$ | -2.830405E+01 | -6.183551E-04 | -2.0125079E-03 |
| $q_8$ | -2.047906E-03 | 0.0 | -1.7757055E-08 |
| $q_9$ | 2.627796E-06 | -9.914615E-12 | 0.0 |
| $q_{10}$ | 1.948423E+03 | -2.048634E-02 | -2.9326904E-02 |
| $q_{11}$ | -1.201087E-04 | 0.0 | 0.0 |
| $q_{12}$ | 1.327249E-02 | 0.0 | 0.0 |
| $q_{13}$ | 3.539774E-07 | 0.0 | 0.0 |
| $q_{14}$ | -3.546851E-10 | 0.0 | 0.0 |
| $q_{15}$ | 1.263778E-01 | 0.0 | 0.0 |

Table III. Parameters for Equation 20 for $W_{1,MX}$ and $U_{1,MX}$ for NaCl and KCl Solutions Evaluated from Osmotic Coefficients up to 350°C at a Reference Pressure of 200 Bars

|  | $W_{1,NaCl}$ | $U_{1,NaCl}$ | $W_{1,KCl}$ | $U_{1,KCl}$ |
|---|---|---|---|---|
| $q_1$ | 1.0620140E+03 | 2.2783431E+03 | 1.870602E+03 | 3.999606E+03 |
| $q_2$ | −2.3263776E+04 | −5.5141510E+04 | −5.071750E+04 | −1.063245E+05 |
| $q_3$ | −2.0149086E+02 | −4.2036595E+02 | −3.326169E+02 | −7.145462E+02 |
| $q_4$ | 6.1763652E−01 | 1.1574638E+00 | 7.159419E−01 | 1.610570E+00 |
| $q_5$ | −2.9491559E−04 | −5.3640257E−04 | −2.458959E−04 | −6.069845E−04 |
| $q_6$ | 0.0 | 2.5441911E+01 | −1.634611E+01 | 0.0 |

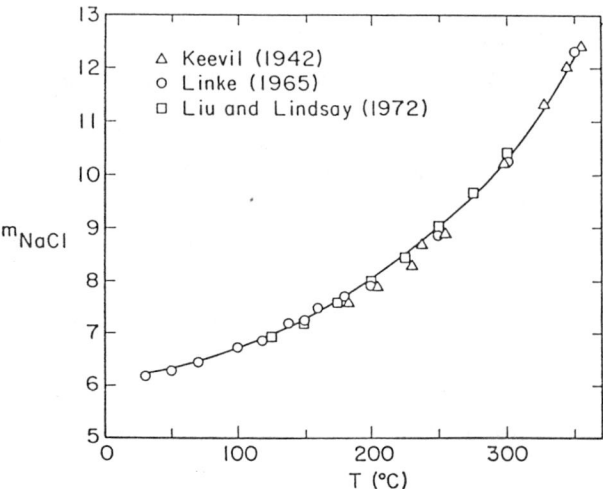

Figure 4. Solubilities of halite (NaCl) in water to 350°C. The curve represents values calculated using the Margules expansion model for activity coefficients (infinite dilution reference state), and standard state Gibbs energies for NaCl(aq) derived from the equations of Pitzer et al. (8) to 300°C, and of Tanger and Helgeson (50) above 300°C.

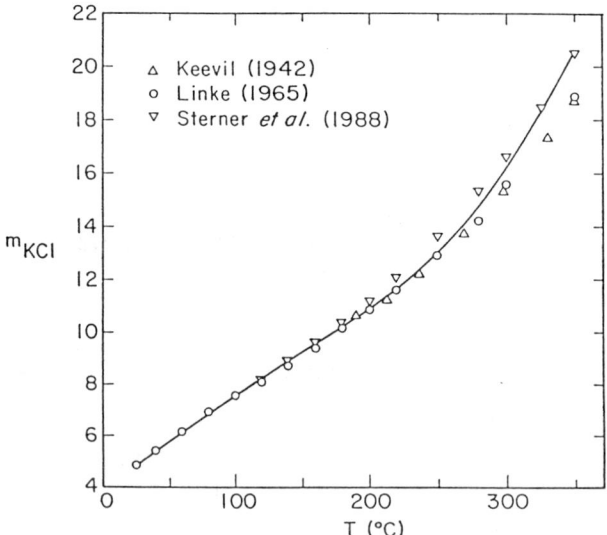

Figure 5. Solubilities of sylvite (KCl) in water to 350°C. The curve represents values calculated using the Margules expansion model for activity coefficients (infinite dilution reference state), and standard state Gibbs energies for KCl(aq) derived from the equations of Pabalan and Pitzer (9) to 300°C, and of Tanger and Helgeson (50) above 300°C.

where the total mole fraction of ions $x_I=(1-x_1)$, whereupon $x_M = Fx_I/2$, $x_N = (1-F)x_I/2$, and $x_X = x_I/2$. The corresponding equation for $(\gamma_N^* \gamma_X^*)$ can be obtained from Equation 21 by interchanging subscripts and replacing F with (1-F). In the general case of a ternary system MX-NX-H$_2$O, all but one of the parameters, namely $Q_{1,MX,NX}$, can be determined from the binary systems. For NaCl-KCl-H$_2$O, however, the temperatures of interest in this study are below the melting points of NaCl and KCl. Thus $W_{NaCl,KCl}$ and $U_{Na,K}$, together with $Q_{1,NaCl,KCl}$ have to be evaluated from activity data in the ternary system. Values of $W_{1,NaCl}$, $W_{1,KCl}$, $U_{1,NaCl}$, and $U_{1,KCl}$ have been previously determined from binary NaCl and KCl solutions.

Robinson (57) provided precise isopiestic data for the NaCl-KCl-H$_2$O at 25°C, while Holmes et al. (52) provided data at 110, 140, 172, and 201°C. These measurements, however, extend only to ionic strengths of about 7.5 m, whereas the maximum solubility in the system exceeds this value at T < 50°C, and reaches more than 30 m at 350°C. It is apparent that values of $W_{NaCl,KCl}$, $U_{Na,K}$, and $Q_{1,NaCl,KCl}$ could be best determined from solubility data in mixtures of NaCl and KCl solutions. These are available from the tabulation of Linke (17) and the extensive evaluation of Sterner et al. (35). By assuming a zero value for $U_{Na,K}$, a constant $Q_{1,NaCl,KCl}$ equal to 0.9547 was derived, as well as a temperature dependent equation for $W_{NaCl,KCl}$ given by Equation 20 for which $q_1$, $q_2$, $q_3$, $q_4$, $q_5$, and $q_6$ have values of 1.778998E+03, -3.19436E+02, -4.64165E+04, 7.33463E-01, -2.80959E-04, and 0.0, respectively. Solubilities calculated from these parameters and those given in Tables II and III are compared to experimental data in Figure 6. These calculated values agree very well with those of Sterner et al. (35) mostly within 2% and not exceeding 5%. Standard deviations between osmotic coefficients calculated using these parameters and experimentally determined values are 0.004, 0.004, 0.005, 0.007, and 0.008 for temperatures of 25, 110, 140, 172, and 201°C, respectively.

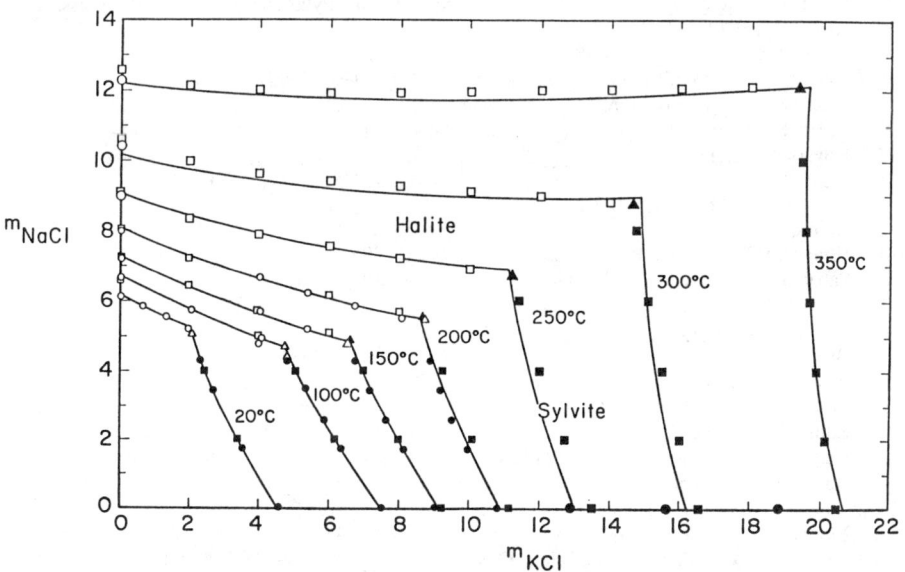

Figure 6. Solubilities of halite and/or sylvite in the NaCl-KCl-H$_2$O system. The squares and circles are experimental data taken from Sterner et al. (35) and Linke (17), respectively (open symbols for halite, closed symbols for sylvite). The open triangles are triple point (halite+sylvite+saturated solution) data from Linke, and the closed triangles are triple point data from Sterner et al. The curves are calculated as in Figures 4 and 5.

## SUMMARY

Two models based on general equations for the excess Gibbs energy of aqueous electrolyte mixtures were shown to be applicable for evaluating and predicting electrolyte solution thermodynamic properties over wide ranges of temperature, pressure, and concentration. Both models make use of empirical parameters which can be derived from various experimentally determined electrolyte solution properties. While the first model [the ion-interaction approach of Pitzer (1) and coworkers] is the most commonly used model for systems of moderate concentration, a second model based on a Margules expansion proposed by Pitzer and Simonson (5) was demonstrated to be a useful alternative for systems which may extend to very high concentrations, even to fused salt systems. Examples of parameter evaluations to high temperatures and pressures and applications of the models to solubility calculations were given for some geologically relevant systems.

## ACKNOWLEDGMENTS

We thank the three anonymous reviewers for their helpful comments and suggestions. This work was supported by the Director, Office of Energy Research, Office of Basic Energy Sciences, Division of Engineering and Geosciences of the U.S. Department of Energy under Contract No. DE-AC03-76SF00098. Final editing was supported by the Nuclear Regulatory Commission under Contract No. NRC-02-88-005.

## LITERATURE CITED

1. Pitzer, K.S. J. Phys. Chem. 1973, 77, 268.
2. Pitzer, K.S. Activity Coefficients in Electrolyte Solutions; Pytkowicz, R., Ed.; CRC Press: Boca Raton, FL, 1979; Vol. 1.
3. Pitzer, K.S. Rev. Mineralogy 1987, 17, 97.
4. Harvie, C.E.; Weare, J.H. Geochim. Cosmochim. Acta 1980, 44, 981.
5. Pitzer, K. S., Simonson, J. M., J. Phys. Chem. 1986, 90, 3005.
6. Pitzer, K.S.; Kim, J.J. J. Amer. Chem. Soc. 1974, 96, 5701.
7. Rogers, P.S.Z.; Pitzer, K.S. J. Phys. Chem. Ref. Data 1982, 11, 15.
8. Pitzer, K.S.; Peiper, J.C.; Busey, R.H. J. Phys. Chem. Ref.Data 1984, 13, 1.
9. Pabalan, R.T.; Pitzer, K.S. J. Chem. Eng. Data 1988, 33, 354.
10. Holmes, H.F.; Mesmer, R.E. J. Phys. Chem. 1983, 87, 1242.
11. Holmes, H.F.; Mesmer, R.E. J. Soln. Chem. 1986, 15, 495.
12. Pabalan, R.T.; Pitzer, K.S. Geochim. Cosmochim. Acta 1987, 51, 2429.
13. Pabalan, R.T., Pitzer, K. S., Geochim. Cosmochim. Acta 1988, 52, 2393.
14. Moller, N. Geochim. Cosmochim. Acta 1988, 52, 821.
15. Brodale, G.E.; Giauque, W.F. J. Amer. Chem. Soc. 1958, 80, 2042.
16. Brodale, G.E.; Giauque, W.F. J. Phys. Chem. 1972, 76, 737.
17. Linke, W.F. Solubilities of Inorganic and Metal Organic Compounds; Amer. Chem. Soc., 1965; 4th Edition, Vols. 1 and 2.
18. Harvie, C.E.; Eugster, H.P.; Weare, J.H. Geochim. Cosmochim. Acta 1982, 46, 1603.
19. Harvie, C.E.; Moller, N.; Weare, J.H. Geochim. Cosmochim. Acta 1984, 48, 723.
20. Meijer, J.A.M.; Van Rosmalen, G.M. Desalination 1984, 51, 255.
21. Monnin, C.; Schott, J. Geochim. Cosmochim. Acta 1984, 48, 571.
22. Langmuir, D.; Melchior, D. Geochim. Cosmochim. Acta 1985, 49, 2423.
23. Felmy, A.R.; Weare, J.H. Geochim. Cosmochim. Acta 1986, 50, 2771.
24. Reardon, E.J.; Armstrong, D.K. Geochim. Cosmochim. Acta 1987, 51, 63.
25. Reardon, E.J.; Beckie, R.D. Geochim. Cosmochim. Acta 1987, 51, 2355.
26. Filippov, V.K.; Nochrin, V.I. Z. Phys. Chem. Leipzig 1988, 269, 50.
27. Schroeder, W.C.; Gabriel, A.; Partridge E.P. J. Amer. Chem. Soc. 1935, 57, 1539.
28. Ananthaswamy, J.; Atkinson, G. J. Chem. Eng. Data 1985, 30, 120.
29. Prausnitz, J.M.; Lichtenthaler, R.N.; Azevedo, E.G. de Molecular Thermodynamics of Fluid-Phase Equilibria; Prentice-Hall: N.J.,1986.

30. Adler, S.B.; Friend, L.; Pigford, R.L. J. A.I.Ch. E. 1966, 12, 629.
31. Pitzer, K.S. Physics and Chemistry of Electrons and Ions in Condensed Matter; Acrivos, J. V.; Mott, N. F.; Yoffe, A. D., Eds.; D. Reidel: Boston, 1984.
32. Pitzer, K.S.; Li, Y. Proc. Nat. Acad. Sci. USA 1983, 80, 7689.
33. Simonson, J.M.; Pitzer, K.S. J. Phys. Chem. 1986, 90, 3009.
34. Weres, O.; Tsao, L. J. Phys. Chem. 1986, 90, 3014.
35. Sterner, S.M.; Hall, D.L.; Bodnar, R.J. Geochim. Cosmochim. Acta 1988, 52, 989.
36. Urusova, M.A. Russ. J. Inorg. Chem. 1975, 20, 1717.
37. Grant-Taylor, D.F. J. Soln. Chem. 1981, 10, 621.
38. Gehrig, M.; Lentz, H.; Franck, E.U. Ber. Bunsenges. Phys. Chem. 1983, 87, 597.
39. Liu, C-T.; Lindsay, W.T. Jr. J. Soln. Chem. 1972, 1, 45.
40. Wood, S.A.; Crerar, D.A.; Brantley, S.L.; Borcsik, M. Amer. J. Sci. 1984, 284, 668.
41. Keevil, N.B. J. Amer. Chem. Soc. 1942, 64, 841.
42. Olander, A.; Liander, H. Acta Chem. Scand. 1950, 4, 1437.
43. Sourirajan, S.; Kennedy, G.C. Amer. J. Sci. 1962, 260, 115.
44. Khaibullin, I. Kh.; Borisov, N.M. High Temp. 1966, 4, 489.
45. Urusova, M.A.; Ravich, M.I. Russ. J. Inorg. Chem. 1971, 16, 1534.
46. Mashovets, V.P.; Zarembo, V.I.; Fedorov, M.K. Zh. Prikl. Khim. 1973, 46, 650.
47. Parisod, C.J.; Plattner, E. J. Chem. Eng. Data 1981, 26, 16.
48. Zarembo, V.I.; Antonov, N.A.; Gilyarov, V.N.; Fedorov, M.K. J. Appl. Chem. USSR 1976, 49, 1259.
49. Chase, M.W. Jr.; Davies, C.A.; Downey, J.R. Jr.; Frurip, D.J.; McDonald, R.A.; Syverud, A.N. J. Phys. Chem. Ref. Data 1985, 14, suppl. no. 1.
50. Tanger, J.C. IV; Helgeson, H.C. Amer. J. Sci. 1988, 288, 19.
51. Robinson, R.A. J. Phys. Chem. 1961, 65, 662.
52. Holmes, H.F.; Baes, C.F. Jr.; Mesmer, R.E. J. Chem. Thermodyn. 1979, 11, 1035.

RECEIVED August 24, 1989

## Chapter 5

# Solubility of Volatile Electrolytes in Multicomponent Solutions with Atmospheric Applications

### Simon L. Clegg[1] and Peter Brimblecombe[2]

[1]Plymouth Marine Laboratory, Citadel Hill, Plymouth PL1 2PB, United Kingdom
[2]School of Environmental Sciences, University of East Anglia, Norwich NR4 7TJ, United Kingdom

> Thermodynamic Henry's law constants are given for the strong acids $HNO_3$, $HCl$, $HBr$ and $HI$ ($K_H$/$mol^2 kg^{-2} atm^{-1}$), and for weak electrolytes $HF$, $HCOOH$, $CH_3COOH$ and $NH_3$ ($K'_H$/$mol\ kg^{-1} atm^{-1}$) from 0 - 40°C. Use of the Pitzer thermodynamic model for the calculation of gas solubilities is evaluated, contrasting the behavior of HCl with new work on $NH_3$ (as an example of a weak electrolyte). Agreement between measurements and model calculations is good for both gases. The effects of dissolved salts, association equilibria, and temperature variations on gas partial pressure are illustrated. The incorporation of non-electrolytes into the model and the estimation of salt effects on neutral species are briefly discussed.

To calculate gas solubility in natural geochemical systems, basic thermodynamic properties such as the Henry's law constant and, in the case of weak electrolytes the dissociation constant, must be combined with a thermodynamic model of aqueous solution behavior. An analogous approach has been used to predict mineral solubilities in concentrated brines (1). Such systems are also relevant to the atmosphere where very concentrated solutions occur as micrometer sized aerosol particles and droplets, which contain very small amounts of water relative to the surrounding gas phase. The ambient relative humidity (RH) controls solute concentrations in the droplets, which will be very dilute near 100% RH, but become supersaturated with respect to soluble constituents (such as NaCl) below about 75% RH. The chemistry of the aerosol is complicated by the non-ideality inherent in concentrated electrolyte solutions.

**THEORY**

The equilibrium of a strong acid HX between aqueous and gas phases is represented by:

$$HX_{(g)} = H^+_{(aq)} + X^-_{(aq)}, \quad K_H = mH^+\ mX^-\ \gamma_\pm^2/pHX \qquad (1)$$

where $K_H$ ($mol^2\ kg^{-2}\ atm^{-1}$) is the thermodynamic Henry's law constant, prefix 'm' represents molality, $\gamma_\pm$ is the stoichiometric mean activity coefficient of $H^+$ and $X^-$ ions in solution and pHX the equilibrium partial pressure of HX. While the undissociated acid molecule must exist in solution, the equilibrium is expressed in the form of Equation 1 in order to be consistent with the thermodynamic convention of treating all strong electrolytes as completely dissociated (2).

A different treatment of the solubility of weak electrolytes is required, where two clearly separable reactions are involved:

0097–6156/90/0416–0058$06.00/0
© 1990 American Chemical Society

$$HY_{(g)} = HY_{(aq)}, \quad K'_H = \gamma_{HY} \, mHY/pHY \tag{2}$$

$$HY_{(aq)} = H^+_{(aq)} + Y^-_{(aq)}, \quad K_{diss} = \gamma_{\pm}^2 \, mH^+ \, mY^-/\gamma_{HY} \, mHY \tag{3}$$

where $\gamma_{\pm}$ is the mean activity coefficient of the *free* aqueous ions and $\gamma_{HY}$ that of the neutral molecule. Here the overall reaction is represented by a Henry's law constant ($K'_H$ / mol kg$^{-1}$ atm$^{-1}$), which describes the equilibrium between the gas phase weak electrolyte and aqueous neutral molecule, and a dissociation constant $K_{diss}$ (mol kg$^{-1}$). Note that we adhere to the convention of using the 'atmosphere' as the unit of pressure (for conversion to S.I. units, atm = 101325 Pa), and also that $K_H$ is equivalent to $K'_H K_{diss}$ where the dissociation reaction is considered separately from that of dissolution.

The term 'thermodynamic Henry's law constant' is used to distinguish $K_H$ and $K'_H$ from the purely empirical constants that are often employed (3). Strictly, gas phase concentrations should be expressed as fugacities. However, for the range of conditions encountered in the earth's atmosphere, deviations from ideal behavior in the gas phase are not significant. Partial pressures are therefore used throughout.

To calculate the partial pressures of volatile electrolytes above solutions of known composition, values of the activity coefficients of the dissolved components are needed in addition to the appropriate Henry's law constants. In this work activity coefficients are calculated using the ion-interaction model of Pitzer (4). While originally formulated to describe the behavior of strong electrolytes, it is readily combined with explicit recognition of association equilibria (1,5), and may be extended to include neutral solutes (4,6). The model has previously been used to describe vapor-liquid equilibria in systems of chiefly industrial interest (7).

In Pitzer's model the Gibbs excess free energy of a mixed electrolyte solution and the derived properties, osmotic and mean activity coefficients, are represented by a virial expansion of terms in concentration. A number of summaries of the model are available (1,4,8). The equations for the osmotic coefficient ($\phi$), and activity coefficients ($\gamma$) of cation (M), anion (X) and neutral species (N) are given below:

$$(\phi - 1) = \left(\frac{2}{\sum_i m_i}\right)\left[\frac{-A^\phi I^{3/2}}{(1+1.2\sqrt{I})} + \sum_c\sum_a m_c m_a \left(B^\phi_{ca} + ZC_{ca}\right)\right.$$

$$+ \sum_{c<c'}\sum m_c m_{c'}\left(\Phi^\phi_{cc'} + \sum_a m_a \Psi_{cc'a}\right) + \sum_{a<a'}\sum m_a m_{a'}\left(\Phi^\phi_{aa'} + \sum_c m_c \Psi_{caa'}\right)$$

$$+ \sum_n\sum_c m_n m_c \lambda_{nc} + \sum_n\sum_a m_n m_a \lambda_{na} + \sum_{n<n'}\sum m_n m_{n'}\lambda_{nn'} + \frac{1}{2}\sum_n m_n^2 \lambda_{nn}$$

$$+ \sum_n m_n^3 \mu_{nnn} + 3\sum_{n<n'}\sum m_n^2 m_{n'}\mu_{nnn'} + 6\sum_{n<n'<n''}\sum\sum m_n m_{n'} m_{n''}\mu_{nn'n''}$$

$$+ 3\sum_n\sum_c m_n^2 m_c \mu_{nnc} + 3\sum_n\sum_c m_n m_c^2 \mu_{ncc} + 6\sum_n\sum_{c<c'}\sum m_n m_c m_{c'}\mu_{ncc'}$$

$$+ 3\sum_n\sum_a m_n^2 m_a \mu_{nna} + 3\sum_n\sum_a m_n m_a^2 \mu_{naa} + 6\sum_n\sum_{a<a'}\sum m_n m_a m_{a'}\mu_{naa'}$$

$$+ 6\sum_{n<n'}\sum\sum_a m_n m_{n'} m_a \mu_{nn'a} + 6\sum_{n<n'}\sum\sum_c m_n m_{n'} m_c \mu_{nn'c}$$

$$\left. + 6\sum_n\sum_c\sum_a m_n m_c m_a \mu_{nca}\right] \tag{4}$$

$$\ln \gamma_M = z_M^2 F + \sum_a m_a (2B_{Ma} + ZC_{Ma}) + \sum_c m_c \left(2\Phi_{Mc} + \sum_a m_a \Psi_{Mca}\right)$$

$$+ \sum_{a<a'} \sum m_a m_{a'} \Psi_{Maa'} + z_M \sum_c \sum_a m_c m_a C_{ca} + 2\sum_n m_n \lambda_{nM} + 3\sum_n m_n^2 \mu_{nnM}$$

$$+ 6\sum_{n<n'} \sum m_n m_{n'} \mu_{nn'M} + 6\sum_n \sum_c m_n m_c \mu_{nMc} + 6\sum_n \sum_a m_n m_a \mu_{nMa} \tag{5}$$

$$\ln \gamma_X = z_X^2 F + \sum_a m_a (2B_{cX} + ZC_{cX}) + \sum_a m_a \left(2\Phi_{Xa} + \sum_c m_c \Psi_{cXa}\right)$$

$$+ \sum_{c<c'} \sum m_c m_{c'} \Psi_{cc'X} + |z_X| \sum_c \sum_a m_c m_a C_{ca} + 2\sum_n m_n \lambda_{nX} + 3\sum_n m_n^2 \mu_{nnX}$$

$$+ 6\sum_{n<n'} \sum m_n m_{n'} \mu_{nn'X} + 6\sum_n \sum_c m_n m_c \mu_{ncX} + 6\sum_n \sum_a m_n m_a \mu_{nXa} \tag{6}$$

$$\ln \gamma_N = 2\sum_n m_n \lambda_{Nn} + 2\sum_c m_c \lambda_{Nc} + 2\sum_a m_a \lambda_{Na} + 3\sum_n m_n^2 \mu_{Nnn}$$

$$+ 6{\sum_{n<n'}}' m_n m_{n'} \mu_{Nnn'} + 6{\sum_n}' m_N m_n \mu_{NNn} + 3\sum_c m_c^2 \mu_{Ncc} + 3\sum_a m_a^2 \mu_{Naa}$$

$$+ 6\sum_{c<c'} \sum m_c m_{c'} \mu_{Ncc'} + 6\sum_{a<a'} \sum m_a m_{a'} \mu_{Naa'} + 6\sum_n \sum_c m_n m_c \mu_{Nnc}$$

$$+ 6\sum_n \sum_a m_n m_a \mu_{Nna} + 6\sum_c \sum_a m_c m_a \mu_{Nca} \tag{7}$$

The water activity ($a_w$), equivalent to ambient RH for solutions at equilibrium with the atmosphere, is given by $\ln(a_w) = -0.018 \phi \Sigma m_i$. In these equations $A^\phi$ is the Debye-Huckel parameter (0.391$_5$ at 25°C), I (mol kg$^{-1}$) is ionic strength, $m_i$ is the molality of species 'i' (cation, anion or neutral), $z_i$ is the charge on ion 'i', and subscripts c, a and n represent cations, anions and neutrals respectively.

Terms $B_{ca}$, $B^\phi_{ca}$, Z, $C_{ca}$, F, $\Phi_{cc'}$ and $\Phi^\phi_{cc'}$ (and corresponding terms for anion pairs) are functions given in full by Pitzer ([4]) and others ([1]) and therefore not reproduced here. The 'B' and 'C' functions incorporate ion interaction parameters obtained from pure electrolyte data, ($\beta^{(0)}_{ca}$, $\beta^{(1)}_{ca}$, $\beta^{(2)}_{ca}$ and $C^\phi_{ca}$). Functions $\Phi^\phi_{cc'}$, $\Phi_{cc'}$, and those for anions, contain parameters ($\theta_{cc'}$ and $\theta_{aa'}$) for interactions between ions of like sign and an unsymmetrical mixing term. The parameters $\psi_{cc'a}$ and $\psi_{caa'}$ account for interactions between one ion of one sign, and two dissimilar ions of opposite sign.

Double summation indices c<c', a<a' and n<n' denote sums over all distinguishable pairs of cations, anions or neutrals. For the primed summations $\Sigma'$ in Equation 7, n (or n' as indicated) cannot equal N. All interactions are assumed to be symmetrical, thus $\theta_{ij}$ is equal to $\theta_{ji}$, and $\lambda_{ij}$ is equal to $\lambda_{ji}$ with corresponding relations applying to other binary and all triplet interactions. In Equations 4-7 there are a large number of possible triplet parameters involving neutral species. Although relatively few have been determined ([5,9]), in systems of environmental interest neutral species such as $CO_2$ or $NH_3$ may only be present at very low concentration. Hence only $\lambda_{Ni}$ parameters are likely to be significant, and the application of Equations 4-7 is greatly simplified.

The Debye-Huckel $A^\phi$ parameter is a function of temperature, as are the pure electrolyte model parameters ([8,10]). Over the temperature range covered in this study the change in $\theta_{ij}$ and $\psi_{ijk}$ is small ([11]). Work involving neutral species, described below, suggests that $\lambda_{ij}$ is also a function of temperature, while the parameter $\mu_{ijk}$ was not found to vary significantly.

Pure electrolyte parameter values ($\beta^{(i)}$, $C^\phi$) at 25°C for each acid, together with first derivatives with respect to temperature ($\beta^{(i)L}$, $C^{\phi L}$), and second derivatives ($\partial \beta^{(i)L}/\partial T$, $\partial C^{\phi L}/\partial T$), are given in Table I for the volatile strong electrolytes being treated here. Superscript 'L' is used to denote $\partial \beta^{(i)}/\partial T$ and $\partial C^\phi/\partial T$ because these derivatives are obtained from apparent relative molal enthalpy data ($\phi L$). Values of the second derivatives are determined from apparent molal heat capacities, e.g. ([12]).

Values of the pure electrolyte parameters at temperature T, required for the calculation of osmotic and activity coefficients, can be obtained from the following equation, given here for $\beta^{(0)}$:

$$^T\beta^{(0)} = {}^{298}\beta^{(0)} + (T-298.15)(\beta^{(0)L} - 298.15\frac{\partial \beta^{(0)L}}{\partial T}) + 0.5(T^2 - 298.15^2)\frac{\partial \beta^{(0)L}}{\partial T} \qquad (8)$$

where $^T\beta^{(0)}$ is the value of $\beta^{(0)}$ at temperature T, and $^{298}\beta^{(0)}$ its value at 25°C. Analogous relationships apply to the other parameters.

Table I. Pitzer Model Parameters for the HNO$_3$, MSA and the Hydrohalic Acids, Valid to 6 mol kg$^{-1}$

| Acid | $\beta^{(0)}$ | $\beta^{(1)}$ | $10^3 C^\phi$ | $10^4 \beta^{(0)L}$ | $10^4 \beta^{(1)L}$ | $10^5 C^{\phi L}$ |
|---|---|---|---|---|---|---|
| HCl[a,d] | 0.1775 | 0.2945 | 0.80 | -3.784 | 4.228 | -3.592 |
| HBr[b] | 0.1984 | 0.3353 | 6.99 | -2.049 | 4.467 | -5.685 |
| HBr[c] | 0.2085 | 0.3477 | 1.517 | " | " | " |
| HI[b] | 0.2211 | 0.4907 | 4.815 | -0.230 | 8.860 | -7.320 |
| HNO$_3$[b,d] | 0.1168 | 0.3546 | -5.39 | 2.345 | 11.39 | -9.223 |
| MSA[e] | 0.1544 | 0.4775 | -4.09 | - | - | - |

| Acid | $10^6 \frac{\partial \beta^{(0)L}}{\partial T}$ | $10^6 \frac{\partial \beta^{(1)L}}{\partial T}$ | $10^7 \frac{\partial C^{\phi L}}{\partial T}$ | Acid | $10^6 \frac{\partial \beta^{(0)L}}{\partial T}$ | $10^6 \frac{\partial \beta^{(1)L}}{\partial T}$ | $10^7 \frac{\partial C^{\phi L}}{\partial T}$ |
|---|---|---|---|---|---|---|---|
| HCl | -1.93 | -17.93 | 4.81 | HI[f] | -5.12 | -29.10 | 6.20 |
| HBr | -4.09 | -15.78 | 6.72 | HNO$_3$ | -19.65 | 24.20 | 25.90 |

Weak acid HF not fitted to Pitzer model, however activity coefficients are available (0 - 35°C), ([13]). Mean stoichiometric activity coefficients at 25°C are also given by Hamer and Wu ([14]) to the following maximum concentrations: HF - 20.0, HCl - 16.0, HBr - 11.0, HI - 10.0, HNO$_3$ - 28.0 mol kg$^{-1}$. All second derivatives listed above were obtained in this study, from data compiled by Parker ([15]). *(a)* Exact fitting equation ([16]) gives parameters over temperature range 0 - 250°C. *(b)* Parameters at 25°C fitted to 6 mol kg$^{-1}$, data of Hamer and Wu ([14]). *(c)* Parameters at 25°C given by Macaskill and Bates ([17]). *(d)* First temperature derivatives obtained in this study, from data compiled by Parker ([15]). *(e)* Parameters fitted to data of Covington et al. ([18]). *(f)* Valid to 2.75 mol kg$^{-1}$.

## THE HENRY'S LAW CONSTANTS

Where possible, Henry's law constants presented here were evaluated directly from partial pressure data and activity coefficients at 25°C. The temperature dependence of $K_H$ and $K'_H$ was obtained thermodynamically.

**THE STRONG ACIDS.** Henry's law constants at 25°C have been evaluated by Brimblecombe and Clegg ([19]). The variation of $K_H$ with temperature was determined from thermodynamic data ([20,21]), by assuming a constant $\delta C^\circ_p$ for the dissolution reaction, and using the following expression obtained by integration of the van't Hoff equation:

$$R \ln(K_{H,2}/K_{H,1}) = \delta H^\circ_1(1/T_1 - 1/T_2) + \delta C^\circ_p(T_1/T_2 - (1 + \ln(T_1/T_2))) \qquad (9)$$

where subscript 1 refers to values at 298.15K; and 2 to the temperature of interest. Values of the Henry's law constants at 25°C are given in Table II, together with fitting equations valid over the range 0-40°C. Although HF is a weak acid ($K_{diss}$ equal to 0.000671 at 25°C) ([13]), its Henry's law constant is presented in Table II both on a stoichiometric basis, and in the more appropriate form for a weak acid described by Equation 2.

Table II. Henry's Law Constants ($K_H$ and $K'_H$) from 0 - 40°C. Values at 25°C determined from partial pressure data. Estimates calculated from available free energy data (22) are also shown. Temperature dependence (T/Kelvin) given by: $\ln(K_H, K'_H) = a + b/T + cT$

Part 1: $K_H$ ($HX_{(g)} = H^+_{(aq)} + X^-_{(aq)}$):

| Species | $K_H(25°)$ | $e^{(-\Delta G/RT)}$ | a | b | c |
|---|---|---|---|---|---|
| HF[a] | 9.61 | 9.54 | -5.91265 | 4918.0344 | -0.02790 |
| HCl[b] | 2.04x10$^6$ | 1.97x10$^6$ | 4.6187 | 5977.5014 | -0.03401 |
| HBr[c] | 1.32x10$^9$ | 7.06x10$^8$ | 7.60095 | 7117.0552 | -0.03512 |
| HI | 2.5x10$^9$ | 2.15x10$^9$ | 9.70617 | 6689.1418 | -0.03522 |
| HNO$_3$ | 2.45x10$^6$ | 2.51x10$^6$ | 2.70104 | 6137.3984 | -0.02876 |

Part 2: $K'_H$ ($HY_{(g)} = HY_{(aq)}$):

| Species | $K'_H(25°)$ | $e^{(-\Delta G/RT)}$ | a | b | c |
|---|---|---|---|---|---|
| HF | 1.43x10$^4$ | 1.37x10$^4$ | 6.61712 | 3360.464 | -0.02789 |
| HCOOH[d] | 5.2x10$^3$ | 5.39x10$^3$ | -10.3069 | 5634.802 | 0.0 |
| CH$_3$COOH | 5.24x10$^3$ | 8.61x10$^3$ | -25.6721 | 8322.372 | 0.02121 |
| NH$_3$[e] | 60.72 | 57.64 | -8.09694 | 3917.507 | -0.00314 |

Value of $K_H$ for MSA is 6.5x10$^{13}$ at 25°C. *(a)* HF included here for calculation of pHF using stoichiometric activity coefficients and total concentrations. *(b)* $K_H$ of 2.04x10$^6$ also derived thermodynamically (23), although the lower 1.97x10$^6$ is marginally preferred. *(c)* Due to uncertainty in $\gamma_{HBr}$ at high concentrations $K_H$ may be lower than that given here. For calculations of pHBr at low concentrations the value 7.064x10$^8$ may be more accurate, (for which 'a' is 6.9759, and other parameters are unchanged). *(d)* Value of $K'_H$ determined from pHCOOH data subject to large uncertainty. *(e)* $K'_H$, including temperature variation, based upon values given by Chen et al. (7).

Henry's law constants calculated directly from thermodynamic data (22) are in generally good agreement with those derived from partial pressure data (Table II), except in the case of HBr. Macaskill and Bates (17) used EMF measurements to obtain values of $\gamma_{HBr}$ at 25°C which are up to 17% lower than those of Hamer and Wu (14). This suggests that the value of $K_H$ derived at 25°C is too great. Although the activity coefficients of Hamer and Wu (14) together with $K_H$ equal to 1.32x10$^9$ should yield reliable partial pressures of HBr over the concentration range used to derive $K_H$ (>7 mol kg$^{-1}$), there may be considerable errors at lower concentrations. The thermodynamically calculated value of 7.06x10$^8$ mol$^2$ kg$^{-2}$ atm$^{-1}$ may therefore be preferred for such systems, particularly if the activity coefficients of Macaskill and Bates (17) are used.

Similar considerations apply to HNO$_3$. Recently Tang et al. (24) have made pH$_2$O and pHNO$_3$ measurements at 25°C and have used these to derive activity coefficients which differ from those of Hamer and Wu (14). These were used, together with other published partial pressure data, to calculate $K_H$ equal to 2.66x10$^6$ mol$^2$kg$^{-2}$atm$^{-1}$. This is greater than the value derived by Brimblecombe and Clegg (19) by approximately one standard deviation. The Henry's law constant of Tang et al. (24) should only be used with the set of corresponding activity coefficients. In terms of absolute accuracy (predicting pHNO$_3$) there appears to be little to choose between the two studies. However, work in progress by the authors using mole fraction based activity coefficient models of Pitzer (25), shows that at concentrations >10 mol kg$^{-1}$ there are systematic errors in the mean osmotic and activity coefficients derived by both Hamer and Wu (14) and Tang et al. (24). This is largely due to the nature of the fitting equations used (extended polynomials in molality), and results in a poorer correlation of the data at high concentrations than might otherwise be achieved.

**WEAK ELECTROLYTES.** The Henry's law constants of these solutes (Table II) have been estimated in the same way as those of the strong acids, relying mainly upon partial pressure data.

Dissociation is slight at the aqueous concentrations for which partial pressures have been measured, and where it has been calculated, a simple Debye-Huckel expression was used to obtain ion activity coefficients.

**Formic and Acetic Acids.** The solubility equilibrium of these weak acids is treated as the two stage process described by Equations 2 & 3. The dissociation constants of both acids are well known and are given as functions of temperature in Table III. While there are several studies of the thermodynamics of aqueous acetic acid, e.g. (26), and of formic acid (27), there are relatively few data for dilute aqueous solutions at 25°C (28-31). The chemistry of these acids is complicated by dimerisation (31-33) and higher association reactions (34) in both aqueous and gas phases.

Using an aqueous phase dimerisation constant of 0.149 kg mol$^{-1}$ (32), a gas phase constant of 1.26x10$^3$ atm$^{-1}$ (35), and assuming unit activity coefficients for uncharged species the quotient mCH$_3$COOH/pCH$_3$COOH was calculated at 25°C for available partial pressure data, see Figure 1. At low aqueous phase concentrations values decrease sharply. Extrapolating the higher concentration data back to zero yields K'$_H$ for CH$_3$COOH of 5.24x10$^3$ mol kg$^{-1}$ atm$^{-1}$, lower than that calculated from thermodynamic data (8.6x10$^3$) (22). Note that the vapor phase dimerisation constant of Chao and Zwolinski (35) is greater than that obtained by earlier workers. Use of a value of 0.000484 atm$^{-1}$ (28) yields K'$_H$ equal to 5.8x10$^3$ mol kg$^{-1}$ atm$^{-1}$ at 25°C. The temperature variation of K'$_H$ was estimated using Equation 9 and published enthalpies (22) and heat capacities (22,36).

Table III. Dissociation Constants of Weak Electrolytes from 0 - 40°C, given by:
$\ln(K_{diss}) = a + b/T + cT$ where T/Kelvin.

| Species | K$_{diss}$(25°) | a | b | c | Ref. |
|---|---|---|---|---|---|
| HF[a] | 6.71x10$^{-4}$ | -12.535 | 1558.334 | 0 | (13,37) |
| NH$_3$ | 1.774x10$^{-5}$ | 16.9732 | -4411.025 | -0.0440 | (38) |
| HCOOH | 1.772x10$^{-4}$ | 13.2948 | -3258.651 | -0.03691 | (39) |
| CH$_3$COOH | 1.754x10$^{-5}$ | 7.4850 | -2724.347 | -0.03118 | (40) |

(a) Hamer and Wu (13) adopt a value of 0.000684 at 25°C, as did the authors in recent calculations of HF solubility (41).

For formic acid an aqueous phase dimerisation constant of 0.00775 kg mol$^{-1}$ (32), and a gas phase value of 345.0 atm$^{-1}$ (33) were used together with three partial pressure measurements (29) to estimate a K'$_H$ of 5.2x10$^3$ mol kg$^{-1}$ atm$^{-1}$ at 25°C, Table II. While the data are scattered, this estimate is in good agreement with the thermodynamically calculated value.

It should be noted that the treatment presented here is not exhaustive, and data for very high aqueous phase concentrations and/or high temperatures have not been included.

**Ammonia.** The solubility of NH$_3$ is described in two stages, as dissolution (Equation 2) followed by base dissociation:

$$NH_{3(aq)} + H_2O = NH_4^+_{(aq)} + OH^-_{(aq)} \tag{10}$$

Henry's law constants of NH$_3$ are readily available over a range of temperatures (6,7), so partial pressure data were used to evaluate $\gamma_{NH_3}$ and the neutral interaction parameters from 0° to 40°C. Fits were restricted to mNH$_3$ ≤20 mol kg$^{-1}$ and the single parameter $\lambda_{NN}$ determined (subscript N henceforth indicates NH$_3$), see Table IV. For pure aqueous NH$_3$ where dissociation is negligible the activity coefficient $\gamma_{NH_3}$ is equal to exp(2mNH$_3\lambda_{NN}$). An independent estimate of the temperature differential $\partial\lambda_{NN}/\partial T$ at 25°C, obtained from apparent molal enthalpy data (6), agrees to within 10% with the value obtained from the partial pressure data.

## CALCULATIONS OF GAS SOLUBILITY IN ELECTROLYTE SOLUTIONS

An understanding of gas solubility in electrolyte solutions is necessary in studies of gas-liquid processes in a wide range of media as diverse as: biological fluids, seawater, brines, scrubbing solutions, extractive solvents and industrial production systems. Our own work concentrates on gas solubility in electrolyte solutions in the atmosphere.

**ATMOSPHERIC CONTEXT.** Concentrated solutions occur in the atmosphere in the form of aqueous aerosols. Seasalt is the largest component of particulate material cycled through the troposphere (42), thus aerosols of marine origin consist largely of brines of approximately seawater composition. Although they are present in smaller quantity, sub-micron sized aerosols consisting of $H_2SO_4$ partially neutralized by $NH_3$ are important because of their role in cloud nucleation. Models of atmospheric aerosols with very simple chemistry give some insight into general aspects of their behavior. However, a more complete chemical description is often required, and there remain many obstacles to achieving this including a number of problems specific to atmospheric systems.

Hysteresis effects during the evaporation of aerosols lead to very high degrees of supersaturation being achieved (>100%) before crystallization occurs (43). Thermodynamic data for such solutions are available for relatively few solutes, e.g. (44), and existing models do not appear adequate to describe solution mixtures at such high concentrations. In addition, the behavior of supercooled and very small (sub-micron) solution droplets, particularly relevant to aerosol behavior in the stratosphere, is another largely unexplored area which has become accessible only recently (45). More mundane, but equally important problems arise from the lack of thermodynamic data to parameterize existing models at normal temperatures and low concentrations.

Temperature and pressure variations of thermodynamic quantities must also be considered. Pressure variations in the atmosphere are relatively small, and are not treated here. As has already been made clear, temperature effects are incorporated within the Pitzer model, and have been investigated over a wide range for some of the major components of brines (46).

The impact of different dissolved salts on gas solubility, the effects of association equilibria and the temperature variation of solubility are briefly outlined below.

**SOLUBILITY OF A WEAK ELECTROLYTE IN SALT SOLUTIONS.** Calculation of the solubility of a volatile strong electrolyte, such as HCl, in aqueous salt solutions is straightforward. However, solubilities of weak electrolytes are more difficult to model accurately, since the dissolved speciation must frequently be determined in addition to the activity of the component of interest. Thus, in the case of $NH_3$, the relevant ionic interactions involving $NH_4^+$ and $OH^-$ must be known in addition to parameters for the interaction of dissolved salts with the neutral $NH_3$ molecule. See, for example, the work of Maeda et al. (47) on the dissociation of $NH_3$ in LiCl solutions.

Few parameters for ion-neutral interactions are available as yet (1,5,9). However, we have recently derived parameters for $NH_3$-salt interactions, using partial pressure, liquid phase partitioning and salt solubility data (6), which may serve as a model for the treatment of other weak and non-electrolytes. The results of this work, and the application of the Pitzer equations to the calculation of neutral species solubility, are discussed below.

**The Activity Coefficient of $NH_3$ in Aqueous Salt Solutions.** The activity coefficient of $NH_3$ in a solution of salt $M_{\nu_+}X_{\nu_-}$ is given by:

$$\ln(\gamma_{NH3}) = 2mNH_3\lambda_{NN} + 3mNH_3^2\mu_{NNN} + 2mM_{\nu_+}X_{\nu_-}(\nu_+\lambda_{NM} + \nu_-\lambda_{NX})$$

$$+ 6mM_{\nu_+}X_{\nu_-}mNH_3(\nu_+\mu_{NNM} + \nu_-\mu_{NNX})$$

$$+ 3(mM_{\nu_+}X_{\nu_-})^2(\nu_+^2\mu_{NMM} + 2\nu_+\nu_-\mu_{NMX} + \nu_-^2\mu_{NXX}) \qquad (11)$$

Note that the final term in Equation 11 may be written as the parameter $\zeta_{NMX}$, proposed by Felmy and Weare ($\underline{9}$). If individual ion molalities are used in place of salt concentration we then obtain:

$$\ln(\gamma_{NH3}) = 2mNH_3\lambda_{NN} + 3mNH_3^2\mu_{NNN} + 2mM\,\lambda_{NM} + 2mX\,\lambda_{NX}$$
$$+ 6mNH_3(mM\,\mu_{NNM} + mX\,\mu_{NNX}) + mM\,mX\,\zeta_{NMX} \qquad (12)$$

where: $\quad \zeta_{NMX} = 6\mu_{NMX} + 3(|z_X|/z_M)\mu_{NMM} + 3(z_M/|z_X|)\mu_{NXX}$

and $z_i$ is the charge on ion 'i'. If the triplet parameters in Equations 11 & 12 are neglected, then the similarity to the various formulations of the empirical Setchenow equation ($\underline{48}$) becomes clear, there being only a self interaction term ($\lambda_{NN}$) and a salting coefficient ($v_+\lambda_{NM} + v_-\lambda_{NX}$).

Where the neutral species is present at sufficiently high concentration, its influence on the activity coefficients of dissolved salts and the osmotic coefficient of the solution must be considered. The equation for the mean activity coefficient of salt $M_{v+}X_{v-}$ dissolved in an aqueous solution containing $NH_3$ (and neglecting dissociation) is given by:

$$v\ln(\gamma_\pm) = \{\text{ion terms}\} + 2mNH_3(v_+\lambda_{NM} + v_-\lambda_{NX}) + 3mNH_3^2(v_+\mu_{NNM} + v_-\mu_{NNX})$$
$$+ 6mNH_3.mM_{v+}X_{v-}(v_+^2\mu_{NMM} + 2v_+v_-\mu_{NMX} + v_-^2\mu_{NXX}) \qquad (13)$$

where the bracketed 'ion terms' are the appropriate multiples of the summations for charged species given in Equations 5 & 6. The constraint of electrical neutrality means that only the parameter sums bracketed together in Equations 11-13 are observable. In order to assign values to individual parameters for $NH_3$ interactions we have set $\lambda_{N,Cl}$ and $\mu_{N,N,Cl}$ to zero. Although the choice of convention does not affect calculated values of $\gamma_{NH3}$ or mean ion activity coefficients, inconsistencies could arise when estimating conventional single ion activity coefficients using neutral-ion parameters from different sources. For example Harvie et al. ($\underline{1}$) set $\lambda_{CO2,H}$ equal to zero in their investigations of the carbonate system.

Ion-$NH_3$ parameters of geochemical interest determined by the authors are listed in Table IV. First derivatives of $\lambda_{Ni}$ with respect to temperature at 25°C are also given for a few ions. Some parameters, in particular those of the alkaline earths, are subject to considerable uncertainty. Further details are given in the original text ($\underline{6}$). Triplet parameters listed in Table IV are chiefly of the form $\mu_{NNi}$ (where i is an ion). Because of the high degree of separation of ions of like sign, it was thought that triplet parameters involving two such ions would be negligible and could be set to zero. Hence the final term in Equations 11 & 13 involves only $\mu_{NMX}$. There is some observational justification for this assumption. When fitting solubilities of NaCl and KCl in the systems Na/Ca/Cl/$NH_3$/$H_2O$ and K/Ca/Cl/$NH_3$/$H_2O$, it was found that the only additional term required (to those determined in systems not containing $Ca^{2+}$) was $\mu_{N,Ca,Cl}$. Its value was found to be the same in both cases, implying that the parameters $\mu_{N,Na,Ca}$ and $\mu_{N,K,Ca}$ were either very similar or negligible. In general the term $\mu_{NMX}$ was rarely required for $NH_3$.

The $NH_3$ - ion interaction parameters listed in Table IV have been obtained from a variety of different data types. Do the results form a coherent whole? In Figure 2 all the available experimental data for KCl/$NH_3$ aqueous solutions at 25°C are compared in terms of the quantity $mK^+\lambda_{N,K}$, obtained by the manipulation of the appropriate equations for partial pressure (p$NH_3$ data), activity (aqueous phase partitioning data) and activity product (KCl solubility data). While there is some scatter, particularly at low salt concentrations, agreement in terms of $\lambda_{N,K}$ (slope) is generally good. There is also consistency in $\mu_{N,N,K}$, since the value of this parameter, found to be necessary for the Baldi and Specchia p$NH_3$ data ($\underline{49}$) in Figure 2, was obtained from salt solubility measurements ($\underline{6}$). These results confirm the utility of the approach adopted here. More importantly, they illustrate the advantage of a thermodynamically based model, when treating different aspects of the system properties, over the purely empirical expressions that are often used.

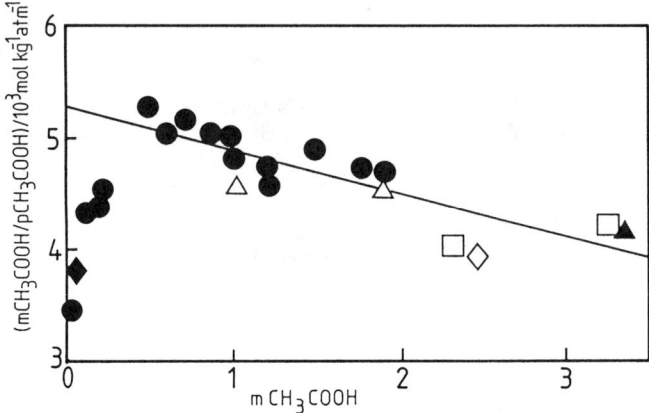

Figure 1. The quotient mCH$_3$COOH/pCH$_3$COOH calculated using partial pressure data from the following sources: dots - Fredenhagen and Liebster (28); triangles - Kaye and Parks (29); squares - Campbell et al. (31); open diamonds - Zarakhani and Vorob'eva (57); closed diamond - Wolfenden (30).

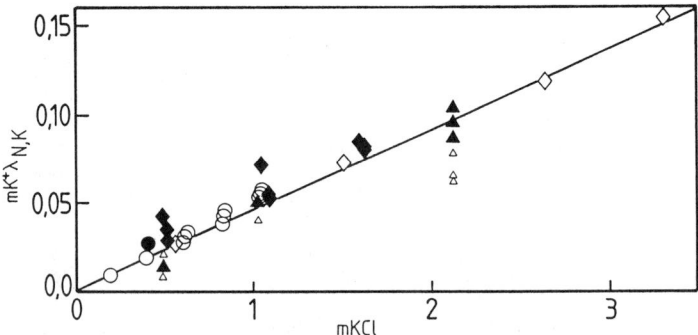

Figure 2. Comparison of the salt effect of KCl on NH$_3$ obtained from different data sets at 25°C: circle - H$_2$O/CHCl$_3$ partitioning data (58); open diamonds - KCl solubility in NH$_{3(aq)}$ (59); closed diamonds pNH$_3$ (60); dots - pNH$_3$ (56); triangles - pNH$_3$ for systems containing >10 mol kg$^{-1}$ NH$_3$ (49), and including $\mu_{N,N,K}$ in Equation 11; small open triangles - as previous data set but neglecting $\mu_{N,N,K}$. The slope of the full line is equal to 0.0454 (Table IV). The influence of $\mu_{N,N,K}$ was significant only for pNH$_3$ data (49), and salt solubility data (59).

TABLE IV. Interaction parameters for $NH_3$ (N) at 25°C (6)

| Species i | $\lambda_{N,i}$ | $\mu_{N,N,i}$ | Species i | $\lambda_{N,i}$ | $\mu_{N,N,i}$ |
|---|---|---|---|---|---|
| [a]$NH_3$ | 0.01472 | - | $F^-$ | 0.091 | - |
| [b]$Mg^{2+}$ | -0.21 | - | $Cl^-$ | 0.0 | 0.0 |
| [b]$Ca^{2+}$ | -0.081 | - | $Br^-$ | -0.022 | - |
| [c]$Na^+$ | 0.0175 | -0.000311 | $I^-$ | -0.051 | - |
| [d]$K^+$ | 0.0454 | -0.000321 | $OH^-$ | 0.103 | - |
| [e]$NH_4^+$ | 0.0 | -0.00075 | $NO_3^-$ | -0.01 | -0.000437 |
| | | | [f]$SO_4^{2-}$ | 0.140 | - |
| | | | $CO_3^{2-}$ | 0.180 | 0.000625 |

*(a)* Temperature variation over the range 0 - 40°C given by the equation: $\lambda_{NN}$ = 0.033161 - 21.12816/T + 4665.1461/$T^2$ where T/K. *(b)* Derived from single data point only. *(c)* Partial pressure and solubility data suggest $\lambda_{N,Na}$ is equal to about 0.031. *(d)* $\partial\lambda_{N,K}/\partial T$ is equal to -0.000141. *(e)* $\partial\mu_{N,N,NH4}/\partial T$ estimated equal to 2.3x$10^{-5}$. *(f)* $\partial\lambda_{N,SO4}/\partial T$ estimated to lie in the range -0.0005 to -0.00095. Additional parameter values: $\mu_{N,Ca,Cl}$, -0.00134; $\mu_{N,K,OH}$, 0.00385; $\mu_{N,NH4,SO4}$, -0.00153.

**COMPARISON OF SALT EFFECTS.** In order to illustrate the effect of dissolved salts on the solubility of both volatile strong electrolytes and neutral species, conditional Henry's law constants $K_H^*$, $K'_H^*$ are defined for the two solute types such that:

$$pHX = mH^+ \, mX^- / K_H^*, \text{ and } pN = mN / K'_H^* \qquad (14)$$

Figure 3a shows values of $K_H^*$ for trace HCl in solutions of the alkali metal chlorides, systems that are all fully parameterized within the Pitzer model. It is clear that both concentration and compositional influences are large. Figure 3b illustrates the variation of $K'_H^*$ for 1.0 mol $kg^{-1}$ $NH_3$ in solutions of the sodium halides, and Li and K chloride. The high concentration of $NH_3$ was adopted to avoid the region where significant dissociation occurs, thus here $NH_3$ behaves purely as a neutral solute. It can be seen firstly that the variation of $K'_H^*$ is much less than $K_H^*$ for HCl, and secondly that the characteristic 'Debye-Huckel' influence on $K_H^*$ (due to the variation of $\gamma_{HCl}^2$) is absent. For both solutes a regular trend is seen in the effects of ions within each periodic group.

The strongest interactions in an electrolyte solution occur between ions of opposite sign. Within the Pitzer model these are accounted for by the $B_{ca}$ and $C_{ca}$ functions, which are known for most solutes. The mixing parameters $\theta_{ij}$ and $\psi_{ijk}$, while having relatively small effect in dilute solutions such as seawater, are important in the much more concentrated mixtures typical of the atmospheric aerosol. Further examples of the effects of individual ions on partial pressures can be seen in partial pressure measurements, e.g. (6,50).

**ASSOCIATION EQUILIBRIA.** There are many solutes which undergo significant ion pairing or association reactions, for example bivalent and higher valency metals with the $SO_4^{2-}$ ion, $Mg^{2+}$ and $Ca^{2+}$ with $F^-$, and the components of the carbonate system. Even conventional strong electrolytes such as HCl and $HNO_3$ can be treated in this way (2,7). While the Pitzer model was originally formulated for strong electrolytes and assumes complete dissociation, many reactions need not be recognized explicitly, for example metal-sulphate and metal-fluoride ion pair formation (41,51). The choice of approach in such cases is essentially pragmatic, but for more strongly associating species the equilibria must be recognized explicitly (1).

For atmospheric chemical calculations one of the most important association equilibria is that involving $HSO_4^-$ formation in acidified sulphate systems. The ions $H^+$, $HSO_4^-$ and $SO_4^{2-}$ are treated separately by the model, and free ion concentrations are determined iteratively (1,52). Harvie et al. (1) have shown that this approach works successfully when predicting mineral solubilities, and Clegg and Brimblecombe (50,53) have verified it in measurements of pHCl and $pHNO_3$ over acid solutions

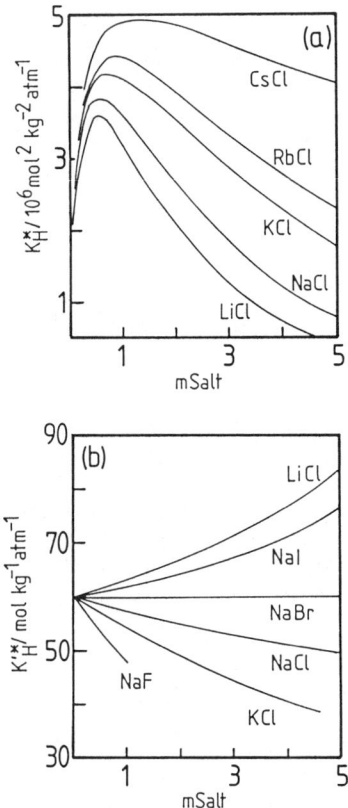

Figure 3. Salt effects on gas solubility: (a) $K_H^*$ for trace HCl in solutions of the alkali metal chlorides; (b) $K'_H{}^*$ for 1.0 mol kg$^{-1}$ NH$_3$ in solutions of the sodium halides and alkali metal chlorides.

containing the $SO_4^{2-}$ anion. Figure 4 shows measured and predicted pHCl for $H/Na/Cl/ClO_4/H_2O$ and $H/Na/Cl/SO_4/H_2O$ systems, together with the concentration of free $H^+$ calculated by the model. It is clear that formation of $HSO_4^-$ considerably reduces pHCl compared to the system containing no $SO_4^{2-}$. The measurements are reproduced to within experimental error. Interestingly, in solutions where $ClO_4^-$ is being substituted for $Cl^-$, the mean activity coefficient of HCl is only slightly affected, thus pHCl is found to be proportional to $mCl^-$.

**THE VARIATION OF TEMPERATURE.** Values of the Henry's law constants of the gases treated in this study can be obtained from the fitting equations given in Table II. Pitzer (4) lists equations giving the pure electrolyte solution parameters for HCl and some alkali metal and alkaline earth chlorides and sulphates as functions of temperature. For other solutes, the first temperature derivatives of the pure electrolyte solution parameters at 25°C are available (10). In addition we have determined values for $HNO_3$, not listed in (10), see Table I. Variations of (ionic) mixture parameters $\theta_{ij}$ and $\psi_{ijk}$ are likely to be small over the temperature range 0-40°C (11). In the case of $NH_3$ it was possible to estimate the temperature variation of $\lambda_{Ni}$ for only a few ions. However, work by the authors on salt effects on $NH_3$ solubility confirms that such variations are small, showing only a slight reduction in strength with increasing temperature. Figure 5 shows calculated and measured total vapor pressures ($pH_2O + pNH_3$) over $Na/SO_4/NH_3$ and $K/OH/NH_3$ aqueous solutions from 20-50°C. (Water vapor pressures were estimated using Equation 4, the equation for water activity given in the text, and values of $pH_2O$ over pure water listed in the *CRC Handbook* (54)). Agreement is good, despite the use of constant $\lambda_{N,Na}$ and $\lambda_{N,SO4}$ parameters determined at 25°C.

When calculating gas partial pressures it quickly becomes apparent that the principal temperature effect lies in the change of $K_H$ rather than in the activity coefficients. In Equation 9 a constant $\delta C°_p$ for the solubility equilibrium is assumed. While this is unlikely to result in appreciable errors over the temperature range 0-40°C, further from 25°C the variation of $\delta C°_p$ has a significant effect on $K_H$, requiring a more complex form of Equation 9 (55). However, the necessary data are not always available, particularly for $C°_p$ of the aqueous solute which generally accounts for most of the change.

Variations of aqueous heat capacities with respect to temperature should also be taken into account, where possible, when estimating activity coefficients. However, errors arising from lack of knowledge of the temperature variation of solute activity coefficients may be less than the uncertainty in $K_H$ or $K'_H$, which remains high for many gases.

## CONCLUSIONS

This work and others (50,53) have shown how the Pitzer model, together with appropriate Henry's law constants, can be used to calculate the solubility of volatile strong electrolytes in multicomponent solutions. The treatment of $NH_3$ summarized above shows that Pitzer formalism can also be used to describe the solubility of weak and non-electrolytes. We have noted how, for low concentrations of $NH_3$, the Pitzer equations reduce to a series of binary interaction terms similar in form to those of the well known Setchenow equations. However, the thermodynamically based approach constitutes a significant improvement over the use of purely empirical equations to predict individual thermodynamic properties because it is equally applicable to both electrolytes and uncharged species, and provides a unified description of a number of important solution properties.

A point that has only been alluded to briefly is that the ion-neutral interactions appear to be quite simply related to the size and charge of ions, something recognized even in early experiments (56). It may therefore be possible to estimate parameters for species for which there are no data. In the field of atmospheric chemistry this might be useful for modeling the thermodynamics of complex reactions, involving free radicals for example, in aerosols or evaporating clouds. There is much work yet to be done.

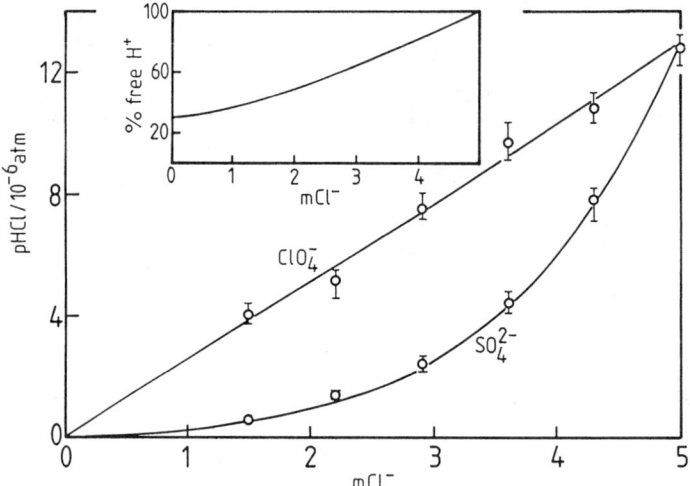

Figure 4. pHCl over H/Na/Cl/ClO$_4$ and H/Na/Cl/SO$_4$ aqueous solutions of constant stoichiometric ionic strength 5.0 mol kg$^{-1}$, showing effect of progressive replacement of Cl$^-$ by ClO$_4^-$, and SO$_4^{2-}$. In both systems total mH$^+$ is equal to 1.5 mol kg$^{-1}$. Symbols - measurements; lines - calculated values. Inset: calculated free H$^+$ ion concentration in the solutions containing SO$_4^{2-}$.

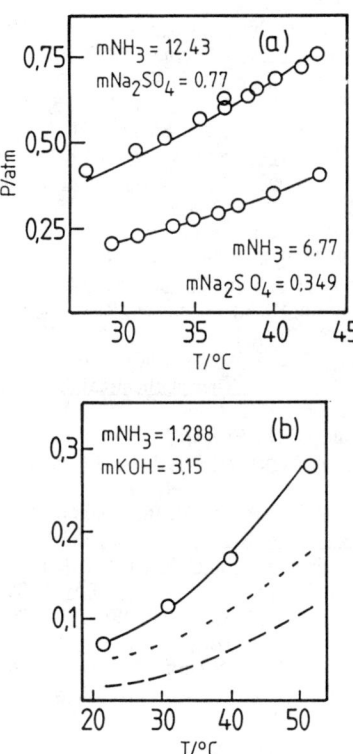

Figure 5. Total vapor pressure ($P = pH_2O + pNH_3$) above (a), aqueous Na/SO$_4$/NH$_3$ (61); and (b), aqueous K/OH/NH$_3$ (62) as a function of temperature. Symbols - measured values; full lines - calculated total pressures; dashed line - pNH$_3$; dotted line - pH$_2$O.

## LITERATURE CITED

1. Harvie, C. E.; Moller, N.; Weare, J. H. Geochim. et Cosmochim. Acta 1984, 48, 723-751.
2. Clegg, S. L.; Brimblecombe, P. Atmos. Env. 1986, 20, 2483-2485.
3. Clegg, S. L.; Brimblecombe, P. Atmos. Env. 1988, 22, 2231-2233.
4. Pitzer, K. S. Rev. Mineral. 1987, 17, 97-142.
5. Pitzer, K. S.; Silvester, L. F. J. Soln. Chem. 1976, 5, 269-277.
6. Clegg, S. L.; Brimblecombe, P. J. Phys. Chem. 1989, 93, xxx.
7. Chen, C.; Britt, H. I.; Boston, J. F.; Evans, L. B. AIChE J. 1979, 24, 820-831.
8. Pitzer, K. S. In Activity Coefficients in Electrolyte Solutions; Pytkowicz, R. M., Ed.; CRC Press: Boca Raton, 1979; Vol. 1.
9. Felmy, A. R.; Weare, J. H. Geochim. et Cosmochim. Acta 1986, 50, 2771-2783.
10. Silvester, L. F.; Pitzer, K. S. J. Soln. Chem. 1978, 7, 327-337.
11. Phutela, R. C.; Pitzer, K. S. J. Soln. Chem. 1986, 15, 649-662.
12. Pabalan, R. T.; Pitzer, K. S. J. Chem. Eng. Data 1988, 33, 354-362.
13. Hamer, W. J.; Wu, Y.-C. J. Res. Nat. Bur. Stand. 1970, 74A, 761-768.
14. Hamer, W. J.; Wu, Y.-C. J. Phys. Chem. Ref. Data 1972, 1, 1047-1099.
15. Parker, V. B. Thermal Properties of Aqueous Uni-univalent Electrolytes; National Bureau of Standards, U.S. Gov. Printing Office: Washington, 1965.
16. Holmes, H. F.; Busey, R. H.; Simonson, J. M.; Mesmer, R. E.; Archer, D. G.; Wood, R. H. J. Chem. Thermo. 1987, 19, 863-890.
17. Macaskill, J. B.; Bates, R. G. J. Soln. Chem. 1983, 12, 607-619.
18. Covington, A. K.; Robinson, R. A.; Thompson, R. J. Chem. and Eng. Data 1973, 18, 422-423.
19. Brimblecombe, P.; Clegg, S. L. J. Atmos. Chem. 1988, 7, 1-18.
20. Stull, D. R.; Prophet, H. JANAF Thermochemical Tables, NSRDS-NBS-37; US Gov. Printing Office: Washington, 1971.
21. Rossini, F. D.; Wagman, D. D.; Evans, W. H.; Levine, S.; Jaffe, I. Selected Values of Chemical Thermodynamic Properties, Part 1, NBS Circular 500; US Gov. Printing Office: Washington, 1961.
22. Wagman, D. D.; Evans, W. H.; Parker, V. B.; Schumm, I. H.; Bailey, S. M.; Churney, K. L.; Nuttall, R. L. J. Phys. Chem. Ref. Data 1982, 11, 392pp.
23. Fritz, J. J.; Fuget, C. R. Chem. and Eng. Data 1956, 10-12.
24. Tang, I. N.; Munkelwitz, H. R.; Lee J. H., Atmos. Env. 1988, 22, 2579-2585.
25. Pitzer, K. S.; Simonson, J. M. J. Phys. Chem. 1986, 90, 3005-3009.
26. Sebastiani, E.; Lacquaniti, L. Chem. Eng. Sci. 1967, 22, 1155-1162.
27. Zarakhani, N. G.; Vorob'eva, N. P.; Vinnik, M. I. Russ. J. Phys. Chem. 1971, 45, 48-50.
28. Fredenhagen, K.; Liebster, H. Z. phys. chem. 1932, A162, 449-453.
29. Kaye, W. A.; Parks, G. S. J. Chem. Phys. 1934, 2, 141-142.
30. Wolfenden, R. J. Amer. Chem. Soc. 1976, 98, 1987-1988.
31. Campbell, A. N.; Kartzmark, E. M.; Gieskes, J. M. T. M. Can. J. Chem. 1963, 41, 407-429.
32. Salomon, M. J. Soln. Chem. 1986, 15, 237-241.
33. Ramsperger, H. C.; Porter, C. W. J. Amer. Chem. Soc. 1926, 48, 1267-1273.
34. Hansen, R. S.; Miller, F. A.; Christian, S. D. J. Phys. Chem. 1955, 59, 391-395.
35. Chao, J.; Zwolinski, B. J. J. Phys. Chem. Ref. Data 1978, 7, 363-377.
36. Allred, G. C.; Woolley, E. M. J. Chem. Thermo. 1981, 13, 155-164.
37. Hepler, L. G.; Hopkins, H. P. Rev. Inorg. Chem. 1979, 1, 303-332.
38. Bates, R. G.; Pinching, G. D. J. Res. Nat. Bur. Standards 1949, 42, 419-430.
39. Harned, H. S.; Embree, N. D. J. Am. Chem. Soc. 1934, 56, 1042-1044.
40. Harned, H. S.; Ehlers, R. W. J. Am. Chem. Soc. 1933, 55, 652-656.
41. Clegg, S. L.; Brimblecombe, P. J. Chem Soc. Dalton Trans. 1988, 705-10.
42. Blanchard, D. C.; Woodcock, A. H. Ann. New York Acad. Sci. 1980, 338, 330-347.

43. Pruppacher, H. R.; Klett, J. D. Microphysics of Clouds and Precipitation; D. Reidel: Dordrecht, 1978.
44. Cohen, M. D.; Flagan, R. C.; Seinfeld, J. H. J. Phys. Chem. 1987, 91, 4563-4574.
45. Davis, E. J. Aerosol Sci. Technol. 1983, 2, 121-141.
46. Pabalan, R. T.; Pitzer, K. S. Geochim. et Cosmochim. Acta 1987, 51, 829-837.
47. Maeda, M.; Hisada, O.; Ikedo, K.; Masuda, H.; Ito, K.; Kinjo, Y. J. Phys. Chem. 1988, 92, 6404-6407.
48. Gordon, J. E. The Organic Chemistry of Electrolyte Solutions; Wiley: New York, 1975.
49. Baldi, G.; Specchia, V. Chim. Ind. (Milan) 1971, 53, 1022-1027.
50. Clegg, S. L.; Brimblecombe, P. Atmos. Env. 1988, 22, 91-100.
51. Pitzer, K. S.; Mayorga, G. J. Soln. Chem. 1974, 3, 539-546.
52. Pitzer, K. S.; Roy, R. N.; Silvester, L. F. J. Am. Chem. Soc. 1977, 99, 4930-4936.
53. Clegg, S. L.; Brimblecombe, P. Atmos. Env. 1988, 22, 117-129.
54. CRC Handbook of Chemistry and Physics; Weast, R. C., Ed.; CRC Press: Boca Raton, 1983.
55. Klotz, I. M.; Rosenberg, R. M. Chemical Thermodynamics. Basic Theory and Methods; Benjamin/Cummings: Menlo Park, 1986.
56. Gaus, W. Z. Anorg. Allgem. Chem. 1900, 25, 236264.
57. Zarakhani, N. G.; Vorob'eva, N. P. Russ. J. Phys. Chem. 1972, 46, 1392-1393.
58. Dawson, H. M.; McCrae, J. J. Chem. Soc. 1901, 79, 493-511.
59. Solubilities of Inorganic and Organic Compounds; Silcock, H. L., Ed.; Pergamon: Oxford, 1979; Vol. 3.
60. Abegg, R.; Reisenfeld, H. Z. Phys. Chem. 1902, 40, 84-108.
61. Perman, E. P. J. Chem. Soc. 1901, 79, 725-729.
62. Sorina, G. A.; Miniovich, V. M.; Efremova, G. D. Zh. Obshchei Khimii 1967, 37, 2150-2154.

RECEIVED August 31, 1989

# Chapter 6

# Modeling Solid–Solution Reactions in Low-Temperature Aqueous Systems

### Pierre D. Glynn

**U.S. Geological Survey, 432 National Center, Reston, VA 22092**

The effect of substitutional impurities on the stability and aqueous solubility of a variety of solids is investigated. Stoichiometric saturation, primary saturation and thermodynamic equilibrium solubilities are compared to pure phase solubilities. Contour plots of pure phase saturation indices (SI) are drawn at minimum stoichiometric saturation, as a function of the amount of substitution and of the excess-free-energy of the substitution. SI plots drawn for the major component of a binary solid-solution generally show little deviation from pure phase solubility except at trace component fractions greater than 1%. In contrast, trace component SI plots reveal that aqueous solutions at minimum stoichiometric saturation can achieve considerable supersaturation with respect to the pure trace-component end-member solid, in cases where the major component is more soluble than the trace.
Field or laboratory observations of miscibility gaps, spinodal gaps, critical mixing points or distribution coefficients can be used to estimate solid-solution excess-free-energies, when experimental measurements of thermodynamic equilibrium or stoichiometric saturation states are not available. As an example, a database of excess-free-energy parameters is presented for the calcite, aragonite, barite, anhydrite, melanterite and epsomite mineral groups, based on their reported compositions in natural environments.

Past studies of solid-solution aqueous-solution (SSAS) systems have focused on measuring the partitioning of trace components between solid and aqueous phases. The effect of solid-solution formation on mineral solubilities was rarely studied. Recently however, Lippmann ([1,2](#)), Thorstenson and Plummer ([3](#)) and Plummer and Busenberg ([4](#)) have enriched our understanding of SSAS systems with their theoretical and experimental descriptions of solid-solution dissolution and component distribution reactions. The objectives of this paper are: 1) to describe and to compare the concepts presented by the above authors, 2) to present some techniques which may help estimate the effect of SSAS reactions on the chemical evolution of natural waters.

DEFINITIONS AND REPRESENTATION OF THERMODYNAMIC STATES

Several thermodynamic states are of interest in the study of SSAS systems. The following sections discuss the concepts of thermodynamic equilibrium, primary saturation and stoichiometric saturation states.

## Thermodynamic Equilibrium States

Thermodynamic equilibrium in a system with a binary solid-solution $B_{1-x}C_xA$ can be defined by the law-of-mass-action equations:

$$[B^+][A^-] = K_{BA}a_{BA} = K_{BA}X_{BA}\gamma_{BA} \quad (1)$$

$$[C^+][A^-] = K_{CA}a_{CA} = K_{CA}X_{CA}\gamma_{CA} \quad (2)$$

where $[A^-]$, $[B^+]$ and $[C^+]$ are the activities of $A^-$, $B^+$ and $C^+$ in the aqueous-solution. $a_{BA}$ and $a_{CA}$, $X_{BA}$ and $X_{CA}$, and $\gamma_{BA}$ and $\gamma_{CA}$ are the activities, mole fractions and activity coefficients of components BA and CA in the equilibrium solid-solution. $K_{BA}$ and $K_{CA}$ are the solubility products of pure BA and pure CA solids.

Using equations 1 and 2, phase diagrams can be constructed which display the series of possible equilibrium states for any given binary SSAS system. By analogy to the pressure versus mole fraction diagrams used for binary-liquid vapor systems, Lippmann (1, 2, 5) defines a variable $\Sigma\Pi = [A^-]([B^+] + [C^+])$, such that adding together equations 1 and 2 yields the following relation, known as the "solidus" equation:

$$\Sigma\Pi_{eq} = K_{BA}X_{BA}\gamma_{BA} + K_{CA}X_{CA}\gamma_{CA} \quad (3)$$

where $\Sigma\Pi_{eq}$ is the value of the $\Sigma\Pi$ variable at thermodynamic equilibrium.

To completely describe thermodynamic equilibrium, a second relation must be derived from equations 1 and 2 (2; Glynn and Reardon, Am. J. Sci., in press). The "solutus" equation expresses $\Sigma\Pi_{eq}$ as a function of aqueous solution composition:

$$\Sigma\Pi_{eq} = 1 / \left( \frac{X_{B,aq}}{K_{BA}\gamma_{BA}} + \frac{X_{C,aq}}{K_{CA}\gamma_{CA}} \right) \quad (4)$$

where the aqueous activity fractions $x_{B,aq}$ and $x_{C,aq}$ are defined as $x_{B,aq} = [B^+]/([B^+]+[C^+])$ and $x_{C,aq} = [C^+]/([B^+]+[C^+])$.

Solid-solution activity-coefficients can be fitted using the following equations (6), derived from Guggenheim's expansion series for the excess-free-energy of mixing (7):

$$\ln \gamma_{BA} = X_{CA}^2[a_0 - a_1(3X_{BA} - X_{CA}) + a_2(X_{BA} - X_{CA})(5X_{BA} - X_{CA}) + ...] \quad (5)$$

$$\ln \gamma_{CA} = X_{BA}^2[a_0 + a_1(3X_{CA} - X_{BA}) + a_2(X_{CA} - X_{BA})(5X_{CA} - X_{BA}) + ...] \quad (6)$$

The first two terms of equations 5 and 6 are generally sufficient to accurately represent the dependence of $\gamma_{BA}$ and $\gamma_{CA}$ on composition (Glynn and Reardon, Am. J. Sci., in press). Indeed, in the case of a solid-solution with a small difference in the size of the substituting ions (relative to the size of the non-substituting ion), the first parameter, $a_0$, is usually sufficient (8). Equations 5 and 6 then become identical to those of the "regular" solid-solution model of Hildebrand (9). For the case where both $a_0$ and $a_1$ parameters are needed, equations 5 and 6 become equivalent to those of the "subregular" solid-solution model of Thompson and Waldbaum (10), a model much used in high-temperature work. Equations 5 and 6 can also be shown equivalent to Margules activity coefficient series (11).

Lippmann's solidus and solutus curves can be plotted and used to predict the solubility of any binary solid-solution at thermodynamic equilibrium, as well as the distribution of components between solid and aqueous phases, if solid-phase and aqueous-phase activity coefficients are known. Figure 1 shows an example of a Lippmann phase diagram for the Ag(Cl,Br) - $H_2O$ system, modeled using the distribution coefficient data of Vaslow and Boyd (12) for trace AgBr in AgCl and trace AgCl in AgBr. Aqueous solutions which plot below the solutus curve are undersaturated with respect to all solid phases, including the pure end-member solids, while solutions plotting above the solutus are supersaturated with respect to one or more solid-solution compositions.

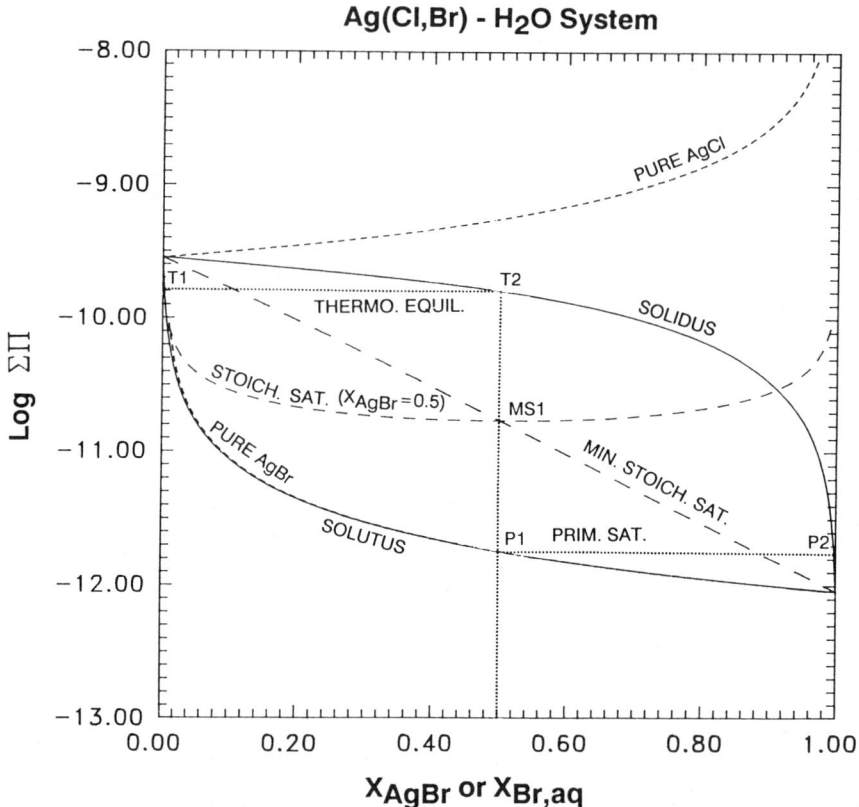

Figure 1. Lippmann diagram (with stoichiometric and pure-phase saturation curves) for the Ag(Cl,Br) - $H_2O$ system at 30° C. Calculated $a_0$ and $a_1$ values are 0.30 and -0.18 respectively. $pK_{AgCl}$ = 9.55 (16). $pK_{AgBr}$ = 12.05 (12). T1 and T2 give the aqueous and solid phase compositions, respectively, of a system at thermodynamic equilibrium with respect to an $AgCl_{.5}Br_{.5}$ solid. P1 and P2 describe the state of a system at primary saturation with respect to the same solid. MS1 gives the composition of an aqueous phase at congruent stoichiometric saturation with respect to that solid.

## 6. GLYNN   *Modeling Solid—Solution Reactions in Low-Temperature Systems*

### Primary Saturation States

Primary saturation is the first state reached during the congruent dissolution of a solid-solution, for which the aqueous-solution is saturated with respect to a secondary solid-phase (13, 14, Glynn and Reardon, Am. J. Sci., in press). This secondary solid will usually have a composition different from that of the dissolving solid. At primary saturation, the aqueous phase is at thermodynamic equilibrium with respect to this secondary solid but remains undersaturated with respect to the primary dissolving solid. The series of possible primary-saturation states for a given SSAS system is represented by the solutus curve on a Lippmann diagram.

In the specific case of a "strictly congruent" dissolution process occurring in an aqueous phase with a $[B^+]/[C^+]$ activity ratio equal to the $B^+/C^+$ ratio in the solid, primary-saturation can be approximately found by drawing a straight vertical line on the Lippmann diagram from the solid-phase composition to the solutus (see figure 1). For an exact calculation, the following relations may be used to determine the primary saturation state:

$$X_{B,aq} = f_B X_{BA} \qquad (7) \qquad\qquad X_{C,aq} = f_C X_{CA} \qquad (8)$$

where $f_B$ and $f_C$ are factors correcting for a possible difference in the aqueous speciation and activity coefficients of $B^+$ and $C^+$.

The equation used to calculate the value of $\Sigma\Pi$ at primary saturation as a function of solid composition, for a strictly congruent dissolution process, may be found by combining the Lippmann solutus equation (4) with equations 7 and 8:

$$\Sigma\Pi_{ps} = 1 / \left( \frac{X_{BA} f_B}{K_{BA} \gamma_{BA,y}} + \frac{X_{CA} f_C}{K_{CA} \gamma_{CA,y}} \right) \qquad (9)$$

where $\gamma_{BA,y}$ and $\gamma_{CA,y}$ refer to the activity coefficients of BA and CA in the secondary solid $B_{1-y}C_yA$ with respect to which the aqueous solution (at primary saturation) is in temporary thermodynamic equilibrium. $\Sigma\Pi_{ps}$ refers to the value of the $\Sigma\Pi$ variable as specifically defined at primary saturation. The composition of the $B_{1-y}C_yA$ phase will generally not be known. By equating $\Sigma\Pi_{ps}(x)$ (equation 9) to $\Sigma\Pi_{eq}(y)$ (equation 3), the relation between the initial solid composition $B_{1-x}C_xA$ and the secondary solid $B_{1-y}C_yA$ may be obtained:

$$1 / \left( \frac{X_{BA} f_B}{K_{BA} \gamma_{BA,y}} + \frac{X_{CA} f_C}{K_{CA} \gamma_{CA,y}} \right) = X_{CA,y} \gamma_{CA,y} K_{CA} + X_{BA,y} \gamma_{BA,y} K_{BA} \qquad (10)$$

In the case of a non-ideal solid-solution series, equation 10 must be solved graphically or by an iterative technique, because $\gamma_{BA,y}$ and $\gamma_{CA,y}$ are typically exponential functions of $x_{CA,y}$.

### Stoichiometric Saturation States

Stoichiometric saturation was formally defined by Thorstenson and Plummer (3). These authors argued that solid-solution compositions typically remain invariant during solid aqueous-phase reactions in low-temperature geological environments, thereby preventing attainment of thermodynamic equilibrium. Thorstenson and Plummer defined stoichiometric saturation as the pseudo-equilibrium state which may occur between an aqueous-phase and a multi-component solid-solution, "in situations where the composition of the solid phase remains invariant, owing to kinetic restrictions, even though the solid phase may be a part of a continuous compositional series".

The stoichiometric saturation concept assumes that a solid-solution can under certain circumstances behave as if it were a pure one-component phase. In such a situation, the dissolution of a solid-solution $B_{1-x}C_xA$ can be expressed as:

$$B_{1-x}C_xA \rightarrow (1-x)B^+ + xC^+ + A^- \tag{11}$$

Applying the law of mass action then gives the defining condition for stoichiometric saturation states:

$$IAP_{ss} = K_{ss} = \frac{[C^+]^x[B^+]^{1-x}[A^-]}{1} = \exp\left(\frac{-\Delta G_r^0}{RT}\right) \tag{12}$$

where x and (1-x) are equal to $x_{CA}$ and $x_{BA}$ respectively and where $\Delta G_r^0$ is the standard free energy change of reaction 11.

According to Thorstenson and Plummer's (3) definition of stoichiometric saturation, an aqueous-solution at thermodynamic equilibrium with respect to a solid $B_{1-x}C_xA$ will always be at stoichiometric saturation with respect to that same solid. The converse statement, however, is not necessarily true: stoichiometric saturation does not necessarily imply thermodynamic equilibrium.

Stoichiometric saturation states can be represented on Lippmann phase diagrams (figure 1) by relating the total solubility product variable $\Sigma\Pi_{ss}$ (defined specifically at stoichiometric saturation with respect to a solid $B_{1-x}C_xA$) to the $K_{ss}$ constant (equation 12) and to the aqueous activity fractions $x_{B,aq}$ and $x_{C,aq}$:

$$\Sigma\Pi_{ss} = \frac{K_{ss}}{\chi_{B,aq}^{1-x}\chi_{C,aq}^{x}} \tag{13}$$

In contrast to thermodynamic equilibrium, for which a single ($x_{B,aq}$, $\Sigma\Pi_{eq}$) point satisfies equations 1 and 2, stoichiometric saturation with respect to a given solid composition is represented by a series of ($x_{B,aq}$, $\Sigma\Pi_{ss}$) points, all defined by relation 13. As shown in figure 1, stoichiometric saturation states never plot below the solutus curve. This is consistent with the fact that stoichiometric saturation can never be reached before primary saturation in a solid-solution dissolution experiment. The unique point at which a stoichiometric saturation curve (for a given solid $B_{1-x}C_xA$) joins the Lippmann solutus represents the composition of an aqueous solution at thermodynamic equilibrium with respect to a solid $B_{1-x}C_xA$.

Saturation curves for the pure BA and CA end-member solids can also be drawn on Lippmann diagrams (2,5):

$$\Sigma\Pi_{BA} = \frac{K_{BA}}{\chi_{B,aq}} \tag{14} \qquad \Sigma\Pi_{CA} = \frac{K_{CA}}{\chi_{C,aq}} \tag{15}$$

These equations define the families of $(x_{BA}, \Sigma\Pi_{BA})$ and $(x_{CA}, \Sigma\Pi_{CA})$ conditions for which a solution containing $A^-$, $B^+$ and $C^+$ ions will be saturated with respect to pure BA and pure CA solids. Thermodynamic equilibrium with respect to a mechanical mixture of the two pure BA and CA solids, in contrast to a solid-solution of BA and CA, will be represented on a Lippmann diagram by a single point, namely the intersection of the pure BA and pure CA saturation curves. The coordinates of this intersection are:

$$\chi_{B,aq}^{int} = \frac{K_{BA}}{K_{BA} + K_{CA}} \tag{16} \qquad \Sigma\Pi^{int} = K_{CA} + K_{BA} \tag{17}$$

## COMPARISON OF SOLID-SOLUTION AND PURE PHASE SOLUBILITIES

In predicting solid-solution solubilities, one of two possible hypotheses must be chosen. In the first, the solid-solution is treated as a one-component or pure-phase solid, given that the equilibration time is sufficiently short, the solid to aqueous-solution ratio is sufficiently high and the solid is relatively insoluble. These requirements are needed to ensure that no significant recrystallisation of the initial solid or precipitation of a secondary solid-phase occurs. For such situations, the stoichiometric saturation concept may apply.

The second hypothesis considers the solid as a multi-component solid-solution, capable of adjusting its composition in response to the aqueous solution composition, given the long equilibration period, the relatively high solubility of the solid and relatively low solid to solution ratio. In this case, the assumption of thermodynamic equilibrium may apply. If the equilibration period is too short for thermodynamic equilibrium to have been achieved, but if an outer surface layer of the solid has been able to recrystallise (because of the high solubility of the solid), the concept of primary saturation may apply.

There are currently insufficient data to determine the exact conditions for which each of these assumptions may apply, especially in field situations. In many instances, neither one of these assumptions will explain the observed solubility of a solid-solution, which may lie between the "maximum" stoichiometric saturation solubility and the "minimum" primary saturation solubility. Nonetheless, these solubility limits can often be estimated.

Stoichiometric Saturation Solubilities

The case of stoichiometric saturation states attained after "strictly congruent" dissolution is examined here. The hypothetical $B_{1-x}C_xA$ solid-solutions considered are 1) calcite ($pK_{sp}$ = 8.48, 15) with a more soluble trace $NiCO_3$ component ($pK_{sp}$ = 6.87, 16) and 2) calcite with a less soluble trace $CdCO_3$ component ($pK_{sp}$ = 11.31, 17). Contour plots of saturation indices (SI=log[IAP/$K_{sp}$]) with respect to major and trace end-member components are drawn in figure 2 as a function of the $a_0$ value (assuming a regular solid-solution model) and of the log of the mole-fraction of the trace component (where $10^{-6} \leq x_{CA} \leq 0.5$). SI values are calculated for major (BA) and trace (CA) components using the relations:

$$SI_{BA} = -\log(K_{BA}) + \log\left[K_{ss}\left(\frac{x_{B,aq}}{x_{C,aq}}\right)^x\right] \quad (18)$$

$$SI_{CA} = -\log(K_{CA}) + \log\left[K_{ss}\left(\frac{x_{C,aq}}{x_{B,aq}}\right)^{(1-x)}\right] \quad (19)$$

$K_{ss}$ values are evaluated from Thorstenson and Plummer's (3) equation 22, modified assuming a regular solid-solution model:

$$K_{ss} = K_{BA}^{(1-x)} K_{CA}^x (1-x)^{(1-x)} x^x \exp[a_0 x(1-x)] \quad (20)$$

Assuming that the dissolution to stoichiometric saturation takes place in initially pure water and that the aqueous activity ratio of the major and minor ions is equal to their concentration ratio, the relation $x_{B,aq}/x_{C,aq} = (1-x)/x$ will apply. Using this relation, applicable only at "minimum stoichiometric saturation" (Glynn and Reardon, Am. J. Sci., in press), the following equations may be derived from equations 18, 19 and 20:

$$SI_{BA} = x \log\left(\frac{K_{CA}}{K_{BA}}\right) + \log(1-x) + \frac{a_0 x(1-x)}{\ln(10)} \quad (21)$$

$$SI_{CA} = (1-x)\log\left(\frac{K_{BA}}{K_{CA}}\right) + \log x + \frac{a_0 x(1-x)}{\ln(10)} \quad (22)$$

The SI contour plots drawn using equations 21 and 22 show the miscibility gap and spinodal gap lines separating intrinsically stable, metastable and unstable solid-solutions (11). In natural environments, while metastable solid-solution compositions may in some cases persist on a geological timescale (depending on the solubility of the solid), unstable solid-solutions formed in low-temperature environments are not likely to do so. Busenberg and Plummer (18) in their study of magnesian calcite solubilities observed that the highest known magnesium contents in modern natural biogenic calcites correspond to the predicted spinodal composition. Metastable compositions will probably not persist, however, in solid-solutions with higher solubilities or which have been reacted for a longer

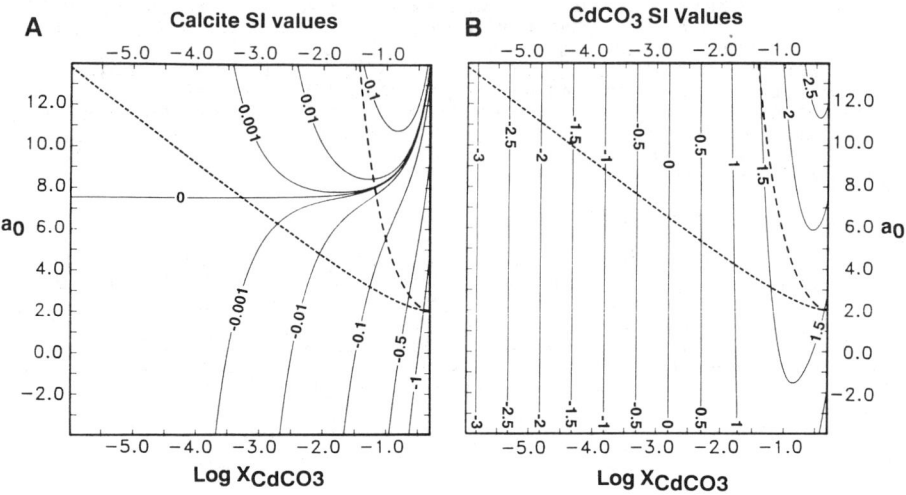
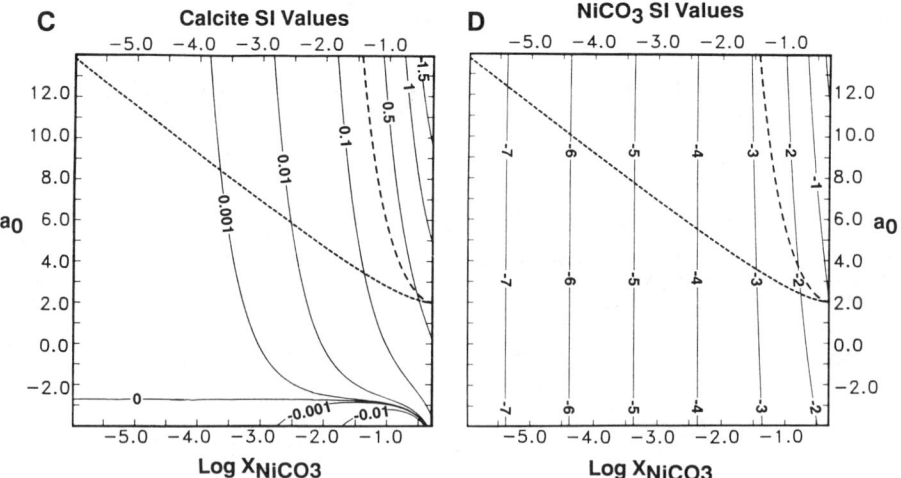

Figure 2. Major and trace component saturation index values (in solid lines) for $(Ca,Cd)CO_3$ (insets A, B) and $(Ca,Ni)CO_3$ (insets C, D) solids at congruent stoichiometric saturation. Miscibility gap lines (short-dashed) and spinodal gap lines (long-dashed) are also shown.

period of time (or at higher temperatures) than the biogenic magnesian calcites. While intrinsically unstable solid-solutions, formed at low-temperatures, may not be found in geologic environments, they can often be synthesized in the laboratory (eg. strontian aragonites (4); barian strontianites, Glynn, unpublished).

Figures 2A and 2B show the case of a solid-solution series, $(Ca,Cd)CO_3$, with a much less soluble trace end-member. If the mole-fraction of trace component is sufficiently high ($x_{CdCO_3} > 10^{-2.8}$), the aqueous phase at stoichiometric saturation will be supersaturated with respect to the trace end-member (except at unrealistic negative $a_0$ values not shown on the plot). The lower solubility of the trace component will generally cause negative SI values for the major component, except at high $a_0$ values (higher than 7.5 in the $(Ca,Cd)CO_3$ case) for which the solid-solutions will generally be metastable or unstable. Calcite SI values drawn in figure 2A show that the mole-fraction of trace component must be sufficiently high ($x_{CdCO_3} > 10^{-2.5}$) for this effect to be measurable in the field (typical uncertainty ≈ 0.01) or in the laboratory.

A laboratory example of the above principle is given by the "strictly-congruent" dissolution experiment of Denis and Michard (14) on a 3.5% Sr-anhydrite. Analysis of their results shows that maximum $SI_{celestite}$ and $SI_{gypsum}$ values of 0.37 and -0.04 respectively were attained after 4 days. Their last sample (after 6 days) gave $SI_{celestite}$ and $SI_{gypsum}$ values of 0.35 and -0.04 respectively. While these results show that stoichiometric saturation was not obtained with respect to the original anhydrite phase (probably because of back-precipitation of a gypsum phase), supersaturation did occur with respect to the less soluble celestite component.

In the case of solid-solutions with a more soluble trace component, aqueous solutions at "minimum stoichiometric saturation" will generally be supersaturated with respect to the major component and undersaturated with respect to the more soluble trace end-member (figures 2C and 2D). Aqueous solutions at minimum stoichiometric saturation with respect to $(Ca,Ni)CO_3$ solid-solutions will be supersaturated with respect to pure calcite at $a_0$ values greater than -2.7. $SI_{calcite}$ values greater than +0.01, however, will only be found at mole fractions of $NiCO_3$ greater than approximately $10^{-2.5}$. In contrast, $SI_{NiCO_3}$ values will exhibit significant undersaturation even at high $x_{NiCO_3}$ mole fractions ($-2 < SI_{NiCO_3} < -1$ at $x_{NiCO_3} = 0.5$).

## Primary Saturation and Thermodynamic Equilibrium Solubilities

A detailed discussion of solid-solution solubilities at primary saturation states and at thermodynamic equilibrium states is given by Glynn and Reardon (Am. J. Sci., in press). The fundamental principles governing these thermodynamic states are given below.

An aqueous solution at primary saturation or at thermodynamic equilibrium with respect to a solid-solution will be undersaturated with respect to all end-member component phases of the solid-solution (see figure 1).

A positive excess-free-energy of mixing (that is a positive $a_0$ value in the case of a regular solid-solution) will raise the position of the solutus curve relative to that of an ideal solid-solution system. A positive excess-free-energy of mixing will therefore increase the solubility of a solid-solution at primary saturation or at thermodynamic equilibrium. The pure end-member saturation curves on a Lippmann diagram offer a upward limit on the position of the solutus. Conversely, a negative excess-free-energy of mixing will lower the position of the solutus relative to that of an ideal solid-solution series.

The solutus curve, in binary SSAS systems with ideal or positive solid-solution free-energies of mixing and with large differences (more than an order of magnitude) in end-member solubility products, will closely follow the pure-phase saturation curve of the least soluble end-member (except at high aqueous activity fractions of the more soluble component, eg. figure 1). In contrast, ideal solid-solutions with very close end-member solubility products (less than an order of magnitude apart) will have a solutus curve up to 2 times lower in $\Sigma\Pi$ than the pure end-member saturation curves. The factor of 2 is obtained for the case where the two end-member solubility products are equal and can be derived from equations 4, 16 and 17.

The composition of a SSAS system at primary saturation or at stoichiometric saturation will be generally independent of the initial solid to aqueous-solution ratio, but will depend on the initial aqueous-solution composition existing prior to the dissolution of the solid. In contrast, the final thermodynamic equilibrium state of a SSAS system attained after a dissolution or recrystallisation process will generally depend not only on the initial composition of the system but also on the initial solid to aqueous-solution ratio (Glynn *et al.*, submitted to Geochim. Cosmochim. Acta).

## ESTIMATION OF THERMODYNAMIC MIXING PARAMETERS

There are two main applications for SSAS theory in the chemical modeling of aqueous systems: 1) the prediction of solid-solution solubilities, 2) the prediction of the distribution of trace components between solid and aqueous phases. Currently, a big problem with both types of predictions is the lack of low-temperature data on solid-solution excess-free-energy functions, and therefore on solid-phase activity coefficients. The two-parameter Guggenheim expansion series (the "subregular" model) has been successfully used to fit laboratory solubility data for the $(Sr,Ca)CO_3$ - $H_2O$ (4), $(Ba,Sr)CO_3$ -$H_2O$ (Glynn, unpublished data), $(Ca,Mg)CO_3$ - $H_2O$ (18) and $K(Cl,Br)$ - $H_2O$ systems (Glynn *et al.*, submitted to Geochim. Cosmochim. Acta). The one-parameter Guggenheim series (the "regular" model) has also been frequently used (2, 19). More laboratory determinations of thermodynamic mixing parameters are needed, not only to acquire data on binary and multi-endmember solid-solutions, but also to further confirm the validity of the regular and sub-regular models, and to compare them with other excess-free-energy models.

In the meantime, as a better approximation than the commonly used assumption of "ideal" solid-solutions, $a_0$ and $a_1$ parameters can often be estimated. A computer code has been written (Glynn, unpublished), which uses either observed or estimated 1) miscibility gap data, 2) spinodal gap data, 3) critical temperature and critical mole-fraction of mixing data, 4) distribution coefficient data, 5) alyotropic point data (2) or 6) activity coefficient data to calculate $a_0$ and $a_1$ parameters. Depending on whether one or two datum points are given, the program assumes either a regular or subregular model.

### Miscibility Gap Data

Miscibility gaps, determined from mineral compositions observed in the field, can be used to estimate thermodynamic mixing parameters in the absence of more accurate laboratory data. This approach suffers from several problems.

The maximum mole-fraction of trace component found may not correspond to the miscibility gap fraction. In this case, if the solid-solution composition is stable, then the excess-free-energy calculated will be overestimated. If the mineral was formed at much higher temperatures than the temperature of interest and if the mineral is fairly unreactive at the lower temperature, the maximum solid-solution mole-fraction observed may well be metastable or even unstable.

The temperature of formation (and of equilibrium) of the solid-solution may not be known. If a lower temperature is assumed, the excess-free-energy will be underestimated.

The extrapolation to 25° C of excess-free-energies estimated at higher temperatures will introduce an additional error, because of the lack of excess enthalpy and excess entropy data.

A partial solid-solution series may not be isomorphous (i.e. the end-members may not have the same structures). In that case the excess-free-energy parameters should be calculated only on a single side of the miscibility gap. On the other side of the miscibility gap, a different model will apply.

Despite the above problems, mixing parameters estimated from miscibility gap information will still be an improvement over the assumption of an ideal solid-solution model. $a_0$ parameters estimated from data in Palache *et al.* (20) and Busenberg and Plummer (21) are presented in table I for a few low-temperature mineral groups. Because of the large uncertainties in the data and in the estimation procedure, a sub-regular model is usually unwarranted. As a result, these estimated $a_0$ values presented should be used only for solid-solution compositions on a single side of the miscibility gap, i.e. only up to the given miscibility fraction.

Table I. Estimated Mixing Parameters for Binary Solid-Solutions at 25° C

| System (main, trace) | Misc. Frac. ($X_{trace}$) | Temp. (°C) | Structure | Estimated $a_0$ at 25 C | Ref. |
|---|---|---|---|---|---|
| $(Mg,Ni)SO_4.7H_2O$ | complete | 25 | orth. | < 2 | 20 |
| $(Mg,Zn)SO_4.7H_2O$ | complete | 25 | orth. | < 2 | 20 |
| $(Mg,Fe)SO_4.7H_2O$ | 16.7% | 25 | orth.-mono. | 2.41 | 20 |
| $(Mg,Mn)SO_4.7H_2O$ | 28.6% | 25 | orth.-mono. | 2.14 | 20 |
| $(Zn,Cu)SO_4.7H_2O$ | 22.2% | 25 | orth.-mono. | 2.26 | 20 |
| $(Zn,Fe)SO_4.7H_2O$ | 22.4% | 25 | orth.-mono. | 2.25 | 20 |
| $(Zn,Mn)SO_4.7H_2O$ | 25.6%? | 25 | orth.-mono. | 2.22 | 20 |
| $(Ni,Fe)SO_4.7H_2O$ | 16.7% | 25 | orth.-mono. | 2.41 | 20 |
| $(Ni,Cu)SO_4.7H_2O$ | 2% | 25 | orth.-mono. | 4.05 | 20 |
| $(Fe,Cu)SO_4.7H_2O$ | 65.4% | 25 | mono. | subregular model needed | 20 |
| $(Fe,Mn)SO_4.7H_2O$ | 8.2% | 25 | mono. | 2.89 | 20 |
| $(Fe,Co)SO_4.7H_2O$ | complete | 25 | mono. | < 2 | 20 |
| $(Fe,Mg)SO_4.7H_2O$ | 58% | 25 | mono.-orth. | subregular model needed | 20 |
| $(Fe,Zn)SO_4.7H_2O$ | 30.7% | 25 | mono.-orth. | 2.11 | 20 |
| $Ag(Cl,Br)$ | complete | 25 | isom. | < 2 | 20 |
| $K(Cl,Br)$ | complete | 25 | isom. | $a_0 = 1.40$, $a_1 = -0.08$ | † |
| $(Ba,Pb)SO_4$ | 20% | 75 | orth. | 2.70 | 20 |
| $(Ba,Sr)SO_4$ | complete | 75 | orth. | 2.34 | 20 |
| $(Ba,Ca)SO_4$ | 7.7% | 75 | orth. | 3.43 | 20 |
| $(Sr,Ca)SO_4$ | 8.3% | 50 | orth. | 3.12 | 20 |
| $(Ca, Mn)CO_3$ | complete ? | 25 | rhomb. | < 2? | 20 |
| $(Ca, Fe)CO_3$ | max. 2.5% | 50 | rhomb. | 4.18 | 21 |
| $(Ca, Zn)CO_3$ | min. 4% | 50 | rhomb. | 3.74 | 20 |
| $(Ca, Co)CO_3$ | min. 2% | 50 | rhomb. | 4.39 | 20 |
| $(Ca, Mg)CO_3$ | 2% | 25 | rhomb. | 4.05 | 21 |
| $(Ca, Mg)CO_3$ | 3.2% | 50 | rhomb. | 3.95 | 21 |
| $(Ca, Mg)CO_3$ | 4.5% | 150 | rhomb. | 4.76 | 21 |
| $(Ca,Ca_{.5}Mg_{.5})CO_3$ "non-defective" | $2.0\% \leq x_{Mg} \leq 49.96\%$ | 25 | rhomb. | $a_0 = 5.08$, $a_1 = 1.90$ | 18 |
| $(Ca,Ca_{.5}Mg_{.5})CO_3$ "defective" | $10.7\% \leq x_{Mg} \leq 47.5\%$ | 25 | rhomb. | $a_0 = 2.54$, $a_1 = 0.71$ | 18 |
| $(Ca, Pb)CO_3$ | min. 10.5% | 50 | rhomb.-orth. | 2.94 | 20 |
| $(Mg,Fe)CO_3$ | complete | 25 | rhomb. | < 2 | 20 |
| $(Mg,Mn)CO_3$ | min. 9.7% | 25 | rhomb. | 2.77 | 20 |
| $(Mg,Ca)CO_3$ | min. 10.1% | 25 | rhomb. | 2.74 | 20 |
| $(Mg,Co)CO_3$ | 6.2% | 25 | rhomb. | 3.10 | 20 |
| $(Fe,Mn)CO_3$ | complete | 25 | rhomb. | < 2 | 20 |
| $(Fe,Ca)CO_3$ | max. 22.7% | 25 | rhomb. | 2.24 | 20 |
| $(Fe,Co)CO_3$ | 13.0% | 25 | rhomb. | 2.57 | 20 |
| $(Mn,Mg)CO_3$ | min. 34.1% | 75 | rhomb. | 2.42 | 20 |
| $(Mn,Zn)CO_3$ | min. 44.6% | 75 | rhomb. | 2.34 | 20 |
| $(Co,Ca)CO_3$ | min. 7% | 50 | rhomb. | 3.26 | 20 |
| $(Zn,Ca)CO_3$ | 24% | 25 | rhomb. | 2.22 | 20 |
| $(Zn,Fe)CO_3$ | min. 38.6% possibly 54.0% | 25 | rhomb. | 2.04 | 20 |
| $(Zn,Co)CO_3$ | 15.9% | 25 | rhomb. | 2.44 | 20 |
| $(Zn,Cu)CO_3$ | 9.7% | 25 | rhomb. | 2.77 | 20 |
| $(Zn,Mn)CO_3$ | min. 16.1% possibly complete | 25 | rhomb. | 2.44 | 20 |
| $(Zn,Cd)CO_3$ | 2.7% | 25 | rhomb. | 3.79 | 20 |
| $(Zn,Mg)CO_3$ | 20.8% | 25 | rhomb. | 2.29 | 20 |
| $(Zn,Pb)CO_3$ | < 0.3% | 25 | rhomb.-orth. | 5.84 | 20 |
| $(Ca,Sr)CO_3$ | 3.8% | 25 | orth. | 3.50 | 20 |
| $(Ca,Sr)CO_3$ | $0.48\% \leq x_{Sr} \leq 85.7\%$ | 25 | orth. | $a_0 = 3.43$, $a_1 = -1.82$ | 4 |
| $(Sr,Ca)CO_3$ | 18.2% | 75 | orth. | 2.76 | 20 |
| $(Sr,Ca)CO_3$ | $2.32\% \leq x_{Sr} \leq 79.0\%$ | 76 | orth. | $a_0 = 2.66$, $a_1 = -1.16$ | 4 |
| $(Ca,Pb)CO_3$ | 7.7% | 25 | orth. | 2.94 | 20 |
| $(Pb,Sr)CO_3$ | 8.7% | 25 | orth. | 2.84 | 20 |

† (Glynn, Reardon, Plummer and Busenberg, submitted to <u>Geochim. Cosmochim. Acta</u>)

## Spinodal Gap Data

Spinodal gaps can be used to estimate low temperature solid-solution mixing parameters. This method is considerably less accurate than the miscibility gap estimation technique. The estimation assumes that unstable compositions are not likely to form during a precipitation process, although metastable compositions may do so. There are at least two cases, where calculated spinodal gaps have been consistent with laboratory or field observations. Using a subregular Guggenheim model to fit the solubilities of "highly-defective" magnesian calcites, Busenberg and Plummer (18) calculated a spinodal composition of 19.8 % Mg. They concluded that this result was in agreement with the magnesium content of modern biogenic magnesian-calcites (which range from 1 to 20% Mg, coralline algae excepted). Similarly, from stoichiometric saturation measurements on the $(Sr,Ca)CO_3 - H_2O$ system, Plummer and Busenberg (4) determined spinodal gaps at 25° and 76° C of $0.065 \leq x_{Sr} \leq 0.620$ and $0.103 \leq x_{Sr} \leq 0.585$ respectively. Synthesis of solid-solutions in these compositional ranges proved to be much more difficult than that of stable and metastable solids (4, Busenberg, pers. commun.). In general however, it is very difficult to distinguish between intrinsically unstable and metastable solid-solution compositions, and therefore the spinodal gap $a_0$ estimation technique can not usually be recommended.

## Critical Mixing Point Data

Critical mixing points can also be used to calculate the non-ideality of a solid-solution series. The critical temperature $T_c$ of a SSAS system refers to the lowest possible temperature at which a complete solid-solution series may form (11). In a regular solid-solution series, a miscibility gap will occur only if $a_0$ is greater than 2 and the following relation may be derived between $a_0$, $T_c$ and T the current temperature of the solid-solution system (2):

$$a_0 = 2T_c/T \qquad (23)$$

Thus, if the critical temperature of a binary solid-solution is known or can be estimated, a regular solid-solution model can be assumed as a first approximation, and the value of $a_0$ can be calculated from equation 23. If the critical mole fraction of mixing is also known, another set of relations can be derived and both the $a_0$ and $a_1$ parameters in the subregular model can be calculated. The use of critical temperatures and critical mole-fractions of mixing to estimate $a_0$, $a_1$ data at low temperatures usually assumes, because of the lack of excess enthalpy and excess entropy data, that the free energy of mixing of a solid-solution decreases with increasing temperature because of the increase in the temperature ideal-entropy of mixing $TS^{m,id}$. At the critical temperature, this term becomes sufficiently large to overcome the repulsive energy exerted between the solid-solution components; the miscibility gap disappears.

An estimate of the maximum possible $a_0$ value for ionic crystals can be obtained by substituting the highest temperature of melting of ionic solids for the critical temperature. The alkaline-earth sulfates and carbonates, for example, melt at temperatures as high as 1700°C. The calculated maximum $a_0$ value for these ionic solids is therefore approximately 13-14. The justification for this estimate depends on two postulates: 1) most ionic melts are completely miscible with each other, 2) the short-lived structures of an ionic melt often closely resemble that of the solid-solution from which it was formed (22). According to Samoilov (22), the last postulate can be justified by comparing the lattice energy of an ionic solid with its heat of fusion. Most commonly, the heat of fusion is a small fraction of the solid's lattice energy, and consequently the lattice energies (and presumably the structures) of the solid and the melt do not differ considerably. If the above reasoning giving a maximum value of 13-14 for $a_0$ is correct, then a minimum miscibility mole-fraction of $10^{-6}$ can be calculated. This estimate agrees with the observation that the purest chemical reagents are at best "five nines" pure (23).

### Distribution Coefficient Measurements

Distribution coefficients observed in the field or in laboratory experiments have often been used to estimate excess-free-energies of mixing. The Ag(Cl,Br) phase diagram shown in figure 1, for example, was constructed from distribution coefficient data (12). Glynn and Reardon (Am. J. Sci., in press) using Crocket and Winchester's (24) (Ca,Zn)CO$_3$ distribution coefficient data confirmed the value of a$_0$ estimated from solid-solution compositions given in Palache et al. (20). There are two main problems in using this technique to calculate solid-solution mixing parameters: 1) the ratio of the component activities in the adjoining aqueous (or non-aqueous) phase must be known, 2) the distribution coefficients measured must be representative of thermodynamic equilibrium between the two phases. The large variations in coefficients reported for SSAS systems (eg. the magnesian calcites, 25), are evidence of the problems in obtaining thermodynamic equilibrium coefficients (4).

### Stoichiometric Saturation Solubilities

Stoichiometric saturation measurements in carefully controlled laboratory experiments offer perhaps the most promising technique for the estimation of thermodynamic mixing parameters (3; Glynn and Reardon, Am. J. Sci., in press). Unfortunately, the results obtained can usually not be verified by a second independent and accurate method, such as reaction calorimetry or measurement of thermodynamic equilibrium solubilities (4). The conditions necessary in obtaining good stoichiometric saturation data (as opposed to thermodynamic equilibrium data) were discussed earlier.

### Thermodynamic Equilibrium Solubilities

Thermodynamic equilibrium measurements of solid-solution solubilities would be the ideal method for obtaining solid-solution excess-free-energies. Unfortunately, thermodynamic equilibrium states commonly can not be obtained in the laboratory, as solid-solution recrystallisation experiments have indicated (4). SSAS systems with highly soluble solid-solutions capable of rapid recrystallisation, such as the K(Cl,Br) - H$_2$O system, are a significant exception to this rule (Glynn and Reardon, Am. J. Sci., in press).

CONCLUSIONS

1) Lippmann phase diagrams can be used to describe and compare thermodynamic equilibrium (equations 3, 4), primary saturation (equations 9, 10), stoichiometric saturation (equation 13) and pure end-member saturation states (equations 14, 15) in binary SSAS systems.
2) Dissolution of a solid-solution to stoichometric saturation may result in either supersaturation or undersaturation with respect to the pure end-member phases. Commonly, solid-solution formation will have only a minor to insignificant effect on the saturation indices of the major component, as long as the mole fraction of trace component in the solid-solution is less than $\approx 1\%$ (equation 21). For large differences in end-member solubility products, this limiting mole-fraction may be lower. Trace component SI values (equation 22) generally show a greater departure from pure phase saturation (compared to major component SI values). If the trace component is much more more insoluble than the major component, a high degree of supersaturation may result. In the case of the (Ca,Cd)CO$_3$ system, supersaturation is predicted as long as the mole-fraction of CdCO$_3$ is higher than $\approx 10^{-2.75}$.
3) Aqueous solutions at primary saturation or at thermodynamic equilibrium with respect to a solid-solution will always be undersaturated with respect to the pure end-member solids. SSAS systems, with a zero or positive solid-solution excess-free-energy of mixing and with more than one order of magnitude difference between end-member solubility products, will be only slightly undersaturated with respect to the least soluble end-member. At most, an ideal solid-solution will be undersaturated by a factor of 2 ($\Sigma\Pi$) from pure end-member saturation. Solid-solutions with negative excess-free-energies of mixing may show a greater degree of undersaturation, although no examples have been found so far in ionic solids.

4) Techniques are presented for the estimation of excess-free-energies. These techniques should only be used, as an alternative to an ideal mixing model, in systems for which no laboratory stoichiometric saturation or thermodynamic equilibrium data is available.
5) Even though little thermodynamic data are available for low temperature SSAS systems, the equations and techniques presented in this paper can be used to estimate the importance of solid-solution aqueous-solution interactions on the chemical evolution of natural waters.

ACKNOWLEDGMENTS

I am very grateful to Ed Busenberg, Don Thorstenson and Niel Plummer for many valuable discussions throughout the preparation of this manuscript. I would also like to thank my three anonymous reviewers and my editors, Randy Bassett and Dan Melchior, for their help in improving this paper.

LITERATURE CITED

1. Lippmann, F. N. Jb. Miner. Abh. 1977, 130, no. 3, 243.
2. Lippmann, F. N. Jb. Abh. 1980, 139, no. 1, 1-25.
3. Thorstenson, D. C.; Plummer, L. N. Am. J. Sci. 1977, 277, 1203-23.
4. Plummer, L. N.; Busenberg, E. Geochim. Cosmochim. Acta 1987, 51, 1393-1411.
5. Lippmann, F. Bull. Mineral. 1982, 105, 273-79.
6. Redlich, O.; Kister, A. T. Ind. Eng. Chem. 1948, 40, no. 2, 345-48.
7. Guggenheim, E. A., Trans. Faraday Soc. 1937, 33, 151-59.
8. Urusov, V. S. Bull. Soc. fr. Mineral. Cristallogr. 1974, 97, 217-22.
9. Hildebrand, J. H. Solubility of Non-Electrolytes; Reinhold Publ. Co.: New York, 1936.
10. Thompson, J. B. Jr.; Waldbaum, D. Geochim. Cosmochim. Acta 1969, 33, 671-90.
11. Prigogine, I.; Defay, R. Chemical Thermodynamics; Longmans, Green & Co Ltd: London, 1954.
12. Vaslow, F.; Boyd, G. E. J. Amer. Chem. Soc. 1952, 74, 4691-4695.
13. Garrels, R. M.; Wollast, R. Am. J. Sci. 1978, 278, 1469-74.
14. Denis, J.; Michard, G. Bull. Mineral. 1983, 106, 309-19.
15. Plummer, L. N.; Busenberg, E. Geochim. Cosmochim. Acta 1982, 46, 1011-40.
16. Smith, R. M.; Martell, A. E. Critical Stability Constants, Volume 4: Inorganic Complexes; Plenum Press, 1976.
17. Davis, J. A.; Fuller, C. C.; Cook, A. D. Geochim. Cosmochim. Acta 1987, 51, 1477-90.
18. Busenberg, E.; Plummer, L. N. Geochim. Cosmochim. Acta 1989, 53, no. 6, 1189-1208.
19. Kirgintsev, A. N.; Trushnikova L. N. Russian J. Inorg. Chem. 1966, 11, 1250-55.
20. Palache, C.; Berman, H.; Frondel, C. Dana's System of Mineralogy; J. Wiley & Sons: New York, 1951.
21. Busenberg, E.; Plummer, L. N. U.S.G.S Open-File Report 83-863, 1983.
22. Samoilov, O. Ya. Structure of Aqueous Electrolyte Solutions and the Hydration of Ions; (Monograph originally pub. in Russian in 1957), Translation by Consultants Bureau, N.Y., 1965.
23. Frye, K. Modern Mineralogy; Prentice-Hall, 1974.
24. Crocket, J. H.; Winchester, J. W. Geochim. Cosmochim. Acta 1966, 30, 1093-1109.
25. Mucci, A. Geochim. Cosmochim. Acta 1987, 51, 1977-84.

RECEIVED July 20, 1989

# Chapter 7

# Effect of Pressure on Aqueous Equilibria

**Pradeep K. Aggarwal[1], William D. Gunter[2], and Yousif K. Kharaka[3]**

[1]Battelle Memorial Institute, 505 King Avenue, Columbus, OH 43201
[2]Oil Sands and Hydrocarbon Recovery, Alberta Research Council, P.O. Box 8330, Postal Station F, Edmonton, Alberta T6H 5X2, Canada
[3]Water Resources Division, U.S. Geological Survey, Mail Stop 427, 345 Middlefield Road, Menlo Park, CA 94025

> A model based on the density of the solvent together with the "iso-Coulombic" form of reactions has been used to estimate the pressure dependence of ionization constants of aqueous complexes. This model reproduces experimentally determined pressure dependencies of several ionization constants of aqueous complexes. Calculated saturation states of calcite in seawater are ~500 cal (100°C, 0.5 kbar) and ~700 cal (100°C, 1.0 kbar) greater, and the temperatures of calcite saturation are ~80°C (0.5 kbar) and ~40°C (1.0 kbar) higher, than the values obtained when the effect of pressure on ionization constants is neglected.

The equilibrium between aqueous solutions and minerals changes significantly with changes in pressure (1,2). To fully characterize the pressure dependence of such equilibria, we need to estimate the effect of pressure on the saturation state ($\Delta G_{diff}$) of minerals:

$$\Delta G_{diff} = RT \ln (Q/K) \tag{1}$$

where K and Q are, respectively, the solubility constant and activity quotient of the mineral; R is the gas constant; and T is temperature in kelvin. For example, the solubility reaction and Q for calcite are:

$$CaCO_3 \text{ (solid)} = Ca^{++} + CO_3^{--} \tag{2}$$

$$\ln Q_2 = \ln a_{Ca^{++}} + \ln a_{CO_3^{--}} - \ln a_{calcite} \tag{3}$$

where $a_i$ is the thermodynamic activity of the subscripted species; the activity of calcite is equal to one for a pure phase in its standard state. Equations 1 and 3 imply that estimates of the effect of pressure on the saturation state should evaluate the pressure dependencies of both the solubility constants and the thermodynamic activities of single ions.

Thermodynamic activities of ionic species in aqueous solutions with ionic strength (I) < 0.01 molal (m) commonly are calculated using the ion-pair model (3), which is valid also for solutions with I < 0.1 m. In dominantly NaCl solutions, the ion-pair model can be used for I ≤ 3 m with appropriate adjustments to the activity coefficients (4). The specific ion interaction model (5) may be more appropriate for solutions of high ionic strengths. The effect of pressure on the thermodynamic activities of single ions in this model can be estimated from the stoichiometric partial molal volume and compressibility data (1). However, a complete data set for all the ion-interaction parameters is not yet available for this model to be used in complex geochemical solutions.

Ion-ion interactions in the ion-pair model are treated by considering the formation of ion-pairs (or aqueous complexes). The effect of pressure on ion-pair formation is considered negligible for most geochemical calculations (6). This approach can lead to an erroneous estimate of the effect of pressure on the saturation states of minerals. A simple relationship based on the density of the solvent (7) which can be used to characterize completely the effects of pressure on aqueous solution - mineral equilibria is presented here.

## ESTIMATION OF PRESSURE EFFECTS

### Effect of Pressure on Solubility Constants

The pressure dependence of the equilibrium or solubility constant (K) for the congruent dissolution of a mineral, as in reaction 2 above, is given by:

$$\left(\frac{d \ln K}{dP}\right)_T = \frac{-\Delta V^o}{RT} \tag{4}$$

$\Delta V^o$ is the molal volume change of the reaction in the standard state:

$$\Delta V^o = \Sigma\ V_P^o - \Sigma\ V_R^o \tag{5}$$

where the subscripts refer to the products (P) or the reactants (R). The standard state for mineral species is the pure phase of unit activity at the temperature and pressure of interest. Standard state for the ionic species is the hypothetical ideal solution of unit molality referenced to infinite dilution.

Standard molal volumes of minerals together with the coefficients of isobaric expansion and isothermal compressibility are available in several compilations (8,9). Standard partial molal volumes of uncomplexed ions are available at 25°C and 1 atm. (1); however, data are sparse for the coefficients of isobaric expansion and isothermal compressibility.

Partial molal volumes of aqueous ionic species vary considerably with changes in pressure and temperature. The molar volume of an aqueous species can be split into a Coulombic and a non-Coulombic term. The non-Coulombic term consists of the intrinsic volume of the ion. The Coulombic term consists of the volume of solvation and the volume of collapse. The volume of solvation is related to the orientation of water dipoles around the aqueous species, and the volume of collapse is the component of the partial molal volume related to the collapse of the water structure in the vicinity of the aqueous species. The pressure and temperature dependence of the molar volume of aqueous species arises from a similar change in the electrostatic properties of the solvent.

Helgeson and co-workers (10,11) have proposed an equation of state based on the electrostatic properties of the solvent that can be used to estimate standard partial molal volumes of uncomplexed ions at elevated temperatures and pressures.

### Effect of Pressure on Ionization Constants

To calculate the effect of pressure on the formation or ionization constants of aqueous complexes, the partial molal volume change of the ionization reaction must be known. Standard partial molal volumes (25°C, 1 bar) of some aqueous complexes are known, and can be used together with the molar volumes of uncomplexed ions to calculate $\Delta V$. The Fuoss equation can also be used to estimate the standard molal volume change in ionization reactions (20,21):

$$K_f = 4\Pi a^3 e^b N / 3000 \tag{6}$$

$$b = |Z_+ Z_-| e_o / a\varepsilon\kappa T \tag{7}$$

where $K_f$ = molar formation constant of an aqueous complex; N = Avogadro's number; $\kappa$ = Boltzmann constant; a = distance of closest approach of ions; $\varepsilon$ = dielectric constant; $e_o$ = permittivity of the vacuum; e = electron charge; and Z = charge on the cation or anion in the ion pair. For a neutral complex, Equations 4, 6 and 7 result in:

$$\Delta V = RT \left[ b \left( \frac{d \ln \varepsilon}{dP} \right)_T - \beta \right] \tag{8}$$

β is the compressibility of the solvent and is used to convert Equation (6) from the molar to molal scale. To calculate the value of ΔV from Equation (8), b is first calculated from Equations (6) and (7) using the known formation constant of an aqueous complex. Estimates of volume change calculated from Equations 6-8 are in reasonably good agreement with measured values when the complexing dominantly is of outer-sphere type, e.g. $MgSO_4^0$. However, large uncertainties result between estimated and measured values when a significant proportion of the ion-pairs are of inner-sphere type, e.g. $CaSO_4^0$ (22).

The molar volume change in ionization reactions at higher temperatures and pressures cannot be calculated for most of the aqueous complexes because of a lack of data on isobaric expansion and isothermal compressibility coefficients. Entropy and heat capacity correlations have recently been used to generate equation of state parameters for estimating molal volumes of aqueous complexes at elevated temperatures and pressures (Sverjensky, 12). These coefficients are available for aqueous complexes only of univalent anions and, therefore, the pressure dependence of ionization constants at elevated temperatures cannot be estimated using Equation 4.

Marshall and Mesmer (7) have used a relationship based upon the density of the solvent for calculating the pressure dependence of ionization constants:

$$\ln K_{P,T} = \ln K^0 - \frac{\Delta V^0}{RT \beta^0} \ln \frac{\rho_{P,T}}{\rho^0} \tag{9}$$

where P is the applied pressure and the superscript (⁰) indicates temperature T and the saturation vapor pressure of water; ρ and β are, respectively, the density and coefficient of isothermal compressibility of water; and ΔV is the molar volume change in the reaction.

Equation 9 also cannot be used to estimate the pressure dependence of ionization constants when the term ($\Delta V^0/RT\beta^0$) shows a temperature dependence (Fig. 1). The following technique can be used to alleviate this problem.

The heat capacity change of "iso-Coulombic" reactions (reactions which are symmetrical with respect to the number of ions of each charge type) is nearly independent of temperature (13,14). Similarly, the molar volume change of "iso-Coulombic" reactions will be expected to display a relatively minor temperature dependence because most of the temperature-dependent changes in the Coulombic and non-Coulombic contributions to the volumes of individual ions will cancel out in the ΔV term. Thus, the 25°C value of ΔV can be used in Equation 9 if the reaction is "iso-Coulombic", or is made so by the addition or subtraction of an appropriate number of water dissociation reactions. For example, the dissociation reaction of the aqueous complex $H_2CO_3^0$ can be written in the "iso-Coulombic" form (Reaction 12) by combining Reactions 10 and 11:

$$H_2CO_3^0 = HCO_3^- + H^+ \tag{10}$$

$$H_2O = H^+ + OH^- \tag{11}$$

$$H_2CO_3^0 + OH^- = HCO_3^- + H_2O \tag{12}$$

The isothermal pressure dependence of the equilibrium constant for (12) is given by:

$$\ln K_{P,13} = \ln K_{13}^0 - \frac{n}{RT \beta^0} \ln \frac{\rho_{P,T}}{\rho^0} \tag{13}$$

where n = $\Delta V_{10}$(25°C,1bar) - $\Delta V_{11}$(25°C,1bar). Using available values of ln $K_{P,11}$, the pressure dependence of the ionization constant of reaction 10 can be obtained:

$$\ln K_{P,10} = \ln K_{P,12} + \ln K_{P,11} \tag{14}$$

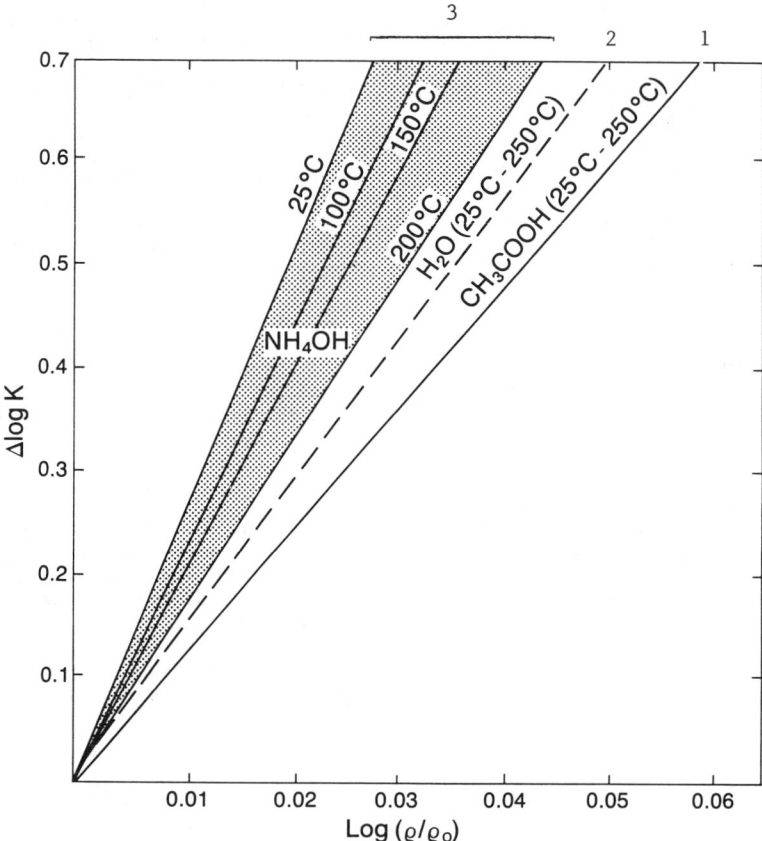

Figure 1. Plot of the change in log ionization constant versus change in log density of water for the following reactions: (1) $H_2O = H^+ + OH^-$ (24); (2) $CH_3COOH = CH_3COO^- + H^+$ (25); (3) $NH_4OH = NH_4^+ + OH^-$ (26). The numbers in parenthesis are for the references from which the data were taken.

## COMPARISON WITH EXPERIMENTAL DATA

Equation 9 has been successfully used to reproduce experimental values of the ionization constants of water and several aqueous complexes at higher temperatures and pressures (7). Eugster and Baumgartner (15) have used a relationship similar to Equation 9 for estimating the pressure dependence of aqueous complexes in supercritical fluids.

Ionization constants at higher pressures for several aqueous complexes calculated using the "iso-Coulombic" form of Equation 9 (as for reaction 12) are shown in Figures 2 through 5 along with the experimental data. As in Equation 14 above, the log K values plotted in Figures 2 to 5 were retrieved from the log K values of iso-Coulombic reactions by subtracting log K for the ionization of water. Estimated pressure dependence of the ionization constants of $H_2CO_3$, $H_3BO_3$, and $HPO_4^{--}$ are in agreement with the experimental data; minor deviations between the calculated and experimental values are within the error of measurement (±0.06 log units).

Calculated pressure dependencies of the ionization constants of $NH_4OH$ and $H_3PO_4$ are generally higher than the measured values. Larger deviations occur at higher temperatures (>100°C). However, the calculated values below 200°C are generally within the error of measurement. Above 200°C, the deviations are higher than the experimental error. These deviations occur probably because the temperature dependence of $\Delta V$ is not completely eliminated by using the iso-Coulombic form of the reaction.

Estimates of pressure dependence of the ionization constant of $PbCl^+$ based on Equation 9 are compared in Table 1 with those obtained by using Equation 4 (12). Both approaches result in similar values for pressures up to 1000 bars. At higher pressures, estimated pressure dependence based on Equation 9 is lower, especially at higher temperatures. As mentioned earlier, disagreement at higher temperatures may be due to the temperature dependence of the reaction volume. In addition, the error of estimation in the data of Sverjensky (12) is unknown.

Table 1. Pressure dependence of the ionization constants of $PbCl^+$ calculated using Equation 9 (from the data in Sverjensky, 12)

| Temp(°C) | $\log K_p - \log K°$ | | |
|---|---|---|---|
| | 500 bars | 1000 bars | 1500 bars |
| 25 | 0.01 (0.00) | 0.01 (0.00) | 0.03 (0.03) |
| 75 | 0.03 (0.01) | 0.06 (0.01) | 0.11 (0.01) |
| 175 | 0.17 (0.12) | 0.29 (0.22) | 0.50 (0.23) |

## APPLICATIONS IN GEOCHEMISTRY

The equilibrium or solubility constants for mineral dissolution reactions increase with increasing pressure, resulting in an increase in the solubility of the minerals (Fig. 6). The ionization constants for the aqueous complexes also increase with increasing pressure (Figs. 2-5), which results in an increase in the activity of the uncomplexed ions. Owing to the increased activity of the uncomplexed ions, the magnitude of the pressure correction for the saturation state of minerals is lower when the pressure effects on the ionization of aqueous complexes are taken into account as compared with no correction for changes in the ionization constants. These are general trends and the saturation state of individual minerals may change differently with increasing pressure, depending upon temperature and solution composition.

To illustrate the effect of pressure on aqueous equilibria, the speciation and saturation states of minerals at higher pressures were calculated for a modified seawater and an oil-field brine from the Texas Gulf Coast (Table 2). The calculations were performed using the computer program SOLMINEQ.88 (16). SOLMINEQ.88 computes the distribution of species and saturation states of minerals using chemical composition, pH, and Eh at 25 °C. A description of the methodology for computing a high temperature pH is given in (17).

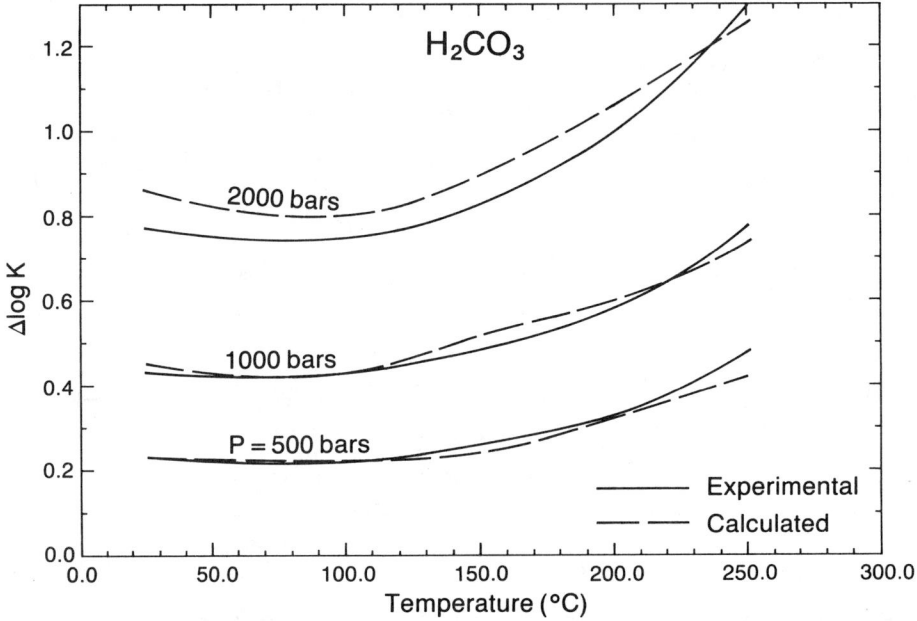

Figure 2. The effect of pressure on the ionization constants of the reaction $H_2CO_3 = HCO_3^- + H^+$. The experimental data were taken from ref. 27. The calculations were performed using the "iso-Coulombic" form of Equation 9.

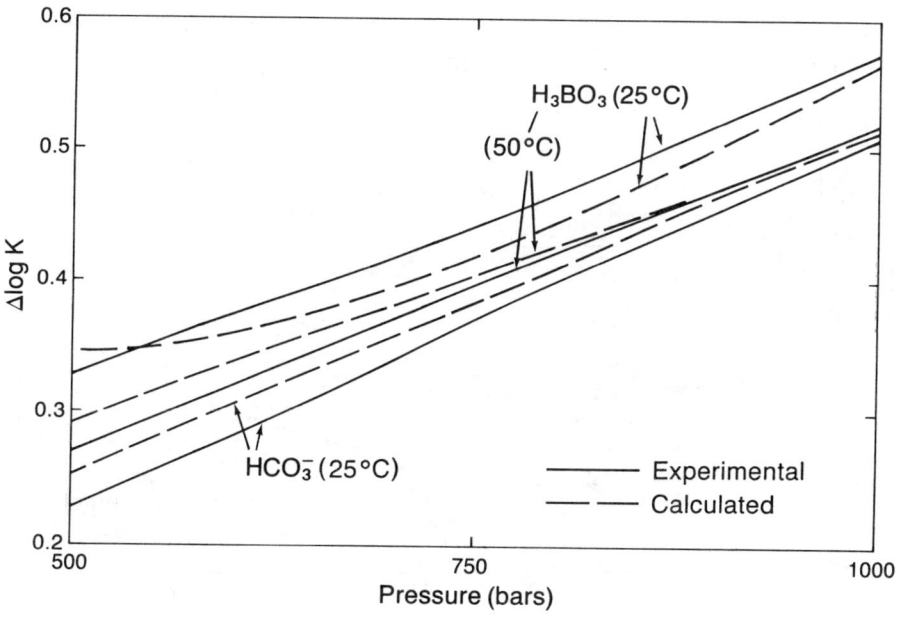

Figure 3. The effect of pressure on the ionization constants of the reactions $H_3BO_3 = H_2BO_3^- + H^+$ (30) and $HCO_3^- = CO_3^{--} + H^+$ (31). The numbers in parentheses are for the references from which the experimental data were taken. The calculations were performed using the "iso-Coulombic" form of Equation 9.

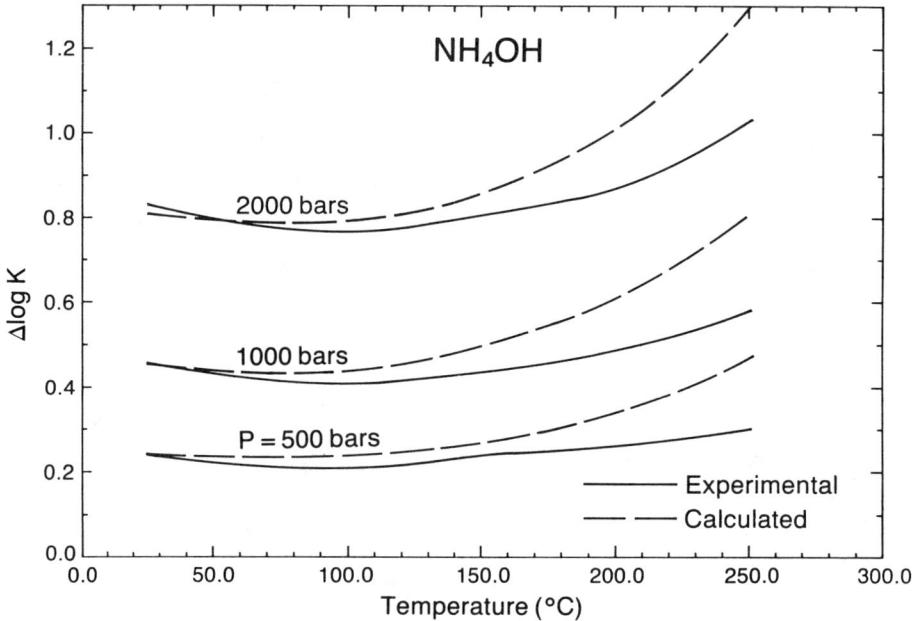

Figure 4. The effect of pressure on the ionization constants of the reaction $NH_4OH = NH_4^+ + OH^-$. The experimental data were taken from ref. 26. The calculations were performed using the "iso-Coulombic" form of Equation 9.

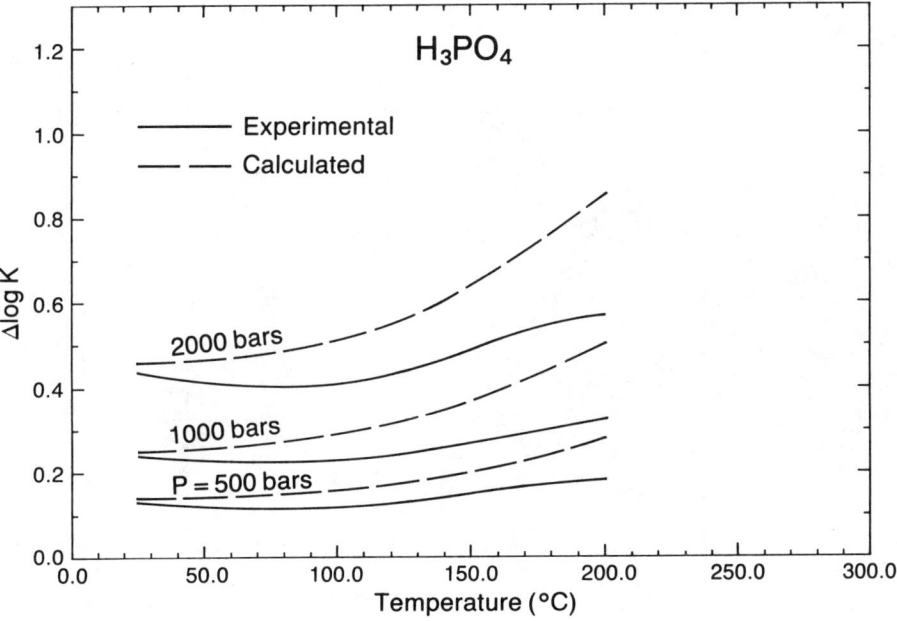

Figure 5. The effect of pressure on the ionization constants of the reaction $H_3PO_4 = H_2PO_4^- + H^+$. The experimental data were taken from ref. 28. The calculations were performed using the "iso-Coulombic" form of Equation 9.

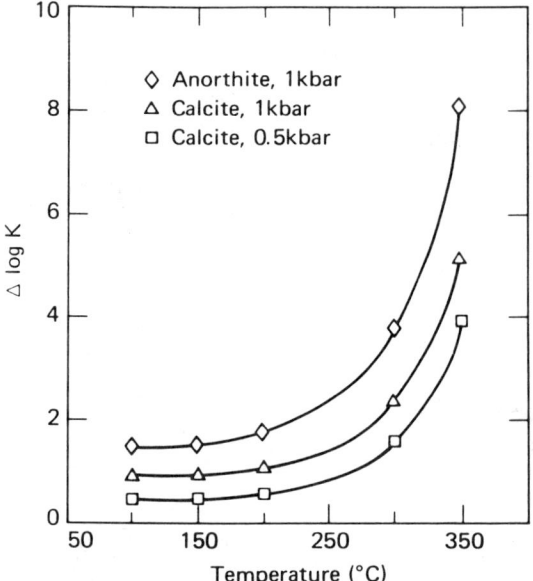

Figure 6. Effect of pressure on the solubility constants of anorthite and calcite at higher temperatures based on data from (8).

Table 2. Compositions of modified seawater and Texas Gulf Coast water used for illustrating the effect of pressure on aqueous equilibria (all concentrations are in mg/L, except pH)

| Component | seawater | Gulf Coast water (Pleasant Bayou #2) |
|---|---|---|
| Ca | 422 | 9100 |
| Mg | 1322 | 660 |
| Na | 11019 | 38000 |
| K | 408 | 840 |
| Cl | 19806 | 80600 |
| $SO_4$ | 2775 | 5.4 |
| $SiO_2$ | 4.28 | 120 |
| Ba | 0.02 | 760 |
| Fe | 0.002 | 62 |
| Li | 0.19 | 39 |
| Pb | | 1.1 |
| Sr | 8.33 | 1020 |
| F | 1.42 | 1.4 |
| B | 4.45 | |
| $PO_4$ | 0.06 | |
| Al* | 0.001 | 0.001 |
| $H_2S$ | 10.23 | 0.5 |
| Total Inorg. Carbon | 24.85 | 216.9 |
| pH(25 °C) | 8.2 | 4.91 |
| Temp (°C) | 25.0 | 150.0 |
| Pressure(bar) | 1.0 | 800.0 |

* assumed for modeling purposes.

Table 3. Standard molal volume change of ionization reactions of aqueous complexes used in SOLMINEQ.88

| Complex | ΔV° | Complex | ΔV° | Complex | ΔV° |
|---|---|---|---|---|---|
| HCO3 - | -28.7 | H4SiO4 | -37.9 | H2S | -13.0 |
| Al(OH)++ | -22.4 | Al(OH)2+ | -43.2 | Al(OH)4- | -64.3 |
| CH3COOH | -11.2 | BaCO3 | -7.5 | BaHCO3 + | -34.3 |
| Ba(OH)+ | -19.7 | BaSO4 | -7.1 | CaCO3 | -10.6 |
| CaHCO3 + | -31.6 | Ca(OH)+ | -19.7 | CaSO4 | -25.0 |
| Cu(OH)+ | -19.7 | CuSO4 | -7.1 | Fe(OH) + | -19.7 |
| Fe(OH)2 | -5.7 | FeSO4 | -7.1 | FeCl ++ | -4.6 |
| Fe(OH)++ | -22.4 | Fe(OH)2+ | -43.5 | Fe(OH)3 | -64.6 |
| Fe(OH)4- | -85.7 | B(OH)4 - | -35.5 | H2CO3 | -27.2 |
| HPO4 -- | -36.0 | H2PO4 - | -25.9 | HSO4 - | -14.8 |
| KCO3 - | -19.8 | MgCO3 | -9.8 | MgHCO3 + | -31.5 |
| Mg(OH) + | -19.7 | MgSO4 | -7.4 | MnHCO3 + | -34.0 |
| MnSO4 | -7.1 | NaCl | -13.7 | NaCO3 - | -23.1 |
| NaHCO3 | -13.8 | Na2SO4 | -7.1 | NaSO4 - | -15.8 |
| NH4OH | -28.8 | Sr(OH) + | -19.7 | SrCO3 | -9.1 |
| SrHCO3 + | -31.5 | SrSO4 | -7.1 | ZnSO4 | -7.6 |

The pressure dependence of mineral dissolution reactions in SOLMINEQ.88 is calculated using data from (8,11,18). The ionization constants at higher pressures for aqueous complexes are calculated using the "iso-Coulombic" form of Equation 9. The coefficient of isothermal compressibility and density of water are obtained from the data in (19). Molal volume change in the ionization reaction of aqueous complexes (Table 3) are obtained from experimental studies, when available, or are estimated based on Equations 6 to 8.

The effect of pressure on calculated pH, activities of several aqueous species, and saturation states of selected minerals for the seawater are listed in Table 4. The calculated pH at higher temperatures and 500 bars pressure is ~0.1 unit lower than that calculated for pressures along the vapor saturation curve; at 1000 bars pressure, the pH is lower by ~0.2 units (Table 4). At 500 bars pressure, the predicted saturation states of calcite are greater by ~200 cal ($\Delta SI=0.15$) at 25°C, and 1800 cal ($\Delta SI=0.75$) at 250°C, compared to those calculated without considering the effect of pressure on the ionization constants (Figure 7). Similar, but numerically larger, differences are seen for the saturation state of anhydrite.

Table 4. Calculated activities of ions and saturation states of minerals in modified seawater at higher pressures. "Sat." is the pressure of steam saturation; other pressures are in bars

A. activities

| | - log activity | | | | | |
|---|---|---|---|---|---|---|
| | T = 100 °C | | | T = 250 °C | | |
| | sat. | 500 | 1000 | sat. | 500 | 1000 |
| $H^+$ | 7.10 | 6.99 | 6.90 | 6.02 | 5.87 | 5.74 |
| $Ca^{++}$ | 2.74 | 2.70 | 2.67 | 3.24 | 3.06 | 2.95 |
| $Mg^{++}$ | 2.00 | 1.97 | 1.95 | 2.38 | 2.27 | 2.18 |
| $Na^+$ | 0.54 | 0.53 | 0.51 | 0.70 | 0.65 | 0.61 |
| $Cl^-$ | 0.50 | 0.49 | 0.48 | 0.68 | 0.63 | 0.59 |
| $SO_4^-$ | 2.73 | 2.66 | 2.59 | 3.23 | 2.97 | 2.82 |
| $CO_3^-$ | 6.22 | 6.04 | 5.88 | 9.53 | 8.95 | 8.54 |

B. saturation states

| | $\Delta G_{diff}$ (cal) | | |
|---|---|---|---|
| Mineral | a | b | c |
| T = 100 °C, P = 500 bars | | | |
| Calcite | 689 | -165 | 243 |
| Anhydrite | -192 | -886 | -699 |
| Albite | -7234 | -8252 | -8556 |
| T = 100 °C, P = 1000 bars | | | |
| Calcite | 689 | -881 | -160 |
| Anhydrite | -192 | -1449 | -1107 |
| Albite | -7234 | -9172 | -9754 |
| T = 250 °C, P = 500 bars | | | |
| Calcite | 766 | -2700 | -900 |
| Anhydrite | 4775 | 3023 | 4082 |
| Albite | -21967 | -23754 | -24361 |
| T = 250 °C, P = 1000 bars | | | |
| Calcite | -766 | -3059 | -1028 |
| Anhydrite | 4775 | 1828 | 3502 |
| Albite | -21967 | -25095 | -26200 |

pressure effects on  a: none;  b: solubility constants;
c: ionization and solubility constants.

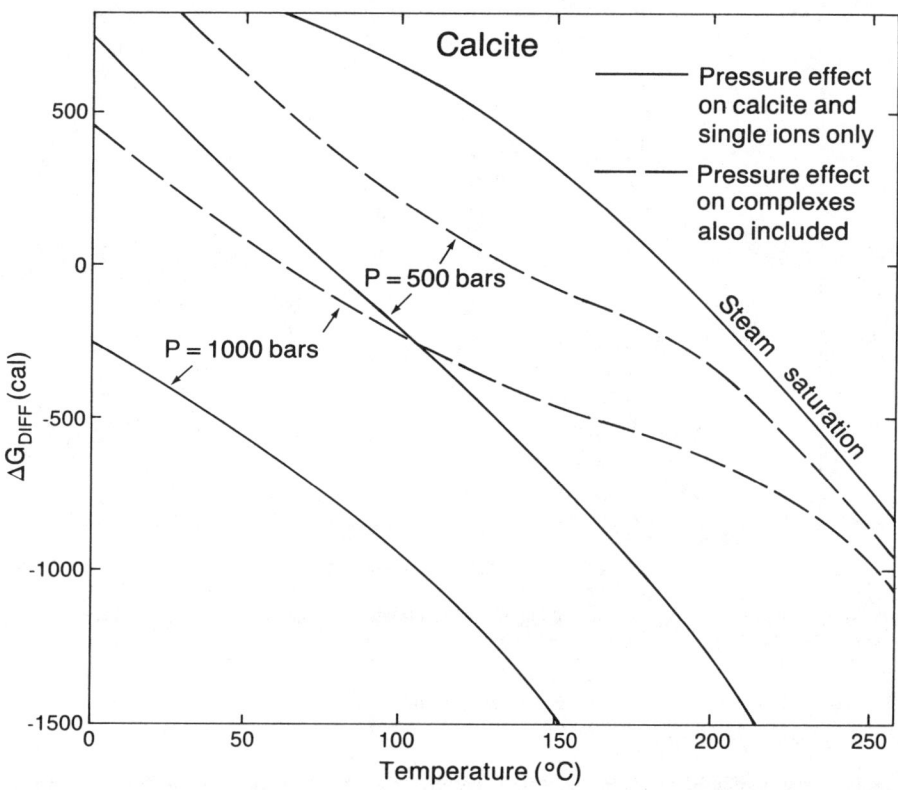

Figure 7. The effect of pressure on the saturation state of calcite in modified seawater. The composition of seawater used in the calculations is given in Table 2.

The pressure dependence of the saturation state of albite is different than that of calcite or anhydrite. The magnitude of pressure correction for the saturation state of albite increases when the pressure correction for the ionization constants is taken into account, in contrast with a decrease observed in the case of calcite and anhydrite (Table 4). The calculated solubility of albite is a function of pH and the activities of aqueous $Na^+$, $Al^{+++}$, and $SiO_2(aq)$. The activities of $Na^+$, $Al^{+++}$, and $SiO_2$ remain nearly unchanged because (1) $Na^+$ and $Al^{+++}$ complex mainly with $Cl^-$ and/or $OH^-$. The fraction of these complexes does not change measurably with a change in pressure; and (2) the effect of pressure on the ionization of silica species is relatively minor. Therefore, the change in the saturation state of albite upon considering the pressure dependence of ionization constants is due mainly to a change in pH. In the case of calcite and anhydrite, the effect of change in pH is countered by a corresponding change in the distribution of carbonate or sulfate species. This change in the distribution of species is responsible for the observed differences in the pressure dependence of the saturation states of albite, calcite, and anhydrite.

An additional argument for considering the effect of pressure on ionization equilibria is presented in Figure 8. This figure plots calculated temperature and pressure conditions at which the modified seawater becomes saturated with calcite or anhydrite. At a constant pressure, the predicted temperature of mineral saturation is significantly different (higher for calcite and lower for anhydrite) when the effect of pressure on ionization equilibria are taken into account. This observation has important consequences for calculated geochemical models of water-rock interactions and ore deposition (23,24).

The effects of pressure on the speciation and saturation states of minerals in the Texas Gulf Coast water are similar to those described above for the seawater.

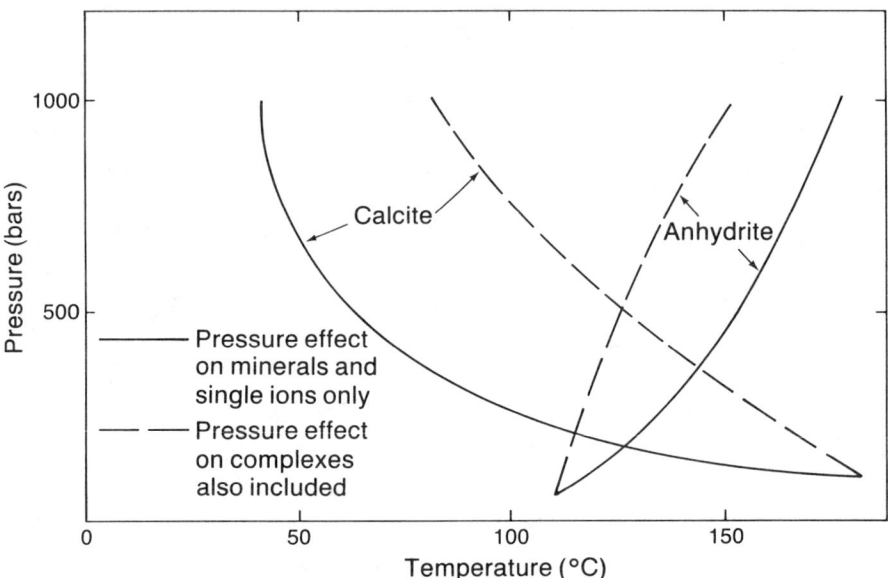

Figure 8. A plot of the pressure-temperature conditions at which calcite and anhydrite are calculated to be in equilibrium with seawater.

CONCLUDING REMARKS

The technique presented in this paper to estimate the pressure dependence of ionization constants is simple to use and requires minimal data. Excellent agreement is observed between the calculated and measured values of ionization constants of several aqueous complexes at higher pressures, at least up to 200°C. The discussion presented above shows that the effect of pressure on ionization constants must be considered in geochemical calculations of mineral-solution equilibria at low and high temperatures.

ACKNOWLEDGMENTS

The authors thank G.W. Bird, J. Hovey, F.J. Millero, E.H. Perkins, and P.R. Tremaine for meaningful discussions, and J. Hovey, M. McKibben, D. Melchior, and J.D. DeBraal for reviews. Financial support for this work was provided in part by the Alberta Research Council.

REFERENCES

1. Millero, F.J. Geochim. Cosmochim. Acta, 1982, 46, 11-22.
2. Hemley, J.J., Cygan, G.L., d'Angelo, W.M. Geology, 1986, 14, 377-379.
3. Garrels, R.M., Thompson, M.E. Am. J. Sci. 1962, 260, 57-66.
4. Helgeson, H.C. Am. J. Sci. 1969, 267, 729-804.
5. Pitzer, K.S. J. Phys. Chem., 1973, 77, 268-277.
6. Seward, T.M. In Chemistry and Geochemistry of Solutions at High Temperatures and Pressures; Rickard, D. Wickman, F. Eds., Pergamon:New York, 165-181.
7. Marshall, W.L., Mesmer, R.E. J. Soln. Chem., 1981, 10, 121-127.
8. Helgeson, H.C., Delany, J., Nesbitt, H.W., Bird, D. Am. J. Sci. 1978, 278A, 1-229.
9. Robie, R.A., Hemingway, B.S., Fisher, J.R. Bulletin U.S. Geological Survey, 1979, 1452, 1-456.
10. Helgeson, H.C., Kirkham, D. Am. J. Sci. 1974, 274, 1199-1261.
11. Tanger, J., Helgeson, H.C. Am. J. Sci., 1988, 288, 19-98.

12. Sverjensky, D.A. In Thermodynamic Modeling of Geological Materials: Minerals, Fluids and Melts; Carmichael, I.S.E., Eugster, H.P., Eds., Mineralogical Society of America: Washington D.C., 1987, Vol. 17, pp 177-210.
13. Lindsay, W.T. Proc. 41st Intl. Water Conf.; Engineering Society of Western Pennsylvania, 1980, 284-294.
14. Murray, R.C., Cobble J.W. Proc. 41st Intl. Water Conf.; Engineering Society of Western Pennsylvania, 1980, 295-310.
15. Eugster, H.P. and Baumgartner, L., In Thermodynamic Modeling of Geological Materials: Minerals, Fluids and Metls; Carmichael, I.S.E., Eugster, H.P., Eds., Mineralogical Society of America: Washington D.C., 1987, Vol. 17, pp 367-404.
16. Kharaka, Y.K., Gunter, W.D., Aggarwal, P.K., Perkins, E.H., and DeBraal, J.D., SOLMINEQ.88: A computer program for geochemical modeling of water-rock interactions; U.S. Geological Survey, Water Resources Investigations Report, under review.
17. Aggarwal, P.K., Hull, R.W., Gunter, W.D., Kharaka, Y.K. In Proc. Third Canadian/American Conference on Hydrogeology; Hitchon, B., Bachu, S. Sauveplane C.M., Eds., National Water Well Association: Dublin, Ohio, 1986, pp 196-203.
18. Helgeson, H.C. and Kirkham, D., Flowers, G. Am. J. Sci. 1981, $\underline{281}$, 1249-1516. 1974
19. Haar, L., Gallagher, J.S., Kell, G.S. NBS/NRC Steam Tables; Hemisphere: Washington, D.C., 1984; p318
20. Hemmes, P., J. Phys. Chem., 1972, $\underline{76}$, 895-900.
21. Fuoss, R. J. Am. Chem. Soc. 1958, $\underline{80}$, 5059-64.
22. Millero, F.J., Gombar, F., Oster, J., J. Soln. Chem., 1977, $\underline{6}$, 269-280.
23. Wolery, T.J., Calculation of Chemical Equilibrium Between Aqueous Solution and Minerals: the EQ3/6 Software Package, Lawrence Livermore Laboratory, 1979, Report UCRL-52658, p41.
24. Reed, M., Geochim. Cosmochim. Acta, 1982, $\underline{46}$, 513-528.
25. Marshall, W.L., Franck, E.V. J. Phys. Chem. Ref. Data, 1981, $\underline{10}$, 295-304.
26. Lown, D.A., Thirsk, M.R., Wynn-Jones, L. Trans. Farad. Soc. 1970, $\underline{66}$, 51-73.
27. Read, A.J. J. Soln. Chem., 1982, $\underline{11}$, 649-664.
28. Read, A.J. J. Soln. Chem., 1975, $\underline{4}$, 53-70.
29. Read, A.J. J. Soln. Chem., 1988, $\underline{17}$, 213-224.
30. Kreamer, T.F., Kharaka, Y.K. Geochim. Cosmochim. Acta, 1986, $\underline{50}$, 1233-1238.
31. Ward, G.K., Millero, F.J. J. Soln. Chem., 1974, $\underline{3}$, 417-430.
32. Disteche, A., Disteche, S. J. Electrochem. Soc., 1968, $\underline{114}$, 330.

RECEIVED October 6, 1989

# CODE DEVELOPMENT AND DOCUMENTATION

# Chapter 8

# Current Status of the EQ3/6 Software Package for Geochemical Modeling

**Thomas J. Wolery, Kenneth J. Jackson, William L. Bourcier, Carol J. Bruton, Brian E. Viani, Kevin G. Knauss, and Joan M. Delany**

**Lawrence Livermore National Laboratory, L-219, P.O. Box 808, Livermore, CA 94550**

> EQ3/6 is a software package for modeling chemical and mineralogic interactions in aqueous geochemical systems. The major components of the package are EQ3NR (a speciation-solubility code), EQ6 (a reaction path code), EQLIB (a supporting library), and a supporting thermodynamic data base. EQ3NR calculates aqueous speciation and saturation indices from analytical data. It can also be used to calculate compositions of buffer solutions for use in laboratory experiments. EQ6 computes reaction path models of both equilibrium step processes and kinetic reaction processes. These models can be computed for closed systems and relatively simple open systems. EQ3/6 is useful in making purely theoretical calculations, in designing, interpreting, and extrapolating laboratory experiments, and in testing and developing submodels and supporting data used in these codes. The thermodynamic data base supports calculations over the range 0-300°C.

EQ3/6 (1-3) is a software package consisting of modeling codes, supporting thermodynamic data files, and some data base management software. Its purpose is to calculate models of chemical interactions in aqueous geochemical systems. It was originally written in the period 1975-1978 by T.J. Wolery (1) to compute reaction path models of seawater/basalt interactions under hydrothermal conditions, and was patterned in function after the PATHI code developed by Helgeson et al. (4, 5). It has subsequently been used in a number of other studies (e.g., refs. 6-19) of rock/water or other solid/water interactions at both low and high temperatures. EQ3/6 is presently being developed and applied as part of the U.S. program for disposal of high level nuclear waste (20; 13-19). Various versions of EQ3/6 have been distributed to the scientific community in the past and work is continuing to improve and extend this software package. EQ3/6 was covered by the modeling software survey of Nordstrom et al. (21) in the 1979 Symposium Series predecessor to this volume. A more recent and more detailed review of EQ3/6 has since appeared (22). The most recent release of EQ3/6 is the "3245" version, general distribution of which began in February, 1988. The "3270" version is presently being developed. These versions will be discussed in the present communication.

The EQ3/6 software package consists of several principal components. These are the EQ3NR and EQ6 codes, the EQLIB library, and the thermodynamic data base. The EQLIB library and the thermodynamic data base support both of the main modeling codes. EQLIB contains math routines, routines that perform various computer system functions, and routines that evaluate scientific submodels, such as for activity coefficients of aqueous species, that are common to both EQ3NR and EQ6. The data base covers a wide range of chemical elements and nominally allows calculations in the temperature range 0-300°C at a constant pressure of 1.013 bar from 0-100°C and the steam-liquid water equilibrium pressure from 100-300°C.

EQ3NR (2) is a speciation-solubility code. It computes a static thermodynamic model of an aqueous solution, given inputs such as pH and analytical (total) concentrations. The typical output includes the distribution of aqueous species, given both as molal concentrations and thermodynamic activities of individual species and as percentage contributions to total mass balances. It also includes saturation indices, defined as $SI = \log Q/K$, where Q is the ion activity product and K the equilibrium constant of a reaction, typically for the congruent dissolution of a mineral phase. EQ3NR is very flexible, and some important parameters that are normally outputs can be made inputs, and vice versa. For example, one might analyze a solution and input the data to compute the saturation index for a given mineral. But one might also want to ask the question, if the same mineral is assumed to be in equilibrium with the solution, what should be the value of one of the analytical data, given that values for the others are known and correct. EQ3NR can make such calculations.

EQ6 (1, 2; Wolery, T.J., and Daveler, S.A., EQ6 User's Guide, in prep.) is a reaction-path program which operates in a few distinct modes. Typically, it is used to compute the consequences of reacting an aqueous solution, previously defined by running EQ3NR, with a set of specified "reactants." The objective is to model the changing chemistry of the solution and the appearance and disappearance of any product phases. This can be done in a titration mode, a closed system mode, or a pseudo-one-dimensional "flow-through" mode which follows the evolution of a packet of solution moving through a reactant medium. There is no provision in this code for modeling actual flow through a column; the "flow through" mode presently in the code only models what happens to the "first" packet of water. All of the above models can be modified to impose fixed values of fugacity for specified gases (23), corresponding to systems that are open with respect to these species and connected to external reservoirs. This capability is useful in modeling systems which are open to the atmosphere, which contains such highly reactive gases as molecular oxygen and carbon dioxide.

The functions which describe the rates at which the "reactants" react with the aqueous solution are chosen by the user and may represent arbitrary relative rates or actual kinetic rate models. If arbitrary relative rates are employed, no time frame is involved in the calculation and progress is measurable only in the change in the remaining amounts of reactants. This is an "equilibrium step calculation," representing a sequence of equilibrium states corresponding to a system whose gross composition is being changed according to some formula. If the rate functions involved are actual rate laws, such as the one proposed for quartz dissolution and growth by Rimstidt and Barnes (24), then the calculation includes a time frame. The word "reactant" as used here refers to a substance which is not in equilibrium with the system composed of the aqueous solution and any "products" which are in equilibrium with that solution. The rate at which a "reactant" dissolves in the aqueous solution can be negative, in which case the "reactant" is really a product whose formation is governed by a kinetic rate law instead of an instantaneous response to satisfy solubility equilibrium. This is the principle behind an option to compute simulations including precipitation kinetics (25).

The temperature along a reaction path may be constant or variable. In non-kinetic mode, it may be treated as a function of the reaction progress variable. In kinetic mode, it may be treated as a function of time. Therefore, EQ6 can also be used to compute the consequences, such as pH shift and mineral precipitation, of heating or cooling an aqueous fluid (1, 26).

MODELING PITFALLS

Software such as EQ3/6 may be a useful tool in geochemical modeling, but there are a number of potential pitfalls of which the user should be keenly aware before trusting the computed results. EQ3/6 is not a "computerized chemical model" or "black box" whose output is always correct (even approximately) or necessarily meaningful, even if correct. It is really a geochemical "calculator" which allows one to compute models which combine or integrate various submodels. One submodel might describe the aqueous species which are thought to be present in a given system, and any equilibrium relations that should pertain among them. Another might describe the thermodynamic activity of water and the activity coefficients of the aqueous species. A third might be a kinetic rate law for a dissolving mineral. Certain other specifications about the nature of the reacting system also pertain in the combination of these submodels, such as whether the system is closed or open with respect to certain gas species, and whether or not the temperature is constant.

Generally speaking, the submodels are tied in some fashion to experimental data. The user of modeling software such as EQ3/6 needs to be aware of the nature and limitations of such submodels and how well they are connected to actual observations. Submodels are best constrained by studies of simple, carefully defined systems. Unfortunately, it is often difficult or impossible to define experimental systems that completely isolate the effects that one wishes to study. For example, in studies of the dissolution rates of minerals, one would like to obtain results which would pertain to the dissolution behavior of minerals in natural systems. However, artifact effects related to sample preparation may be present in the absence of special precautions (27). Thus, many of the data reported in the older literature are unsuitable for building appropriate rate models, and new studies in which such effects are eliminated or ameliorated (e.g., 27-29) are required.

Because of the difficulty in isolating specific effects, a submodel may be constructed by assuming the validity of a second submodel. For example, solubility product constants may be determined from solubility measurements by assuming a specific model for the activity coefficients of the species in the fluid phase. An inconsistency may result if these constants are later employed in conjunction with a different model of the activity coefficients. The nature of speciation and activity coefficient submodels is such that they should really be considered one model and be developed concurrently. An outstanding example of this approach is the "sea salt" system model of Harvie, Moller, and Weare (30). Inconsistencies may also result by building a submodel, as for speciation, by combining various reported results. This problem is commonly encountered in building large thermodynamic data bases.

It is important to recognize that the use of geochemical modeling software such as EQ3/6 will almost always involve extrapolating submodels beyond the bounds of the experimental data on which they were constructed. For example, in putting together the basis for a calculation simulating the reaction of silicate minerals with ground water, one might choose to utilize the model for quartz dissolution and growth proposed by Rimstidt and Barnes (24). Their model was proposed on the basis of experiments in which the aqueous solution was always neutral to slightly acidic. It reflected no dependence of rate on solution pH. Later experiments by Knauss and Wolery (29) showed such a dependence in the mildly alkaline to alkaline pH range. Further experiments might show other dependencies that could be significant under conditions of interest to users of modeling codes.

A given set of experimental data may be explainable by different models. For example, Reddy et al. (31) measured the growth rate of calcite and fit the data to a rate law model proposed by Plummer et al. (32). It was later shown (22) that a very simple "transition state theory" rate law, quite different in form, fit the data about as well (neither model fit the data perfectly). One should keep in mind that such models may extrapolate differently.

Models are often developed to explain certain kinds of data, ignoring other kinds that also might be pertinent. The initial development of Pitzer's equations (33, 34) for activity coefficients in concentrated solutions was focused on explaining measurements of vapor pressure equilibrium and of electromotive force (emf). The data could be explained by assuming that the electrolytes examined were, at least in a formal sense, fully dissociated. Later work using these equations to explain solubility data required the formal adoption of a few ion pair species (30). Even so, no speciation/activity coefficient model based on Pitzer's equations is presently consistent with the picture of much more extensive ion-pairing based on other sources, such as Smith and Martell's (35) compilation of association constants. This compilation is a collective attempt to explain other kinds of data, such as electrical conductance, spectrophotometry, and acoustic absorption.

Other problems may arise if the modeler's objective is to explain or predict the results of an "applications level" experiment (one involving a relatively complex system) that is carried out in the field or, more commonly, the laboratory. First, the conditions assumed to prevail in the experiment may not be the actual ones. For example, the experiment may be thought to be a closed system, when in fact there is loss or gain of volatiles such as carbon dioxide. For another, the walls of the experimental vessel may be thought to be not a factor in the course of reaction when in fact they are via such mechanisms as diffusive absorption or corrosion.

A more difficult problem is that the user of a modeling code such as EQ6 must specify whether a given reaction is controlled by partial equilibrium, is governed by a rate law, or simply does not proceed on the time scale of interest. This issue can not be conveniently addressed by reference to some data base, but requires user cognizance of what sort of behavior has been observed in similar systems. Often, a number of such experiments must be done and analyzed to build up the necessary experience or "lore" to make successful predictions in an a priori sense. Field observations are an equally important part of such experience if modeling results are to be applied to field scenarios.

A point that often receives insufficient appreciation is that a system in the laboratory often behaves differently in sometimes critical ways from what is nominally the same system in the field. If one is studying rocks that are the products of natural hydrothermal alteration, the partial equilibrium assumption is more likely to be valid than it is in a laboratory hydrothermal apparatus in which one attempts to recreate such alteration. Such differences are commonly manifested in the appearance of different mineral assemblages, though changes in fluid chemistry may be very similar. An example is hydrothermal reaction of seawater and basalt, a process which occurs naturally at mid-ocean ridges (see refs. 1, 36, and many sources cited therein). The naturally altered basalts become rich in chlorite or chlorite plus epidote. In experimental systems, smectite clays appear instead. Time appears to be the limiting factor.

A further problem facing the user of geochemical modeling software is that no code, even EQ3/6, currently includes provision for all of the possible phenomena that may occur in reacting systems of interest. Beyond that, the fundamental scientific data base to support adequate modeling of all of these phenomena under all possible conditions of interest is incomplete. New fundamental measurements may be necessary to model many systems and scenarios of interest.

The preceding discussion is not intended to leave the reader with the feeling that little can be done with geochemical modeling codes. Much can now be done with them. In general, though, the user must be knowledgeable about the submodels included in the code, the related science base, the options offered by the code, and most importantly, the lore or base of experience concerning the problem. The computer code can then be useful in refining knowledge and testing hypotheses.

THE EQ3/6 SOFTWARE PACKAGE

The original version of EQ3/6 was written in FORTRAN 66 on CDC 6400 and 6600 computers at Northwestern University (1). In 1978, Wolery brought EQ3/6 to Lawrence Livermore National Laboratory (LLNL), where it was adapted and further developed on CDC 7600 machines. Some of this work shifted for a time onto Cray computers. A few years ago, almost all development work shifted to 32-bit machines with UNIX-based operating systems. A concurrent shift was made in the programming language, from FORTRAN 66 to FORTRAN 77. A deliberate effort has been made to maintain a high degree of portability of EQ3/6 on all 32- and 64-bit machines which offer full ANSI FORTRAN 77. This includes essentially all 32- and 64- bit UNIX-based machines, VAX machines, and Cray machines. The recent product is not very portable to CDC machines or IBM mainframes. To obtain reasonable run times on smaller problems, EQ3/6 requires a machine at least as powerful as a Sun 3/50 or a VAX 11/780. For larger EQ6 problems, a much faster machine, such as a Cray, Convex, Alliant, or high-end DEC machine is more suitable.

The guiding principle of EQ3/6 development has been to offer the user a a number of modeling options. For each major type of submodel, the approach has been to offer a menu of possibilities. For example, the user chooses the available model for activity coefficients of aqueous species that best fits his current problem. This reflects the belief of the code developers that there is usually no single "best" model to use as the submodel of a given type; rather, there are various models with different strengths and weaknesses. This approach to code development also lets the user investigate the consequences of using alternative models for a given phenomenon.

Generally speaking, the strong points of EQ3/6 include flexibility in terms of allowed input in speciation-solubility calculations, automatic decision-making in reaction path calculations (constrainable a priori by the user by means of the input file), a large thermodynamic data base, capability to treat dissolution and precipitation growth kinetics, a variety of models for treating the activity coefficients of aqueous species, and a significant and growing capability to treat the thermodynamics of solid solution. The principal weak points presently are that no capability is included for explicit treatment of reactions on mineral surfaces (sorption), for even one-dimensional fully coupled transport and chemistry, and pressure corrections for dealing with pressures off the 1.013 bar-steam saturation curve. Sorption modeling is a strong point of other codes, prominent examples being MINEQL (37) and its derivative MINTEQ (38). Plans exist (20) to extend EQ3/6 capabilities to cover such phenomena, where the current scientific base exists.

There is a published user's guide (3) for the EQ3NR code, written to correspond to the 3230 version. The only really significant difference between that and the presently available 3245 version is that the newer code allows the use of Pitzer's equations to describe the activity coefficients of aqueous species (39). The main user's guide for the EQ6 code (written to correspond to the 3245 version) is now in the final stages of preparation. Until it becomes available, the reader is best referred to ref. 2. Several supplementary user's guides have been already published. Ref. 23 describes the option for modeling reaction paths by imposing fixed fugacities of selected gases. Ref. 25 addresses the options for dealing with kinetic rate laws.

<u>Activity Coefficients of Aqueous Species</u>. The original version of EQ3/6 followed Helgeson et al. (5) in using the "B-dot" equation to describe the activity coefficients of aqueous solutes and a recommended approximation for the activity of water. The "B-dot" equation represents a simple extension of the Debye-Hückel equation and is only useful in relatively dilute solutions (deviations from precise measurements can be seen at ionic strengths below 0.1 molal, and become severe above 1.0 m). Beginning with version 3245, EQ3/6 offers two alternatives, the Davies (40) equation and Pitzer's equations (33, 34, 30, 39).

The Davies equation is another simple extended Debye-Hückel model and is generally about as good or bad as the "B-dot" model. The attractive aspect of the "B-dot" and Davies equations is that they are universally applicable to any aqueous species, real or imagined. However, they can not be applied to concentrated solutions. Pitzer's equations can be applied in dilute or concentrated solutions, but require special parameters (virial coefficients) for the species present. There is presently no reliable way to estimate the parameters; instead, they must be fit to experimentally determined quantities, such as the osmotic coefficient. Furthermore, the speciation model one employs using this option must be consistent with that employed in the fitting process. At present, the usage of Pitzer's equations is largely restricted to systems containing strong electrolytes, such as the "sea salt" system studied by Harvie, Moller, and Weare (30).

Three other options are presently being studied, with an eye on including them in the forthcoming 3270 version. One of these is the equations of Helgeson, Kirkham, and Flowers (41), for which further model development is required. The second (42) is based on the hydration theory concept of Stokes and Robinson (43). This also requires further model development. The third set is a model (44) recommended by the European Nuclear Energy Agency for obtaining equilibrium constants for the formation of aqueous complexes of interest in nuclear waste disposal, such as of uranium and plutonium.

In older versions of EQ3/6, no special effort was made to ensure consistency between the activity coefficient equations, which give results for single ions, and any special pH scale. The pH scale governing standard measurements is the NBS scale (45), which is defined in terms of a simple Debye-Hückel model for the activity coefficient of the chloride ion. If one uses the "B-dot" equation for the ions, and makes no corrections, one is essentially defining pH on the "B-dot" scale. Numerically, the inconsistency is small in dilute solutions, and has traditionally been ignored in perhaps all geochemical modeling codes. Beginning with the 3245 version, EQ3/6 now corrects all activity coefficients of aqueous ions to either an extended NBS scale or a "rational" scale, depending on instructions from the user. The NBS scale in the code is said to be extended because the formal limit of 0.1 m ionic strength is ignored. There is an ambiguity in the NBS scale because the correction depends on the ionic strength, which is a model dependent parameter whose value depends on what species one takes to be present. The "rational" scale defines the activity coefficient of the hydrogen ion as unity. The correction for this scale is independent of any model dependent parameters.

**Thermodynamic Data Base.** The thermodynamic data base was originally a reformatted version of the data base for the PATHI program of Helgeson and coworkers (4, 5; these references say little about this data base). The aqueous speciation model included many ion pairs and aqueous complexes. The PATHI data base had just been updated to conform with the contemporaneous efforts of Helgeson et al. (46-48) to provide better thermodynamic data. This affected the equilibrium constants for mineral dissociation reactions and a few aqueous redox reactions, but left the data for ion-pairing and complexation reactions unchanged. The mineral data on the first EQ3/6 data base were supplemented by Wolery (1) with some estimates of the constants for a number of clay mineral compositions.

Since EQ3/6 was brought to LLNL, the data base has been periodically updated and expanded. A particular effort has been made to respond to the subsequent efforts of Helgeson's group, which has done much work to correlate and predict the thermodynamic properties of the major rock-forming minerals (mostly exclusive of evaporite minerals) and many of the major aqueous solutes found in natural geochemical systems. The data for evaporite minerals were taken from the work of Harvie, Moller, and Weare (30). Other data base development has taken place in response to needs in the areas of radionuclide migration and disposal of high level nuclear waste. These efforts include Rard's work on the thermodynamics of species of technetium (49) and ruthenium (50). A complete documentation of all these changes is beyond the scope of the present communication.

The most recently released version of the EQ3/6 data base (3245; Feb. 1988) contains data for 47 chemical elements, 686 aqueous species, 638 aqueous reactions, 713 pure minerals, 11 gases, and 15 solid solutions. The data base itself is generally known by the file name of DATA0. The modeling codes do not read DATA0, but an unformatted equivalent known as DATA1. In the process of writing DATA1, a preprocessor performs various checks (such as to ensure that all reactions satisfy mass and charge balance) and replaces equilibrium constants on a temperature grid with equivalent interpolating polynomials. The preprocessor makes a special DATA1 to support the use of Pitzer's equations. An additional data file containing the Pitzer interaction parameters is read and all aqueous species that appear on DATA0 but not this file are stripped out to maintain consistency between the activity coefficient and speciation models. The interaction parameters on the additional data file are all of the thermodynamically observable type; these are broken down following a set of arbitrary conventions to non-observable theoretical equivalents to facilitate calculation of single-ion activity coefficients.

The DATA0 file was formerly maintained by a combination of hands-on editing, the use of a data block editing code, and the use of several data-generating codes, each of which had one or more of its own data files. This patchwork system has been replaced by a system based on INGRES, a commercial relational data base. There is a unified master data file, which, unlike DATA0, contains thermodynamic data in different forms (e.g., Gibbs energies, enthalpies, entropies, and heat capacities). The previously used data base codes have been integrated into the system as INGRES "applications." Internal documentation of the data and how it has been processed is an integral part of the system. The application code for adding to and updating the master data file generates audit records of all changes. Additional modifications in progress include provisions for the correction of apparent equilibrium constants to standard state values, for consistent treatment of multiple activity coefficient formalisms, and for ensuring compatibility with values recommended by CODATA and the NEA. The problem of multiple activity coefficient/speciation formalisms will be treated in future releases by generation of separate DATA0 files. Also in preparation is a major update in response to recent work by Helgeson and colleagues (51, 52; Schock, E.L., and Helgeson, H.C., Geochim. Cosmochim. Acta, submitted; Schock, E.L., Helgeson, H.C., and Sverjensky, D.A., Geochim. Cosmochim. Acta, submitted). Beginning in late 1988, LLNL will start issuing, on a regular basis, updated versions of DATA0 files.

**Solid Solutions.** EQ3/6 includes provision for dealing with solid solutions, both as "reactant" and "product" phases. Both ideal and non-ideal solid solution models have been incorporated. The compositions of "product" solid solutions continually readjust to remain in equilibrium with the changing fluid composition. In the "flow-through" open system mode, "product" solid solutions are removed from the reacting system as they form, resulting in zoning along the flow path. In the other modes, the entire mass of such phases reequilibrates with the fluid. The treatment of solid solution "reactants" is straightforward; the user defines the specific compositions on the input file. The treatment of "product" solid solutions is complicated by the fact that the saturation index for a solid solution is a function of its composition. Until the solid solution actually forms, however, the value of the saturation index is indeterminate. The approach taken in EQ3/6 is to use the maximum value over the composition space of the phase.

Application of EQ3/6 to many important problems will depend critically on the ability to model compositional variation in clays and zeolites. Thermodynamic data for 2:1 clays, and to a lesser extent zeolites, are not sufficiently abundant or of high enough quality to construct definitive solid solution models for these phases. Nevertheless, incorporation of reasonable models into EQ3/6 is a prerequisite to assessing the sensitivity of the geochemical modeling results to different solid solution approaches and for testing predictions against experimental and field observations.

In the 3245 version of EQ3/6, the solid solution models are restricted to the molecular-mixing type. Included for example is a 12 component model for dioctahedral smectite, partly based on thermodynamic estimation techniques ([46], [53]). Models are also included for some zeolites, based on similar methods ([54]).

Work is in progress to include site-mixing models in the forthcoming 3270 version. In site-mixing models, the ions are allowed to mix randomly on distinct kinds of sites, independently of the mixing on other kinds of sites. Molecular mixing is mathematically identical to site mixing over a single site, excluding mixing of vacancies. Both molecular- and site-mixing models can be treated as thermodynamically ideal or non-ideal, but there is little data in the 0-300°C temperature range to support the construction of non-ideal models. Work is presently focused on better models for smectites, illite, chlorite, and various zeolites. For example, to extend the compositional range of the smectite solid solution and to utilize a more appropriate model, an ideal site-mixing model for dioctahedral smectites ([55]) is being incorporated into EQ3/6.

An ion exchange approach is also being worked on. To allow reactant smectites and/or zeolites that are not in overall equilibrium with the solution phase to vary their exchange composition, fictive aqueous species, i.e. exchange `complexes', are defined and used to model cation exchange ([56]). The form of the equilibrium relationship between exchange complexes and basis species determines the exchange convention, i.e. mole-fraction or equivalent fraction representation of ideal exchange. The association reactions are formulated to preserve equivalent fraction formalism because this is numerically equivalent to the treatment of exchange using ideal site-mixing.

Rate Laws. The principal kinetic rate laws included in the EQ6 code are the transition-state theory form (e.g., [28]) and the Plummer et al. ([32]) rate law proposed for the dissolution and growth of carbonate minerals. Less important forms are discussed in ref. [25]. Generally speaking, these models may include an implicit model of speciation on the surface of the dissolving or growing mineral. However, no explicit models for speciation on mineral surfaces are presently accounted for in EQ3/6. Further development of kinetics theory may require the inclusion of such models for coupling with future rate law models.

Redox Disequilibrium in the Aqueous Phase. Aqueous redox disequilibrium modeling is permitted in the 3245 version of EQ3NR, but not the corresponding version of EQ6. It will be permitted in the 3270 version of EQ6. This will allow, for example, calculation of seawater/basalt reaction models in which total sulfate is conserved instead of partially reduced to sulfide.

Numerical Methods and Data Structure. Both EQ3NR and EQ6 make extensive use of a combined method, using a "continued fraction" based "optimizer" algorithm, followed by the Newton-Raphson method, to make equilibrium calculations. The method uses a set of master or "basis" species to reduce the number of iteration variables. Mass action equations for the non-basis species are substituted into mass balance equations, each of which corresponds to a basis species.

In the 3245 version of EQ3/6, the Newton-Raphson algorithm was modified to treat activity coefficients of aqueous species as known constants during a Newton-Raphson step. The activity coefficients are adjusted by a "double update" method of back-substitution between such steps. This simplifies the coding of the Jacobian matrix and gives performance comparable to the older method in which these activity coefficients were adjusted along with other variables using a pure Newton-Raphson method.

The "double" update is employed to expand the description of the system from the new concentrations of the basis species because, strictly speaking, new values of the activity coefficients are also required to calculate new concentrations of the non-basis species, and vice versa. The code first recomputes the concentrations of dependent species using the old values of the activity coefficients, recomputes the activity coefficients, uses them to recompute the non-basis species, and recomputes the activity coefficients. The 3270 version will similarly treat the activity coefficients of species in solid solutions.

The operational basis set can be changed by a process known as "basis switching." Thus, a basis species which contributes little to a mass balance can be exchanged with a non-basis species, preferably one which dominates that balance. Such an operation tends to lead to improved numerical behavior. EQ3/6 has allowed users to specify basis switches on the input file. It has also had an option (not much used) to use basis switching as part of the "optimizer" to reduce the value of residuals prior to Newton-Raphson iteration. For version 3270, EQ6 will be modified to insure after each step of reaction progress that the basis set employed is the "best" one. Note that this adjustment will take place after Newton-Raphson iteration has converged. This change is being implemented to improve numerical behavior in calculations involving very large numbers of chemical components.

C.M. Bethke (57) has shown that significant numerical advantages in such calculations can be realized by switching into the basis set a mineral species that is in partial equilibrium with the aqueous phase. This avoids expansion of the size of the Jacobian matrix and reduces computation time. A method based on this concept is being developed for use in the 3270 version of EQ3/6. The concept appears to show promise for improvement of the "optimizer" algorithm as well as the Newton-Raphson one.

The data structure used in EQ6 through version 3245 did not permit implementation of these ideas. The basis set was defined as a subset of the aqueous species, so a mineral species could not be switched into the basis. The properties of the various types of species (aqueous, mineral, gas) were distributed over separate, analogous arrays. The old structure had also become inefficient in terms of storage when the data base became large. A new data structure is being implemented in EQ3/6 version 3270. This has a single list of species of all types. The basis set is defined by a pointer array. There is a single list of phases, and the species belonging to a given phase appear contiguously on the list of species. The arrays for stoichiometric composition and reaction coefficients have been restructured to eliminate the storage of many zero values. Parallel arrays define the elements or species that correspond to the coefficient values, and pointer arrays determine where in these arrays the entries for a given species begin and end.

Through version 3245, mass balances in EQ6 were defined by stoichiometric numbers based on elemental composition. In the 3270 version, they will be based upon reaction coefficients. To maintain a maximum degree of correspondence of mass balance totals defined in this manner with physically measurable quantities, the mass balances and associated stoichiometric factors will be defined and maintained in terms of the original (data file) basis set (or more strictly, the reactions written using this set).

An example shows how non-physical mass balances arise. Dissociation of the non-basis species $HgCl_3^-$ is written on the data file as $HgCl_3^- = Hg^{2+} + 3Cl^-$. Similarly, for $HgBr_3$, the reaction is written as $HgBr_3^- = Hg^{2+} + 3Br^-$. The mercuric, chloride, and bromide ions are all basis species. The stoichiometric relationships are all obvious; e.g., the mercuric trichloride complex carries a weight of three in calculating the total chloride. However, if this complex is switched into the basis set in place of mercuric ion, then the reaction for the dissociation of the bromide complex must be rewritten as $HgBr_3^- + 3Cl^- = HgCl_3^- + 3Br^-$. Looking only at the reaction coefficients, one would infer that this species carries a weight of -3 in the balance for chloride ion. Curiosities of this sort have routinely been dealt with in other modeling codes, such as Reed's codes (58) and MINTEQ

(38). The totals are meaningful only in terms of the basis set employed. The only problem with the use of non-physical totals is that code users may confuse such quantities with those reported from chemical analysis. Except for three special cases discussed below, the procedure used in the new EQ3/6 code keeps the mass balance totals physically meaningful.

In the new code, the mass balance for water replaces one for elemental oxygen, the mass balance for hydrogen ion replaces one for elemental hydrogen, and the mass balance for oxygen gas replaces the charge balance. There is no general way to write the reactions to guarantee physical meaning to these new total quantities. The mass balance for water calculated by this method is generally numerically very nearly equal to the physical amount of water. However, in high pH solutions, a mass balance of hydrogen ion usually yields a negative number. In reducing solutions, the same sort of result is seen for oxygen gas. EQ3NR deals with these quantities only as calculated outputs, so their usage is significant primarily to EQ6. A principal advantage for EQ6 of redefining mass balances in this way is that the existing "continued fraction" based "optimizer" algorithm can be modified to apply to all the basis species; through version 3245, it could not be applied to these three special species.

Rate Law Integration. Kinetic rate laws are ordinary differential equations (ODEs) that are numerically integrated in EQ6 when such rate laws are invoked by the user. In version 3245, the method of integration is a simple "predictor" function modeled after the predictor-corrector method of Gear (59, 60). Integration accuracy is maintained by cutting the step size if an accuracy test is failed. In version 3270, a corrector function is being added to increase the robustness of the code. In version 3245, some problems have been observed to get caught running at small steps sizes for long periods of time due to the lack of an ODE corrector.

CURRENT DIRECTIONS IN USING EQ3/6 IN GEOCHEMICAL MODELING

Current directions in code usage are illustrated by two examples, one dealing with pH buffers, the other with reaction of ground water with spent nuclear reactor fuel. The former case deals with small, relatively simple systems, the latter with a system that is about as large and complex as a geochemist is ever likely to see.

pH Buffer Calculations. Knauss and Wolery performed monomineralic dissolution rate experiments on albite (28) and quartz (29) in various pH buffer solutions at 70°C. The buffer compositions chosen were fairly standard ones taken from the literature (see ref. 28 for sources). The pH of these solutions was given for 25°C, not 70°C. It was necessary to obtain the values at the temperature of the experiments. This was done using the EQ3NR code. The method of calculation is illustrated for the case of the pH 4 buffer 0.05 m potassium hydrogen phthalate plus 0.0001 m HCl. The solution composition was constrained to satisfy 0.05 m total dissolved potassium, 0.05 m total dissolved phthalate, 0.0001 m total dissolved chloride. pH was input as 4.0 and the code was instructed to adjust the concentration of hydrogen ion as required to satisfy electrical balance. For a temperature of 25°C, the calculated pH was 4.00, in excellent agreement with expectations and close to a measured value of 4.06. For 70°C, the pH was calculated to be 4.11.

It was subsequently found that the high level of potassium in this buffer interfered with the measurement of dissolved sodium and aluminum in the albite experiments. Such high levels of buffering agent were not required to maintain the pH. In a subsequent study on the dissolution kinetics of muscovite (Knauss, K.G., and Wolery, T.J., Geochim. Cosmochim. Acta, in press), the concentrations of the buffering agents were reduced tenfold (e.g., the concentration of potassium hydrogen phthalate in the "pH 4" buffer was reduced from 0.05 m to 0.005 m). The code calculation was constrained to satisfy 0.005 m total dissolved potassium, 0.005 m dissolved phthalate, and pH 4.00 at 25°C. The total dissolved chloride was set to 0.0001 m and the code was instructed to adjust this to achieve electrical neutrality. The total dissolved chloride after adjustment was 0.0003 m, indicating that 0.0003 m HCl is required to make up the buffer. From this, the 70°C pH was calculated to be 4.06. The new buffer could as easily have been calculated to satisfy pH 4.00 at 70°C by, as before, adjusting the chloride to satisfy charge balance. This would have yielded a slightly different concentration of HCl in the buffer.

The use of EQ3NR to design custom pH buffer solutions represents a useful and one of the "safer" applications of EQ3/6. It is safer because there are no issues of kinetics or metastability to deal with. It is not absolutely safe, however, because the results do depend on the thermodynamic data used. On the other hand, many buffer compositions have been well studied and provide useful but limited tests of validation of the code and the thermodynamic data base.

Reaction of Spent Fuel and Groundwater. At the other extreme end of code usage is the work of Bruton and Shaw (19) to model reaction of spent nuclear fuel with ground water. These calculations include about forty chemical elements, employing almost the whole of the EQ3/6 thermodynamic data base. This modeling is currently only in the early stages, requiring a number of simplifying assumptions and proceeding in the likely absence of important thermodynamic data. Nevertheless, the modeling is proving useful in pushing for better characterization of experiments and in identifying thermodynamic data needs. A resonance between modeling and experimentation is expected to lead to models sufficiently dependable to make predictions of behavior of waste emplaced in the field.

One of the major outputs of these model simulations is a prediction of a sequence of secondary phases. These phases are important because they can act as sinks for radionuclides and other toxic components of the waste, limiting their migration away from the point of emplacement. Even phases not containing such components affect the overall course of reaction via effects on pH, redox conditions, etc. Figure 1 shows the results of one simulation for reaction of spent fuel and J-13 well water at 25°C. Haiweeite (a calcium uranyl silicate), soddyite (a uranyl silicate), and schoepite (a uranyl hydroxide) are successive (and overlapping) sinks for uranium, the dominant constituent of spent fuel.

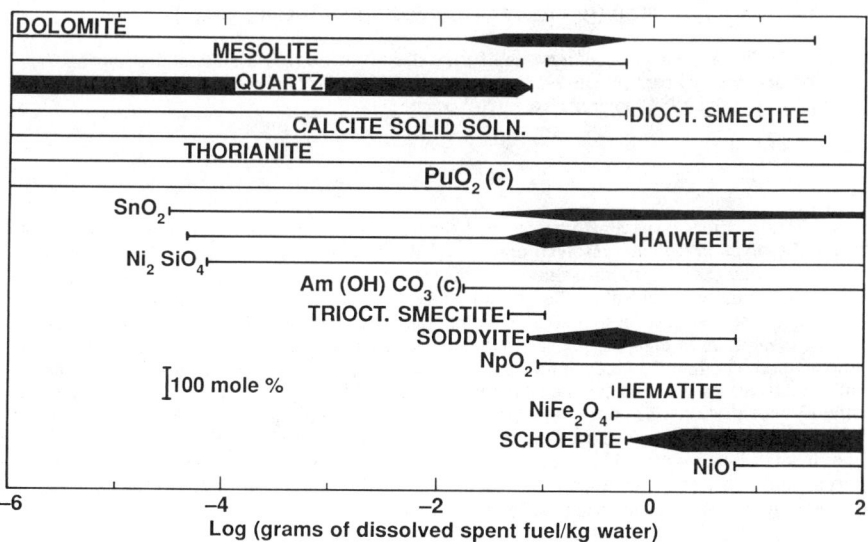

Figure 1. The predicted sequence of solid reaction products of reaction between spent nuclear fuel and J-13 well water at 25°C. The width of the bars represents the percentage of the number of moles of all precipitates at any given point of reaction progress. There is a break in the appearance of mesolite. There is no break in the appearance of $PuO_{2(c)}$. (Reproduced with permission from Ref. 19. Copyright 1988 Materials Research Society.)

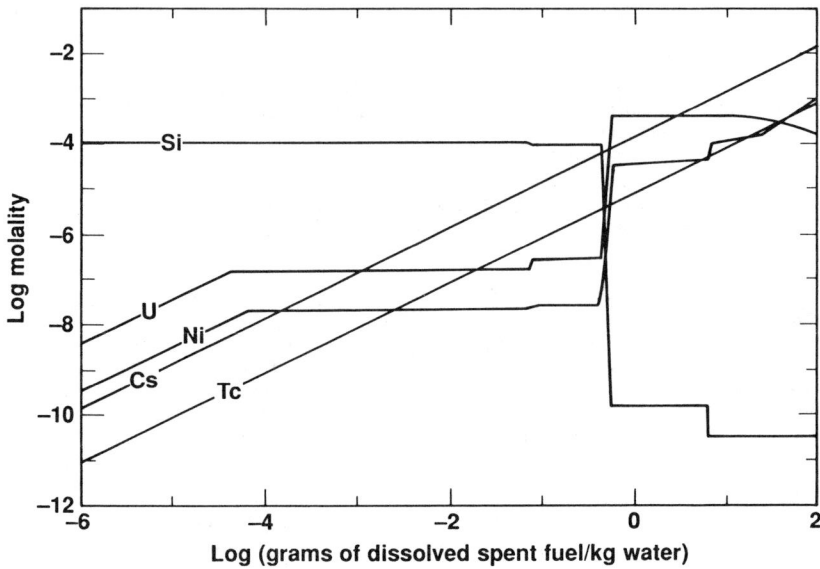

Figure 2. The predicted concentrations of total dissolved Cs, Ni, Si, Tc, and U during the reaction of spent nuclear fuel and J-13 well water at 25°C. (Reproduced with permission from Ref. 19. Copyright 1988 Materials Research Society.)

The solubility constraints imposed by haiweeite and soddyite create an inverse relationship between dissolved silica and dissolved uranium. This is illustrated in Figure 2, which shows the evolution of total concentration of uranium, silica, and other selected elements. The silica in the system is essentially provided by the ground water. As reaction proceeds, uranium released by dissolution of the fuel is sequestered by silica into haiweeite. When a large fraction of the silica in the system is tied up in this phase, soddyite begins to form. In this phase, a given amount of silica can tie up more uranium than it can in haiweeite. The haiweeite redissolves as more soddyite forms. Finally, the released uranium has tied up nearly all the silica in the system in soddyite and the uranium concentration rises until it is limited by the precipitation of another uranium phase, schoepite, which contains no silica. The silica at the end of the reaction path is extensively tied up in the nickel silicate phase. The sequestering of uranium in uranyl silicates has been observed experimentally (see ref. 19), although uranophane, another uranyl silicate, is found instead of soddyite. Further work is planned to resolve this difference.

ACKNOWLEDGMENTS

Prepared by Yucca Mountain Project (YMP) participants as part of the Civilian Radioactive Waste Management Program. The YMP project is managed by the Yucca Mountain Project Office of the U.S. Department of Energy, Nevada Operations Office. YMP Project work is sponsored by the DOE Office of Civilian Radioactive Waste Management.

This work was performed under the auspices of the U.S. Department of Energy by Lawrence Livermore National Laboratory under contract No. W-7405-Eng48. The authors thank Paul Cloke, Miki Moore, and three anonymous reviewers for their helpful comments.

## Literature Cited

1. Wolery, T.J. Ph.D. thesis, Northwestern University, Evanston, IL, 1978; University Microfilms International, Ann Arbor, MI, Order No. 7907954.
2. Wolery, T.J. Calculation of Chemical Equilibrium between Aqueous Solution and Minerals: The EQ3/6 Software Package. UCRL-52658, Lawrence Livermore National Laboratory, Livermore, CA, 1979.
3. Wolery, T.J. EQ3NR, A Computer Program for Geochemical Aqueous Speciation-Solubility Calculations: User's Guide and Documentation. UCRL-53414, Lawrence Livermore National Laboratory, Livermore, CA, 1983.
4. Helgeson, H.C. Geochim. Cosmochim. Acta 1968, 32, 853-877.
5. Helgeson, H.C.; Brown, T.H.; Nigrini, A.; Jones, T.A. Geochim. Cosmochim. Acta 1970, 34, 569-592.
6. Taylor, R.W.; Jackson, D.D.; Wolery, T.J.; Apps, J.A. In Howard, J.H., et al., eds., Geothermal Resource and Reservoir Investigations of U.S. Bureau of Reclamation Leaseholds of East Mesa, Imperial Valley, California. LBL-7094, Lawrence Berkeley Laboratory, Berkeley, CA, 1978; Section 5.
7. Sverjensky, D. Econ. Geol. 1984, 79, 23-27.
8. Brimhall, G.H., Jr. Econ. Geol., 1980, 75, 384-409.
9. Garven, G. Ph.D. thesis, University of British Columbia, Vancouver, British Columbia, 1982.
10. Janecky, D.R.; Seyfried, W.E., Jr. Geochim. Cosmochim. Acta 1984, 48, 2723-2738.
11. Gitlin, E. Geochim. Cosmochim. Acta 1985, 49, 1567-1579.
12. Bowers, T.S.; von Damm, K.L.; Edmond, J.M. Geochim. Cosmochim. Acta 1985, 49, 2239-2252.
13. Kerrisk, J. Reaction-Path Calculations of Ground-Water Chemistry and Mineral Formation of Rainier Mesa, Nevada. LANL-9912-MS, Los Alamos National Laboratory, Los Alamos, NM, 1984.
14. Knauss, K.; Oversby, V.M.; Wolery, T. In Scientific Basis for Nuclear Waste Management VII; McVay, G.L., Ed.; Mat. Res. Soc. Symp. Proc. 1984, Vol. 26; pp. 301-308.
15. Knauss, K.; Delany, J.; Beiriger, W.; Peifer, D. In Scientific Basis for Nuclear Waste Management VIII; Jantzen, C.M., Stone, J.A., and Ewing, R.C., Eds.; Mat. Res. Soc. Symp. Proc. 1985, Vol. 44; pp. 539-546.
16. Delany, J.M. Reaction of Topopah Spring Tuff with J-13 water: A Geochemical Modeling Approach Using the EQ3/6 Reaction Path Code. UCRL-53631, Lawrence Livermore National Laboratory, Livermore, CA, 1985.
17. Wolery, T.J. Chemical Modeling of Geologic Disposal of Nuclear Waste: Progress Report an a Perspective. UCRL-52748, Lawrence Livermore National Laboratory, Livermore, CA, 1980.
18. Bruton, C.J. In Scientific Basis for Nuclear Waste Management XI; Apted, M.J., and Westerman, R.E., Eds.; Mat. Res. Soc. Symp. Proc. 1987, Vol. 112; pp. 607-619.
19. Bruton, C.J.; Shaw, H.F. In Scientific Basis for Nuclear Waste Management XI; Apted, M.J., and Westerman, R.E., Eds.; Mat. Res. Soc. Symp. Proc. 1988, Vol. 112; pp. 485-494.
20. McKenzie, W.F.; Wolery, T.J.; Delany, J.M.; Silva, R.J.; Jackson, K.J.; Bourcier, W.L.; Emerson, D.O. Geochemical Modeling (EQ3/6) Plan Office of Civilian Radioactive Waste Management Program. UCID-20864, Lawrence Livermore National Laboratory, Livermore, CA, 1986.
21. Nordstrom, D.K.; Plummer, L.N.; Wigley, T.M.L.; Wolery, T.J.; Ball, J.W.; Jenne, E.A.; Bassett, R.L.; Crerar, D.A.; Florence, T.M.; Fritz, B.; Hoffman, M.; Holdren, G.R., Jr.; Lafon, G.M.; Mattigod, S.V.; McDuff, R.E.; Morel, F.; Reddy, M.M.; Sposito, G.; Thrailkill, J. In Chemical Modeling in Aqueous Systems; Jenne, E.A., Ed.; Amer. Chem. Soc. Symp. Series 1979, Vol. 79; pp. 857-892.
22. Wolery, T.J.; Isherwood, D.J.; Jackson, K.J.; Delany, J.M.; Puigdomenech, I. In Proceedings of the Conference on the Application of Geochemical Models to High-Level Nuclear Waste Repository Assessment, Oak Ridge, Tennessee, October 2-5, 1984, Jacobs, G.K., and Whatley, S.K., Eds.; NUREG/CP-0062 and ORNL/TM-9585, 1985; pp. 54-65.
23. Delany, J.M.; Wolery, T.J. Fixed-Fugacity Option for the EQ6 Geochemical Reaction Path Code. UCRL-53598, Lawrence Livermore National Laboratory, Livermore, CA, 1984.

24. Rimstidt, J.D.; Barnes, H.L. Geochim. Cosmochim. Acta 1980, 44, 1683-1699.
25. Delany, J.M.; Puigdomenech, I.P.; Wolery, T.J. Precipitation Kinetics Option for the EQ6 Geochemical Reaction Path Code. UCRL-53642, Lawrence Livermore National Laboratory, Livermore, CA, 1986.
26. Wolery, T.J. Geothermal Resources Council Transactions 1979, 3, 793-795.
27. Holdren, G.R., Jr.; Berner, R.A. Geochim. Cosmochim. Acta 1979, 43, 1161-1171.
28. Knauss, K.G.; Wolery, T.J. Geochim. Cosmochim. Acta 1986, 50, 2481-2497.
29. Knauss, K.G.; Wolery, T.J. Geochim. Cosmochim. Acta 1988, 52, 43-53.
30. Harvie, C.E.; Moller, N.; Weare, J.H. Geochim. Cosmochim. Acta 1984, 48, 723-758.
31. Reddy, M.M.; Plummer, L.N.; Busenberg, E. Geochim. Cosmochim. Acta 1981, 45, 1281-1289.
32. Plummer, L.N.; Wigley, T.M.L.; Parkhurst, D.L. Am. J. Sci. 1978, 278, 179-216.
33. Pitzer, K.S. J. Phys. Chem. 1973, 77, 268-277.
34. Pitzer, K.S. In Activity Coefficients in Electrolyte Solutions; Pytkowicz, R.M., Ed.; CRC Press: New York, 1979; pp. 157-208.
35. Smith, R.M.; Martell, A.E. Critical Stability Constants, Volume 4: Inorganic Complexes; Plenum Press: New York and London, 1976.
36. Mottl, M.J. Bull. Geol. Soc. Amer. 1983, 94, 161-180.
37. Westall, J.C.; Zachary, J.L.; Morel, F.M.M. MINEQL, A Computer Program for the Calculation of Chemical Equilibrium Composition of Aqueous Systems; Water Quality Laboratory, Ralph M. Parsons Laboratory for Water Resources and Environmental Engineering, Department of Civil Engineering, Massachusetts Institute of Technology; Tech. Note 18, 1976.
38. Felmy, A.R.; Girvin, D.C.; Jenne, E.A. MINTEQ--A Computer Program for Calculating Geochemical Equilibria; PB84-157148; Battelle Pacific Northwest Laboratories, Richland, WA, 1984.
39. Jackson, K.J.; Wolery, T.J. Extension of the EQ3/6 computer codes to geochemical modeling of brines. Mat. Res. Soc. Symp. Proc. 44, 507-514, 1985.
40. Davies, C.W. Ion Association; Butterworths: London; 1962.
41. Helgeson, H.C.; Kirkham, D.H.; Flowers, G.C. Am. J. Sci. 1981, 281, 1249-1516.
42. Wolery, T.J.; Jackson, K.J. This volume.
43. Stokes, R.H.; Robinson, R.A. J. Am. Chem. Soc. 1948, 70, 1870-1878.
44. Grenthe, I.; Wanner, H. Guidelines for the extrapolation to zero ionic strength, OECD Nuclear Energy Agency, Data Bank Report TDB-2, 1988.
45. Bates, R.G.; Guggenheim, E.A. Pure Appl. Chem. 1960, 1, 163-168.
46. Helgeson, H.C.; Delany, J.M.; Nesbitt, H.W.; Bird, D.K. Am. J. Sci. 1978, 278A, 1-229.
47. Helgeson, H.C.; Kirkham, D.H. Am. J. Sci. 1974, 274, 1089-1198.
48. Helgeson, H.C.; Kirkham, D.H. Am. J. Sci. 1976, 276, 97-240.
49. Rard, J.A. Critical Review of the Chemistry and Thermodynamics of Technetium and Some of Its Inorganic Compounds and Aqueous Species, UCRL-53440, Lawrence Livermore National Laboratory, Livermore, CA, 1983.
50. Rard, J.A. Chem. Rev. 1985, 85, 1-39.
51. Tanger, J.C.; Helgeson, H.C. Am. J. Sci. 1988, 288, 19-98.
52. Schock, E.L.; Helgeson, H.C. Geochim. Cosmochim. Acta 1988, 52, 2009-2036.
53. Tardy, Y.; Garrels, R. Geochim. Cosmochim. Acta 1974, 40, 1051-1056.
54. La Iglesia, A.; Aznar, A.J. Zeolites 1986, 6, 26-29.
55. Aagaard, P.; Helgeson, H.C. Clays Clay Miner. 1983, 31, 207-217.
56. Mattigod, S.V.; Sposito, G. In Chemical Modeling in Aqueous Systems; Jenne, E.A., Ed.; Amer. Chem. Soc. Symp. Series 1979, Vol. 79; pp. 837-856.
57. Bethke, C.M. In Collected Abstracts for the Workshop on Geochemical Modeling, Sept. 14-17, 1986, Fallen Leaf Lake, CA, Lawrence Livermore National Laboratory, Livermore, CA.
58. Reed, M.H. Geochim. Cosmochim. Acta 1982, 46, 513-528.
59. Gear, C.W. Comm. ACM 1971, 14, 176-179.
60. Gear, C.W. Comm. ACM 1971, 14, 185-190.

RECEIVED October 6, 1989

# Chapter 9

# Geochemical Modeling of Water—Rock Interactions Using SOLMINEQ.88

**Ernest H. Perkins[1], Yousif K. Kharaka[2], William D. Gunter[1], and Jeffrey D. DeBraal[2]**

[1]Oil Sands and Hydrocarbon Recovery, Alberta Research Council, P.O. Box 8330, Postal Station F, Edmonton, Alberta T6H 5X2, Canada
[2]Water Resources Division, U.S. Geological Survey, Mail Stop 427, 345 Middlefield Road, Menlo Park, CA 94025

> SOLMINEQ.88 is an ANSI standard FORTRAN-77 geochemical modeling program, based on the 1973 version of SOLMNEQ and various updated versions. The algorithms have been improved, resulting in faster program execution and tighter convergence. SOLMINEQ.88 has a revised thermodynamic data base, including over 270 inorganic and 80 organic aqueous species, and 214 minerals. The program calculates the distribution of mass among aqueous species and complexes and calculates the saturation indices of minerals at temperatures up to 350°C and pressures up to one kilobar. Several examples of typical SOLMINEQ.88 calculations are given in the text.
>
> SOLMINEQ.88 has a number of new options, including the ability to calculate the effects of boiling, mixing of solutions and partitioning of gases between water, oil and a vapour phase. It has a mass transfer option to predict the effects of dissolution and/or precipitation of minerals. An ion exchange and an ion adsorption options have also been included. A number of different options, including the use of the Pitzer activity coefficient model, the printing of a comprehensive set of geothermometers, activity ratios and mass abundances, can be specified at run time. Additional minerals and components (both anions and cations, including their complexes) can be added at run time.
>
> A user-friendly, interactive input program, SOLINPUT, is available and can be used to update input files. SOLMINEQ.88 PC/SHELL, a user-friendly interactive, full screen environment for IBM personal computers (clones and compatibles) which controls input, program execution and examination of output is also available.

Water-rock interactions in both natural systems and systems perturbed by man are of significant interest because of their potential to concentrate or remove minerals or elements of economic and social interest, and because of their potential to modify the transport and thermal processes in a system. The solid phases can generally be examined in fossil systems and given enough samples, an

This chapter not subject to U.S. copyright
Published 1990 American Chemical Society

overview of the chemical processes which occurred can be obtained. In currently active systems, the fluids may be the only phase available for examination.

The composition of the fluid phase is the result of many different physical and chemical processes which typically occur at temperatures and pressures different than the sampling conditions, which in turn are often different from the analytical conditions. The processes which can control or modify the chemical composition of the water include mineral dissolution and precipitation, ion exchange and adsorption/ desorption, mixing of different fluids, gas separation by boiling or pressure draw-down, and organic/inorganic interactions. Other complications which must be addressed before interpretation, include analytical error, loss of volatiles after sampling but before analysis, failure to perform a complete analysis, as well as poor thermodynamic data. All of these processes and problems must be kept in mind during the interpretation of any water sample.

Many different geochemical programs exist which address some of the processes and complications previously mentioned. A few of the programs currently available and commonly used include PATH and it's modified versions (1-2), SOLMNEQ and it's modified versions (3-4), WATEQ and it's modified versions (5-6-7), MINEQL (8), EQ3/EQ6 (9-10-11), PHREEQE (12), GEOCHEM (13); SOLVEQ (14), MINTEQ (15), DYNAMIX (16).

These programs and many others are general and comprehensive, however, each program is more suited and particularly useful for a special process and/or environment. SOLMINEQ.88 is particularly useful for modeling interactions in sedimentary basins and in thermally stimulated oil reservoirs where petroleum, organic species, and high temperatures, pressures and salinities prevail.

The original version of SOLMNEQ (3) was written in PL/1 for the IBMs 360 series of computers, however, the present version of SOLMINEQ.88 is written in ANSI standard FORTRAN-77 using improved numerical methods to force convergence. It has been tested on computers ranging from IBM pc's (compatibles and clones) to mini-computers (VAXes and PRIMEs) to large mainframe computers (AMDAHLs, IBM). The only portions of the code which may have to be modified between the various architectures and compilers are the file opening statements and these are clearly commented. On machines which are severely limited in memory (less than 500 K), overlaying sections of code may be necessary; the current structure of the code has been written to allow this.

SOLMINEQ.88 has a revised thermodynamic data base, containing more than 270 inorganic aqueous species, 80 organic species and 210 minerals, with easy methods of adding more. The user now has the option of using the Pitzer equations to calculate the activity coefficients of aqueous species. In addition to the options that existed in previous version, new modeling options have been added which can be used to study the effects of boiling, mixing of different fluids, partitioning of gases between water, oil and gas phases. Also included are several mass transfer options that can be used to predict the effects of ion exchange, adsorption, desorption and the effects of dissolution and precipitation of solids on the aqueous chemistry.

SOLMINEQ.88 has an interactive input program available for the mainframe use, and a complete user friendly, full screen interactive shell which controls the data input, running of SOLMINEQ.88 and examination of the output available for the pc version.

COMPUTATIONAL DETAILS

The computational basis currently used for SOLMINEQ.88 is very similar to that discussed in previous papers (3-17-4); details are given in Kharaka et al., (1988). Essentially, given a water composition, pH, Eh, temperature and pressure, the distribution of species in the water is calculated by solving the set of mass action and mass balance equations using a back-substitution method. This method has been improved from the original by the addition of an "optimal scaler" which speeds up convergence considerably.

Once convergence has been reached, the "hydronium" balance for the system (the sum of all of the ionizable hydrogen ion minus the sum of all of the ionizable hydroxide) is calculated. Even if the distribution of species in the aqueous solution is modified by changes in temperature and/or pressure, this value is constant. If chemical components are added to or removed from the aqueous solution, the change in the hydronium balance is known.

The convergence on the difference on hydronium balance is used as a basis for all of the program options, including changes in temperature and pressure, precipitation or dissolution of any components, or loss of volatiles and adsorption of ions on surfaces. As an example, consider the loss of carbon dioxide as a volatile into a co-existing gas phase during boiling. The water is initially at 25°C with the pH and total mass of each of the elements known. The distribution of species is obtained and the initial hydronium balance calculated. The distribution of species is then calculated at the temperature of boiling by using the same hydronium balance as that at 25°C, and modifying the pH to obtain convergence. The mass of all of the components in the water is then concentrated by the boiling factor (80 percent quality steam would result in concentrating the components in solution by a factor of 5). The initial amount of $CO_2$ lost as a volatile is guessed and the hydronium balance recalculated accordingly. The pH is varied (and the distribution of species continually recalculated) until convergence on the old and new hydronium balances are obtained. The amount of lost $CO_2$ is then recalculated. If this agrees with the previous amount, the iterative process is terminated. Otherwise a new value for the amount of $CO_2$ fractionated into the vapour phase is guessed and the iterative process continued.

One new feature added to SOLMINEQ.88 is the Pitzer method of calculating the activity coefficients of aqueous species as discussed by Kharaka et al. (1987), and detailed in Kharaka et al. (1988). The Pitzer activity coefficient formulation expands the activity coefficients of the charged species using second and third virial parameters, but generally denies the existance of complexed species. The SOLMINEQ.88 modifications use the calculated Pitzer activity coeffients to force the activity of the corresponding component species, but also includes all known species and complexes. For aqueous species without published virial coefficients the deviation function is obtained from the Pitzer activity coefficient for Na (for monovalent species) and Ca (for multivalent species). Although this approach is clearly an operational approach, it does allow activity coeffients to be calculated in high ionic strength solutions.

Another set of features added to SOLMINEQ.88 is the ability to include additional anions and cations in the user defined data set at run time. Up to 14 complexes of each can be included if their equilibrium constants are known. The addition of two user defined anions is particularly useful in the study of the role of organic ligands because SOLMINEQ.88 incorporates thermochemical data for aqueous species of only acetate, oxalate and succinate. These are the dominate organic ligands in sedimentary basins, but others may be important in these and other systems ([20]). In addition, up to five user defined minerals can be included with the necessary information to calculate both low and high temperature saturation indices.

MODELING CAPABILITIES

Once the distribution of mass among the various aqueous species in solution has been calculated, there are a number of different geochemical modeling options available (Fig. 1). The most commonly used option is to increase the temperature and pressure and re-calculate the distribution of species and saturation states of minerals, as initially discussed by Kharaka and Mariner (1977). The effects of increasing pressure on the complexation of the aqueous species was not included at that time but has now been added based on the correlations of Aggarwal et al. (this issue). Including this correction is important because the dissociation constant of complexes can vary by a factor of three over 1 kilobar. The pressure correction for aqueous complexes results in reducing the enhanced solubility of minerals from that predicted without this correction ([22]).

Both the lost gas option and the carbon dioxide option were discussed by Aggarwal et al. (1986). The lost gas option allows a known amount of carbon dioxide, of hydrogen sulfide and/or of

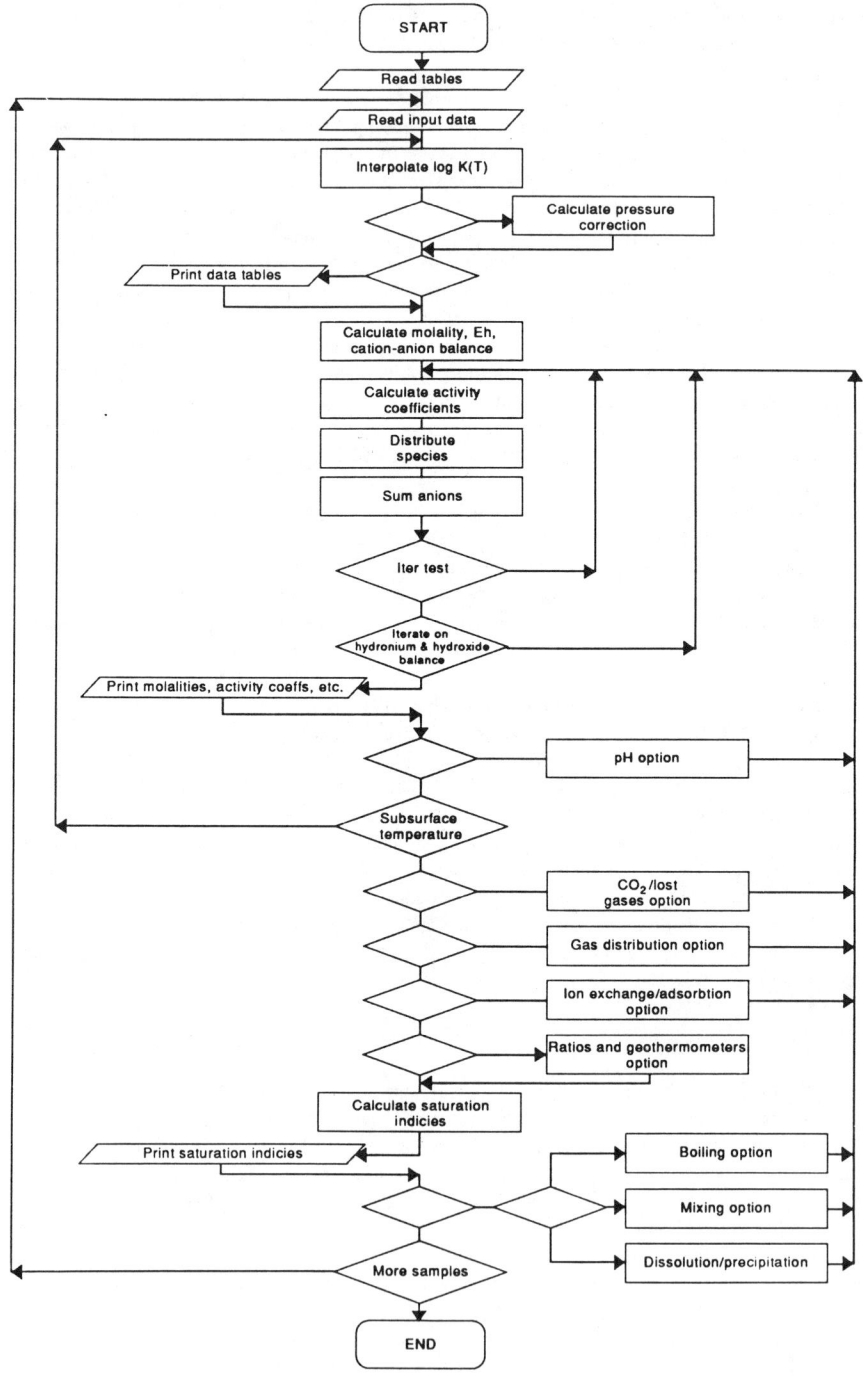

Figure 1. Flow chart for SOLMINEQ.88.

ammonia to be added/removed from the aqueous phase, while the carbon dioxide option adds or removes carbon dioxide until one of six criteria are satisfied. These criteria can be any one of either saturation with calcite, dolomite or siderite, or attainment of a fixed pH, of a fixed partial pressure of carbon dioxide gas or of a fixed molality of carbonic acid in solution. Another option is to modify the pH of the solution such that the solution is in equilibrium with calcite, dolomite or siderite. This option, called the pH option, was also discussed by Aggarwal et al. (1986). The convergence technique for both of these has been improved.

The mixing option in SOLMINEQ.88 is particularly useful in modeling deep well disposal of industrial waters. This option allows two different aqueous fluids to be mixed in varying proportions. At a constant or varying temperature, the variables typically monitored in the mixed fluids, are the saturation indices of the various minerals in order to determine if there will be problems with precipitation of various phases in intermediate mixtures. A detailed example of the use of this option is reported in Gunter et al. (1986), who modeled the effects of waste water disposal from enhanced oil recovery operations (EOR) in two different disposal formations. In cases where mineral precipitation is predicted, the code can be used to indicate a remedial action in the form of chemical additives, or by mixing with water of different composition (Fig. 2).

The dissolution/precipitation option in SOLMINEQ.88 can be used to carry out:
1. Dissolve/precipitate a given amount of mineral;
2. dissolve/precipitate a mineral to saturation;
3. dissolve/precipitate a mineral to saturation with another mineral;
4. add/subtract aqueous components to saturation with a mineral; and;
5. add/subtract components from a congruent, incongruent, or net reaction.

In previous versions of SOLMINEQ, boiling operated only as a concentration mechanism and did not allow fractionation of any of the volatile components into the gas phase. This option has been extended in SOLMINEQ.88 and now includes fractionation of carbon dioxide, ammonia, hydrogen sulfide and methane into the gas phase. When combined with the mineral dissolution and precipitation option, the effects of additives and steam quality on condensate pH can be calculated. Steam quality is defined by the mass of water in the vapour phase divided by the total mass of water times 100°. Figure 3 illustrates the effect on the condensate pH when a surface water is boiled. The solid curve on figure 3 is the locus of condensate pH and steam quality values for 250°C. If the condensate is separated from the steam and cooled to 25°C, the pH of the resulting solution is represented by the dashed line. For this curve, the abscissa refers to the steam quality that the condensate was derived from. This figure clearly shows that under moderate to high steam quality, waters with extreme pH values can be generated. These waters will have significant potential to react with and damage either the production equipment and the formation mineralogy. A more detailed example showing the effects of an ammonium additive has been discussed by Perkins and Gunter (1988).

An option which shares many features of the boiling option is the gas fractionation option. In most oil wells, oil and water are produced up the tubing string while the gas (carbon dioxide, methane, hydrogen sulfide) is produced up the annulus. The annulus gas separates from the oil and water at the sand face and is present either because a three phase system (oil, water and gas) exists in the reservoir or because it was formed due to the pressure drop in the near well bore region. If the compositions of the annulus gas and of the co-existing water have been determined, the relative masses of gas, water and oil have been measured, and the near well temperature known, the change in water chemistry due to gas separation can be calculated. In addition, the composition of the co-existing gas phase (if one is present) and the amount of dissolved volatiles in the oil phase are also calculated, as they exist in the reservoir. If calcite is present and is though to be in equilibrium with the water, the reservoir pressure can be estimated with confidence.

Tables 1, 2 and 3 illustrate a typical set of data and results when using the gas fractionation option. Table 1 gives the water composition, the annulus gas composition and the oil-water ratio.

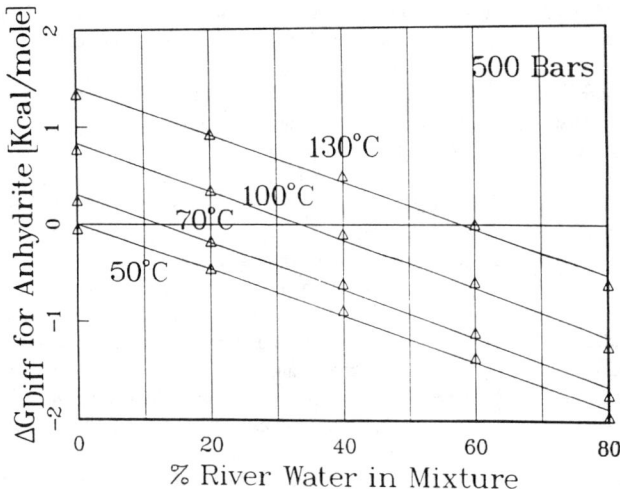

Figure 2. Saturation state of anhydrite in a mixture of Dolores river water and Paradox Valley brine, western Colorado, at likely aquifer temperatures of 50°C to 130°C and 500 bars. The chemical composition of Paradox Valley brines and other pertinent information are given by Flak and Brown (1988). The percentage of river water required to stop precipitation of anhydrite in the injection aquifer is obtained at the point where $\Delta G_{diff} = 0$.

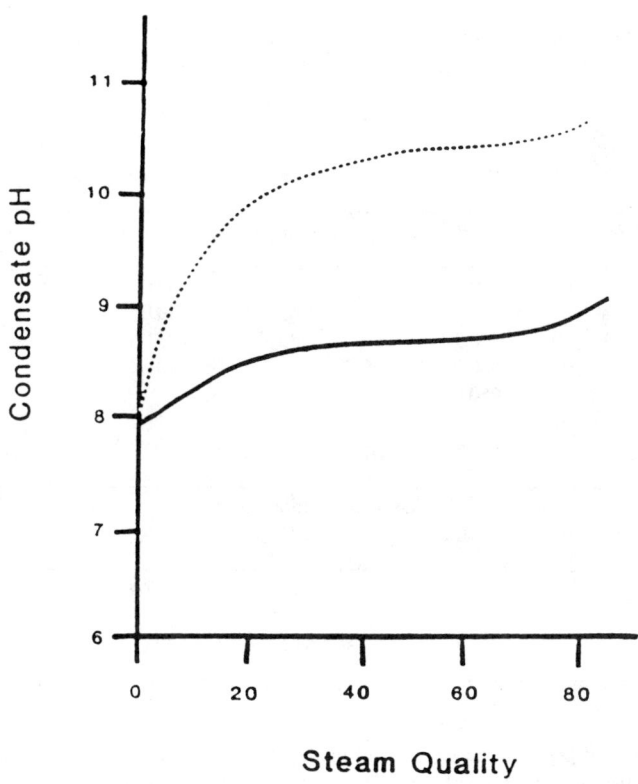

Figure 3. pH of condensate versus the steam quality. The solid curve is for the condensate at 250°C and dashed curve is the 25°C equivalent value.

Table 2 gives the temperatures, the pH (measured at 25°C, calculated at higher temperatures) and the saturation indices of calcite for laboratory conditions, well head conditions and sandface conditions. Of immediate interest is that calcite may be forming scale from the sand face up through the well tubulars, but by the time that the water has reached wellhead conditions, there is no potential for scale formation. The high saturation index for calcite at the sand face is the result of the formation and separation of a gas up the annulus. Table 3 gives the saturation indices of calcite, calculated for a variety of pressures. Because calcite is present in significant quantities in the reservoir, it probably is in equilibrium with the water. Examining table 3, this occurs for total pressures of approximately 9 bars and above. As noted for this particular pressure, no gas phase exists. Thus, in the near well region, the pressure must be greater than 9 bars and a gas phase has not evolved. Other samples from different wells at the same pilot indicated a gas phase was present, but that total pressures were lower. Knowledge of the presence of a gas and of the near well pressure field is important in developing an optimal production strategy. Without using a geobarometer, this information is extremely difficult to obtain.

Table 1. Water and gas composition and oil/water ratio

| Water Composition (mg/L) | | | |
|---|---|---|---|
| Na | 3500.0 | K | 99.0 |
| Ca | 121.5 | Mg | 40.0 |
| Al | 0.3 | $SiO_2$ | 191.0 |
| Cl | 5680.0 | $SO_4$ | 40.0 |
| TIC | 250.0 | | |

Measured pH at 25°C was 6.34

| | Gas Composition (mcf/d) | | |
|---|---|---|---|
| Stream | $CO_2$ | $CH_4$ | $N_2$ |
| 222.0 | 19.0 | 0.5 | 7.4 |

Oil/water ratio = 0.42

Minerals and oil have the potential to modify the composition of fluids by the release and absorption of material on their surfaces. These processes which are relatively fast, and dependant upon the available surface area of the minerals, may significantly buffer the composition of water. SOLMINEQ.88 has two options which can be used to simulate either an ion exchange or an ion adsorption process. The ion exchange option simulates the exchange of either anions or cations on a surface of constant charge while the adsorption option simulates the exchange process on an amphoteric surface, developed due to the ionization of surfaces sites. The adsorption option is based on the Gouy-Chapman theory as described by Healy and White (1978). The prime difference between the ion exchange and the ion adsorption options, is the way in which activity of aqueous complexes vary as they approach the surface. The ion exchange option assumes the surface activity of the ions are the same as the bulk solution, while the adsorption option uses an electrical potential term to calculate the surface activity of ions from the activity of the ions in the bulk solution. SOLMINEQ.88

does not assume parameters for a particular surface type, thus these models are general, and can be used to simulate the interaction of any of the aqueous species in the data base with a surface (19).

Table 2. Measured and calculated values of the pH and the saturation index of calcite at several points in a production well

| Conditions | Temperature (°C) | pH | Calcite (SI) |
|---|---|---|---|
| Ambient | 25 | 6.34 | -0.631 |
| Well Head | 105 | 6.29 | 0.025 |
| Sand Face | 150 | 6.31 | 0.267 |

Table 3. Calculated values for the pH and calcite saturation index, when including the effects of the annulus gas

| Temperature (°C) | Pressure (bars) | pH | Calcite (SI) |
|---|---|---|---|
| 150 | 5. | 6.07 | .552 |
| 150 | 7. | 6.09 | .214 |
| 150 | 9. | 6.04 | .066 |
| 150 | 11. | 6.01 | .039 |
| 150 | 13. | 6.01 | .039 |

No gas was present at pressures above approximately 10 bars.

## LIMITS OF MODELING

A clear understanding of the limits of applicability of geochemical models and of the uncertainties and potential errors in the analytical and thermodynamic data is essential to the correct use of SOLMINEQ.88 and the interpretation of the results. These limits can be primarily divided into four different groups, which are: errors and uncertainties in the physical parameters; the chemical analysis; and in the thermodynamic data; and extrapolation of the equations and formulas beyond their range of applicability.

The physical parameters are properties such as temperature, pressure, oil-water ratio, amount of gas produced up the annulus, etc. Although the potential to accurately measure these parameters is high, frequently they are not measured at all or the values reported represents a temporal average often taken over a pad comprised of 20 or more wells. In addition, large discrepancies often exist as different methods of measurement were used at different times.

Errors in chemical analysis are the most obvious source of uncertainty in the modeling of water-rock interactions. The uncertainty may result from incomplete analysis, unreliable sample collection and preservation, or from incorrect interpretation of the data. Incomplete analysis are unfortunately prevalent, elements like aluminum are typically present in very low concentrations and are difficult to measure. Concentrations of the dissolved organic species are often not measured due to the analytical expense, even though some of the mono- and dicarboxylic acid anions may be present in high concentrations (26).

Poor sample preservation will often lead to results considerably in error. If pH, alkalinity, $H_2S$ and $NH_3$ are not determined in the field, their values must be considered suspect (27-28). Sample preservation may require filtering in the absence of air by pressurizing the vessels with an inert gas (29). Dependent on which cations and anions are to be analysed for, the samples may have to be either acidified or diluted with distilled water.

The reliability of thermodynamic data used to calculate the equilibria between the minerals and aqueous species, and the equilibria between the aqueous species and complexes is variable and often the range of data is considerable. Many of the more common minerals are non-stoichiometric and show large variations in surface characteristics, crystallinity and structure. Especially for aqueous species, the data may have been extrapolated over a considerable temperature range. Fortunately, thermodynamic data bases are continually being refined. The SOLMINEQ.88 thermodynamic data base has been prepared considering the reliability of the available data. Recognizing that new and better data will become available, the SOLMINEQ.88 user has the option of modifying the thermodynamic data for the equilibria between the minerals, the aqueous species components and/or the remaining aqueous species and complexes at run time.

Some of the limits mentioned above can be relaxed by carefully using various SOLMINEQ.88 options. For example, if the formation temperature is not known, the calculated value from one of the chemical geothermometers can be used (21). If an element has not been analysed for, an approximate value can be obtained by assuming equilibrium with the appropriate formation mineral using the mineral dissolution and precipitation option. If carbon dioxide was known to have been lost between the time of sampling and analysis but the pH was measured in the lab and at the time of sampling, the amount of carbon dioxide in the water at the time of sampling can be recalculated. Other techniques and options in SOLMINEQ.88 can be used, but each assumption of this type should reduce confidence in the results.

CONCLUSION

SOLMINEQ.88 is a complete rewrite of SOLMNEQ. This is reflected in the programing language, the numerical techniques used and by a variety of new modeling options. SOLMINEQ.88 has been tested on a wide variety of FORTRAN-77 compilers and meets the ANSI Fortran-77 language standard. The speed of program execution has been increased and considerable effort has gone into program verification.

Major new options in SOLMINEQ.88 include a boiling option, volatile fractionation between oil, gas and water, an adsorption and ion exchange option and a dissolution/precipitation option. Interactive programs have been written which facilitate input, running and examination of the output.

SOLMINEQ.88 with complete program documentation describing the input, output, program options and equations, and sample input and output examples is available from the authors (19).

LITERATURE CITED

1. Helgeson, H. C.; Brown, T. H.; Nigrini, A.; Jones, T. A. Geochim. et Cosmochim. Acta, 1970, 34, 569-592.
2. Perkins, E. H. University of British Columbia, unpublished M.Sc. thesis, 1980.
3. Kharaka, Y. K; Barnes, I. U.S. Geol. Surv. Computer Contributions, Nat'l. Technical Information Service # PB-215 899, 1973, 81 pp.

4. Aggarwal, P. K.; Hull, R. W.; Gunter, W. D.; Kharaka, Y. K. Proc. of the 3rd. Canadian-American Conf. on Hydrogeology of Sedimentary Basins. Nat'l. Water Well Asso. 1986, 3, 196-203.
5. Truesdell, A. H.; Jones, B. F. J. of Research, U.S. Geol. Surv. 2, 1974, p. 233-248.
6. Plummer, L. N.; Jones, and Busenberg, E. U.S. Geol. Surv. Water Resources Investigations Report 76-13, 1984, 70 pp.
7. Ball, J. W.; Nordstrom, D. K.; Zachmann, D. W. U.S. Geol. Surv. Open-File Report 87-50, 1987, 108 pp.
8. Westall, J.; Zachary, J. L.; Morell, F. M .M. Mass. Inst. Tech. Dept. Civil Eng. Tech. Note 18, 1976, 91 p.
9. Wolery, T. J. Lawrence Livermore Laboratory (UCRL-52658), 1979, 41 p.
10. Wolery, T. J. Lawrence Livermore Laboratory (UCRL-5314), 1983, 191 pp.
11. Wolery, T. J. Lawrence Livermore Laboratory (UCRL-51) (draft), 1984, 251 pp.
12. Parkhurst, S. L.; Thorstenson, C.; Plummer, L. N. U.S. Geol. Surv. Report No. PB81-167801, 1980,
13. Sposito, G.; Mattigod, S. V. Report, Kearney Foundation of Soil Science, University of California. 1980.
14. Reed, M. H. Geochim. et Cosmochim. Acta, 46, 1982, 513-528.
15. Felmy, A. R.; Girvin, D.; Jenne, E. A. U.S. Environmental Protection Agency, Athens, Georgia. 1984
16. Narasimhan, T. N.; White, A. F.; Tokunaga, T. Water Resources Research, 22, 1986, p. 1820-1834.
17. Kharaka, Y. K.; Mariner, R. H. Proc. 2nd International Symposium on Water-Rock Interaction, 1977.
18. Kharaka, Y. K.; Maest, A. S.; Carothers, W. W.; Law, L. M.; Lamothe, P. J.; Fries, T. L. Applied Geochemistry, 2, 1987, p. 543-561.
19. Kharaka, Y. K.; Gunter, W. D.; Aggarwal, P. K.; Perkins, E. H.; DeBraal, J. D. U.S. Geol. Surv.Water Resources Investigation Report 88-4227, 1988, 430 pp.
20. Lundegard, P. D.; Kharaka, Y. K. 1989. Origin and significance of dissolved organic acid anions in subsurface waters (this issue).
21. Kharaka, Y. K.; Mariner, R. H. In Thermal History of Sedimentary Basins. (eds. N. D. Naeser and T. H. McCulloh) Springer, 1988, p. 99-117..
22. Aggarwal, P. K.; Gunter, W. D.; Kharaka, Y. K. Effect of Pressure on Aqueous Equilibria, 1989, (this issue).
23. Gunter, W. D.; Fuhr, B. J.; Young, B. Proc. of the 3rd. Canadian-American Conference on Hydrogeology of Sedimentary Basins, Nat'l. Water Well Asso., 1986, 3, 233-249.
24. Perkins, E. H.; Gunter, W. D. In Proc. of the 4th Internat'l. Conf. on Heavy Crude and Tar Sands, 1988, v. 5, paper 11 (preprint).
25. Healy, T. W.; White, L. R. Advances in Colloid and Interface Science, 1978, 9, 303-345.
26. Carothers, W. W.; Kharaka, Y. K. Amer. Assoc. of Petrol. Geol. Bull., 1978, 62, 2441-2453.
27. Presser, T. S.; Barnes, I. 1974. U.S. Geol. Surv. Water Resources Investigation Report 22-74, 1974, 11 pp.
28. Lico, M. S.; Kharaka, Y. K.; Carothers, W. W.; Wright, V. A. U.S. Geol. Surv. Water-Supply Paper 2194, 1982, 21 pp.
29. Hull, R. W.; Kharaka, Y. K.; Maest, A. S.; Fries, T. L. Proc. of 1st Canadian/American Conference on Hydrogeology. Nat'l. Water Well Assoc., Ohio, 1984.
30. Flak, L. H.; Brown, J. Soc. of Petrol. Engineers LADS/SPE 17222, 1988, 24 pp.

RECEIVED October 6, 1989

## Chapter 10

# Application of the Pitzer Equations to the PHREEQE Geochemical Model

**L. Niel Plummer[1] and David L. Parkhurst[2]**

[1]U.S. Geological Survey, 432 National Center, Reston, VA 22092
[2]U.S. Geological Survey, Denver Federal Center, Mail Stop 418, Lakewood, CO 80225

A computer program is presented that simulates geochemical reactions in brines and other concentrated electrolyte solutions using the ion-interaction virial-coefficient approach for activity-coefficient corrections developed by Pitzer. Reaction-modeling capabilities include calculation of (1) aqueous speciation and mineral saturation indices, (2) mineral solubilities, (3) mixing or titration of aqueous solutions, (4) irreversible reactions, and (5) reaction paths, including evaporation. The program's data base of Pitzer interaction parameters includes a partially validated data base at 25 °C for the system $Na-K-Mg-Ca-H-Cl-SO_4-OH-HCO_3-CO_3-CO_2-H_2O$, and largely untested literature data for Fe(II), Mn(II), Sr, Ba, Li, and Br, with provision for calculations at temperatures other than 25 °C. The need to maintain an internally consistent data base of interaction parameters and equilibrium constants is emphasized through an example of the calculated solubility of nahcolite in $Na_2CO_3$ solutions. The choice of activity-coefficient scale has particular significance to thermodynamic calculations in brines if the measured pH is introduced. Several examples are presented to demonstrate applications of the code.

PHRQPITZ (1) is a Fortran 77 computer program that simulates geochemical reactions in brines and other electrolyte solutions to high concentrations. PHRQPITZ has been adapted from the U.S. Geological Survey geochemical simulation computer code PHREEQE (PH-REdox-EQuilibrium-Equations) (2) in which the aqueous model of PHREEQE has been replaced with the Ion-Interaction virial-coefficient approach (3-7). The PHRQPITZ code contains most of the reaction-modeling capabilities of the original PHREEQE code including calculation of (1) aqueous speciation and mineral-saturation indices, (2) mineral solubility, (3) mixing or titration of aqueous solutions, (4) irreversible reactions and mineral-water mass transfer, and (5) reaction path, including evaporation. Computed results for each aqueous solution include the osmotic coefficient, activity of water, mineral-saturation indices, mean-activity coefficients, total-activity coefficients, and scale-dependent values of pH, individual-ion activities and individual-ion activity coefficients. The capability of PHREEQE to simulate redox reactions is retained in the PHRQPITZ software, but is not currently accessible due to lack of an internally consistent data base of Pitzer interaction parameters for important redox couples. Full documentation for PHRQPITZ is given elsewhere (1).

The Pitzer treatment of the aqueous model is based largely on the equations and data as presented by (8-9). A data base of Pitzer interaction parameters is provided that is identical to the data base of (9) at 25 °C for the system $Na-K-Mg-Ca-H-Cl-SO_4-OH-HCO_3-CO_3-CO_2-H_2O$. This data base has been partially validated (8-9) using previously published laboratory solubility measurements for evaporite minerals in water

This chapter not subject to U.S. copyright
Published 1990 American Chemical Society

and mixed-salt solutions, including seawater. The PHRQPITZ data base has been extended to include largely untested literature data for Fe(II), Mn(II), Sr, Ba, Li, and Br with provision for calculations at temperatures other than 25 °C. Some new data for the temperature dependence of mineral equilibrium constants accompanies the additional (untested) data. As with PHREEQE, the aqueous model and thermodynamic data of PHRQPITZ are user-definable and external to the code.

An interactive input code for PHRQPITZ called PITZINPT (1) is analogous to the PHREEQE input code, PHRQINPT (10). The reader is referred to the PHREEQE documentation (2) and the PHRQPITZ documentation (1) for further background and modeling information.

## IMPLEMENTATION OF THE PITZER EQUATIONS IN PHRQPITZ

Details of the Pitzer equations and definitions of the notations utilized in this paper and in PHRQPITZ are given in the literature (3-10). As the focus of this report is on the capabilities and limitations of PHRQPITZ in relation to its application to geochemical problems, only selected aspects of the implementation of the Pitzer equations in PHRQPITZ are presented.

### THE DEBYE-HUCKEL PARAMETER, $A^\phi$

In PHRQPITZ values of Debye-Huckel parameter $A^\phi$ are computed over the temperature range 0-350 °C. The total pressure is taken to be 1 atm between 0 and 100 °C and that of the vapor pressure curve for pure water (11) beyond 100 °C. The dielectric constant of pure water is calculated from (12). Values of $A^\phi$ are reported in (12) to three significant figures between 0 and 350 °C and are identical (to at least three significant figures) to those calculated in PHRQPITZ. Between 0 and 100 °C at 1 atm total pressure, $A^\phi$ calculated in PHRQPITZ agrees with values calculated in (13) within 0.00004 or better. The computed value of $A^\phi$ in PHRQPITZ at 25 °C is 0.39148 which compares with 0.39145 reported by (13) and 0.391 from (12). To be consistent with (9), $A^\phi$ is defined to be 0.392 at 25 °C and 1 atm total pressure in PHRQPITZ (8). Therefore, small inconsistencies in calculations with PHRQPITZ may be observed between results at 25 °C and those very near 25 °C.

### HIGHER-ORDER ELECTROSTATIC TERMS

The terms $^E\theta_{ij}(I)$ and $^E\theta'_{ij}(I)$ account for electrostatic mixing effects of unsymmetrical cation-cation and anion-anion pairs (7). These higher-order electrostatic terms are calculated routinely in PHRQPITZ for all unsymmetrical pairs of cations or unsymmetrical pairs of anions using the Chebyshev approximation to the integrals $J_0(x)$ and $J_1(x)$ (7-8). Test calculations showed little difference from more simplified approximations to $J_0(x)$ and $J_1(x)$ given in (7). Values of $^E\theta_{ij}(I)$ and $^E\theta'_{ij}(I)$ depend only on ion charge and total ionic strength and are zero when cation or anion pairs have the same charge.

Caution should be exercised in using literature values of $\theta_{ij}$ (14). Values of $\theta_{ij}$ must be compatible with the same single-salt data ($\beta^0$, $\beta^1$, $\beta^2$, and $C^\phi$) used in the model and, for use in PHRQPITZ, their determination from mixed-salt solutions must include the higher-order electrostatic terms discussed above. Both types of $\theta_{ij}$ are reported in the literature (i.e., determined with and without provision for higher-order electrostatic terms). In PHRQPITZ, the higher-order electrostatic terms are always included. This precaution applies to cation or anion pairs such as $Ca^{2+}$-$Na^+$ or $SO_4^{2-}$-$Cl^-$, but not to interactions such as $Cl^-$-$F^-$ or $Ca^{2+}$-$Mg^{2+}$ where, because of the identical charge, the higher-order electrostatic terms are zero. Values of $\theta_{ij}$ given in (5, 15) do not include the higher-order electrostatic terms for unsymmetrical cation pairs and unsymmetrical anion pairs, whereas the higher-order electrostatic terms are included in the data of (8-9). There is significant improvement in modeling the system Na-K-Mg-Ca-Cl-$SO_4$-$H_2O$ when the higher-order electrostatic terms were included (8).

Values of the parameters $\psi_{ijk}$ are included for all different combinations of two cations and an anion or two anions and a cation. $\psi$ is usually determined from the same two-salt mixture used to define $\theta_{ij}$ and is therefore internally consistent with that value of $\theta_{ij}$ as well as the individual single-salt interaction parameters.

## DATA BASE

Two data files are required to run PHRQPITZ. The first file, PHRQPITZ.DATA, is analogous to the thermodynamic data base of PHREEQE (though much smaller) and contains data under the keywords ELEMENTS, SPECIES, LOOK MIN and MEAN GAM. The first three of these keywords are identical to their usage in (2). The keyword MEAN GAM is used in PHRQPITZ to define the stoichiometries of selected salts for calculation of the mean-activity coefficient.

Equilibrium constants at 25 °C for (1) the formation of $OH^-$, $HCO_3^-$, $H_2CO_3^*$, $HSO_4^-$, $CaCO_3°$, $MgOH^+$, and $MgCO_3°$, and (2) dissolution of minerals listed under LOOK MIN in PHRQPITZ.DATA are computed from the free energies of (9).

Expressions for the temperature dependence of log K of the form,

$$\log K = A_1 + A_2 T + A_3/T + A_4 \log T + A_5/T^2, \tag{1}$$

where $A_1$ - $A_5$ are constants and T is temperature in Kelvins, are given for the available data. The form of Equation 1 corresponds to a model (16) for the heat capacity as used in PHREEQE. Expressions of this form were either taken from the literature or fitted to solubility data (17) after speciation in PHRQPITZ. The $A_1$ term has been adjusted to agree with (9) at 25 °C. In some cases the temperature dependence of log K is calculated from the van't Hoff equation using a value of $\Delta H_r°$ at 25 °C. No data for the temperature dependence of the equilibrium constants are available for many minerals of interest. In these cases, PHRQPITZ uses the 25 °C value for all temperatures.

The second data file read by PHRQPITZ (PITZER.DATA) contains values of the Pitzer ion interaction parameters $\beta^0$, $\beta^1$, $\beta^2$, $C^\phi$, $\theta$, $\lambda$, $\psi$, including a limited amount of data on the temperature derivatives of $\beta^0$, $\beta^1$, $\beta^2$, and $C^\phi$. For the chemical system Na-K-Mg-Ca-H-Cl-$SO_4$-OH-$HCO_3$-$CO_3$-$CO_2$-$H_2O$ at 25 °C, the data of PHRQPITZ are identical to those of (9), and in verification procedures, calculations have reproduced results reported in (9). Details of the PHRQPITZ data base are given elsewhere (1).

## ADDITIONS TO THE DATA BASE

Extensions beyond the data base of (9) are largely untested and include additions to PITZER.DATA for (1) the calculation of the thermodynamic properties of aqueous solutions containing Fe(II), Mn(II), $Sr^{2+}$, $Ba^{2+}$, $Li^+$, and $Br^-$; (2) the estimation of the temperature dependence of many of the single-salt parameters from selected literature data; and (3) the calculation of the thermodynamic properties of NaCl solutions to approximately 300 °C along the vapor pressure curve of water beyond 100 °C (18). Except for the NaCl-$H_2O$ system, the PHRQPITZ aqueous model should not be used outside the temperature range 0 to 60 °C. Several recent evaluations of the temperature dependence of Pitzer interaction parameters to relatively high temperatures (19-21) have not yet been incorporated in the PHRQPITZ data base.

Extensions to PHRQPITZ.DATA beyond that of (9) include estimates of $\Delta H_r°$ for the formation of aqueous non-master species such as $OH^-$, $HCO_3^-$, $MgOH^+$, etc., and calculation of equilibrium constants for mineral-dissolution reactions at temperatures other than 25 °C. If changes in the temperature-dependence of the Pitzer interaction parameters are made, the appropriate mineral equilibrium constants and their temperature-dependence will need to be reexamined for consistency (see (1) for an extensive compilation and summary of published Pitzer interaction parameters).

## MAINTAINING INTERNAL CONSISTENCY

It is important to maintain the internal consistency of the data base given by (9). As an example, consider the solubility of nahcolite in $Na_2CO_3$ solutions (see Figure 7c, p.734 of (9)). The solubility of nahcolite in $Na_2CO_3$ solutions calculated from PHRQPITZ using the data base of (9) is shown in Figure 1 (curve 1). The results are identical to those of (9) and, of course, adequately reproduce the original experimental data. We will now examine the consequences of changing the data base of (9) with regards to the solubility of nahcolite. The $Na^+$-$HCO_3^-$ interaction parameters in the data base of (9) were originally taken from (22) based on an evaluation of the original electrochemical measurements of (23-24) in $NaHCO_3$-NaCl aqueous solutions. Alternatively, Pitzer interaction parameters for $NaHCO_3$ have been determined from isopiestic measurements of mixed aqueous solutions of $NaHCO_3$ and $Na_2CO_3$ at 25 °C (25). The resulting calculated solubility of nahcolite in $Na_2CO_3$ solutions if the single-salt parameters of $NaHCO_3$ are arbitrarily changed from those of (22) to those of (25) is shown in curve 2 of Figure 1.

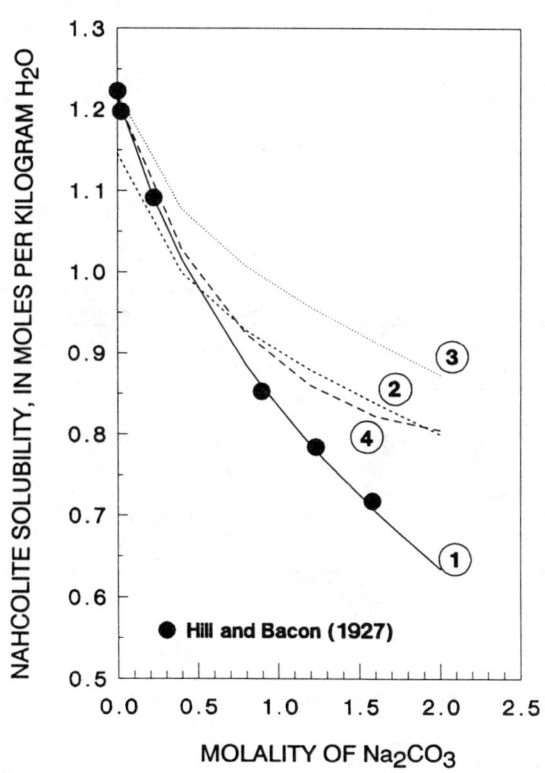

Figure 1. Comparison of calculated solubility of nahcolite in aqueous solutions of $Na_2CO_3$ to 2 molal (from (1)). See text for explanation of curves.

Curve 3 of Figure 1 retains the parameters for NaHCO$_3$ of (25) and uses an equilibrium constant of nahcolite which was adjusted to be internally consistent with (25). As shown, the agreement is still poor with the data base (9) and experimental data.

One final adjustment is to change $\theta$ and $\psi$ for CO$_3$-HCO$_3$-Na interactions. Because no other values of $\theta$ and $\psi$ are known for NaHCO$_3$-Na$_2$CO$_3$ mixtures, we have substituted values of $\theta$ and $\psi$ determined by (26) for KHCO$_3$-K$_2$CO$_3$-KCl aqueous salt mixtures. This adjustment shows a marked improvement (curve 4, Figure 1), but further adjustments in these parameters are clearly needed to obtain a revised data set for the NaHCO$_3$-Na$_2$CO$_3$ system consistent with the data of (25).

New values of $\theta$ and $\psi$ for CO$_3$-HCO$_3$-Na interactions that are internally consistent with the NaHCO$_3$ data of (25) and the known solubility of nahcolite in Na$_2$CO$_3$ solutions (27) are compared with values from (9) in Table I. Both sets of parameters in Table I reproduce the experimental data for the solubility of nahcolite in Na$_2$CO$_3$ solutions (curve 1, Figure 1). However, the $\beta^1$ parameter for NaHCO$_3$ of (25) falls well outside the range of reasonable values demonstrated for a large number of single salts (5, 15), while the NaHCO$_3$ values of (22) are consistent with previously recognized single-salt parameters. Further errors and inconsistencies are likely if the alternate data set of Table I is applied beyond the NaHCO$_3$ system. This example demonstrates the need for both accuracy and internal consistency among all Pitzer interaction parameters and thermodynamic data for the given chemical system when changes to the data base of (9) are contemplated.

Table I. Comparison of Two Internally Consistent Sets of Pitzer Interaction Parameters for the System NaHCO$_3$-Na$_2$CO$_3$-H$_2$O at 25 °C. $A^\phi$ = .392. Higher-order electrostatic terms included in calculation of $\theta$ and $\psi$

| Parameters from (9) | Alternate Parameter Set |
|---|---|
| NaHCO$_3$ (22) | NaHCO$_3$ (25) |
| $\beta^\circ$ = .0277 | $\beta^\circ$ = -.04096 |
| $\beta^1$ = .0411 | $\beta^1$ = .5062 |
| $C^\phi$ = 0 | $C^\phi$ = .005250 |
| Na$_2$CO$_3$ (9) | Na$_2$CO$_3$ (9) |
| $\beta^\circ$ = .0399 | $\beta^\circ$ = .0399 |
| $\beta^1$ = 1.389 | $\beta^1$ = 1.389 |
| $C^\phi$ = .0044 | $C^\phi$ = .0044 |
| $\theta_{CO3-HCO3}$ = -.04 | $\theta_{CO3-HCO3}$ = .111 |
| $\psi_{Na-CO3-HCO3}$ = .002 | $\psi_{Na-CO3-HCO3}$ = -.025 |
| Log $K_{Nahcolite}$ = -10.742 | Log $K_{Nahcolite}$ = -10.696 |

## SCALE CONVENTION OF ACTIVITY COEFFICIENTS

It has long been recognized that individual-ion activities and individual-ion activity coefficients cannot be measured independently. Single-ion activity coefficients are valid only when combined to define properties of neutral combinations of ions such as in the calculation of mean-activity coefficients, saturation indices, solubility, etc. (9, 15). Therefore, values of individual-ion activities and individual-ion activity coefficients have meaning only in a relative sense and individual values depend on the particular choice of scale convention. The subject has received recent attention as it applies to interpretation of pH in seawater (28-36).

PHRQPITZ offers two scaling conventions based on (9). In the first case no scaling is performed and individual-ion activity coefficients are computed directly from the equations of (9). In the second case, all individual-ion activity coefficients calculated from the equations of (9) are scaled according to the MacInnes convention (37). In this case the activity coefficient of Cl$^-$ is defined to be equal to the mean-activity coefficient of KCl in a KCl solution of equivalent ionic strength, $\gamma_{Cl^-(Mac)} \equiv \gamma_{\pm KCl}$. The scaling factor for the $i$th ion is computed from the term $(\gamma_{Cl^-}/\gamma_{\pm KCl})^{z_i}$ and is multiplied through all other calculated individual-ion activity coefficients (9). That is

$$\gamma_{i(Mac)} = \gamma_i (\gamma_{Cl^-}/\gamma_{\pm KCl})^{z_i} \qquad (2)$$

where $\gamma_{i(Mac)}$ is the individual-ion activity coefficient of the $i$th ion converted to the MacInnes scale, $\gamma_i$ is the activity coefficient of the ion consistent with some other convention, $\gamma_{Cl^-}$ is the activity coefficient of $Cl^-$ according to the same alternate convention, $\gamma_{\pm KCl}$ is the mean-activity coefficient of KCl in a pure KCl solution of equivalent ionic strength and $z_i$ is the charge of the $i$th ion (+ for cations, - for anions). The activity coefficients could be placed on other scales by substituting for $\gamma_{\pm KCl}$ or $\gamma_{Cl^-}/\gamma_{\pm KCl}$ in Equation 2. Further details are given in (9).

As an example, Table II compares log values of the molality, activity, and activity coefficient of individual ions computed for seawater in equilibrium with aragonite at 25 °C and a $CO_2$ partial pressure of $10^{-3}$ atm. The results are presented for the problem computed on the MacInnes scale and without scaling. In the problem, pH was calculated from the various equilibria and is therefore internally consistent with the aqueous model and the respective scale. When all individual-ion activity coefficients are consistent with a single scale convention, the individual-ion molalities are independent of choice of scale, but the individual-ion activities and individual-ion activity coefficients are scale dependent. For example, the pH of seawater in equilibrium with aragonite at 25 °C and $10^{-3}$ atm $P_{CO_2}$ (using the data base of (9)) is 7.871 on the MacInnes scale and 7.828 without scaling, as shown in Table II.

Table II. Seawater In Equilibrium with Aragonite at $PCO_2$ of $10^{-3}$ atm. and 25 °C

| Species | MacInnes Scale | | | Unscaled | | |
|---|---|---|---|---|---|---|
| | Log Molality | Log Activity | Log Gamma | Log Molality | Log Activity | Log Gamma |
| $H^+$ | -7.694 | -7.871 | -0.178 | -7.694 | -7.828 | -0.134 |
| $H_2O$ | -0.008 | -0.008 | 0.000 | -0.008 | 0.008 | 0.000 |
| $Ca^{2+}$ | -1.981 | -2.795 | -0.814 | -1.982 | -2.708 | -0.726 |
| $Mg^{2+}$ | -1.259 | -2.034 | -0.775 | -1.259 | -1.946 | -0.687 |
| $Na^+$ | -0.313 | -0.552 | -0.239 | -0.313 | -0.508 | -0.195 |
| $K^+$ | -1.973 | -2.248 | -0.274 | -1.973 | -2.204 | -0.230 |
| $Cl^-$ | -0.246 | -0.363 | -0.118 | -0.246 | -0.407 | -0.162 |
| $CO_3^{2-}$ | -4.513 | -5.424 | -0.911 | -4.513 | -5.512 | -0.999 |
| $SO_4^{2-}$ | -1.532 | -2.419 | -0.887 | -1.532 | -2.507 | -0.975 |
| $OH^-$ | -5.938 | -6.134 | -0.196 | -5.938 | -6.178 | -0.240 |
| $HCO_3^-$ | -2.781 | -2.956 | -0.176 | -2.780 | -3.000 | -0.219 |
| $H_2CO_3$ | -4.535 | -4.482 | 0.053 | -4.535 | -4.482 | 0.053 |
| $HSO_4^-$ | -8.204 | 8.312 | -0.108 | -8.204 | -8.356 | -0.152 |
| $CaCO_3$ | -5.068 | -5.068 | 0.000 | -5.068 | -5.068 | 0.000 |
| $MgOH^+$ | -5.890 | -5.980 | -0.100 | -5.880 | -5.936 | -0.056 |
| $MgCO_3$ | -4.530 | -4.530 | 0.000 | -4.529 | -4.529 | 0.000 |

Further inspection of Table II indicates that when different scales are used, which are in themselves internally consistent, all mean quantities such as $\gamma_\pm$, $a_\pm$, $m_\pm$ and other neutral salt combinations such as ion-activity products, saturation indices, and therefore solubility, are independent of scale. Clearly then, the same thermodynamic conclusion will be reached for a given data base regardless of scale when all individual-ion values are internally consistent with a single scale convention. However, no significance can be assigned to the individual-ion activities, individual-ion activity coefficients, and pH in comparing results on differing scales.

The problem of activity coefficient scale is more important when the <u>measured</u> pH is introduced in geochemical calculations. The measured pH is not likely to be on the same activity coefficient scale as the aqueous model because the buffers used to define pH are conventional (28). Even if the pH is on the same scale as the aqueous model, uncertainties in its measurement in brines, such as due to liquid-junction potentials (28,38), will always introduce inconsistencies. Consequently, it is unlikely that the measured pH will be consistent with the particular scale used for the individual ions.

When the measured pH is used, the following conditions prevail: (1) the individual-ion activities and individual-ion activity coefficients again depend on choice of scale (regardless of the inconsistency of the measured pH), (2) the mean-activity coefficients are independent of the measured pH and, thus, independent of scale, and (3) the calculated mean activities, mean molalities, ion-activity products, saturation indices and $CO_2$ partial pressure are not independent of scale. This dependency of thermodynamic properties on scale when the measured pH is used in calculations is particularly acute in the carbonate system and other chemical systems where the equilibria being considered depend significantly on pH. However, it is not likely to be important to calculations involving chloride and sulfate minerals in most natural waters.

Uncertainties in the calculated saturation index due to inconsistency of the measured pH with the adopted scale increase with increasing ionic strength, but are relatively unimportant at ionic strengths less than that of seawater (I=0.7). At relatively high ionic strengths, such as those of formation brines, the inconsistency in pH scale leads to differences in the calculated calcite saturation index (SI) approaching 0.5 SI units or greater between unscaled and scaled (MacInnes) calculations (see Example 1, below). Therefore, in addition to the relatively formidable task of measuring pH in brines, it is important to recognize the magnitude of error that can result simply from differences in activity-coefficient scale.

## PRECAUTIONS AND LIMITATIONS

Because of its adaptation from PHREEQE, PHRQPITZ retains some of the limitations of the original code. These are discussed in (2) and reviewed here as they apply to geochemical reactions in brines.

All calculations are made relative to one kg $H_2O$. As there is no mass balance for the elements H and O, there is no formal provision for keeping track of the amount of water used in reactions such as hydration and dehydration of solids. This is a source of error, for example, in simulation of the evaporation of brines when hydrated minerals are precipitated and remove water from solution.

PHRQPITZ retains the original logic of PHREEQE concerning oxidation-reduction reactions, but because an internally-consistent data base of Pitzer interaction parameters for multiple oxidation states is not currently available, geochemical redox reactions may not be attempted in PHRQPITZ. All redox equilibria used with PHREEQE have been removed from the PHRQPITZ data base. Because of the lack of Pitzer interaction parameters for aqueous aluminum and silica species, calculations with aluminosilicates are not possible in PHRQPITZ.

Many of the original convergence problems of PHREEQE were redox-related. Improvements to the convergence criteria of PHREEQE (noted in the January, 1985 version) have been incorporated in PHRQPITZ. Precautions and comments on the use of ion exchange, and titration/mixing reactions in PHREEQE, and the lack of uniqueness of modeled reaction paths (2, 39) also apply to PHRQPITZ.

The activity of water in PHRQPITZ is computed from the osmotic coefficient, and represents a substantial improvement over water activity calculations in PHREEQE (2).

The temperature range for equilibria in PHRQPITZ is variable and is generally 0 - 60 °C if $\Delta H_r^\circ$ is known. However, the NaCl system is valid to approximately 350 °C. The temperature dependence of the solubility of many of the minerals in PHRQPITZ is not known and large errors could result if calculations are made at temperatures other than 25 °C for these solids. Limited temperature-dependent data for single salt parameters are included, but are probably not valid outside the interval 0 to 60 °C.

## EXAMPLES

Several problems involving brines and evaporation of seawater are now examined as a means of demonstrating some of the geochemical applications of the PHRQPITZ code.

### Brine Sample Speciation on Different Activity-Coefficient Scales

The speciation of a brine solution is used to determine mineral-water saturation state and demonstrate the significance of uncertainties in activity-coefficient scale on the predicted saturation state, particularly as applied to the carbonate system. The brine analysis is sample T-93 from (40) and has the following properties: Temperature (18 °C); pH (5.0); density (1.204 g/cc); ionic strength (8.59); water activity (0.721); and in mg/L: Ca (64,000); Mg (5,100); Na (45,000); K (199); Sr (1,080); Cl (207,000); $HCO_3$ (19); $SO_4$ (284); Br (1,760); $SiO_2$ (4.8).

The problem has 2 parts. In the first part the sample T-93 is speciated on the MacInnes, and unscaled conventions, respectively, using the measured pH. In the second part an attempt is made to resolve uncertainties in the carbonate system by assuming the brine is in equilibrium with calcite, on the MacInnes and unscaled conventions, respectively. Use of calcite equilibrium is preferable to alternate means of defining pH, such as through charge balance, owing to uncertainties in the analytical data.

The results of part 1 show that the saturation indices of calcite and dolomite vary because of choice of scales from -0.31 and -1.08 (MacInnes scale) to +0.16 and -0.13 (unscaled), respectively. The computed total concentration of inorganic carbon and log $PCO_2$, which is based on the pH and alkalinity measurements, varies from 32.43 mmol/kg $H_2O$ and 0.379 (MacInnes scale) to 10.25 mmol/kg $H_2O$ and -0.13 (unscaled), respectively. These represent changes of more than 300 percent for these variables. Because total inorganic carbon, total alkalinity, and $PCO_2$ can be measured within ten percent or better in many brines, the thermodynamics of the carbonate system in this brine would be better defined by measuring total alkalinity and either total inorganic carbon or $PCO_2$. If such data are available, the carbonate system is unambiguously defined and independent of the measured pH and choice of scale convention. A value of pH would then be defined by the aqueous model according to any particular choice of activity-coefficient scale.

In part 2, as a means of examining possible variations in the pH of brine T-93, and in the absence of total inorganic carbon or $PCO_2$ data, the solution was equilibrated with calcite on both activity-coefficient scales in part 2 by adjusting the total inorgainc carbon content. This caused outgassing (MacInnes scale) and ingassing (unscaled) of $CO_2$ to reach calcite saturation. The results suggest that pH could vary as much as 0.5 between the two scales (pH 5.328, MacInnes scale; pH 4.828, unscaled). By equilibrating with calcite, the carbonate system is more closely defined, as evidenced by the similarity in the calculated $PCO_2$ (1.086 atm, MacInnes scale; 1.123 atm unscaled) and total inorganic carbon (0.0149 molal, MacInnes scale; 0.0154 molal, unscaled). The total inorganic carbon and $PCO_2$ are not in complete agreement because of the initial inconsistency of the measured pH (5.00) with either scale. This inconsistency can be resolved only if total inorganic carbon or $PCO_2$ is measured in conjunction with the total alkalinity.

<u>Invariant Point in the Evaporation of Seawater</u>

Using the computer code SNORM (41) and the mineral stability data of (9), the final equilibrium mineral assemblage upon evaporation of seawater to dryness in an environment open to air (log $PCO_2$ = -3.5) at 25 °C is anhydrite ($CaSO_4$), bischofite ($MgCl_2 \cdot 6H_2O$), carnallite ($KMgCl_3 \cdot 6H_2O$), halite (NaCl), kieserite ($MgSO_4 \cdot H_2O$), and magnesite ($MgCO_3$). According to the mineralogic phase rule this is an invariant system. That is, once evaporation of seawater reaches saturation with these phases, further evaporation of that solution or other inputs of calcium, magnesium, sodium, potassium, carbon, chloride or sulfate change only the masses of the solids formed without altering the composition of the equilibrium solution. In this problem PHRQPITZ is used to test this reasoning and, at the same time, evaluate the internal consistency of SNORM and PHRQPITZ. Because both the SNORM and PHRQPITZ codes use the data base of (9) (though in different ways) we expect to find with PHRQPITZ that no other phases than those listed above could precipitate from the final equilibrium solution; that is, that the saturation indices of all other minerals in the data base of (9) are less than zero in the equilibrium solution.

In this problem an irreversible reaction (the addition of 0.1 and 1.0 moles of "sea salt") is used as one means of modeling evaporation (of seawater) while maintaining equilibrium with the 7 phases listed above.

In the first part, pure water is equilibrated with the six minerals in contact with air, and the resulting mass transfer indicates that all six minerals dissolve accompanied by a small loss of $CO_2$ to the atmosphere. The saturation indices show that all other minerals in the data base of (9) are undersaturated in the equilibrium solution. This confirms the prediction from SNORM and constitutes a partial validation of the two codes. The final equilibrium solution is predominantly a $MgCl_2$ solution of ionic strength 17.3873, water activity of 0.3382 and pH of 6.01 (unscaled) and 7.48 (MacInnes scale) with the following (molal) composition: Ca (0.00093075); Mg (5.7357); Na (0.093130); K (0.021516); Cl (11.4608); $SO_4$ (0.063396); C (0.000057242).

The invariant system is confirmed through PHRQPITZ by irreversibly adding 0.1 and 1.0 moles of "sea salt" to the solution and again equilibrating with the same set of minerals resulting in identical solution compositions to that given above.

<u>Evaporation of Seawater</u>

In the previous problem PHRQPITZ was used to define the final invariant composition of the aqueous phase in the evaporation of seawater. Some of the possibilities and limitations in modeling the initial stages of the evaporation of seawater are now considered.

Usually when reaction paths are simulated, the irreversible reactant is an unstable mineral or a suite of unstable minerals; that is, the stoichiometry of the irreversible reaction is fixed. Evaporation poses a special problem in reaction path simulation because the stoichiometry of the irreversible reaction (defined by the aqueous solution composition) continually changes as other minerals precipitate (or dissolve). In the second problem (above) evaporation of seawater was simulated by irreversible addition of "sea salt", that is, a hypothetical solid containing calcium, magnesium, sodium, potassium, chloride, sulfate and carbon in stoichiometric proportion to seawater. The approach used was valid as long as intermediate details of the reaction path are not required. The reaction path during evaporation could be solved in PHRQPITZ by changing the stoichiometry of the irreversible reactant (altered "sea salt") incrementally between phase boundaries, but this method would be extremely laborious.

Full consideration of the evaporation of seawater is too complex to be considered here. For example purposes, the evaporation of seawater was taken to the halite phase boundary only. The starting solution (seawater) was first equilibrated with dolomite in contact with air. This equilibration step causes precipitation of dolomite and outgassing of $CO_2$. This thermodynamically modified seawater is now undersaturated with all other minerals. The modified seawater was used as the starting point in evaporation.

An evaporation factor of 3.4878 is required to make seawater just saturated with gypsum (while maintaining equilibrium with dolomite in contact with air). Of the initial 1000 grams of water in the starting solution, 286.7 grams remain. The modified seawater is saturated with gypsum and dolomite at a $PCO_2$ of $10^{-3.5}$ atm and undersaturated with all other minerals in the data base of (9).

It is assumed that the next phase encountered in evaporation will be halite and an evaporation factor of 3.0572 is required to reach halite saturation. The cumulative evaporation factor is the product of both evaporation factors, that is, 3.4878 x 3.0572 = 10.6629 and the cumulative water remaining at the halite phase boundary is 1,000/10.6629 = 93.8 grams $H_2O$.

The problem becomes more complicated now because the final saturation indices show that at the halite phase boundary the solution is oversaturated with anhydrite and magnesite. The calculations have not been carried further, but if the goal were to follow rigorously the thermodynamic path, it would be necessary to back up and locate at least the magnesite phase boundary and evaluate again whether anhydrite saturation would be reached before halite saturation. At the magnesite phase boundary the system would be treated either open or closed to re-reaction of dolomite previously formed. In an open system it would be necessary to locate the point in evaporation where all the dolomite previously formed had reacted to form magnesite. After locating the magnesite phase boundary the point where dolomite vanishes must be located. Beyond this point in evaporation the solution will be undersaturated with dolomite while maintaining equilibrium with magnesite.

There are many other considerations, if a geologically meaningful reaction path is to be simulated. For example, if for kinetic reasons it is assumed that magnesite would not precipitate in this environment, magnesite equilibrium would not be included and subsequent computed supersaturation with respect to magnesite in solution would be disregarded. Seawater evaporation reaction paths could also be examined more realistically if formation of protodolomite or magnesian calcites were considered rather than dolomite which is well known for its irreversible behavior in low-temperature environments. In simulation of geochemical reactions it is often necessary to recognize minerals that react reversibly in the environment and those that do not. Although PHRQPITZ (and PHREEQE) can treat both equilibrium and nonequilibrium conditions, logical decisions regarding kinetically-controlled processes require guidance from the modeler.

ACKNOWLEDGMENTS

We have benefitted from discussions and correspondence with (the late) M.W. Bodine, B.F. Jones (USGS, Reston), C.E. Harvie and J.H. Weare (University of California, San Diego), E.J. Reardon (University of Waterloo), R.K. Stoessell (University of New Orleans) and R.J. Spencer (University of Calgary). G.W. Fleming and E.C. Prestemon contributed computer calculations. We thank S.A. Dunkle for compilations of Pitzer interaction parameters. The manuscript was improved by review comments from O.P. Bricker and D.C. Thorstenson (USGS, Reston).

## REFERENCES

1. Plummer, L. N.; Parkhurst, D. L.; Fleming, G. W.; Dunkle, S. A. U.S. Geological Survey, Water-Resources Investigations Report, 88-4153, 1988; p 310.
2. Parkhurst, D. L.; Thorstenson, D. C.; Plummer, L. N. U.S. Geological Survey, Water-Resources Investigations Report 80-96, 1980; p 210.
3. Pitzer, K. S. J. Phys. Chem. 1973, 77, 268-277.
4. Pitzer, K. S.; Mayorga, G. J. Phys. Chem. 1973, 77, 2300-2308.
5. Pitzer, K. S.; Mayorga, G. J. Sol. Chem. 1974, 3, 539-546.
6. Pitzer, K. S.; Kim, J. J. J. Am. Chem. Soc. 1974, 96, 5701-5707.
7. Pitzer, K. S. J. Sol. Chem. 1975, 4, 249-265.
8. Harvie, C. E.; Weare, J. H. Geochim. Cosmochim. Acta 1980, 44, 981-997.
9. Harvie, C. E.; Moller, N.; Weare, J. H. Geochim. Cosmochim. Acta 1984, 48, 723-751.
10. Fleming, G. W.; Plummer, L. N. U.S. Geological Survey, Water-Resources Investigations Report 83-4236 1983, p 108.
11. Haar, L.; Gallagher, J. S.; Kell, G. S. NBS/NRC STEAM TABLES 1984, Hemisphere Pub. Corp., Washington, DC, p 320.
12. Bradley, D. J.; Pitzer, K. S. J. Phys. Chem. 1979, 83, 1599-1603.
13. Ananthaswamy, J.; Atkinson, G. J. Chem. Eng. Data 1984, 29, 81-87.
14. Reardon, E. J.; Armstrong, D. K. Geochim. Cosmochim. Acta 1987, 51, 63-72.
15. Pitzer, K. S. In Activity Coefficients in Electrolyte Solutions; Pytkowicz, R. M., Ed.; CRC Press, Boca Raton, Florida, 1979, 157-208.
16. Maier, C. G.; Kelly, K. K. J. Am. Chem. Soc. 1932, 54, 3243-3246.
17. Linke, W. F. Solubilities of Inorganic and Metal Organic Compounds; American Chemical Society: Washington, DC, 1965, 1, p 1487, 2, p 1914.
18. Pitzer, K. S.; Peiper, J. C.; Busey, R. H. J. Phys. Chem. Ref. Data 1984, 13, 1-102.
19. Pabalan, R. T.; Pitzer, K. S. Geochim. Cosmochim. Acta 1987, 51, 829-837.
20. Pitzer, K. S. In Reviews in Mineralogy; Thermodynamic Modeling of Geological Materials; Minerals, Fluids and Melts; Carmichael, I. S. E.; Eugster, H. P., Eds.; Mineralogical Society of America: Washington, DC, 1987.
21. Moller, N. Geochim. Cosmochim. Acta 1988, 52, 821-837.
22. Pitzer, K. S.; Peiper, J. C. J. Phys. Chem. 1980, 84, 2396-2398.
23. Harned, H. S.; Davis, R., Jr. J. Am. Chem. Soc. 1943, 65, 2030-2037.
24. Harned, H. S.; Bonner, F. T. J. Am. Chem. Soc. 1945, 67, 1026-1031.
25. Sarbar, M.; Covington, A. K.; Nuttal, R. L.; Goldberg, R. N. J. Chem. Therm. 1982, 14, 967-976.
26. Roy, R. N.; Gibbons, J. J.; Williams, R.; Godwin, L.; Baker, G.; Simonson, J. M.; Pitzer, K. S. J. Chem. Therm. 1984, 16, 303-315.
27. Hill, A. E.; Bacon, L. R. J. Am. Chem. Soc. 1927, 49, 2487-2495.
28. Bates, R. G. Determination of pH, Theory and Practice; John Wiley & Sons: New York, 1973, p 479.
29. Bates, R. G. In The Nature of Seawater; Goldberg, E. D., Ed.; Dahlem Workshop Report, Abakon Verlagsgesellschaft: Berlin, 1975, 313-338.
30. Bates, R. G.; Culberson, C. H. In The Fate of Fossil Fuel $CO_2$ in the Oceans; Anderson, N. R.; Malahoff, A., Eds.; Plenum: New York, 1977; 45-61.
31. Millero F. J. Geochim. Cosmochim. Acta 1979, 43, 1651-1661.
32. Millero, F. J.; Schreiber, D. R. Am. J. Sci. 1982, 282, 1508-1540.
33. Plummer, L. N.; Sundquist, E. T. Geochim. Cosmochim. Acta 1982, 46, 247-258.
34. Millero, F. J. Geochim. Cosmochim. Acta 1983, 47, 2121-2129.
35. Dickson, A. G. Geochim. Cosmochim. Acta 1984, 48, 2299-2308.
36. Covington, A. K.; Bates, R. G.; Durst, R. A. Pure Appl. Chem. 1985, 57, 531-542.
37. MacInnes, D. A. J. Am. Chem. Soc. 1919, 41, 1086-1092.
38. Westcott, C. pH Measurements; Academic: New York, 1978, p 172.
39. Plummer, L. N.; Parkhurst, D. L.; Thorstenson, D. C. Geochim. Cosmochim. Acta 1983, 47, 665-686.
40. Frape, S. K.; Fritz, P.; McNutt, R. H. Geochim. Cosmochim. Acta 1984, 48, 1617-1627.
41. Bodine, M. W., Jr.; Jones, B. F. U.S. Geological Survey, Water-Resources Investigations Report 86-4086, 1985, p 130.

RECEIVED October 6, 1989

# APPLICATIONS TO MODELING: EQUILIBRIUM AND MASS TRANSFER

## Chapter 11

# Reconstruction of Reaction Pathways in a Rock–Fluid System Using MINTEQ

### Hannah F. Pavlik[1,3] and Donald D. Runnells[2]

[1]F. G. Baker Associates, 2970 Howell Road, Golden, CO 80401
[2]Department of Geological Sciences, University of Colorado, Boulder, CO 80309–0250

The mass transfer model MINTEQ was used to evaluate probable reaction pathways driving the reaction of processed oil shale leachate (Lawrence Livermore L2 leachate) with a sandstone of the Uinta Formation collected near Parachute, Colorado. The purpose of the study was to model the chemical interactions associated with the accidental discharge of oil shale leachate into the subsurface. The geochemical interaction of leachate and sandstone was reconstructed in consecutive steps that were considered additive toward the attainment of final equilibrium. The outcome of each hypothetical equilibration step was used as the starting point for successive chemical mass balance calculations. Modeling results suggest that the approach to chemical equilibrium is controlled by the recarbonation of the leachate, during which $Ca^{2+}$ activity and pH are driven by precipitation of calcite. Recarbonation is accompanied by the apparent precipitation of sepiolite from the leachate, dissolution of magnesium from an inferred magnesium carbonate mineral in the sandstone, and partitioning of lithium and fluoride between liquid and solid phases. The adsorption isotherms for Li and F were linear and insensitive to the system parameters, and therefore a constant Kd was used to model surface reactions involving Li and F. Modeling results compare favorably with laboratory studies of the interaction between sandstone and leachate.

This paper is the result of research conducted to evaluate the use of distribution coefficient (Kd) values and mineral solubility data to simulate migration of contaminants from disposal sites of oil shale waste planned in the Piceance Creek Basin of western Colorado. In a broad context, the purpose of the research was to simulate geochemical impacts of the accidental discharge of oil shale leachate into the subsurface. The aquifer of concern was a sandstone member of the Uinta Formation, the bedrock that underlies potential disposal sites for oil shale waste near Parachute, Colorado. The contaminant fluid was a leachate derived from the Lawrence Livermore L2 modified in-situ processed shale.

The interaction of leachate and sandstone was studied in a series of batch experiments in which crushed samples of Uinta Sandstone were reacted with the leachate for five days. The contact time was determined in previous screening experiments which showed that major chemical constituents in the leachate had attained a steady-state composition during this period. During the experiments, the initial and final compositions of the leachate were determined. Characterization of the Uinta Sandstone and the L2 leachate and details of the analytical methods used in these studies have been described elsewhere (1).

MINTEQ (2) was the equilibrium mass-transfer code chosen to model the geochemical interaction between leachate and sandstone. The model was selected to perform three functions for which it is well-suited:
(1) To compute the activity of major inorganic species in the L2 leachate before and after reaction with the sandstone;
(2) To calculate and evaluate the solubility controls imposed on the geochemical system during recarbonation; and,

[3]Current address: EBASCO Services, 143 Union Boulevard, Lakewood, CO 80228

(3) To model the partitioning of lithium and fluoride between solid and liquid phases.

Hypothetical reaction pathways were evaluated in four general stages. First, the chemical composition of the L2 leachate before and after reaction with the sandstone was examined and plausible reactions that might represent stages of partial equilibrium were selected. Second, MINTEQ was used to compute the equilibrium composition of the leachate at each reaction step, given the precipitation or dissolution of plausible minerals and assuming that the system was closed to $CO_2$. Third, the outcome of each hypothetical equilibration step was used as input for subsequent mass balance calculations. And fourth, after each equilibration step, progress toward equilibrium was reevaluated through comparison of predicted and observed leachate chemistry.

Because of the complex mineralogy of the Uinta Sandstone and the complex chemistry of the leachate, it was impractical to measure all of the parameters needed to simulate surface reactions using the complexation algorithms in MINTEQ. Therefore, the more generalized approach of using empirical Kd values for Li and F was tested and found to be appropriate. The Kd parameter represents a general descriptor of the surface reactions occurring in this rock-fluid system. As is well known, the Kd is not a universal parameter but must be tested and verified for each system of interest.

Composition and Properties of the Uinta Sandstone

The Uinta Formation is the stratigraphic unit that overlies the Tertiary Eocene oil shale deposits of the Green River Formation in western Colorado. This stratigraphic unit contains interbedded fluviatile and lacustrine beds marking the margins of ancient Lake Uinta (3-6). It is characterized by tuffaceous siltstones and dolomitic limestones which grade into medium to fine-grained orange sandstones. Tuff beds containing diagenetic zeolite minerals are abundant in both the Green River and Uinta Formations (7).

The suite of minerals found in the Uinta Sandstone collected from Parachute Creek reflects the unique geochemical environment provided by Lake Uinta. The dominant minerals in the fresh sandstone are quartz, sodium plagioclase feldspars, and potassic feldspars as indicated by X-ray diffraction analysis (1). Important accessory minerals are calcite, nahcolite, dolomite-ankerite, and talc. Trace minerals include iron-enriched magnesite in the siderite-magnesite series, pyrite, pyrrhotite, and hematite. Chemical properties of representative samples of the Uinta Sandstone are summarized in Table I.

Characterization of L2 Oil Shale Leachate

The fluid generated by agitating the L2 processed shale in water in closed containers for 30 days attained a pH of 11.8. The geochemistry of high-pH oil shale leachates has been studied by a number of investigators as part of the U.S. Department of Energy Oil Shale Program (for example, 8-10). The high pH has generally been attributed to the hydrolysis of calcium and magnesium oxides produced by the decomposition of carbonates during retorting (11,12).

The L2 leachate is essentially a sodium-sulfate solution containing significant concentrations of K, Ca, Mo, F, Li, Si, inorganic carbon, OH, and Cl. The ionic strength ranges from 0.08 to 0.09. Arsenic is present in trace concentrations. The organic fraction of the L2 leachate was not characterized in this study; however, the concentration of total dissolved organic carbon in the leachate was 8 mg/L. In view of the high concentrations of sulfate (2765 mg/L) and carbonate (428 mg/L) as complexing ligands, we do not believe that the dissolved organic carbon in the leachate will play a significant role in complexation.

Speciation of the fluid was modeled using MINTEQ. The major inorganic species in the L2 leachate are listed in Table II. The thermodynamic data base for MINTEQ was adapted from that of WATEQ3 (13-15). The VAX version of the model used in this study was obtained from Battelle Pacific Northwest Laboratory and implemented at the University of Colorado by Davis (16).

PRELIMINARY SCREENING OF PLAUSIBLE REACTIONS

Batch reaction experiments were conducted and the data obtained were examined to establish changes in aqueous speciation that had occurred in the L2 leachate during contact with the Uinta Sandstone. The rock-fluid system was assumed to have attained a state of "apparent equilibrium" during the batch experiments; that is, it was assumed that both equilibrium and non-equilibrium

**TABLE I. Chemical Properties of the Uinta Sandstone**

| Analysis | Results |
|---|---|
| pH (1:1 distilled $H_2O$ after 2 hr) | 8.4 ± 0.02 |
| Specific Conductance (1:1 distilled $H_2O$ after 2 hr) | 18 ± 2 $\mu$S |
| Oxid. org. C | 0.18 ± 0.03 wt % |
| Total carbonate C | 0.97 ± 0.64 wt % |
| Total free Fe | 2.1 ± 0.3 wt % |
| Amorphous Fe | 0.30 ± 0.04 wt % |
| Total free Si | 0.5 ± 0.1 wt % |
| Exchangeable Ca | 1.2 ± 0.1 wt % |
| Exchangeable Mg | 0.5 ± 0.1 wt % |
| Zero net surface charge ($Na_2SO_4$) | pH = 6.7 |
| Cation exchange capacity (Na saturation) | 10.4 ± 0.3 meq/100g |
| Anion exchange capacity (Cl exchange) | 0.07 ± 0.02 meq/100g |
| BET Surface Area Crushed bedrock <2mm | mean 12. $m^2$/g range 11. - 14. $m^2$/g |

TABLE II. Principal Aqueous Species of the L2 Leachate Before and After Reaction with the Uinta Sandstone *

| Species | (-)log Concentration (mol/Kg) | |
|---|---|---|
| | Before Reaction | After Reaction |
| Total Na | 1.44 | 1.39 |
| ($Na^+$) | (1.47) | (1.41) |
| ($NaSO_4^-$) | (2.82) | (2.76) |
| ($NaCO_3^-$) | (2.92) | (5.44) |
| Total K | 2.00 | 2.26 |
| ($K^+$) | (2.02) | (2.28) |
| ($KSO_4^-$) | (3.25) | (3.50) |
| Total Li | 2.65 | 2.86 |
| ($Li^+$) | (2.66) | (2.88) |
| Total Ca | 2.20 | 2.33 |
| ($CaCO_3°aq$) | (2.62) | (5.09) |
| ($Ca^{2+}$) | (2.68) | (2.58) |
| ($CaSO_4°aq$) | (2.80) | (2.68) |
| ($CaOH^+$) | (3.71) | (7.54) |
| Total Mg | 5.70 | 2.59 |
| ($Mg^{2+}$) | (6.19) | (2.83) |
| ($MgSO_4°aq$) | (6.36) | (2.98) |
| ($MgOH^+$) | (6.40) | (6.96) |
| Total $SO_4$ | 1.54 | 1.53 |
| ($SO_4^{2-}$) | (1.60) | (1.61) |
| ($CaSO_4°aq$) | (2.80) | (2.68) |
| ($NaSO_4^-$) | (2.82) | (2.76) |
| Total Al | 4.47 | -- |
| ($Al(OH)_4^-$) | (4.47) | -- |
| Total $CO_3$ | 2.07 | 2.74 |
| ($CO_3^{2-}$) | (2.31) | (4.90) |
| ($CaCO_3°aq$) | (2.62) | (5.09) |
| ($NaCO_3^-$) | (2.92) | (5.44) |
| ($HCO_3^-$) | (4.12) | (2.77) |
| Total Cl | 3.49 | 3.36 |
| ($Cl^-$) | (3.49) | (3.36) |
| Total F | 3.98 | 4.28 |
| ($F^-$) | (3.98) | (4.30) |
| Total $H_4SiO_4$ | 3.44 | 3.83 |
| ($H_4SiO_4°$) | (6.05) | (3.83) |
| ($H_2SiO_4^{2-}$) | (3.58) | (9.24) |
| ($H_3SiO_4^-$) | (4.02) | (5.74) |
| pH | 11.8 | 7.9 |
| Specific conductance ($\mu S$) | 7280. | 5490. |
| Total dissolved solids (mg/L) | 4563. | 4319. |
| Ionic strength | (0.09) | (0.08) |

\* Laboratory-measured data are reported as total concentrations; speciated data indicated in parentheses were computed using MINTEQ.

processes had attained steady-state levels of saturation with respect to mineral phases in the system (17).

Interaction of the fluid with the sandstone minerals lowered the concentrations of Li, K, Ca, Mo, Si, F, $SO_4$, and $CO_3$ but increased the concentrations of Na, Mg, and Cl. The chemical compositions of the L2 leachate before and after reaction with the sandstone are compared in Table II. By far the greatest changes observed in fluid chemistry reflect changes in the activity of the chemical components Mg, $CO_3$, and $H_4SiO_4$ as a function of pH. Based on these observations, the batch experiments suggested that three types of reactions are important. These include: (1) leachate recarbonation accompanied by precipitation of calcite, (2) reaction of the leachate with potassium and magnesium silicates and magnesium carbonate minerals in the sandstone, and (3) partitioning of Li and F between dissolved and solid phases. Plausible reactions were evaluated by modeling the equilibrium chemistry and solubility of minerals in the fluid and by evaluating experimentally-derived batch Kd values for Li and F.

Calcium Solubility Controls and Recarbonation of the L2 Leachate

Recarbonation was inferred to be a key process affecting the mass balance of the H, Ca, and $CO_3$ components of the system. During recarbonation, carbon dioxide dissolves in the fluid and calcite precipitates with a decrease in pH according to the following reaction:

$$Ca^{2+} + CO_2 (g) + H_2O \rightarrow CaCO_3 (s) + 2H^+ \tag{1}$$

Carbon dioxide partial pressure was not directly measured during the batch experiments. However, it was observed that in the absence of Uinta Sandstone, when the fluid is open to atmospheric carbon dioxide (theoretical log $pCO_2$ = -3.57 atm. at the University of Colorado), it gradually recarbonates and the pH drops from 11.8 to 9.6 after 36 days.

Based on a knowledge of the mineralogy of the Uinta Sandstone, the mineral phases most likely to be controlling the solubility of Ca in the sandstone - L2 leachate system are calcite, dolomite, gypsum, and fluorite. Stability lines and saturation indices calculated for these minerals are presented in Figure 1 and Table III, respectively. The observed data point (black circle) plotted in Figure 1 represents the measured pH and log $Ca^{2+}$ activity in the L2 leachate after reaction with the Uinta Sandstone. The log $CO_2$ gas partial pressure of -2.95 atmosphere is based on the measured pH and alkalinity of the reacted solution. The open circle represents the log $Ca^{2+}$ activity and pH calculated by MINTEQ for the raw leachate recarbonated to a log $CO_2$ partial pressure of -2.95 atmosphere. The leachate apparently developed a $CO_2$ gas overpressure because equilibrium with calcite was attained in sealed containers at the relatively low pH of 7.91 (18). The calcite-dolomite line shown in the figure represents the pH-dependent activity of $Ca^{2+}$ in equilibrium with both calcite and dolomite.

The data presented in Figure 1 and Table III illustrate that upon recarbonation and exposure to the sandstone, the activity of $Ca^{2+}$ in the leachate most closely approaches equilibrium with calcite. Both before and after reaction with the sandstone, $Ca^{2+}$ activity is undersaturated with respect to gypsum and fluorite and oversaturated with respect to dolomite. The gypsum and fluorite solubility lines in Figure 1 were developed using $SO_4^{2-}$ and $F^-$ activities calculated in the raw leachate.

A summary of the response of the leachate-sandstone system to recarbonation is presented in Figure 2. The figure illustrates that the approach to equilibrium with calcite (according to Equation 1) can be simulated by the progressive dissolution of $CO_2$ gas into the leachate and by equilibration of the leachate with calcite in the Uinta Sandstone.

The recarbonation pathway for the system was simulated by MINTEQ as follows (Figure 2). First, MINTEQ was used to calculate the partial pressure of $CO_2$ gas in equilibrium with the two fluids listed in Table III. Based upon measured alkalinity and pH, the raw L2 leachate (point 1, Figure 2) was computed to have a $CO_2$ pressure of $10^{-8.23}$ atmosphere. After 5-day reaction with the Uinta Sandstone (point 4, Figure 2), the measured pH and alkalinity of the fluid yielded a calculated $CO_2$ pressure of $10^{-2.95}$ atmosphere. Leachate recarbonation was simulated by fixing the $CO_2$ gas partial pressure of the unreacted fluid in the absence of sandstone at the theoretical value of $10^{-3.57}$ atmosphere (point 2, Figure 2), and in the presence of sandstone at the value of $10^{-2.95}$ atmosphere calculated from leachate chemistry (point 3, Figure 2). It is assumed that, when open to the atmosphere, the system will ultimately approach equilibrium with ambient $CO_2$ pressure (in Boulder) of $10^{-3.57}$ atmosphere (point 5, Figure 2).

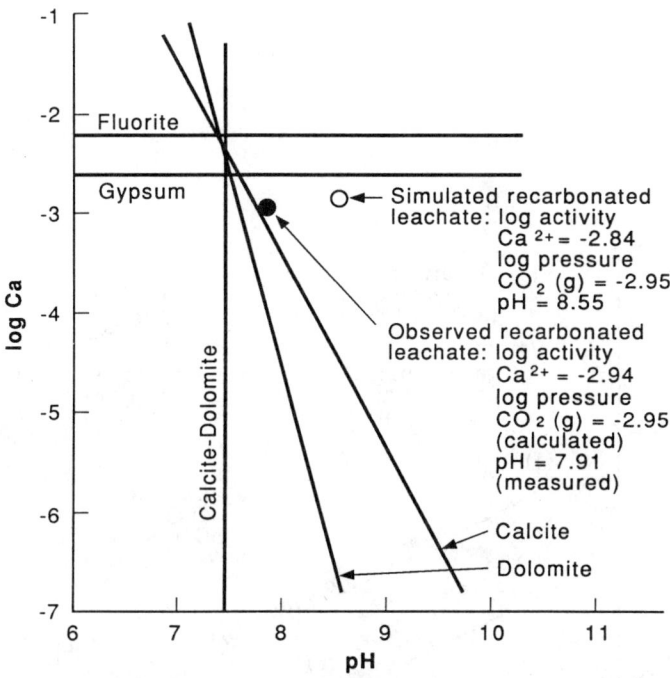

Figure 1. Mineral stability lines for solid phases potentially controlling Ca activity at log $pCO_2$ = -2.95 (error bars fall within the areas of the circles).

TABLE III. MINTEQ-Calculated Saturation Indices for Selected Minerals in the L2 Leachate Before and After Reaction with the Uinta Sandstone

| Mineral | Log SI Before Reaction | Log SI After Reaction |
| --- | --- | --- |
| Calcite | 2.71 | 0.25 |
| Gypsum | -0.26 | -0.14 |
| Fluorite | -0.30 | -0.81 |
| Dolomite | 1.97 | 0.31 |
| Talc | 4.15 | -0.45 |
| Sepiolite | 0.21 | -2.12 |
| Magnesite | -1.24 | -0.44 |
| Hydromagnesite | -11.17 | -12.45 |
| Nesqehonite | -3.65 | -2.85 |
| Brucite | 0.34 | -4.15 |
| Diopside | 5.78 | -2.02 |

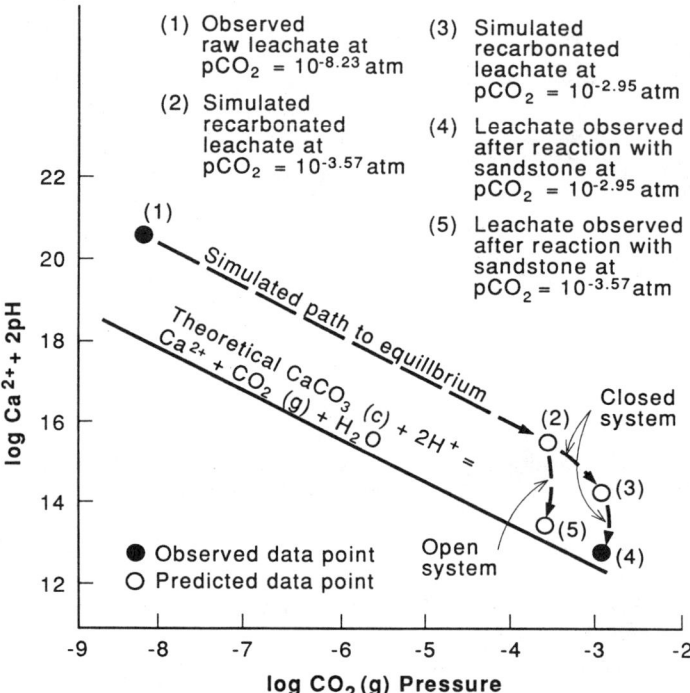

Figure 2. Simulated recarbonation of the L2 leachate and the approach to system equilibrium with calcite (error bars fall within the area of the black circles).

## Magnesium and Silica Solubility Controls

Recarbonation of the L2 leachate affects the activity of both $Ca^{2+}$ and $Mg^{2+}$ by lowering the pH. Key mineral phases in the sandstone-leachate system which might control the activity of $Mg^{2+}$ are shown in Figures 3 and 4. In the computations for these diagrams, dissolved silica activity and pH were fixed at the values observed after reaction with the sandstone. Calculated saturation indices for selected Mg-bearing minerals are given in Table III.

As shown by the raw L2 leachate data point (empty circle) in Figure 3, the leachate is initially supersaturated with respect to (at least) calcite, dolomite, talc, magnesite, sepiolite, hydromagnesite, nesquehonite, and brucite. After reaction with the Uinta Sandstone (indicated by the solid circle, point L2-USS in Figure 3), $Mg^{2+}$ activity appears to be most nearly controlled by the solubility of magnesite.

To better understand the magnesium and silicate chemistry of the rock-fluid system, the solubilities of talc, sepiolite, and diopside are compared in Figure 4. These three minerals were plotted because they have been reported to be stable in high silica, alkaline systems similar to the one under study (19-23). The mineral stability lines were developed using the following equations:

$$3Mg^{2+} + 4H_4SiO_4^° \rightarrow Mg_3Si_4O_{10}(OH)_2 + 4H_2O + 6H^+ \tag{2}$$
$$(\text{talc})$$

$$4Mg^{2+} + 6H_4SiO_4^° \rightarrow Mg_4Si_6O_{15}(OH)_2 \cdot 6H_2O + 8H^+ \tag{3}$$
$$(\text{sepiolite})$$

$$Mg^{2+} + Ca^{2+} + 2H_4SiO_4^° \rightarrow MgCaSi_2O_6 + 4H^+ + 2H_2O \tag{4}$$
$$(\text{diopside})$$

Inspection of Figure 4 indicates that after reaction with the Uinta Sandstone for five days (see data point L2-USS in the figure), the activity of $Mg^{2+}$ in solution most closely approaches apparent equilibrium with talc or sepiolite. However, these results must be interpreted with caution. The published literature contains large discrepancies in the thermodynamic data available for the stabilities of talc and sepiolite (24,25). Garrels and Mackenzie (23), among others, have reported that sepiolite precipitates preferentially to talc in silica-rich, alkaline fluids. Similarly, Nordstrom and Munoz (24) noted that error in the existing thermodynamic data is sufficiently large to reverse the relative stabilities of these two magnesium silicate minerals. On the basis of the data plotted in Figure 4, and the field evidence for occurrences of sepiolite (20,22), it seems reasonable to assume that sepiolite is the solid phase most likely to control the chemistry of magnesium in the presence of dissolved silica in this system.

## Partitioning of Lithium and Fluoride

Information on the partitioning of dissolved lithium and fluoride species between liquid and solid phases was obtained from sorption isotherms derived for the rock-fluid system in a parallel research effort (1,26). These studies showed that the distribution coefficient parameter successfully integrated the surface interactions of lithium and fluoride with the Uinta Sandstone; this approach provided a useful first-approximation for modeling sorption in the system. Isotherm data indicated that for both lithium and fluoride: (1) equilibrium, reversible partitioning occurred, (2) the slope of the partitioning function was constant and linear, (3) the choice of solid/solution ratio (from 1:5 to 1:20) had no statistically significant effect on the determination of the distribution coefficient, and, (4) the Kd value was independent of concentration, pH, and ionic strength effects after equilibrium was attained. Experimental Kd values obtained for lithium and fluoride in the Uinta Sandstone-L2 leachate system were $0.6 \pm 0.1$ mL/g and $2.2 \pm 0.5$ mL/g, respectively.

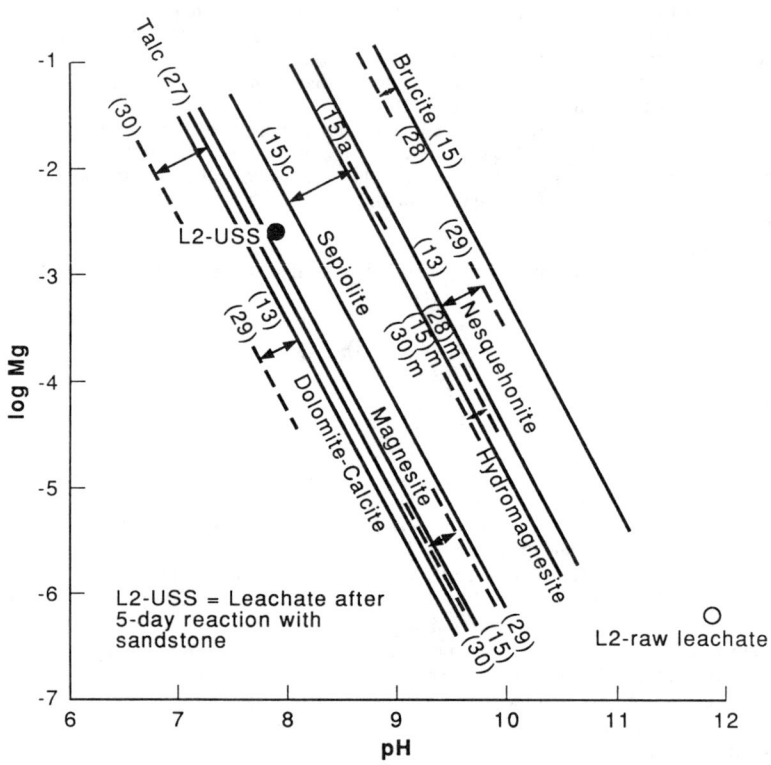

**Thermodynamic Data Sources**

(13) Ball et al., 1980
(15) Truesdell and Jones, 1974
(15)a Ibid (amorphous sepio.)
(15)c Ibid (crystalline sepio.)
(15)m Ibid ($Mg_5$ hydromag.)
(27) Plummer et al., 1976
(28) Stumm and Morgan, 1981
(28)m Ibid ($Mg_4$ hydromag.)
(29) Lindsay, 1979
(30) Robie et al., 1979
(30)m Ibid ($Mg_5$ hydromag.)

**Figure 3. Mineral stability lines for solid phases potentially controlling Mg activity at log $pCO_2$ = -2.95 (error bars fall within the area of the data points).**

## MINTEQ SIMULATION OF SYSTEM BEHAVIOR

Hypothetical reaction pathways chosen to model the L2 leachate-Uinta Sandstone system are illustrated in Figure 5. As a first approximation, dissolution/precipitation reactions affecting the mass balance of Na, K, Mo, $SO_4$, and Cl were not considered. Instead, based upon the solubility controls discussed in the previous sections of this paper, the working hypothesis for the simulations is that the recarbonation of L2 leachate drives the reactions toward equilibrium. Along the path toward equilibrium, recarbonation is accompanied by the precipitation and dissolution of sepiolite, calcite, and an inferred hydrated magnesium carbonate mineral such as hydromagnesite.

Dissolution of hydromagnesite appears to be one reasonable mechanism for releasing $Mg^{2+}$ into the L2 leachate for several reasons. First, small intensity peaks for hydromagnesite were detected during X-ray diffraction of the Uinta Sandstone ([1]). Although these were not conclusive, the presence of a magnesium carbonate mineral of similar composition can be inferred in the sandstone. A mass balance calculation on magnesium indicates that the increase in Mg in solution can be explained by dissolution of approximately 1 percent hydromagnesite from the sandstone. Data presented in Figure 3, supported by MINTEQ mineral saturation index calculations (Table III), show that hydromagnesite should dissolve in the leachate. However, saturation indices given in Table III indicate that other dissolution and precipitation reactions involving Mg-bearing minerals are also possible, such as dissolution of magnesite and precipitation of either dolomite or talc.

The results from one reaction-path simulation of the L2 leachate-Uinta Sandstone system are presented in Figure 6. The series of reactions chosen for this simulation are as follows:

(1) Sepiolite precipitates from the silica-rich alkaline fluid according to the reaction:

$$2Mg^{2+} + 3H_4SiO_4^\circ \rightarrow Mg_2Si_3O_6(OH)_4 + 2H_2O + 4H^+ \tag{5}$$

(2) Recarbonation of the fluid occurs, causing calcite to precipitate, and driving the pH of the system down according to the reaction:

$$Ca^{2+} + CO_2(g) + H_2O \rightarrow CaCO_3 + 2H^+ \tag{6}$$

(3) Continued recarbonation lowers pH further causing hydromagnesite in the Uinta Sandstone to dissolve as follows:

$$Mg_5(CO_3)_4(OH)_2 \cdot 4H_2O + H^+ \rightarrow 5Mg^{2+} + 4CO_3^{2-} + 6H_2O \tag{7}$$

(4) Lithium and fluoride partitioning between liquid and solid phases reaches a steady-state.

In Figure 6, column (1) represents the initial chemical composition of the leachate observed prior to contact with the sandstone. Column (5) summarizes the solution composition observed after reaction with the sandstone for five days. Columns (2)-(4) represent intervening steps in the reaction-path simulation. The major changes in chemistry observed between columns (1) and (5) are an increase of three orders of magnitude in the concentration of Mg and significant decreases in total dissolved carbonate, fluoride, and silica.

The data presented in columns (3) and (4) of Figure 6 illustrate the effect of lithium and fluoride Kd mass action equations on the simulation of system equilibrium. In column (3) of Table IV, where the Li-Kd constant was used to simulate adsorption, predicted and observed concentrations of lithium match quite closely. However, fluoride levels are overestimated. In column (4), where the F-Kd constant was used to simulate adsorption, the match between predicted and observed fluoride in solution is improved. In this case, however, the predicted concentration of lithium in the leachate is overestimated.

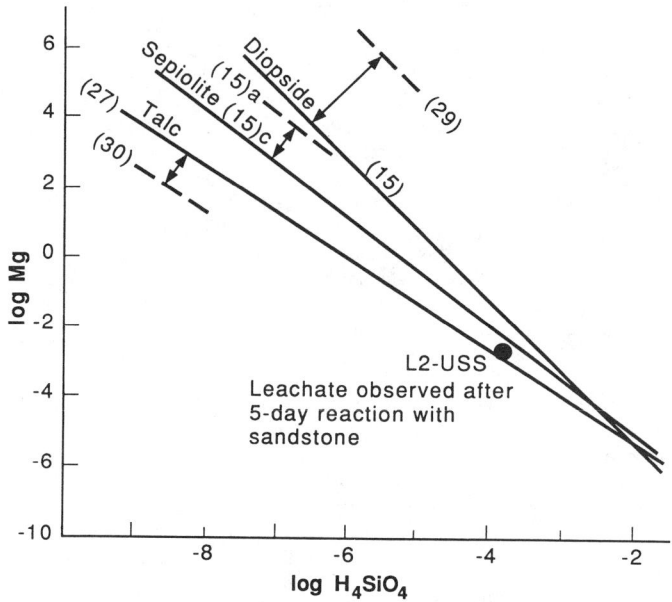

**Thermodynamic Data Sources**

(15)  Truesdell and Jones, 1974  (27)  Plummer et al., 1976
(15)a Ibid (amorphous sepio.)   (29)  Lindsay, 1979
(15)c Ibid (crystalline sepio.) (30)  Robie et al., 1979

**Figure 4.** Mineral stability lines for talc, sepiolite, and diopside after the 5-day reaction period (error bars fall within the area of the black circle).

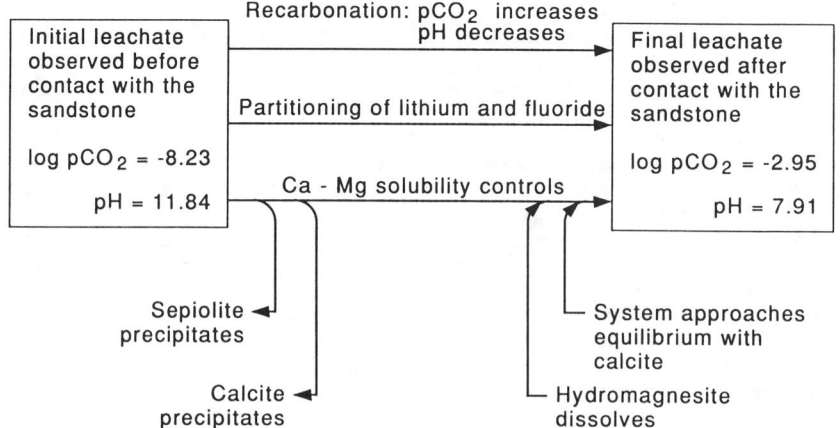

**Figure 5.** Schematic diagram showing the modified reaction-path simulation of the system.

11. PAVLIK & RUNNELLS  *Reconstruction of Reaction Pathways*  151

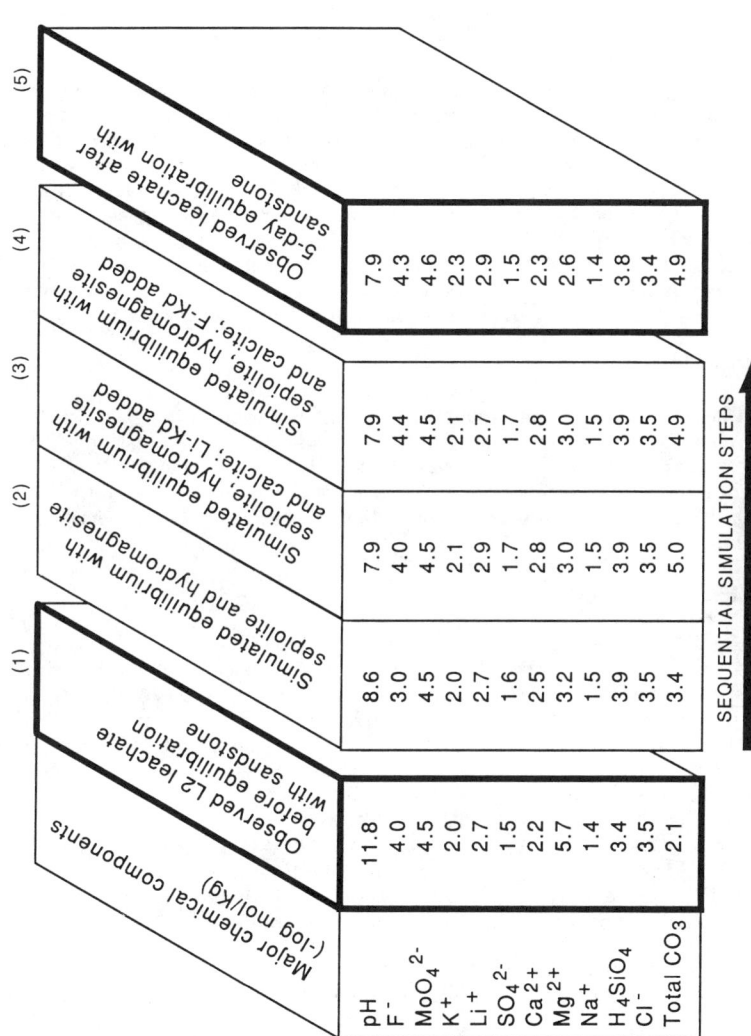

Figure 6. Results of the MINTEQ reaction-path simulation.

CONCLUSIONS

MINTEQ reaction-path simulations supported by mineral saturation data suggest that the chemical equilibrium of the Uinta Sandstone-L2 leachate system is controlled by recarbonation of the leachate, during which calcite precipitation largely controls $Ca^{2+}$ activity and pH. Recarbonation is accompanied by the apparent precipitation of sepiolite from the leachate, dissolution of magnesium from an inferred magnesium carbonate mineral in the sandstone, and partitioning of lithium and fluoride between liquid and solid phases. The modeling results are reasonable in light of the observed mineralogy, measured Kd values, and observed changes in chemical composition of the leachate.

In this modeling effort, the complex interaction of oil shale leachate and Uinta Sandstone was reconstructed in consecutive steps that were considered additive toward the attainment of final equilibrium. The predicted outcome of each equilibration step was used as the starting point for successive mass balance calculations. In this way, plausible reaction pathways along which the leachate and sandstone approach chemical equilibrium were hypothesized and evaluated. It should be noted that given the number of plausible phases that may be reacting in the system, predictions of solution composition along the path to heterogeneous equilibrium are likely to be non-unique. Despite this fact, and as illustrated in this work, the value of the reaction-path modeling approach lies in its use of equilibrium concepts to distill a complex set of hypothetical geochemical processes into a few major reactions that are likely to improve our understanding of system behavior.

ACKNOWLEDGMENTS

The research in this paper was performed in partial fulfillment of a Ph.D dissertation completed by the senior author at the University of Colorado-Boulder. Financial support from the U.S. Department of Energy, Atlantic Richfield Foundation, Inc., and the University of Colorado Department of Geological Sciences are gratefully acknowledged. We also wish to thank Al Burnham and Bill Miller of Lawrence Livermore National Laboratory for providing samples of L2 retorted oil shale.

LITERATURE CITED

1. Pavlik, H. F. Ph.D. Thesis, University of Colorado, Boulder, 1987; p 207.
2. Felmy, A. R.; Girvin, D. C.; Jenne, E. A. MINTEQ, a Computer Program for Calculating Aqueous Geochemical Equilibria; Final Project Report, Contract 68-03-3089, U.S. Environmental Protection Agency: Athens, GA, 1983; p 87.
3. Miknis, F. P.; McKay, J. F., Eds. Geochemistry and Chemistry of Oil Shales; ACS Symposium Series No. 230; American Chemical Society: Washington, DC, 1983; p 565.
4. Bradley, W. H. U.S. Geol. Survey Prof. Paper 168, 1931; p 58.
5. Juhan, J. P. The Mountain Geologist 1965, 2, 123-128.
6. Cashion, W. B.; Donnell, J. R. U.S. Geol. Survey Bull. 1394-G, 1974; p 9.
7. Surdam, R. C.; Parker, R. B. Geol. Soc. of Amer. Bull. 1972, 83, 689-700.
8. Runnells, D. D.; Glaze, M.; Saether, O.; Stollenwerk, K. In Trace Elements in Oil Shale; Chappell, W.R., Ed.; Progress Report 1978-79, Center for Environmental Sciences, University of Colorado: Denver, CO, 1979; p 134.
9. Runnells, D. D.; Esmaili, E. In Trace Elements in Oil Shale; Chappell, W.R., Ed.; Progress Report 1980-81, Center for Environmental Sciences, University of Colorado: Denver, CO, 1981; p 163.
10. Peterson, E. J.; Henicksman, A.; Wagner, P. Investigations of Occidental Oil Shale, Inc., Retort 3E Spent Shales; Report LA-8792-MS, Los Alamos National Laboratory: Los Alamos, NM, 1981; p 39.
11. Park, W. C.; Lindemanis, A. E.; Raab, G. A. In Situ 1980, 3, 353.
12. Burnham, A. K.; Subblefield, C. T.; Campbell, J. H. Effects of Gas Environment on Mineral Reactions in Colorado Oil Shale; Report UCRL-81951, Lawrence Livermore National Laboratory: Livermore, CA, 1978.

13. Ball, J. W.; Nordstrom, D. K.; Jenne, E. A. U.S. Geol. Survey Water Resources Invest. 78-116, 1980; pp. 78-116.
14. Ball, J. W.; Jenne, E. A.; Cantrell, M. W. U.S. Geol. Survey Open File Report 81-1183, 1981; p 81.
15. Truesdell, A. H.; Jones, B. F. U.S.Geol. Survey Jour. Res. 1974, 2, 233-248.
16. Davis, A. O. Ph.D. Thesis, University of Colorado, Boulder, 1985; p 214.
17. Plummer, L.N.; Parkhurst, D.L.; Thorstenson, D.C. Geochim. Cosmochim. Acta 1983, 47, 665-686.
18. Drever, J. I. The Geochemistry of Natural Waters; 2nd ed.; Prentice-Hall: New Jersey, 1988; Figure 4.4, pp. 61.
19. Reddy, K. J.; Lindsay, W. L. Jour. Environ. Qual. 1986, 15, 1-4.
20. Wollast, R.; Mackenzie, F. T.; Bricker, O. P. Am. Mineral 1968, 53, 1645-1662.
21. Bricker, O. P.; Nesbitt, H. W.; Gunter, W. D. Am. Mineral 1973, 58, 64-72.
22. Hathaway, J. C.; Sachs, P. L. Am. Mineral 1965, 50, 852-867.
23. Garrels, R. M.; Mackenzie, F. T. In Equilibrium Concepts in Natural Water Systems; Jenne, E. A., Ed.; Advances in Chemistry Series No. 67; American Chemical Society: Washington, DC, 1967; pp. 222-242.
24. Nordstrom, D. K.; Munoz, J. Geochemical Thermodynamics; Benjamin/Cummings: San Francisco, CA, 1985; p 477.
25. Bassett, R. M.; Kharaka, Y. K.; Langmuir, D. In Chemical Modeling in Aqueous Systems; Jenne, E. A., Ed.; ACS Symposium Series No. 93; American Chemical Society: Washington, DC, 1979; pp. 389-400.
26. Pavlik, H. F.; Runnells, D. D. "Evaluation of the Validity and Statistical Variability of Lithium and Fluoride Distribution Coefficient Values in a Complex Geochemical System"; Research Highlights, DOE/ER-0261, U.S. Department of Energy: Washington, DC, 1986; p 16.
27. Plummer, L. N.; Jones, B. F.; Truesdell, A. H. U.S. Geol. Survey Water Resources Invest. 73-13, 1976; p 61.
28. Stumm, W.; Morgan, J. J. Aquatic Chemistry; 2nd ed.; Wiley-Interscience: New York, 1981; pp. 273.
29. Lindsay, W. L. Chemical Equilibria in Soils; Wiley-Interscience: New York, 1979; pp. 107-108.
30. Robie, R. A.; Hemingway, B. S.; Fisher, J. R. U.S. Geol. Survey Bulletin No. 1452, 1978; p 456.

RECEIVED July 20, 1989

## Chapter 12

# Hydrogeochemical Interactions and Evolution of Acidic Solutions in Soil

**Patrick Longmire[1,2], Douglas G. Brookins[2], and Bruce M. Thomson[2]**

[1]Roy F. Weston, Inc., Albuquerque, NM 87108
[2]University of New Mexico, Albuquerque, NM 87131

>Leachate generated from surface disposal of acidic uranium mill tailings in New Mexico and Colorado significantly alters hydrogeochemical characteristics of subjacent sediments including pH, Eh, mineralogical transformation, and acid neutralizing capacity. Experimental investigations and thermodynamic equilibrium modeling with the geochemical code PHREEQE show that the relatively oxidizing tailings pore water is in near equilibrium with jurbanite (AlOHSO$_4$), gypsum (CaSO$_4$·2H$_2$O), strengite (FePO$_4$·2H$_2$O), and lepidocrocite ($\gamma$-FeOOH), and is oversaturated with alunite (KAl$_3$(SO$_4$)$_2$(OH)$_6$), goethite ($\alpha$-FeOOH), and jarosite (KFe$_3$(SO$_4$)$_2$(OH)$_6$). Ions concentrated in tailings pore water include Mg, Na, Mn, V, Ni, Al, Fe, Ca, K, SO$_4$, NO$_3$, PO$_4$, Mo, Se, As, and U. Leach experiments on tailings material demonstrated that As, Cr, Mo, U, and V are associated with clay minerals, jarosite, and ferric oxyhydroxide coatings. The enrichment factors (clay/sand abundance) for these solutes are greater than unity, which may be the result of anion adsorption below pH$_{zpc}$ literature values for ferric oxyhydroxide, silica gel, and montmorillonite. The concentrations and mobilities of several species and elements follow the order SO$_4$>NH$_4$>Al>Mn>NO$_3$>U>Fe>Se>PO$_4$>Ni>As>Cd at pH 4.0. Sulfate-dominated leachate reacts with tailings subsoil calcite producing gypsum, which results in a continued decrease in SO$_4$ concentrations. Dissolved concentrations of U, NO$_3$, SO$_4$, and other major ions remain elevated above background concentrations downgradient from the tailings impoundment.

The safe disposal of uranium-mill tailings is one of the major environmental challenges currently facing the front end of the nuclear fuel cycle (1). It is estimated that there are 300 million metric tons of tailings associated with U.S. uranium mills which are operating, on standby basis, inactive, or abandoned. Most uranium tailings in the United States and Canada are characterized by pore waters with a low pH, a moderately oxidizing Eh, high total dissolved solids content, and high activities of radionuclides including $^{226}$Ra, $^{210}$Pb, and $^{230}$Th (2,3). Acid-leach uranium tailings contain large amounts of complex soluble SO$_4$ salts resulting from the use of sulfuric acid during the ore extraction process (4). Surface impoundments containing uranium tailings are commonly in areas of structurally complex geology and overlie recharge zones of alluvial and bedrock aquifers. The Grants Mineral Belt, New Mexico was selected for part of this investigation because of the extensive amount of uranium mining and milling that has taken place over the past three decades. Uranium milling at Maybell, Colorado continued over a fifteen-year period and this uranium locality is one of the largest in that state. Consequently, degradation of ground-water quality has resulted from tailings seepage in the Grants Mineral Belt and at Maybell, Colorado.

The objectives of this investigation are to characterize tailings-pore water solutions and tailings mineralogy, and to evaluate hydrogeochemical processes occurring within acid-leach tailings impoundments and tailings subsoils. Experimental investigations, computer modeling, and field studies are used in this investigation to assess the long-term impact of tailings leachate and to evaluate the mobility of Al, Fe, $SO_4$, Mo, Se, As, and U concentrated in tailings leachate.

## SAMPLING AND ANALYTICAL METHODS

Acid-leach tailings samples from five surface tailings impoundments and pore-water samples from lysimeters were collected from 1982 to 1989. Tailings samples were collected from one operating and four inactive mills in New Mexico and Colorado. Details of sample collection and routine chemical analysis procedures are contained in Thomson et al. (2). The major cations generally were analyzed by inductively coupled plasma emission spectroscopy and the trace elements were determined by graphite-furnace atomic absorption and instrumental neutron activation analysis. The major anions were analyzed by ion chromatography and titration. The accuracy and precision of the tailings pore water data were evaluated through charge-balance calculations as well as through analysis of blank, spike, and replicate samples. Field analyses included measurements of temperature, pH (Orion Ross), Eh (Orion 96-78 platinum electrode), dissolved oxygen, alkalinity, and specific conductance. Ground water has been shown not to have a "system" Eh because internal disequilibrium exists between the redox couples (5). Under acidic and relatively oxidizing conditions, however, the $Fe^{2+}/Fe^{3+}$ redox couple has been shown to be representative of measured Eh values in surface waters contaminated by acid mine drainage (6). Measured Eh values for highly oxidizing, acid leach pore waters within uranium tailings impoundments may approximate a "system" Eh. In this investigation, field Eh measurements were made on tailings raffinate samples and lysimeter solutions. Laboratory Eh values were recorded for tailings subsoils, geologic samples, and tailings used for leach experiments. The precision of Eh measurements was 5 mV and the precision on pH measurements was 0.05 pH units.

Batch leach experiments were performed on tailings material to determine the nature of contaminants distributed on sand and silt and clay-sized fractions. For the batch leach experiments, a mixture of tailings material was prepared using a chemical dispersant (sodium hexametaphosphate $(NaPO_3)_6$). The mixture was shaken and allowed to settle in covered beakers for sufficient time such that no particles with a diameter > 50 $\mu$m remained in suspension. The fines, which remained in suspension, represented the silt and clay-sized fraction of the samples. At the end of the settling period, the liquid was decanted from the beaker. The remaining sand-sized tailings were dried and transferred to a sealed vial. The decanted solution was passed through a 0.45 $\mu$m membrane filter, previously washed with distilled water, and the filtrate was collected and placed in a sealed container. The solids were dried and transferred to a sealed vial. The samples were then analyzed by instrumental neutron activation analysis.

## TAILINGS-PORE-FLUID CHEMISTRY AND MINERALOGY

Characterization of the tailings solids and associated pore waters is an important aspect of tailings geochemistry. Geochemical conditions within a tailings impoundment provides the basis for precipitation-dissolution and adsorption-desorption reactions which control the chemical composition of tailings leachate. Due to the nature of the milling process, which mainly uses sulfuric and nitric acids for leaching of uranium minerals, aqueous solutions associated with tailings impoundments are of extremely poor quality. Concentrations of selected species in tailings raffinates (the aqueous solution remaining after uranium has been extracted by a solvent) from four uranium mills in the Grants Mineral Belt, New Mexico are presented in Table I. The raffinates contain both chemicals added during the leaching process and elements from the ore and gangue, which have been solubilized during the milling process. The composition of the raffinates is dependent on: (1) type of ore; (2) type of leach circuit; (3) byproduct recovery; (4) initial quality of process water; and (5) surface evaporation. These factors account for the concentration ranges of constituents reported in Table I. For example, an ammonium nitrate solution was used in the mills to elute U from the loaded resin, and the U was then precipitated from the eluate with anhydrous ammonia, generating an ammonia-rich tailings pore water (4). Additional analyses of tailings raffinates from the Grants Mineral Belt, New Mexico are provided by Longmire (7). Contaminants of concern generally include As, Mo, $NO_3$, $SO_4$, Se, and U due to their relatively high concentrations in uranium-tailings impoundments, tailings subsoils, and ground water.

Analytical results of tailings pore water samples from Maybell, Colorado are shown in Table II. These two lysimeter samples are enriched in Al, F, Fe, Mn, Ni, $PO_4$, $NO_3$, U, Se, and major ions. Elevated concentrations of Al probably result from surface reaction dissolution of aluminosilicate minerals under the strongly acidic conditions of the tailings pore water. A major source of acidity in the tailings impoundment is the reprecipitation of Al in hydroxide and hydroxy-sulfate phases, which consume base (OH) and decrease the pH. In some acidic-sulfate waters, aluminum solubility may be controlled by several minerals including alunite, jurbanite, k-alum, and basaluminite (8). Iron is an important species because of this element's ability to flocculate, adsorb, and coprecipitate other trace elements (9-14). Jarosite is the most plausible secondary mineral controlling Fe concentrations in the Maybell tailings pore water, based on X-ray diffraction (XRD) analysis and scanning electron microscope (SEM) studies. Jarosite is stable under strongly oxidizing, acidic conditions (15). Possible sources of K include K-rich salts in the initial mill reagents and K derived from dissolution of K-feldspar, montmorillonite, and illite present as silicate gangue minerals. Sulfate is the dominant species in the Maybell tailings constituting 70 to 73% of the total dissolved solids (TDS). Sulfate is an important solute because it competes strongly with $PO_4$ and other anions for adsorption sites (9). The solution obtained from lysimeter 082 contains higher concentrations of major ions, and most of the trace elements relative to lysimeter sample 083. Lysimeter 082 is completed in a fine-grained zone within the Maybell tailings impoundment; this zone is characterized by a low saturated hydraulic conductivity ($10^{-8}$ cm sec$^{-1}$). Lysimeter 083 is completed in a sand zone characterized by lessor amounts of clay-sized material and a resultant higher saturated hydraulic conductivity ($10^{-5}$ cm sec$^{-1}$).

The chemical and mineralogical properties of tailings salts are determined by the composition of the tailings raffinate. In this investigation, the most common soluble solids found in acid leach uranium tailings impoundments identified from XRD analysis included gypsum ($CaSO_4 \cdot 2H_2O$), alunite [$(K,Na)Al_3(SO_4)_2(OH)_6$], jurbanite ($Al(SO_4)(OH) \cdot 5H_2O$), basaluminite ($Al_4(SO_4)(OH)_{10} \cdot 5H_2O$), jarosite ($KFe_3(SO_4)(OH)_6$), natrojarosite ($NaFe_3(SO_4)(OH)_6$), calcium sulfate, thenardite ($Na_2SO_4$), and gibbsite ($Al(OH)_3$) (7). Potassium jarosite was the most abundant form of jarosite observed in all of the tailings impoundments; this mineral forms a fine-grained precipitate, which gives a good X-ray pattern. Subordinate amounts of natrojarosite were also identified by XRD analysis. From XRD analysis, quartz, chlorite, magnetite, K-feldspar, Na-feldspar, montmorillonite, mixed-layer illite-montmorillonite, and a poorly crystalline Fe-rich compound were identified as the gangue minerals.

During chemical oxidation of uranium ore and gangue minerals in the presence of sulfuric acid and excess free oxygen, the precipitation of jarosite and alunite is indicated by the reactions

$$K^+ + 3Fe^{3+} + 2SO_4^{2-} + 6H_2O = KFe_3(SO_4)_2(OH)_6 + 6H^+ \quad (1)$$

$$K^+ + 3Al^{3+} + 2SO_4^{2-} + 6H_2O = KAl_3(SO_4)_2(OH)_6 + 6H^+ \quad (2)$$

Table I. Concentrations of Selected Constituents of Acid-Leach Uranium Mill Tailings Pond Raffinates, Grants Mineral Belt, New Mexico. Sampled 1982-1985 (2).
(all concentrations in mg/L unless noted)

| Constituent | Four acid leach mills (14 samples) | | |
|---|---|---|---|
| | Minimum | Median | Maximum |
| Gross radioactivity (pCi/L) | 3200 | 38000 | 73000 |
| Ra-226 (pCi/L) | 15 | 70 | 1800 |
| As | 0.18 | 1.3 | 5.6 |
| Mo | 0.20 | 0.90 | 29.5 |
| Se | 0.006 | 0.21 | 6.97 |
| $SO_4$ | 300 | 29700 | 56000 |
| U | 1.1 | 15 | 69 |
| V | 39.0 | 74 | 107 |
| $NH_3$ | 4.0 | 486 | 4809 |
| TDS | 17900 | 39800 | 72800 |
| pH | 0.30 | 1.05 | 2.15 |

Table II. Analyses of Tailings Pore Water, Maybell, Colorado. Sampled March 17, 1989.
(all concentrations in mg/L unless noted)

| Species and Parameter | Lysimeter 082 | Lysimeter 083 |
|---|---|---|
| Mg | 440 | 310 |
| Ca | 470 | 460 |
| Na | 114 | 118 |
| K | 46 | 61 |
| $SO_4$ | 11100 | 8100 |
| Cl | 13.10 | 21.10 |
| Al | 720 | 350 |
| Mn | 870 | 710 |
| V | 1.95 | 1.15 |
| Fe | 380 | 560 |
| Ni | 1.35 | 1.04 |
| $NO_3$ | 420 | 400 |
| $PO_4$ | 3.3 | 1.30 |
| Cd | 0.03 | 0.03 |
| Mo | 0.10 | 0.05 |
| Se | 0.95 | 1.27 |
| As | 0.08 | 0.09 |
| U | 1.1 | 0.44 |
| TDS | 15300 | 11500 |
| Temp (C) | 17.0 | 17.0 |
| pH | 2.87 | 3.26 |
| Eh (field, mV) | +550 | +480 |

Equation (1) suggests that Fe and $SO_4$ are consumed and H is liberated at a 3:2:6 ratio. Precipitation of jarosite and alunite liberates 6 moles of H, which contributes to the generation of an acidic solution. These reactions depend on the relative concentrations of dissolved Fe and Al produced from minerals undergoing dissolution within tailings pore water. The dominant source of Al, based on scanning electron microscope (SEM) analysis, is probably from Na- and K- feldspars, montmorillonite, and illite. The feldspars have undergone surface reaction dissolution as evidenced by pronounced crystallographically controlled etching (7). Iron concentrations within tailings pore water are the result of dissolution of gangue minerals and the addition of $Fe^{2+}$ from the oxidation of uraninite to $UO_2^{2+}$ during the milling process (4).

THERMODYNAMIC CONSIDERATIONS

The major environmental problem associated with acid-leach tailings is the potential for long-term acid generation and the increased mobility of $NO_3$, $SO_4$, U, Fe, Mn, Cr, and other ions. It is important to understand the geochemistry of the major ions because chemical reactions between the major ions are controlled by precipitation-dissolution processes, which may control the distribution of some trace elements including As, Se, Mo, Cr, and U. Thermochemical data for the system $K_2O$-$Fe_2O_3$-$SO_3$-$SiO_2$-$Al_2O_3$-$O_2$-$H_2O$ were compiled from the literature (Table III) for minerals of the jarosite group, alunite, gibbsite, K-feldspar, and goethite. This system is used to model the chemical evolution of pore waters within tailings impoundments, tailings subsoil, and ground water.

Table III. Equilibrium Constants Used in Calculations for the System $K_2O$-$Fe_2O_3$-$SO_3$-$SiO_2$-$Al_2O_3$-$O_2$-$H_2O$

| Reaction | Log K (298 K) |
|---|---|
| K-feldspar[a] $KAlSi_3O_8 + 4H_2O + 4H^+ = K^+ + Al^{3+} + 3H_4SiO_4$ | 1.29 |
| Gibbsite[a] $Al(OH)_3 = Al^{3+} + 3(OH)^-$ | -34.04 |
| Alunite[b] $KAl_3(SO_4)_2(OH)_6 + 6H^+ = K^+ + 3Al^{3+} + 2SO_4^{2-} + 6H_2O$ | -1.54 |
| Goethite[c] $FeOOH + 3H^+ = Fe^{3+} + 2H_2O$ | 2.25 |
| Jarosite[c] $KFe_3(SO_4)_2(OH)_6 + 6H^+ = K^+ + 3Fe^{3+} + 2SO_4^{2-} + 6H_2O$ | -7.12 |

(References)

[a] Helgeson (16)
[b] Knight (17)
[c] Bladh (18)

The stabilities of alunite, jarosite, gibbsite, goethite, and K-feldspar as functions of activities of $Fe^{3+}$, Al, H, and $SO_4$ in solution are shown in Figure 1. Alunite is a soluble mineral and is typically found in a low pH, high $SO_4$, and Fe-poor environment (6). When the concentrations of dissolved $SO_4$, Al, and K exceed $10^{-3}$ m, alunite is more stable than gibbsite below pH 3. Alunite has been observed in some uranium tailings impoundments in New Mexico (7) and at Maybell, Colorado. Jarosite, however, is more abundant than alunite in most of the uranium tailings impoundments characterized in this investigation. From figure 1, tailings pore waters are characterized by a high ratio of log $(a(Fe^{3+})/a(H^+)^3)$. An increase in $\log(a(SO_4^{2-}) \cdot a(H^+)^2)$, caused by the dissociation of sulfuric acid used in the ore-extraction process, shifts the phase boundaries between alunite and K-feldspar, and between jarosite and K-feldspar to less positive values of log $(a(Al^{3+})/a(H^+)^3)$ and log $(a(Fe^{3+})/a(H^+)^3)$, respectively (Figure 1). Tailings-pore waters from New Mexico apparently are in equilibrium with jarosite, whereas Maybell tailings pore waters are in equilibrium with alunite. Additional analyses of tailings pore water from New Mexico, shown in Figure 1, are from Longmire (7).

Conversely, a median ground water composition (19) consisting of a moderately low TDS (413 mg/L), low concentrations of Al (0.07 mg/L), Fe (0.84 mg/L), and $SO_4$ (36.97 mg/L), a near neutral pH (7.3), and an oxidizing Eh (+164 mV) is defined by relatively high log $(a(Al^{3+})/a(H+)^3)$ (10.4) and log $(a(Fe^{3+})/a(H+)^3)$ (7.6) ratios. Ground water exceeds gibbsite and goethite saturation where alunite and jarosite are predicted to be unstable with a low log $(a(SO_4^{2-}) \cdot a(H^+)^2)$ (-18.2) ratio. Potassium feldspar is also predicted to be the dominant Al-rich mineral stable in subtailings sediments. With reaction development occurring as tailings leachate approaches the median ground water composition, the leachate follows a path on the stability diagram that is away from the alunite and jarosite fields (Figure 1). Values of log $(a(Al^{3+})/a(H^+)^3)$ and log $(a(Fe^{3+})/a(H^+)^3)$ increase when tailings leachate reacts with soil solution or ground water, because the decrease in H activity does not exceed the decrease in $Fe^{3+}$ or Al activities.

Another important system to consider is $CaO-SO_3-CO_2-H_2O$, which includes gypsum and calcite. Gypsum is the most abundant sulfate mineral observed in New Mexico and Colorado uranium tailings impoundments. This observation is in agreement with other investigators who have conducted mineralogical studies at other uranium tailings impoundments in Wyoming (21). Ion speciation and solubility calculations using the geochemical code PHREEQE (20) suggest that Maybell tailings pore water is in near equilibrium with gypsum. The modeling results suggest that gypsum could be precipitating and/or dissolving without kinetic constraints to control concentrations of Ca and $SO_4$. Precipitation of gypsum results when leachate containing sulfuric acid reacts with calcium carbonate, and is given by the following overall reaction:

$$H_2O + CaCO_3 + H_2SO_4 = CaSO_4 \cdot 2H_2O + CO_2(g) \qquad (3)$$

Precipitation of gypsum, and possibly barite, in tailings pore water and tailings subsoil may result in removal of other group II metals, including Ba and Ra through coprecipitation (22).

RESULTS OF LEACH STUDIES

Results of batch leach tests performed on tailings from the Grants Mineral Belt, New Mexico are summarized in Tables IV and V. These results represent the average of two analyses on the acid leach tailings samples. Table IV contains the elemental concentrations of the sand and silt and clay-sized fractions, as well as the soluble concentration of each element found when 100 g of material is leached with 250 mL of sodium hexametaphosphate solution. The right-most column presents the ratio of the elemental concentration associated with the silt and clay-sized fraction to that of the sand-sized fraction, and it is presented as an enrichment factor. The high enrichment factors for As, Mo, Cr, Se, and V may be due to the fact that these elements are adsorbed onto jarosite, montmorillonite, and ferric oxyhydroxides (9-14). Speciation

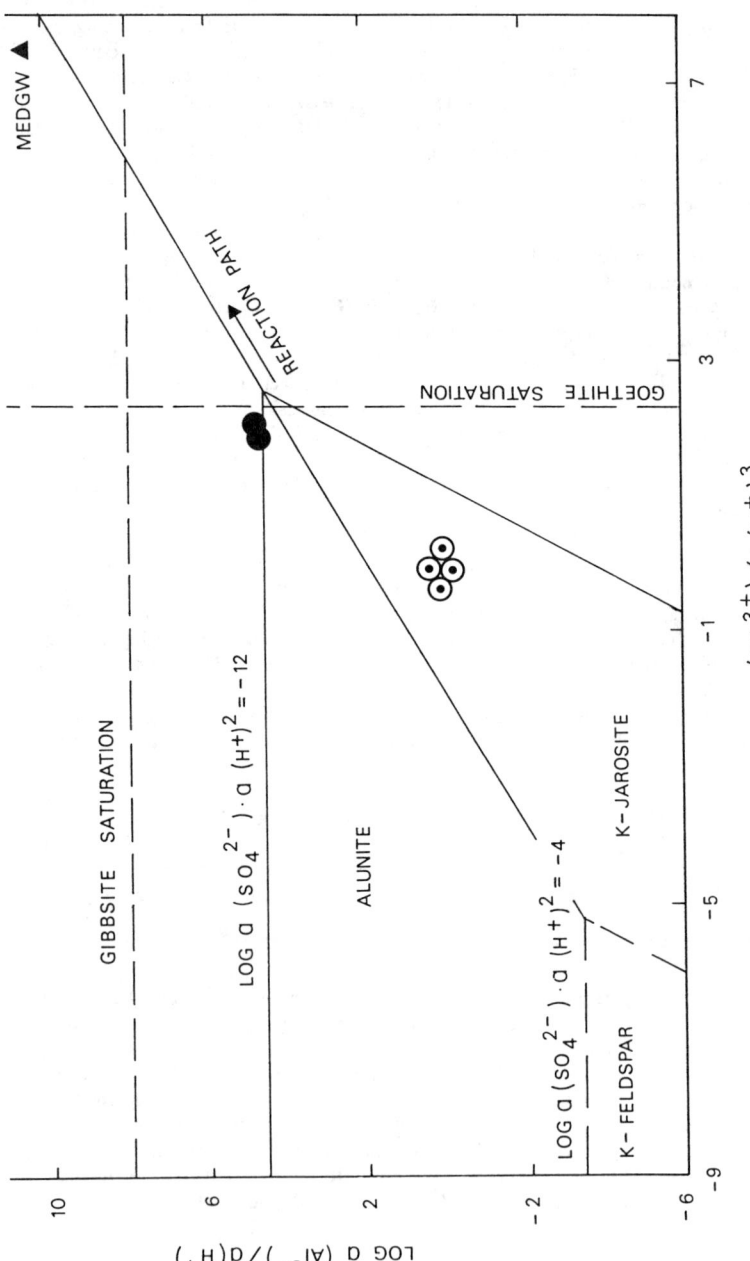

Figure 1. Activity diagram of jarosite-alunite-potassium feldspar-gibbsite-goethite system at 1 bar and 298 K. Modified from Bladh (18). Tailings pore water from Grants Mineral Belt, New Mexico and Maybell, Colorado are denoted by ⊙ and ●, respectively. Median ground water composition (19) is represented by ▲.

Table IV. Average Concentrations in Each Fraction of Acid Leach Uranium Tailings by INAA[1], Grants Mineral Belt, New Mexico (2 Samples)

| Element | Sand (ppm) | Silt and Clay (ppm) | Water[2] (mg/L) | Enrichment Clay/Sand |
|---|---|---|---|---|
| As | 2.51 | 20.44 | 0.17 | 8.2 |
| Cr | 9.88 | 157.00 | 0.01 | 15.9 |
| Fe | 2387.00 | 30370.00 | 11.50 | 12.7 |
| Mo | 5.39 | 216.00 | 0.50 | 40.1 |
| Se | 8.17 | 124.00 | 0.13 | 15.2 |
| Th | 1.17 | 2.49 | 0.00 | 2.1 |
| U | 19.04 | 118.00 | 0.10 | 6.2 |
| V | 169.00 | 954.00 | 0.69 | 5.6 |
| pH(1.2) | | | | |
| Eh(+600 mV) | | | | |

[1] Instrumental neutron activation analysis.
[2] Concentrations in 250 mL of solution containing 100 g of acid leach uranium tailings.

Table V. Distribution of the Total Mass of Each Element from a 100 g Sample of Acid Leach Uranium Tailings, Grants Mineral Belt, New Mexico

| Element | Sand (g) | Silt and Clay (g) | Soluble (g) |
|---|---|---|---|
| As | 1.88x10-04 | 5.11x10-04 | 5.20x10-05 |
| Cr | 7.41x10-04 | 3.93x10-03 | 0.00 |
| Fe | 0.1791 | 0.7593 | 0.0035 |
| Mo | 4.04x10-04 | 5.40x10-03 | 0.00 |
| Se | 6.13x10-04 | 3.10x10-03 | 3.88x10-05 |
| Th | 8.79x10-05 | 6.22x10-05 | 0.00 |
| U | 1.43-03 | 2.96x10-03 | 3.08x10-05 |
| V | 1.27x10-02 | 2.39x10-02 | 2.03x10-04 |

calculations using PHREEQE (20) show that these solutes are stable as $H_2AsO_4^-$, $MoO_4^{2-}$, $Cr^{3+}$, $HSeO_3^-$, and $VO^{2+}$, respectively. At a pH value of 1.2, the adsorbents are characterized by a net-positive surface charge (montmorillonite, $pH_{zpc}$ 2.5; $Fe(OH)_3$ am, $pH_{zpc}$ 8.5; $SiO_2$ gel, $pH_{zpc}$ 1.0-2.5) enhancing anion adsorption (23-25).

Using the data from Table IV, together with the results of the particle size analysis, the mass of each element associated with each fraction of a 100 g soil sample is presented in Table V. Thus, in a 100 g sample of tailings there was 0.179 g of Fe in the sand-sized fraction, 0.759 g in the silt and clay-sized fraction, and 0.003 g in aqueous solution (Table V). The leach experiments show the decreasing order of soluble constituents listed in Table V is Fe > V > As > Se > U > Th = Mo = Cr at pH 1.2. All of the above elements show higher distributions on the silt and clay-sized material, except for Th which is distributed throughout all fractions. Thorium has the smallest enrichment factor (Table IV), which may be due to the occurrence of detrital rather than adsorbed Th.

Column leach tests were performed to assess the mobility of elements concentrated in the Maybell tailings. Three columns were loosely packed with composite tailings to increase the movement of solution through the columns. This resulted in a bulk density of 1.26 g cm$^{-3}$ for the tailings. Deionized water was used as the influent for the column leach tests, where one pore volume was passed through every 30 hours. The recycled solution percolated through each corresponding column two more times for a total of three pore volumes. Element concentrations, pH, and temperature values were measured for the third pore volume. The final effluent was passed through a 0.45 $\mu$m membrane filter membrane, separated into three aliquots, and preserved according to the type of analysis required. Results of the column leach tests are shown in Table VI. The concentrations and mobilities of several species and elements generally follow the order $SO_4 > NH_4 > Al > Mn > NO_3 > U > Fe > Se > PO_4 > Ni > As > Cd$ at pH 4.0. The mobility of these elements after they are released by sulfuric acid dissolution is controlled by precipitation-dissolution reactions, coprecipitation reactions, solid solution substitutions, and adsorption-desorption reactions.

Table VI. Results of Column Leach Tests on Tailings Samples, Maybell, Colorado[1]
(all concentrations in mg/L unless noted)

| Element | Minimum | Average | Maximum |
| --- | --- | --- | --- |
| Al | 49.2 | 54.2 | 59.4 |
| As | 0.035 | 0.038 | 0.045 |
| Cd | 0.0017 | 0.0020 | 0.0023 |
| Fe | 0.78 | 0.94 | 1.11 |
| Mn | 34.8 | 43.3 | 49.0 |
| Mo | <0.001 | <0.001 | <0.001 |
| $NH_4$ | 131 | 142 | 151 |
| Ni | 0.11 | 0.13 | 0.15 |
| $NO_3$ | 5.3 | 5.9 | 6.2 |
| $PO_4$ | 0.2 | 0.3 | 0.3 |
| Se | 0.7 | 0.7 | 0.8 |
| $SO_4$ | 2371 | 2448 | 2511 |
| U | 1.83 | 2.72 | 3.25 |
| V | <0.01 | <0.01 | <0.01 |
| pH | 3.31 | 4.00 | 5.37 |
| Temperature (C) | 23 | 23 | 23 |

[1]Triplicate analyses

## IMPACT OF TAILINGS LEACHATE ON SOIL, MAYBELL, COLORADO

At Maybell, Colorado, tailings leachate has migrated through tailings subsoil and the vadose zone of the Brown Park Formation. The Browns Park Formation of late Tertiary age consists of unconsolidated material with minor amounts of carbonaceous matter, pyrite, and calcite (26). Bulk mineralogy of these sediments consists of quartz, Na-feldspar, K-feldspar, montmorillonite, and illite. Depths to unconfined ground water vary from 18 m beneath the tailings impoundment, to over 91 m southwest of the impoundment. Samples from tailings subsoil and the Browns Park Formation were collected to evaluate the geochemical impact of tailings leachate on subsurface materials.

Table VII lists the physical and chemical properties of tailings subsoil and the underlying Browns Park Formation. The sediments were characterized using standard soil-testing procedures (27,28). The effects of acidic leachate on the Browns Park Formation are apparent. The tailings subsoil has an average pH value of 3.75, the measured Eh is relatively oxidizing, the $CaCO_3$ content is depleted, and the acid neutralizing capacity has been exhausted. Precipitation of gypsum has occurred beneath the tailings impoundment. Acid-soluble Fe is more abundant in the tailings subsoil, compared to the Browns Park Formation, which indicates that Fe, possibly in the form of ferric oxyhydroxide, has precipitated from solution during the neutralization of tailings leachate. The cation exchange capacity is similar for both sample types, and the clay minerals include illite (15 percent) and Ca-montmorillonite (85 percent). Calcium is the predominant exchangeable cation, suggesting that the bulk of the montmorillonite may be Ca saturated. Exchangeable $NH_4$ was less than 0.01 meq/100g, indicating that $NH_4$ derived from tailings leachate is not attenuated by cation exchange reactions, which is supported by the column leach experiments. Some $NH_4$ could reprecipitate in jarosite. The anion exchange capacity of the sediments ranged between 0.5 and 5.7 meq/100g.

## GEOCHEMICAL MODELING

Ion-speciation and solid-phase solubility calculations have been widely used to develop a better understanding of geochemical processes controlling the chemical composition of natural waters and waters contaminated from anthropogenic sources. In this investigation, geochemical modeling involving ion-speciation and solubility calculations was compared to observed mineralogical assemblages to determine if precipitation reactions were thermodynamically possible.

The geochemical code PHREEQE (20) was used in this analysis. Tailings pore water was equilibrated with jarosite, gypsum, and jurbanite under open conditions ($PCO_2$ $10^{-3.5}$) to evaluate the long-term acidity within the tailings impoundment. The program solves simultaneous equations describing the equilibrium-chemical reactions that may occur in a given water. From the input of the solution analyses, PHREEQE computes the activities of complexed and free ionic species, neutral ion pairs, and the distribution of ionic species. The model then calculates ion activity products (IAP) and compares the ion activity products to the solubility products ($K_T$) for the minerals and solid compounds contained in the data base. The relative degree of saturation is measured by the saturation index (SI) (29), which is defined as the $\log_{10}$ (IAP/KT).

For meaningful model simulations, it is important that the thermochemical data base is accurate and internally consistent. The data base contained in PHREEQE is the same as the data base developed for WATEQ2 (30). This data base has been critically reviewed (31) and compared with WATEQFC (32). The values of Log K for solids included in this simulation, along with appropriate reactions from which they were determined, are listed in Table VIII.

Table VII. Characterization of Tailings Subsoil and the Browns Park Formation, Maybell, Colorado[1]

| Property | Tailings Subsoil | Browns Park Formation |
|---|---|---|
| Water content (g/g)(%) (after air drying) | 0.35 | 0.08 |
| Particle density (g/cm3) | 2.63 | 2.63 |
| Particle size distribution (weight %) | | |
| Sand (50-2000 um) | 13.0 | 18.0 |
| Silt (2-50 um) | 72.0 | 66.0 |
| Clay (< 2 um) | 15.0 | 16.0 |
| pH of saturated paste | 3.75 | 7.41 |
| Eh of saturated paste (millivolts) | +520 | +276 |
| Organic matter (g/g)(%) | 0.05 | 0.22 |
| Acid-soluble iron (g/g)(%) | 1.51 | 0.16 |
| $CaCO_3$ (g/g)(%) | 0.04 | 25.5 |
| Gypsum (g/g)(%) | 1.26 | <0.01 |
| Acid neutralizing capacity (tons $CaCO_3$ equivl/1000 tons) | -0.84 | 320.0 |
| Cation exchange capacity (CEC)(meq/100g) | 13.4 | 13.8 |

[1]Average of six measurements for tailings subsoil and Browns Park Formation.

Table VIII. Equilibrium Constants Used in PHREEQE for Several Controlling Minerals in Tailings Pore Water, Maybell, Colorado

| Reaction | Log K (298 K) |
|---|---|
| Jarosite $3e^- + KFe_3(SO_4)_2(OH)_6 + 6H^+ = K^+ + 3Fe^{2+} + 2SO_4^{2-} + 6H_2O$ | 24.30 |
| Jurbanite $AlOHSO_4 + H^+ = Al^{3+} + SO_4^{2-} + H_2O$ | -3.23 |
| Gypsum $CaSO_4 \cdot 2H_2O = Ca^{2+} + SO_4^{2-} + 2H_2O$ | -4.85 |

Dissolved concentrations from lysimeter sample 082 (Table II) were used as input for the modeling simulations, and the results are presented in Table IX. Hematite exhibits a high degree of oversaturation, and tailings pore water is moderately oversaturated with respect to jarosite, goethite, and alunite. These solubility calculations for jarosite, goethite, and alunite indicate that attainment of equilibrium is probably inhibited by kinetic and/or other constraints. The modeling results indicate that jurbanite, gypsum, lepidocrocite, and strengite are in near equilibrium, and these minerals could be controlling the concentrations of some of their dissolved components. The solubility calculations using the PHREEQE code for tailings pore water are in reasonable agreement with minerals identified by XRD analysis. Mineralogical characterization studies verify that gypsum is the dominate $SO_4$ mineral controlling the dissolved concentrations of Ca and $SO_4$ for the Maybell tailings pore water. The solution is undersaturated with respect to ferrihydrite, amorphous $Fe(OH)_3$, and gibbsite. The ferric ion may first precipitate as $Fe(OH)_3$ and poorly crystalline goethite. These more soluble compounds preferentially dissolve and reprecipitate into crystalline lepidocrocite and goethite, which are thermodynamically more stable. Speciation calculations using PHREEQE show that the dominant free ions and complexes for Al, Fe, Mn, Mo, Se, As, and U include $Al(SO_4)_2^-$, $Fe^{2+}$, $Mn^{2+}$, $MoO_4^{2-}$, $HSeO_3^-$, $H_2AsO_4^-$, and $UO_2(SO_4)_2^{2-}$, respectively (Table X).

Table IX. Saturation Indices (SI = log $(IAP/K_T)$) for Acid Leach Tailings Raffinate, Maybell, Colorado. Computed by PHREEQE ([20])

| Mineral phase | Formula | Acid leach Raffinate SI |
|---|---|---|
| Jarosite | $KFe_3(SO_4)_2(OH)_6$ | 1.97 |
| Hematite | $Fe_2O_3$ | 7.41 |
| Lepidocrocite | $\gamma$-FeO(OH) | 0.66 |
| Alunite | $KAl_3(SO_4)_2(OH)_6$ | 1.89 |
| Gypsum | $CaSO_4 \cdot 2H_2O$ | 0.38 |
| Strengite | $FePO_4 \cdot 2H_2O$ | 0.96 |
| Ferrihydrite | $5Fe_2O_3 \cdot 9H_2O$ | -2.88 |
| Goethite | $\alpha$-FeO(OH) | 1.22 |
| $Fe(OH)_3$ amorphous | $Fe(OH)_3$ | -0.66 |
| Jurbanite | $AlOHSO_4 \cdot 5H_2O$ | 0.96 |
| Gibbsite | $Al(OH)_3$ | -3.98 |

These results show how geochemical modeling can be used with mineralogical studies to evaluate chemical reactions occurring in tailings pore water and subjacent soils. The identification of gypsum, jarosite, and alunite support the predictions made by PHREEQE that these secondary minerals precipitated from tailings pore water.

Table X. Speciation Calculations for Tailings Pore Water, Maybell, Colorado. Computed by PHREEQE (20)

| Element | Computed Species | Log Molality | Molal Percentage |
|---------|------------------|--------------|------------------|
| Al | total | -1.57 | |
|    | $Al(SO_4)_2^-$ | -1.97 | 39.81 |
|    | $AlSO_4^+$ | -2.03 | 34.67 |
|    | $Al^{3+}$ | -2.16 | 25.70 |
| Fe | total | -2.16 | |
|    | $Fe^{2+}$ | -2.39 | 58.88 |
|    | $FeSO_4^0$ | -2.56 | 39.81 |
| Mn | total | -1.79 | |
|    | $Mn^{2+}$ | -2.03 | 57.54 |
|    | $MnSO_4^0$ | -2.17 | 41.69 |
| Mo | total | -5.97 | |
|    | $MoO_4^{2-}$ | -5.97 | 100.00 |
| U  | total | -5.33 | |
|    | $UO_2(SO_4)_2^{2-}$ | -5.69 | 43.65 |
|    | $UO_2SO_4^0$ | -5.87 | 28.84 |
|    | $UO_2^{2+}$ | -6.14 | 15.49 |
|    | $UO_2(HPO_4)_2^{2-}$ | -6.29 | 10.96 |
| Se | total | -4.91 | |
|    | $HSeO_3^-$ | -5.05 | 72.44 |
|    | $H_2SeO_3$ | -5.48 | 26.92 |
| As | total | -5.97 | |
|    | $H_2AsO_4^-$ | -6.02 | 89.13 |
|    | $H_3AsO_4^0$ | -6.88 | 11.30 |

Additional simulations were made with PHREEQE in which Maybell tailings-pore water was set in equilibrium with jarosite, gypsum, and jurbanite at constant Eh in an open system ($PCO_2$ $10^{-3.5}$) (Table VIII). Jurbanite was selected as a solid controlling dissolved Al concentrations because the SI value for this mineral was calculated to be closer to equilibrium (SI = 0.96) than the SI value for alunite (SI = 1.89) (Table IX). In addition, the Maybell data plot close to the jurbanite stability line defined by a pAl + 3 pH vs 2 pH + p $SO_4$ diagram presented by Nordstrom (8). The calculated pH value of 2.17 for the equilibrated solution was slightly lower than the measured pH value of 2.87 for lysimeter 082. The lower pH value results from the addition of protons to solution through the reactions listed in Table VIII. The solution is undersaturated with calcite (SI = -11.89) under open-system conditions, suggesting that the long-term buffering capacity of the gangue material in the tailings impoundment is exhausted. Because large amounts of acidic-tailings pore water have already migrated from the impoundment into the tailings subsoil and Browns Park Formation (Table VII) with concurrent depletion of the acid neutralizing capacity, continued depletion of the carbonate buffer appears to be likely in the future. In deeper sections of the Browns Park Formation, the carbonate buffer has not been exhausted. As the acid neutralizing capacity decreases, the acidic tailings leachate will migrate through the vadose zone. Dissolved species including $SO_4$, $NO_3$, Cl, and U have migrated to the water table. Ground-water monitoring data collected at Maybell, Colorado confirm the mobility of these four solutes and in addition to other major ions.

CONCLUSIONS

1. Significant soil contamination has resulted from seepage from uranium tailings impoundments in the United States and Canada. Pore waters associated with tailings commonly are highly acidic, relatively oxidizing, and contain very high concentrations of TDS, $NO_3$, Fe, As, Mo, Se, U, V, and major ions.

2. Results of batch leach experiments on tailings samples have shown that As, Se, Mo, Cr, and V, stable as neutral ion pairs, ions, and oxyanions, are enriched in the silt and clay-sized fraction. This enrichment may contribute to a contaminant source lasting for several decades. From column leach experiments, the concentrations and mobilities of several species and elements follow the order $SO_4 > NH_4 > NO_3 > Al > Mn > U > Fe > Se > PO_4 > Ni > As > Cd$ at pH 4.0.

3. Results of geochemical modeling suggest that the tailings pore water is in near equilibrium with respect to strengite, jurbanite, lepidocrocite, and gypsum, and is oversaturated with respect to jarosite, alunite, and goethite. Those minerals in equilibrium control concentrations of Al, Fe, K, $SO_4$, $PO_4$ and Ca in acidic, sulfate-rich tailings impoundments.

4. A major geochemical process occurring in tailings subsoil includes precipitation-dissolution reactions with calcite and gypsum as tailings leachate becomes neutralized. X-ray diffraction analysis, SEM studies, and soil characterization investigations confirm the depletion of acid neutralizing capacity in tailings subsoil at Maybell, Colorado. This acid neutralizing capacity, however, has not been exhausted in the underlying Browns Park Formation aquifer.

ACKNOWLEDGMENTS

Partial funding for the water analysis was provided by the New Mexico Energy Research and Development Institute under contract no. EMD-2-69-1107 and partial funding for the soil collection and analysis were provided by the U.S. Bureau of Mines, Spokane Research Center, contract no. J0225002. Dr. William Downs and Mr. Armando Groffman conducted the column tests on the Maybell tailings. Assistance in collection of the pore-water samples was provided by the Jacobs Engineering Group personnel. Dr. Bimal Mukhopadhyay critically reviewed an earlier draft of the paper. Ms. J. Binder and Ms. R. Landers prepared the manuscript.

LITERATURE CITED

1. Crawford, M. Science 1985, 229, 537-538.
2. Thomson, B. M.; Longmire, P.; Brookins, D. G. Applied Geochemistry 1986, 1, 335-343.
3. Morin, K. A.; Cherry, J. A.; Dav, N. K.; Lim, T. P.; Vivyurka, A. J. Jour. of Contaminant Hydrology 1988, 2, 305-322.
4. Merritt, R. C. The Extractive Metallurgy of Uranium: Colorado School of Mines: Golden, Colorado, 1971, p. 576.
5. Lindberg, R. D.; Runnells, D. D. Science 1984, 225, 925-927.
6. Ball, J. W.; Nordstrom, D. K. Water Resource Investigation No. 85-4169: U.S. Geological Survey: Denver, CO, 1985.

7. Longmire, P. M.S. Thesis, University of New Mexico, New Mexico, 1983, p. 182.
8. Nordstrom, D. K. Geochim. Cosmochim. Acta. 1982, 46, 681-692.
9. Rai, D.; Zachara, J. M. Electric Power Research Institute, Report No. EA-3256, Palo Alto, California, 1984.
10. Borovec, Z. Chem. Geol. 1981, 32, 45-58.
11. Tripathy, V. J. Ph.D. Thesis, Stanford University, California, 1984. p. 297.
12. Hsi, C. D.; Langmuir, D. Geochim. Cosmochim. Acta. 1985, 49, 1931-1941.
13. Benjamin, M. M.; Bloom, N. S. In Adsorption from Aqueous Solutions; Plenum Press: New York, 1981; Chapter 3.
14. Leckie, J. O.; Benjamin, M. M.; Hayes, K.; Kaufman, G.; Altmann, S. Electric Power Research Institute Report No. CS-1513, Palo Alto, California, 1980. p. 197.
15. van Breeman, N. Spec. Publ. 10. SSSA 1982, 95-108.
16. Helgeson, H. C. Am. Jour. Sci., 1969, 267, 155-186.
17. Knight, J. E. Econ. Geol., 1977, 72, 1321-1336.
18. Bladh, K. W. PhD Thesis, University of Arizona, Arizona, 1978, p. 98.
19. Lindberg, R. D. Ph.D Thesis, University of Colorado, Colorado, 1983.
20. Parkhurst, D. L.; Thorstenson, D. C.; Plummer, L. N. Water Resource Investigation No. 80-96; U.S. Geological Survey: Denver, CO, 1980, 210.
21. Peterson, S. R.; Serne, R. J.; Felmy, A. R.; Erikson, R. L.; Gee, G. W. Proc. 1st Can./Amer. Conf. on Hydrogeology 1985, p 211.
22. Langmuir, D.; Melchior, D. Geochim. Cosmochim. Acta. 1985, 49, 2423-2432.
23. Parks, G. A. Chem. Rev. 1965, 65, 177-198.
24. Leckie, J. O.; James, R. O. In Aqueous-Environmental Chemistry of Metals: Ann Arbor Science Publishers: Michigan, 1974; Chapter 1.
25. James, R. O.; MacNaughton, M. G. Geochim. Cosmochim. Acta. 1977, 41, 1549-1555.
26. Chenoweth, W. L. Rocky Mountain Association of Geologists, 1986, 289-292.
27. Black, C. A. Methods of Soil Analysis. Part 1. Physical and Mineralogical Properties Including Statistics of Measurement and Sampling; American Society of Agronomy, Monograph 9. p. 770.
28. Black, C. A. Methods of Soil Analysis, Part 2. Chemical and Microbiological Properties; American Society of Agronomy, Monograph 9. p. 1572.
29. Barnes, I.; Clark, F. E. Professional Paper No. 498-D; U.S. Geological Survey: Denver, CO, 1969. p. 8.
30. Ball, J. W.; Nordstrom, D. K.; Jenne, E. A. Water Resource Investigation No. 78-116; U.S. Geological Survey: Denver, CO, 1980. p. 109.
31. Noronha, C. J.; Pearson, F. J., Jr.; INTERA Inc. report, 1983.
32. Runnells, D. D.; Lindberg, R. D. J. Geochem. Exp., 1981, 15, 37-50.

RECEIVED August 31, 1989

Chapter 13

# Geochemistry of Organic Acids in Subsurface Waters

## Field Data, Experimental Data, and Models

Paul D. Lundegard[1] and Yousif K. Kharaka[2]

[1]Unocal Science and Technology, 376 South Valencia Avenue, Brea, CA 92621
[2]Water Resources Division, U.S. Geological Survey, Mail Stop 427, 345 Middlefield Road, Menlo Park, CA 94025

>   Integration of field observations, experimental data, and computer simulations provides valuable insights into the origin and inorganic interactions of organic acids and organic acid anions in subsurface waters. Present-day concentrations show significant variations with respect to temperature, basin, and reservoir age, reflecting complex interaction of biological, geological, and geochemical factors that influence the production or destruction of these species. Hydrous pyrolysis experiments demonstrate that most dissolved organic acids are produced by thermal degradation of kerogen, and that variation in kerogen chemistry can cause relative generating capacities to vary by a factor of two or more between samples of equal thermal rank. A mathematical model of the kinetics of organic acid generation from kerogen shows that variations in thermal gradients can cause significant variations in maximum organic acid concentrations, and in trends of organic acid concentrations with depth and temperature. Experimental and field studies of acetate decarboxylation rates indicate that they are strongly affected by temperature and catalysis. In nature, halflives of acetate are on the order of 20 to 60 million years at 100° C. Geochemical models predict that major organic acid species 1) are important proton sources in subsurface waters, 2) buffer pH, but increases in $pCO_2$ are still likely to favor calcite undersaturation in most subsurface waters, 3) do not form important complexes with Pb, Zn, and U, and 4) do form important complexes with Fe, Ca and possibly Al.

In recent years there has been considerable interest in the origin and significance of dissolved organic acids in natural subsurface waters (1). Geochemical interest in these species stems from their potential influence as proton donors for a variety of pH-dependent reactions, as pH buffering agents, and as complexers of metals such as aluminum, iron, lead, and zinc (2-7). This paper reviews the occurrence of dissolved organic acids in subsurface waters, and discusses their origin and geochemical significance. Present-day concentrations of organic acids are influenced by factors that control the competing processes of organic acid production and destruction. Together, field and experimental data allow identification of the important factors and provide insight into the processes they control. A kinetic model of organic acid production from kerogen is described which illustrates the impact of thermal and burial history on patterns of organic acid concentrations. Finally, several examples of geochemical speciation and reaction path models are given which illustrate the importance of organic acids to specific diagenetic processes.

## OCCURRENCE OF DISSOLVED ORGANIC ACIDS

Data on the concentration of dissolved organic acids in natural formation water come predominantly from analysis of samples from Cenozoic reservoirs (Fig. 1), although sampled strata range in age from Permo-Triassic to Pleistocene. These data show that: 1) concentrations in subsurface waters are highly variable, both for a given temperature range and age, 2) concentrations greater than 3000 mg/l (0.05 m) acetate equivalent are apparently rare, an important observation for geochemical modeling, and 3) maximum concentrations vary with temperature, generally showing a peak in the range 80-120°. Studies of individual basins show that average organic acid concentration in the temperature range 80° - 120° C is related to the age of the reservoir sampled. While variations in source material influence concentrations, this relationship between geologic time and organic acid concentration is apparently caused by the kinetics of organic acid decarboxylation and will be discussed later.

As shown by figure 1, maximum organic acid concentration increases from less than a few hundred mg/l at temperatures less than 40° C, to a peak of a few thousand mg/l at approximately 100° C, and then declines rapidly again with further increases in temperature. Available data suggest that maximum concentrations probably do not exceed 100 mg/l at temperatures greater than 200° C. Water samples from two geothermal wells (temperatures greater than 250° C) at Salton Sea, California, showed no detectable organic acids (Y.K. Kharaka, unpublished data).

Organic acid anions can dominate the total alkalinity of subsurface waters, and the anions of short-chain aliphatic acids contribute the great majority of the organic alkalinity. Acetate is by far the most important individual species (4, 8-10). The abundance of other aliphatic acids generally decreases with increasing carbon number (propionate > butyrate > valerate).  In low temperature waters (less than 80° C), where bacterial alteration has presumably been active and total organic acid concentrations are low, propionate is sometimes more abundant than acetate (8, 11). Recent data indicate that propionate dominance at low temperature is less common than previously thought (5, 10, 12, 13-14) and may be a product of primary organic acid generation (14). At temperatures less than 80° C, the low and variable concentrations of organic acids are controlled by several factors. These include 1) bacterial consumption of organic acids, especially acetate (8), 2) reduced generation of organic acids, and 3) dilution by mixing with meteoric water (10, 13).

Few data are available on the concentration of dicarboxylic acid anions in subsurface waters. $C_2$ through $C_{10}$ saturated acid anions have been reported in addition to maleic acid (*cis*-butenedioic acid) (5, 15-16). Oxalic acid (ethanedioic) and malonic acid (propanedioic) appear to be the most abundant. Reported concentrations range widely from 0 to 2540 mg/l but mostly are less than a few 100 mg/l. Concentrations of these species in formation waters are probably limited by several factors, including the very low solubility of calcium oxalate and calcium malonate (5), and the susceptibility of these dicarboxylic acid anions to thermal decomposition (16). This paper will focus on the monocarboxylic acids because they are much more abundant and widespread, and stability constants for their complexes with metals are better known.  We do recognize that dicarboxylic acid anions may be locally important, especially for complexing metals.

## ORIGIN OF MAJOR SPECIES

### Field Data.

Variations in the organic acid concentrations of diverse waters with respect to temperature suggest a strong thermal influence (Fig. 1). In Cenozoic age reservoirs maximum organic acid

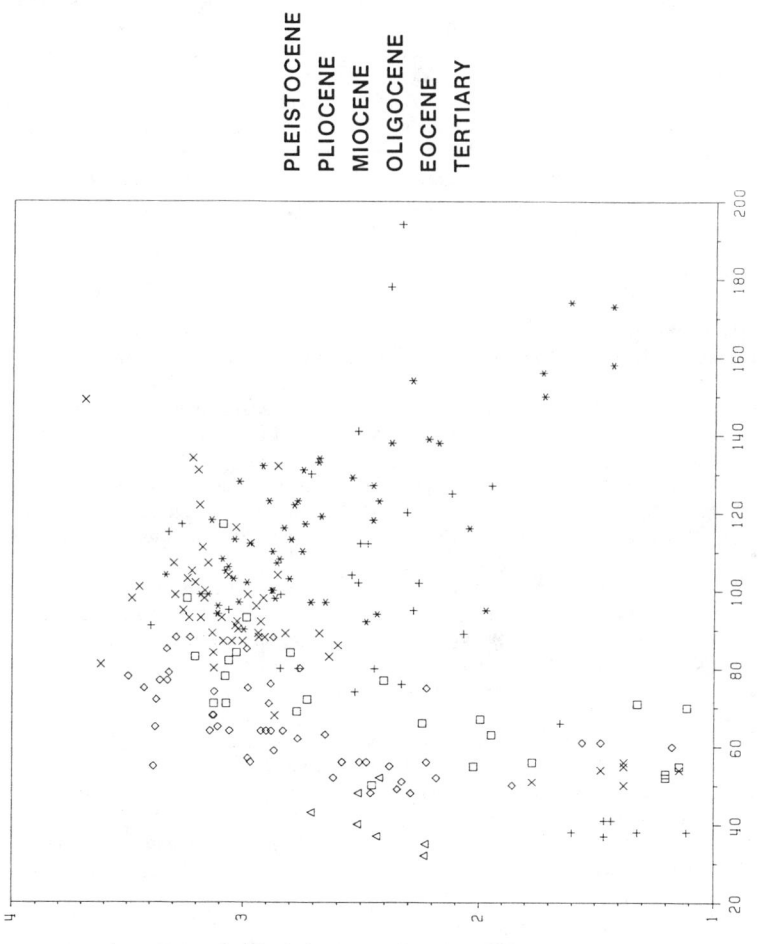

Figure 1. Compilation of organic alkalinity analyses (n=263) for waters produced from Cenozoic reservoirs versus reservoir temperature (5, 8-10).

concentrations consistently occur in the temperature range 80° to 120° C. If, as is conventionally believed, bacteria can not survive in the subsurface at temperatures greater than approximately 80° C (17), the distribution of organic acids with respect to temperature suggests that bacterial processes are not important in their production. Bacterial processes dependent on meteoric water invasion for an energy source or for transport into the subsurface appear unimportant in organic acid production. Many waters with high organic acid concentrations are produced from deep overpressured reservoirs, where oxygen isotopic data indicate no mixing with meteoric water has occurred (9). Fisher (10) reported that in the Pliocene of the Eastern Venezuelan Basin total aliphatic acid concentrations are lower in waters associated with biodegraded oils). High organic acid concentrations found in association with non-biodegraded oils suggest that biodegradation is not a major acid-producing process. Carothers and Kharaka argued that at temperatures less than 80° C, bacterial consumption of acetate is an active process because in that environment propionate sometimes shows a dominance over acetate (8). Fisher, however, found no propionate dominance in waters associated with biodegraded oils in the Eastern Venezuelan Basin (10). Organic acid concentration in formation water also bears no relationship to whether samples are collected from oil or gas wells. Field data show, therefore, that thermal energy is required to produce the major organic acid species. Biological processes are unimportant in the production of organic acids but are important in their destruction at temperatures less than 80° C.

Experimental Data.

Laboratory experiments provide convincing evidence for a thermal origin for the major dissolved organic acids. Heating kerogen or rocks containing kerogen in the presence of water under laboratory conditions is commonly referred to as hydrous pyrolysis (18). While experimental artifacts exist (14), due to the catalytic effects of reactor materials and the high temperatures commonly used to reduce reaction times, hydrous pyrolysis has proven a useful method for studying the general controls on organic acid synthesis (14, 19).

Thermally immature kerogen contains appreciable oxygen, from about 8 to 25 weight percent, depending on type (20). Variations in the atomic O/C ratios of kerogen and its infrared adsorption spectra as a function of thermal maturity (Fig. 2), indicate that oxygen is liberated from kerogen during the thermal stress associated with burial (21, 22). Most of the oxygen is liberated at thermal ranks lower than those needed for major hydrocarbon generation. Further, carboxyl and carbonyl groups are the oxygen functions that are liberated earliest. Hydrous pyrolysis experiments demonstrate that low molecular weight organic acids and carbon dioxide are among the oxygen-containing products of thermal degradation of kerogen (14, 15, 19).

Unless total organic acid yields are specifically of interest, experimental data are most usefully normalized to the amount of organic matter in the unreacted rock. Quantitative comparison of the organic acid generating capacities of different types of organic matter further requires that thermal maturity be held constant. This requirement is not always easy to meet when comparing samples from different localities. Small differences in thermal maturity can result in substantial differences in organic acid generating capacity (Fig. 2).

In experiments with terrestrial mudrocks (250°-350° C, 72 h), Lundegard and Senftle (14) produced $C_1$ through $C_4$ aliphatic acids, but no oxalic acid. These experiments show that as much as 1.2 weight % of the organic carbon in kerogen of low thermal rank can be converted to short-chain aliphatic acids. Acetic and propionic acids were the dominant acids produced. Propionic was the more abundant acid in the lower temperature experiments and acetic was the more abundant acid in the higher temperature experiments (Fig. 3). Propionic acid was destroyed in higher temperature experiments, while acetic acid yields simply increased.

Initial studies suggest that higher relative yields of low molecular weight organic acids correlate with high vitrinite content and Rock-Eval oxygen indices, which are proportional to kerogen oxygen content. However, these bulk geochemical parameters do not account for all the variations observed in hydrous pyrolysis experiments (Table I; 14). Further, some samples

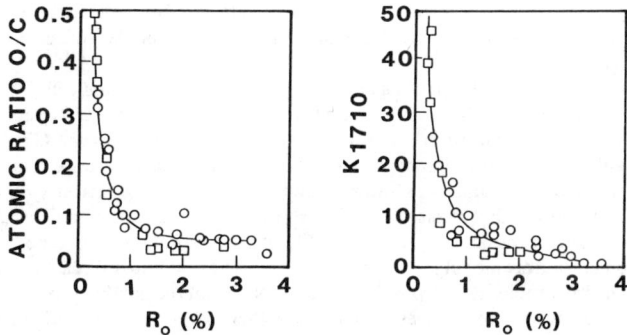

Figure 2. Plots of atomic O/C ratio and the infrared adsorption peak characteristic of carboxyl and carbonyl groups in Type III kerogen (circles) and coals (squares) as a function of vitrinite reflectance. (Reproduced with permission from Ref. 22. Copyright 1980 Editions Technip).

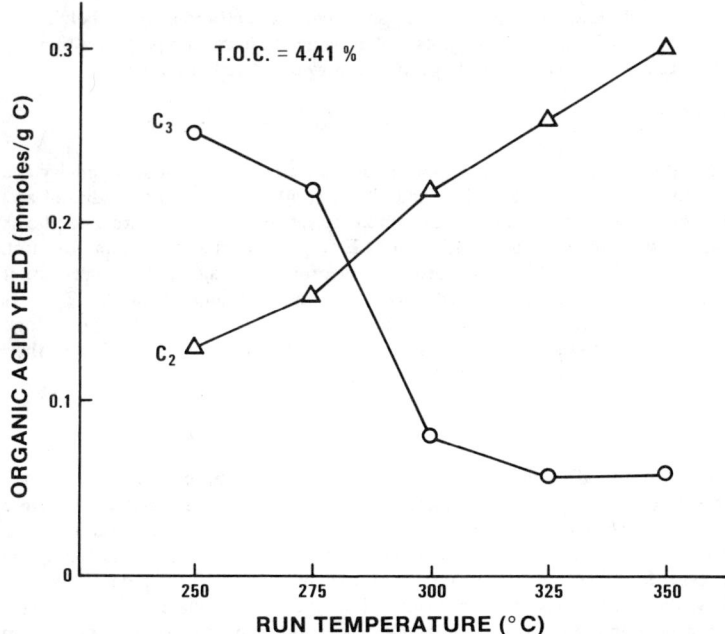

Figure 3. Relative yields of acetate ($C_2$) and propionate ($C_3$) for (72 h) hydrous pyrolysis experiments with a terrestrial mudrock containing Type II and Type III kerogen. (Reproduced with permission from Ref. 14. Copyright 1987 Pergamon Journals).

like the Kishenehn oil shale (samples 3-1, 3-2, and 3-3 in Table I) generate abundant organic acids in spite of low oxygen indices. Evidently, the oxygen present in these samples has a strong tendency to be liberated as acids. The low relative yield of the Monterey sample in Table I (sample 2-1) suggests that the anomalously high concentrations of organic acids in California oil field waters (8) are controlled more by the high organic content of the source rocks than the quality of the organic material.

Experiments with closely spaced samples in a single core show that relative organic acid generating capacities can vary by a factor of two or more in narrow stratigraphic intervals. Samples 1-1, 1-2, and 1-3 came from a three meter interval of a single core (Fig. 4, Table I). This variation is attributable soley to properties of the organic matter, because mineral matrices and thermal rank were very similar in all three samples. Since organic matter type and content of shaly rocks vary considerably, these experiments predict the primary organic acid generating capacity will show similar variation. The probability of such variation should be considered in any attempt to quantitatively model the effects of organic acids on mineral reactions (4, 6).

Liquid hydrocarbons are a possible intermediate source in the production of some organic acids, but this has yet to be investigated. Since their oxygen content is significantly lower than most immature kerogens (20, 23), it seems likely that the organic acid generating capacity of liquid hydrocarbons is also lower. Organic acids are also abundant in both oil-productive and non-productive reservoirs suggesting that oil is not an intermediate source of universal importance.

SURVIVABILITY OF ORGANIC ACIDS

The data in figure 1 show that maximum organic acid concentrations occur between $80^\circ$ and $120^\circ$ C in sedimentary basins. This suggests that at higher or lower temperatures they are either produced in lesser abundance or that they are somehow destroyed or diluted.

Experimental Data.

Acetic acid and other organic acids are thermodynamically unstable at sedimentary conditions and will eventually decarboxylate to $CO_2$ and alkanes (24). Experimental studies of acetic acid decarboxylation show that the rate is extremely sensitive to temperature and the types of catalytic surfaces available (Table II; 25-26). Extrapolated rate constants for acetic acid decarboxylation at $100^\circ$ C differ by more than 14 orders of magnitude between experiments conducted in stainless steel and catalytically less active titanium (Table II; 26). Inherent (uncatalyzed) decarboxylation rates are similar for acetic acid and acetate (26). However, in catalytic environments their rates of decarboxylation differ markedly (25-26), and therefore a pronounced pH effect on total decarboxylation rate is observed.

Field Studies.

While experimental studies of organic acid decarboxylation have established some of the controls, the relevance experimentally determined rate constants for natural systems, where potential catalysts of many types abound, is questionable. Field calibrations clearly are advantageous from the standpoint of eliminating the effects of kinetic artifacts, but other limitations exist. In trying to relate the effects of time and temperature on decarboxylation rates in natural systems, the effects that variations in the type and abundance of organic matter can have on the production of organic acids, and therefore on their primary concentration, must be minimized.

In the Gulf Coast Basin, broadly similar sedimentary facies are represented throughout the Cenozoic section. As a result, organic matter type and abundance is reasonably similar. Using data from Eocene through Pleistocene reservoirs in the Gulf Coast Basin, Lundegard and Land (7) showed that the average organic acid concentration in formation water in the

Table I. Comparison of Relative Acetate and Propionate Yields in Hydrous Pyrolysis Experiments (325° C, 72 h)

| Sample | Kerogen type | H index | O index | $V_r$ (%) | Acetate (mmole/g C) | Propionate (mmole/g C) |
|---|---|---|---|---|---|---|
| 1-1 | II/III | 320 | 242 | .55 | .351 | .060 |
| 1-2 | II/III | 498 | 91 | .55 | .261 | .057 |
| 1-3 | II/III | 433 | 192 | .55 | .167 | .044 |
| 2-1 | II/I | 611 | 31 | .41 | .162 | .042 |
| 3-1 | I | 459 | 51 | .37 | .335 | .047 |
| 3-2 | I | 420 | 62 | .28 | .417 | .059 |
| 3-3 | I | 1048 | 26 | .39 | .201 | .048 |

Samples: 1- Tertiary of Cook Inlet, Alaska; 2- Monterey Fm., California; 3- Kishenehn Fm., Montana. For explanation of kerogen types see Tissot and Welte (20). Yields are per gram of carbon in the unreacted rock.

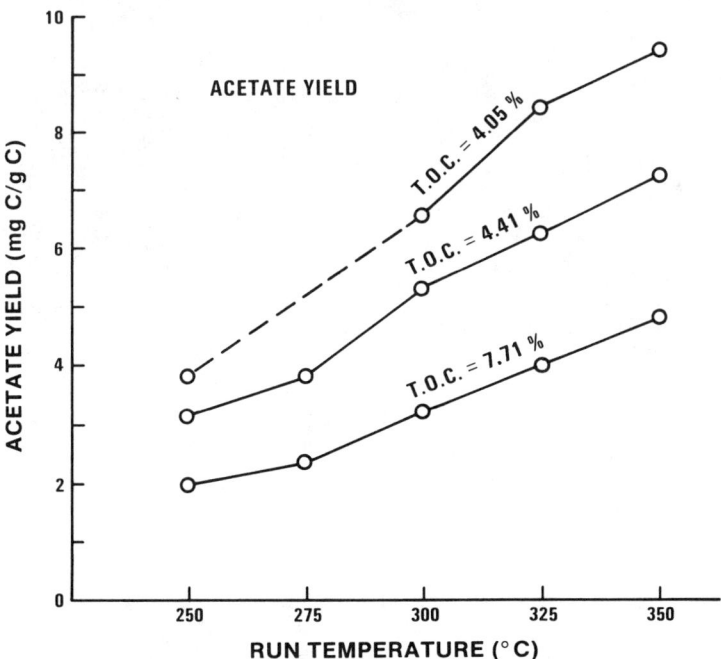

Figure 4. Relative yields of acetate for (72 h) hydrous pyrolysis experiments with three terrestrial mudrocks of identical thermal rank and similar mineral matrices. (Reproduced with permission from Ref. 14. Copyright 1987 Pergamon Journals).

Table II. Experimental Decarboxylation of Acetic Acid and Acetate. Data from Kharaka et al. (25) and Palmer and Drummond (26)

| Solution | Reactor material | Halflife at $100^\circ$ C (years) |
|---|---|---|
| acetic acid | Treated Ti | $3.7 \times 10^{14}$ |
|  | Ti | $4.0 \times 10^{8}$ |
|  | SS | $1.2 \times 10^{1}$ |
| sodium acetate | SS | $3.9 \times 10^{14}$ |
|  | Au | $1.6 \times 10^{14}$ |
|  | Ti | $2.6 \times 10^{12}$ |

Ti = titanium; SS = stainless steel; Au = gold

temperature range 80-120° C decreased with increasing age of the reservoir (Fig. 5). This trend represents the existence of fewer samples of low concentration in the younger reservoirs. These data indicated an approximate halflife of 60 million years for total acetate. In the Gulf Coast Basin maximum organic acid concentrations are not related to reservoir age in the same way as the average concentrations. Older reservoirs can have higher maximum concentrations than younger reservoirs. This fact probably reflects local variations in the primary generating capacities of the source rocks. Combining data from three different basins, Kharaka estimated halflives of 27 to 51 million years at 100° C (27). Field calibrations of these types are approximations and probably give maximum halflives. The greatest unknown is the residence time of the dissolved organic acids at a particular temperature. This is a function of the thermal and hydrologic history of the basin, and is difficult to reconstruct. Nevertheless, the field data demonstrate a substantially shorter halflife than indicated by experimental determinations of the inherent stability of acetate (Table II). However, at 100° C, concentrations in Cenozoic waters have probably been reduced by at most a factor of two or three, and a much lower factor for late Cenozoic waters.

Low concentrations of organic acids at temperatures less than 80° C occur for several reasons. One reason is the decreased organic acid generation from kerogen as a result of low thermal stress. The second reason is that dilution can occur where upward-moving acid-rich, formation waters mix with acid-poor waters of meteoric or other origins. This situation may exist in the Palo Duro Basin of west Texas (13) and the Pleistocene of offshore Louisiana (5). The third reason is that bacterial consumption of organic acids can occur at temperatures less than 80° C (8).

MODELING ORGANIC ACID GENERATION

By analogy with the first-order kinetic behavior of oil production from kerogen (28), organic acid production from kerogen should be affected by variations in geothermal gradient and burial rate. These factors will induce spatial (with respect to depth of burial) and temporal variations in the rates of organic acid production. Primary organic acid concentration will vary directly with these variations in production rate and inversely with porosity at the time of production. Primary concentration is considered here to be the concentration that exists before significant organic acid destruction takes place. High geothermal gradients should increase production rates. But, because shale porosity decreases exponentially with depth (29), it is not obvious what net effect high geothermal gradients will have on the dissolved concentration of organic acids. To evaluate this, a mathematical model is next discussed.

Although activation energies and Arrhenius frequency factors for organic acid production from kerogen are poorly known, rough estimates can be made that are compatible with known relative rates of oxygen and hydrogen loss from kerogen, and field data on the temperature distribution of organic acids in basins. Relative effects of time, temperature, and porosity on organic acid concentration can then be incorporated into a simple mathematical model.

Oxygen functional groups in kerogen are lost at thermal ranks lower than those of principal hydrocarbon production (Fig. 2; 21-22). Furthermore, carboxyl and carbonyl groups are the oxygen functions that are most readily liberated from kerogen. Based on these observations an activation energy at the low end of the spectrum of activation energies observed for hydrocarbon generation (28-29) was used. An activation energy of 23 kcal/mole and an Arrhenius frequency factor of $5 \times 10^{13}$ (30) was used. Because these values are only rough estimates, calculated yields and concentrations are plotted with respect to an arbitrary relative scale (Fig. 6). The emphasis here is on the relative impact of variable thermal histories.

The model simulates the history of shaly source rock for an assumed burial and thermal history. The results discussed here are based on a constant sedimentation rate of 500 m/m.y., and a geothermal gradient of either 30° C/km or 60° C/km. A Gulf Coast-type porosity gradient for shale is used, where porosity = $0.39\ e^{-.000095 \times depth\ (ft)}$ (31). For each increment

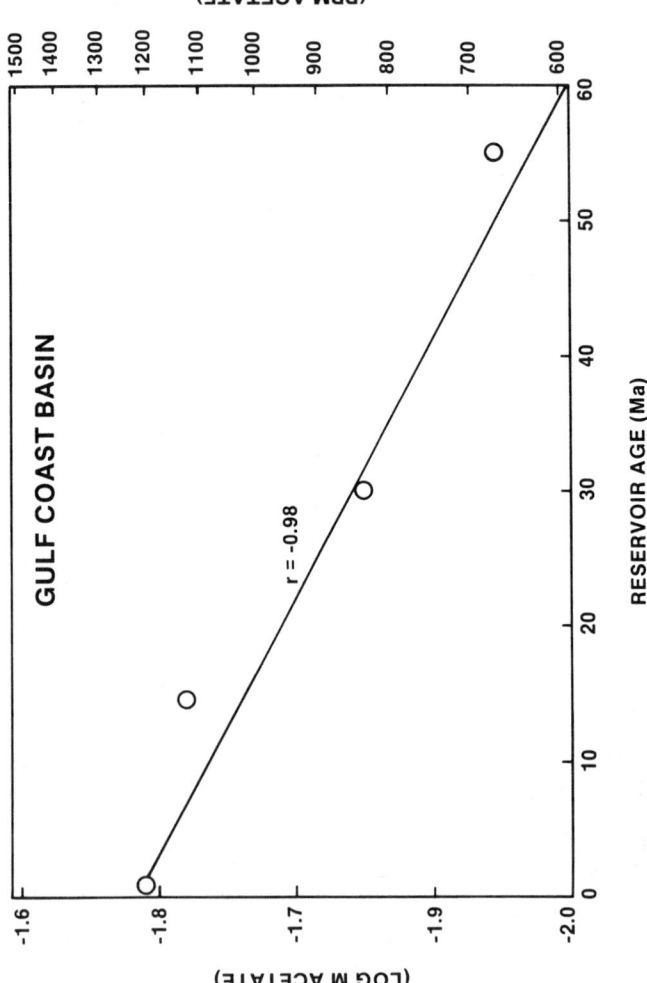

Figure 5. Mean organic alkalinity in temperature range 80-120 C versus reservoir age for Gulf Coast formation water. Trend suggests an acetate halflife of approximately 60 m.y. (Reproduced with permission from Ref. 7. Copyright 1989 Elsevier).

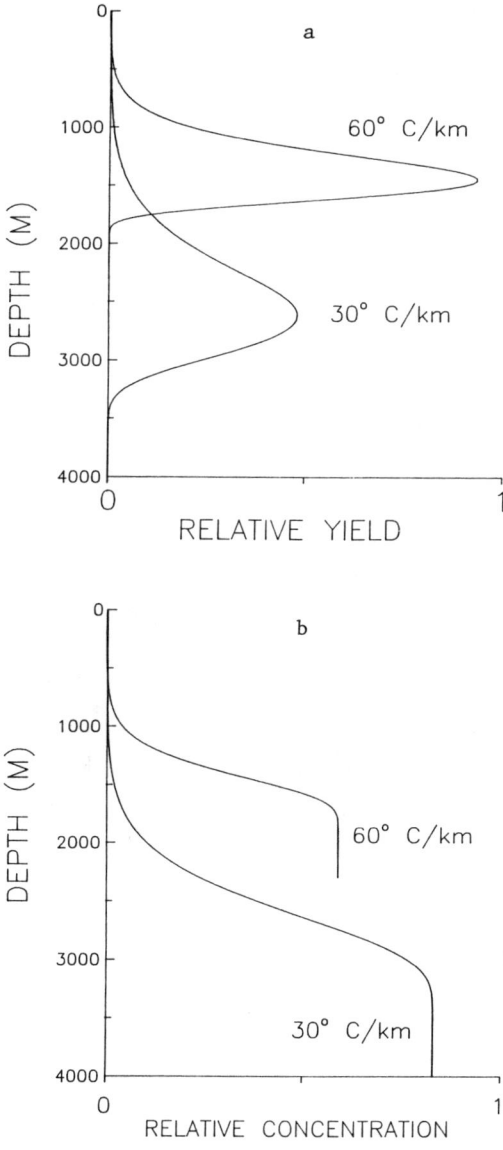

Figure 6. Results of a mathematical model of organic acid generation from kerogen for two geothermal gradients (30 and 60°C/km). Sedimentation rate equals 500 m/m.y. in both cases. An arbitrary scale is used on X axes because of uncertainties in the accuracy of chosen kinetic parameters (see text). a, Organic acid yield versus depth; b, organic acid concentration versus depth. *Continued on next page.*

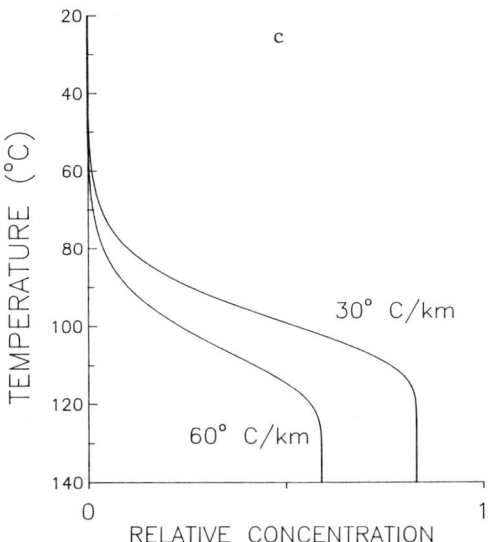

Figure 6. *Continued.* c, Organic acid concentration versus temperature. Note that the higher geothermal gradient produces a higher peak yield but a lower maximum organic acid concentration. *See* text for details.

of time the average depth, temperature, porosity, and rate constant are calculated. From these values, the incremental yield of organic acid is calculated from the rate constant and the time interval. Organic acid concentration at any depth is equal to the sum of the organic acid concentrationa at the beginning of the time interval, plus the incremental yield of organic acid divided by the porosity. For both the high and low thermal gradient models, the time-integrated or cumulative yield is the same and is controlled by initial oxygen content of kerogen. However, in the presence of the higher thermal gradient, oxygen is liberated from kerogen at shallower depths and over a narrower range of depths (Fig. 6a). Peak yield is greater in the case of the higher thermal gradient, and occurs after less time at a slightly higher temperature. Accounting for the effect of ambient porosity in the source rock, predicted organic acid concentrations were calculated. In the case of the higher thermal gradient, higher acid concentrations are expected at shallow depths, but higher maximum concentrations are obtained when the thermal gradient is lower (Fig. 6b). At all temperatures, the higher geothermal gradient actually results in lower concentrations, in spite of the higher production rates (Fig. 6c). The explanation lies in the competing effects of production rate and porosity. Peak production in the high thermal gradient model occurs at higher porosity conditions, which mediates the effect of the greater peak yield on acid concentration. Conversely, when the geothermal gradient is lower, the lower porosity at the time of peak yield results in higher maximum acid concentrations. This analysis shows that high organic acid concentrations may, surprisingly, be favored by low thermal gradients. Maximum concentrations would presumably be achieved when geothermal gradients increase with time at a rate that results in rapid generation at depths where porosities are low.

The model results discussed above express only the effects time and temperature on organic acid production from kerogen, and the effect of porosity on organic acid concentrations. They do not account for the effects of time and temperature on organic acid decarboxylation. Therefore, the results are best applied to Neogene and source rocks where the field data discussed earlier indicate that decarboxylation effects on total acetate concentrations are small, where temperatures do not exceed approximately $150^{\circ}$ C.

ROLE OF ORGANIC ACIDS IN DIAGENESIS

The concentration of organic acid anions in formation water can be high, especially between $80^{\circ}$ and $120^{\circ}$ C, where acetate is usually the dominant anion (9, 27). Because acetate and other organic acid anions are proton donors or acceptors, depending on the pH of water as well as the anion, they play an important role in all pH-dependent homogeneous and heterogenous reactions (32). Also, these organic anions form strong complexes with some metals and other cations and have been suggested as the mechanism for the transport of many metals, including Pb and Zn in Mississippi Valley-type ore deposits (2), and Al in sedimentary basins (4, 16). Thus, organic acid anions and their complexes with inorganic species should be included in geochemical codes to study mineral diagenesis.

Metal Complexation.

The role of organic acid anions in the transport of U, Pb, and Zn in sedimentary basins was examined using detailed inorganic and organic compositions of formation waters from several oil and gas fields. Using the computer code SOLMNEQ, Kraemer and Kharaka (33) showed that in Gulf Coast geopressured-geothermal waters the portion of U complexed with organic anions is negligible. In the case of Pb and Zn, Kharaka et al. (34) showed that the concentrations of organic complexes of Pb and Zn in metal-rich brines from the central Mississippi Salt Dome basin form a very small percentage of the total metal. Their calculations indicated that at subsurface conditions, 65 to greater than 99 percent of the dissolved Pb is present as chloride complexes, and more than 99 percent of the Zn is present as chloride complexes.

The role of metal complexes with organic acid anions in mineral diagenesis in sedimentary basins was investigated using the geochemical code SOLMINEQ.88 (35). This code has thermochemical data for various metal complexes with acetate, oxalate, and succinate, and an option to add the complexes of two additional (user specified) organic anions for a total of 80 organic species (see also, Perkins et al., this volume). At present, stability constants for calcium and aluminum oxalate have only been measured at 25° C and thermochemical data are insufficient to accurately extrapolate these constants to higher temperatures (35). As a first approximation, the 25° C constants are currently used for all temperatures. The chemical composition of water from Pleasant Bayou #2 well (Texas Gulf Coast Tertiary) was used as a test case (33). The runs were made for subsurface temperature (150° C) and pressure (700 bars) conditions. The pH of the formation water measured at the well site was 6.50; the pH calculated for subsurface conditions from detailed inorganic chemistry (including gases) and organic chemistry of the fluid was 4.92. To test the effects of organic-metal complexing in the presence of extreme amounts of organic acid anions, 10,000 mg/l acetate and 2,000 mg/l oxalate were added. These concentrations are well beyond those normally found in natural waters and were chosen to investigate maximum complexing capabilities. Simulations were made at pH 4.92 +/- 2 to investigate the role of pH in complex formation and mineral diagenesis. Selected results of runs with and without the added organics at pH 4.92 are compared below.

Results showed that a significant percentage of iron and calcium were complexed with acetate and oxalate (Table III). The percentage of calcium complexed by oxalate is relatively low, perhaps because the stability constants used are for 25° C. No stability constant data for calcium-oxalate are currently available at higher temperatures (35). Only a small portion of sodium and other monovalent metals form complexes with acetate and oxalate. The percentage of aluminum complexed with these organic species is only 10 percent of total aluminum. The fact that much more of the aluminum is complexed with acetate than with oxalate may reflect the availability of stability constants for aluminum-acetate at elevated temperatures (S. E. Drummond, 1989, pers. communication) and lack of such data for aluminum-oxalate complexes. The amount of aluminum complexed with these organic species generally increases as the pH of the water decreases. The increase results even though more of the organics are protonated, mainly because the percentage of $Al^{+3}$ is higher at lower pH values. In the case of the sample from the Pleasant Bayou #2 well (Table III), the percentage of aluminum complexed with these organics at pH 3.92 and 2.92 is 46 and 81, respectively; at pH 5.92 and 6.92 the percentages are less than 0.1. These simulation results for aluminum should be contrasted with experimental studies which suggest greatly enhanced aluminum concentrations in the presence of these organics (4, 16). The discrepancy may be related to the fact that thermochemical data for Al are incomplete and unreliable. Or, it is possible that the experiments were terminated before equilibrium with aluminosilicates was reached.

The saturation states of selected minerals with and without the added organics in the Pleasant Bayou #2 water at subsurface conditions and pH 4.92 are shown in Table IV. The saturation state of halite is unchanged because the Na-organic complexes are insignificant. On the other hand, albite, kaolinite, anhydrite and calcite are significantly more undersaturated in the presence of the organics because an important proportion of the Ca and Al are complexed. The saturation state of siderite shows the largest change, from -1.9 to -2.6 kcal/mole. This is a result of the particularly strong complex that Fe forms with acetate (36), and suggests that high concentrations of Fe can be retained in solution at temperatures of 80° to 120° C, where organic acid anions are in highest concentration. At higher temperatures, where decarboxylation destroys organic acid anions (Fig. 1), complexed Fe may be released to solution to promote siderite or ankerite formation. Ankerite is a common late-stage cement in Eocene sandstones of the Texas Gulf Coast (37-38).

In the computation of metal-organic complexes and mineral saturation states discussed above, the aim is not to quantify the changes in all natural waters, but to show when organic complexing can be significant. The results warrant the effort to include dissolved organics in the

## pH Effects.

As proton sources, organic acids have an important influence on a variety of pH dependent reactions, especially between 80° and 120° C. Using the geochemical algorithm PHREEQE (39), the effect of acetic acid addition on calcite solubility was investigated. PHREEQE was used in preference to SOLMINEQ because it allows the user to specify saturation with respect to specific mineral phases and gases. Addition of 0.04 m (2400 mg/l) acetic acid to a calcite-saturated solution in equilibrium with a fixed $pCO_2$ of 10 atm at 100° C, will increase calcite solubility by nearly 3 times (Table V).

Buffering of pH by organic acid anions has potential consequences for various pH-dependent reactions, such as carbonate precipitation and dissolution. Organic acid anions commonly dominate the total alkalinity of formation waters, especially in the 80-120° C range (Fig. 7). Some workers have suggested that an increase in the partial pressure of carbon dioxide ($pCO_2$) may promote the precipitation of carbonate minerals in waters whose alkalinity is dominated by organic acid anions (15). Their reasoning was that if pH is fixed by organic acid buffers, then an increase in $pCO_2$ should cause carbonate precipitation by increasing the activity of dissolved carbonate ions. This hypothesis was quantitatively evaluated by Lundegard and Land (6-7) using the reaction path modeling program PHREEQE (39) and including thermochemical data for complexing of Ca and Na by acetate (40). They modeled the response of acetic acid solutions, initially at calcite saturation and a specified $pCO_2$, to changes in the partial pressure of carbon dioxide. Most of the simulations were done for conditions of 100° C, 0.5 m NaCl concentration, and a change in $CO_2$ pressure from 0.1 to 1.0 atm, however these conditions were varied to determine the sensitivity of results. The simulations showed that at 100° C, initial acetic acid concentrations must exceed approximately 0.06 m (3600 mg/l) in order for increases in $pCO_2$ to cause calcite precipitation in initially calcite-saturated solutions (Fig. 8). The minimum initial concentration of acetic acid needed to cause precipitation instead of dissolution (crossover cencentration) decreases with increasing temperature as a result of the decreasing $pK_d$ of acetic acid (Fig. 9). However, this effect should largely be compensated by the rapidly decreasing concentrations of organic acids at temperatures above 100° C (Fig. 1). The position of the crossover acetic acid concentration is relatively insensitive to modest changes in ionic strength (0.5 to 1.0 m NaCl), initial calcium concentration (0 to 0.01 m), and the magnitude of the change in $pCO_2$ (0.1 to 1.0 atm versus 1.0 to 10.0 atm) (7). Given the existing empirical data on organic acid concentrations (Fig. 1), increases in $pCO_2$ should promote calcite dissolution in the majority of natural waters.

## CONCLUSIONS

Concentrations of organic acids in formation waters are controlled by a number of factors that influence their production and/or destruction. These factors include organic matter type and abundance in source rocks, temperature, and burial history. Hydrous pyrolysis experiments confirm the thermal generation of short-chain aliphatic acids from kerogen, and indicate that natural concentrations will be strongly influenced by the abundance and nature of organic matter in the reacting system. Thermal and burial history of source rocks influences the primary concentration of dissolved organic acids because of their effect on the timing and rate of organic acid production from kerogen, and on the porosity at the time of organic acid production. A mathematical model, incorporating the effects of time and temperature on organic acid generation and depth-dependent porosity changes, illustrates that while high geothermal gradients promote rapid generation of organic acids, the consequent generation at shallower depths (higher porosities) will tend to reduce maximum dissolved concentrations of organic acids, and therefore their diagenetic impact. In the temperature range 80-120° C the rate of

Table III. Concentrations of Selected Cations and Anions from Pleasant Bayou #2 Well and the Percent of Cations Complexed with Added Organics. Calculated with SOLMNEQ.88

| Temp. | 150° C | Acetate 10,000 mg/l |
|---|---|---|
| Press. | 700 bars | Oxalate 2,000 mg/l |
| pH = 4.92 | | |

| Constituent | mg/l | | |
|---|---|---|---|
| | | % Complexed With | |
| | | Acetate | Oxalate |
| TDS | 132,000. | | |
| $HCO_3$ | 365. | | |
| $SO_4$ | 5.4 | | |
| $SiO_2$ | 120. | | |
| Cl | 80,000. | | |
| Na | 38,000. | 1. | 0.4 |
| Ca | 9,100. | 21. | 3. |
| Fe | 62. | 7. | 54. |
| Al | 0.01 | 10. | <0.1 |

Table IV. Saturation States of Selected Minerals With and Without Added Organic Species to the Water from Pleasant Bayou #2 Well. See Table III for Chemical Analysis

| Mineral | $\Delta G$ (without) | $\Delta G$ (with org.) |
|---|---|---|
| | kcal/mole | |
| Halite | -2.9 | -2.9 |
| Albite | -4.1 | -4.2 |
| Kaolinite | -6.6 | -6.8 |
| Anhydrite | -4.0 | -4.1 |
| Calcite | -0.6 | -0.7 |
| Siderite | -1.9 | -2.6 |

Table V. Effect of Acetic Acid Addition on Calcite Solubility. Modeled with PHREEQE for 100° C, 0.5 m NaCl

| Acetic Acid | | Log $pCO_2$ | Calcite Dissolved |
|---|---|---|---|
| (m) | mg/l | (atm) | (m) |
| 0.00 | 0 | 0 | .0035 |
| 0.02 | 1200 | 0 | .0117 |
| 0.04 | 2400 | 0 | .0208 |
| 0.06 | 3600 | 0 | .0300 |
| 0.00 | 0 | 1 | .0080 |
| 0.02 | 1200 | 1 | .0151 |
| 0.04 | 2400 | 1 | .0228 |
| 0.06 | 3600 | 1 | .0306 |

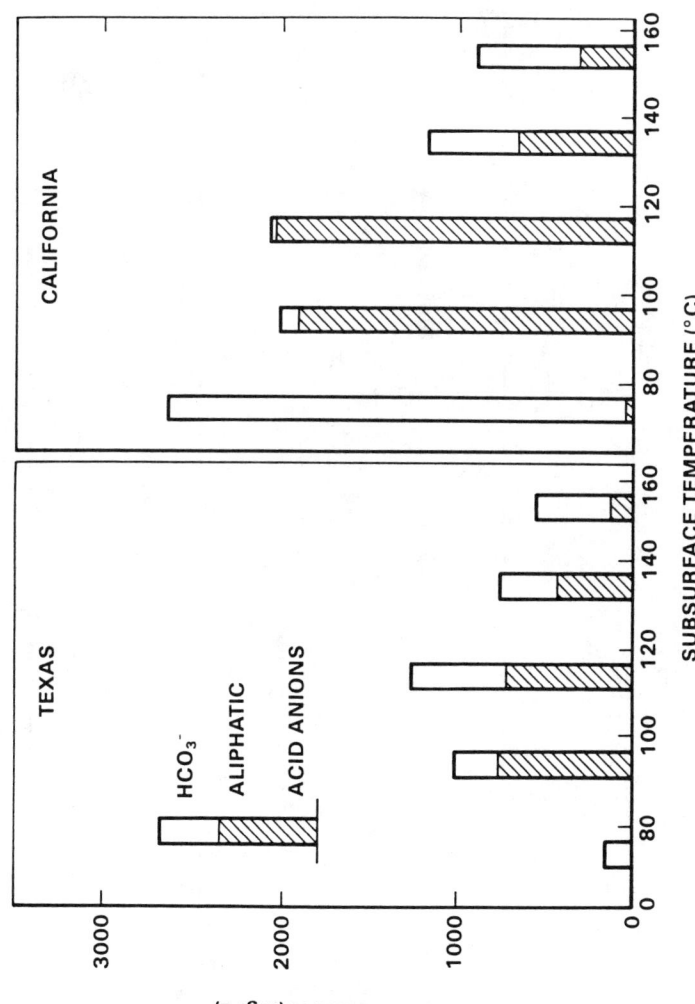

Figure 7. Relative contributions of organic acid anions and bicarbonate to total alkalinity of formation waters. After Carothers and Kharaka (8). Note that organic acid anions can dominate alkalinity between 80° and 120° C.

186                    CHEMICAL MODELING OF AQUEOUS SYSTEMS II

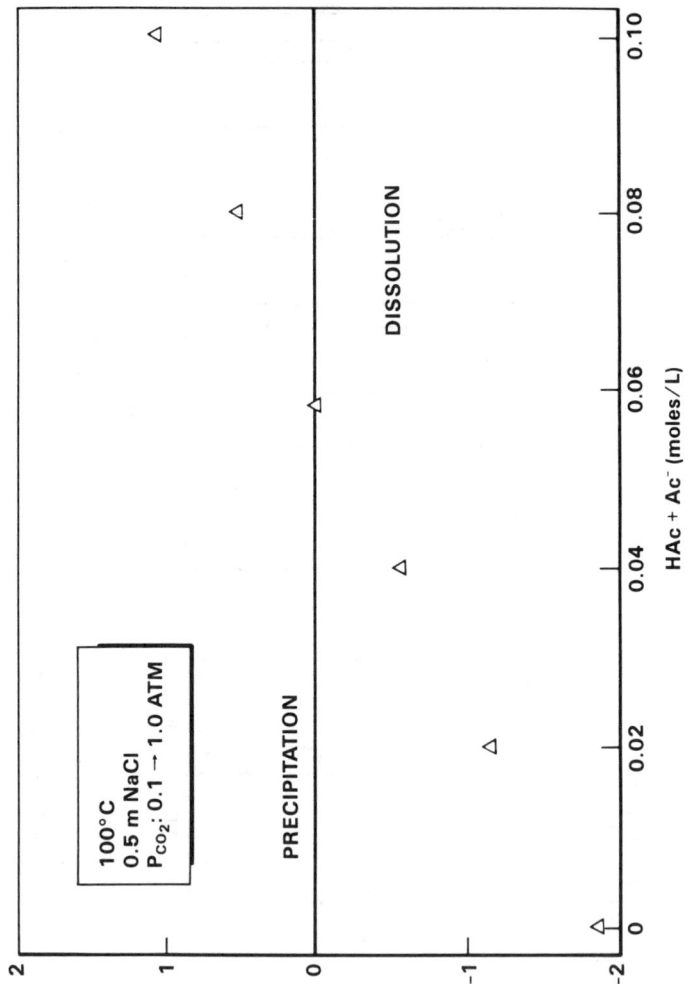

Figure 8. Results of a geochemical model investigating the effects of pH buffering by acetic acid-acetate on the response of the carbonate system to changes in $pCO_2$. Increases in the $pCO_2$ of calcite-saturated solutions will promote calcite undersaturation for initial concentrations of acetic acid up to approximately 0.06 m (3600 mg/l). This is called the crossover concentration. (Reproduced with permission from Ref. 7. Copyright 1989 Elsevier).

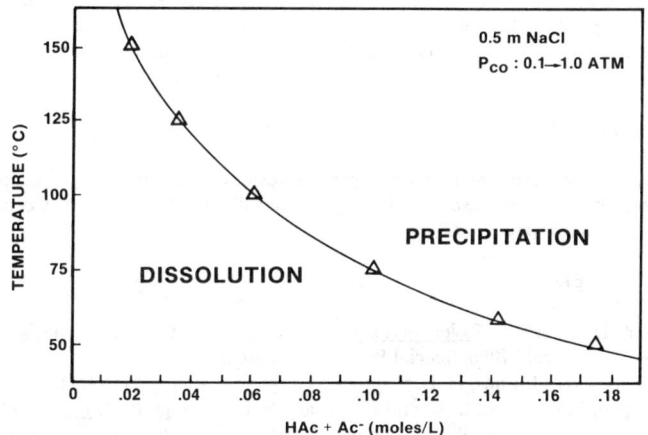

Figure 9. Temperature dependence of the acetic acid crossover concentration for calcite, 0.5 m NaCl concentration, and a $pCO_2$ change from 0.1 to 1.0 atm. Crossover concentration (see Fig. 8) decreases with increasing temperature as a result of the decrease in the $pK_d$ of acetic acid. (Reproduced with permission from Ref. 7. Copyright 1989 Elsevier).

organic acid synthesis from kerogen greatly exceeds the rate of thermal decarboxylation in sedimentary basins. At higher temperatures, the acid generating capacity of kerogen is diminished, while rates of decarboxylation greatly increase. Consequently, observed concentrations of organic acids are lower at temperatures greater than 120° C.

Assessment of the geochemical significance of cation complexing by organic acid anions is preliminary due to uncertainities in the thermochemical data for complexes at elevated temperatures. Complexing of Pb and Zn by organic acid anions in brines appears to be less important than chloride complexing. Significant organic complexing of calcium and iron probably occurs in sedimentary basins and may have an important impact on carbonate diagenesis. The role of organic complexing of aluminum in formation waters is still incompletely resolved due the absence of reliable thermochemical data for organic complexes at elevated temperatures.

As proton sources, organic acids are likely to have an important influence on a variety of pH dependent reactions, especially between 80° and 120° C. Geochemical models predict that addition of typical amounts of acetic acid to calcite saturated solutions at 100° C, can increase calcite solubility by several times. Organic acid anions also have an appreciable impact on the alkalinity and pH buffering capacity of formation waters. However, in most formation waters it is unlikely that calcite precipitation will result from simply increasing the $pCO_2$ of calcite-saturated solutions, unless pH is buffered by systems other than the acetate and carbonate systems.

ACKNOWLEDGMENTS

Dan Melchior, Bob Sweeney, and two anonymous reviewers provided many comments that substantially improved this manuscript. We also would like to thank Unocal Corporation for permission to publish this paper.

LITERATURE CITED

1. Gautier, D. L., Ed., Roles of Organic Matter in SedimentDiagenesis; Soc. Econ. Paleontol. Mineral.: 1986; Special Publ. No. 38, 203p.
2. Giordano, T. H.; Barnes, H. L. Economic Geology 1981, 76, 2200-2211.
3. Lundegard, P. D.; Land, L. S.; and Galloway, W. E. Geology 1984, 12, 399-402.
4. Crossey, L. J.; Surdam R. C.; Lahann, R. In Roles of Organic Matter in Sediment Diagenesis; Gautier, D. L., Ed.; Soc. Econ. Paleontol. Mineral., 1986; Special Publ. No. 38, p 147.
5. Kharaka, Y. K.; Law, L. M.; Carothers, W. W.; Goerlitz, D. F. In Roles of Organic Matter in Sediment Diagenesis; Gautier, D. L., Ed.; Soc. Econ. Paleontol. Mineral., 1986; Special Publ. No. 38, p 111.
6. Lundegard, P. D.; Land, L. S. In Roles of Organic Matter in Sediment Diagenesis; Gautier, D. L., Ed.; Soc. Econ. Paleontol. Mineral., 1986; Special Publ. No. 38, p 129.
7. Lundegard, P. D.; Land, L. S. Chemical Geology 1989, 74, 277-287.
8. Carothers, W. W.; and Kharaka, Y. K. Amer. Assoc. Petrol. Geol. Bull. 1978, 62, 2441-53.
9. Lundegard, P. D. Ph.D. Thesis, The University of Texas at Austin, Austin, Texas, 1985.
10. Fisher, J. B. Geochim. Cosmochim. Acta 1987, 51, 2459-2468.
11. Kharaka, Y. K.; Hull, R. W.; Carothers, W. W. In Relationships of Organic Matter and Mineral Diagenesis; Soc. Econ. Paleontol. Mineral. Short Course No. 17, 1985; Chapter 2.
12. Hanor, J. S.; Workman, A. L. Applied Geochemistry 1986, 1, 37-46.
13. Means, J. L.; and Hubbard, N. Org. Geochemistry 1987, 11, 177-92.
14. Lundegard, P.D.; and Senftle, J.T. Applied Geochem. 1987, 2, 605-612.

15. Surdam, R. C.; Boese, S. W.; Crossey, L. J. In Clastic Diagenesis; McDonald, D. A. and Surdam, R. C., Ed.; Amer. Assoc. Petrol. Geol.; Tulsa, 1984; Memoir 37, p. 127-49.
16. McGowen, D. B.; Surdam, R. C. Org. Geochem. 1988, 3, 245-259.
17. Davis, J. B. Petroleum Microbiology; Elsevier: Amsterdam, 1967; 604 p.
18. Lewan M. D.; Winters J. C.; McDonald J. H. Science 1979, 203, 897-99.
19. Kawamura K.; Tannenbaum E.; Huizinga B. J.; Kaplan I. R. Geochem. Jour. 1986, 20, 51-59.
20. Tissot, B. P.; Welte, D. H. Petroleum Formation and Occurrence; Springer-Verlag: Berlin, 1978; p 538.
21. Robin P. L.; Rouxhet P. G. Geochim. Cosmochim. Acta. 1978, 42, 1341-49.
22. Rouxhet P. G.; Robin P. L.; Nicaise G. In Kerogen; Durand, B., Ed.; Editions Technip: Paris, 1980; p 163-90.
23. Seifert, W. K., In Progress in the Chemistry of Organic Natural Products; Springer-Verlag: New York, 1975; 32, p. 1-49.
24. Shock, E. L. Geology 1988, 16, 886-890.
25. Kharaka, Y. K.; Carothers, W. W.; and Rosenbauer, R. J. Geochim. Cosmochim. Acta. 1983, 47, 397-402.
26. Palmer, D. A.; and Drummond, S. E. Geochim. Cosmochim. Acta 1986, 50, 813-23.
27. Kharaka, Y. K. In Proc. Third Canadian/American Conference on Hydrogeology; Hitchon, B., Bachu, S., and Sauveplane, Eds.; National Water Well Assoc., 1986, 173-195.
28. Tissot, B. P.; Espitalie, J. Revue de l'Institut Francais du Petrole 1975, 30, 743-77.
29. Tissot, B. P.; Pelet, R.; Ungerer, PH. Amer. Assoc. Petrol. Geol. Bull. 1987, 71, 1445-1466.
30. Abelson, P. H. Sixth World Petroleum Congress, 1963; 397-407.
31. Magara, K. Compaction and fluid migration; Elsevier: Amsterdam, 1978; p 319.
32. Willey, L. M.; Kharaka, Y. K.; Presser, T. S.; Rapp, J. B.; Barnes, I. Geochim. Cosmochim. Acta 1975, 39, 1707-10.
33. Kraemer, T. F., Kharaka, Y. K. Geochim. Cosmochim. Acta 1986, 50, 1440-1455.
34. Kharaka, Y. K., Maest, A. S., Carothers, W. W., Law, L. M., Lamothe, P. J., Fries, T. L. Applied Geochem. 1987, 2, 543-561.
35. Kharaka, Y. K., Gunter, W. D., Aggaral, P. K., Perkins, E. H., Debraal, J. D. U. S. Geological Survey Water Resources Investigation Report 88-4227 1988, 420p.
36. Drummond, S. E., Palmer, D. A. Geol. Soc. Amer. Abs. with Prog. 1985, 17, 567.
37. Boles, J. R. Contrib. Mineral. Petrol. 1978, 68, 13-22.
38. Fisher, R. S., Land, L. S. Geochim. Cosmochim. Acta 1986, 50, 551-561.
39. Parkhurst, D. L.; Thorstenson, D. C.; and Plummer, L. N. U. S. Geol. Survey Water Res. Invest. 80-96 1982, p 210.
40. Martell, A. E.; Smith, R. M. Critical stability constants, volume 3: other organic ligands; Plenum Press:, New York, 1977; p 495.

RECEIVED August 24, 1989

# Chapter 14

# Carbon Isotope Mass Transfer as Evidence for Contaminant Dilution

## Laura Toran

### Environmental Sciences Division, Oak Ridge National Laboratory, P.O. Box 2008, Oak Ridge, TN 37831-6036

Carbon isotope data provided evidence for dilution of sulfate-contaminated groundwater near an underground mine; this result would not be predicted from thermodynamic reaction path modeling alone. The effect of $CO_2$ outgassing and carbonate precipitation and dissolution on $\delta^{13}C$ of dissolved inorganic carbon was modeled with an integrated form of Rayleigh distillation. The observed carbon isotope ratios were 2-4 $°/_{oo}$ lighter than those modeled for most samples, indicating an additional source of light carbon. Mixing with uncontaminated water surrounding the mine is hypothesized to explain the discrepancies. Alternative hypotheses include sulfide oxidation and $CO_2$ outgassing at pH less than or equal to 5 and siderite precipitation which preferentially removes heavy carbon.

Mixing of different groundwaters can be difficult to recognize when dissolved constituents are controlled by similar processes and sources in the mixed waters. Modeling of isotope ratios can provide an additional source of information to indicate mixing. Carbon isotope ratios were measured and calculated for groundwater that has been contaminated with sulfate and undergone $CO_2$ outgassing and carbonate dissolution and precipitation. Geochemical modeling can help explain how sulfate contamination occurred and provide information that will help predict or prevent future contamination. However, additional data on carbon isotopes was needed to try to better understand the observed carbon mass transfer. Carbon isotope modeling suggests additional processes such as dilution that may be important. Although the lack of chemical data from before and, in particular, during mining makes it difficult to obtain a unique reaction path, the modeling in this study points out alternative mechanisms and new data collection needs.

Sulfide oxidation in a carbonate environment involves two sets of reactions: oxidation and neutralization. Sulfide minerals exposed to the atmosphere can oxidize to produce sulfate and acidity (Equation 1):

$$FeS_2 + \frac{7}{2}O_2 + H_2O \rightarrow Fe^{2+} + 2H^+ + 2SO_4^{=}. \tag{1}$$

In a carbonate environment, dissolution of minerals such as dolomite neutralizes the acidity and increases the $HCO_3^-$, $Mg^{2+}$, and $Ca^{2+}$ in solution (Equation 2):

$$2H^+ + 2SO_4^= + MgCa(CO_3)_2 \rightarrow 2SO_4^= + Mg^{2+} + Ca^{2+} + 2HCO_3^-. \qquad (2)$$

These reactions can cause the system to become saturated with respect to calcite, iron hydroxide, siderite, and some sulfate minerals. Furthermore, if the neutralization takes place in an open system, $CO_2$ outgassing may occur as the dissolved inorganic carbon (DIC) increases the partial pressure of $CO_2$.

## SITE DESCRIPTION

The Shullsburg mines in southwestern Wisconsin present an example of these reactions. The mines are part of the upper Mississippi Valley zinc-lead district and were operated from the 1920s until 1979. While the mines were exposed to air, sulfide minerals in the Galena-Platteville Formation oxidized to sulfate. When groundwater re-entered the mined formation after mine closure, the high sulfate concentrations spread to the local groundwater system (Figure 1) and forced closure of 11 wells (1,2). However, because the surrounding carbonate rocks neutralized the pH, acidity and trace metals were not an environmental problem.

Contaminated water is defined here as having sulfate concentration greater than 1 mmol/L. The contaminated water had higher $Ca^{2+}$, $Mg^{2+}$ and $SO_4^=$ than uncontaminated water from the surrounding area (Figure 2). However, contaminated and uncontaminated samples had similar pH and $HCO_3^-$ values.

## REACTION PATH MODELING

The computer program PHREEQE (3) was used for chemical equilibrium and reaction path modeling. PHREEQE does speciation and mass transfer calculations to find the distribution of aqueous complexes and the saturation indices of potential mineral phases present.

The PHREEQE reaction path model of sulfide oxidation in the Shullsburg mines includes five chemical changes.

(1) Sulfide oxidation increases $SO_4^=$ and $H^+$ according to the stoichiometry of Equation 1.

(2) Dolomite plus or minus calcite dissolution buffers pH, increases $Ca^{2+}$ and $Mg^{2+}$, and initially increases $HCO_3^-$ in the water.

(3) Calcite precipitation occurs when the dissolution of dolomite causes the solution to become saturated with respect to calcite (as it was in all cases modeled).

(4) Iron concentration can be reduced by precipitation of an iron mineral such as siderite ($FeCO_3$), jarosite ($NaFe_3(SO_4)_2(OH)_6$), or amorphous iron hydroxide ($Fe(OH)_{3a}$). It was unclear what mineral or phase controls iron concentration, especially since precipitation was sensitive to oxidation potential (pe), and the field redox electrode measurements used in the model are not always relevant for specific redox couples. Since the mine workings flooded before the study was begun, samples of the solid phases could not be collected.

(5) To reduce the DIC concentration, $CO_2$ outgassing is fixed by selecting an appropriate log $P_{CO2}$ to match the observed DIC and pH.

The amount of carbon observed in the Shullsburg waters was less than the modeled carbon in a closed system. An increase in the amount of sulfide oxidation implies an increase in the amount of carbonate dissolution in a 1:1 stoichiometric ratio (Equation 2). When calcite precipitation, driven by dolomite dissolution, is included in reaction modeling, the relationship between sulfate and carbon becomes nonlinear (Figure 3). Even for this nonlinear model, there is a discrepancy between measured carbon concentrations and modeled closed system carbon concentrations.

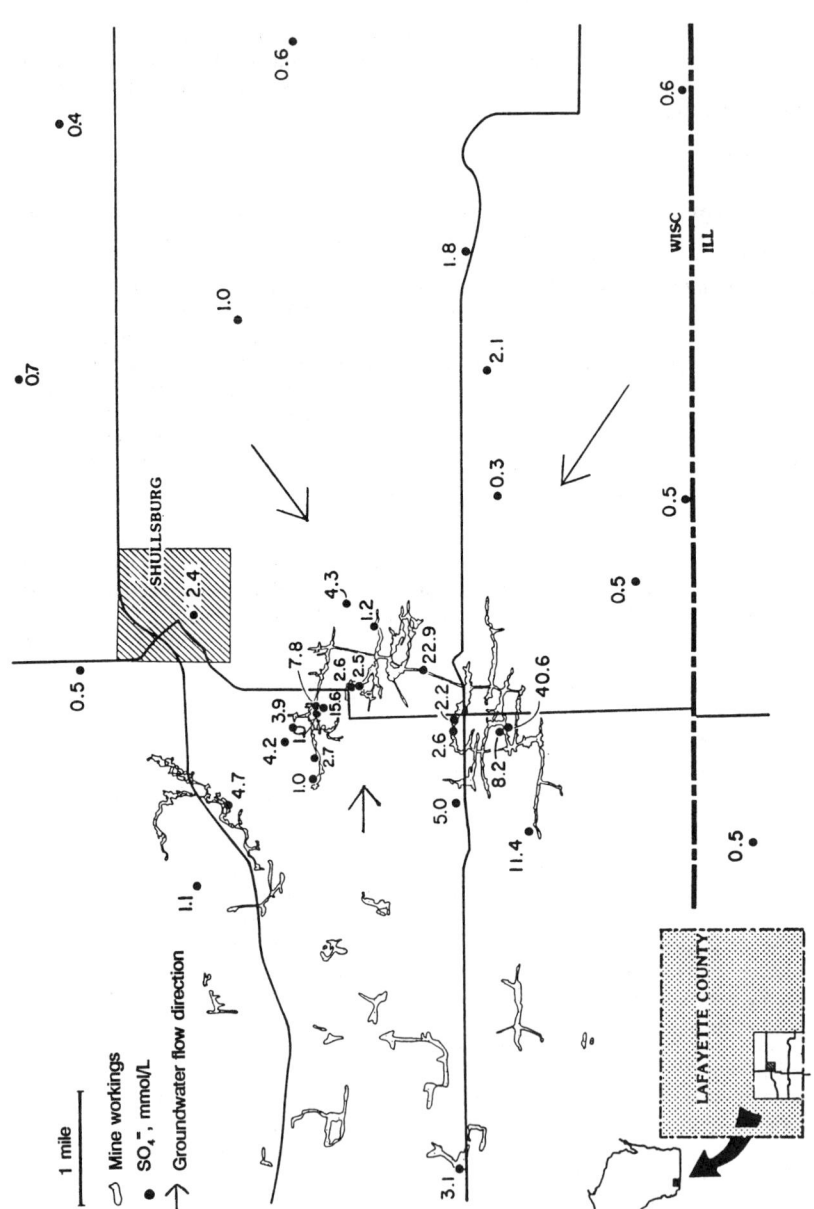

Figure 1. Location map, average sulfate concentrations near the Shullsburg mines measured between 1983 and 1985 in the Galena-Platteville formation (11), and groundwater flow directions measured in 1983 (19).

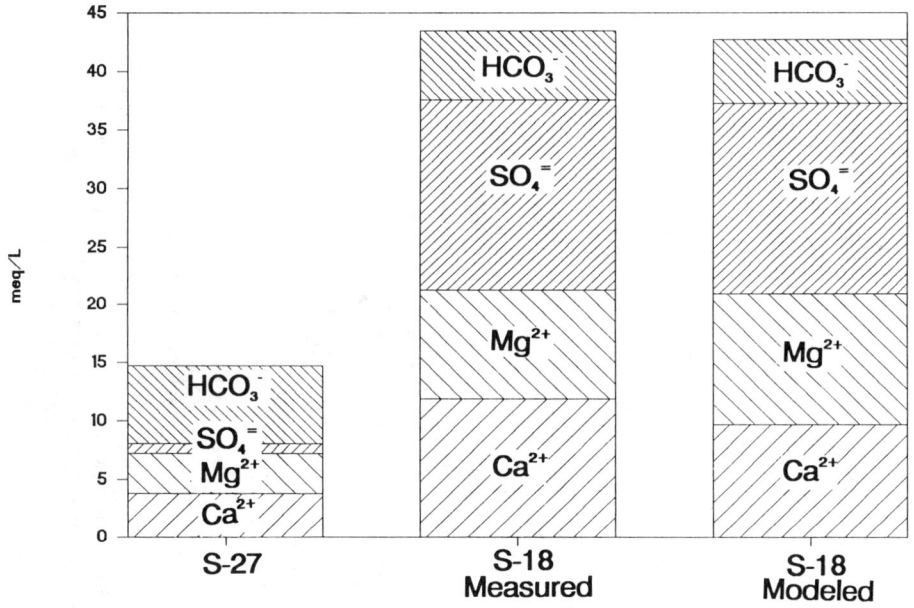

Figure 2. Histogram of major cations and anions for an uncontaminated sample used in modeling (S-27), a contaminated sample (S-18), and results of PHREEQE reaction path model.

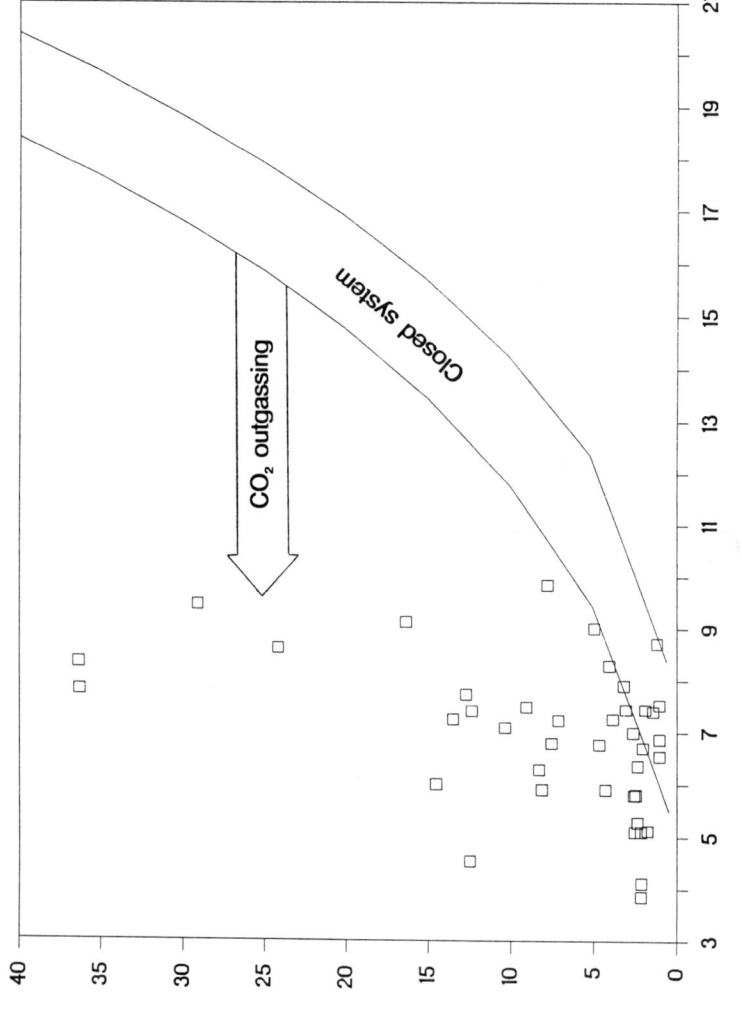

Figure 3. Sulfate versus alkalinity for contaminated Shullsburg samples. Diagonal curves show PHREEQE calculations for total carbon production during closed system oxidation and neutralization using high and low initial DIC. Samples to the left of the closed system region have undergone $CO_2$ outgassing at nearly constant pH.

Loss of carbon could be accounted for by $CO_2$ outgassing in an open or "partially" open system. The predicted amount of carbon lost is not large enough to correspond to a system in equilibrium with a partial pressure of $CO_2$ ($P_{CO2}$) in the air at $10^{-3.5}$ atm. However, the system would not have been completely closed, since oxidation took place in open mine workings, exposed to air. "Open" and "closed" are end members for what is actually a variable system in open caves (4), and the $P_{CO2}$ in water is difficult to predict (5). The calculated $P_{CO2}$ in contaminated water varies from $10^{-1.5}$ to $10^{-2.5}$ atm, which is typical of soil gas $P_{CO2}$ or streams from carbonate caves (6). To model the $CO_2$ outgassing along with the sulfide oxidation and carbonate neutralization, the carbon concentration was matched by selecting a log $P_{CO2}$, different for each sample modeled. Fixing the amount of $CO_2$ outgassing converts the PHREEQE model to a mass balance calculation for carbon since this constraint is not based on equilibration with a known phase. Although the kinetic control on outgassing was not known, alkalinity seemed to have an upper limit of around 10 mmol/L in this carbonate-pH buffered system. The $P_{CO2}$ in the models varied from $10^{-1.6}$ to $10^{-2}$ atm, the range being close to observed values, but differing slightly because of differences in modeled and measured pH. Matching the carbon rather than the $P_{CO2}$ provides the input data needed for carbon isotope modeling.

The modeled and observed cation concentration could be matched closely (Figure 2). The small differences between observed and calculated concentrations can be attributed to differences in starting water, mixing, analytical error, or lack of information about the controlling iron phase.

## METHODS OF CARBON ISOTOPE MODELING

Carbon isotope effects of outgassing and carbonate dissolution and precipitation can be modeled to gain more insight into the reaction mechanism and environment of oxidation. These effects were modeled using an integrated form of Rayleigh distillation derived by Wigley et al. (7). Their equation expresses the isotope consequences of equilibrium fractionation between multiple inputs and outputs. The input parameters for the Rayleigh model are mass transfer coefficients for carbon, fractionation factors for each output, initial carbon isotope ratio, and the carbon isotope ratio of any input carbon.

Carbon isotope modeling has been used to distinguish closed and open carbonate systems (8), calculate the $\delta^{13}C$ of $CO_2$ dissolved in water for radiocarbon age determinations (9), investigate different modes of calcite deposition in caves (4), and determine sources of water discharging through an arid canyon (10). For Shullsburg groundwater, the inputs and outputs that may affect carbon isotope ratios included $CO_2$ outgassing, dolomite dissolution, calcite precipitation, siderite precipitation, and dilution during water level recovery.

The isotopic composition of carbon in these reactions depend on two factors that can no longer be measured in the Shullsburg mines: the pH during sulfide oxidation and the composition of incoming water that dilutes the contaminated mine water. These factors may be quite variable. The pH in an analogous mine that is still open (measured with pH paper) varied from 5.5 on damp mine walls to 7 in drips from the wall and ceiling (11). During sulfide oxidation, it is possible that acidic microenvironments occur, and the outgassing and precipitation might take place in these microenvironments. In any case, the carbon isotope fractionation factors are constant below a pH of about 5. In the carbon isotope modeling, the pH was varied between 5 and 7 to study the $\delta^{13}C$ effects.

The chemical and isotope composition of the water that may dilute contaminated water is needed to calculate both the sulfate concentration and the $\delta^{13}C$ of the mixture. If sulfate was diluted, the amount of $CO_2$ outgassing was larger than the sampled composition indicates because increased sulfate production requires increased carbonate dissolution, and the DIC gets reduced by $CO_2$ outgassing. An additional effect of dilution is to alter the carbon isotope ratio by the addition of lighter carbon. The carbon isotope ratio of three background samples with low sulfate concentration was measured; the $\delta^{13}C$ varied from -8.6 to -13.7‰. Some of the contaminated samples had even lighter ratios (-14 and -14.7‰), which may indicate that even a lighter range is possible for background water. Siegel (12) estimated the regional carbon isotope ratio to be -12‰, which would be typical of dissolved carbon from mixing of soil gas (-24‰) and carbonate rocks (0‰). The average carbon isotope ratio of the background samples measured was -12‰, and this estimate was used as the initial carbon isotope ratio.

Despite some unknown factors, the isotope modeling shows several trends and points out the possible significance of several reactions that would not be indicated by reaction path modeling alone.

## RESULTS OF CARBON ISOTOPE MODELING

The overall trend calculated in carbon isotopes is toward heavier isotope ratios for carbonate in solution after sulfide oxidation and neutralization. Dolomite dissolution introduces heavier carbon ($\delta^{13}C$ = -1‰, 13). Calcite precipitation reverses this trend by preferentially removing $^{13}C$ ($\epsilon_{ps}$ = 11.8 to 2‰ at 10°C, where $\epsilon_{ps}$ is the isotopic fractionation between calcite precipitated (p) and (s) the solution), but the mass transfer for this step tended to be small. The effect of $CO_2$ outgassing above pH = 5.5 is to make the $\delta^{13}C$ in solution heavier by preferentially removing light carbon ($\epsilon_{gs}$ = 0.5 to -9.0‰ at 10°C, where $\epsilon_{gs}$ is the isotopic fractionation between $CO_2$ outgassed (g) and (s) the solution).

Even when assuming a conservative pH value of 5, most of the calculated $\delta^{13}C$ values were 2 to 4‰ heavier than observed values (Table I), except one sample that matches and one sample that is lighter than observed. If the pH was 7 during the reactions, the model predicts values 4 to 9‰ heavier than observed except for one sample which matches (Table I).

TABLE I: Calculated and observed $\delta^{13}C$ ‰ for different pHs assuming iron hydroxide precipitation. Calculated assuming initial carbon = 8 mmol/L, initial $\delta^{13}C$ = -12‰, dolomite = -1‰, pe = 2.8, starting water S-27. Listed in order of decreasing sulfate concentration

| ID | OBS | CALC pH 5 | CALC pH 6 | CALC pH 7 | $SO_4^=$ mmol/L |
|---|---|---|---|---|---|
| S-2c | -7.3 | -4.8 | -2.9 | 1.6 | 10.8 |
| S-18 | -8.8 | -6.0 | -4.2 | 0.4 | 8.2 |
| S-2a | -10.6 | -6.3 | -4.4 | 0.4 | 7.3 |
| S-19 | -9.5 | -8.0 | -6.9 | -4.1 | 4.7 |
| S-21 | -10.3 | -10.2 | -9.0 | -5.9 | 2.6 |
| S-15 | -14.7 | -10.4 | -9.1 | -6.0 | 2.6 |
| S-16 | -14.0 | -10.1 | -9.0 | -6.5 | 2.5 |
| S-22 | -13.0 | -9.9 | -9.1 | -7.2 | 2.4 |
| S-20 | -6.4 | -10.9 | -9.6 | -6.5 | 2.2 |

Mixing of isotopically light carbon from groundwater refilling the mined area is one explanation for the observed carbon isotope ratios. The amount of carbon in the surrounding groundwater was similar to contaminated water, so mixing would not alter the carbon concentrations significantly. In contrast, introduction of light carbon from dissolution of organic matter would require an additional carbon sink. No information is available on organic matter in the formation, so this hypothesis in not quantified further.

To check for other parameters that might vary enough to explain the observed isotope ratios, a sensitivity analysis was performed. A possible range for each parameter was selected, holding the other parameters constant and using the mass transfer coefficients as calculated by PHREEQE.

DIC of starting water ranged from 6.4 to 11.4 mmol/L with an average value of 8.0. The initial carbon isotope ratio was varied from -6 to -18‰ with the average at -12‰. The $\delta^{13}C$ of the carbonate host rock in the mines varied between -2 and 0‰, so a mean of -1‰ was used in the basic model. In addition, calculations were made using 1 and -3‰. The pe, which affects iron mineral precipitation, was varied from 2 to 4 with the measured value of the mean carbon sample at pe = 2.8 used in the basic model. These variations either had a small effect on the carbon isotope ratio (especially considering the range of values selected) or made it even heavier than the basic model (Table II).

TABLE II: Sensitivity analysis for $\delta^{13}C$ calculations for sample S-18

| MODEL | $\delta^{13}C$ ‰ |
|---|---|
| Basic[+] | -6.0 |
| Initial DIC = 6.4 mmol/L | -7.3 |
| Initial DIC = 11.4 mmol/L[*] | -7.0 |
| Siderite ppt. | -11.0 |
| pH = 7 | 0.4 |
| Dolomite $\delta^{13}C$ = -3‰ | -7.3 |
| Dolomite $\delta^{13}C$ = 1‰ | -4.7 |
| Initial $\delta^{13}C$ = -18‰ | -8.0 |
| Initial $\delta^{13}C$ = -6‰ | -4.0 |
| pe = 4 | -5.9 |
| pe = 2 | -6.5 |
| Kinetic $\epsilon_{gs}$ | 2.4 |
| Kinetic $\epsilon_{ps}$ | -5.1 |
| Observed | -8.8 |

[+]See Table I and text for parameter values.
[*]Calculation based on modified equation for $C \approx C_o$ ([14](#)).

Precipitation of siderite can explain the observed light carbon isotope ratios if the reaction takes place between pH 5 and 6 (Table III). Some of the samples are presently close to saturation with respect to siderite (saturation index, SI = -0.27 to -2.2), and may have been saturated during pre-dilution conditions. Siderite precipitation has a larger fractionation factor than calcite (5.1‰ larger, [15](#)) and leaves the remaining solution lighter. The use of siderite instead of iron hydroxide as a control on the iron concentration creates one of the stronger shifts in the modeled carbon isotope ratio. However, the calculated ratios at pH 7 are still heavier than observed (Table III). These siderite calculations should be considered end members since siderite precipitation has been maximized by omitting iron hydroxide or jarosite from the reaction path. This example demonstrates the importance of determining the appropriate solid phase fractionation factor.

TABLE III: Observed and calculated $\delta^{13}C$ ‰ for different pHs assuming siderite precipitation. Other input parameters listed in Table I and text. Listed in order of decreasing sulfate concentration

| ID | OBS | CALC pH 5 | CALC pH 6 | CALC pH 7 |
|---|---|---|---|---|
| S-2c | -7.3 | -10.7 | -8.7 | -4.8 |
| S-18 | -8.8 | -11.0 | -8.8 | -4.5 |
| S-2a | -10.6 | -11.4 | -9.1 | -4.6 |
| S-19 | -9.5 | -11.1 | -9.7 | -7.0 |
| S-21 | -10.3 | -12.5 | -11.2 | -8.1 |
| S-15 | -14.7 | -12.6 | -11.1 | -8.1 |
| S-16 | -14.0 | -12.1 | -10.9 | -8.5 |
| S-22 | -13.0 | -11.6 | -10.7 | -8.8 |
| S-20 | -6.4 | -12.9 | -11.5 | -8.5 |

The equilibrium constants for mass transfer and isotope fractionation factors used in the models are based on macrocrystalline, pure minerals, under equilibrium conditions, which may not be representative of the complexities of the real system. The assumption that these conditions hold could introduce error in the calculations. Some alternative, nonequilibrium values for the calcite precipitation and $CO_2$ outgassing fractionation factors have been suggested. Michaelis et al. (16) observed kinetically controlled $CO_2$ outgassing in springs that yield an equivalent fractionation factor of -11.1‰ in bicarbonate solutions. This is a larger fractionation than the equilibrium value, so the remaining solution is heavier than during equilibrium fractionation, and kinetic control will not explain the observed lighter values (Table II). Turner (17) and Usdowski et al. (18) measured kinetic isotope fractionation between calcite and bicarbonate as low as 0.35‰ during rapid precipitation. However, the increased enrichment of the solid phase is not sufficient to explain the observed light isotope ratios in solution (Table II).

HYPOTHETICAL MIXING MODEL

If dilution is an important process in producing the observed water chemistry, can the amount of dilution be predicted? To model dilution, the composition of the incoming water must be known; this chemical data is not available, and could vary considerably for different locations. By using a hypothetical dilution water composition, the method for calculating dilution can be explained, and an estimate of dilution can be suggested.

Both the sulfate concentration and the $\delta^{13}C$ of dilution water are needed to calculate dilution. The average carbon isotope ratio of the background samples measured was -12‰, and this estimate was used in the mixing model. Some of the contaminated samples with lower sulfate concentration also had light carbon isotope ratios (Table I). The sulfate concentration of the dilution water is also unknown. However, groundwater flow modeling of the Shullsburg mines (19) indicates that water level rise after mining is dominated by lateral flow. For example, a typical water balance in the model obtained 98% of water level rise from the horizontal flow, as opposed to direct recharge from above. Thus, it is likely that the high sulfate water closest to the mines would be diluted by somewhat contaminated sources that are adjacent (Figure 1). The mode of sulfate concentrations in contaminated samples throughout the study area was between 2 and 3 mmol/L, so an observed sample with 2.6 mmol/L was selected to represent the dilution water.

The mixing model requires simultaneous solutions to $\delta^{13}C$ and $SO_4^=$ dilution because additional outgassing occurs at higher pre-mixing sulfate concentrations. These solutions were obtained graphically by finding the intersection of mixing curves for both sulfate concentration and carbon isotope composition (Figure 4a). The relationship between $\delta^{13}C$ and $SO_4^=$ needed to scale the $\delta^{13}C$ can be obtained with a curve of carbon isotope composition versus sulfate composition calculated by PHREEQE and the Wigley et al. (14) equation for a given final carbon concentration and pH (Figure 4b). The three curves create a triple "point" which provides a solution to the mixing problem.

Three of the samples with heavy calculated carbon isotope ratios (Table IV) could be matched with the selected hypothetical dilution water. The calculated dilution ranged from 45 to 85%. For a given pH, the range was smaller (Table IV). The dilution was also calculated for siderite precipitation at pH 7, and the dilution values were similar to those calculated for pH 5 without siderite precipitation (Table IV). For three samples whose calculated isotope ratios were not greatly different from background, and one sample whose mixing curves did not intersect, a different background water composition needs to be selected. One sample (S-20) had an observed carbon isotope ratio that was heavier than observed. This sample may have been diluted by a more contaminated solution with heavier carbon.

TABLE IV: Calculated % dilution water for different pHs. Calculated using intersection of mixing curves for $\delta^{13}C$ vs. dilution water, $SO_4^=$ vs. dilution water, and $\delta^{13}C$ vs. $SO_4^=$ calculated from PHREEQE and Wigley et al. equation. Sample S-15 assumed composition for dilution water, with $\delta^{13}C$ of $-12°/_{oo}$

| ID | pH 5 | pH 6 | pH 7 | SIDERITE pH 7 |
|---|---|---|---|---|
| S-2c | 44 | 57 | 73 | 51 |
| S-18 | 63 | 72 | 81 | 67 |
| S-19 | 63 | 72 | 84 | 62 |

CONCLUSIONS

Carbon isotope ratios were measured and calculated for groundwater that has been contaminated with sulfate and undergone $CO_2$ outgassing and carbonate dissolution and precipitation. The observed ratios can be explained by some combination of carbon mass transfer calculated from PHREEQE, dilution by isotopically light carbon from background water, siderite precipitation, introduction of light carbon from organic matter, and varied pH and initial water. These possible controls were not predicted by thermodynamic modeling alone.

Of the mechanisms suggested, dilution is a likely influence. Dilution certainly occurred during water level recovery after the mines closed. The amount and influence of dilution cannot be precisely quantified, but isotope modeling suggests dilution factors of 45 to 85%. These dilution factors could saturate the water with respect to siderite, gypsum, anhydrite and cerussite, but not smithsonite or anglesite (2). The pH during oxidation is unknown, but pH 5 or below in the mine environment is not an unrealistic hypothesis, and low pH makes the calculated carbon isotope values shift toward the observed ratios. Siderite precipitation

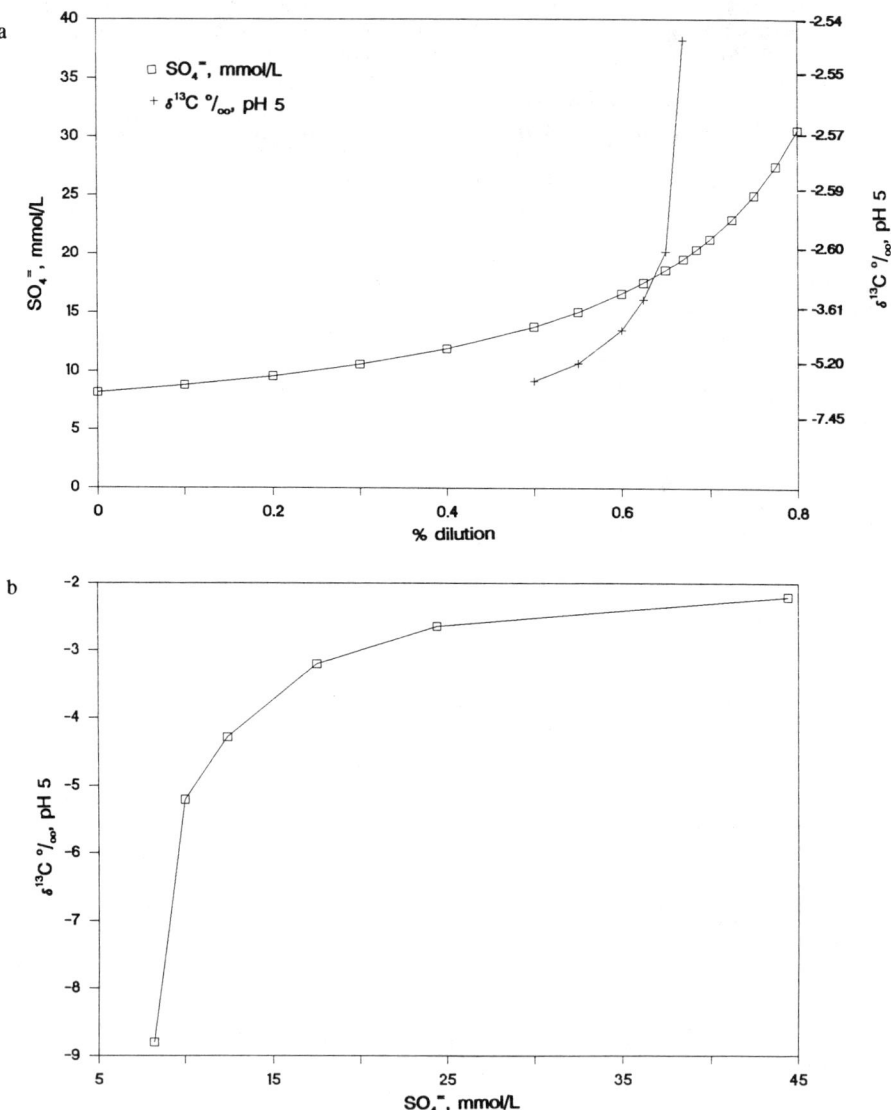

Figure 4. (a) Intersection of mixing curves for sample S-18 used to calculate % dilution water at pH = 5. (b) The scale for $\delta^{13}C$ calculated from mixing curves for $SO_4^=$ versus $\delta^{13}C$ calculated by PHREEQE at different pH levels. Dilution water hypothesized to be sample S-15 ($SO_4^= = 2.6$) with $\delta^{13}C = -12$ o/oo.

may have occurred before the mines flooded, although the water is presently undersaturated according to PHREEQE modeling. This precipitation is probably not the exclusive control on carbon isotope ratios or iron concentrations, and again the amount of precipitation cannot be quantified.

More data on the sulfide oxidizing environment is needed to distinguish the possible carbon isotope models. For example, the pH during oxidation, the extent of siderite precipitation, the composition of incoming dilution water, and the controls on $CO_2$ outgassing are unknown factors that influence the final composition of the contaminated water. Modeling of the carbon isotope data pointed out the importance of these additional factors and the limitations in thermodynamic modeling alone, which did not quantify their influence on the groundwater chemistry.

ACKNOWLEDGMENT

The Environmental Sciences Division is operated by Martin Marietta Energy Systems, Inc., under contract DE-AC05-84OR21400 with the U.S. Department of Energy.

LITERATURE CITED

1. Evans, T.J.; Cieslik, M.J. Wisconsin Geological and Natural History Survey Miscellaneous Paper 85-1 1985, 16 pp.
2. Toran, L. J. Contam. Hydrol. 1987, 2, 1-29.
3. Parkhurst, D.L.; Thorstenson, D.C.; Plummer, L.N. U.S. Geol. Surv. Water Resour. Invest. 80-96 1980, 210 pp.
4. Hendy, C.H. Geochim. Cosmochim. Acta 1971, 35, 801-824.
5. Bogli, A. Karst Hydrology and Physical Speleology; Translated by June C. Schmid; Springer-Verlag: New York, 1980, p 18.
6. Herman, J.S.; Lorah, M.M. Chem. Geol. 1987, 62, 251-262.
7. Wigley, T.M.L.; Plummer, L.N.; Pearson, F.J. Geochim. Cosmochim. Acta 1978, 42, 1117-1139.
8. Deines, P.; Langmuir, D.; Harmon, R.S. Geochim. Cosmochim. Acta 1974, 38, 1147-1164.
9. Reardon, E.J.; Fritz, P. J. Hydrol. 1978, 36, 201-224.
10. Cheng, S.; Long, A.; Adar, E. EOS, Transactions of the Am. Geophys. Union 1984, 65, p 888.
11. Toran, L. Ph.D. Thesis, University of Wisconsin-Madison, Wisconsin, 1986.
12. Siegel, D.I. U.S. Geol. Surv. Prof. Pap. 1405-D 1989, 76 pp.
13. Hall, W.E.; Friedman, I. U.S. Geol. Surv. Prof. Pap. 650-C 1969, 140-148.
14. Wigley, T.M.L.; Plummer, L.N.; Pearson, F.J. Geochim. Cosmochim. Acta 1979, 43, 1395.
15. Golyshev, S.I.; Padalko, N.L.; Pechenkin, S.A. Geochem. Intern. 1981, 18, 85-99.
16. Michaelis, J.; Usdowski, E.; Menschel, G. Am. Jour. Sci. 1985, 285, 318-327.
17. Turner, J.V. Geochim. Cosmochim. Acta 1982, 46, 1183-1192.
18. Usdowski, E.; Hoefs, J.; Menschel, G. Earth Planet. Sci. Lett. 1979, 42, 267-276.
19. Toran, L.; Bradbury, K.R. Ground Water 1988, 26, 724-733.

RECEIVED August 4, 1989

Chapter 15

# Transport of $^{14}CO_2$ in Unsaturated Glacial and Eolian Sediments

**Robert G. Striegl and Richard W. Healy**

**U.S. Geological Survey, Box 25046, Mail Stop 413, Federal Center, Denver, CO 80225**

> Measurements of losses of $CO_2$ to unsaturated sediment-water mixtures indicate that diffusion of $^{14}CO_2$ in the unsaturated zone may be substantially retarded by isotopic exchange of $^{14}C$ to an adsorbed inorganic C phase. Two geochemical models for calculating $^{14}CO_2$ retention in the unsaturated zone were compared. The first accounted only for $^{14}CO_2$ retention caused by $^{14}C$ dilution to dissolved inorganic carbon. The second accounted for additional $^{14}C$ dilution to an adsorbed C phase predicted from $^{14}CO_2$-loss experiments. The geochemical models were separately coupled with a two-dimensional, finite-difference model for gas diffusion to simulate the distribution of $P^{14}CO_2$ in the unsaturated zone near a disposal trench at a low-level radioactive waste-disposal site near Sheffield, Illinois. Comparison of simulated $P^{14}CO_2$ distribution with onsite data supported the presence of the adsorbed C phase.

Carbon exchange among mobile gaseous or aqueous phases and less mobile adsorbed or solid phases can result in alteration of the carbon-isotope composition of $CO_2$ in the unsaturated zone atmosphere and of dissolved inorganic carbon (DIC) in soil water. In locations where gas or water are enriched with $^{14}C$ relative to immobile phases, incomplete accounting of the C-exchange reservoir could result in unrealistically large predictions of $^{14}C$ transport from radioactive-waste disposal sites, or in radiocarbon ages that overestimate actual ages of underlying ground water. In this paper, the transport of $^{14}CO_2$ in the unsaturated zone at a low-level radioactive-waste disposal site is considered. Two geochemical models are used to calculate the loss of $^{14}CO_2$ to isotopically exchangeable phases. The first model uses DIC concentration calculated from calcite equilibria to quantify the exchange reservoir. The exchange reservoir in the second model also includes an adsorbed inorganic C phase determined from measured $CO_2$ isotherms. The geochemical models are separately coupled with a model for gas diffusion in the unsaturated zone to simulate the spatial distribution of $^{14}CO_2$ partial pressure ($P^{14}CO_2$) near a waste-disposal trench. Spatial distributions of $P^{14}CO_2$ predicted by each model are compared to the distribution of mean $P^{14}CO_2$ measured at the site during 1984-1986.

SITE STUDY

The primary gaseous carrier for transport of $^{14}C$ from the low-level radioactive-waste disposal site (lat. 41°20′ N, long. 89°47′ W), near Sheffield, Bureau County, Illinois is $^{14}CO_2$ [1]. Production of the gas is caused by aerobic microbial decomposition of organic waste buried in waste-disposal trenches. This results in steep gradients in $P^{14}CO_2$ in undisturbed glacial and eolian sediments adjacent to the site. To collect samples of unsaturated zone gases, nests of gas piezometers were installed along a cross section in boreholes located at distances of 12, 29, and 46 m from the end wall

This chapter not subject to U.S. copyright
Published 1990 American Chemical Society

of a waste-disposal trench. The piezometers were screened in four lithostratigraphic units at depths of 1.8 to 13.6 m below the land surface (Figure 1).

The uppermost units, the Peoria Loess and the Roxana Silt, are postglacial eolian silts having similar textural composition and porosity. The silts are underlain by the Radnor Till Member of the Glasford Formation, a mottled gray, clayey-silt till that overlies a pebbly sand outwash deposit of the Toulon Member of the Glasford Formation (2). The Toulon Member overlies weathered Pennsylvanian shale of the Carbondale Formation of the Desmoinesian Series (1, 3). Particle-size distributions, porosity values, and surface areas of the sediments are listed in Table I. Surface area was not determined for the Roxana Silt, but was assumed to be similar to that of the Peoria Loess because of the textural similarity of the silts. The water table is generally located at a depth of about 15 m in the Carbondale Formation, near the contact with the Toulon Member. More detailed explanations of the hydrogeology and the water balance at the site are presented in (3) and (4).

Gas samples were collected from the piezometers on 10 occasions during 1984-86. A detailed description of the gas-piezometer installation; gas collection and analytical procedures; and listings of $N_2$, $O_2 + Ar$, $CO_2$, $^{14}CO_2$, $CH_4$ and $^{222}Rn$ partial pressures for specific collection locations and dates are presented in (1).

Partial pressures are products of the total barometric pressure and gas mole fractions. On days when samples were collected, barometric pressures ranged from 97,600 to 99,200 Pa; the average barometric pressure for the period of study was 98,600 Pa. Pressure transducer measurements indicated that differences between barometric pressure at the land surface and barometric pressure at the depths of the piezometer screens were not substantial. Consequently, $PCO_2$ and $P^{14}CO_2$ were calculated using the average barometric pressure as the total pressure. Mole fractions of $CO_2$ were measured directly by gas chromatography, whereas $^{14}CO_2$ mole fractions were a product of the $CO_2$ mole fraction and the measured ratio of $^{14}CO_2$ to $CO_2$ (1). $CO_2$ samples were collected from the Peoria Loess, the Roxana Silt, the Radnor Till Member, and the Toulon Member; $^{14}CO_2$ samples were collected from the Roxana Silt, the Radnor Till Member, and the Toulon Member.

There was little difference in mean $PCO_2$ between boreholes at each depth. Therefore, a single mean $PCO_2$ value was calculated for each lithostratigraphic unit (Table II). Steep horizontal gradients in $P^{14}CO_2$ were observed within lithostratigraphic units, therefore a mean $P^{14}CO_2$ value was calculated for each piezometer sampled (Table III, column B). Variance in mean $PCO_2$ and $P^{14}CO_2$ was greatest at locations nearest to the gas sources. The source for $CO_2$ was root and microbial respiration. This resulted in annual peaks of increased $PCO_2$ near the land surface during the warm growing season; the amplitudes of annual cycles decreased with depth as $CO_2$ diffused downward (1). The predominant source of $^{14}CO_2$ was microbial decomposition of buried organic waste; greatest variance in $P^{14}CO_2$ occurred in the Toulon Member at the piezometer that was closest to the disposal trench. Releases of $^{14}CO_2$ did not exhibit temporal cycles and were apparently determined by the availability of substrate for decomposition.

## GEOCHEMICAL AND TRANSPORT MODELING

Coupling of a geochemical model of the phase distribution of carbon isotopes with a model of gas transport in unsaturated porous media is required for prediction of the spatial distribution of $P^{14}CO_2$ in the unsaturated zone. Because $P^{14}CO_2$ is very small relative to $PCO_2$, seemingly large increases in $P^{14}CO_2$ caused by decomposition of low-level radioactive waste have negligible effect on the total concentration of inorganic carbon. Retardation of $^{14}CO_2$ transport is therefore controlled chiefly by isotopic exchange of gaseous $^{14}C$ to less mobile phases. Quantification of isotope exchange requires knowledge of the initial isotopic ratio and the concentration of C in all exchangeable phases. Assuming that isotopic equilibrium is attained, equilibrium concentrations can then be calculated.

Two geochemical models were used to quantify the exchangeable C reservoir: (1) a theoretical model based on calcite equilibrium control (calcite equilibrium model), and (2) an empirical model based on measured losses of $CO_2$ from a surrogate unsaturated zone atmosphere to unsaturated water-sediment mixtures ($CO_2$ retention model).

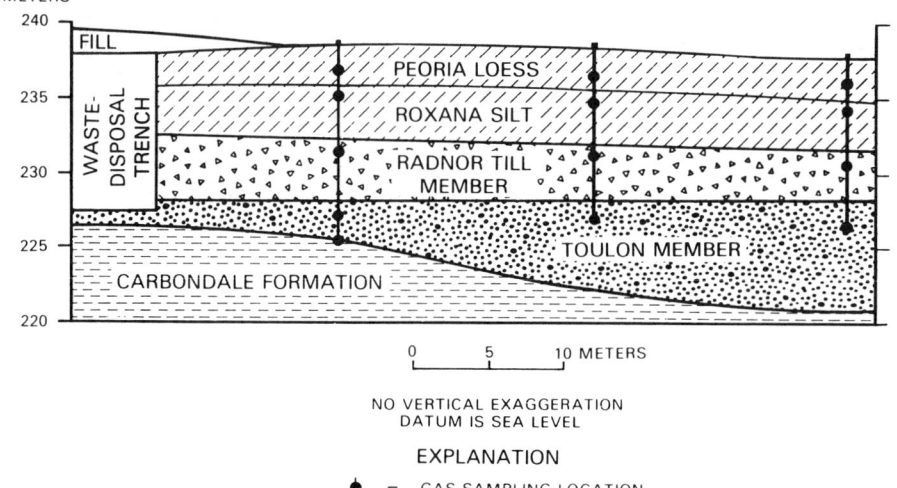

Figure 1. Lithology of the cross section.

Table I. Physical Properties of Unsaturated Sediments

| | Particle size (5) (weight percent) | | | Total porosity (6) $\theta_T$ (volume fraction) | Air-filled porosity (6) $\theta_D$ (volume fraction) | Surface area (7) $SA_s$ ($m^2/g$) |
|---|---|---|---|---|---|---|
| | Sand >62 μm | Silt 4-62 μm | Clay 4 μm | | | |
| Peoria Loess | 4 | 81 | 15 | 0.43 | 0.30 | 9.82 |
| Roxana Silt | 2 | 81 | 17 | .43 | .30 | - - |
| Radnor Till | 18 | 53 | 29 | .32 | .02 | 26.40 |
| Toulon Member | 84 | 10 | 6 | .35 | .29 | 2.14 |

Table II. Mean $PCO_2$ Values (in pascals ± 1 standard deviation)

| Depth (m) | Location | $PCO_2$ ($P_a$) |
|---|---|---|
| 0 | Atmosphere | 33 ± 3% |
| 1.8 | Peoria Loess | 2100 ± 48% |
| 3.6 | Roxana Silt | 2310 ± 31% |
| 7.3 | Radnor Till Member | 3420 ± 19% |
| 11.6, 13.6 | Toulon Member | 3800 ± 5% |

Table III. Simulated and Mean $P^{14}CO_2$ Values For Locations Where $^{14}CO_2$ Samples Were Collected (in pascals)

| Depth, (m) | A | B | C |
|---|---|---|---|
| | Horizontal distance from trench = 12 meters. | | |
| 3.6 | * 5.79 x $10^{-6}$ | 5.80 x $10^{-6}$ ± 92% | * 4.91 x $10^{-6}$ |
| 7.3 | * 1.69 x $10^{-5}$ | 1.19 x $10^{-5}$ ± 97% | * 1.10 x $10^{-5}$ |
| 11.6 | * 4.17 x $10^{-5}$ | 2.54 x $10^{-5}$ ± 103% | * 2.93 x $10^{-5}$ |
| 13.6 | * 1.54 x $10^{-5}$ | 2.03 x $10^{-5}$ ± 120% | * 5.07 x $10^{-6}$ |
| | Horizontal distance from trench = 29 meters. | | |
| 3.6 | + 8.04 x $10^{-7}$ | 3.54 x $10^{-7}$ ± 98% | * 1.16 x $10^{-7}$ |
| 7.3 | + 8.49 x $10^{-6}$ | 3.89 x $10^{-6}$ ± 73% | * 1.83 x $10^{-6}$ |
| 11.6 | + 2.21 x $10^{-5}$ | 7.88 x $10^{-6}$ ± 85% | * 5.85 x $10^{-6}$ |
| | Horizontal distance from trench = 46 meters. | | |
| 3.6 | ++ 4.76 x $10^{-7}$ | 5.90 x $10^{-9}$ ± 70% | * 9.50 x $10^{-9}$ |
| 7.3 | ++ 7.75 x $10^{-6}$ | 7.45 x $10^{-7}$ ± 24% | * 4.56 x $10^{-7}$ |
| 11.6 | ++ 1.58 x $10^{-5}$ | 7.48 x $10^{-7}$ ± 28% | + 1.12 x $10^{-6}$ |

A    $P^{14}CO_2$ simulated using calcite equilibrium model.
B    Mean $P^{14}CO_2$ ± 1 standard deviation (S.D.).
C    $P^{14}CO_2$ simulated using $CO_2$ retention model.
\*    Modeled $P^{14}CO_2$ is in the range of mean $P^{14}CO_2$ ± 1 S.D.
\+    Modeled $P^{14}CO_2$ is greater than mean $P^{14}CO_2$ ± 1 S.D.
++    Modeled $P^{14}CO_2$ is more than 10 times greater than mean $P^{14}CO_2$

### Calcite Equilibrium Model.

Estimates of $^{14}CO_2$ transport are simplified if concentrations of exchangeable inorganic C phases can be calculated from thermodynamic constants and a minimum of site-specific measurements. This is possible if calcite equilibrium control accurately quantifies the exchangeable C reservoir in the unsaturated zone; similar calcite equilibrium control in water-saturated systems is well documented (8, 9, 10). Because isotope exchange to mineral lattices is slow with respect to gas residence times (11, 12), DIC is regarded as the exchange reservoir for gaseous $^{14}C$. For moderately alkaline locations where the DIC concentration is approximately equal to the concentration of $HCO_3^-$, the DIC reservoir can be quantified from the calcite equilibrium reaction:

$$CO_2 + H_2O + CaCO_{3(s)} = 2\ HCO_3^-{}_{(aq)} + Ca^{++}{}_{(aq)}. \tag{1}$$

In terms of $HCO_3^-$ and $PCO_2$, the equilibrium constant for equation 1 is:

$$K_{eq} = [HCO_3^-]^3/2\ PCO_2, \tag{2}$$

which re-arranges to

$$[HCO_3^-] = (2\ K_{eq}\ PCO_2)^{1/3} \tag{3}$$

Equation 3 was previously applied by Thorstenson et al. (13) to describe the relation between $[HCO_3^-]$ and $PCO_2$ in an unsaturated zone in the Western Great Plains. It has direct application to the site near Sheffield where unsaturated zone pH is about 7.5 and calcite coatings on sediment particles are present (1).

### Carbon Dioxide Retention Model.

To simulate onsite conditions, batch experiments were conducted that measured losses of $CO_2$ and $^{14}CO_2$ from a surrogate atmosphere similar to the unsaturated zone atmosphere at the cross section ($N_2$, 0.752; $O_2$, 0.200; $A_r$, 0.009; $CO_2$, 0.039; and enriched with 48 dpm/mL of $^{14}CO_2$) to unsaturated sediments (10 percent water content by mass) collected from the site. Measured $CO_2$ losses were 8 to 17 times larger than losses predicted from Equation 3 (14). The majority of $CO_2$ loss is thought to be dominated by adsorption of bicarbonate (15, 16) or carbonate (16) anions on metal oxide surfaces. Ratios of the relative losses of $CO_2$ and $^{14}CO_2$ in the batch experiments indicate bicarbonate formation.

The reactive surfaces of carbonate minerals are potentially an additional reservoir for $^{14}C$ dilution (12, 17, 18). Exchange of $^{14}C$ to carbonate surfaces was estimated to represent 5 to 15 percent of total $^{14}CO_2$ losses in several batch experiments (14).

For the purpose of transport modeling, measured $CO_2$ losses to unsaturated sediments were quantified by Freundlich isotherms for a $PCO_2$ range commonly present in unsaturated zones (400 to 4000 Pa). The isotherms define the moles of dissolved plus adsorbed $CO_2/m^2$ of sediment surface ($\widetilde{C}O_2$) associated with the deposits in which onsite measurements were made:

eolian silts (Peoria Loess and Roxana Silt),

$$\widetilde{C}O_2 = 2[(1.514 \times 10^{-8})\ P_{CO_2}{}^{0.61}]; \tag{4a}$$

glacial till (Radnor Till Member),

$$\widetilde{C}O_2 = 2[(4.365 \times 10^{-10})\ P_{CO_2}{}^{0.80}]; \tag{4b}$$

and outwash sand (Toulon Member),

$$\widetilde{C}O_2 = 2[(1.698 \times 10^{-10})\ P_{CO_2}{}^{0.97}]. \tag{4c}$$

## CONCENTRATIONS OF EXCHANGEABLE C PER VOLUME OF UNSATURATED ZONE

Coupling of geochemical and gas diffusion models requires common units of concentration for all interacting phases. For the modeled example, the units are moles per m³ of unsaturated zone (gas-water-solid matrix) at *in-situ* conditions.

Gas Phase.

Regarding $CO_2$ and $^{14}CO_2$ as ideal gases at *in-situ* temperatures and pressures, their concentrations in the unsaturated zone can be calculated by:

$$\hat{C}_A = (\theta_D)(P_A)/RT \qquad (5)$$

where
- $\hat{C}_A$ = concentration of gas A in the unsaturated zone, mol/m³;
- $\theta_D$ = air-filled porosity, dimensionless;
- $P_A$ = partial pressure of gas A, Pa;
- $R$ = gas constant, 8.314 Pa m³/K mol; and
- $T$ = mean *in-situ* temperature, 283 K.

Aqueous Phase (Calcite Equilibrium Model)

For *in-situ* conditions, the concentration of $HCO_3^-$ per m³ of unsaturated zone ($\overline{HCO_3^-}$) is:

$$\overline{HCO_3^-} = (\theta_T - \theta_D)(2\ K_{eq}\ PCO_2)^{1/3} \qquad (6)$$

where $\theta_T$ = total porosity, dimensionless.

Assuming that isotopic equilibrium is attained between the gaseous and aqueous phases, the concentration of $H^{14}CO_3^-$ per m³ of unsaturated zone ($\overline{H^{14}CO_3^-}$) is:

$$\overline{H^{14}CO_3^-} = (\theta_T - \theta_D)(2\ K_{eq}\ PCO_2)^{1/3}\ (P^{14}CO_2/PCO_2). \qquad (7)$$

Aqueous plus Adsorbed Phases ($CO_2$ Retention Model)

According to the $CO_2$ retention model the total concentration of aqueous plus adsorbed $CO_2$ per m³ of unsaturated zone ($\overline{CO_2}$) is:

$$\overline{CO_2} = (1-\theta_T)(\tilde{C}O_2)(SA_s)(\rho_s) \qquad (8)$$

where
- $1-\theta_T$ = volume fraction of solids, unitless;
- $SA_s$ = surface area of sediment, m²/g; and
- $\rho_s$ = average density of solids, g/m³.

Assuming that isotopic equilibrium is attained between the gaseous and aqueous plus adsorbed phases, the concentration of aqueous plus adsorbed $^{14}CO_2$ per m³ of unsaturated zone ($\overline{^{14}CO_2}$) is:

$$\overline{^{14}CO_2} = (\hat{C}_{14CO_2})(\overline{CO_2}/\hat{C}_{CO_2}). \qquad (9)$$

GAS TRANSPORT

Ordinary molecular diffusion is generally recognized as the primary mechanism for gas transport in the unsaturated zone ([19, 20, 21]). Fundamental theory for ordinary diffusion according to Fick's first and second laws is presented in ([22]); more extensive theoretical discussions of gas transport in porous media are presented in ([23, 24, 25]).

Transient one-dimensional diffusion of gas A into gas B is given by Fick's second law:

$$D_{AB} \frac{\partial^2 C_A}{\partial x^2} = \frac{\partial C_A}{\partial t} \quad (10)$$

where  $D_{AB}$ = molecular diffusion constant for diffusion of gas A into gas B, m² /s;
$C_A$ = concentration of gas A in the gas mixture, mol/m³ of gas;
x = dimension in the direction of diffusion, m; and
t = time, s.

Development of $^{14}CO_2$ Transport Equations

According to the calcite equilibrium model, Fick's second law for $^{14}CO_2$ diffusion in the unsaturated zone can be generalized as:

$$\tau D_{AB} \frac{\partial^2 \hat{C}_{14CO_2}}{\partial x^2} = \frac{\partial \hat{C}_{14CO_2}}{\partial t} + \frac{\partial \overline{H^{14}CO_3^-}}{\partial t} + \lambda \left( \hat{C}_{14CO_2} + \overline{H^{14}CO_3^-} \right) \quad (11)$$

where  $\tau$ = tortuosity factor for resistance to diffusion caused by the physical structure of the porous medium, dimensionless; and
$\lambda$ = radioactive decay constant, t⁻¹.

For the $CO_2$ retention model, the transport equation is generalized as:

$$\tau D_{AB} \frac{\partial^2 \hat{C}_{14CO_2}}{\partial x^2} = \frac{\partial \hat{C}_{14CO_2}}{\partial t} + \frac{\partial \overline{C_{14CO_2}}}{\partial t} + \lambda \left( \hat{C}_{14CO_2} + \overline{C_{14CO_2}} \right). \quad (12)$$

Because the $^{14}C$ decay constant (1.21 x 10⁻⁴ yr⁻¹) is very small, it was possible to disregard the last term on the right side of Equations 11 and 12 for the waste site example. Substitution of Equations 5 and 7 into Equation 11 gives:

$$\tau D_{AB} \frac{\partial^2 [\theta_D (P^{14}CO_2)/RT]}{\partial x^2} = \frac{\partial [\theta_D (P^{14}CO_2)/RT]}{\partial t} + \frac{\partial [(\theta_T - \theta_D)(2 K_{eq} PCO_2)^{1/3} (P^{14}CO_2/PCO_2)]}{\partial t} \quad (13)$$

for $CO_2$ transport according to the calcite equilibrium model.

The transport equation for the $CO_2$ retention model is similarly obtained by substitution of Equations 5, 8, and 9 into Equation 12:

$$\tau D_{AB} \frac{\partial^2 [\theta_D (P^{14}CO_2)/RT]}{\partial x^2} = \frac{\partial [\theta_D (P^{14}CO_2)/RT]}{\partial t} + \frac{\partial [(1-\theta_T)(\widetilde{C}O_2)(SA_s)\rho_s](P^{14}CO_2)/PCO_2}{\partial t} \quad (14)$$

If it is assumed that $\frac{\partial PCO_2}{\partial t} = 0$, Equation 13 reduces to:

$$\tau D_{AB} \frac{\partial^2 [\theta_D (P^{14}CO_2)/RT]}{\partial x^2} = [\theta_D/RT + (\theta_T - \theta_D)(2 K_{eq})^{1/3} PCO_2^{-2/3}] \frac{\partial P^{14}CO_2}{\partial t} \quad (15)$$

and equation 14 reduces to:

$$\tau D_{AB} \frac{\partial^2 [\theta_D (P^{14}CO_2)/RT]}{\partial x^2} = [\theta_D/RT + (1-\theta_T)(\widetilde{C}O_2)(SA_a)(\rho_s/PCO_2)] \frac{\partial P^{14}CO_2}{\partial t} \quad (16)$$

If $\tau$ is to represent only physical resistance to diffusion, then it needs to be applicable to diffusion of all gases in a medium. To aid in selection of $\tau$, three theoretical (26, 27, 28) and three empirical

(29, 30, 31) formulations were tested to determine the $\tau$ that produced the best numerical fit to $P\text{CH}_4$ distribution at the site near Sheffield (32). $\text{CH}_4$ is chemically more conservative than $^{14}\text{CO}_2$ and, like $^{14}\text{CO}_2$, is produced by decomposition of the buried waste. Computer simulations that applied the tortuosity equations of (27) and (28) produced very good numerical fit of the data. The formulation of Millington and Quirk (28):

$$\tau = \theta_D^{1/3} (\theta_D/\theta_T)^2 \qquad (17)$$

was selected for use in solving Equations 15 and 16.

The molecular diffusion constant ($D_{AB}$) for diffusion of $CO_2$ into air at 273.15 K and 101,325 Pa is $1.39 \times 10^{-5}$ m$^2$/s (33). That value is corrected to $1.53 \times 10^{-5}$ m$^2$/s according to (22, equation 16.3-1) for the mean temperature and atmospheric pressure at the cross section (283 K; 98,600 Pa).

## NUMERICAL MODELING

The purpose of numerical modeling was to simulate the movement of $^{14}\text{CO}_2$ in the unsaturated zone so that the geochemical model that best represents *in-situ* $^{14}\text{CO}_2$ retardation could be determined. This was accomplished by solving the two-dimensional numerical equivalents of Equations 15 and 16 for the movement of $^{14}\text{CO}_2$ in the cross section, and comparing the results of those computer simulations to the measured onsite $P^{14}\text{CO}_2$ distribution.

The simulations were performed using a finite-difference model for solution of the diffusion equation in porous media (34). For modeling purposes, the cross section (Figure 1) was divided into a 3-layered, 36 by 53 block-centered grid. The layers represented the Toulon Member, the Radnor Till Member, and eolian silts (Roxana Silt and Peoria Loess). Physical properties of the unsaturated zone that were used for modeling were identical to those listed in Table I.

The lowermost boundary of the grid was the water table, which was considered to be impermeable to gas diffusion (relative to diffusion in air-filled pore spaces), and was located at the contact of the Toulon Member and the Carbondale Formation. The right boundary of the grid was located about 10 m beyond the borehole that is farthest from the trench. The upper boundary of the grid was the atmosphere, having a constant $P^{14}\text{CO}_2$ of $3.96 \times 10^{-11}$ Pa. The left boundary was the trench wall. A constant source of $P^{14}\text{CO}_2$ of $8.0 \times 10^{-5}$ Pa was assumed at the trench wall. This source is about 3.5 times greater than the mean $P^{14}\text{CO}_2$ in the Toulon Member at the piezometer 12 m to the east of the trench. A time period of 15 years was simulated. This was the approximate length of time that waste was buried.

The results of the numerical simulations and the onsite mean $P^{14}\text{CO}_2$ data are listed in Table III. $P^{14}\text{CO}_2$ values that were calculated using the calcite equilibrium model (column A) were within 1 standard deviation (S.D.) of onsite mean $P^{14}\text{CO}_2$ data (column B) only at the piezometers located in the borehole nearest the trench. Simulated $P^{14}\text{CO}_2$ values for piezometers at the second and third borehole were substantially larger than measured values. Decreasing the source term for $^{14}\text{CO}_2$ at the trench produced a fit to the onsite data at either the second or third borehole, but simulated $P^{14}\text{CO}_2$ values at the two remaining boreholes were unlike the onsite data.

Column C lists simulated $P^{14}\text{CO}_2$ values that were calculated using the $CO_2$ retention model. Simulated $P^{14}\text{CO}_2$ values were within 1 S.D. of the onsite data at 9 of 10 sampling locations.

## CONCLUSIONS

Laboratory measurements of the losses of $CO_2$ and $^{14}\text{CO}_2$ from a surrogate unsaturated zone atmosphere to unsaturated sediments indicate the presence of an adsorbed C phase that can retard $^{14}\text{CO}_2$ transport in the unsaturated zone. Measured losses of $CO_2$ from the atmosphere were 8 to 17 times greater than those predicted by calcite equilibrium calculations. Modeled predictions of $^{14}\text{CO}_2$ transport in a cross section near buried low-level radioactive waste support the presence of the adsorbed C phase; distribution of $P^{14}\text{CO}_2$ was more accurately simulated using a model of $^{14}\text{CO}_2$ retention based on measured $CO_2$-loss isotherms than with a model based on calcite equilibrium control. Failure to account for the adsorbed C phase can lead to substantial errors when using models to estimate $^{14}C$ transport and exchange in the unsaturated zone.

LITERATURE CITED

1. Striegl, R.G. U.S. Geol. Surv. Water-Resour. Inv. Rep. 88-4025, 1988, 69 p.
2. Willman, H.B.; Frye, J.C. Illinois State Geol. Survey Bull. 94, 1970, 204 p.
3. Foster, J.B.; Erickson, J.R.; Healy, R.W. U.S. Geol. Surv. Water-Resour. Inv. Rep. 83-4125, 1984, 83 p.
4. Healy, R.W.; Gray, J.R.; deVries, M.P.; Mills, P.C. Water Resour. Bull., 1989, 25, 381-90.
5. Day, P.R. In Methods of Soil Analysis; Black, C.A., Ed.; Amer. Soc. of Agronomy: Madison, Wis., 1965; pp. 545-66.
6. Vomocil, J.R. In Methods of Soil Analysis, Black, C.A., Ed.; Amer. Soc. of Agronomy: Madison, Wis., 1965; pp. 299-314.
7. Busenberg, E. Dynamic nitrogen adsorption, written commun., 1986.
8. Plummer, L. N.; Busenberg, E. Geochim. Cosmochim. Acta, 1982, 46, 1011-40.
9. Butler, J.N. Carbon Dioxide Equilibria and their Applications; Addison-Wesley Publ. Co.: Reading, Mass., 1982; Chapter 4.
10. Stumm, W.; Morgan, J.J. Aquatic Chemistry; John Wiley and Sons: New York, 1981; Chapter 4.
11. Mozeto, A.A. Ph.D. Thesis, Univ. of Waterloo, Ont., 1981.
12. Mozeto, A.A.; Fritz, D.; Reardon, E.J. Geochim. Cosmochim Acta, 1984, 48, 495-504.
13. Thorstenson, D.C.; Weeks, E.P.; Haas, H.; Fisher, D.W. Radiocarbon, 1983, 25, 315-46.
14. Striegl, R.G. Ph.D. Thesis, Univ. of Wisconsin-Madison, 1988.
15. Aylmore, L.A.G. Clays and Clay Minerals, 1974; 22, 175-83.
16. Russell, J.D.; Paterson, E.; Fraser, A.R.; Farmer, V.C. J. Chem. Soc.: Faraday Trans. I, 1975, 5, 1623-30.
17. Wendt, I. Earth and Plan. Sci. Letters, 12, 1971, 439-42.
18. Garnier, J-M. Geochim. Cosmochim. Acta, 1985, 49, 683-93.
19. Keen, B.A. The Physical Properties of the Soil; Longmans Green: Toronto, Ont., 1931; 380 pp.
20. Evans, D.D. In Methods of Soil Analysis, Black, C.A., Ed; Amer. Soc. of Agronomy: Madison, Wis., 1965; pp. 319-30.
21. Weeks, E.P.; Earp, D.E.; Thompson, G.M. Water Resour. Res., 1982, 18, 1365-78.
22. Bird, R.B.; Stewart, W.E.; Lightfoot, E.N. Transport Phenomena; John Wiley and Sons: New York, 1960; 780 p.
23. Cunningham, R.E.; Williams, R.J.J. Diffusion in Gases and Porous Media; Plenum Press: New York, 1980; 275 p.
24. Mason, E.Z., Malinuskas, A.P. Chem. Eng. Monogr. 17; Elsevier Press: New York, 1983; 194 p.
25. Thorstenson, D.C.; Pollock, D.W. Water Resour. Res., 1989, 25, 477-507.
26. Marshall, T.J. J. Soil Sci., 1959, 10, 79-82.
27. Millington, R.J. Science, 1959, 130, 100-02.
28. Millington, R.J.; Quirk, J.M. Trans. Faraday Soc., 1960, 57, 212-27.
29. Wesseling, J. Neth. J. Agric. Sci., 1962, 10, 109-17.
30. Grable, A.R.; Siemer, E.G. Soil Sci. Soc. Amer. Proc., 1968, 32, 180-86.
31. Lai, S-H.; Tiedje, J.M.; and Erickson, A.E. J. Soil Sci. Soc. Amer., 1976, 40, 3-6.
32. Striegl, R.G.; Ishii, A.L. J. Hydrol., In Press.
33. CRC Handbook of Chemistry and Physics, 60th Ed.; Weast, R.C., Ed., CRC Press: Boca Raton, Flor., 1981; p. F-62.
34. Ishii, A.L.; Healy, R.W.; Striegl, R.G. U.S. Geol. Surv. Water-Resour. Inv. Report 89-4027, 1989.

RECEIVED August 31, 1989

# APPLICATIONS TO MODELING: TRANSPORT AND COUPLED CODES

# Chapter 16

# Evolution of Dissolution Patterns

## Permeability Change Due to Coupled Flow and Reaction

### Carl I. Steefel and Antonio C. Lasaga

### Department of Geology and Geophysics, Yale University, P.O. Box 6666, New Haven, CT 06511

The effects of coupling chemical reactions and fluid flow on the space-time evolution of rock dissolution patterns have been investigated using two dimensional numerical simulations. The simulations focus on the nonlinear, positive feedback between chemical reactions and flow which arises through the permeability of the medium. The feedback may lead to channelization of flow where regions of higher permeability capture sufficient reactive flow that their own propagation accelerates at the expense of adjacent areas. This paper investigates the importance of the rate of fluid flow, the dispersion coefficient, and the rate of reaction on channel formation.

The results of the simulations indicate that the most effective channel propagation occurs when the reactions are transport-controlled (reaction rate constants faster than transport) since in these cases, the channel maximizes its own rate of growth by restricting permeability change to its walls and tip. In this reaction regime, the existence and amplitude of channels depends primarily on the ratio of the flow velocity to the dispersion coefficient and is independent of the reaction rate constant. Large ratios of flow velocity to dispersion result in large channel amplitudes and more finely spaced channels. In the kinetic rate-controlled reaction regime, permeability change is more diffuse, resulting in a decrease in how effectively an individual channel may be propagated and an increase in channel branching. These generalizations are scale-dependent, however, since a given reactive flow regime can produce pervasive permeability change at the pore scale while causing coherent channels at a regional scale.

Reaction-induced or "secondary porosity" has been widely recognized and described in the geological literature (1) and yet relatively little research has been devoted to investigating this phenomenon quantitatively. One of the most interesting and potentially geologically important examples of reaction-induced porosity change occurs where the reaction is coupled to fluid flow, since in this case a positive, nonlinear feedback arises between the flow and the reaction through the permeability (2,3). Where the reaction results in a net molar volume decrease of the solids (in the simplest case, by dissolution) the resulting increase in permeability may divert flow into this area, thus accelerating the rate of permeability increase at the expense of adjacent regions. This instability leads to channeling of flow as the reaction front develops "fingers" rather than propagating as a planar front. The most extreme example of this phenomenon is the karstification of limestone (4), but the process is important on the pore size scale as well since the permeability structure of the rock may be controlled by it. Moreover, channeling may occur on much larger scales and has simply not been recognized as such. To the extent that channeling does occur, it is clearly an important process since it provides, unlike pervasive permeability change, a means of establishing higher permeability zones over larger distances given a limited volume of reactive fluid.

Two fundamentally different regimes can exist: 1) those characterized by transport-controlled reaction and 2) those characterized by kinetic rate-controlled reaction (3). In the case of transport-controlled reaction, the reaction rate constant is much faster than any of the transport processes involved so that the length scale over which a moving fluid comes to equilibrium is small. In this regime, therefore, the walls of a dissolution channel are essentially discontinuities in permeability while in the kinetic rate-controlled case, where equilibrium between the fluid and the reacting mineral occurs over some distance, the boundaries of a channel are blurred by a more gradual permeability change.

The purpose of this paper is to investigate the importance of the rate of fluid flow, the dispersion coefficient, and the rate of reaction on channel formation. The analysis of Ortoleva et al. (2), who focussed primarily on a linear stability analysis of the problem, will be extended by examining the full range of non-linear effects involved in reaction-induced channelization of flow.

## MATHEMATICAL FORMULATION OF THE CONSERVATION EQUATIONS

The conservation equations for flow, mass transport, and reaction in porous media are formulated using a continuum approach in which a representative elemental volume (REV) is chosen which is assumed to be smaller than the length scale of the phenomena being monitored and within which the various properties are assumed to be constant.

### Continuity Equation

The continuity equation, which describes the conservation of fluid mass in the system, can be written

$$\frac{\partial \phi}{\partial t} = -\nabla \bullet \mathbf{q} \qquad (1)$$

where $\phi$ is the porosity, $\mathbf{q}$ is the fluid velocity vector, and where we have neglected changes in fluid density due to either temperature or to concentration changes.

### Conservation of Momentum

For the conservation of momentum, Darcy's Law is used

$$\mathbf{q} = -\frac{\mathbf{k}}{\mu}(\nabla P - \rho_f \mathbf{g}) \qquad (2)$$

where $\mathbf{k}$ is the permeability tensor, $\mu$ is the fluid viscosity, $P$ is the fluid pressure, $\rho_f$ is the fluid density, and $\mathbf{g}$ is the gravity vector. The Darcian velocity $\mathbf{q}$, which is a volume flux in units of volume of fluid per unit area of porous medium per unit time, is related to the true fluid velocity $\mathbf{v}$ by the expression $\mathbf{q} = \phi \mathbf{v}$. In practice the fluid pressure $P$ is solved for by substituting the Darcian velocity $\mathbf{q}$ from equation 2 into equation 1.

A relationship between porosity and permeability based on the Kozeny-Carman equation (5) is used in which porosity is modelled as a bundle of capillary tubes of equal length. The Kozeny-Carman equation is given by

$$\mathbf{k} = C_k \frac{\phi^3}{(1-\phi)^2} \frac{1}{S^2} \qquad (3)$$

where $S$ is the total surface area of solids per unit volume of solids and $C_k$ is a shape factor for which the value 0.2 is generally used (5,6). With this formulation it is possible to calculate both the total surface area of solids and the permeability as they evolve with time. The total surface area can also be related to the reactive surface area of the rock if the simplifying assumption is made that the average radius of cylinders of the reactive mineral is the same as the other minerals in the rock.

Conservation of Solute Mass

A differential equation expressing conservation of mass of the species $i$ in solution must contain terms which account for changes in the concentration of $i$ in the fluid phase due to chemical reaction. We also need to account for the flux of $i$ into and out of the REV. The conservation of the $i^{th}$ species, in units of moles per unit volume of porous medium per unit time, can be written

$$\frac{\partial(\phi C_i)}{\partial t} = -\sum_\alpha \nabla \bullet \mathbf{J}_\alpha + R_i \tag{4}$$

where $C_i$ (in units of moles/m³ fluid) refers to the concentration of some species in solution, $\mathbf{J}_\alpha$ is the sum of the fluxes described below, and $R_i$ is the net rate (moles $i$/m³ rock/s) for all the chemical reactions involving $i$ that result from mineral-solution interaction (7).

The dispersive flux (moles $i$/unit area rock/s) is given by

$$\mathbf{J}_{disp} = -\phi \mathbf{D_h} \bullet \nabla C_i \tag{5}$$

where $\mathbf{D_h}$ is the dispersion tensor defined as the sum of the mechanical dispersion $\mathbf{D}$ and of the molecular diffusion in a porous medium, $\mathbf{D_d^*}$ (8)

$$\mathbf{D_h} = \mathbf{D} + \mathbf{D_d^*} \tag{6}$$

The advective flux (moles $i$/m² rock/s) is given by

$$\mathbf{J}_{adv} = \mathbf{q} C_i \tag{7}$$

It is through this term that the coupling between the flow and reaction arises since the Darcian flow velocity $\mathbf{q}$ depends on the permeability of the medium, which is in turn affected by the chemical reactions.

If we restrict the treatment here to an idealized system involving a single chemical component and a single reacting phase, the term $R_i$ can be written simply as

$$R_i = -\nu_i \phi A_\theta k (C - C_{eq}) \tag{8}$$

where, $\nu_i$ is the number of moles of $i$ in mineral $\theta$, $A_\theta$ is the reactive surface area of the mineral (m²/m³ rock), $k$ is the reaction rate constant (m/s), and where the reaction is assumed to follow first-order kinetics (7). In the numerical simulations which follow, the reactive surface area is held constant to simplify the interpretation of the results, although in practice it can be continually recomputed. Since the reaction term $R_i$ represents the number of moles of the species $i$ per second precipitated as or dissolved from mineral $\theta$, the reaction term $R_i$ can be written

$$R_i = -\nu_i V_\theta^{-1} \frac{\partial \phi_\theta}{\partial t} \tag{9}$$

where $\phi_\theta$ and $V_\theta$ are the volume fraction and molar volume respectively of mineral $\theta$. In this case where mineral $\theta$ is the only one reacting and therefore the only mineral whose volume fraction changes, equation 4 can be written as (for $\nu_i = 1$)

$$V_\theta^{-1}\frac{\partial \phi}{\partial t} = \frac{\partial(\phi C_i)}{\partial t} + \sum_\alpha \nabla \bullet \mathbf{J}_\alpha \qquad (10)$$

where the porosity is expressed by $\phi = 1 - \phi_\theta$.

## INITIAL AND BOUNDARY CONDITIONS

The problem considered here is a geometrically simple one in two dimensions designed to simulate isothermal flow and reaction in a medium in which some percentage of the rock is reactive (e.g, a carbonate cement) while the remainder is treated as inert (e.g, a quartz sandstone at low temperature). The analysis presented here should apply in most respects to the case where the entire rock is reactive (e.g., a pure limestone), although in this instance the flow can no longer be treated as Darcian. The problem as formulated here is essentially the same as that considered by Ortoleva et al. (2). Although our formulation is based on a one-component system, the results should be broadly applicable to relatively simple multi-component reactions (e.g., calcite dissolution).

In order to investigate the full range of non-linear effects, the governing equations must be solved numerically. A rectangular grid is used, two ends of which are fixed at some constant fluid pressure and the sides of which have no flow boundary conditions (Figure 1). Only the net pressure drop across the region of interest is important since the density of the fluid is assumed constant, therefore the outlet of the system is fixed at $P = 0$. The inlet of the system is fixed at a pressure which gives a pressure gradient of 0.007MPa/m resulting in a Darcian flow velocity of about 0.3 m/yr for the permeabilities of 1 to 10 millidarcies chosen here. The inlet concentration is fixed at $C_o$ such that the fluid is undersaturated with respect to the reactive mineral phase, while the outlet is fixed at the equilibrium concentration, $C_{eq}$. As initial conditions, we consider a reaction front with a "scallop" or irregularity in it, on the upstream side of which is inert porous medium and on the downstream side of which is a lower permeability and lower porosity zone which contains 5% by volume reactive cement. With this formulation, it is possible to follow the evolution of both the reaction front as it moves downstream and the flow field as the permeability changes in the aquifer.

## NUMERICAL METHODS

The coupled set of nonlinear differential equations (equations 1 and 4) are solved by the alternating direction implicit (ADI) method (9-10) on an evenly spaced grid. The advective transport of a solute species was solved using the Lax-Wendroff two-step method (10). To ensure that numerical dispersion is avoided, a grid spacing was chosen such that the grid Peclet number (defined by $\frac{v\Delta X}{D}$) $\leq 2$ (11). The computational expense involved in using a fine grid spacing required that only relatively small ratios of the advection to the dispersion term be considered.

The quasi-steady state approximation was used to solve Equation 4 (12,2). This approximation holds where the time required to change the porosity of the rock is long compared to the time required to achieve steady-state with respect to solute concentrations. Therefore, we can write

$$\frac{\partial C}{\partial t} \simeq 0 \qquad (11)$$

which allows us to take much larger time steps in calculating changes in porosity and permeability by using a series of quasi-steady state solutions to the conservation of solute mass equation. The approximation is generally valid for simple cases where the concentration in solution of some species is much less than the density of the mineral phase with which it is reacting (2). In more complicated problems where additional transient terms enter (e.g., temperature or porosity change due to precipitation of a second mineral), the quasi-steady state assumption with respect to concentration in the fluid phase may not be justified.

The steady-state form of the continuity equation is solved. The reaction front is tracked using the enthalpy method (13) in which the location of the front remains uncertain to within the grid spacing $\Delta x$.

## NONDIMENSIONAL PARAMETERS

It is often useful to non-dimensionalize a differential equation, not only because the resulting equation is generally both simpler in form and more general in its applicability, but also because it provides a convenient way of comparing the relative importance of terms in the equation (e.g., the importance of diffusive versus advective flux). Two important non-dimensional parameters which appear in the conservation of solute mass equation are the Peclet number and the Damkohler number. The Peclet number is given by

$$Pe = \frac{v\ell^*}{D_h} \tag{12}$$

where $v$ is the true velocity, $D_h$ is given by Equation 6, and $\ell^*$ is the length scale of interest. The Damkohler number (14) can be written as

$$Da = \frac{A_\theta k \ell^*}{v} \tag{13}$$

where $A_\theta k$ has same meaning and units as in Equation 8. The Peclet number is a measure of the relative importance of the advective to the dispersive flux over a given length scale while the Damkohler number is a measure of the importance of the reaction rate constant (combined with the reactive surface area) versus the flow velocity for a given length scale. When the Damkohler number is $\gg 1$, for instance, the rate of the chemical reaction is much larger than the rate at which the solute species is transported by flow and the solution remains close to equilibrium with the reacting mineral phase. In contrast, where the Damkohler number is $\leq 1$, the reaction rate cannot keep up with the advection term and local disequilibrium results. It is important to stress here that both the Peclet number and the Damkohler number depend critically on the length scale considered.

Although the problem considered here is both two-dimensional and non-linear, the simplest form of the one-dimensional reaction-transport equation,

$$\frac{\partial C}{\partial t} = D_h \frac{\partial^2 C}{\partial x^2} - v \frac{\partial C}{\partial x} - k_{eff}(C - C_{eq}) \tag{14}$$

where $k_{eff} = A_\theta k$, yields some additional insight into the importance of the Damkohler number. An analytical solution to Equation 14 has been presented by a number of workers (15,8,12). Lichtner (12) presents a steady-state solution for the case where a reaction front is located at the position $\ell$

$$C(x) = C_{eq} - (C_{eq} - C_\ell)e^{-\gamma(x-\ell)} \tag{15}$$

where the important quantity $\gamma$ is defined by

$$\gamma = \frac{v}{2D_h}\left[\left(1 + \frac{4D_h k_{eff}}{v^2}\right)^{1/2} - 1\right] \tag{16}$$

The dimensionless quantity $4D_h k_{eff}/v^2$, a combination of the Damkohler and Peclet numbers discussed above, provides an approximate measure of the relative importance of reaction, dispersion, and advection in determining a solute concentration profile. As pointed out by

Lichtner (12), the quantity $\gamma^{-1}$ which appears in Equation 15 gives the approximate distance from the reaction front at which the fluid will be in equilibrium with the mineral involved in the reaction. Where the reaction rate is infinitely fast, $\gamma^{-1} \to 0$ so that the local equilibrium condition is recovered (12). In the case where $4D_h k_{eff}/v^2 \leq 1$ and therefore the quantity $\gamma^{-1}$, which we might refer to as the equilibration length scale, is greater than zero, the reaction front will be spread out over some finite distance. Where the reaction involves a permeability change, $\gamma^{-1}$ provides a measure of the distance over which this change occurs. As demonstrated below, the equilibration length scale has an important effect on whether a channel of a particular size will be stable or not.

## TRANSPORT-CONTROLLED REGIME

A numerical simulation of the unstable growth of a dissolution channel is shown in Figure 2 as it grows from some initial small amplitude by capture of flow from adjacent areas (indicated by the streamlines). Note that in this particular simulation the boundaries of the channel are very sharp since we have chosen a value of the reaction rate constant which is significantly larger than the rate of advection. Ortoleva et al. (2) have presented a linear stability analysis which applies to such a case of transport-controlled reaction (their analysis assumes an infinitely thin reaction front). They find that the stability of the planar reaction front (i.e., whether a perturbation will grow or decay) depends primarily on the local Peclet number, that is the ratio of the fluid velocity over the given length scale to the dispersion coefficient. Rewriting slightly equation IX.B3 from Ortoleva et al. (2), we obtain

$$Pe_{crit} = \frac{v_{crit} L_y}{D_h} = \frac{\pi(3-\Gamma)(1+\Gamma)}{2(1-\Gamma)} \tag{17}$$

where $v_c$ is the critical velocity evaluated upstream of the reaction front, $\Gamma$ is the quantity $\phi_{in} k_{in}/\phi_f k_f$ where the subscripts $in$ and $f$ refer to the initial and final medium properties respectively, and $L_y$ is the width of the zone from which flow could potentially be captured (i.e., the width of the aquifer or the distance between channels). If $L_y$ is too small relative to the ratio of the flow velocity to the dispersion coefficient, then the undersaturation necessary to drive the unstable growth of the channel is wiped out by dispersion. As pointed out by Ortoleva et al. (2), any value of $v/D_h$ can lead to an instability if the width of the aquifer is large enough.

Since the linear stability analysis applies only to small perturbations from the originally planar geometry of the reaction front, it does not consider the effects of dispersion on the concentration profile as the fluid moves down the channelway. Figure 3 shows the results of a numerical simulation and the linear stability analysis in which the length of the channel is plotted against time. Note that the channel growth in the numerical simulation initially agrees closely with the results of the stability analysis, but at some channel length we see a divergence of the two curves which is due to the effects of dispersion from the sides of the channel.

The point where the divergence between the stability analysis and the numerical results occurs can be predicted roughly if we make the approximation that the walls of the channel represent two plane sheets. For diffusion between two infinite plane sheets, the results depend on the dimensionless ratio $D_h t/a^2$ where $a$ is the half-width between the two planes (16). The calculations by Crank show that the concentration of a species in solution at the center of the channel (i.e., a distance $a$ from the walls) will begin to be affected when the parameter $D_h t/a^2$ approaches the value of about 0.1. Making the substitution

$$t = \frac{L_z}{\bar{v}} \tag{18}$$

we can calculate an approximate channel length, given by $L_z$, at which dispersion from the sides will become important for some average flow velocity $\bar{v}$ down the channel. In the simulations presented in Figure 3, the value of the parameter $D_h L_z/\bar{v} a^2$ was found to be

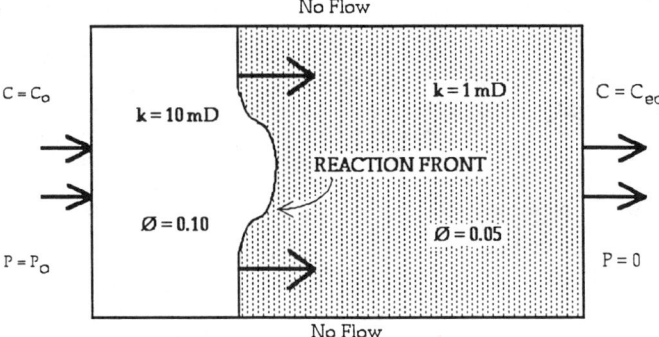

Figure 1. Initial and boundary conditions for the model problem. Undersaturated flow enters from the left, causing the reaction front to migrate gradually downstream. Shaded portion of the aquifer contains 5% reactive cement, resulting in a porosity of 5% and a permeability of 1 millidarcy. Upstream of the reaction front the porosity is 10% and the permeability is 10 millidarcies.

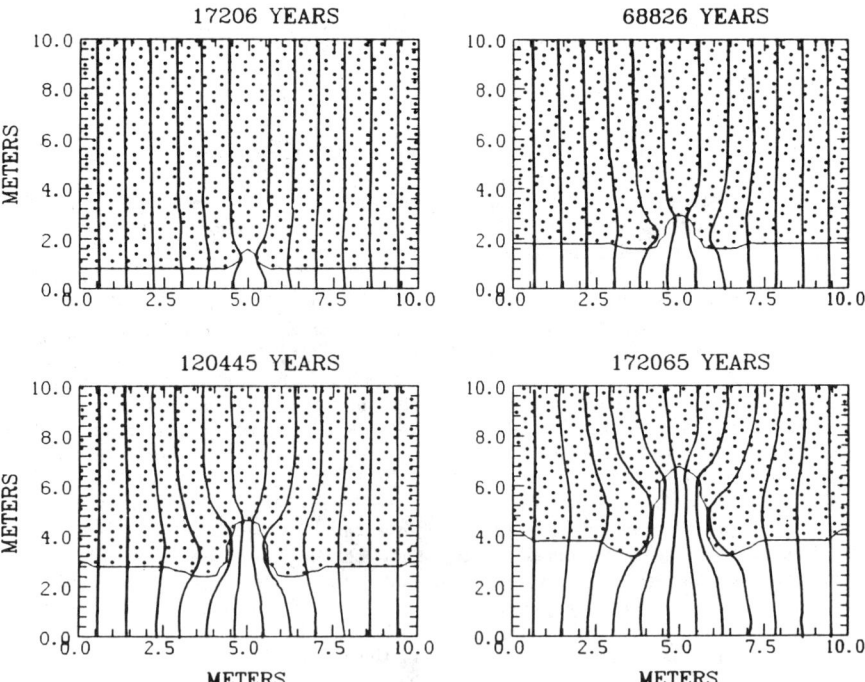

Figure 2. Unstable growth of a dissolution channel in the transport-controlled reaction regime. Flow is from bottom to top, with the cement-bearing portion of the aquifer shown as shaded. Note the convergence of flow into the dissolution channel. Times are based on the dissolution of calcite using an undersaturation of $1 \times 10^{-4} M$ calcite. The Darcian velocity upstream of the front is $1 \times 10^{-8} ms^{-1}$ ($\sim 0.3 myr^{-1}$), the dispersion coefficient is $1 \times 10^{-8} m^2 s^{-1}$.

about 0.1, suggesting that this is a useful method for estimating the channel length at which the linear stability analysis will break down.

The fact that the rate of growth of a channel is slowed by the effects of dispersion from the sides of the channel once it attains some length suggests that there should be a final, steady-state channel length where the effects of dispersion and flow are in balance. This steady-state channel length will never be achieved, however, where the channel continues to capture flow from other regions, since in this case the rate of fluid flow in the channel will continue to rise. The case we consider here is where the channel can capture flow from the full width of the aquifer and the steady-state channel length is attained only once this has occurred. Figure 4 shows the results of a series of simulations using aquifer widths of 3, 5 and 7 meters and carried out at various values of $v^2/k_{eff}D_h$. When the aspect ratio of the channels (the ratio of the channel length to the aquifer width) is plotted versus the nondimensional parameter $v^2/k_{eff}D_h$, a straight line with positive slope is obtained. These results are strictly applicable to a fairly narrow range of Damkohler numbers, however. As shown below, the behavior differs both for very large Damkohler numbers (where the reaction rate appears to have no effect and only the Peclet number is important) and for very small Damkohler numbers (where channels cannot propagate at all).

The dependence of the final channel length on the diameter of the channel is less clear but, in contrast to Hoefner and Fogler (3), the dependence cannot be zero. This is because the length of the channel depends on the ratio of the characteristic time required to diffuse in from the walls of the channel (which depends on $a^2$, the square of the channel half-width) and the characteristic time required for a flow packet to traverse the length of the channel (given by $L_z/v$). Accordingly, one would expect that since the time required for diffusion and dispersion depends on the square of the distance, the length of the channel (which together with the flow velocity determine the time required for advective transport) should follow the square of the channel half-width. This may be true for channels with lengths much greater than their widths, but in the case of channels with small aspect ratios modelled here, the channel length $L_z$ and the channel width $L_y$ are linearly related (Figure 4). This may be due to the pronounced convergence of flow toward the center of the channel which occurs in these small aspect ratio channels. In any event, it is clear that larger aquifer widths will result in larger channel amplitudes.

Channel width may have an important effect where two channels are competing for flow. Given two channels of approximately the same length, fluid in the wider channel will arrive at the tip of the channel less affected by dispersion than the fluid in the narrower channel, resulting in different rates of channel growth. Partly because of this effect, it is unlikely that two separate channels can be maintained indefinitely. Differences in length between competing channels may have a similar effect since the longer channel should eventually capture the flow of the smaller one (3).

The simulations presented in Figures 4 were carried out for a relatively narrow range of ratios of flow velocity to dispersion coefficient because of the difficulties in accurately and stably calculating solute transport in the high Peclet number case. There are several geological situations, however, where this ratio may become very large. One possible example is where dissolution of a relatively pure carbonate with no inert matrix occurs such that the permeability jump across the reaction front is very large. In this situation, the ratio $v/D_h$ increases dramatically because 1) the high permeability in the channels results in high rates of fluid flow and 2), in the absence of an inert matrix, dispersion will be minimal, consisting only of the contribution by molecular diffusion. In a relatively pure limestone, for instance, we expect that channels can attain very large amplitudes while in calcite-cemented sandstone, channels should be of much smaller amplitude because of the importance of dispersion and the slower flow rates.

KINETIC RATE-CONTROLLED REGIME

While the formation of coherent channels is favored in the transport-controlled case because of the ability of the channel to capture flow and restrict permeability change to the walls and tip of the channel, in the case of kinetic rate-controlled reaction, the permeability change is

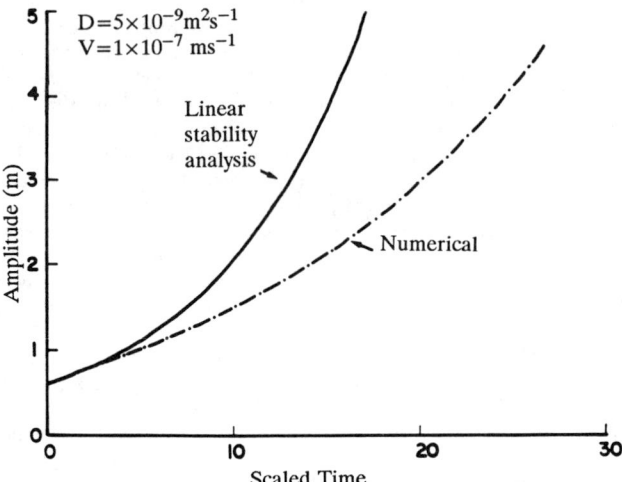

Figure 3. Comparison of results of numerical simulation with the linear stability analysis (2). Note the initial agreement between the numerical and linear stability analysis. The divergence of the nonlinear numerical simulation and the linear stability analysis occurs where dispersion from the sides of the channel affects the degree of undersaturation. Time is in units of $1 \times 10^6$ seconds, scaled by the product of the molar volume and the degree of undersaturation of the fluid (i.e, $V_\theta [C_o - C_{eq}]$).

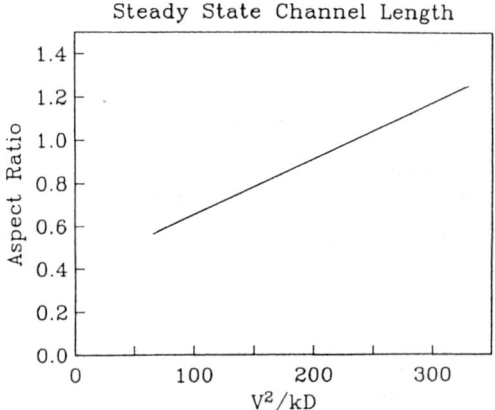

Figure 4. The steady-state aspect ratio (length/width) of the channel versus the nondimensional parameter $v^2/kD$. Results are valid only for a limited range of Damkohler number. Aquifer widths of 3, 5, and 7 meters were used.

spread out over some finite distance. The effect of varying the effective reaction rate constant ($k_{eff} = A_\theta k$) on channel length after a given time (for a constant flow velocity and dispersion coefficient) is shown in Figures 5 and 6. Note that at Damkohler numbers less than about 0.01 (where the length scale is given by the width of the aquifer) the permeability change occurs over a sufficiently large distance that the channel cannot propagate efficiently. In the limiting case where the Damkohler number is very small, any perturbation of the planar reaction front decays away altogether (i.e., channels cannot be propagated at all).

It is interesting to note in Figures 5 and 6, however, that the channel length is greater for a Damkohler number of 0.05 than it is for 0.1. This is because the growth rate of the channel depends on the quantity $4D_h k_{eff}/v^2$ (or Da/Pe) rather than solely on the ratio $v/D_h$. For slower reaction rates, the walls of the channel are not held at the equilibrium concentration (as in the transport-controlled case), thereby decreasing the concentration gradient (and thus the dispersive-diffusive flux) in the channel. Therefore, Figure 6 indicates that the length of the channel will initially increase with decreasing Damkohler number (in the range Da = 0.1 − 0.01).

Another effect of decreasing the reaction rate should be to make the probability of propagating any one channel more nearly equal. In the transport-controlled case, branching of channels may occur, but one of the channels will quickly dominate at the expense of the other because of the efficiency of reactive flow capture. In the kinetic rate-controlled case, any one channel will be less able to capture flow from another. Therefore, in a porous medium with a random distribution of permeability, we expect that the degree of branching should be greater in the kinetic rate-controlled case.

This has been borne out in a series of elegant experiments conducted by Hoefner and Fogler (3). They began by pumping mildly acidic water through both limestone and dolomite cores followed by injection of a low-temperature metal alloy to preserve the reaction-induced channel structure. The remainder of the carbonate was then dissolved in strong acid, leaving only the alloy cast of the dissolution channels. Hoefner and Fogler observed a systematic variation in channel structure with Damkohler number, with injections at high Damkohler number (close to the local equilibrium condition) showing coherent channels with a minimum of branching and with injections at lower Damkohler number showing a high degree of branching and very diffuse to nonexistent channels. In the transport-controlled case represented by the dissolution of calcite, although channel branching occurred, only a single channelway eventually survives. In the limiting case of an extremely low Damkohler number, no single channel is favored at all and dissolution was observed to occur pervasively over the entire rock.

The dependence of the channel structure on the Damkohler number, therefore, leads to two completely different rates of permeability change in the transport-controlled and the kinetic rate-controlled cases and, in fact, to a reversal in the effect of an increase in flow velocity. In both cases, Hoefner and Fogler (3) found in their experiments that the permeability rose dramatically once the dissolution channels reached the other end of the core and a breakthrough occurred. As we would expect in the transport-controlled case, an increase in the flow velocity decreased the time required to achieve the breakthrough. In the kinetic rate-controlled case (represented by dissolution of dolomite), however, increasing the flow velocity had the opposite effect of increasing the time required for the breakthrough. This occurs because an increase in the flow rate (which causes the Damkohler number to decrease) lessens the efficiency with which channels can be propagated.

## CHARACTERISTIC LENGTH SCALES FOR CHANNEL FORMATION

We have tried to stress the importance of the length scale on the phenomena we are considering. Length scale is clearly important in determining both the Peclet and the Damkohler numbers. Moreover, our ability to recognize patterns in the field depends on the characteristic length scale of these patterns. While channeling in a pure limestone should be easily observable on the outcrop scale, channeling in a metamorphic belt characterized by large dispersion coefficients and/or low reaction rate constants may be recognizable only through regional mapping. In the transport-controlled regime, the critical parameter is the Peclet number which ultimately determines whether a channel will be able to propagate at all and,

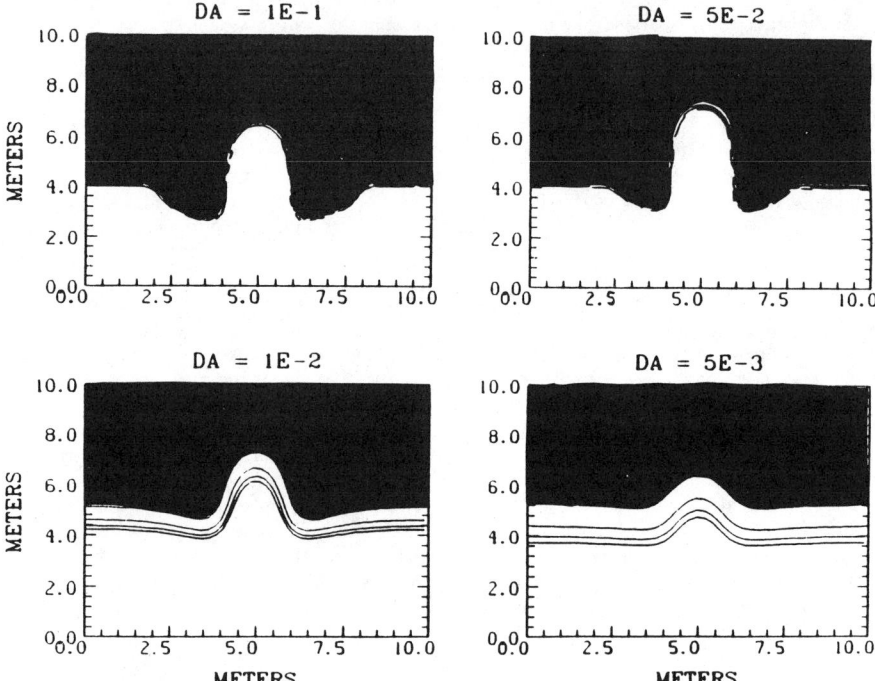

Figure 5. Contours of permeability at the same point in the time evolution of a channel for a variety of Damkohler numbers. Flow is from bottom to top and the shaded region contains the reactive cement. The initial perturbation is a 1 meter high by 0.3 meter wide zone. Permeability contours are equally spaced at increments of 2 millidarcies. True velocity of $1 \times 10^{-7}$ m/s and a dispersion coefficient of $1 \times 10^{-8}$ m$^2$/s. The length scale is taken as 10 meters, the width of the aquifer.

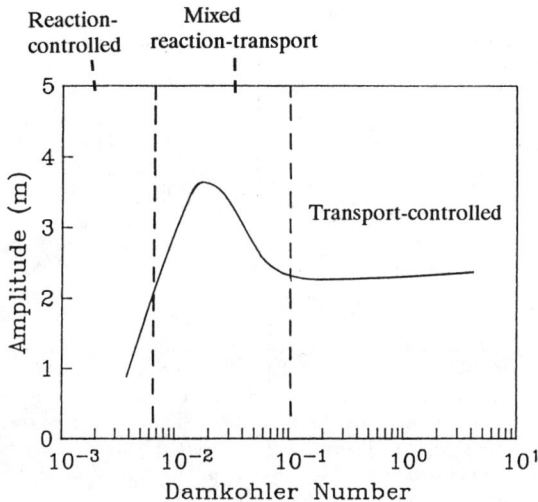

Figure 6. Channel length at the same point in time versus Damkohler number. In the transport-controlled regime, the channel length is independent of the reaction rate. As the Damkohler number decreases from 0.1 to 0.01, the length of channels increase (see text for discussion). At Damkohler numbers below about 0.01, dissolution becomes pervasive and the channel does not propagate.

if so, what the fastest growing wavelength (distance between channels) will be (2). In the case of kinetic rate-controlled dissolution, the critical parameter is the ratio of the equilibration length scale $\gamma^{-1}$ to the size of the channel. If we are considering channels which are much larger than the equilibration length scale, then these channels will propagate coherently even though on the pore scale the permeability change appears pervasive.

To illustrate the importance of the length, consider initially a "fast" dissolution reaction. The rate of dissolution of calcite at $80°C$ is about $3.4 \times 10^{-5}$ moles/m$^2$/s (17). In order to cast the rate constant for use in Equation 8, it must be converted to units of s$^{-1}$ by dividing by the equilibrium concentration which we take as $1 \times 10^{-4}$M Ca$^{+2}$ and the porosity. Assuming a reactive surface of 1 m$^2$/m$^3$ rock, the rate constant is $3.4 \times 10^{-3} s^{-1}$. On the basis of the results of the numerical simulations presented above, channel propagation begins to become inefficient at a Damkohler number less than about 0.01. Therefore, for a Darcian flow velocity of 10 meters per year and a porosity of 0.1 (so $v_{true}$ = 100 m/yr), the length scale corresponding to a Damkohler number of 0.01 is approximately 9 micrometers. Because calcite should be essentially at local equilibrium, limestone makes the ideal rock within which to form coherent channels at all length scales. When we consider some of the major rock-forming minerals, however, which either dissolve or form at much slower rates, the permeability structure which develops will depend on the scale of the observation. As an example, consider the rate of dissolution of quartz at $80°C$ which is about $4.8 \times 10^{-12}$ moles/m$^2$/s (18). An equilibrium concentration of $\sim 5 \times 10^{-4} M$ gives a reaction rate constant of $9.6 \times 10^{-11} s^{-1}$ which, for a Damkohler number of 0.01 (assuming again a flow rate of 10 meters/yr and porosity of 0.1), yields an approximate length scale of 330 meters. This indicates that only channels larger than about 300 meters will propagate coherently in a silicate rock in which the reactions proceed at a rate similar to quartz. Another possible example is given by the transformation of calcite to dolomite which involves a net molar volume decrease of 18% and is notoriously slow at the lower temperatures characterizing sedimentary basins. Accordingly, we expect that where dolomitization forms at less than hydrothermal temperatures, the reaction will be pervasive on the pore level but that at some larger scale (which will depend in detail on the as yet undetermined reaction rate constant) the reaction front may show channeling.

CONCLUSIONS

Both the existence and the geometry of dissolution channels in rocks depend critically on two key non-dimensional parameters: the Peclet number, which gives the ratio of advective to dispersive flux for a given length scale, and the Damkohler number, which measures the relative importance of the reaction rate constant versus advection over some length scale. The Damkohler number determines whether the reaction regime will be transport-controlled or kinetic rate-controlled, for which we observe two fundamentally different kinds of permeability change. In the case of transport control, where the local equilibrium assumption approximately holds, channel formation is favored because reaction-induced permeability change is focused along the walls and tips of discrete channels. In this regime, the growth of channels is independent of the reaction rate constant and depends primarily on the Peclet number. Large ratios of advective to dispersive flux result in the maximum rates of channel propagation and the largest final lengths for the channels. Where the ratio of the flow velocity to the dispersion coefficient is small, as, for example, in most metamorphic systems, channeling may develop only on scales which are recognizable by regional mapping.

In the kinetic rate-controlled regime, the ability of the system to propagate discrete channels is either reduced or eliminated altogether. Since the Damkohler number, however, depends on the length scale involved, we expect that reactions which are too slow to form channels at relatively small length scales may develop channels at larger length scales. A knowledge of how the key parameters, the Peclet and Damkohler numbers, affect channel formation in reactive porous media offer us some hope of understanding its reaction-induced permeability structure.

ACKNOWLEDGMENTS

We are indebted to the Amoco Foundation for a Graduate Fellowship given to C. Steefel. We are grateful to Jianxin Jiang, Philippe Van Cappellen, Edward Bolton, and Neil Ribe, and for a series of valuable discussions about the work presented here. We are also very grateful to J. Jiang for his help with the final production of the manuscript.

LITERATURE CITED

1. Roehl, P.O.; Choquette, P.W. (eds.) Carbonate Petroleum Reservoirs; Springer-Verlag: New York, 1985, p 622.
2. Ortoleva, P.; Chadam, J.; Merino, E.; Sen, A. Am J. Sci. 1987, 287, 1008-40.
3. Hoefner, M.L.; Fogler, H.S. AIChE J. 1988, 34, 45-54.
4. Bögli, A. Karst Hydrology and Physical Speleology; Springer-Verlag: New York, 1980, p 270.
5. Bear, J. Dynamics of Fluids in Porous Media; American Elsevier: New York, 1972; p 764.
6. Lerman, A. Geochemical Processes: Water and Sediment Environments; John Wiley and Sons: New York, 1977; p 481.
7. Lasaga, A.C. J. Geophys. Res. 1984, 89, 4009-4025.
8. Bear, J. Hydraulics of Groundwater; McGraw-Hill: New York, 1979, p 569.
9. Peaceman, D.W.; Rachford, H.H. SIAM J. 1955, 3, p. 28.
10. Press, W.H.; Flannery, B.P.; Teukolsky, S.A.; Vetterling, W.T. Numerical Recipes: The Art of Scientific Computing; Cambridge University Press: Cambridge, 1986, p 818.
11. Patel, M.K.; Cross, M.; Markatos, N.C. Int. J. Numerical Methods in Engineering 1988, 26, 2279-304.
12. Lichtner, P.C. Geochim. Cosmochim. Acta 1988, 52, 143-65.
13. Lichtner, P.C. Geochim. Cosmochim. Acta 1985, 49, 779-800.
14. Bahr, J.M.; Rubin. J. Water Resources Research 1987, 23, 438-52.
15. Ogata, A.; Banks, R.B. United States Geological Survey, Professional Paper no. 411-A, U.S. Department of Interior, U.S. Government Print Office: Washington, D.C., 1961.
16. Crank, J. The Mathematics of Diffusion, Oxford University Press: Oxford, 1975, p 414.
17. Sjöberg, E.L.; Rickard, D.T. Geochim. Cosmochim. Acta 1984, 48, 485-93.
18. Rimstidt, J.D.; Barnes, H.L. Geochim. Cosmochim. Acta 1980, 44, 1683-99.

RECEIVED August 18, 1989

## Chapter 17

# Modeling Dynamic Hydrothermal Processes by Coupling Sulfur Isotope Distributions with Chemical Mass Transfer: Approach

### David R. Janecky

**Los Alamos National Laboratory, Isotope Geochemistry Group, University of California, Los Alamos, NM 87545**

> A computational modeling code (EQPS_S) that couples sulfur isotope distribution and chemical mass transfer reaction calculations has been developed. Isotopic distribution is calculated using standard fractionation factor equations. A post processor approach to EQ6 calculations was chosen so that a variety of isotopic pathways could be examined for each reaction pathway. Two types of major bounding conditions were implemented: (1) equilibrium isotopic exchange between sulfate and sulfide species or exchange only accompanying chemical reduction and oxidation events, and (2) existence or lack of isotopic exchange between solution species and precipitated minerals, parallel to the open and closed chemical system formulations of chemical mass transfer modeling codes. All of the chemical data necessary to calculate isotopic distribution pathways is generated by most mass transfer modeling codes and can be input to the EQPS code. Routines are built in to directly handle EQ6 tabular files.
> Chemical reaction models of seafloor hydrothermal vent processes and accompanying sulfur isotopic distribution pathways illustrate the capabilities of coupling EQPS_S with EQ6 calculations, including the extent of differences that can exist due to the isotopic bounding condition assumptions described above.

Isotopes have long been known to provide unique information about natural hydrothermal systems. Light-stable isotopes of H, O, S, and C, in particular, have received considerable attention due to their measurable fractionation both between minerals and solution and as a function of temperature during hydrothermal processes, and their ubiquitous occurrence and dominant chemical role in such systems ([1]). While isotopic systematics have been integrated into mineral phase and solution speciation diagrams ([2]), general computational chemical mass transfer models have largely ignored this source of information until recently ([3-5]).

For investigating reactions involving high temperature, metal transporting hydrothermal solutions, sulfur isotopic processes have been chosen for systematic examination. The stability of multiple sulfur redox states (predominantly $S^{-2}$ and $S^{+6}$) and a relatively large natural range of isotopic compositions ($\delta^{34}S$ <0 to >20 per mil) provides an excellent probe into important hydrothermal processes of fluid mixing, wall rock interaction, and deep metasomatic reactions. A computer model which examines isotopic processes quantitatively has been implemented. Application to reactions occurring during hydrothermal venting at mid-ocean ridges are presented here to provide examples and an evaluation of the approach.

## APPROACH

Chemical reaction pathways for input into the isotopic model have been computed using the EQ3/6 reaction pathway modeling codes (6). Distribution of sulfur isotopes between aqueous species and minerals are calculated using a new computer code (EQPS_S). Isotopic fractionation factors (7) are used by the code to determine the distribution among components as described below. Thus, this approach does not make or apply any assumptions about the chemical mechanism by which isotopic exchange or transfer occurs. The 'descriptive', rather than 'mechanistic' approach, is due in part to the lack of understanding of such mechanisms and inability of chemical reaction codes to handle kinetics of homogeneous solution reactions.

Hydrothermal flow and reactions often occur at rates faster than that of sulfur isotopic equilibration (8). Therefore, a variety of isotopic distribution pathways are possible due to competition between three processes: isotopic fractionation (*e.g.*, equilibrium-disequilibrium of exchange), chemical reactions (*e.g.*, sulfur oxidation-reduction reactions), and physical processes (*e.g.*, mixing between solutions, conductive cooling-heating, and/or wall rock reactions). To evaluate the consequences and importance of possible pathways without knowledge of explicit mechanisms and kinetics requires examination of the boundary conditions that determine sulfur isotopic distribution and composition attained by the hydrothermal solution and sulfide minerals, rather than only the absolute pathways followed. However, in exploring the effects of some of the boundary conditions and comparing the results to natural products, the relative importance and range of the possible processes involved and the most critical variables can be elucidated.

Chemical reaction constraints of interest are redox equilibrium *vs.* disequilibrium (especially whether sulfate reduction progresses appreciably or not) and open *vs.* closed system processes. These types of constraints are implemented routinely when running chemical reaction models using EQ3/6 (9,5).

Isotopic reaction constraints of interest include parallels to those for chemical reactions: isotopic equilibria or disequilibria between redox species independent of chemical equilibria, and open *vs.* closed isotopic exchange reactions between aqueous species and precipitated minerals. The latter isotopic process is more complex than that considered for chemical reactions, because precipitated and/or saturated reactant phases may partially redissolve to equilibrate with solution, while not simultaneously exchanging isotopically with the solution. In addition, during isotopic fractionation and chemical reaction it is of interest to examine the extent to which mineral precipitation is coupled with redox processes during incongruent dissolution of a reactant phase. For example, if chemical reduction occurs at a mineral surface, rather than homogeneously in solution, the resulting reduced sulfur may have a tendency to be precipitated inhomogeneously, thus affecting the resulting mineral isotopic composition and apparent fractionation factors.

### EQPS Computation Implementation.

The EQPS code is composed of of approximately 3500 lines of FORTRAN statements, including comments and local graphics interface. The code does not involve any machine restrictions. CPU time for each of the examples presented below is approximately 35 seconds on a MicroVAX II when chemical mass transfer pathway data is read from EQ6 tabular files.

Total moles of all sulfur-bearing components present in the system are required by EQPS from data on tabular files output by EQ6 or input from the user. The input modules allow expansion to integrate with other codes as necessary. The choice of a post-processor mode of operation for EQPS, rather than integration into the EQ6 code as for the oxygen and hydrogen isotopic calculations of Bowers and Taylor (4), was made primarily because of the complexities of sulfur chemistry discussed above. By using the post-processor approach the computationally intensive EQ6 code is only run once for each chemical reaction path, while the simpler and much faster EQPS code can be run multiple times to examine the effects of various isotopic reaction constraints.

An isotopic data input file provides EQPS with coefficients to calculate a fractionation factor ($\alpha$) for each solution and mineral component ($i$), relative to a reference solution specie (chosen to be $H_2S$, consistent with previous convention) using:

$$1000 \ln\alpha_i = \frac{A}{T^2} \cdot 10^6 + \frac{B}{T} \cdot 10^3 + C \tag{1}$$

with A, B, and C being constants for each specie or mineral ($\underline{7}$) and temperature ($T$) in Kelvin. The fractionation factors ($\alpha_i$) are then used to calculate the composition ($\delta^{34}S_i$) of each component from:

$$\delta^{34}S_i = \delta^{34}S_{H_2S} + \Delta_i \tag{2}$$

where

$$\Delta_i = 1000(\alpha_i)(1 + \frac{\delta^{34}S_{H_2S}}{1000}) \tag{3}$$

The isotopic composition of $H_2S$ is constrained by the bulk composition ($\delta^{34}S_{\sum s}$) and total mass ($m_{\sum s}$) of sulfur in the system, and the composition ($\delta^{34}S_i$) and mass ($m_i$) of all components:

$$\delta^{34}S_{\sum s} \cdot m_{\sum s} = \sum \delta^{34}S_i \cdot m_i \tag{4}$$

In addition to fractionation coefficients, the isotopic data file contains information on the stoichiometric amount of sulfur in each component and a flag indicating the redox state of each component (sulfide or sulfate in the present treatment). The code has been designed to be sufficiently general such that it is not limited by the present data base for sulfur isotopic fractionation and redox state, or specific to sulfur; by changing the problem input file and the fractionation specification file a variety of other elements including carbon, hydrogen and oxygen could be examined.

The code operates by calculating masses of each sulfur isotopic specie present in the system and in individual components at each reaction progress point. Reactant sulfur is added to the bulk system and distributed among aqueous species and minerals. Sulfur redox isotopic disequilibria can be considered using EQPS by evaluating isotopic distribution separately for oxidized and reduced species using the redox flag from the isotopic data file, with $H_2S$ and $SO_4^=$ as reference species. Chemical oxidation and reduction processes during redox isotopic disequilibria are accounted for by examining the mass balance relations and transferring the required masses and isotopic compositions between the two subsystems. Open system behavior between minerals and solution is computed by tracking the composition and masses of previously precipitated minerals separately from newly precipitated minerals. This results in sequential addition of isotopically evolving sulfide or sulfate to the solids as precipitation proceeds. For computational simplicity, and because minerals seldom dissolve regularly, dissolution of precipitated minerals is treated as a bulk average process. Thus, even minerals which were precipitated in an open isotopic system provide sulfur with the average isotopic composition of the mineral upon dissolution. Inhomogeneous oxidation/reduction reaction processes occurring at a mineral surface have not yet been generalized and included in the code.

To operate EQPS on an EQ6 calculated reaction path, the user selects the boundary constraints affecting a process (open vs. closed system, isotopic equilibrium or disequilibrium between redox subsets, and precipitation pathway accompanying redox reactions), inputs initial solution or system $\delta^{34}S$, reactant(s) $\delta^{34}S$ composition(s), and specifies an input file generated from an EQ6 run. The EQ6 data consists of solution composition and mineral amounts at discrete points on a reaction pathway. EQPS can either be set to calculate at the points produced by EQ6, or use a curve crawler technique to produce pseudo-continuous isotopic pathways at user definable granularity. Accuracy of either computational procedure depends most on the step size executed by EQ6 and only slightly on the step size selected during the

EQPS calculation. Given sufficient output points from EQ6, the isotopic fractionation calculations provide more detailed information than present analyses of sulfur isotopic compositions for several reasons. Firstly, EQPS calculations are sufficiently fast that small step sizes are not a problem, especially when coupled to graphic output routines. Secondly, calculated EQ6 pathways are smooth and continuous between discontinuities, which are forced by EQ6 to be part of the output set of data points. This results in a typical computed $\delta^{34}S$ uncertainty of less than 0.05 per mil, which are sensitive to zoning and varying solution isotopic composition during precipitation. In contrast, sulfur isotopic analyses are limited by sample size requirements and heterogeneous mixtures of phases.

## EXAMPLE: APPLICATION TO SEAFLOOR HYDROTHERMAL PROCESSES

Mid-ocean ridge hydrothermal processes provide an ideal application for geochemical reaction path modeling, involving temperature dependent reactions, fluid mixing, reaction with sulfide products, and reaction with seafloor basalts (5). The solution and solid compositions are well characterized, including that of sulfur isotopes (10,11). However, measured fractionations between solution and solid samples can not be the result of simple equilibrium processes (10-12). The ability to track reactions involving seawater sulfate ($\delta^{34}S = 21$ per mil) and hydrothermal sulfide (assumed to have a relatively consistent $\delta^{34}S \doteq 1$ per mil) provides unique insight into mixing, reaction, and redox processes. Thus, the mechanisms by which components of the system achieve their sulfur isotopic compositions and the resulting constraints which can be placed on processes occurring in the overall hydrothermal system by such modeling efforts are of significant interest. Questions involving the sulfur isotopic results and their relationships to hydrothermal processes in seafloor systems were, in fact, the rationale for developing this sulfur isotopic modeling approach. Two examples which illustrate application of such models and the effects of various constraints on sulfur isotopic model pathways are reactions involved in mixing of seawater into hydrothermal solution, and the consequences of reaction of hydrothermal solution with basalt components and seawater sulfate.

### Hydrothermal Fluid Mixing with Seawater.

Adiabatic-mixing pathways, where seawater (2°C) mixes into hydrothermal fluid (350°C), have been successfully used to model formation of sulfide minerals associated with venting hydrothermal solutions at mid-ocean ridges (9). Sulfate reduction can be quantitatively and isotopically important in such reactions (5). Combinations of three types of isotopic path constraints discussed above have been examined, using the mixing reaction pathways calculated by Janecky and Seyfried (9) for chemical equilibrium and initial sulfur isotopic compositions of 1 per mil for the hydrothermal solution and 21 per mil for seawater (Figure 1).

Complete sulfate-sulfide isotopic exchange equilibrium and chemical equilibrium in a closed system results in significantly decreasing $\delta^{34}S$ values for sulfide minerals and solution sulfide, while sulfate components increase. Such mineral isotopic compositions are not observed in active mid-ocean ridge seafloor systems (10,11). Below approximately 200°C, inorganic sulfate-sulfide isotopic exchange reactions are known to be particularly inhibited (8), and thus the extremely negative values achieved by these models are not likely in natural systems without substantial biological mediation.

Sulfate-sulfide isotopic exchange disequilibrium (where sulfate is reduced to sulfide without isotopic exchange but sulfide minerals and $H_2S$ isotopically equilibrate) in closed and open isotopic systems result in significantly smaller shifts in isotopic compositions (Figure 1). In addition, the calculated shifts are to higher sulfide $\delta^{34}S$ compositions, generally consistent with the measured $\delta^{34}S$ values for seafloor sulfide minerals. Analyses of the natural samples, however, provide no evidence of systematic isotopic differences between pyrite, sphalerite, and chalcopyrite, as predicted by both standard fractionation approaches and these calculated isotopic reaction pathways, possibly due to dynamic variations of mixing and precipitation on a local scale.

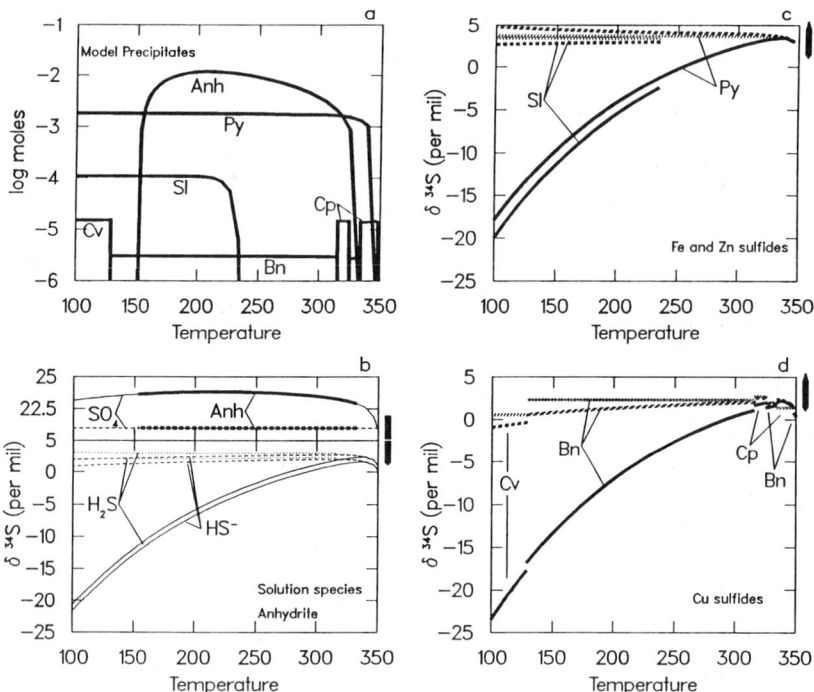

Figure 1. (a) Calculated amounts of mineral precipitates (after Janecky and Seyfried (9)) and (b-d) $\delta^{34}S$ reaction pathways during adiabatic mixing seawater into hydrothermal solution from 350°C to 100°C for sulfate-sulfide chemical equilibrium. The results of three sets of isotopic constraints are compared for (b) solution species and anhydrite, (c) Fe and Zn sulfide minerals, and (d) Cu sulfide minerals. Equilibrium sulfate-sulfide isotopic exchange pathways in an isotopically closed system are shown as solid lines, disequilibrium sulfate-sulfide isotopic exchange pathways (sulfate reduced is quantitatively converted to isotopically heavy sulfide, see text) in an isotopically closed system by dashed lines, and disequilibrium sulfate-sulfide isotopic exchange pathways in an isotopically open system by dotted lines. Mineral isotopic compositions are shown as heavy lines ($Cp$ – chalcopyrite, $Sl$ – sphalerite, $Py$ – pyrite, $Cv$ – covellite, $Bn$ – bornite, and $Anh$ – anhydrite) and solution species as thin lines. For comparison, the range of measured isotopic compositions from seafloor hydrothermal vents is indicated with range bars to the left of each isotopic figure (10,11).

The maximum calculated increase in these models of solution sulfide $\delta^{34}S$ is only about 2 per mil above the initial value of 1 per mil, while mineral compositions can be increased only slightly higher. Such limited reducing potential of the hydrothermal solution is a significant constraint on hydrothermal reaction processes and interpretation of the products of these systems. Measured solution sulfide $\delta^{34}S$ compositions, however, are as high as 4 to 7.5 per mil in samples from both 21°N East Pacific Rise and Juan de Fuca and mineral compositions also cover a wider range than the models (10,11). In addition, the solution composition range extends to higher values than the bulk of sulfide solids measured from the same areas, inconsistent with all relatively simple fractionation models. Such solutions with high $\delta^{34}S$ sulfide apparently represent special processes in seafloor vents, because they are not found to be precipitating significant amounts of sulfide minerals. Shanks and Seyfried (10) and Woodruff and Shanks (11) have argued that the lack of variation in vent fluid chemistry from a given vent site, despite large variations in $\delta^{34}S$ values of $H_2S$ relative to sulfide minerals, points to near surface reactions modifying a relative uniform solution evolved at depth. One reaction which may be involved in producing such solutions is reaction with basalts within the near surface hydrothermal flow zone, coupled with addition of seawater sulfate by mixing or other processes.

## Simplified Model for Sulfate Reduction in a Hydrothermal Stockwork.

Alteration due to reactions between basalt and hydrothermal solution (±ambient seawater) below the seafloor in massive sulfide ore deposits has been studied extensively, and the physical relationship of a hydrothermal flow zone to such deposits is understood in general terms (13). However, the potential influence of chemical reactions in such stockwork zones on venting solutions and processes of massive sulfide deposit formation has been largely unrecognized. In examining affects on sulfur isotopic composition of the hydrothermal solution, the components of such reactions which may result in sulfate reduction are of primary interest.

Crystalline tholeiitic basalt, initially composed of plagioclase, pigeonitic pyroxene, olivine, and minor oxide and sulfide minerals, commonly is replaced in stockwork zones by chlorite-rich assemblages. Alteration of olivine and pyroxene in such rocks releases ferrous iron which may result in enhanced reducing potential (14) beyond that attributable solely to pristine hydrothermal solution. Observations from altered basalts indicate that olivine is generally more reactive than pyroxenes, consistent with experimental and theoretical constraints (15,16). In addition, relative reaction rates between plagioclase and olivine favor olivine alteration. Thus, when examining stockwork alteration with respect to sulfate reduction potential, the reactive component of basalt has been limited to the fayalite component of olivine (5). A variety of reaction paths were calculated for hydrothermal solution + fayalite + anhydrite. Two bounding pathways based on the presence or absence of secondary magnetite illustrate potential effects on both sulfide concentration and isotopic composition. Solution composition changes of most elements during reaction are only subtle. Sulfide concentration and isotopic composition, however, can change significantly (Figure 2).

Comparison of computed reaction pathways to the available data on $H_2S$ concentrations and $\delta^{34}S_{H_2S}$ values from 21°N East Pacific Rise and Juan de Fuca Ridge vent sites (10,11) indicates that the models may represent reasonable extremes in reaction processes (Figure 2). When magnetite is not allowed to precipitate during the reaction, the $\delta^{34}S$ of $H_2S$ dissoved in solution increases by about 3.5 per mil while the dissolved concentration of sulfide decreases significantly. Suppression of magnetite precipitation severly limits iron oxidation and sulfate reduction. In contrast, when magnetite is allowed to precipitate and significant oxidation of iron and reduction of sulfate occurs, $\delta^{34}S$ of $H_2S$ in solution increases exceed 5 per mil and the dissolved sulfide concentration stays relatively high. The latter reaction continues to increase $\delta^{34}S_{H_2S}$ to greater than 15 per mil. Varying the extent or rate of iron oxidation between these two extremes can reproduce the range of samples which have been analyzed and suggests possible differences between active systems which should be investigated using petrographic and geophysical techniques. In addition, the results indicate that the processes that cause large sulfur isotopic variations in vent compositions may be relatively insensitive to

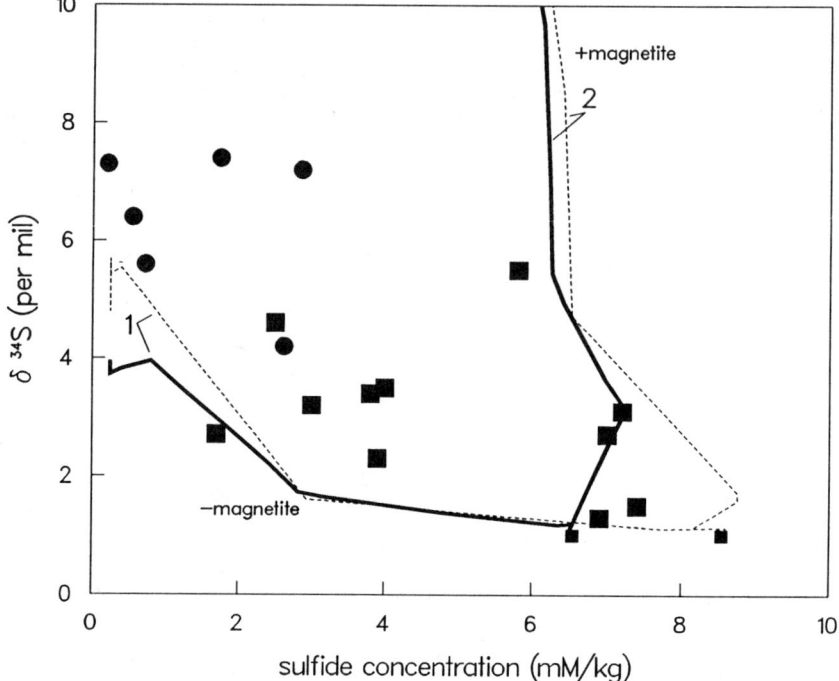

Figure 2. $\delta^{34}S$ of solution sulfide vs. sulfide concentration for data from active 21°N East Pacific Rise (solid squares) and Juan de Fuca (solid circles) hydrothermal vent solution samples (10,11). Pathways from 'analog' stockwork alteration computational models (after Janecky and Shanks (5)) are shown as lines starting from small solid circle symbols. Solid lines represent reaction paths with an initial $H_2S$ concentration in hydrothermal solution of 6.5 mM. Dashed lines represent reaction pathways calculated using hydrothermal solution composition of 8.5 mM $H_2S$. Hydrothermal solution reaction pathways illustrated are (1) anhydrite+fayalite reactants with suppression of magnetite and all silica polymorphs and (2) anhydrite+fayalite reactants with magnetite precipitation allowed and suppression of all silica polymorphs. Restriction of the solution $H_2S$ and $\delta^{34}S_{H_2S}$ data to within the bounds of the calculated reaction pathways strongly suggests that subseafloor basalt alteration is the process which produces the $^{34}S$-enriched $H_2S$ values.

substantially different initial hydrothermal fluid compositions and that such reactions may only cause subtle variations in other solution parameters. Thus, in spite of the inherent limitations of such simplified models, the significant consequences for $\delta^{34}S_{H_2S}$ values are consistent with natural results and provide important insight (Figure 2).

## SUMMARY

A computational modeling approach has been implemented to allow examination of sulfur isotopic distribution during chemical reaction processes. The approach integrates isotopic distribution reaction path calculations based on fractionation factors and the results of chemical mass transfer reaction pathway calculations. Various combinations of chemical and isotopic equilibria bounding conditions can be rapidly evaluated, including redox and isotopic exchange equilibrium or disequilibrium and open or closed system behavior. The results can be readily compared to processes which formed natural samples or occurred during experiments. Application to seafloor hydrothermal systems has resulted in new insights and better constraints on important hydrothermal processes.

The code and isotopic data files are available from the author. Example calculations and results presented in this paper are included in the code release package.

## ACKNOWLEDGMENTS

This work has benefited greatly from discussions with W. C. Shanks, W. E. Seyfried, T. S. Bowers, R. O. Rye, B. R. Erdal, C. J. Duffy, R. W. Charles, and T. J. Wolery, among many others. Reviewer comments improved the manuscript. Initial calculations were supported by NSF grants OCE-8400676 and OCE-8542276 to W. E. Seyfried, University of Minnesota. Preparation of this paper and continued refinement of the approaches to examining water-rock interaction is supported by funding from Geosciences Program, US DOE Office of Basic Energy Science and the Institute of Geophysics and Planetary Physics, Los Alamos National Laboratory, University of California.

## LITERATURE CITED

1. Valley, J. W.; Taylor, H. P.; O'Neil, J. R. Stable isotopes in high temperature geological processes; Mineralogical Society of America, Reviews in Mineralogy. 1986, 16, p570.
2. Ohmoto, H. Econ. Geol., 1972, 67, 551-578.
3. Walshe, J. L.; Solomon, M. Econ. Geol. 1981, 76, 246-284.
4. Bowers, T. S.; Taylor, H. P., Jr. J. Geoph. Res. 1985, 90, 12,583-12,606.
5. Janecky, D. R.; Shanks, W. C., III Can. Mineral. 1988, 26, 805-825.
6. Wolery, T. J. Calculation of chemical equilibrium between aqueous solution and minerals: the EQ3/6 software package; Lawrence Livermore National Laboratory, Livermore, Ca, 1979; UCRL-52658.
7. Ohmoto, H.; Rye, R. O. In Geochemistry of Hydrothermal Ore Deposits, Barnes, H. L., Ed., Wiley: New York, 1979; 509-567.
8. Ohmoto, H.; Lasaga, A. C. Geochim. Cos. Acta 1982, 46, 1727-1745.
9. Janecky, D. R.; Seyfried, W. E., Jr. Geochim. Cos. Acta 1984, 48, 2723-2738.
10. Shanks, W. C., III; Seyfried, W. E., Jr. J. Geoph. Res. 1987, 92, 11,387-11,399.
11. Woodruff, L. G.; Shanks, W. C. J. Geoph. Res. 1988, 93, 4562-4572.
12. Kerridge, J.; Haymon, R. M.; Kastner, M. Earth Planet. Sci. Lett. 1983, 66, 91-100.
13. Franklin, J. M.; Lydon, J. W.; Sangster, D. F. In Econ. Geol. 75th Anniv. Volume, Skinner, B. J., Ed., Econ. Geol. Publ. Co., 1981; 485-627.
14. Shanks, W. C.; Bischoff, J. L.; Rosenbauer, R. J. Geochim. Cos. Acta 1981, 45, 1977-1995.
15. Seyfried, W. E., Jr. Ann. Rev. Earth Planet. Sci. 1987, 15, 317-335.
16. Murphy, W. M.; Helgeson, H. C. Geochim. Cos. Acta 1987, 51, 3137-3154.

RECEIVED August 18, 1989

## Chapter 18

# Coupling of Precipitation−Dissolution Reactions to Mass Diffusion via Porosity Changes

**Chalon L. Carnahan**

**Earth Sciences Division, Lawrence Berkeley Laboratory, University of California, Berkeley, CA 94720**

> Coupling of precipitation/dissolution reactions to diffusive mass transport via porosity changes has been implemented in the computer program THCC, a simulator of reactive chemical transport. The coupling is accomplished without increasing the set of primary unknowns. Porosity is included explicitly in the transport equations and is tracked by accounting for changes of volumes of precipitates. The coupling prevents the volume of a precipitated solid from exceeding available pore volume. Results of calculations are presented for two examples, each done with and without variable porosity.

Computer programs (1-5) that couple chemical reactions, including precipitation/dissolution, to mass transport processes have not, in general, accounted for the effects of precipitation/dissolution reactions on the transport processes. In particular, changes in the sizes of pores or apertures of fractures caused by precipitation/dissolution reactions accompanying mixing of fluids of different compositions can alter the effective mass diffusivities and permeabilities of porous or fractured materials (6,7), and these alterations can affect the subsequent movement of dissolved chemicals. Neglect of these alterations not only might produce inaccurate computational results (8), but might also lead to physically unrealizable consequences such as the calculated volume of a precipitate exceeding available pore space.

An examination of calculated permeability changes associated with chemical changes accompanying simulated acidization of sandstone with HCl/HF mixtures has been reported (9). Calculations of the permeability changes were done separately from the reactive chemical transport simulations; thus, there was not complete coupling between precipitation/dissolution reactions and fluid flow in the simulations.

Reaction-transport feedback in advecting systems with chemical disequilibrium has been described and has been simulated by the computer program REACTRANS to demonstrate the origins of patterned states ("geochemical self-organization") in rocks (10,11).

The reactive chemical transport simulator THCC (12,13) is being used to study effects on mass transport of precipitation/dissolution reactions. This paper sets forth the mathematical and numerical bases for the coupling of these reactions to mass *diffusion* in a porous medium and presents results of calculations to demonstrate consequences of the coupling.

THE COMPUTER PROGRAM THCC

The THCC computer program (12,13) is a thermodynamically based simulator of multicomponent, reactive chemical transport using the direct method of solution. The program simulates transport of reactive chemical species by advection and by hydrodynamic dispersion or mass diffusion in one-dimensional or cylindrically symmetric geometry. Chemical reactions are assumed to be in a state of local equilibrium. The reactions simulated are complexation, oxidation-reduction, and ionization of water in the aqueous phase, reversible precipitation of solid phases, and ion exchange. Chemical reactions are described by mass action relations among thermodynamic activities of participating species. The THCC program has the capability to simulate systems with temporally and spatially variable fields of temperature and to simulate radioactive decay of selected reactants.

VARIABLE POROSITY IN THCC

The variable porosity must be included explicitly in the transport equations for mobile species. Values of porosity are related to volumes of reactive solids.

Transport Equations. In a system with fluid-filled porosity $\epsilon_f$, $N_p$ reactive solid phases, and aqueous-phase species consisting of $N_b$ basis species (the smallest set of species needed to define the chemical system in the aqueous phase) and $N_c$ complexes, the conservation equation for the mass of a component $i$ is:

$$\frac{\partial}{\partial t}\left(\epsilon_f W_i + \sum_{k=1}^{N_p} \nu_{ik} P_k\right) = \nabla \cdot \left(\frac{\epsilon_f D}{\tau} \nabla W_i\right), \quad i = 1, \ldots, N_b. \tag{1}$$

where $t$ is time (seconds), $D$ is the diffusion coefficient in the fluid phase (m²/s), $\tau$ is the tortuosity, $P_k$ is the concentration (mole/dm³ of porous medium) of a precipitate $P_k$, and the $\nu_{ik}$'s are the number of moles of basis species $B_i$ per mole of solid $P_k$. $W_i$ is the total concentration (mole/dm³ of the fluid phase) of component $i$ and is defined by

$$W_i = B_i + \sum_{j=1}^{N_c} \nu_{ij} C_j, \quad i = 1, \ldots, N_c, \tag{2}$$

where $B_i$, $i = 1, \ldots, N_b$, is the concentration (moles/dm³ of the fluid phase) of the basis species $B_i$ containing component $i$ and $\nu_{ij}$ is the number of moles of $B_i$ per mole of complex $C_j$, $j = 1, \ldots, N_c$, having concentration $C_j$ (moles/dm³ of the fluid phase). $C_j$ is formed by reactions among basis species, thus:

$$\nu_{lj} B_l^{z_l} + \nu_{mj} B_m^{z_m} + \ldots = C_j^{\nu_{lj} z_l + \nu_{mj} z_m + \ldots},$$

where the $z_i$ are signed ionic charges of the basis species. The concentration of complex $C_j$ is given by the mass action relation,

$$C_j = \frac{K_{cj}}{\gamma_j} \prod_{l=1}^{N_b} (\gamma_l B_l)^{\nu_{lj}}, \quad j = 1, \ldots, N_c, \tag{3}$$

where $K_{cj}$ is the temperature-dependent thermodynamic equilibrium constant for formation of complex $C_j$ and the $\gamma$'s are activity coefficients. The left side of Equation (1) represents the rate of change in time of the total amount of component $i$ (contained in aqueous species and reactive solids) within an infinitesimal volume of porous medium. The right side of Equation (1) represents the accumulation of component $i$ in the volume due to spatial variations of the diffusive flux of aqueous species migrating through the volume.

Numerical Solution. In the numerical formulation of THCC, Equations (2) and (3) are substituted into Equation (1). The resulting set of $N_b$ partial differential equations is transcribed into $N_b$ finite-difference equations, using central differencing in space and the Crank-Nicolson method to obtain second-order accuracy in time. The set of unknowns consists of $B_i$, $i = 1, \ldots, N_b$, and $P_k$, $k = 1, \ldots, N_p$, at each finite-difference node. Residue equations for the basis species are formed by algebraically summing all terms in the finite-difference forms of the transport equations. The finite-difference analogs of Equation (1) provide $N_b$ residue equations at each node; the remaining $N_p$ residue equations are provided by the solubility products for the reactive solids.

A reactive solid $P_k$ dissolves to form its constituent basis species, thus:

$$P_k = \nu_{lk} B_l^{z_l} + \nu_{mk} B_m^{z_m} + \ldots.$$

The activity product, $Q_{sk}$, for this reaction is given by

$$Q_{sk} = \prod_{l=1}^{N_b} (\gamma_l B_l)^{\nu_{lk}}. \qquad (4)$$

If the the basis species formed by dissolution of solid $P_k$ are in chemical equilibrium with the solid, then the activity product, $Q_{sk}$, is equal to the thermodynamic solubility product, $K_{sk}$. Residue equations for the solids are formed in the following manner. At each finite-difference node the activity product, $Q_{sk}$, is computed and compared to the theoretical solubility product, $K_{sk}$, for the solid. If the solid is present at the node, or if the solid is not present and $Q_{sk} > K_{sk}$, then the residue for the solid at the node is set equal to the algebraic difference, $Q_{sk} - K_{sk}$. On the other hand, if $Q_{sk} \le K_{sk}$ and the solid is not present at the node, the residue is set equal to zero. This procedure provides a residue equation for each solid at each node and eliminates the need to change the number of unknowns at nodes where solids have precipitated or dissolved.

The residues ($N_b + N_p$ at each node) are reduced to "zero" (a small positive number fixed by specifying an error tolerance at input) iteratively by computing corrections to current values of the unknowns using the Newton-Raphson method ([14]). Elements of the Jacobian matrix required by this method are computed from analytical expressions. The system of equations to be solved for the corrections has block tridiagonal form and is solved by use of a published software routine ([15]).

Currently, the tortuosity, $\tau$, and the quotient, $D/\tau$, are assumed constant. However, $\tau$ may depend on the habit and texture of the particular precipitated mineral. Also, the assumption is made that precipitation is uniform over the fluid-porous medium interface, rather than concentrated in small pores or in "bottlenecks" connecting large pores. However, non-uniform deposition may be expected in natural systems. Both assumptions are first approximations and are subject to refinement when experimental data become available that provide more detailed information about effects of precipitation/dissolution on tortuosity and about uniformity of precipitation.

Volumetric Relations. The fluid-filled porosity $\epsilon_f$ is not a primary unknown because it is determined by the $P_k$, $k = 1, \ldots, N_p$. In the iterative Newton-Raphson solution scheme, values of $\epsilon_f$ are updated based on new values of the $P_k$'s at the end of each iteration.

A conservation equation can be written for a unit volume of porous medium:

$$\epsilon_m + \epsilon_p + \epsilon_f + \epsilon_g = 1, \qquad (5)$$

where $\epsilon_m$ is the volume fraction of unreactive solid (constant), $\epsilon_p$ is the volume fraction of reactive solids, $\epsilon_f$ is the volume fraction of fluid phase, and $\epsilon_g$ is the volume fraction of gas phase. $\epsilon_p$ is calculated from the current concentrations of reactive solids by

$$\epsilon_p = \sum_{k=1}^{N_p} P_k \overline{V}_k, \qquad (6)$$

where $\overline{V}_k$ is the molar volume (dm³/mole) of reactive solid $k$ and is assumed constant. It follows from Equation (5) that

$$\Delta(\epsilon_f + \epsilon_g) = -\Delta\epsilon_p. \qquad (7)$$

In the current numerical formulation if $\epsilon_g > 0$, then $\Delta\epsilon_f = 0$ and $\Delta\epsilon_g = -\Delta\epsilon_p$ until $\epsilon_g = 0$; thereafter, $\Delta\epsilon_f = -\Delta\epsilon_p$.

To simulate constant porosity in a system with diffusion and precipitation/dissolution reactions, all $\overline{V}_k$'s are given input values of zero.

FLUX CONDITION AT INNER BOUNDARY

Both examples given in this paper were calculated with the following condition at the boundary $x = 0$:

$$-[D\,\nabla W_i]_{x=0} = J_i\,[\epsilon_f]_{x=0}, \quad i = 1, \ldots, N_b, \qquad (8)$$

where $J_i$ [mole/(m²s)] is the input value of the flux of basis species $i$ referred to unit area of the fluid phase. Equation (8) equates the computed diffusive flux of component $i$ at $x = 0$ (left side of the equation) to the prescribed input flux multiplied by the current value of fluid-filled porosity at $x = 0$ (right side of the equation). This formulation allows the incoming flux to decrease in response to decreasing porosity at the boundary. The quantity $J_i$ is the value of the input flux in the hypothetical case that $\epsilon_f = 1$ at $x = 0$.

EXAMPLES

The examples involve (1) precipitation along a gradient of temperature and (2) isothermal precipitation of a solid by reaction between two diffusing ions.

Transport of Silica along a Temperature Gradient. This example simulates diffusion of silicic acid into a domain where temperature decreases with increasing distance from the inlet at $x = 0$. The temperature field is steady in time. Quartz precipitates according to the reaction

$$SiO_2(\text{quartz}) + 2H_2O = Si(OH)_4^0(\text{aq}),$$

for which the equilibrium constant, $K_s$, is given as a function of absolute temperature, $T$ (K), by (16)

$$\log K_s = 1.881 - \frac{1560.}{T} - 2.028 \times 10^{-3}T. \qquad (9)$$

The molar volume of quartz is 0.0227 dm³/mole (17). The quantity $D/\tau$ is assumed constant and equal to $10^{-9}$ m²/s. The temperature gradient is $-20$ °C/m, and the temperature at the boundary $x = 0$ is 150 °C. The outer boundary is located at $x = 5$ m. Initial conditions in the domain $x \geq 0$ were $\epsilon_f = 0.05$, $\epsilon_g = 0.$, $\epsilon_m = 0.95$, and zero concentrations of $Si(OH)_4^0(\text{aq})$ and $SiO_2(\text{quartz})$. The boundary condition at $x = 0$ is $J(Si(OH)_4^0(\text{aq})) = 10^{-4}$ mole/m²s. Figure 1 is a diagrammatic representation of the initial and boundary conditions used in this simulation.

Because quartz exhibits prograde solubility, it precipitates along the decreasing temperature profile within the domain $x \geq 0$. However, significant changes of porosity occur only at the boundary $x = 0$. (The volume fraction of reactive solid, $\epsilon_p$, at $x = 0$ is $3.47 \times 10^{-2}$ when $t = 10^9$ seconds; values of $\epsilon_p$ at $x > 0$ at this time are less than $10^{-4}$ times the value at $x = 0$.) Figure 2 shows the accumulation of quartz at this boundary at times up to $10^9$ seconds for simulations with variable porosity and constant porosity. In the case of constant porosity the non-physical result that the volume of precipitated quartz exceeds available pore space is obtained at about $0.45 \times 10^9$ seconds, when the solid concentration exceeds 2.2 moles/dm³ of porous medium. On the other hand, in the case of variable porosity the concentration of solid increases gradually with time and does not exceed the critical concentration. This is a result of a decreasing influx of $Si(OH)_4^0(aq)$ at the boundary caused by the decreasing porosity there, as indicated by Equation (8). Also, the concentration of precipitated solid at the adjacent node is smaller in the variable porosity case than in the constant porosity case for all times, a consequence of reduced effective diffusivity at the boundary as reflected in the term $\epsilon_f D/\tau$ in Equation (1).

Figure 3 shows the porosity at the boundary $x = 0$ as a function of time for the variable porosity case. The porosity declines monotonically with time, approaching the asymptotic value of zero; this result was verified by plotting porosity against reciprocal time and extrapolating to $t^{-1} = 0$.

Precipitation of Gypsum at Constant Temperature. This example simulates diffusion of $Ca^{2+}$(aq) and $SO_4^{2-}$(aq) into a domain where the two ions react and gypsum precipitates according to the reaction

$$CaSO_4 \cdot 2H_2O(c) = Ca^{2+}(aq) + SO_4^{2-}(aq) + 2H_2O,$$

for which $\log K_s = -4.50$ at 25 °C (18). The molar volume of gypsum is 0.0747 dm³/mole (17). The quantity $D/\tau$ is assumed constant and equal to $10^{-9}$ m²/s. Initial conditions in the domain $x \geq 0$ were $\epsilon_f = 0.05$, $\epsilon_g = 0.$, $\epsilon_m = 0.95$, and zero concentrations of $Ca^{2+}$(aq), $SO_4^{2-}$(aq), and $CaSO_4 \cdot 2H_2O(c)$. The boundary condition at $x = 0$ is $J(Ca^{2+}(aq)) = J(SO_4^{2-}(aq)) = 10^{-3}$ mole/m²s. The outer boundary is located at $x = 0.5$ m. Figure 4 is a diagrammatic representation of the initial and boundary conditions used in this simulation.

Under the isothermal condition imposed in this example, gypsum precipitates only at the boundary $x = 0$ in both constant porosity and variable porosity cases. Significant changes of porosity occur at this location in the variable porosity case. Figure 5 shows the accumulation of gypsum at this location at times up to $10^7$ seconds for the two cases. The results are qualitatively similar to the results of the previous example. In the case of constant porosity the volume of precipitated gypsum exceeds available pore space at times greater than about $0.14 \times 10^7$ seconds. In this example, the critical concentration of solid is 0.67 mole/dm³ of porous medium. In the variable porosity case the solid concentration rises asymptotically toward the critical value, again as a result of decreasing influxes of reactants at the boundary caused by decreasing porosity there.

Figure 6 shows the porosity at the boundary $x = 0$ as a function of time for the variable porosity case. As in the previous example, the porosity declines steadily and approaches zero asymptotically.

DISCUSSION

The method described here to account for variations of porosity and mass diffusivity associated with precipitation/dissolution of reactive solids provides coupling of chemical reactions to mass transport. The method does not allow the physically impossible exceedance of available pore volume by precipitated solids.

It is important to note that in systems of the type considered here variations of porosity and mass diffusivity *do not affect fluid-phase concentrations of reactants in equilibrium with a reactive solid*; only the rate of accumulation of the solid is affected. In the present examples the

| Boundary $x = 0$ | Initial Conditions | Boundary $x = 5\,\text{m}$ |
|---|---|---|
| $\text{Si(OH)}_4^0(\text{aq})$ flux= $10^{-4}$ mole/(m$^2$s) 150 °C | $[\text{Si(OH)}_4^0(\text{aq})] = 0.$ $[\text{SiO}_2(\text{quartz})] = 0.$ $-20\,°\text{C/m}$ $\epsilon_f = 0.05$ $\epsilon_g = 0.$ $\epsilon_m = 0.95$ | $[\text{Si(OH)}_4^0(\text{aq})] = 0.$ 50 °C |

Figure 1. Initial and boundary conditions for simulation of diffusion of silicic acid along a temperature gradient ($-20\,°\text{C/m}$) with precipitation of quartz. Brackets denote concentrations.

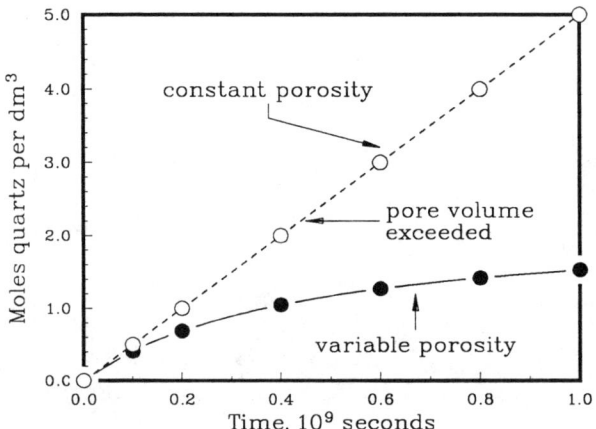

Figure 2. Quartz precipitated at boundary $x = 0$, temperature 150 °C.

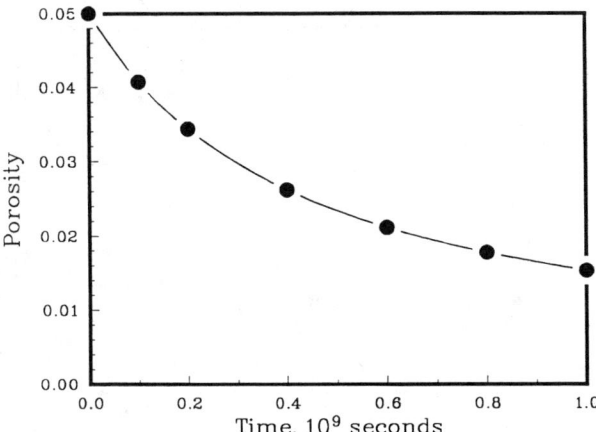

Figure 3. Porosity at boundary $x = 0$ for quartz precipitation, temperature 150 °C.

| Boundary $x = 0$ | Initial Conditions | Boundary $x = 5\,\text{m}$ |
|---|---|---|
| $\text{Ca}^{2+}(\text{aq})$ flux= $\text{SO}_4^{2-}(\text{aq})$ flux= $10^{-3}$ mole/(m$^2$s) 25 °C | $[\text{Ca}^{2+}(\text{aq})] = 0.$ $[\text{SO}_4^{2-}(\text{aq})] = 0.$ $[\text{CaSO}_4 \cdot 2\text{H}_2\text{O}(c)] = 0.$ 25 °C $\epsilon_f = 0.05$ $\epsilon_g = 0.$ $\epsilon_m = 0.95$ | $[\text{Ca}^{2+}(\text{aq})] = 0.$ $[\text{SO}_4^{2-}(\text{aq})] = 0.$ 25 °C |

Figure 4. Initial and boundary conditions for simulation of diffusion of calcium and sulfate ions at constant temperature (25 °C) with precipitation of gypsum. Brackets denote concentrations.

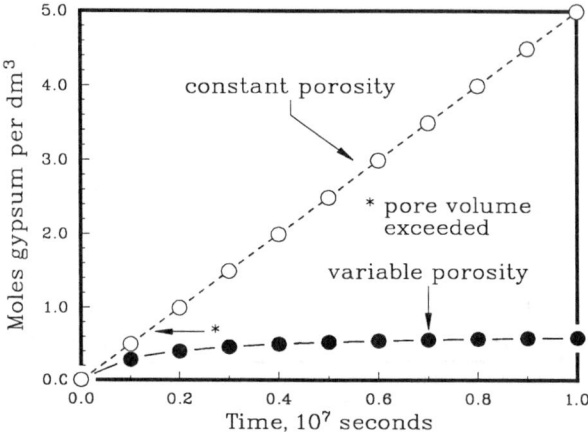

Figure 5. Gypsum precipitated at boundary $x = 0$, temperature 25 °C.

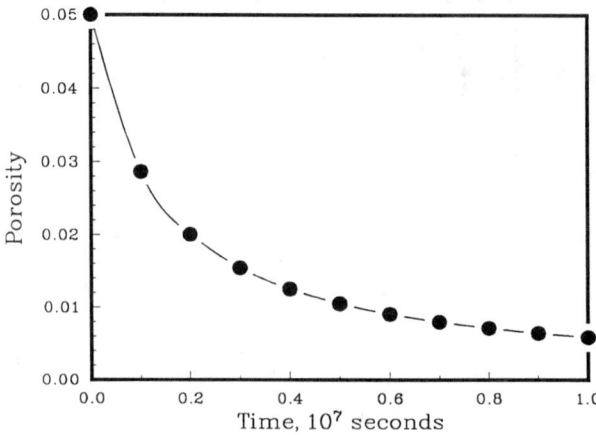

Figure 6. Porosity at boundary $x = 0$ for gypsum precipitation, temperature 25 °C.

profiles of fluid-phase concentrations of reactants were identical in the cases with and without porosity variations *whenever the reactive solid was present*. This is a necessary consequence of the assumption of chemical equilibrium in the precipitation/dissolution process. However, in simulations involving dissolution of a previously precipitated solid and transport of the dissolution products, the limitation imposed by available porosity on quantities of precipitate may have to be considered.

In real porous media, mineral deposition or dissolution would create preferential pathways to solute migration by diffusion and by advective fluid flow. Simulation of the formation of preferential pathways is beyond the capabilities of the present, one-dimensional model. Future development will extend the present model to higher dimensionality in order to examine these interesting processes.

ACKNOWLEDGMENTS

I am grateful for helpful comments by three anonymous reviewers. This work was supported by the Director, Office of Civilian Radioactive Waste Management, Office of Facilities Siting and Development, Siting and Facilities Technology Division, of the U. S. Department of Energy under Contract No. DE-AC03-76SF00098.

LEGEND OF SYMBOLS

$B_i$    concentration of basis species $i$, mole/dm$^3$ of aqueous phase
$C_j$    concentration of complex $j$, mole/dm$^3$ of aqueous phase
$D$     coefficient of hydrodynamic dispersion, m$^2$/s
$J_i$    flux of component $i$ at $x = 0$ for $\epsilon_f = 1$, mole/(m$^2$s)
$K_{cj}$   equilibrium constant for formation of complex $j$, units variable
$K_{sk}$   equilibrium constant (solubility product) for dissolution of solid $k$, units variable
$N_b$    total number of basis species
$N_c$    total number of complexes
$N_p$    number of solids subject to reversible precipitation/dissolution
$P_k$    concentration of solid $k$, mole/dm$^3$
$Q_{sk}$   activity product of basis species constituents of solid $k$, units variable
$t$     time, s
$T$     absolute temperature, K
$\overline{V}_k$    molar volume of solid $k$, dm$^3$/mole
$W_i$    total concentration of element $i$, mole/dm$^3$
$x$     distance, m
$z_i$    signed number of units of electronic charge carried by an ionic species
$\gamma_i$    activity coefficient of aqueous-phase species $i$
$\epsilon_f$    fraction of volume of porous medium occupied by fluid
$\epsilon_g$    fraction of volume of porous medium occupied by gas
$\epsilon_m$    fraction of volume of porous medium occupied by nonreactive solids
$\epsilon_p$    fraction of volume of porous medium occupied by reactive solids
$\nu_{ij}$    number of moles of basis species $i$ per mole of complex $j$
$\nu_{ik}$    number of moles of basis species $i$ per mole of solid $k$
$\tau$     tortuosity

LITERATURE CITED

1. Grove, D. B.; Wood, W. W. Ground Water 1979, 17, 250–257.
2. Miller, C. W.; Benson, L. V. Water Resour. Res. 1983, 19, 381–391.
3. Walsh, M. P.; Bryant, S. L.; Schechter, R. S.; Lake, L. W. Am. Inst. Chem. Eng. J. 1984, 30, 317–328.
4. Bryant, S. L.; Schechter, R. S.; Lake, L. W. Am. Inst. Chem. Eng. J. 1986, 32, 751–764.

5. Carnahan, C. L. In Coupled Processes Associated with Nuclear Waste Repositories; Tsang, C.-F., Ed.; Academic Press: Orlando FL, 1987; Chapter 19.
6. Witherspoon, P. A.; Wang, P. S. Y.; Jwai, K.; Gale, J. E. Water Resour. Res. 1980, 16, 1016–1024.
7. Coudrain-Ribstein, A. Docteur-Ingenieur Thesis, l'Ecole Nationale Superieure de Mines de Paris, 1983.
8. Pearson, F. J., Jr. Proc. Workshop on Fundamental Geochemistry Needs for Nuclear Waste Isolation, US DOE CONF8406134, 1985, p 173.
9. Walsh, M. P.; Lake, L. W.; Schechter, R. S. J. Pet. Tech. September 1982, 2097–2112.
10. Ortoleva, P.; Merino, E.; Moore, C.; Chadam, J. Amer. J. Sci. 1987, 287, 979–1007.
11. Ortoleva, P.; Chadam, J.; Merino, E.; Sen, A. Amer. J. Sci. 1987, 287, 1008–1040.
12. Carnahan, C. L. In Scientific Basis for Nuclear Waste Management X; Bates, J. K.; Seefeldt, W. B., Eds.; Materials Research Society: Pittsburgh PA, 1987; pp 713–721.
13. Carnahan, C. L. In Scientific Basis for Nuclear Waste Management XI; Apted, M. J.; Westerman, R. E., Eds.; Materials Research Society: Pittsburgh PA, 1988.
14. Carnahan, B.; Luther, H. A.; Wilkes, J. O. Applied Numerical Methods; Wiley: New York, 1969; p 319.
15. Hindmarsh, A. C. Solution of Block-Tridiagonal Systems of Linear Algebraic Equations; Lawrence Livermore National Laboratory Report UCID-30150, 1977.
16. Rimstidt, J. D.; Barnes, H. L. Geochim. Cosmochim. Acta 1980, 44, 1683–1699.
17. Robie, R. A.; Bethke, P. M.; Beardsley, K. M. In CRC Handbook of Chemistry and Physics; Weast, R. C., Ed.; CRC Press: Cleveland, 1977; pp B-220-B-252.
18. Phillips, S. L.; Hale, F. V.; Silvester, L. F.; Siegel, M. D. Thermodynamic Tables for Nuclear Waste Isolation; Lawrence Berkeley Laboratory Report LBL-22860 Vol. 1, 1988.

RECEIVED July 20, 1989

# Chapter 19

# Simulation of Molybdate Transport with Different Rate-Controlled Mechanisms

### Kenneth G. Stollenwerk and Kenneth L. Kipp

### U.S. Geological Survey, Box 25046, Mail Stop 413, Federal Center, Denver, CO 80225

> Laboratory column experiments were used to identify potential rate-controlling mechanisms that could affect transport of molybdate in a natural-gradient tracer test conducted at Cape Cod, Mass. Column-breakthrough curves for molybdate were simulated by using a one-dimensional solute-transport model modified to include four different rate mechanisms: equilibrium sorption, rate-controlled sorption, and two side-pore diffusion models. The equilibrium sorption model failed to simulate the experimental data, which indicated the presence of a rate-controlling mechanism. The rate-controlled sorption model simulated results from one column reasonably well, but could not be applied to five other columns that had different input concentrations of molybdate without changing the reaction-rate constant. One side-pore diffusion model was based on an average side-pore concentration of molybdate (mixed side-pore diffusion); the other on a concentration profile for the overall side-pore depth (profile side-pore diffusion). The mixed side-pore diffusion model gave a reasonable correlation with experimental data, and the parameters could be used for a variety of input concentrations. However, the profile side-pore diffusion model gave the most accurate simulations for the largest variety of input concentrations.

Nonequilibrium transport of solutes through porous media occurs when ground-water velocities are sufficiently fast to prevent attainment of chemical and physical equilibrium. Chemical reactions in porous media often require days or weeks to reach equilibrium. For example, Fuller and Davis ([1]) reported that cadmium sorption by a calcareous sand was characterized by multiple reactions, including a recrystallization reaction that continued for a period of days. Sorption of oxyanions by metal oxyhydroxides often occurs at an initially rapid rate; the rate then decreases until steady-state is achieved ([2-4]). Unless ground-water velocity in such a situation is extremely slow, nonequilibrium transport will occur.

Diffusion of a solute through immobile water to a reaction site also is affected by interstitial water velocity. If the diffusion rate is slow compared to the interstitial velocity, physical nonequilibrium occurs ([5-7]). The immobile water can be a layer on the grain surface (film diffusion), in dead-end pores between tightly packed grains (pore diffusion), or within crevices or pits on the grain surfaces (particle diffusion). Calcium and chloride breakthrough curves from column experiments done by James and Rubin ([8]) indicate that nonequilibrium transport occurs unless interstitial velocities are decreased so that the hydrodynamic-dispersion coefficient is of the same order of magnitude as the molecular-diffusion coefficient.

Information about rate-controlling mechanisms that could occur in the field often can be obtained from carefully performed laboratory column experiments. The asymmetric shape of breakthrough curves from column experiments done using field ground-water velocities can indi-

This chapter not subject to U.S. copyright
Published 1990 American Chemical Society

cate the occurrence of nonequilibrium solute transport; numerical simulation of these breakthrough curves, using a solute-transport model that contains the appropriate equations, may provide information about the rate-controlling mechanism.

Results described in this paper are from column experiments designed to evaluate potential rate-controlling mechanisms that could affect transport of Mo(VI) in a Cape Cod natural-gradient tracer test (9). This test was done to study dispersive transport and chemical processes in a glacial-outwash aquifer that was affected by recharge of treated sewage (10); the tracers were superimposed on the sewage plume. The zone of Mo(VI) injection was partially in freshwater above the sewage plume and partially in a transition zone between freshwater and the sewage plume. Molybdate was introduced with other constituents in the initial injection pulse to evaluate transport of a reactive oxyanion ($MoO_4^{2-}$). Knowledge gained from studying the reactions and transport of Mo(VI) is likely to apply to other more toxic oxyanions such as chromate ($CrO_4^{2-}$), selenite ($SeO_3^{2-}$), and arsenate ($AsO_4^{3-}$).

## COLUMN EXPERIMENTS

Columns, with a length of 0.302 m and an inside diameter of 0.025 m, were packed with sediment collected from an uncontaminated part of the aquifer adjacent to the tracer-test plume. Ground water of two different compositions was used in the experiments. One composition, from above the sewage plume, had a pH of 5.7 and no measurable phosphate ($PO_4$). The other composition, from the transition zone between the freshwater and the sewage plume, had a pH of 6.3 and had 0.036 mmol/l $PO_4$. Major element chemistry for both ground waters is listed in Table I.

Molybdate-free ground water was eluted through the columns until pH and specific conductance of the eluent stabilized. A nonreactive tracer, bromide (Br), then was added to determine dispersion characteristics of the packed columns. When Br could no longer be detected in eluent from the columns the influent was changed to Mo(VI)-spiked ground water. Six Mo(VI) concentrations, which ranged from 0.0016 to 0.096 mmol/l, were used for the sewage-contaminated ground water; replicates were run for the 0.0053 and 0.043 mmol/l columns. Sorption of Mo(VI) from ground water not contaminated by sewage was evaluated in replicate columns at one concentration (0.043 mmol/l). The interstitial velocity for each column was about the same as that observed during the tracer test (0.43 m/d). Physical and chemical parameters for the columns are listed in Table II. Porosity and bulk density were determined gravimetrically, and longitudinal dispersivity was obtained from Br breakthrough data.

## MOLYBDATE SORPTION

Molybdate occurs as the $MoO_4^{2-}$ oxyanion at pH values used in these experiments. Complexation with cations is negligible. Batch experiments (11-12) and column experiments showed that less Mo(VI) is sorbed from sewage-contaminated ground water than from the fresh ground water above the sewage plume. There are two causes for this behavior. 1) Alluvial grains were visibly coated with iron oxides. Thus, the net positive charge of the grain surfaces was greater at pH 5.7 than pH 6.3 which resulted in more Mo(VI) sorption at pH 5.7. 2) Phosphate in the sewage plume competed with Mo(VI) for sorption sites. The extent of competition increased with the ratio of $PO_4$ to Mo(VI) in solution.

Sorption of Mo(VI) in these experiments can be described by the Freundlich isotherm. Freundlich constants for sorption from sewage-contaminated ground water were calculated from a linear plot of the concentration of Mo(VI) sorbed by sediment in each column compared to input concentration (Figure 1) using the following equation:

$$\bar{c} = K_f c^n, \qquad (1)$$

where $\bar{c}$ is the concentration of Mo(VI) sorbed, in mmol/kg;
$K_f$ is the Freundlich adsorption equilibrium constant, in $(l/kg)^{-n}$;
c is the concentration of Mo(VI) in solution, in mmol/l; and
n is the Freundlich exponent (slope of line on a log plot of $\bar{c}$ versus c).

Table I. Ground-Water Chemistry

| | pH | Na | K | Ca | Mg | $HCO_3$ | $SO_4$ | Cl | $PO_4$ |
|---|---|---|---|---|---|---|---|---|---|
| | | | | Concentration (mmol/l) | | | | | |
| Uncontaminated ground water | 5.7 | 0.20 | 0.02 | 0.03 | 0.03 | 0.08 | 0.08 | 0.14 | --- |
| Sewage-contaminated ground water | 6.3 | 1.7 | 0.18 | 0.21 | 0.13 | 1.3 | 0.32 | 0.82 | 0.036 |

Table II. Physical and Chemical Parameters for Column Experiment

| Length, m | Porosity | Bulk density, g/cm$^3$ | Interstitial Velocity, m/d |
|---|---|---|---|
| 0.302 | 0.34 | 1.75 | 0.43 |

| Longitudinal dispersivity, m | Time of pulse, d | Input concentration mmol/l |
|---|---|---|
| 0.0027 | 4.19 | 0.0016-0.096 |

| Freundlich isotherm constant, (l/kg)$^{-n}$ | Slope of Freundlich isotherm |
|---|---|
| 0.14[1] | 0.72[1] |
| 4.64[2] | 0.37[2] |

[1] Values are for sewage-contaminated ground water.
[2] Values are for ground water not contaminated by the sewage plume.

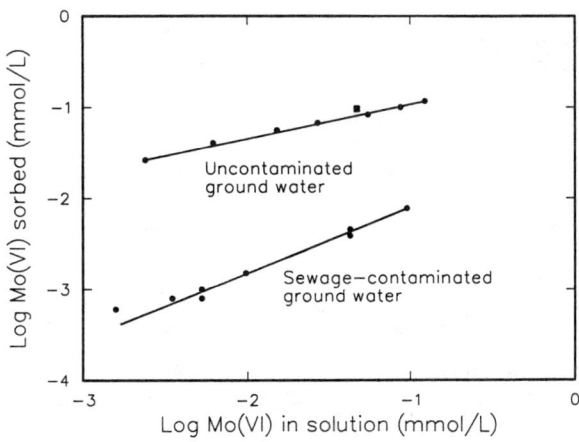

Figure 1. Freundlich plot of Mo(VI) sorption parameters for sewage-contaminated ground water and uncontaminated ground water. The size of plotted experimental data points in all Figures encompasses the analytical error.

Data from batch experiments (11) were used to calculate the Freundlich constants for sorption from ground water that was uncontaminated by the sewage plume (Figure 1). Data from the replicate column experiments (closed square) plot almost on the line, which indicates that the batch-determined sorption constants can be used to describe Mo(VI) sorption in these two columns.

The Freundlich constants in Table II were used in the solute-transport models. The exponent is less than one for both sets of data, so the isotherms are nonlinear and predict a decrease in sorption at larger solution concentrations.

## NUMERICAL SIMULATIONS

Many one-dimensional solute-transport models have been developed and used to analyze column data. For a recent review, see Grove and Stollenwerk (13). Four different models were used in the study discussed in this article to simulate the shape of the column-breakthrough curves. All four models contain a one-dimensional solute-transport equation and use the Freundlich equation to describe sorption. They differ in the rate mechanism that is assumed to control transport of Mo(VI) from flowing phase to solid surface. The essential features of each model are summarized in Table III.

The basic equation that describes the transport of solutes in the flowing-water phase can be expressed as:

$$\frac{\partial c}{\partial t} = -V\frac{\partial c}{\partial x} + D\frac{\partial^2 c}{\partial x^2} + \bar{r}. \qquad (2)$$

Accumulation in the sorbed phase is described by:

$$\frac{\partial \bar{c}}{\partial t} = -\frac{\theta}{\rho_b} \bar{r}, \qquad (3)$$

where  c   is the concentration in the flowing-water phase, in mmol/l;
        t   is time, in d;
        x   is distance along the column, in m;
        V   is the interstitial velocity, in m/d;
        D   is the dispersion coefficient, in m$^2$/d;
        $\theta$   is the porosity;
        $\rho_b$  is the bulk density, in kg/l;
        $\bar{r}$   is the rate of addition of solute to the flowing-water phase from the sorbed phase, in mmol/l/d;
        $\bar{c}$   is the concentration on the solid phase, in mmol/kg.

All parameters are assumed to be constant and uniform along the column.

The boundary conditions for Equation 2 are:

$$\text{at } x = 0 \qquad -D\frac{\partial c}{\partial x} + Vc = Vc_b(t), \text{ and} \qquad (4a)$$

$$\text{at } x = L \qquad \frac{\partial c}{\partial x} = 0, \qquad (4b)$$

where $c_b$ is the inlet concentration and L is the column length. Solute is introduced to the column as a slug of a specified duration, $t_s$. The initial conditions for Equations 2 and 3 are:

$$\text{at } t = 0, \quad c = 0 \quad \text{and} \quad \bar{c} = 0. \qquad (4c,d)$$

The first transport model (local equilibrium sorption) is formulated by assuming that local equilibrium exists along the column. No immobile-water phase is present. The water velocity is assumed slow enough so sorption rapidly attains equilibrium (14). An early example of this model with a finite-difference solution was developed by Gupta and Greenkorn (15).

The second transport model (rate-controlled sorption) is based on the assumption that the sorption mechanism does not have time to reach equilibrium at each point along the column. Mansell et al. (16) used this model to simulate phosphorus transport through sandy soils. Therefore, Equation 1 is replaced by the sorption rate expression:

$$\bar{r} = \frac{\rho_b}{\theta} k_r (K_f c^n - \bar{c}) \qquad (5)$$

where $k_r$ is the sorption-rate constant, $d^{-1}$.

Enfield and Bledsoe (17) used a sorption-rate equation similar to Equation 5 to represent sorption of herbicides by soil.

The third and fourth models are based on the assumption of an immobile-water phase. This phase can be caused by the existence of crevices or pits on the grain surfaces and by dead-end pores created by tight packing of the grains. In this article, we did not consider an immobile-water film that could cover entire grain surfaces. Diffusion of Mo(VI) into the immobile-water phase retards transport because of the capacitance of this phase and also provides a rate-limiting step for sorption on the solid matrix adjacent to the immobile-water phase. For these models, transport of Mo(VI) from the flowing phase can occur by sorption to the solid phase or by diffusion to the immobile-water phase which then is followed by sorption to the solid phase. The sorption from either phase is assumed to be at local equilibrium with the adjacent fluid phase. An additional balance equation for the solid phase adjacent to the immobile-water phase is as follows:

$$\frac{\partial \bar{c}_s}{\partial t} = -\frac{\theta_s}{\rho_{bs}} \bar{r}_s, \qquad (6)$$

where $\rho_{bs}$ is the dry bulk density adjacent to the immobile-water phase, in g/cm ;
  $\theta_s$ is the porosity of the immobile-water phase;
  $\bar{r}_s$ is the rate of addition of solute to the immobile-water phase from the adjacent solid phase, in mmol/l/d;
  $\bar{c}_s$ is the sorbed concentration adjacent to the immobile-water phase, in mmol/kg.

The total dry bulk density and total porosity in Equation 3 must be changed to the values for the flowing and solid phases that are in contact, denoted by $\rho_{bf}$ and $\theta_f$. Sorption sites also need to be partitioned into those adjacent to the flowing phase and those adjacent to the immobile-water phase. This concept of dividing the porous matrix into two regions for sorption was described by van Genuchten and Wierenga (18). We have used their idea, but have divided the sites on the basis of bulk density.

The third model (mixed side-pore diffusion) (18) is based on the assumption of an effective average or mixed concentration of solute in the immobile-water phase. This model has no spatial dependence in the immobile-water phase. The balance equation for the immobile-water phase is:

$$\frac{\partial c_s}{\partial t} = -\frac{\theta_f}{\theta_s} r_s + \bar{r}_s \qquad (7)$$

where $c_s$ is the immobile-water phase concentration, in mmol/l; and
  $r_s$ is the rate of addition of solute to the flowing phase from the immobile-water side pore phase, in mmol/l/d. The initial condition is:

$$\text{at } t = 0, \quad c_s = 0; \qquad (8)$$

and the transfer-rate expression is:

$$r_s = -\frac{\theta_s}{\theta_f} k_s A(c-c_s), \qquad (9)$$

where $k_s$ is a transfer-rate constant, in m/d; and A is the effective interfacial area per unit volume of immobile-water phase, in $m^{-1}$.

Parameters V, $\theta_f$, $\theta_s$, $k_s$, and A in the mixed side-pore model are estimated from the shape of the breakthrough curve for a nonreactive tracer. The parameters k and A can be determined only as a lumped parameter, $k_s A$. The parameters $\rho_{bf}$ and $\rho_{bs}$ are estimated from the shape of the breakthrough curve for the reactive solute.

The transfer rate in the mixed side-pore model is proportional to the difference in concentration between the flowing-water and immobile-water phases. The transfer-rate constant $k_s A$ is a characteristic-rate parameter for diffusion in the immobile-water phase. Without the Freundlich sorption mechanism, this third model is the same as the dead-end pore model developed by Coats and Smith (19). The Freundlich sorption isotherm was included by van Genuchten and Wierenga (18) in their study, but they solved for the linear case only. Grove and Stollenwerk (20) described a similar model but included Langmuir sorption and a continuous immobile-water film phase.

The fourth model (profile side-pore diffusion) is similar to the third but the assumption is made that a concentration profile exists throughout the thickness of the immobile-water phase. Molecular diffusion of solute is the major transport mechanism in the immobile-water phase. The transfer rate of solute from the flowing- to the immobile-water phase is assumed to be the diffusional flux at the interface between these phases. Therefore, Equations 7 and 9 are replaced by:

$$\frac{\partial c_s}{\partial y} = D_m \frac{\partial^2 c_s}{\partial y^2} + \bar{r}_s. \qquad (10)$$

The boundary conditions are:

$$\text{at } y = 0, \qquad c_s = c; \text{ and} \qquad (11a)$$

$$\text{at } y = L_s, \qquad \frac{\partial c_s}{\partial y} = 0, \qquad (11b)$$

where $D_m$ is the effective molecular diffusivity in the immobile-water zone, in $m^2/d$; y is distance into side pore from the entrance, in m; and $L_s$ is length of side pore, in m. The second boundary condition states that no solute leaves through the back end of the side pore. The transfer rate expression at the side-pore entrance is:

$$r_s = AD_m \left.\frac{\partial c_s}{\partial y}\right|_{y=0} \frac{\theta_s}{\theta_f} \qquad (12)$$

The concentration profile in the immobile-water phase is controlled by a diffusional-transport mechanism. The transfer rate from the immobile-water phase to the flowing-water phase is the diffusive flux, which depends on the concentration gradient in the immobile-water phase at the interface. Parameters V, $\theta_f$, $\theta_s$, A, and $L_s$ in the profile side-pore model are estimated from the shape of the breakthrough curve for a nonreactive tracer. Parameters $\rho_{bf}$ and $\rho_{bs}$ are estimated from the shape of the breakthrough curve for a reactive solute. The effective molecular diffusivity $D_m$ is estimated from values published in the literature.

The profile side-pore diffusion model is the most complex and, perhaps, the most realistic of the four models. Without the Freundlich sorption mechanism, the model is the same as that developed by Kipp (21). The case of radial diffusion with linear sorption was considered by van Genuchten et al. (22); whereas, spherical diffusion that had linear sorption was described by

Nkedi-Kizza et al. (23). Even when shallow dead-end pores or thin immobile-water pockets exist, the diffusion rates may be so slow that a concentration profile is more realistic than an effective, average concentration in the immobile-water phase.

Finite-difference techniques were used to compute numerical solutions as column-breakthrough curves because of the nonlinear Freundlich isotherm in each transport model. Along the column, 100 nodes were used, and 10 nodes were used in the side-pore direction for the profile model. A predictor-corrector calculation was used at each time step to account for nonlinearity. An iterative solver was used for the profile model; whereas, a direct solution was used for the mixed side-pore and the rate-controlled sorption models.

## MODELING RESULTS

The four potential rate mechanisms were evaluated by calculating column-breakthrough curves for various parameter sets to obtain the most accurate correlation between observed column-breakthrough curves and calculated concentration data. The parameters $\rho_{bf}$ and $\rho_{bs}$ for the mixed side-pore and profile side-pore diffusion models were estimated from the 0.043 mmol/l breakthrough curves. Simulations at other concentrations were made by changing only the solution concentration value in the Freundlich equation. Physical and chemical parameters common to all four models are listed in Table II. Results are for 0.096-, 0.043-, 0.01- and 0.0016-mmol/l columns.

### Breakthrough Curve for Br

The breakthrough curve for nonreactive Br for one of the columns is shown in Figure 2; it is similar in shape to the Br-breakthrough curves for all columns. An immobile-water phase is indicated by the asymmetrical shape of the curve, and the profile side-pore diffusion model gave the best match to the experimental data.

### Sewage-Contaminated Ground Water

### Equilibrium Sorption Model

Simulation of Mo(VI) breakthrough by the equilibrium sorption model is compared with the experimental data from the 0.043 mmol/l column in Figure 3. Results for the 0.096-, 0.01- and 0.0016-mmol/l columns were similar and are not shown. The model simulates a very steep slope for the adsorption limb of the breakthrough curve and complete site saturation by the second pore volume. Experimental data from the column show that complete breakthrough did not occur until the sixth pore volume, which indicates the effects of a rate process. The equilibrium model also simulated complete rinse-out of Mo(VI) by the 9th pore volume; whereas Mo(VI) in the column effluent did not reach zero until the 15th pore volume, which indicates that desorption also was affected by a rate process.

### Rate-Controlled Sorption Model

The assumption that sorption was controlled by the rate of the sorption reaction resulted in a more accurate simulation of the experimental data from the 0.43 mmol/l column (Figure 4b). A sorption-rate constant of 1.5 $d^{-1}$ resulted in the most accurate visual correlation to the experimental data; however, some discrepancy between the two curves still occurs in the shoulder part of the adsorption curve and in the tail part of the desorption curve. The rate-controlled sorption model also was used to simulate the column breakthrough curves at other input concentrations of Mo(VI) by using the reaction-rate constant of 1.5 $d^{-1}$ (Figures 4a, 4c, and 4d). Simulation of these breakthrough curves became progressively less accurate as concentrations of Mo(VI) decreased from 0.043 mmol/l. Only by using a different rate constant for each concentration could results similar to Figure 4b be achieved. If the sorption rate were truly controlling sorption of Mo(VI), one rate constant would be able to more accurately simulate transport at different concentrations.

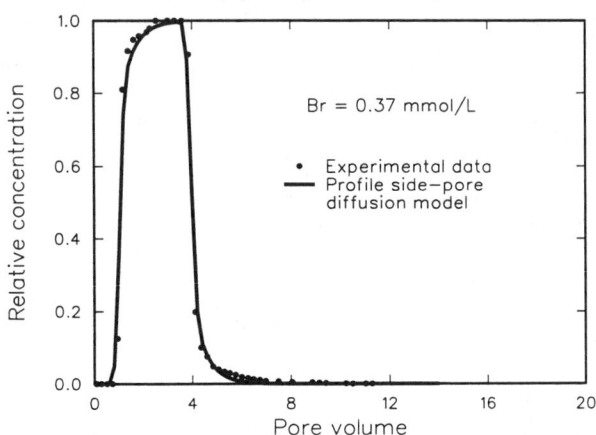

Figure 2. Simulation of Br experimental data, using the profile side-pore diffusion model.

Table III. Comparison of the Four Transport Models Used to Simulate Column Breakthrough of Mo(VI)

| Model | Features |
|---|---|
| Local Equilibrium Model | Assumes chemical and physical equilibrium conditions throughout column. |
| Rate-Controlled Sorption Model | Accounts for the possibility that the rate of sorption reaction may be too slow for equilibrium to be achieved. |
| Mixed Side-Pore Diffusion Model | Diffusion into immobile water in pores is the rate-limiting step. An average concentration is assumed throughout each side pore. |
| Profile Side-Pore Diffusion Model | Diffusion into immobile water in side pores is the rate-limiting step. A concentration gradient exists in each side pore. |

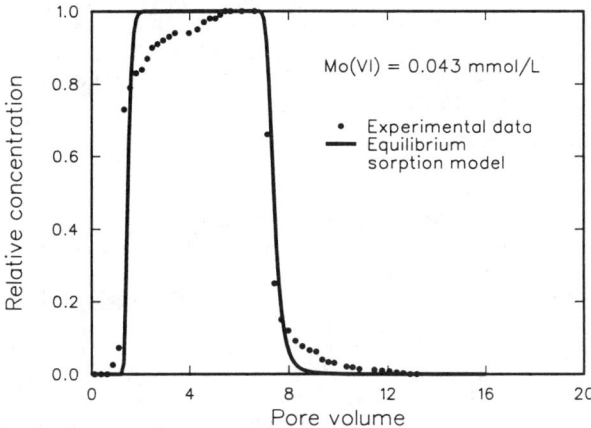

Figure 3. Simulation of Mo(VI) experimental data from sewage-contaminated ground water, using the equilibrium sorption model.

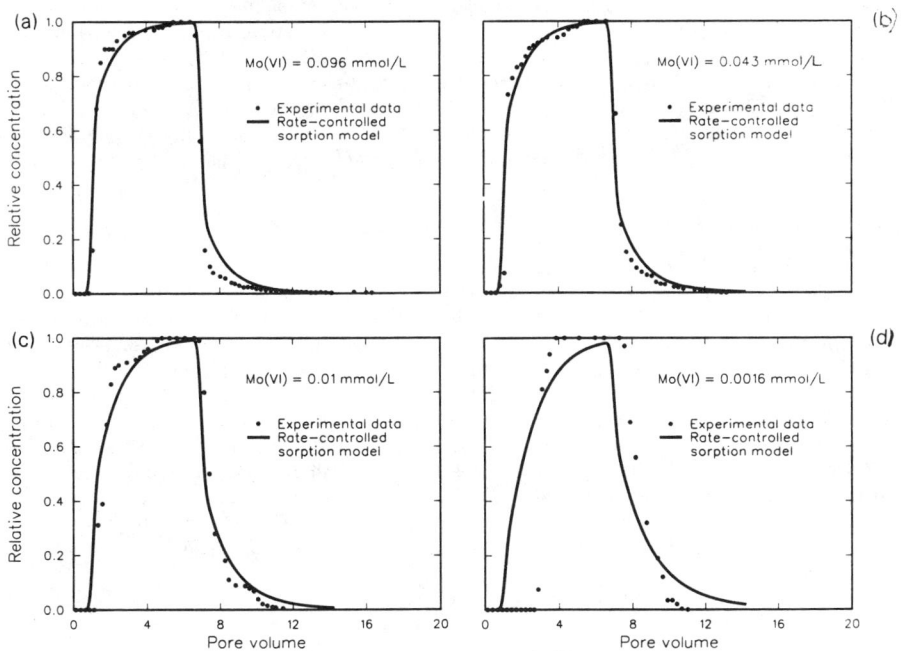

Figure 4. Simulation of experimental data from sewage-contaminated ground water, using the rate-controlled sorption model for four concentrations of Mo(VI).

Mixed Side-Pore Diffusion Model

The mixed side-pore diffusion model also reasonably simulated the experimental data (Figures 5a and 5b). This model was slightly more accurate than the reaction-rate model in simulating breakthrough curves for a range of input concentrations (Figure 5a-5c); however, significant discrepancies also were observed between experimental data and model simulations at concentrations of less than 0.01 mmol/l Mo(VI).

Profile Side-Pore Diffusion Model

The profile side-pore diffusion model simulated the experimental data from the 0.043 mmol/l column almost exactly and was within the accuracy of the breakthrough data (Figure 6b). Based on the best fit simulation of the Br-breakthrough curve (Figure 2), the immobile-water phase was calculated to be about 5 percent of the total porosity. Apparently, diffusion into and out of this volume of immobile water was responsible for the observed shoulder and tail of the curves of the experimental data.

The profile side-pore model also was the most transferable to other input concentrations (Figures 6a and 6c). Even at concentrations of (Mo(VI) less than 0.01 mmol/l, the profile side-pore model simulated experimental data more accurately than the other models (data not shown). Only at 0.0016 mmol/l Mo(VI) did the profile side-pore model fail to correlate with the experimental data.

As the concentration of Mo(VI) in the influent decreased, the shoulder and tail parts of the experimental breakthrough curves became progressively less prominent (compare Figures 6a and 6d). This trend may be explained by a slower rate of Mo(VI) diffusion into side pores caused by a decrease in the concentration gradient between flowing and immobile phases.

A possible explanation for the progressive divergence between experimental data and model simulations as Mo(VI) concentrations decreased, is failure of the Freundlich equation to accurately describe Mo(VI) sorption for large ranges in concentration. Batch experiments have indicated that Mo(VI) sorption decreases as the relative concentration of $PO_4$ increases (11). Because each column contained a different ratio of Mo(VI) to $PO_4$, each data point in Figure 1 (sewage-contaminated ground water) actually may be part of a different isotherm, each having different $K_f$ values.

Uncontaminated Ground Water

Simulations of Mo(VI) breakthrough data for $PO_4$-free ground water are shown in Figures 7a-7d. Experimental data from one column are plotted, results from the replicate column are similar. The effects of increased sorption due to lower pH and the absence of $PO_4$ are indicated by the experimental data. Complete saturation of sites did not occur until pore volume 60, and rinse-out of Mo(VI) from the column was not complete until pore volume 200. The asymmetric shape of the experimental data was not simulated as well by the equilibrium model (Figure 7a) or by the rate-controlled sorption model (Figure 7b). The mixed side-pore diffusion model failed to simulate the rinse-out of Mo(VI) from the column (Figure 7c), but reasonably simulated the sorption part of the breakthrough curve. The sharp break in the adsorption limb of the breakthrough curve is an artifact of the mathematical model used to represent Freundlich sorption. The profile side-pore model had the most accurate simulation of experimental data for ground water without $PO_4$ (Figure 7d) and is a further indication of the robustness of this model for a variety of conditions.

CONCLUSIONS

Transport of Mo(VI) through laboratory columns packed with alluvium was retarded by sorption on iron oxide coatings. The amount of Mo(VI) sorbed decreased as pH and the ratio of $PO_4$ to Mo(VI) increased. Molybdate sorption could be described by the Freundlich isotherm.

The shape of breakthrough curves from column experiments provided evidence that Mo(VI) transport was not equilibrium controlled. The decrease in slope as the adsorption limb of the

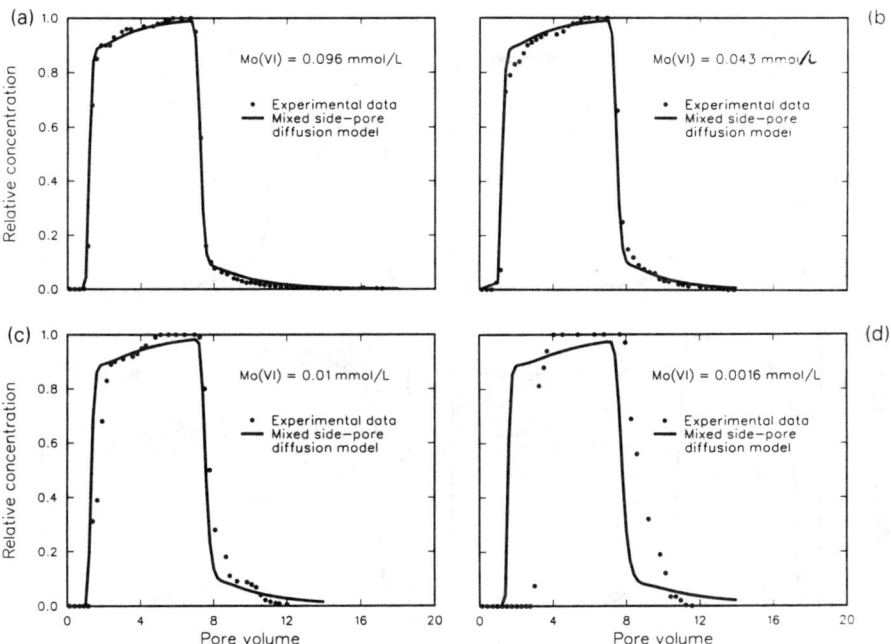

Figure 5. Simulation of experimental data from sewage-contaminated ground water, using the mixed side-pore diffusion model for four concentrations of Mo(VI).

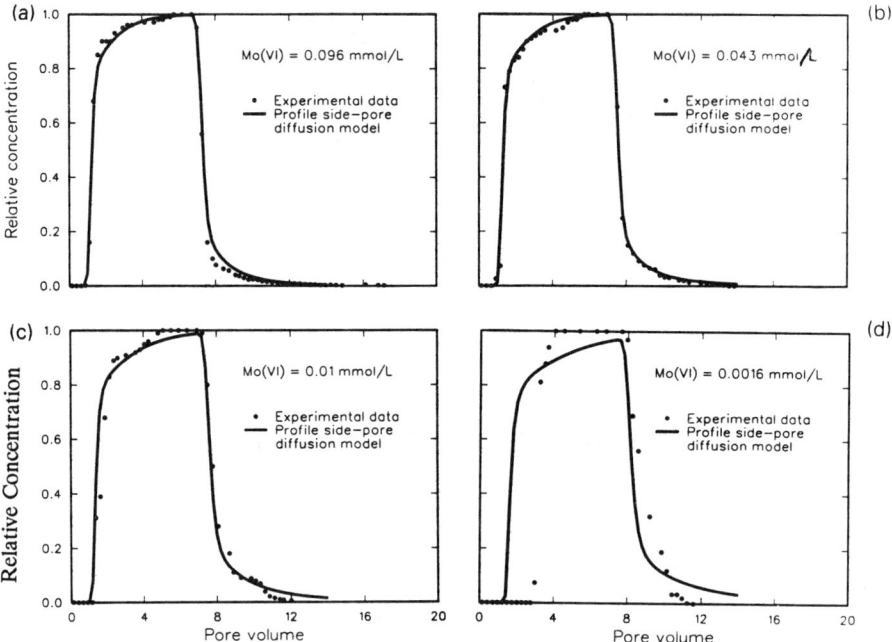

Figure 6. Simulation of experimental data from sewage-contaminated ground water, using the profile side-pore diffusion model for four concentrations of Mo(VI).

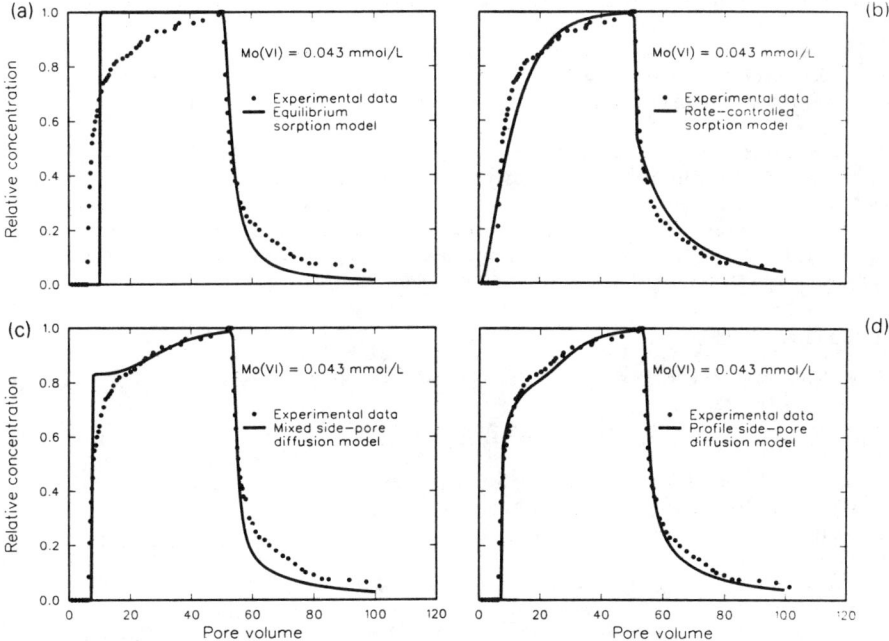

Figure 7. Simulation of Mo(VI) experimental data from uncontaminated ground water, using the equilibrium sorption, rate-controlled sorption, mixed side-pore diffusion, and profile side-pore diffusion models.

breakthrough curve achieved steady state and the decrease in slope of the desorption limb as Mo(VI) was rinsed out of the column could only be simulated adequately by using a solute-transport model that included a rate mechanism. When transport was assumed to be controlled only by the rate of Mo(VI) sorption, experimental data were simulated reasonably well, but the reaction-rate constant used in the model was applicable only for the breakthrough curve to which it was fitted.

The concept of Mo(VI) diffusion into and out of side pores that had an immobile-water phase resulted in a more accurate simulation of experimental breakthrough curves for a wider range of concentrations. The mixed side-pore diffusion model could be used to fit a particular experimental breakthrough curve with about the same degree of accuracy as the reaction rate model; however, the mixed side-pore diffusion model was applicable for a wider range of concentrations.

The most accurate results were achieved by using the more complex and conceptually realistic profile side-pore diffusion model. This model had a fit through all of the experimental data points from one column, could be used for a range of concentrations, and could simulate breakthrough curves for ground water of different compositions.

None of the models could simulate experimental data at very small Mo(VI) concentrations in the presence of $PO_4$. One possible explanation is the failure of the Freundlich isotherm to adequately describe sorption of Mo(VI) over a wide concentration range when a competing anion such as $PO_4$ is present.

Transferability of the results from this study to the Cape Cod natural gradient tracer test will provide information about the validity of laboratory experiments in providing information about onsite processes. Although actual values for some of the physical parameters determined in the laboratory may not apply to an aquifer because of scale differences between laboratory and field, conceptually realistic models such as the profile side-pore diffusion model may be able to simulate onsite transport conditions more accurately.

LITERATURE CITED

1. Fuller, C.C.; Davis, J.A. Geochim. Cosmochim. Acta 1987, 51, 1491-502.
2. Barrow, N.J. Soil Sci. 1970, 109, 282-88.
3. Munns, D.N.; Fox, R.L. Soil Sci. Soc. Am. J. 1976, 40, 46-51.
4. Bolan, N.S.; Barrow, N.J.; Posner, A.M. J. Soil Sci. 1985, 36, 187-97.
5. Biggar, J.W.; Nielsen, D.R. Soil Sci. Soc. Am. Proc. 1962, 26, 125-28.
6. DeSmedt, F.; Wierenga, P.J. J. Hydrol. 1979, 41, 59-67.
7. Rao, P.S.C.; Rolston, D.E.; Jessup, R.E.; Davidson, J.M. Soil Sci. Soc. Am. J. 1980, 44, 1139-146.
8. James, R.V.; Rubin, J. In Chemical Modeling in Aqueous Systems; Jenne, E.A. Ed.; ACS Symposium Series No. 93; American Chemical Society, Washington D.C. 1979, pp 225-35.
9. LeBlanc, D.R.; Garabedian, S.P.; Wood, W.W.; Hess, K.M.; Quadri, R.D. U.S. Geol. Surv. Open-File Report 87-109, 1987, B-13-16.
10. LeBlanc, D.R. U.S. Geol. Surv. Open-File Report 87-109, 1987, B-9-12.
11. Stollenwerk, K.G.; Grove, D.B. U.S. Geol. Surv. Open-File Report 87-109, 1987, B-17-22.
12. Stollenwerk, K.G. EOS, Transactions of the American Geophysical Union 1988, 69(16), 354.
13. Grove, D.B.; Stollenwerk, K.G. Water Resour. Bull. 1987, 23(4), 601-15.
14. Grove, D.B.; Stollenwerk, K.G. U.S. Geol. Surv. Water Resour. Invest. Report 1984, 84-4059, 58 p.
15. Gupta, S.P.; Greenkorn, R.A. Water Resour. Res. 1974, 10(4), 839-46.
16. Mansell, R.S.; Selim, H.M.; Kanchanasnt, P.; Davidson, J.M.; Fiskell, J.G.A. Water Resour. Res. 1977, 13(1), 189-94.
17. Enfield, C.G.; Bledsoe, B.E. Jour. Irrigation and Drainage Division Amer. Soc. of Civil Eng., 1975, 101(IR3), 145-55.
18. van Genuchten, M.Th.; Wierenga, P.J. Soil Sci. Soc. Am. J. 1976, 40, 473-80.
19. Coats, K.H.; Smith, B.D. Soc. Pet. Eng. J. 1964, 4, 73-84.
20. Grove, D.B.; Stollenwerk, K.G. Water Resour. Res. 1985, 21(11), 1703-709.

21. Kipp, K. L. In Proc. Symposium on Uncertainties Associated with the Regulation of the Geological Disposal of High-level Radioactive Waste; Kocher, D.C., Ed.; U.S. Nuclear Regulatory Commission: Washington D.C. 1982; NUREG/CP-0022, CONF-810372, 321-32.
22. van Genuchten, M.Th.; Tang, D.H.; Guennelon, R. Water Resour. Res. 1984, 20(3), 335-46.
23. Nkedi-Kizza, P.; Rao, P.S.C.; Jessup, R.E.; Davidson, J.M. Soil Sci. Soc. Am. J. 1982, 46, 471-76.

RECEIVED August 18, 1989

# APPLICATIONS TO MODELING: SURFACE CHEMISTRY

# Chapter 20

# Numerical Simulation of Coadsorption of Ionic Surfactants with Inorganic Ions on Quartz

### Rebecca L. Rea and George A. Parks

### Department of Applied Earth Science, Stanford University, Stanford, CA 94305-2225

When little organic carbon is present in a sediment or aquifer, sorption of organic solutes must be controlled by mineral surfaces. Adsorption of these solutes depends on pH, electrolyte concentration, and adsorption of unrelated inorganic ions. Interactions among adsorbing species can significantly enhance or inhibit adsorption. The adsorption of anionic and cationic surfactants on quartz and corundum in the presence of a background electrolyte, and the coadsorption of an anionic surfactant enhanced by $Ca^{++}$ have been simulated successfully using HYDRAQL, a chemical speciation program including a triple-layer adsorption model. Surface ionization constants, electrolyte binding constants, and surfactant binding constants were determined from five independent experimental investigations. Surfactants are assumed to adsorb as single ions at low solution concentration and as clusters resulting from hydrophobic interactions between molecules at high concentration. The simulation accounts for experimental adsorption densities, zeta potentials, and for $Ca^{++}$ coadsorption, froth flotation recovery.

Adsorption of organic solutes is commonly controlled by the solid organic matter in soils, sediments, and aquifers ([1-3]). However, when organic matter is absent, mineral surfaces play an important role and interactions with inorganic solutes may significantly influence adsorption of the organic species. Tipping and Heaton ([4]) and Davis ([5]), for example, found that adsorption of natural organic solutes is sensitive to calcium concentration and pH. Gerstl and Mingelgrin ([6]) found that pre-adsorption of a long-chain cationic surfactant on the clay mineral, attapulgite significantly enhanced the adsorption capacity of the clay for parathion. Ainsworth *et al.* ([7]) showed that the adsorption of PCBs on goethite is enhanced in the presence of surfactants and that cationic and anionic surfactants introduced opposite pH dependence. Finally, Gaudin and Chang ([8]) in one of the few direct measurements of simultaneous or co-adsorption of two species, found that adsorption of $Ba^{++}$ enhances adsorption of laurate ion ($C_{11}H_{23}COO^-$) on quartz.

As a first step in the development of methods for dealing with these synergistic effects in hydrogeochemical transport programs, this paper demonstrates that conceptual models developed for adsorption of surfactants on oxides, including enhancement by $Ca^{++}$ can be incorporated into an existing surface complexation model, HYDRAQL ([9]). HYDRAQL is a version of MINEQL, a chemical speciation program originated by Westall *et al.* ([10]) and expanded to deal with adsorption in a triple layer model by Davis *et al.* ([11]).

Adsorption of Surfactants

Conceptual models for adsorption of surfactants are described in detail by Somasundaran and Fuerstenau (12) and Parks (13). Illustrative experimental data for adsorption of dodecylamine ion ($C_{12}H_{25}NH_3^+$) on quartz ($\alpha$-$SiO_2$) are provided by DeBruyn (14) and for adsorption of dodecylsulfate ion ($C_{12}H_{25}SO_4^-$) on corundum ($\alpha$-$Al_2O_3$) by Chandar et al. (15).

At low concentrations, ionic surfactants adsorb chiefly as counterions in response to surface charge. Anionic species adsorb strongly when pH < $pH_{pzc}$, where surface charge is positive, and weakly, if at all, when pH > $pH_{pzc}$. Cations adsorb weakly when pH < $pH_{pzc}$ and strongly when pH > $pH_{pzc}$. In this range, electrokinetic behavior is identical to that of simple monovalent indifferent electrolytes, i.e., the adsorbed surfactant reduces, but does not reverse the sign of the zeta potential, $\zeta$ (16). In keeping with the behavior of indifferent counterions, the surfactant is assumed to adsorb in the outer Helmholtz plane (OHP) and the diffuse layer.

At higher ionic surfactant concentrations, Van der Waals interaction between hydrocarbon chains and hydrophobic bonding results in aggregation, forming clusters called "hemi-micelles" (12,17-21). The aggregation numbers of hemi-micelles are not well established; estimates range from 2 to ~250 (15,22,23). Surface charge is reduced more rapidly than at lower solution concentrations and is ultimately reversed as solution concentration and adsorption increase so the adsorption bond includes a "chemical" or "specific" contribution.

Cation Enhancement of Surfactant Adsorption

The presence of multivalent inorganic cations can initiate or enhance adsorption of otherwise weakly or non-adsorbing anionic surfactants (8,24). Multivalent anions inhibit adsorption of anionic surfactants, but may enhance adsorption of cationic surfactants. There are two conceptual models for cation enhancement (24,25). When a surfactant fails to adsorb because its charge is the same as that of the surface, then: (1) the enhancing ion may adsorb onto the surface, reversing surface charge, with the surfactant adsorbing in response to that charge, or (2) the surfactant and coadsorbing ion may form an aqueous complex with charge opposite that of the surface, and this complex may adsorb. In either case, adsorption is possible because the surfactant and surface are oppositely charged in the presence of the enhancing ion. These mechanisms are thermodynamically indistinguishable.

ADSORPTION MODEL

HYDRAQL (10) treats adsorption as surface complexation with bound hydroxide functional groups, SOH, and their ionization products, $SO^-$ and $SOH_2^+$. The calculations in this paper use HYDRAQL in its triple layer mode. Surface charge and counterionic accumulate in three "layers": (1) at the surface itself, i.e., in the plane of the SOH groups where the surface potential is $\Psi_0$; (2) in the outer Helmholtz plane (OHP), where adsorbed ions retain their inner hydration sheaths (26) and the potential is $\Psi_\beta$; and (3) in the diffuse layer. The triple layer model is ideal for our purposes because of its ability to compute an estimate of $\Psi_\beta$. The computed $\Psi_\beta$ can be compared with experimental measurements of the zeta potential, providing an additional means of constraining models.

In triple layer approximations the location of each adsorbate with respect to the surface must be specified. Protons and all ions assumed bound as inner-sphere complexes (specifically or chemically adsorbed species) are assumed to lose part of their hydration sheaths, bonding directly to sites in the surface itself. Adsorbates assumed to remain hydrated, forming outer-sphere surface complexes, are assigned to the OHP. In the intrinsic equilibrium constants for adsorption reactions, $K^{int}$, the activities of ions transferred from solution to the surface are corrected for the electrical potential they experience, $\Psi_0$ or $\Psi_\beta$ (27).

A variety of parameters must be evaluated in any surface complexation model. Common to all surface reactions are the density of SOH sites, $N_s$, the specific surface area, $A_s$, the solid/liquid ratio, and the near-surface capacities, $C_1$ inside the OHP, and $C_2$ outside the OHP. $N_s$ is a property of the solid; for both quartz and corundum, $N_s$ was taken as 5 sites/nm² (11,28). $A_s$ and solid/liquid ratio are experimental variables; values reported with published adsorption data were used in each case. The value, 0.20 F/m² (Farads per m²) (27), was adopted for $C_2$ throughout this work. Every surface ionization and adsorption reaction requires selection of the stoichiometric reaction itself, the location of the adsorption complex, and an equilibrium constant.

As a first approximation in formulating any reaction, the aqueous species predominant under the conditions of the experiment and the oppositely charged surface species were selected as reactants. These reactants were combined in one-to-one stoichiometry and treated as outer sphere complexes. If the resulting reaction proved inadequate for simulating experimental data, reactants and/or stoichiometry and/or location were adjusted.

The complex systems of interest involve adsorption of $H^+$, $Na^+$, $Cl^-$, a surfactant ion, and the enhancing ion, if any. For each solid, ionization and electrolyte binding reactions and equilibrium constants were estimated for $H^+$, $Na^+$, and $Cl^-$ using an appropriate set of experimental adsorption data. The value of $C_1$ is determined as part of this process. All of these parameters were subsequently held constant and used when the appropriate ions were present in other systems. Having fixed these parameters, only adsorption reactions and their equilibrium constants were used to describe adsorption of surfactants and enhancing ions. Using the same strategy, these were determined using data from the simplest possible experimental systems, then applied without further adjustment in the final effort to simulate the most complex systems containing all interacting solutes. The only new parameters needed to simulate enhanced adsorption were a single reaction describing interaction of the enhancing ion and the surfactant, and its equilibrium constant.

Surface Ionization and $Na^+$ and $Cl^-$ Binding Constants

Reactions and selected intrinsic equilibrium constants describing ionization of SOH and adsorption of $Na^+$ and $Cl^-$ on quartz and corundum are listed in Table I. The formulas of surface complexes indicate the assumed location of constituents of the complex in the electrical triple layer. The entire surface complex is located directly on the surface unless a dash is present in the formula. If a dash is present, everything to the left of the dash is located on the surface, and everything to the right of the dash is located in the OHP.

Table I. Reactions and Selected Intrinsic Equilibrium Constants for Surface Ionization and Electrolyte Adsorption

| Reaction | Quartz | Corundum |
|---|---|---|
| $C_1$ | 1.295 F/m$^2$ | 1.4 F/m$^2$ |
| $SOH_2^+ \leftrightarrows SOH + H^+$ | $pK^{int}_{a1} = 3.2$ | $pK^{int}_{a1} = 5.2$ |
| $SOH \leftrightarrows SO^- + H^+$ | $pK^{int}_{a2} = 7.2$ | $pK^{int}_{a2} = 12.3$ |
| $SOH + Na^+ \leftrightarrows SO\text{-}Na + H^+$ | $p^*K^{int}_{Na^+} = 6.7$ | $p^*K^{int}_{Na^+} = 9.7$ |
| $SOH_2\text{-}Cl \leftrightarrows SOH + H^+ + Cl^-$ | —— | $p^*K^{int}_{Cl^-} = 7.9$ |

As first approximations, equilibrium constants for ionization of SOH and for $Na^+$ and $Cl^-$ binding on quartz and corundum were assumed equal to values selected by Davis et al. (27) for amorphous silica and by James and Parks (28) for $\gamma\text{-}Al_2O_3$, respectively. If necessary, the equilibrium constants were adjusted to fit experimental data for systems more closely approximating those in which surfactant adsorption data were available. The adjusted equilibrium constants are summarized in Table I. Sources of data and details of fitting procedure are described below.

Li and DeBruyn (19) measured the adsorption density of $Na^+$ on ground Brazilian optical quartz (using radio-tracer methods) and zeta potential, both as functions of pH and NaCl concentration. Tabular data were obtained directly from Li's SM thesis (29). Davis et al. provide only $pK^{int}_{a2}$. A reaction forming a positive surface site, $SOH_2^+$, was included, and its equilibrium constant selected so that the model $pH_{pzc}$ matched Li's experimental value of $pH_{pzc} = 2.0$. The values of $pK^{int}_{Na^+}$ and $C_1$, only, were then adjusted to optimize simulation of the sodium adsorption and zeta-potential data (29). Results are shown in Figure 1. The fits are typical of results obtained by others (11,28).

No direct adsorption data were available for $Na^+$ and $Cl^-$ on corundum. Fuerstenau and Modi (18), however, computed the zeta potential, $\zeta$, as a function of pH from experimental measurements of streaming potential in NaCl solutions. The value of $\Psi_\beta$ computed by the triple layer model in HYDRAQL is an approximation of $\zeta$ (11). Because $\zeta$ depends on $H^+$, $Na^+$, and $Cl^-$ adsorption densities, it was possible to use these data to constrain the equilibrium constants and $C_1$. The value of $pK^{int}_{a2}$ was adjusted to force $pH_{pzc} = 9.1$ as observed by Fuerstenau and Modi (18), and $p^*K^{int}_{Na^+}$ was adjusted to fit the zeta-potential for $pH < pH_{pzc}$. Results are shown in Figure 2.

Surfactant adsorption

Dodecylamine on Quartz

DeBruyn (14) measured the adsorption density of dodecylamine on ground Brazilian optical quartz as a function of concentration and pH, using radio-tracer methods. Two adsorption reactions are needed to represent the conceptual model for surfactant adsorption, one describing outer-sphere adsorption of $RNH_3^+$ and one describing hemi-micelle formation. A third reaction, producing an adsorbed $RNH_3Cl$ complex, is suggested by evidence that both adsorbed surfactants (30) and micelles (31) bind counterions. These three reactions and the $pK^{int}$ values derived by fitting deBruyn's data in the context of the appropriate surface ionization and electrolyte binding reactions (Table I), accounting for aqueous hydrolysis and dimerization of the amine (32) are:

$$SOH + RNH_3^+ \leftrightarrows SOH\text{-}RNH_3^+ \qquad\qquad p^*K^{int}_{RNH_3^+} = 3.3 \qquad (1)$$

$$SOH + RNH_3^+ + Cl^- \leftrightarrows SOH\text{-}RNH_3Cl \qquad\qquad p^*K^{int}_{RNH_3Cl} = 4.8 \qquad (2)$$

$$SOH + 5RNH_4^+ \leftrightarrows SO(RNH_4)_2(RNH_3)_3^+ + 4H^+ \qquad\qquad p^*K^{int}_{HM^+} = 12.5 \qquad (3)$$

The resulting simulation is compared with experimental data in Figure 3.

Monomeric surface species were assumed to be outer-sphere complexes located in the OHP. The hemi-micelle was assumed to be an inner-sphere complex located on the surface. These assumptions are supported by the work of Chander, et al. (22). They observed significant ionic strength dependence of the adsorption of dodecylsulfonate on alumina in NaCl solutions at low surfactant concentrations and no ionic strength dependence in the higher concentration region corresponding to hemi-micelle formation. High ionic strength dependence suggests outer-sphere complexes, while lack of ionic strength dependence suggests inner-sphere adsorption complexes (33).

The composition of the hemi-micelle in Reaction 3 was determined by optimizing the fit to the adsorption density at high concentration and high pH, recognizing an experimentally observed reversal in zeta potential (16), and the high-density packing of molecules in hemi-micelles observed by Waterman et al. (34).

Dodecylsulfate on Corundum

Chandar et al. (15) measured the adsorption density of dodecylsulfate on corundum at pH = 6.5 in 0.1 M NaCl. The experimental data are adequately described by a model closely analogous to that used for dodecylamine on quartz. In addition to the surface ionization and NaCl binding reactions (Table I), the reactions and equilibrium constants used were:

$$SOH_2\text{-}RSO_3 \leftrightarrows SOH + H^+ + RSO_3^- \qquad\qquad p^*K^{int}_{RSO_3^-} = 1.1 \qquad (4)$$

$$SOH_2(RSO_3H)_3(RSO_3)_2^- \leftrightarrows SOH + H^+ + 5RSO_3^- \qquad\qquad p^*K^{int}_{HM^-} = 41.1 \qquad (5)$$

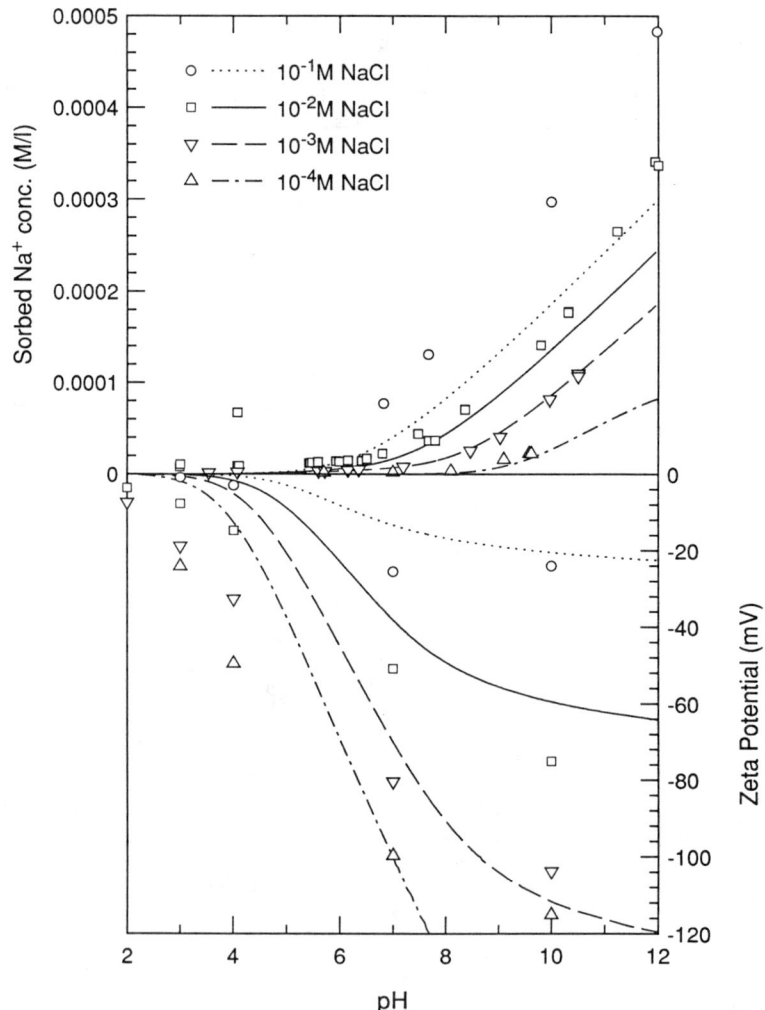

Figure 1. Na⁺ on Quartz: Comparison of experimental (29) and simulated adsorption and zeta-potential data. Modeling parameters for surface ionization and sodium adsorption are given in Table I. Surface site concentration: $\Sigma \text{SOH} = 7.06 \times 10^{-4}$ M.

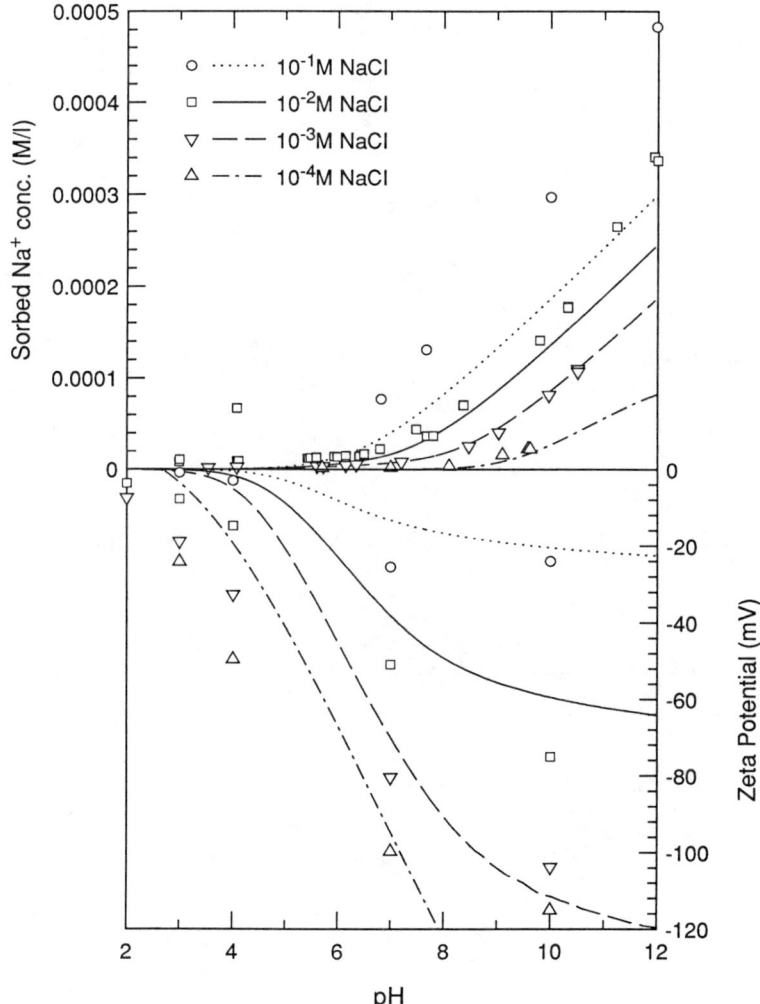

Figure 2. Na⁺ and Cl⁻ on Corundum: Comparison of experimental (18) and simulated zeta-potential data. Modeling parameters for surface ionization and electrolyte adsorption are given in Table I. Surface site concentration: $\Sigma$ SOH = $2.34 \times 10^{-4}$ M.

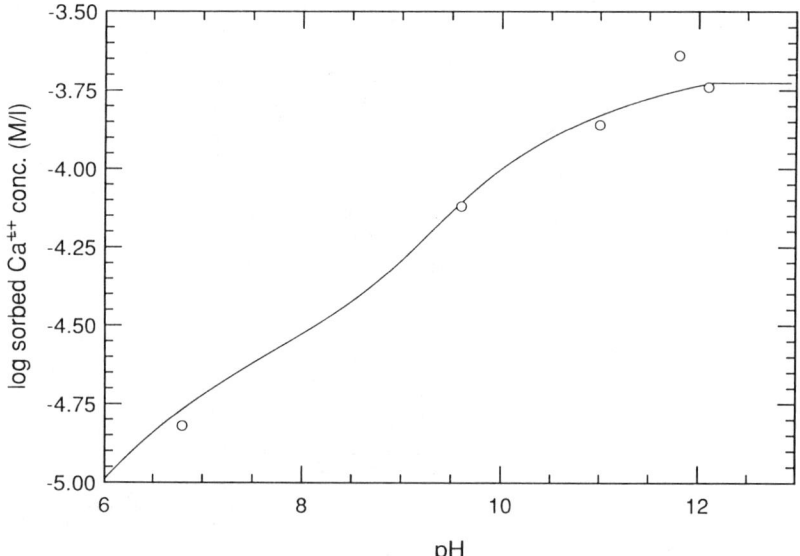

Figure 3. Dodecylamine on Quartz: Comparison of experimental (14) and simulated adsorption data. Modeling parameters for surface ionization and NaCl electrolyte adsorption are given in Table I.  Surface site concentration: $\Sigma SOH = 1.27 \times 10^{-4}$ M.  Dodecylamine adsorption Reactions 1, 2, and 3 were assumed with $p^*K^{int}_{RNH_3^+} = 3.3$, $p^*K^{int}_{RNH_3Cl} = 4.8$, and $p^*K^{int}_{HM^+} = 12.5$, respectively.

Equilibrium constants for Reactions 4 and 5 were estimated by trial and error to fit the experimental dodecylsulfate adsorption data. No counterion binding reaction, equivalent to Reaction 2, was needed. The experimental data and simulation are compared in Figure 4.

Coadsorption

$Ca^{++}$ improves froth flotation recovery of quartz with dodecylsulfate, implying that calcium enhances adsorption of dodecylsulfate on the mineral, probably by coadsorption (35). Direct simultaneous adsorption measurements of both surfactant and enhancing ion are not presently available, but adsorption behavior can be inferred from flotation studies. Froth flotation is a technique used to separate one kind of particulate solid from another through selective attachment to air bubbles in an aqueous suspension. Successful flotation is a complex process, dependent upon both equilibrium chemistry and bubble attachment kinetics. However, there are good empirical correlations between flotation recovery and adsorption density (36,37). Flotation requires adsorption of an appropriate surfactant at a density equivalent to at least 5% surface coverage (14,36). As reported by Fuerstenau and Healy (35) and Fuerstenau et al. (37), calcium-enhanced flotation of quartz by dodecylsulfate commences near pH = 10 (implying surface coverage of at least 5% at that point), peaks near pH = 12, and drops off at higher pH (Figure 5).

To simulate coadsorption of calcium and surfactant, it is first necessary to define reactions and equilibrium constants for adsorption of calcium on quartz in the absence of surfactants. The same two reactions used by other investigators to describe adsorption of divalent cations (38) were adequate to match the limited $Ca^{++}$ adsorption data (Figure 5) published by Cooke and Digre (39). The reactions and equilibrium constants used in addition to those listed in Table I were:

$$SOH + Ca^{++} \leftrightarrows SO\text{-}Ca^+ + H^+ \qquad p^*K^{int}_{Ca^{++}} = 3.8 \qquad (6)$$

$$SOH + Ca^{++} + H_2O \leftrightarrows SO\text{-}CaOH + 2H^+ \qquad p^*K^{int}_{CaOH^+} = 13.5 \qquad (7)$$

Calcium-enhanced surfactant adsorption can be simulated by addition of only one additional reaction, formation of a mixed $Ca^{++}$-surfactant surface complex:

$$SOH + Ca^{++} + RSO_3^- \leftrightarrows SOCa\text{-}RSO_3 + H^+ \qquad p^*K^{int}_{CaRSO_3^+} = 8.7 \qquad (8)$$

Accepting the predetermined surface ionization and NaCl binding reactions (Table I), $Ca^{++}$ binding Reactions 6 and 7, and surfactant adsorption Reactions 4 and 5, it was possible to simulate the necessary 5% surface coverage at pH ≈ 10.3 and adsorption maximum at pH ≈ 12 by setting $p^*K^{int}_{CaRSO_3^+} = 8.7$. Results are shown in Figure 6.

CONCLUSIONS

These results demonstrate the feasibility of simulating the adsorption of monofunctional anionic and cationic alkyl surfactants on mineral surfaces, and the cation enhancement of surfactant adsorption with the electrical triple layer, surface complexation model in HYDRAQL. No changes in the basic model were needed. No adjustment of surface ionization or electrolyte binding constants or of capacities were needed, once these parameters had been derived from independent data obtained from surfactant free systems; in fact, these parameters proved nearly identical to those derived by others for a different silica and a different alumina (27,28). Surfactants were accommodated by adding a set of aqueous solution reactions to account for known solution behavior and a small set of adsorption reactions based on conceptual models for surfactant adsorption promulgated largely by the flotation community. The fact that it was possible to apply reactions and equilibrium constants derived from diverse simple systems, without change, in increasingly complex systems suggests that this modeling approach has wide general application.

The simulation models are assuredly not unique. Adsorption density measurements, electrokinetic data, and even froth flotation data constrain the problem, and more extensive data in these areas would be useful, but direct observation of the compositions and aggregation numbers of

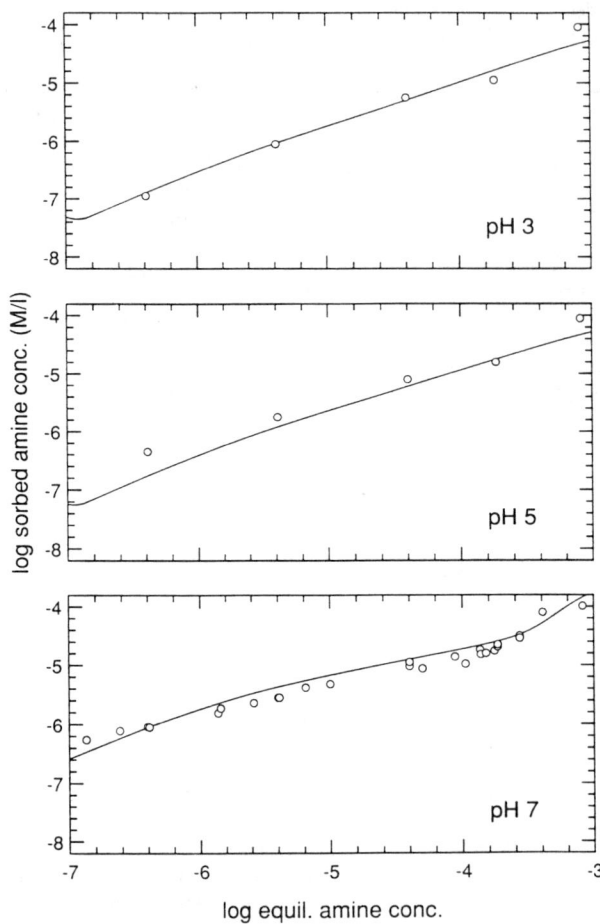

Figure 4. Dodecylsulfate on Corundum: Comparison of experimental (15) and simulated adsorption data. Modeling parameters for surface ionization and NaCl electrolyte adsorption are given in Table I. Surface site concentration: $\Sigma SOH = 2.34 \times 10^{-4}$ M. Dodecylsulfate adsorption Reactions 4 and 5 were assumed with $p^*K^{int}_{RSO_3^-} = 1.1$ and $p^*K^{int}_{HM^-} = 41.1$, respectively.

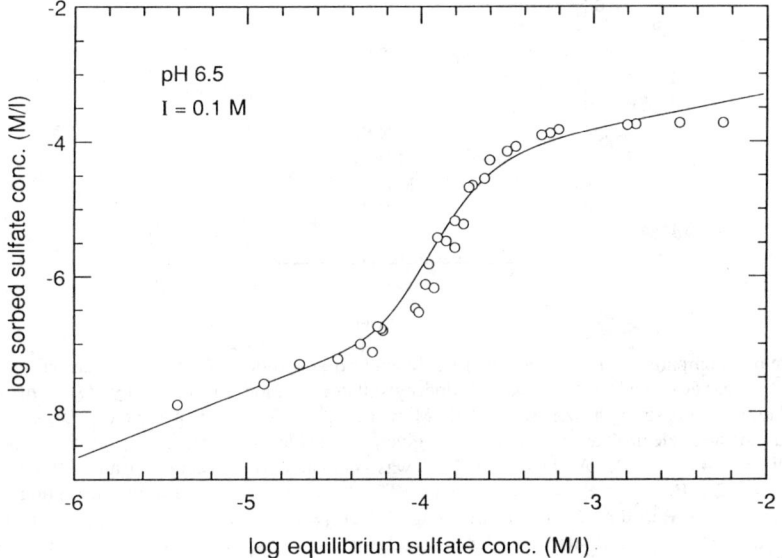

Figure 5. $Ca^{++}$ on Quartz: Comparison of experimental (39) and simulated adsorption data. No background electrolyte was present; $\Sigma CaCl_2 = 7.2 \times 10^{-3}$ M. Modeling parameters for surface ionization and $Cl^-$ adsorption are given in Table I. Surface site concentration: $\Sigma SOH = 2 \times 10^{-4}$ M. $Ca^{++}$ adsorption Reactions 6 and 7 were assumed with $p^*K^{int}_{Ca^{++}} = 3.8$ and $p^*K^{int}_{CaOH^+} = 13.5$, respectively.

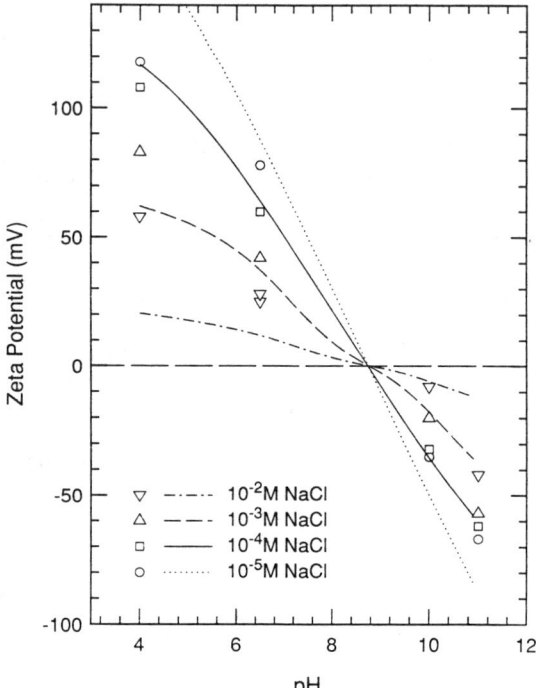

Figure 6. Comparison of simulated $Ca^{++}$ enhanced adsorption of dodecylsulfate on quartz with experimental, $Ca^{++}$-enhanced dodecylsulfate flotation of quartz (35,37). In the simulation, ionic strength was set at 0.01 M with NaCl. Modeling parameters for surface ionization and electrolyte adsorption are given in Table I. Surface site concentration: $\Sigma SOH = 2.34 \times 10^{-4}$ M. Adsorption of $Ca^{++}$ was simulated with Reactions 4 and 5 assuming $\Sigma Ca^{++} = 7.2 \times 10^{-4}$ M, $p^*K^{int}_{Ca^{++}} = 3.8$, and $p^*K^{int}_{CaOH^+} = 13.5$. Dodecylsulfate adsorption was simulated with Reactions 6 and 7 assuming $\Sigma RSO_3 = 2.3 \times 10^{-4}$ M, $p^*K^{int}_{RSO_3^-} = 1.1$, and $p^*K^{int}_{HM^-} = 41.1$. $Ca^{++}$-dodecylsulfate coadsorption Reaction 8 was assumed with $p^*K^{int}_{CaRSO_3^+} = 8.7$.

adsorption complexes and their locations relative to the surface are needed to define reactions uniquely.

LITERATURE CITED

1. Goring, C. A. I. Ann. Rev. Phytopathol. 1967, 5, 285-318.
2. Karickhoff, S. W. Chemosphere 1981, 10, 833-846.
3. Schwarzenbach, R. P.; Westall, J. T. Environ. Sci. Technol. 1981, 15, 1360-1367.
4. Tipping, E.; Heaton, M. J. Geochim. Cosmochim. Acta 1983, 47, 1393-1397.
5. Davis, J. A. Geochim. Cosmochim. Acta 1982, 46, 2381-2393.
6. Gerstl, Z.; Mingelgrin, U. Clays and Clay Minerals 1979, 27, 285-290.
7. Ainsworth, C. C.; Chou, S. F. J.; Griffin, R. A. 24th Hanford Life Sciences Symposium October 1985.
8. Gaudin, A. M.; Chang, M. C. AIME Trans. 1952, 193, 193-201.
9. Papelis, C.; Hayes, K. F.; Leckie, J. O. Technical Report No. 306, Environmental Engineering and Science, Dept. Civil Engineering, Stanford University, 1988.

10. Westall, J. C.; Zachary, J. L.; Morel, F. M. M. Technical Note 18, Ralph M. Parsons Laboratory, M. I. T., Cambridge, MA 1976.
11. Davis, J. A.; James, R. O.; Leckie, J. O. J. Colloid Interface Sci. 1978, 63, 480-499.
12. Somasundaran, P.; Fuerstenau, D. W. J. Phys. Chem. 1966, 70, 90-96.
13. Parks, G. A. In Marine Geochemistry; Riberg, J. P.; Skirrow, G., Eds.; Academic: New York, 1976, p 241-308.
14. deBryun, P. L. Trans. AIME 1955, 202, 291-296.
15. Chandar, S.; Somasundaran, P.; Turro, N. J. J. Colloid Interface Sci. 1987, 117, 31-46.
16. Fuerstenau, D. W. J. Phys. Chem. 1956, 60, 981-985.
17. Gaudin, A. M.; Fuerstenau, D. W. AIME Trans. 1955, 202, 958-962.
18. Fuerstenau, D. W.; Modi, H. J. J. Electrochemical Soc. 1959, 106, 336-341.
19. Li, H. C.; deBruyn, P. L. Surface Sci. 1966, 5, 203-220.
20. Wakamatsu, T.; Fuerstenau, D. W. In Advances in Chemistry Series 79, Adsorption from Aqueous Solution; Gould, R. F., Ed.; American Chemical Society: Washington, DC, 1968; p 161-172.
21. Dick, S. G.; Fuerstenau, D. W.; Healy, T. W. J. Colloid Interface Sci. 1971, 37, 595-602.
22. Chander, S.; Fuerstenau, D. W.; Stigter, D. In Adsorption From Solution; Ottewill, R. H.; Rochester, C. H.; Smith, A. L., Eds.; Academic: London, 1983; p 197-210.
23. Cases, J. M.; Levitz, P.; Poirier, J. E.; Van Damme, H. In Advances in Mineral Processing; Somasundaran, P., Ed.; S.M.E. Pub.: Littleton, CO, 1986; p 171-188.
24. Fuerstenau, M. C., Miller, J. D., Kuhn, M. C. Chemistry of Flotation; AIME: New York, 1964.
25. Davis, J. A.; Leckie, J. O. Environ. Sci. and Tech. 1978, 12, 1309-1315.
26. Bockris, J. O'M.; Reddy, A. K. N. Modern Electrochemistry; Plenum: New York, 1970.
27. Davis, J. A.; James, R. O.; Leckie, J. O. J. Colloid Interface Sci. 1978, 63, 480-499.
28. James, R. O.; Parks, G. A. Surface and Colloid Sci. 1982, 12, 119-216.
29. Li, H. C. S. M. Thesis, M. I. T., Cambridge, MA, 1955.
30. Bitting, D.; Harwell, J. H. Langmuir 1987, 3, 500-511.
31. Adamson, A. W. Physical Chemistry of Surfaces; John Wiley and Sons: New York, 1976.
32. Anamanthpadmabhan, K.; Somasundaran, P.; Healy, T. W. AIME Trans. 1979, 266, 2003-2009.
33. Hayes, K. F.; Leckie, J. O. J. Colloid Interface Sci. 1987, 115, 564-572.
34. Waterman, K. C.; Turro, N. J.; Chandar, P.; Somasundaran, P. J. Phys. Chem. 1986, 90, 6828-6830.
35. Fuerstenau, M. C.; Healy, T. W. In Adsorptive Bubble Separation Techniques; R. Lemlich, Ed.; Wiley: New York, 1972; p 91-131.
36. Fuerstenau, D. W. Trans. AIME 1957, 208, 1365-1367.
37. Fuerstenau, M. C.; Martin, C. C.; Bhappu, R. B. Trans AIME 1963, 226, 449-454.
38. Davis, J. A.; Leckie, J. O. J. Colloid Interface Sci. 1949, 67, 90-107.
39. Cooke, S. R. B.; Digre, Marcus AIME Trans. 1949, 184, 299-305.

RECEIVED August 18, 1989

Chapter 21

# Constant-Capacitance Surface Complexation Model

## Adsorption in Silica—Iron Binary Oxide Suspensions

**Paul R. Anderson[1] and Mark M. Benjamin**

**Department of Civil Engineering, University of Washington, Seattle, WA 98195**

A conceptual and mechanistic model of particle interactions in silica-iron binary oxide suspensions is described. The model is consistent with a process involving partial $SiO_2$ dissolution and sorption of silicate onto $Fe(OH)_3$. The constant capacitance model is used to test the mechanistic model and estimate the effect of particle interactions on adsorbate distribution. The model results, in agreement with experimental results, indicate that the presence of soluble silica interferes with the adsorption of anionic adsorbates but has little effect on cationic adsorbates.

A number of researchers have attempted to model trace element distribution among multi-component solids. For example, Balistrieri et al. ([1]) modeled trace metal scavenging by heterogeneous, deep ocean particulate matter; Oakley et al. ([2]) and Davies-Colley et al. ([3]) modeled trace metal partitioning in marine sediments and estuarine sediments, respectively; and Goldberg and Sposito ([4]) modeled phosphorus adsorption onto soils with the constant capacitance model. Common to all these modeling efforts was the simplifying assumption that the multi-component system could be represented either by some average collective property for the group or as a collection of discrete pure solid phases, a concept which Honeyman ([5]) called "adsorptive additivity". Honeyman tested this concept and demonstrated in several experiments that particle interactions in binary suspensions of oxides can lead to significant deviations from the adsorptive additivity concept. This paper presents an alternative approach which can account for some of these deviations.

The model used to evaluate surface chemistry in these systems is the constant capacitance surface complexation model. This model has been used to describe the adsorption of cations ([6,7]) and anions ([4,8]) onto oxides similar to those used in our experiments. A significant difference between those studies and the present study is that we have adapted the model to simulate some of the interactions that might occur between particles in a binary suspension.

In a previous paper (Anderson and Benjamin; accepted for publication in Environmental Science and Technology), surface and bulk characteristics of amorphous oxides of silica, aluminum, and iron, both singly and in binary mixtures were described. The solids were characterized with an array of complementary analytical and experimental techniques, including scanning electron microscopy, particle size distribution, x-ray photoelectron spectroscopy (XPS),

---

[1]Current address: Pritzker Department of Environmental Engineering, Illinois Institute of Technology, Chicago, IL 60616

and measurement of the pH of the zero point of charge (PZC) and of the specific surface area ($N_2$-BET). Also, batch adsorption experiments were used to characterize the adsorption behavior of Ag, Cd, $PO_4$, $SeO_3$, and Zn. The experiments revealed a number of physical and chemical differences in the particles' properties in the binary versus the single oxide suspensions. These changes resulted from interactions such as aggregation, disaggregation, dissolution, and precipitation among the suspended particles.

Details on the adsorbent preparation and experimental and analytical techniques are presented elsewhere (9). This paper briefly reviews the experimental results for the $Fe(OH)_3$ and $SiO_2$ suspensions and describes a conceptual and mechanistic model for particle interactions which is qualitatively consistent with the experimental observations. Similar results were obtained for binary $Al(OH)_3$ and $SiO_2$ suspensions (9). The constant capacitance surface complexation model is then used to test the mechanistic model and estimate the quantitative influence of the particle-particle interactions on adsorbate distribution.

## CHARACTERISTICS OF THE BINARY SUSPENSIONS

The binary systems were synthesized using two different processes, and the bulk and surface characteristics of suspensions synthesized by these two processes were indistinguishable. One, the mixed suspension, was prepared by mixing an Fe oxide suspension with a suspension of $SiO_2$. In the other method, $Fe(OH)_3$ was precipitated from a solution in which amorphous $SiO_2$ particles were suspended.

The particle size distributions (PSD) for the binary suspensions were much more like that of the pure $SiO_2$ suspension than that of the Fe oxide suspension (Figure 1). The smaller and intermediate size particles from the $Fe(OH)_3$ suspensions were not present as discrete particles in the mixed systems, while the size distribution of the larger (apparently $SiO_2$) particles remained about the same. Other characteristics of the $SiO_2$ binary oxides and the component solids are summarized in Table I. Most of these properties suggest that the surfaces of particles in the binary $SiO_2$ suspensions were dominated by Fe oxides. The most likely process by which this might have occurred is heterocoagulation between the negatively charged $SiO_2$ and positively charged Fe oxide particles. Coverage of the $SiO_2$ particles by the smaller Fe colloidal particles is consistent with the surface charge of the resulting particles being dominated by $Fe(OH)_3$, and the specific surface areas being nearly those expected for a collection of discrete solids. However, the $SiO_2$ was not completely masked because the XPS analyses indicate the presence of $SiO_2$ at the particle surface.

The adsorbent properties of the solids in the binary suspensions were also characterized. Adsorption of Cd (Figure 2), Ag, and Zn, in systems containing $Fe(OH)_3$ was nearly the same whether $SiO_2$ was present or not. In contrast, anion removal was inhibited in the binary systems relative to the pure $Fe(OH)_3$ system. For both $SeO_3$ and $PO_4$, the pH region of the adsorption edge shifted in the acid direction about 1 pH unit when $SiO_2$ was present compared to the corresponding $SiO_2$-free systems. The data for $PO_4$ are shown in Figure 3.

## A MODEL OF INTERACTIONS IN THE BINARY $SIO_2$ SYSTEMS

One mechanism that is consistent with the observed properties of the particles in these suspensions involves the dissolution of amorphous $SiO_2$ and adsorption of soluble silicate on the $Fe(OH)_3$ surface. This process could occur in parallel with the heterocoagulation mentioned earlier. Soluble silicate species might then compete with $SeO_3$ or $PO_4$ for surface sites as suggested by Goldberg (8) for the $PO_4$/silicate/goethite system. Sorption of silicate species onto $Fe(OH)_3$ need not affect cationic adsorbates. Benjamin and Bloom (10) demonstrated that adsorption of cations is often minimally affected by anion adsorption even under conditions where anion-anion competition is severe (11).

The computer programs FITEQL (12) and MICROQL (13) were used to model chemical speciation in this study. FITEQL uses a non-linear, least squares optimization technique to calculate equilibrium constants from chemical data. The program was used here to select

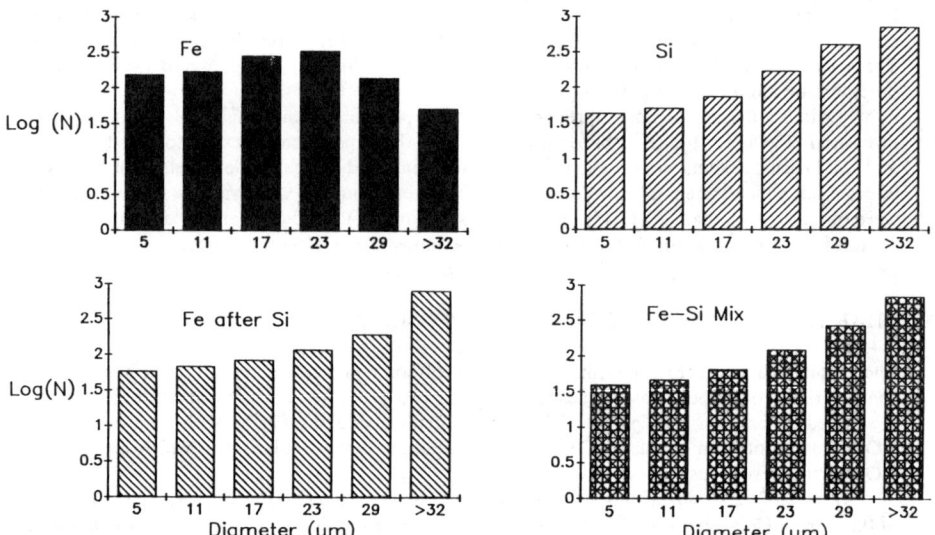

Figure 1. Histograms of the particle size distribution in $Fe(OH)_3$, $SiO_2$, and the binary suspensions.

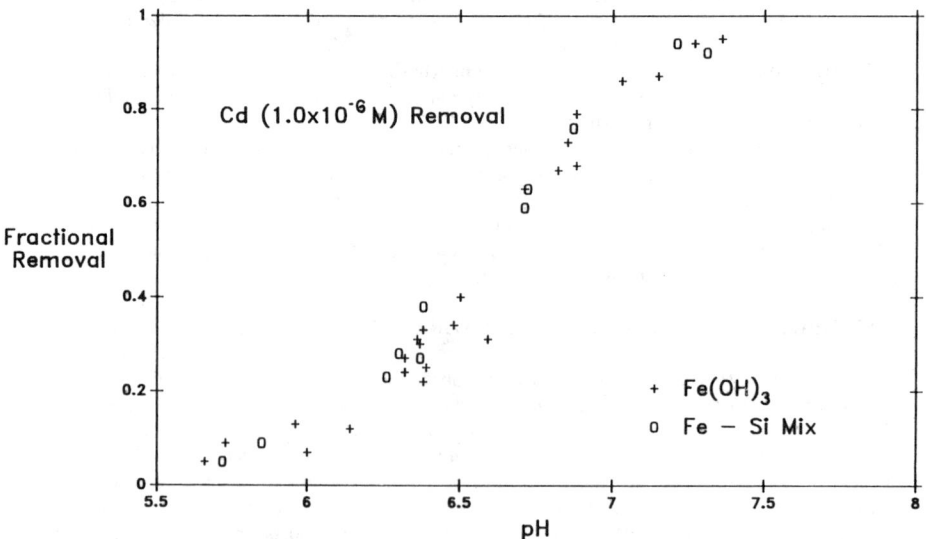

Figure 2. Fractional removal of Cd onto $Fe(OH)_3$ and an Fe-Si binary suspension. Solids concentrations are given in Table I.

Table I. Bulk and Surface Characteristics of Reference Oxides and Binary Solid

| Property | Suspension | | |
|---|---|---|---|
| | $Fe(OH)_3$ | $SiO_2$ | Fe after Si |
| Total Conc. Added[1] (mol/L) | | | |
| Fe | 0.001 | - | 0.001 |
| Si | - | 0.008 | 0.008 |
| Specific Surface Area ($m^2$/g) | | | |
| Measured | 186-201 | 105 | 144 |
| Expected[2] | - | - | 120 |
| PZC[3] | 7.2(S.A.) | <3.0(EM) | 7.2(S.A.) |
| XPS Surface Stoichiometry | | | |
| Si/Fe | - | - | 13.3 |
| Surface Sites | | | |
| Density[4] (sites/$nm^2$) | 27 | 5 | - |
| Expected Ratio[5] Si/Fe | - | - | 10.0 |

(1) Fe and Al are $10^{-4}$ M for systems with $PO_4$.

(2) Expected values based on mass of each oxide added and measured values for pure oxides.

(3) SA = salt addition; EM = electrophoretic mobility

(4) Literature values for solids prepared in the same manner. The value for $Fe(OH)_3$ is based on $9.85 \times 10^{-3}$ mol sites/g $Fe_2O_3 \cdot H_2O$ used by Davis and Leckie (15) from the work by Yates (16). The value for $Al(OH)_3$ is also from Davis and Leckie (15) from the work by Huang (17). The value for $SiO_2$ is taken from the work of Kent (18).

(5) Expected values based on literature values for the single component oxides and the mass of each oxide added to the binary systems.

**276** CHEMICAL MODELING OF AQUEOUS SYSTEMS II

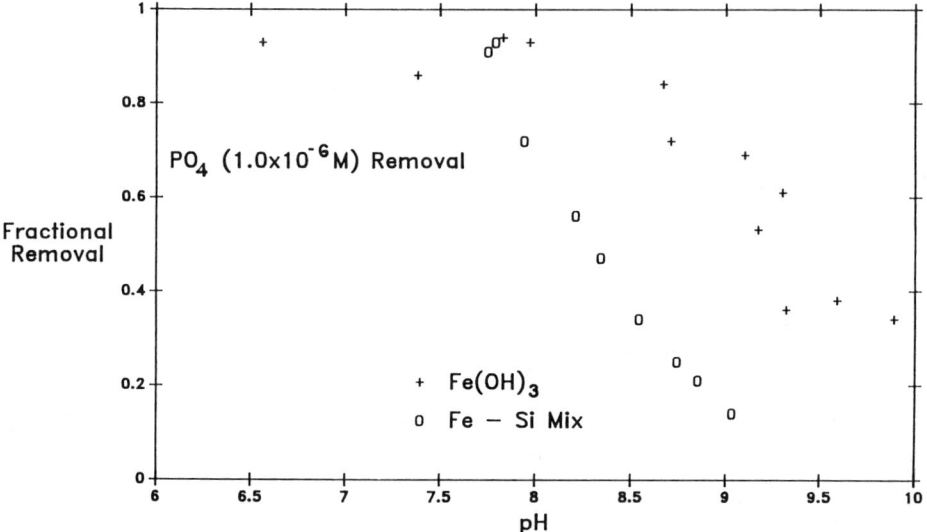

Figure 3. Fractional removal of $PO_4$ onto $Fe(OH)_3$ and an Fe-Si binary suspension. Solids concentrations are given in Table I.

surface complexation constants for the adsorbates in the reference oxide suspensions. MICROQL is a chemical speciation program. Like FITEQL, it includes surface chemical reactions and can be adapted to use the constant capacitance model. A more detailed description of the constant capacitance model can be found elsewhere (4,6).

The first attempt to model sorption in these suspensions was based on the adsorptive additivity approach. The model parameters which best fit the sorption data for pure $Fe(OH)_3$ and $SiO_2$ suspensions were identified (Table II) and using these values the binary systems were modeled by including surface sites from both pure solids in the calculations. However, because this approach overestimated the overall removal of ions from solution in almost all cases, dissolution and sorption of silicate were then incorporated into the model.

MICROQL as modified to include the constant capacitance model was used to evaluate the possible influence of soluble silica in the Fe-Si binary systems. Sorption of soluble silica was represented by the following reactions where SOH represents the surface functional groups.

$$SOH + H_4SiO_4 = S\text{-}H_3SiO_4 + H_2O \tag{1}$$

$$\text{Log } K_{(int)} = 4.15$$

$$SOH + H_4SiO_4 = S\text{-}H_2SiO_4^- + H^+ + H_2O \tag{2}$$

$$\text{Log } K_{(int)} = -3.85$$

The surface complexation constants, $K_{(int)}$, are intrinsic constants which do not incorporate the influence of the surface electrical charge. In the model these are corrected for the effects of surface charge. The values used in this study are average values calculated from the study by Goldberg (8) of silicate adsorption onto goethite. Therefore, the only unknown model parameter was the concentration of soluble silica added to the system by partial dissolution of $SiO_2$. This value, represented by *Si*, was used as an adjustable parameter to fit the model to the experimental data. An additional assumption in the model was that when a surface Si ion dissolved, a new $SiO_2$ surface site was exposed, so that partial dissolution did not lead to any loss of $SiO_2$ sites.

The first binary suspension data to be modeled this way were those for $PO_4$ sorption in the Fe-Si system. The value of *Si* which provided the best fit for these data was $8.0 \times 10^{-5}$ M. At pH 8, the model indicated that 42% of this dissolved Si adsorbed, occupying about 38% of the total $Fe(OH)_3$ sites, while 58% of the silica released from the solid $SiO_2$ remained in equilibrium solution. This is considerably less than the reported equilibrium solubility for amorphous silica of about $2 \times 10^{-3}$ M (14), suggesting that the suspensions were not in true equilibrium, perhaps because partial covering of the surface with Fe oxide inhibited dissolution. The same equilibrium partitioning of Si species (i.e. same dissolved Si concentration at equilibrium and same adsorption density of Si on $Fe(OH)_3$) was assumed to exist in the other systems, and the effects of this redistribution of Si on the behavior of other adsorbates were predicted. These predictions were then compared to the experimental data. Because the $Fe(OH)_3$ concentration (and the total number of $Fe(OH)_3$ sites) in the $PO_4$ adsorption experiments was 10% of the value in the other adsorption experiments, the total *Si* concentrations used in the model differ among the systems.

The results for $PO_4$ adsorption demonstrate that the experimental data for removal in the pure $Fe(OH)_3$ suspension (*Si* = 0.0 M) could be reproduced by the model (Figure 4). As expected, the predicted adsorption edge for $PO_4$ removal shifts to lower pH values as *Si* concentration increases.

When the same equilibrium Si distribution is used to simulate $SeO_3$ removal, a qualitatively similar shift in the adsorption edge is predicted (Figure 5). The results of model simulations at higher and lower *Si* concentrations indicate the sensitivity of the model, but none of these curves adequately matches the slope of the experimental data. The difficulties

Table II. Adsorbent characteristics and parameters used in the constant capacitance model

| | Adsorbent | |
|---|---|---|
| | $Fe(OH)_3$ | $SiO_2$ |
| Solid (g/L) | 0.103 | 0.483 |
| Specific surface area ($m^2/g$) | 193 | 105 |
| Site concentration (mol/L) | $8.8 \times 10^{-4}$ | $4.2 \times 10^{-4}$ |
| Capacitance ($F/m^2$) | 1.25 | 1.25 |
| Log $K_{+\,(int)}$ | 5.0 | - |
| Log $K_{-\,(int)}$ | -9.4 | -7.5 |

| | Log $K_{(int)}$ | |
|---|---|---|
| Species | | |
| SO-Ag | -4.5 | - |
| SO-Cd$^+$ | -3.14 | -7.35 |
| SO-Zn$^+$ | -2.68 | -4.17 |
| S-SeO$_3^-$ | 12.9 | - |
| S-PO$_4^{2-}$ | 7.27 | - |

Notes:

In all batch experiments with $PO_4$, the solid and site concentrations for $Al(OH)_3$ and $Fe(OH)_3$ were 10% of these values.

Reactions are of the form:

$$SOH + M^{x+} = SOM^{(x-1)+} + H^+$$

$$SOH + A^{y-} + H^+ = SA^{y-} + H_2O$$

where $A^{y-}$ is $HPO_4^{2-}$ or $SeO_3^{2-}$

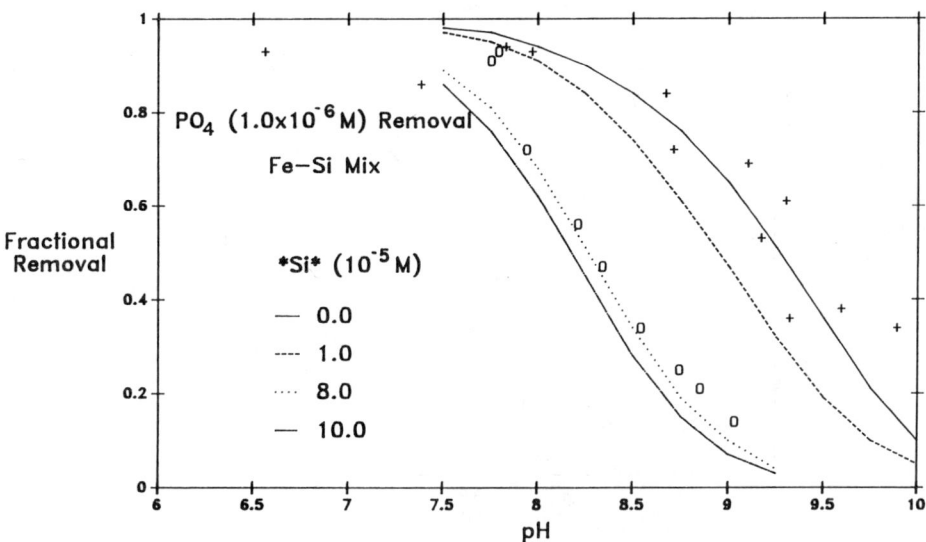

Figure 4. Fractional removal of $PO_4$ onto $Fe(OH)_3$ (crosses) and an Fe-Si binary suspension (open symbols) compared to results from the constant capacitance model. *Si* represents the total silicate which dissolved from the pure solid (some of which adsorbs onto the $Fe(OH)_3$). Solids concentrations are given in Table I.

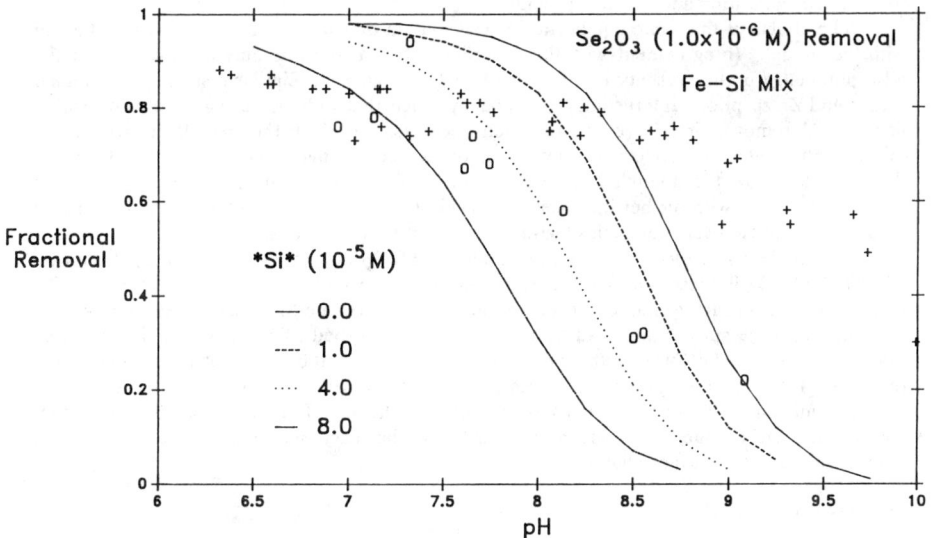

Figure 5. Fractional removal of $SeO_3$ onto $Fe(OH)_3$ (crosses) and an Fe-Si binary suspension (open symbols) compared to results from the constant capacitance model. *Si* represents the total silicate which dissolved from the pure solid (some of which adsorbs onto the $Fe(OH)_3$). Solids concentrations are given in Table I.

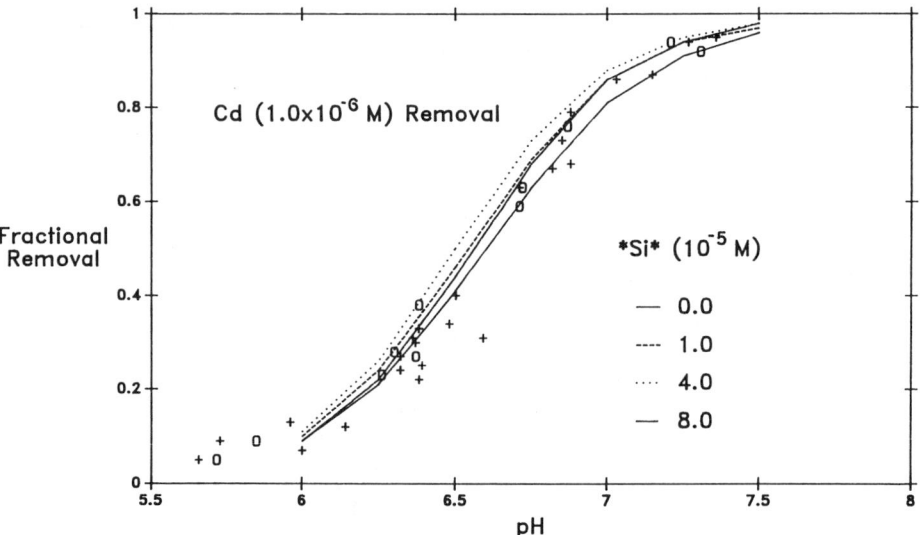

Figure 6. Fractional removal of Cd onto $Fe(OH)_3$ (crosses) and an Fe-Si binary suspension (open symbols) compared to results from the constant capacitance model. *Si* represents the total silicate which dissolved from the pure solid (some of which adsorbs onto the $Fe(OH)_3$). Solids concentrations are given in Table I.

encountered in modeling $SeO_3$ adsorption may be related to the model assumptions regarding total site concentration and surface speciation (9).

The effects of this same equilibrium Si distribution on Cd removal were also investigated with the model. In agreement with the experimental data, there is only a slight shift of the adsorption edge in the presence of adsorbing $H_4SiO_4$ (Figure 6). Similar results were obtained for Ag and Zn sorption. Interestingly, the model predicts that with increasing silicate adsorption density, Cd removal first increases and then decreases, although the net effect is relatively minor. This behavior reflects the net effect of two competing trends. At low adsorption densities of silicate, the dominant effect is that the surface is made more negative by adsorbed silicate. However, with further increases in the silicate adsorption density, the loss of available binding sites for Cd overwhelms this factor, and Cd sorption decreases.

The modeling results indicate that dissolution of $SiO_2$ may explain the decreased removal of $SeO_3$ and $PO_4$ in these binary suspensions and the absence of a significant effect on Ag, Cd, or Zn removal. In fact, to model the $PO_4$ removal data in the Fe-Si binary suspension, 38% of the total $Fe(OH)_3$ surface sites had to be occupied by adsorbed silica. It seems that covering that fraction of the $Fe(OH)_3$ with silica should have affected the PZC, but no such shift was observed (Table I). One possible explanation is that the Si adsorbed without altering surface charge significantly. This explanation was supported by the model simulation which showed that 85% of the surface bound silica adsorbed according to the reaction described by equation 1 and did not influence the surface charge.

## SUMMARY

Experimental data for adsorption of Ag, Cd, and Zn in two reference systems, $Fe(OH)_3$ and $SiO_2$, were reproduced using MICROQL and the constant capacitance adsorption model. Model results for the reference systems were used to simulate the effects of soluble silica on sorption in $Fe(OH)_3$ + $SiO_2$ binary suspensions. Adsorptive additivity is not observed in such systems. However, the experimental results are consistent with a scenario involving partial $SiO_2$

dissolution and subsequent sorption of silicate on the Fe(OH)$_3$. Soluble silica apparently competes for surface sites with PO$_4$ and SeO$_3$, but has little influence on the adsorption of Ag, Cd, or Zn. The constant capacitance model was used to simulate adsorption of soluble silicate, significantly improving the correlation between model and experimental results.

ACKNOWLEDGEMENTS

Funding for this study was provided by the Valle Scandinavian Exchange Program at the University of Washington, the United States Environmental Protection Agency through grant number R810902-01-2, and the University of Washington Graduate School Research Fund. Special thanks are expressed to NESAC/BIO at the University of Washington and McCrone Associates for their help with the surface analyses. We are also grateful for the comments and suggestions offered by John Ferguson.

LITERATURE CITED

1. Balistrieri, L.; Brewer,P.G.; Murray,J.W. Deep-Sea Res. 1981, 28A, 101-121.
2. Oakley, S.M.; Nelson, P.O.; Williamson, K.J. Environ. Sci. Technol. 1981, 15, 474-480.
3. Davies-Colley, R.J.; Nelson, P.O.; Williamson, K.J. Environ. Sci. Technol. 1984, 18, 491-499.
4. Goldberg, S.; Sposito, G. Soil Sci. Soc. Am. J. 1984, 48, 779-783.
5. Honeyman, B.D. Ph.D. Thesis, Stanford University, Stanford, CA. 1984.
6. Huang, C.P.; Stumm, W. J. Colloid Interface Sci. 1973, 43, 409-420.
7. Schindler, P.W.; Furst, B.; Dick, R.; Wolf, P.U. J. Colloid Interface Sci. 1976, 55, 469-475.
8. Goldberg, S. Soil Sci. Soc. Am. J. 1985, 49, 851-856.
9. Anderson, P.R. Ph.D. Thesis, University of Washington, Seattle, WA. 1988.
10. Benjamin, M.M.; Bloom, N.S. In Adsorption from Aqueous Solutions; P.H. Tewari, ed.; Plenum Publishing Co., 1981; pp 41-60.
11. Leckie, J.O.; Benjamin, M.M.; Hayes, K.; Kaufman, G. Electric Power Research Institute Report CS-1513, RP910-1 Final Report, Palo Alto, CA, 1980.
12. Westall, J.C. FITEQL, A program for the determination of chemical equilibrium constants from experimental data. User's guide version 1.2. Chemistry Department, Oregon State University, Corvallis, Oregon, 1982.
13. Westall, J.C. MICROQL I. A chemical equilibrium program in BASIC. Report 86-02, Department of Chemistry, Oregon State University, Corvallis, Oregon, 1986.
14. Kittrick, J.A. Clays Clay Min. 1969, 17, 157-167.
15. Davis, J.A.;Leckie, J.O. J. Colloid Interface Sci. 1978, 67, 90-107.
16. Yates, D.E. Ph.D. Thesis, University of Melbourne, Melbourne, Australia, 1975.
17. Huang, C.P. Ph.D. Thesis, Harvard University, Cambridge, MA. 1975.
18. Kent, D.B.; Kastner, M. Geochim. Cosmochim. Acta, 1985, 49, 1123-1136.

RECEIVED August 18, 1989

# Chapter 22

# Influence of Temperature on Ion Adsorption by Hydrous Metal Oxides

## Michael L. Machesky

**Geosciences Department, Pennsylvania State University, 208 Deike Building, University Park, PA 16802**

Adsorption processes are most often studied and modeled at room temperatures. At other temperatures, however, several factors act to perturb adsorption phenomena. These include shifts in equilibria among solution species, changes in the zero point of charge (pHzpc) of the sorbent, and changes in the ratio of adsorbed to solution phase ions. Fortunately, the magnitude of these factors can be predicted with standard thermodynamic relationships if solution and adsorption enthalpy data are available. Comparative adsorption studies at several temperatures, and calorimetry can be used to obtain adsorption enthalpies. Relatively few studies have been performed but these do suggest several general trends apply. First, the pHzpc of hydrous metal oxides appear to decrease with increasing temperature and consequently, proton adsorption enthalpies are exothermic. Typical values at the pHzpc are -20 to -45 kJ/mole which corresponds to pHzpc changes of less than ±0.5 pH units for 25 ±20°C. Second, anion adsorption decreases with increasing temperature while metal cation adsorption increases and as a result, anion adsorption is exothermic and metal cation adsorption endothermic. Available data suggest adsorption enthalpies are often ±20 kJ/mole or greater. This implies 20°C temperature gradients, which are rather common in temperate climates, could more than double or halve the adsorbed to solution ratio of a particular ion. Thus, temperature should be considered an important variable when investigating or modeling adsorption processes.

Cation and anion adsorption by hydrous metal oxides influence several processes of environmental concern including contaminant transport, nutrient availability, and mineral dissolution rates ([1,2]). Various factors influence the amount of a particular ion adsorbed including solution pH, type of oxide and its surface area and crystallinity, time, ionic strength, properties and concentration of the adsorbing species, and competing species. These factors have received various degrees of scrutiny in previous studies. Temperature is another potentially important variable but has not to date received as much attention. Most laboratory studies have been conducted at room temperatures (20 to 30°C) and few investigations have obtained data at other temperatures for comparison. Extrapolating these room temperature data to actual environmental situations is difficult for a number of reasons and one of these may be the temperature differences involved. For example, average surface and shallow ground water temperatures are less than 25°C in polar and temperate climates and, seasonal fluctuations occur. In addition, thermoclines divide surface and deeper waters in many lakes and the oceans. Conversely, temperatures in deep sedimentary basins and hydrothermal systems are greater than 25°C and the importance of adsorption processes in these environments is largely unknown. Finally, the utility of metal oxides (e.g., alumina) as chromatographic supports or to help decontaminate waste process streams may be enhanced by manipulating temperature.

The purpose of this paper is to outline a general framework with which the importance of temperature to ion adsorption processes can be evaluated. This will be accomplished by examining ion adsorption enthalpy data from several calorimetric and variable temperature studies. Hopefully, this information will stimulate further discussion and study concerning the temperature dependence of the adsorption process and result in more accurate predictions when modeling adsorption processes in natural systems is of concern.

IMPORTANT FACTORS

Several factors require consideration to assess the importance of temperature to adsorption processes. First, equilibria between solution forms of an adsorbing ion are temperature dependent. Second, the zero point of charge (pHzpc) will change with temperature and this alters the pH range of positive and negative surface charge. Third, adsorbed to solution phase ratios of cations and anions will vary with temperature. A fourth factor concerns adsorption kinetics but this is beyond the scope of this discussion which deals only with trends which can be defined on the basis of thermodynamic (i.e., adsorption equilibrium) relationships.

Temperature Effects on Solution Equilibria

Changes in solution species distributions with temperature must be considered because amounts adsorbed depend on the predominant form of a species present. For example, adsorption of a divalent anion onto a positively charged oxide surface would be favored over the monovalent form of the same species according to electrostatic effects. The simplest thermodynamic expression available for estimating changes in solution species distributions is the integrated form of the van't Hoff equation with the reaction enthalpy ($\Delta H_r$) taken to be independent of temperature,

$$\log(K_2/K_1) = (\Delta H_r/2.303R) \cdot \left( \frac{1}{T_1} - \frac{1}{T_2} \right) \tag{1}$$

where $K_2$ is the reaction equilibrium constant at temperature, $T_2$ (Kelvin); $K_1$ is the reaction equilibrium constant at temperature, $T_1$; R is the gas constant. Thus, if the reaction enthalpy can be determined and the reaction equilibrium constant is known at one temperature, the equilibrium constant at a second temperature can be calculated. If the reaction enthalpy is temperature dependent (invariably the case over a large enough temperature range), more complicated expressions involving heat capacity terms must be used (<u>3</u>).

Fortunately, equation (1) is adequate for most solution reactions near room temperature, and several computer equilibrium models make these corrections if the required enthalpy values are available. Unfortunately, enthalpy data for many important solution species (e.g., metal ion species and ion pairs) have not been determined. In a few instances the temperature dependence of a reaction is very well known. A particularily relevant example for aquatic chemistry is log $K_w$ for water which is given by,

$$\log K_w = \frac{-4470.99}{T} + 6.0875 - 0.01706T \quad (2)$$

where T is the temperature in Kelvin (3).

## Temperature Effects on the pHzpc of Oxides

This effect refers to how the affinity of the potential determining ions for hydrous oxide surfaces ($H^+$, $OH^-$) change with temperature. Shifts in the pHzpc with temperature will change the magnitude of the surface charge at a particular pH and hence the electrostatic influence on ion adsorption processes will vary. Two methods have been used to evaluate this dependence; potentiometric acid-base titrations performed at various temperatures and isoperibol solution calorimetry adapted to suspension titrations. Procedural details for both these methods are available elsewhere (4,5). A simple modification of the various temperature technique termed a '$\Delta T$ titration' has also been developed (6). Briefly, a hydrous metal oxide suspension is equilibrated at the pHzpc, the temperature is changed slightly, and the new pH equated to the pHzpc at that temperature. Good agreement was found for rutile and hematite between pHzpc values determined with '$\Delta T$ titrations' and complete acid-base titrations at several temperatures (6).

With isoperibol solution calorimetry, proton adsorption and desorption enthalpies are obtained and these can be used to predict how the pHzpc varies with temperature. The '1 $pK_a$' model is especially useful for this purpose. This model has been gaining increasing popularity relative to the '2 $pK_a$' model for describing the surface charging characteristics of hydrous metal oxide surfaces (7,8). The basic charging equation for this model is,

$$[SO^{-1/2}] + [H^+] = [SOH^{+1/2}] \quad (3)$$

where SOH represents a surface hydroxyl group which has a formal charge of ±1/2 because of crystallographic considerations (7). Since at the pHzpc,

$$[SO^{-1/2}] = [SOH^{+1/2}] \quad (4)$$

it follows that,

$$\log K_H = pHzpc \quad (5)$$

where $K_H$ is the equilibrium constant for equation (3).

Assuming the validity of the '1 $pK_a$' model, proton adsorption enthalpies at the pHzpc can be used to predict the temperature variation of $K_H$ (and therefore the pHzpc) with the van't Hoff relation (6),

$$\frac{d \log K_H}{dT} = \Delta H_{ads}/2.303 \, RT^2 \quad (6)$$

where $\Delta H_{ads}$ is the proton adsorption enthalpy at the pHzpc which is assumed to be independent of temperature. Conversely, if the pHzpc variation with temperature is known (the dlog $K_H$/dT term), the proton adsorption enthalpy can be calculated.

Table I lists the pHzpc values determined at various temperatures for several hydrous metal oxides along with proton adsorption enthalpy values calculated using equation (6). Except for one instance, pHzpc values decrease with increasing temperature and therefore calculated proton adsorption enthalpies are exothermic. The first $\gamma$-$Al_2O_3$ entry is clearly anomalous in sign and magnitude and can not be considered representative of hydrous metal oxides in general. Table II lists proton adsorption enthalpies determined using calorimetry and there is reasonable agreement between Tables I and II for similar oxides.

Proton adsorption free energy ($\Delta G_{ads}$) and entropy values ($T\Delta S_{ads}$) are also included in Table II. Free energy values are obtained at the pHzpc using the standard thermodynamic relation,

$$\Delta G = -2.303 \text{ RT log } K_H \tag{7}$$

assuming the validity of the '$1pK_a$' model and, adsorption entropies are determined by difference.

Enthalpy and entropy both contribute to the favorable proton adsorption free energy values with enthalpy being relatively more important. Reaction enthalpies reflect bond formation energetics while net solvation changes are large contributors to reaction entropies in aqueous solutions (17). Thus, the energetics of the bond formation process appears to be more important than net solvation changes in the interfacial region for proton adsorption. Also, proton adsorption free energies and enthalpies for ferric and Al-oxides are similar and larger than those for rutile. This is consistent with the larger formal charge of the Ti cation (+4) which would tend to form weaker bonds with protons (6). However, other factors must also influence the strength of the proton-surface bond since some metal oxides with lower cation formal charges (e.g., NiO) have relatively small proton adsorption enthalpies.

Proton adsorption enthalpy data at the pHzpc and equation (6) permit estimation of expected changes in pHzpc with temperature. Ferric and Al-oxides, for example, are characterized by a proton adsorption enthalpy in the pHzpc region of about -40 kJ/mole. Thus, we can predict that the pHzpc will decrease about 0.024 pH units per degree temperature increase or, a pHzpc of 8.0 at 25°C would be 8.5 at 5°C and 7.5 at 45°C. It has also been suggested that an 'upper bound' for proton adsorption enthalpies is the enthalpy of water formation which is -55.83 kJ/mole at 25°C (6). Most of the enthalpy values in Tables I and II are less than this value but a much larger data base is needed to confirm this hypothesis. For example, no enthalpy data are available for $SiO_2$ although zeta potential values at pH 4 become more negative as temperature increases which suggests the pHzpc is also decreasing (18). The change in pHzpc with temperature although predictable for many hydrous metal oxides is rather slight given the variability in the measured pHzpc for a particular oxide. For example, pHzpc values for goethite range from 7.0 to 9.0 (19). Thus, changes in pHzpc over earth surface temperature ranges are expected to only slightly influence cation and anion adsorption. At higher temperatures, however, changes in the pHzpc will become relatively more important.

Table I. Changes in pHzpc with temperature obtained from the literature and corresponding proton adsorption enthalpy values calculated using equation (6)

| Oxide | Temp(°C) | pHzpc | $\Delta H_{ads}$(kJ/mole) | Source |
|---|---|---|---|---|
| α-$Fe_2O_3$ | 5 | 9.50 | -45.1 (5 to 60°C) | (6) |
| | 20 | 8.60 | -62.4 (5 to 20°C) | |
| | 60 | 7.80 | -37.4 (20 to 60°C) | |
| $TiO_2$(rutile) | 5 | 5.80 | -26.7 (5 to 50°C) | (6) |
| | 20 | 5.60 | -17.6 (20 to 50°C) | |
| | 50 | 5.10 | | |
| $TiO_2$(rutile) | 25 | 6.00 | -19.9 (25 to 75°C) | (9) |
| | 75 | 5.50 | | |
| $Fe_3O_4$ | 25 | 6.55 | -36.6 (25 to 90°C) | (10) |
| | 90 | 5.40 | | |
| $Fe_3O_4$ | 30 | 6.80 | -32.8 (30 to 50°C) | (11) |
| | 50 | 6.45 | -32.8 (30 to 80°C) | |
| | 80 | 6.00 | | |
| γ-$Al_2O_3$ | 10 | 4.45 | +197 (10 to 50°C) | (12) |
| | 50 | 8.95 | | |
| γ-$Al_2O_3$ | 30 | 9.06 | -24.6 (30 to 90°C) | (10) |
| | 90 | 8.36 | | |
| γ-$Al_2O_3$ | 25 | 8.00 | -44 (25°C)[1] | (13) |
| $Co_3O_4$ | 25 | 10.35 | -26.7 (25 to 80°C) | (14) |
| | 80 | 9.62 | | |
| NiO | 25 | 9.85 | -28.9 (25 to 80°C) | (14) |
| | 80 | 9.06 | | |

[1] Determined from the change in surface potential with temperature and the '2p$K_a$' model. The listed enthalpy is the average of those for the individual surface hydrolysis constants (-54 and -34.8 kJ/mole).

Table II. Proton adsorption enthalpies determined using calorimetry and calculated adsorption free energies and entropies at the pHzpc and 25°C

| Oxide | pHzpc | $\Delta H_{ads}$ | $\Delta G_{ads}$ | $T\Delta S_{ads}$ | Source |
|---|---|---|---|---|---|
| | | | kJ/mole | | |
| $TiO_2$(rutile) | 6.0 | -22 | -34 | +12 | (5) |
| $TiO_2$(rutile) | 5.5 | -22 | -31 | +9 | (6) |
| $\alpha$-$Fe_2O_3$ | 8.7 | -38 | -50 | +12 | (6) |
| $\alpha$-FeOOH | 8.1 | -39 | -46 | +7 | (5) |
| $\gamma$-$Al_2O_3$ | 8.5 | -42 | -49 | +7 | (15) |
| $\alpha$-$Al_2O_3$ | 8.8 | -44[1] | -50 | +6 | (16) |

[1] Estimated from heat of immersion data.

## Temperature Effects on Cation and Anion Adsorption

The specific adsorption of anions and cations are also influenced by temperature changes and adsorption studies at various temperatures and isoperibol solution calorimetry have been used to investigate this influence. However, relatively few of these studies have been conducted. Residual solution concentrations at two temperatures can be used with a form of the Clausius-Clayperon equation to calculate ion adsorption enthalpies (20),

$$\Delta H_{ads} = [2.303R \log (C_2/C_1)]/(\frac{1}{T_2} - \frac{1}{T_1}) \qquad (8)$$

where $\Delta H_{ads}$ is the isosteric heat of ion adsorption (assumed to remain constant over the temperature range considered) at a given surface coverage; R is the gas constant; $C_2$ is the equilibrium solution concentration of the ion at temperature, $T_2$ and the given surface coverage; and $C_1$ is the equilibrium concentration of the ion at temperature $T_1$ and the given surface coverage. Complete adsorption isotherm data (low to high surface coverage) at several temperatures is most useful since this also provides information about the variation in adsorption enthalpy with surface coverage (20). If residual solution concentrations increase with increasing temperature (adsorption decreases) calculated adsorption enthalpies will be exothermic and residual solution concentrations will decrease with increasing temperature (adsorption increases) for endothermic adsorption enthalpies. Conversely, ion adsorption enthalpies determined using calorimetry can be used to predict the expected ratio of residual solution concentrations at two temperatures. Strictly speaking, comparison of isosteric and calorimetric enthalpies requires use of the relation,

$$\Delta H_{isosteric} = \Delta H_{differential} + RT \qquad (9)$$

where $\Delta H_{differential}$ refers to adsorption enthalpy measurements determined incrementally from low to high surface coverage. In actual practice, however, adsorption enthalpies determined using calorimetry are probably intermediate between true differential and isosteric enthalpies (21). In any case, the RT factor (+2.48 kJ/mole at 25°C) makes only a slight contribution to the measured differential enthalpies. For ion adsorption studies, much more significant corrections are associated with proton uptake and release during anion and cation adsorption, respectively (6,22).

Representative anion isosteric adsorption enthalpies at several surface coverages for goethite are compiled in Table III.

Table III. Isosteric adsorption enthalpies for representative anions and several fractional surface coverages ($\Theta$) of goethite

| Anion | pH | $\Theta$ | $\Delta H_{ads}$(kJ/mole) | Method | Source |
|---|---|---|---|---|---|
| $H_2PO_4^-$ | 4 | 0.1 | -24 | Calorimetry | (22) |
|  |  | 0.5 | -19 |  |  |
|  |  | 0.8 | +3 |  |  |
| $IO_3^-$ | 4 | 0.1 | -28 | " | " |
|  |  | 0.5 | -17 |  |  |
|  |  | 0.8 | -9 |  |  |
| $C_7H_5O_3^-$ | 4 | 0.1 | -27 | " | " |
|  |  | 0.5 | -7 |  |  |
|  |  | 0.8 | -7 |  |  |
| $F^-$ | 4 | 0.1 | -10 | " | " |
|  |  | 0.5 | -7 |  |  |
|  |  | 0.8 | 0 |  |  |
| $HSeO_3^-$ | 6.7 | 0.074 | -82 | Two | (20) |
|  |  | 0.15 | -29 | Temperatures |  |
|  |  | 0.22 | -22 |  |  |

In all instances adsorption enthalpies are exothermic at surface coverages < 80% of maximum for each anion. Adsorption enthalpies are more exothermic at low coverages with values as large as -82 kJ/mole for selenite at surface coverages < 10% of maximum. These data indicate the goethite surface is energetically heterogeneous with a relatively small fraction of the available surface sites (< 10%) being of highest energy. Also, phosphate, salicylate and selenite have similar adsorption enthalpies at low (10 to 20%) surface coverages which suggests bonding mechanisms are similar. This has been confirmed with in-situ spectroscopic techniques which indicate these anions adsorb in an inner-sphere bidentate or chelate type fashion (23,24). Fluoride, however, must bind in a monodentate fashion and consequently, lower adsorption enthalpies result. It can also be suggested that outer-sphere complex formation would result in relatively low adsorption enthalpies (< -10 kJ/mole at low surface coverages).

Fewer data, particularily calorimetric, are available for evaluating the influence of temperature on metal cation adsorption. The enthalpy of Cd(II) adsorption onto rutile was determined using isoperibol solution calorimetry and a value of +10 kJ/mole was found (6). A recent variable temperature study (25) allows enthalpies for Cd(II), Zn(II), and Ni(II) adsorption onto hematite (synthesized in the presence of 0.86% Si) to be calculated using equation (8). These data are summarized in Table IV.

Table IV. Residual solution concentrations at 5, 20, 25 and 35°C and average isosteric adsorption enthalpies calculated using equation (8) for cadmium, zinc and nickel onto hematite at pH=6 and a total metal concentration of 10$\mu$M. Concentration data from (25)

| Metal ion | 5°C | 20°C | 25°C | 35°C | $\Delta H_{ads}$(kJ/mole) |
|---|---|---|---|---|---|
| | solution concentration ($\mu$M) | | | | |
| Cd(II) | 6.9 | 5.7 | 5.0 | 3.9 | +13 |
| Zn(II) | 4.1 | 2.0 | 1.1 | 0.5 | +49 |
| Ni(II) | 8.1 | 5.9 | 3.8 | 2.2 | +30 |

The calculated enthalpies vary depending on which solution concentration data are used but they are always endothermic since residual solution concentrations decrease with increasing temperature. Calculated and measured adsorption enthalpies for Cd(II) agree very well even though two different oxides and techniques were used. Furthermore, since adsorption enthalpies are endothermic, the adsorption free energy is dominated primarily by a positive entropy contribution. Partial desolvation of the metal cation upon adsorption is most likely a primary contributing factor to this entropy increase (6). The paucity of available data does not permit an assessment how metal cation adsorption enthalpies and entropies vary as surface coverage increases. If previous calorimetric cation exchange studies prove to be analogous, however, enthalpies would become more endothermic and entropies more positive since an increasing fraction of hydration waters would be shed as surface coverage increases (26).

Table V illustrates the expected influence of temperature on ion adsorption processes for the range of exothermic adsorption enthalpies listed in Table III.

Table V. Calculated solution concentration ratios for the range of exothermic adsorption enthalpies listed in Table III. $C_{25}/C_5$ and $C_{25}/C_{45}$ refer to the ratios between 25 and 5°C and 25 and 45°C, respectively

| $\Delta H_{ads}$(kJ/mole) | $C_{25}/C_5$ | $C_{25}/C_{45}$ |
|---|---|---|
| -80 | 10.2 | 0.13 |
| -40 | 3.2 | 0.36 |
| -20 | 1.8 | 0.59 |
| -10 | 1.3 | 0.77 |
| -5 | 1.2 | 0.87 |

Residual solution concentrations are greater at 25 compared to 5°C and less at 25 compared to 45°C and both ratios approach unity as the enthalpy becomes less exothermic. Thus, at a given total concentration, a temperature decrease from 25 to 5°C is expected to result in increased adsorption while a temperature increase to 45°C should result in anion desorption. For cation adsorption, the expected concentration ratios are the inverse of those listed since cation adsorption enthalpies are endothermic.

The predicted temperature influence is indeed significant since for an adsorption enthalpy of ±20 kJ/mole (less than that found for many species in Tables III and IV), a temperature ±20°C removed from 25°C would nearly double or halve residual solution concentrations. Thus, adsorption experiments should be conducted at the temperature of interest. In addition, temperature gradients or fluctuations of ±20°C are fairly common in natural systems. Thermoclines separate warm surface water from colder deeper waters in oceans and many lakes and it is conceivable that the scavenging of trace species by settling inorganic particulate matter is perturbed by this temperature gradient. In this case enthalpy data suggest cations would be released and anions scavenged more efficiently as particulates settle through the thermocline although pH decreases could augment or even dominate the temperature effect. As another example, seasonal temperature fluctuations in vadose zones or shallow ground waters could perturb adsorption processes affecting contaminant transport.

CONCLUSIONS

The influence of temperature on ion adsorption processes can be predicted using adsorption enthalpy data and standard thermodynamic relationships. Comparative studies at several temperatures and calorimetry can be used to obtain adsorption enthalpies and those data that are available suggest certain generalizations may apply. First, the pHzpc of hydrous metal oxides decreases as temperature increases and vice-versa. With one exception all data in Tables I and II follow this trend. Furthermore, the magnitude of the calculated (from variable temperature studies) and measured (using isoperibol solution calorimetry) proton adsorption enthalpies generally fall between -20 and -45 kJ/mole and there is reasonable agreement among different studies for the same solid. Still, more data is needed to confirm these trends. In any event, the magnitude of the resulting pHzpc change is rather slight and so this influence on the specific adsorption of ions is not great provided the pHzpc and the 'adsorption edge' of a particular ion do not coincide. Second, specific adsorption of anions appears to increase and metal cation adsorption to decrease with decreasing temperature and vice-versa. This generalization is advanced on the basis of a very limited data set which suggests metal cation adsorption is endothermic and anion adsorption exothermic although it is far from certain that all adsorbed cations and anions will follow this trend. The reasons underlying this trend are still unclear but are probably strongly dependent on the net solvation changes of the adsorbing ion and displaced surface groups. In addition, the shift in pHzpc and anion or cation adsorption compliment each other. That is, as temperature decreases the pHzpc increases which enlarges the domain of positive surface charge and should enhance anion adsorption. Conversely, a temperature increase would decrease the pHzpc thereby promoting cation adsorption which is observed to increase as temperature increases.

LITERATURE CITED

1. Kinniburgh, D.G.; Jackson, M.L. In "Adsorption of Inorganics at Solid-Liquid Interfaces"; Anderson, M.A., and Rubin, A.J., Eds.; Ann Arbor Science: Ann Arbor, MI, 1981; Chapter 3.
2. Sigg, L.; Stumm, W. Colloids and Surfaces. 1981, 2, 101-117.
3. Stumm, W.; Morgan, J.J. "Aquatic Chemistry"; Wiley-Interscience: New York, 1981, 2nd ed.; p. 68-72.
4. Huang, C.P.; In "Adsorption of Inorganics at Solid-Liquid Interfaces"; Anderson, M.A., Rubin, A.J., Eds.; Ann Arbor Science: Ann Arbor, MI, 1981; Chapter 5.
5. Machesky, M.L.; Anderson, M.A. Langmuir. 1986, 2, 582-587.
6. Fokkink, L.G.J. Ph.D. Dissertation, Agricultural Univ., Wageningen, The Netherlands, 1987.
7. Van Riemsdijk, W.H.; De Wit, J.C.M.; Koopal, L.K.; Bolt, G.H. J. Colloid Interface Sci. 1987, 116, 511-522.
8. Westall, J.C. In "Aquatic Surface Chemistry"; Stumm, W., Ed.; Wiley-Interscience: New York, 1987; Chapter 1.
9. Berube, Y.G.; DeBruyn, P.L. J. Colloid Interface Sci. 1968, 27, 305-318.
10. Tewari, P.H.; McLean, A.W. J. Colloid Interface Sci. 1972, 40, 267-272.
11. Blesa, M.A.; Figliolia, N.M.; Maroto, A.J.G.; Regazzoni, A.E. J. Colloid Interface Sci. 1984, 101, 410-418.
12. Akratopulu, K. Ch.; Vordonis,L.; Lycourghiotis, A. J. Chem. Soc., Faraday Trans. I. 1986, 82, 3697-3708.
13. Vlekkert, H.V.D.; Bousse, L.; Rooij, N.D. J. Colloid Interface Sci. 1988, 122, 336-345.
14. Tewari, P.H.; Campbell, A.B. J. Colloid Interface Sci. 1976, 55, 531-539.
15. Machesky, M.L.; Jacobs, P.A., Colloids and Surfaces, in press.
16. Griffiths, D.A.; Fuerstenau, D.W. J. Colloid Interface Sci. 1981, 80, 271-283.
17. Nancollas, G.H. "Interactions in Electrolyte Solutions"; Elsevier: Amsterdam, 1966; Chapter 5.
18. Alekhin, Yu.V.; Sidorova, M.P.; Ivanova, L.I.; Lakshtanov, L.Z. Koll. Zhur. USSR (Engl. Trans.). 1984, 46, 1195-1198.
19. Zeltner, W.A.; Anderson, M.A. Langmuir. 1988, 4, 469-474.
20. Balistrieri, L.S.; Chao, T.T. Soil Sci. Soc. Am. J. 1987, 51, 1145-1151.

21. Hayward, D.O.; Trapnell, B.M.W. "Chemisorption"; Buttersworths: Washington, D.C., 1964, Chapter 6.
22. Machesky, M.L.; Bischoff, B.L.; Anderson, M.A., Environ. Sci. Technol. 1989, 23, 580-587.
23. Zeltner, W.A.; Yost, E.C.; Machesky, M.L.; Tejedor-Tejedor, M.I.; Anderson, M.A. In "Geochemical Processes at Mineral Surfaces"; Davis, J.A.; Hayes, K.F., Eds.; American Chemical Society Symposium Series No. 323: Washington, D.C., 1986, p. 142-161.
24. Hayes, K.F.; Roe, A.L.; Brown, G.E., Jr.; Hodgson, K.O.; Leckie, J.O.; Parks, G.A. Science. 1987, 238, 783-786.
25. Bruemmer, G.W.; Gerth, J.; Tiller, K.G. J. Soil Sci. 1988, 39, 37-52.
26. Clearfield, A.; Tuhtar, D.A. J. Phys. Chem. 1976, 80, 1302-1305.

RECEIVED August 31, 1989

# Chapter 23

# Coagulation of Iron Oxide Particles in the Presence of Organic Materials

## Application of Surface Chemical Model

### Liyuan Liang[1] and James J. Morgan

### Environmental Engineering Science, W. M. Keck Laboratories, California Institute of Technology, Pasadena, CA 91125

> Experiments using hematite particles (70 nm in diameter) in the presence of organics (phthalic acid, fatty acids, polyaspartic acid, fulvic and humic acid) reveal important features of particle coagulation dynamics. A light scattering technique was used to determine quantitatively the initial coagulation rates of hematite particles. Electrokinetic measurements were taken to obtain the sign and magnitude of electrical potential at the oxide/aqueous solution interface. Adsorption experiments were carried out to evaluate affinities of aqueous molecules for the metal oxide surface. Intrinsic equilibrium constants for surface complexes are derived from a Surface Complex Formation/Diffuse Layer Model (SCF/DLM) accounting for interfacial electrostatic charge and potential. Small organic molecules, such as phthalate, show specific chemical reaction with the hematite surface and influence colloidal hematite coagulation kinetics by altering interfacial charge and potential. For fatty acids, hydrophobic interaction, in addition to covalent and electrostatic interaction, offers a plausible explanation for observed systematic changes in hematite stability and electrokinetic data. In the presence of polyelectrolytes, such as polyaspartic acid, fulvic and humic acids, a combination of specific chemical, electrostatic, and hydrophobic energies of carboxyl segments favors adsorption, and these materials have a relatively great impact on particle coagulation and stability.

Suspended solids play an important role in the geochemical cycles of metals, such as iron and manganese, in lake, river, estuary and ocean waters. Small particles in the range of 10 nm to 1 μm possess large specific surface areas and many physical, chemical processes readily take place at the solid-aqueous interface. Observed non-conservative behavior of iron in estuarine waters has been attributed to "flocculation of dissolved iron" ([1-4]). Rates of coagulation (i.e., the aggregation of colliding particles to form larger particles) may determine the fate of naturally occurring particles. Hence, an understanding of the process is desirable in natural water environments and in improving water quality in engineering facilities. Colloids are said to be "stable" when they experience slow coagulation, to be "unstable" when they experience transfer-controlled coagulation. A "stability ratio" for coagulation is defined as the ratio of the transfer-controlled coagulation rate to the actual one.

Earlier experimental and theoretical studies of systems in which particles interact through Van der Waal's attraction and electrostatic repulsion led to the development of Derjaguin-Landau-Verwey-Overbeek theory (DLVO). This theory has been successful in providing a quan-

---

[1]Current address: Department of Environmental Health Sciences, School of Public Health, Room 311, University of South Carolina, Columbia, SC 29208

titative understanding of the relationship between critical coagulation concentrations and the valences of simple electrolytes (5). Multivalent ions are more effective in causing coagulation than monovalent ions, and polymeric species often are more effective than monomeric species.

In a complex aqueous system, ionic species may interact specifically with particle surfaces in addition to electrostatic interaction. As a result of specific adsorption, the particle surface charge and potential may be altered, which in turn may influence the coagulation rate of particles. Ali et al. (6) showed that dissolved organic matter increases the stability of lake particles while magnesium and calcium are able to decrease the stability. Adsorption of dissolved organic matter on metal oxides changes the particle surface characteristics (7), and the aggregation process is affected by the chemical interaction of bivalent ions with surface-adsorbed humics (8). Further evidence of the importance of adsorption on coagulation kinetics comes from the experiments of Gibbs (9), who studied the coagulation of river-borne particles under natural conditions and also with their organic coatings removed. Gibbs observed that naturally coated particles coagulate more slowly than uncoated ones. Surface chemistry influences the surface charge on particles and hence has an important effect on particle coagulation. To model coagulation of particles in the presence of a specifically adsorbed species (such as organic solutes), it is necessary to apply surface chemical models to adsorption results, and thereby relate the adsorption to coagulation.

## OBJECTIVES

This study examines the influence of organic species on particle stability with respect to coagulation. Hematite particles were chosen because hematite is one of the naturally occurring iron oxides. From previous work and this work, the zero point of charge ($pH_{zpc}$) of hematite particles is found to be around 8.5, which allows the study of chemical interactions on both positively and negatively charged surfaces. Aqueous iron(III) is capable of forming complexes with many organic and inorganic species. A knowledge of relevant aqueous Fe(III) equilibrium constants should aid in interpretating iron oxide surface speciation.

Several organic compounds were chosen to study electrostatic, specific chemical, hydrophobic effects on particle stability. The organics include phthalic acid as an example of small organic materials, polyaspartic acid as an example of polymeric organics and fatty acids with different numbers of carbons in the hydrophobic chains. Naturally occurring organic materials, such as Suwannee River fulvic and humic acids were also used in the study. To identify the factors that influence the rate of coagulation in natural waters, adsorption, electrokinetic and coagulation rate data are combined and interpreted on the basis of surface chemical and colloid stability models.

## EXPERIMENTAL

### General Remarks.

Details of experimental procedures are given elsewhere by Liang (14). Deionized distilled water was used to prepare all solutions. All reagents were analytical grade and were used without further treatment. Fatty acids and polyaspartic acids were obtained from Sigma and were at least 99% pure. Fulvic and humic acids were supplied by the International Humic Substances Society.

Pyrex glassware was used and was cleaned first in a detergent solution (Linbro) soaked in concentrated hydrochloric acid, rinsed with deionized distilled water, and then oven dried. A suction pump was used to facilitate the cleaning of syringes and electrophoresis cells.

The solution pH was monitored in all coagulation, titration, electrophoresis and adsorption experiments. These measurements were made using a Radiometer glass combination electrode (GK2401C) and a pH meter (Radiometer Model PHM84 research pH meter). The electrode was calibrated by NBS buffers and, if extended measurements were made, calibration was checked against buffers every two hours.

### Particle Preparation and Characterization.

Hematite particles were synthesized according to the method of Matijević and Scheiner (15). The characterization of the bulk and surface properties of the particles has been described by Liang (14). Table I summarizes the properties of hematite particles used in this study.

Table I: Hematite Surface Physical and Chemical Properties

| Sample Preparation | $pH_{zpc}$ | Electrolyte | Specific Surface Area | Surface Equilibrium Constants |
|---|---|---|---|---|
| Hydrolysis of $Fe(ClO_4)_3$ at $100°C$ | 8.50 | NaCl | BET:<br>Batch 1: $40.0 m^2/g$<br>Estimated from averaged diameter:<br>Batch 1: $38.3 m^2/g$<br>Batch 2: $16.3 m^2/g$ | Diffuse Layer Model<br>$pK_{a1}^{int}=7.25$<br>$pK_{a2}^{int}=9.75$<br>Site density=$4.8 nm^{-2}$ |

### Adsorption Measurement.

Adsorption of organic solutes, such as phthalate, lauric (fatty acid, C12) and Suwannee River fulvic acid, on hematite particles was studied using a total organic carbon analyzer. Adsorption isotherms at constant pH and adsorption as a function of pH at fixed organic concentration were obtained. The desired pH was achieved by adding a small amount of NaOH or HCl and ionic strength was adjusted by adding NaCl. An organic compound was introduced into a particle suspension of known solid concentration. Stirring was maintained for 24 hours and at the end of the period, pH measurements were taken. Particles were filtered out of suspension following adsorption.

Because the solid concentration was low in the sample, the uptake of organic acid by particles was low and the difference between the initial amount of organics and the amount in the filtrate could not yield an accurate measure of the adsorbed acid by particles. Therefore, the amount of adsorbed organics was determined as follows. After filtration the particles collected on the filter were then resuspended in a small volume (~10ml) of $0.02N\ H_2SO_4$ solution. The carbon present in the $H_2SO_4$ suspension was determined using a Dorhmann DC-80 Total Organic Carbon Analyzer.

### Electrophoretic Mobility Measurement.

A Mark II particle micro-electrophoresis apparatus (Rank Brothers, London) was used to measure hematite electrophoretic mobility. These studies were performed at $25°C$ in a KCl solution. In each measurement at least 20 particles were timed in each direction of movement (16).

### Coagulation Experiments.

Light scattering measurements were applied to determine the initial rate of hematite coagulation. The details of the procedure and interpretation are described in Liang (14).

To perform a coagulation kinetics experiment, 3 ml of colloidal suspension of desired pH and solid concentration was placed in a cell of 1 cm light path length. Ten to 100 $\mu l$ of aqueous organic solution (measured by a Hamilton syringe) were introduced to an adder-mixer (17), and then mixed with a particle suspension by plunging the mixer into the cell. Spectrophotometric measurements were begun immediately after the initial mixing time of approximately 0.4 sec. Extinction as a function of time was measured at wavelength $\lambda = 546nm$ and recorded automatically.

## RESULTS AND DISCUSSION

Effects of inorganic species, such as phosphate, sulfate, magnesium and calcium were investigated, and have been discussed in Liang (14). Hematite surface chemical equilibria were studied by acid-base titration and surface acidity equilibrium constants were derived using a surface chemical model. The model used was surface complex formation in which electrostatics were described with a Gouy-Chapman relationship for interfacial charges and potentials. All surface-bound species were assumed to be located at a surface mean plane. Thus, the model is a two layer model (a surface layer and a diffuse layer) and the fitting parameters are the surface species formation constants, e.g., for $\equiv\text{FeOH}_2^+$, $\equiv\text{FeOMg}^+$, $\equiv\text{FeL}$, etc. We refer to this simple model as SCF/DLM, surface complex formation plus diffuse layer model. Similar models with different electrostatic description for interfaces have been used to interpret titration and adsorption data (18-22).

The estimation of equilibrium constants with a SCF/DLM model can be facilitated by using computer programs, such as SURFEQL and FITEQL (23-24). Hematite surface equilibrium acidity constants are listed in Table I, together with other surface physical parameters. Here we summarize the effects of organic solutes on kinetics of hematite coagulation and discuss the results on the basis of surface chemical concepts.

### Effects of Small Organic Adsorbates.

Phthalate adsorbs specifically on iron oxide surfaces (14, 25). The agreement of experimental results on adsorption, electrokinetics and rate of coagulation demonstrates the effect of chemical interaction on colloid stability. In Figure 1 adsorption density, electrokinetic mobility and stability ratio are compared as a function of total phthalate concentration. At low concentration (total phthalate concentration of 1 micromolar), the adsorption density is of the order $8 \times 10^{-11}$ moles/cm$^2$. This corresponds to a tenth of the total surface site density. Most surface sites are occupied by surface hydroxyl groups in protonated or neutral species, so that surface charge and potential are similar to values in the absence of organics, and mobility data show that at $1\mu$M phthalate concentration the mobility is unaffected by the presence of phthalate ions. Consequently, particles have a high stability ratio, on the order of 100. At higher organic concentration, adsorption of phthalate ions is able to reduce the interfacial potential, and at a total phthalate concentration of $2 \times 10^{-4}$M the potential is near zero, as is reflected in the mobility measurements in Figure 1(b). At this phthalate concentration the stability ratio is a minimum, and the coagulation rate is close to diffusion-controlled. As phthalate concentration is further increased beyond $2 \times 10^{-4}$M, mobility data show that surface charge reverses sign. If phthalate did not specifically adsorb on the hematite surface, the mobility would be expected to level off, approaching zero when the phthalate concentration is increased beyond $2 \times 10^{-4}$M. The increase in adsorption density indicates that the chemical bonding between phthalate and surface hydroxyl groups is sufficiently strong, for phthalate ions to overcome a negative surface potential and be adsorbed. Specific adsorption results in a negative interfacial potential and a rise in the stability ratio. Although the tendency of increased stability ratio is evident when phthalate concentration exceeds the critical coagulation concentration, the absolute values of stability ratio are small. The stability ratio is not expected to increase substantially when phthalate concentration is greater than $10^{-3}$M, for two reasons. First, using Langmuir's treatment, the maximum adsorption density corresponds to $\sim 40$Å$^2$ per molecule adsorbed on the surface. This specific area is of the same order of magnitude as the cross-sectional area calculated for phthalate ions. Hence, further adsorption of phthalate is restricted by lack of available surface to accommodate additional molecules. Consequently, surface charge will be approximately constant when the phthalate concentration exceeds $1 \times 10^{-3}$M. Second, an increase in phthalate ion concentration is unavoidably accompanied by an increase in the concentration of counterions, such as Na$^+$. In the presence of a high concentration of positive ions, stability is reduced through electrostatic interactions.

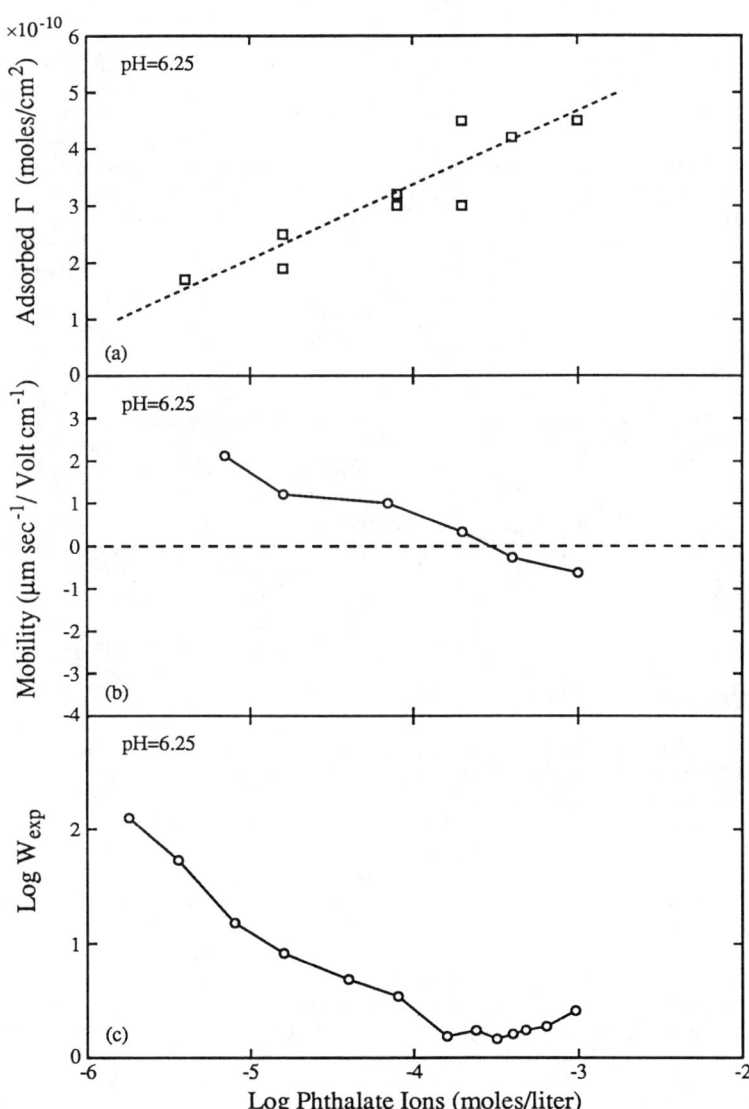

Figure 1: Comparison of hematite adsorption density, $\Gamma$, mobility, and stability ratio, $W_{exp}$, as a function of total phthalate concentration at pH 6.25. The typical variations of the data are less than 10% of the average plotted.

A quantitative treatment of surface speciation can be achieved through the equilibria

$$\equiv\text{FeOH} + 2\text{H}^+ + \text{L}^{2-} \rightleftharpoons \equiv\text{FeHL} + \text{H}_2\text{O} \qquad K_{HL} \qquad (1)$$

$$\equiv\text{FeOH} + \text{H}^+ + \text{L}^{2-} \rightleftharpoons \equiv\text{FeL}^- + \text{H}_2\text{O} \qquad K_L \qquad (2)$$

Where $L^{2-}$ represents phthalate ion, which are considered here as potential determining ions. $K_{HL}$ and $K_L$, are the equilibrium constants, obtained by fit to adsorption data. The adsorption density as a function of phthalate aqueous concentration is illustrated in Figure 2, where the solid line shows SCF/DLM model calculation with $\log K_{HL}=16.45$, $\log K_L=11.28$, in addition to the constants for hematite surface hydroxyl species. The model yields a good estimation at higher adsorption densities, but it overestimates the adsorbed organics at lower solution concentration.

The surface complex equilibrium constants derived above can be used in calculating surface potentials. Figure 3 illustrates the surface potential from the SCF/DLM model (solid line) and $\zeta$-potential evaluated from the mobility measurements using the treatment of Wiersema et. al. (26). The modeled potential values in the neighborhood of zero agree well with the ones derived experimentally.

The pH dependence of $\zeta$-potential, surface speciation and colloid stability ratios is presented in Figure 4. In panel (a) of Figure 4, the solid line represents the surface potential from the model, and the data points correspond to $\zeta$-potential from electrokinetic measurements. The dashed line represents a reduced potential calculated at a distance of 4.0 nm from the surface. In panel (b), the surface species distribution is calculated by the SCF/DLM model. Under these conditions, the observed zero interfacial potential is in the range pH 7.2 to 7.5, which suggests that the isoelectric point ($pH_{iep}$) is reduced from 8.5 to near 7, as a result of phthalate adsorption. In panel (c) the stability ratio is plotted, and shows consistency with the potential data. At pH around 7 the potential is approximately zero as a result of adsorption of phthalate, and the corresponding stability is also a minimum. At pH 4, although the adsorption is substantial, the particles still carry a net positive charge, which results in relative stability. At pH 10, a relatively high stability ratio can be attributed to the predominant neutral and negative species, $\equiv\text{FeOH}$ and $\equiv\text{FeO}^-$, and the relatively high negative surface potential.

## Effects of Higher Molecular Weight Organics.

In studies by Hunter and Liss (27) and Tipping (7) adsorption of dissolved natural organic matter is said to account for the surface properties of various types of naturally occurring particles. The presence of natural organic matter changes interfacial potential and charge, and consequently electrostatic interaction between particles is altered. Hence, physical-chemical interaction between natural organic matter and particle surfaces plays a key role in determining observed coagulation rates. Results for electrophoretic mobility and coagulation kinetic studies presented in this research demonstrate the influence of adsorption on coagulation.

Carboxyl groups have been shown to have a strong affinity for hematite surfaces (adsorption of phthalate on hematite is due primarily to carboxyl group bonding to Fe centered surface sites). A polymer molecule with a large number of carboxyl groups, such as polyaspartic acid, is expected to exert a stronger effect than simple organic molecules on hematite stability, as proposed by Lyklema (16). Lyklema pointed out that polymer adsorption is driven by the free energy of bonding polymer segments to the adsorbent. As a result, adsorption of polymers is widely observed, since only modest segment adsorption free energy is needed. The effect of polyaspartic acid (PAA) on the coagulation rate of hematite is shown in Figure 5(a). The plot, in part, resembles Figure 1(c), in which the stability ratio is plotted as a function of phthalate concentration. However, the concentration corresponding to the minimum stability is much lower for PAA ($\sim 2 \times 10^{-5}$ g/l, or $2 \times 10^{-7}$ M of $-\text{COOH}$) than for phthalate ($\sim 5 \times 10^{-2}$ g/l, or $6 \times 10^{-4}$ of $-\text{COOH}$). At the pH of the study, the side chain carboxyl groups in PAA are extensively dissociated (intrinsic $pK_a$=3.9). The polymer structure is illustrated on p 302.

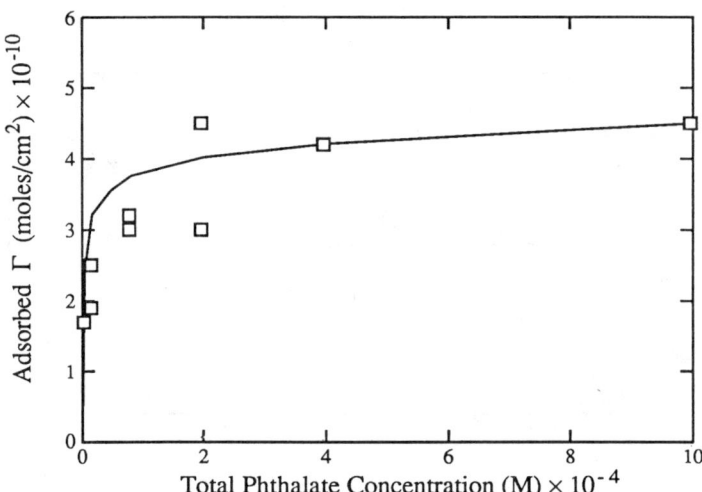

Figure 2: Adsorption density, $\Gamma$, of hematite suspensions as a function of total phthalate concentration. Adsorption was under the following conditions: hematite solid concentration=17.6mg/l, pH=6.2 and $(NaClO_4)$=5 millimolar. Solid line corresponds to a SCF/DLM model calculation.

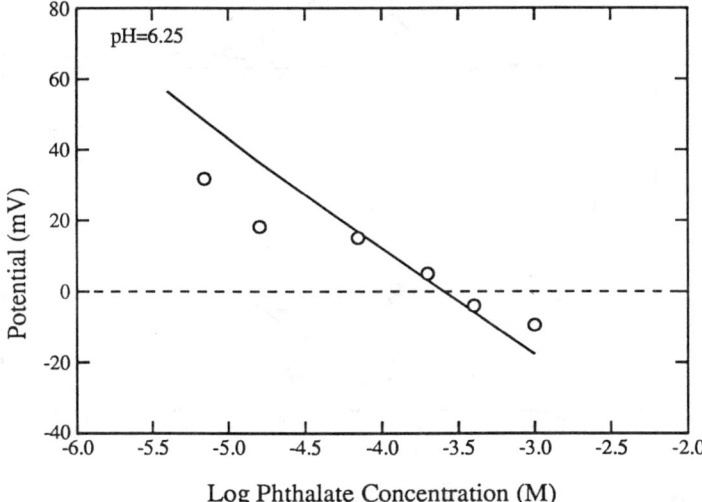

Figure 3: Potentials in millivolts plotted as a function of phthalate concentration under the following conditions: pH=6.25, hematite solid concentration=8.6 mg/l, and (KCl)=5 millimolar. The solid line represents the surface potential calculated by the SCF/DLM model, and the data points are the $\zeta$-potential evaluated from the mobility data using Wiersema's treatment (26).

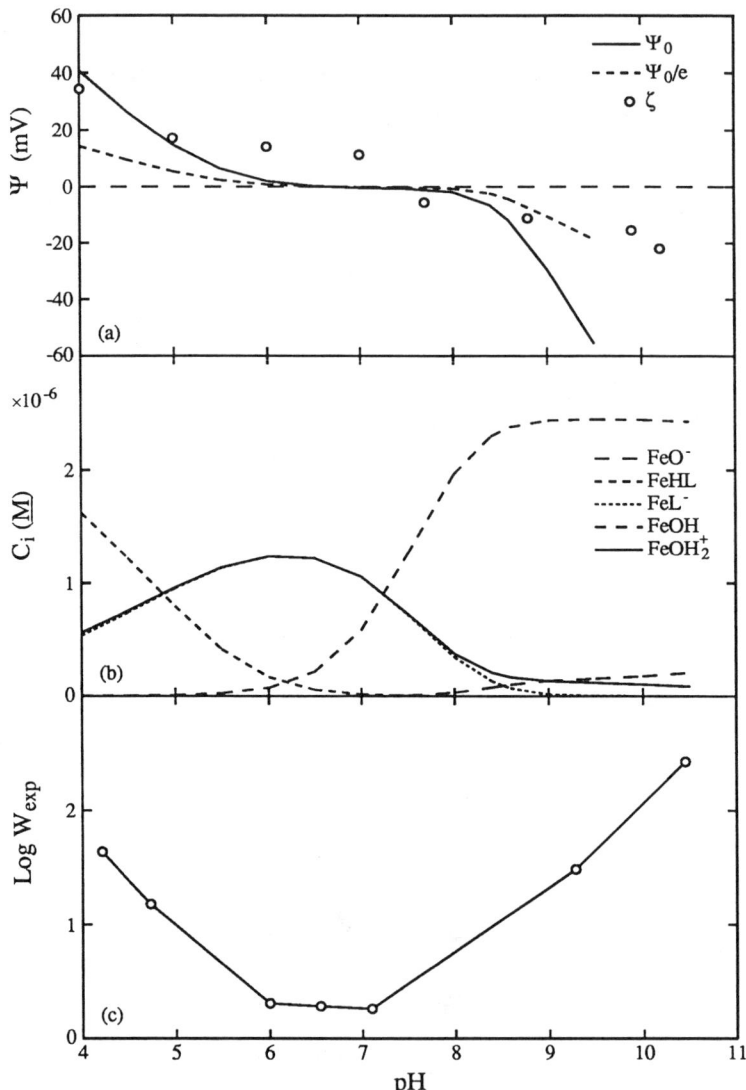

Figure 4: Comparison of potentials, $\Psi$ and $\zeta$, surface species distribution, $C_i$, and experimental stability ratio, $W_{exp}$, as a function of pH in the presence of 0.2 millimolar total phthalate species. Hematite solid concentration is 17.0 mg/l, and ionic strength is 5 millimolar. The diffuse layer "thickness", $\kappa^{-1}$, is 4.0 nm.

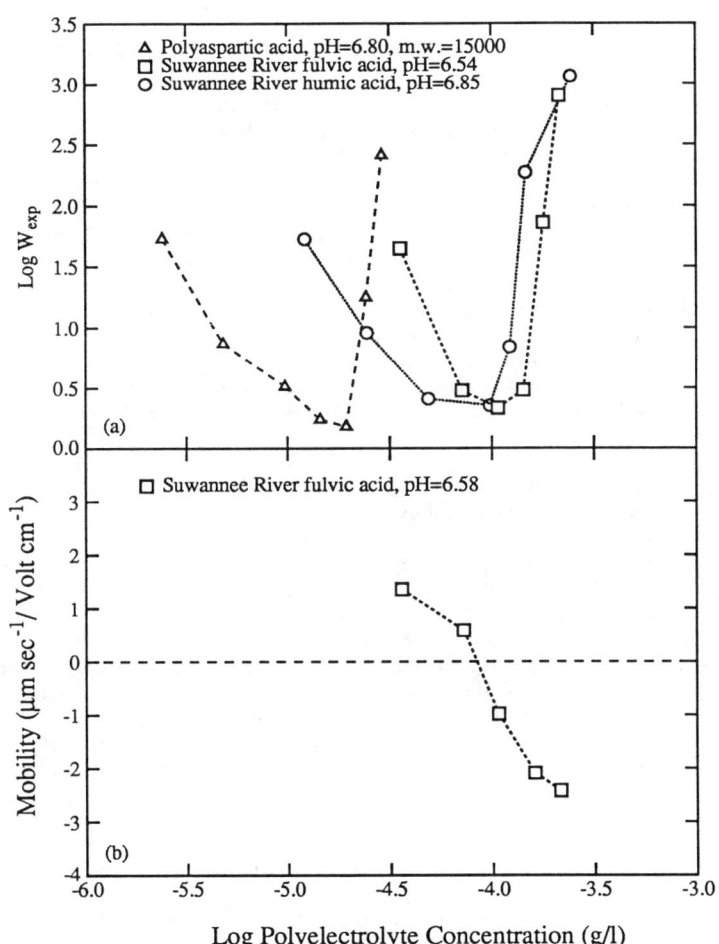

Figure 5. (a) Experimentally derived stability ratio, $W_{exp}$, of a hematite suspension, plotted as a function of polyelectrolyte concentration in the presence of 1 millimolar NaCl. Hematite concentration is 17.2 mg/l. (b) Electrophoretic mobility of a hematite suspension as a function of fulvic acid concentration at pH 6.58. Hematite concentration is 8.6 mg/l and ionic strength as KCl is 5 millimolar.

$$\begin{array}{c}\text{H}_2\text{N}-\text{CHCO}-(\text{NHCHCO})_n-\text{NHC}-\text{COO}^-\\|\quad\quad\quad\quad|\quad\quad\quad\quad\quad|\\\text{CH}_2\quad\quad\text{CH}_2\quad\quad\quad\text{CH}_2\\|\quad\quad\quad\quad|\quad\quad\quad\quad\quad|\\\text{C}=\text{O}\quad\quad\text{C}=\text{O}\quad\quad\quad\text{C}=\text{O}\\|\quad\quad\quad\quad|\quad\quad\quad\quad\quad|\\\text{O}^-\quad\quad\quad\text{O}^-\quad\quad\quad\quad\text{O}^-\end{array}$$

As the PAA concentration increases from zero to $2 \times 10^{-5}$ g/l, the binding of carboxylate anionic groups to the hematite surface effectively reduces the positive surface potential, and a decreased stability ratio is observed. As the PAA concentration is increased beyond $2 \times 10^{-5}$ g/l, the continued adsorption of negative carboxylate groups is able to reverse the interfacial potential, resulting in an increase in the stability ratio.

Coagulation of hematite in the presence of fulvic (FA) and humic acid (HA) follows the same trend shown for polyaspartic acid. Because carboxyl groups are dominant in natural organics (29) the organic substances behave as negatively charged polyelectrolytes in the usual pH range of natural systems. Thurman and Malcolm reported that Suwannee River fulvic acid contains 6 millimoles/g of –COOH functional groups. The critical coagulation concentration of fulvic acid is $10^{-4}$ g/l (Figure 5(a)), which corresponds to $6 \times 10^{-7}$ M of –COOH. Thus, the critical coagulation concentrations of fulvic acid and PAA are roughly the same order of magnitude, in terms of –COOH groups. The molecular weight of Suwannee River humic acid is 3000–5000, greater than that reported for fulvic acid (1000–1700). However, both appear to contain approximately the same number and type of functional groups. Humic acid and fulvic acid, therefore, are expected to have similar effects on hematite stability. They do, as seen in Figure 5(a).

Preliminary adsorption experiments revealed that the adsorption of FA on hematite surfaces is of the "high affinity" type. Electrophoretic mobility data (Figure 5(b)) indicate that adsorption of fulvic acid on hematite is not only a consequence of electrostatic interaction between polyanions and positive surface sites, but depends on specific chemical interactions between surface iron ($\equiv$Fe) and carboxyl (–COO$^-$) groups as well. If electrostatic interaction were the sole driving force for adsorption, the mobility would be expected to be reduced to zero as the FA concentration increases, and would remain at zero as the concentration of FA is further increased. The strong reversal in mobility observed arises from the chemical interaction in addition to the electrostatic force. Consequently, the mobility data are in accord with the observed stability ratio, where at (FA)$< 1 \times 10^{-4}$ g/l the stability results from net repulsions of positive surfaces, and at (FA)$> 1 \times 10^{-4}$ g/l the stability arises from repulsion between negative surfaces. Similar conclusions can be drawn with respect to the driving forces for HA adsorption on hematite particles.

### Effects of Fatty Acids.

Coagulation experiments were performed with the following linear aliphatic acids: propionic (3 carbon atoms, C3), caprylic (C8), capric (C10), and lauric (C12) acid. The supporting electrolyte concentration was maintained at 0.05 molar with NaCl. Figure 6(a) shows the effect of concentration of the different acids on colloidal hematite stability. It can be seen that there is little or no effect on stability when the fatty acid concentration is low. As the concentration is increased, an influence on hematite coagulation rate becomes noticeable. First, the increase in acid concentration makes hematite less stable, and a minimum stability ratio is reached at the critical coagulation concentration. The stability ratio then increases as the fatty acid concentration is increased beyond the critical coagulation concentration. Although the plots of the coagulation rate versus acid concentration for C8, C10 and C12 are similar to one another, the critical coagulation concentration values differ significantly, with successive critical concentrations differing by a factor of about 10. For the C3 data set the stability ratio flattens out when the C3 concentration is increased beyond the critical coagulation concentration. This behavior differs sharply from that of fatty acids with a higher number of carbon atoms, where we

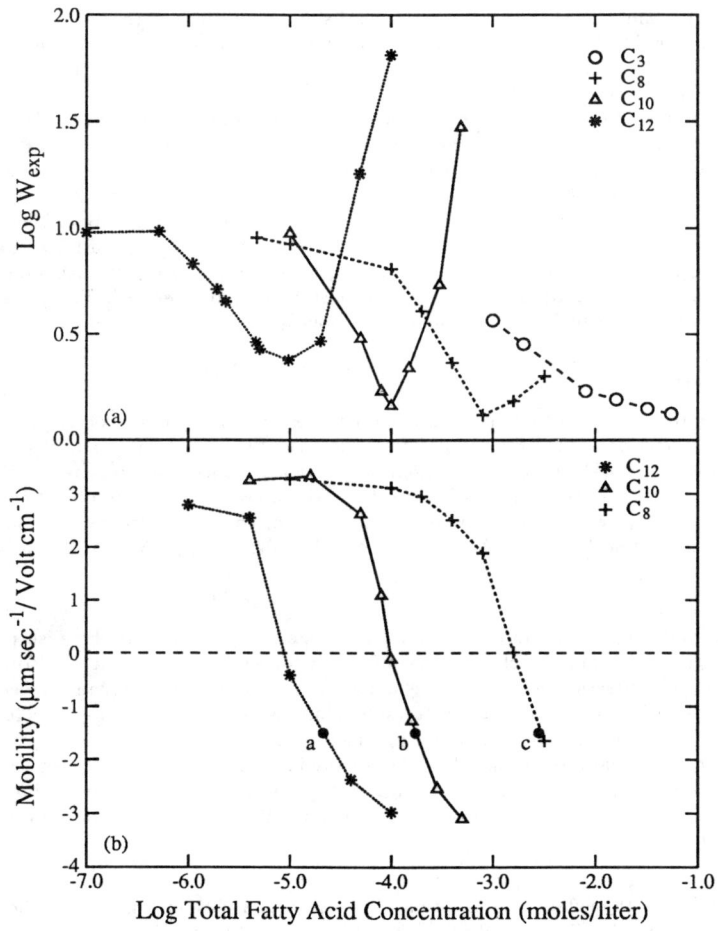

Figure 6: (a) Experimentally derived stability ratio, $W_{exp}$, of a hematite suspension, plotted as a function of fatty acid concentration at pH 5.2. The ionic strength is 50 millimolar of NaCl and hematite concentration is 34.0 mg/l. (b) Electrophoretic mobility of a hematite suspension as a function of fatty acid concentration at pH 5.15. Hematite concentration is 8.6 mg/l and the ionic strength is 5 millimolar of KCl. Lauric acid is denoted by $C_{12}$, capric acid by $C_{10}$, caprylic acid by $C_8$, and propionate by $C_3$.

observe colloid restabilization as concentrations are increased beyond the critical coagulation concentrations.

The corresponding electrophoretic mobilities of hematite as a function of total acid concentration are plotted in Figure 6(b). The supporting ionic strength is 5 millimolar KCl and the pH value is approximately 5.2. The plots show that the mobility changes most rapidly when it is close to zero, while at larger absolute mobility values the slope is smaller. The three curves in Figure 6(b) resemble each other in their general shape, but differ in the total concentration at which zero mobility is attained. It can be seen that the fatty acid concentration required to achieve zero mobility is less for a acid molecule with a longer carbon chain length, and differs by a factor of 10 between C8, C10, and between C10 and C12.

A key to understanding the colloidal stability and electrokinetic mobility of hematite in the presence of different fatty acid species is that their adsorption alters the electrostatic forces between particles. An adsorption isotherm was constructed for lauric acid to obtain a quantitative estimate of the affinity for the hematite surface. Figure 7 shows the amount of lauric acid (C12) adsorbed per unit surface area as a function of solution concentration. The hematite solid concentration was 35.5 mg/l, the ionic strength was 0.05M NaCl and pH 5.2. Unlike the adsorption isotherm for phthalate (Figure 2), the plot does not show typical Langmuir adsorption behavior. Instead, an exponential increase in adsorbed C12 is observed. To understand the underlying energetics, a surface equilibrium model was applied to the following reaction:

$$\equiv\text{FeOH} + \text{H}^+ + \text{L}^- \rightleftharpoons \equiv\text{FeL} + \text{H}_2\text{O} \tag{3}$$

where $\text{L}^-$ represents the ligand, laurate ion. The surface equilibrium model, incorporating electrostatic corrections to Equation (3), did not yield good fit to Figure 7. The observed rapid rise in the adsorption of lauric acid as the solution fatty acid concentration increases (Figure 7) is not explained by surface complexation, electrostatic interaction, or a combination of both. There are at least three contributions to the equilibrium constant: electrostatic interaction, $\Delta G_{ele}$, intrinsic chemical reaction, $\Delta G_{chem}$, and a hydrophobic energy, $\Delta G_{hyd}$. In mathematical terms:

$$-RT \ln K_{HL}^{app} = \Delta G_{ele} + \Delta G_{chem} + \Delta G_{hyd} \tag{4}$$

Specific and electrostatic interactions are both operative throughout the adsorption process; an additional lateral interaction energy appears to depend on the extent of adsorption. At low fatty acid concentration, a large surface area is able to accommodate organic molecules in a tangential orientation. Entropy change as a result of fatty acids leaving the aqueous solution contributes to the total adsorption energy. At high laurate ion concentrations the surface becomes crowded. A radial orientation of the adsorbate is more likely in that it allows a larger number of anions to be accommodated on the surface. Interactions between the hydrophobic tails appears to yield additional energy for adsorption. To summarize, the adsorption equilibrium constant is represented by:

$$K_{HL} = f(\theta) \tag{5}$$

where $\theta = \Gamma/\Gamma_m$ is the extent of adsorption. The adsorption data in Figure 7 can be modeled by a diffuse layer model. The equilibrium constants are fitted by a polynomial as follows:

$$\log K_{HL} = 9.868 + 0.9\theta - 6.551 \times 10^{-1}\theta^2 + 1.766\theta^3 \tag{6}$$

Because the electrostatic interaction is modeled through an equilibrium calculation the equilibrium constant given by Equation (6) is the result of specific interaction and hydrophobic energy.

The mobility data and stability ratios from coagulation experiments both show systematic change with variations in concentration and in the carbon number of the molecules. In general, the adsorption of fatty acid is represented by Equation (4), where $K_{HL}^{app}$ is proportional to the adsorption density and inversely proportional to the aqueous anion concentration. If $K_{HL}^{app} = A \cdot \Gamma/L_T$ ($A$ is a constant), and it is assumed that each $-\text{CH}_2$ group contributes an energy $\phi$ to the total hydrophobic interaction energy, then $\Delta G_{hyd} = n\phi$, where $n$ is the number of carbon atoms in the fatty acid molecule. Using Fuerstenau's approach (30), Equation (4) can be modified to:

$$\frac{A \cdot \Gamma}{L_T} = \exp[-(\Delta G_{elec} + \Delta G_{chem} + n\phi)/RT] \qquad (7)$$

When the mobility reaches zero, the energy contribution due to electrostatic interactions vanishes; this in turn corresponds to a minimum stability, and hence $L_T$ equals the critical coagulation concentration. The slope of a plot of log($c.c.c.$) versus $n$ yields the energy $\phi$ for each $-CH_2$ group. Using the mobility data in Figure 6(b), $\phi$ is determined to be $(1.2\pm0.1)$RT (Figure 8). This value is approximately that associated with the hydrophobic effect (31).

The C12, C10 and C8 acids differ only in the number of carbon atoms in their hydrophobic tails, so the specific chemical contribution through the carboxylic group should remain approximately the same for all of these molecules. At the same mobility, the electrostatic contribution is also similar. Hence, differences in overall equilibrium constants are due essentially to hydrophobic interactions. With the experimentally determined values of $\phi$, $K_{HL}$ for C10 and C8 is easily modeled. For example, at point $a$ in Figure 6(b), log$K_{HL}$ is determined to be 10.39 for C12 by fitting the corresponding adsorption data. The constant for C10 at point $b$ differs from $K_{C12}$ by $\exp(2\phi/RT)$ or a numerical value of log $K_{C10}$ = 9.27. Similarly, $K_{C8}$ at point $c$ is given by $K_{C8} = K_{C12}/\exp(4\phi/RT)$ and log $K_{C8}$ = 8.15.

Hematite stability and mobility data in the presence of a series of fatty acids (Figure 6) demonstrate the importance of hydrophobic interaction. Each 10-fold decrease in $c.c.c.$ as the number of carbon atoms in the chain increases by 2 can be attributed to the extra hydrophobic contribution of two ($-CH_2$) groups.

## SUMMARY AND CONCLUSIONS

Small organic molecules, for example phthalate, show specific interaction with hematite surface hydroxyl groups. Phthalate forms strong complexes on the hematite surface, thereby reducing the initially positive surface potential. When sufficient phthalate is present in the suspension, hematite surface potential can be reversed as a result of phthalate adsorption.

In the presence of fatty acids, hydrophobic interactions are operative in addition to the specific chemical and electrostatic interactions. As a consequence molecules, with longer carbon chains have a greater effect on the stability of a hematite suspension than molecules with shorter carbon chains and the same functional groups. The influence of organic molecules on hematite stability can be explained qualitatively by recognizing the influences of hydrophobic contributions on binding of organic molecules to the hematite surface. At pH 6, the order of effectiveness of organic anions in causing hematite coagulation is,

$$\text{laurate} > \text{caprate} > \text{caprylate} \gg \text{propionate}$$

Macromolecules, such as polyaspartic acid, humic and fulvic acid, are observed to have the strongest effects on the stability of hematite among all the organics studied. At pH<pH$_{zpc}$, the critical coagulation concentrations for organic species in terms of *carboxyl groups* are summarized in Table II. At pH about 6.5, the concentrations (in terms of carboxyl group) required to achieve the minimum stability are at least 100 times less for PAA, FA and HA than those of simple organics, such as oxalic and phthalic acids. The strong effect of polyelectrolytes on hematite stability can be accounted for by the adsorption characteristics of polyelectrolytes. Macromolecules and polymers are adsorbed on surfaces through bonding individual segments to the surface sites. Since many segments are linked together (linearly or branched) in a polymer, the overall interaction energy can be very high, even if the adsorption energy of an individual segment is small. The carboxyl groups in FA and HA are the important segments in accounting for the adsorption of these organic molecules on the hematite surface. The critical coagulation concentrations in terms of –COOH at pH 6.5 are similar for both PAA and FA. The combined effects of the electrostatic energy, the specific chemical energy, and the hydrophobic energy of each segment favor adsorption, and consequently the naturally occurring fulvic and humic acids are the most effective materials in influencing hematite particle stability.

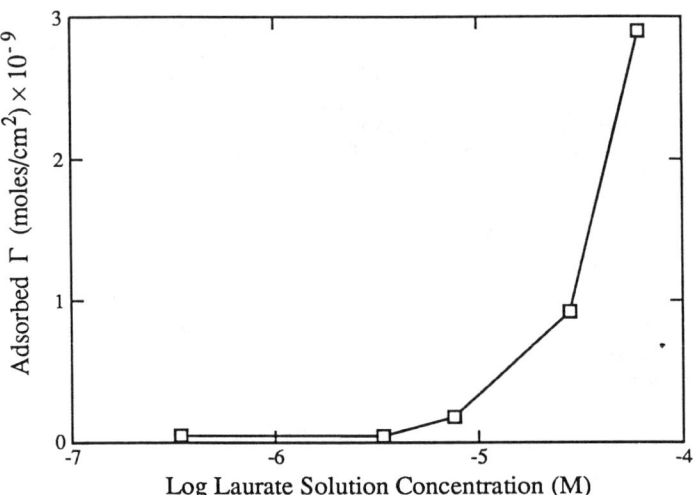

Figure 7: Lauric acid adsorption on the surface of hematite particles as a function of lauric acid concentration. Hematite concentration is 35.5 mg/l, NaCl concentration is 50 millimolar and pH is 5.2.

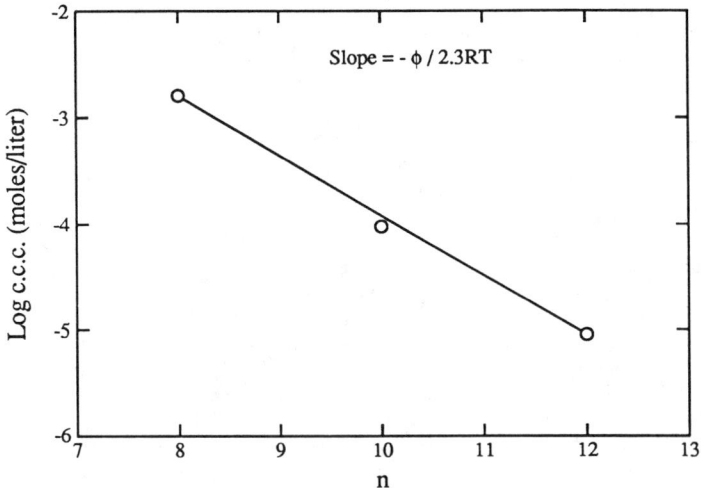

Figure 8: Critical coagulation concentration for linear alphatic acids as a function of the number, $n$, of carbon atoms.

Table II: Critical Coagulation Concentration (c.c.c.) of Organics in Terms of Moles of Carboxyl Functional Groups

| pH | Organics | Log c.c.c. M | Log c.c.c. as carbon | Log c.c.c. as carboxyl |
|---|---|---|---|---|
| 6.2 | Oxalic | -3.9 | -3.4 | -3.6 |
| 6.2 | Phthalic | -3.8 | -2.8 | -3.5 |
| 6.8 | PAA | -4.7* | -6.2 | -6.7 |
| 6.5 | FA | -4.0* | -5.4 | -6.2 |
| 6.8 | HA | -4.0* | -5.4 | -6.2 |
| 5.2 | $C_{12}$ | -5.0 | -3.9 | -5.0 |
| 5.2 | $C_{10}$ | -4.0 | -3.0 | -4.0 |
| 5.2 | $C_8$ | -3.1 | -2.1 | -3.1 |
| 5.2 | $C_3$ | -1.2 | -0.5 | -1.2 |

* Concentrations are expressed in mg/l.

The experimental results for coagulation of colloidal hematite particles in the presence of aqueous organic species demonstrate the importance of specific chemical adsorption at the surface in influencing oxide particle coagulation rates. A chemical model (SCF/DLM) provides a quantitative description of variations in interfacial charge and potential resulting from the addition of organic electrolytes to hematite suspensions. The laboratory observations on the particle coagulation behavior in the presence of fulvic and humic acids are in accord with observations by other workers on naturally-occurring particles and polymeric species (32). Laboratory measurements of adsorption, electrokinetic mobility and coagulation rates for oxides can be used to understand mechanisms of particle coagulation and stabilization in modeling a complex geological system.

LITERATURE CITED

1. Sholkovitz, E.R. Geochim. Cosmochim. Acta 1976, 40, 831.
2. Boyle, E.A.; Edmond, J.M.; Sholkovitz, E.R. Geochim. Cosmochim. Acta 1977, 41, 1313.
3. Mayer, L.M. Geochim. Cosmochim. Acta 1982, 46, 2527.
4. Fox, L.; Wofsy S.C. Geochim. Cosmochim. Acta 1983, 47, 211.
5. Verwey, E.J.W.; Overbeek, J.Th.G. Theory of the Stability of Lyophobic Colloids; Elsevier: Amsterdam; 1948.
6. Ali, W.; O'Melia, C.R.; Edzwald, J.K. Water Sci. Technol. 1984, 17, 701.
7. Tipping, E. Geochim. Cosmochim. Acta 1981, 45, 191.
8. Tipping, E.; Ohnstad, M. Nature 1984, 308, 266.
9. Gibbs, R.J. Environ. Sci. Technol. 1983, 17, 237.
10. Breeuwsma, A.; Lyklema, J. J. Colloid Interface Sci. 1973, 43, No.2, 437.
11. James, R.O.; Parks, G.A. Surface and Colloid Science; Matijević, E., Ed.; Plenum: New York; 1982; Vol. 12.
12. Parks, G.A.; DeBruyn, P.L. J. Phys. Chem. 1962, 66, 967.
13. Penners, N.H.G.; L.K. Koopal Colloids and Surfaces 1987, 28, 67.
14. Liang, L. Ph.D. Thesis, California Institute of Technology, Pasadena, CA, 1988.
15. Matijević, E.; Scheiner, P. J. Colloid Interface Sci. 1978, 63, No.3, 509.
16. Bales, R. Ph.D. Thesis, California Institute of Technology, Pasadena, CA, 1986.
17. Ottewill, R.H.; Sirs, J.A. Photoelectric Spectrometry Group, Bulletin 1957, 10, 262.
18. Davis, J.A.; J.O. Leckie J. Colloid Interface Sci. 1978, 67, No.1, 90.
19. Davis, J.A.; J.O. Leckie J. Colloid Interface Sci. 1980, 74, No.1, 32.
20. Dzombak, D.A.; F.M.M. Morel J. Hydraulic Engineering 1987, 113, No.4 430.
21. Sigg, L.; W. Stumm Colloid and Surfaces 1980, 2, 101.

22. Hohl, H.; Sigg, L.; Stumm, W. ACS Advances in Chemistry Series No. 189; Kavanaugh, M.C.; Leckie, J.O., Eds.; 1980.
23. Westall, J.C.; Zachary, J.L.; Morel, F.M.M. Technical Note, No.18, Parsons Labs, MIT, Cambridge, MA, 1976.
24. Westall, J.C. Report 82-01, Oregon State University, Corvallis, OR, 1982.
25. Balistrieri, L.S.; Murray, J.W. Geochim. Cosmochim. Acta 1987, 51, 1151.
26. Wiersema, P.H.; Loeb, A.L.; Overbeek, J.Th.G. J. Colloid Interface Sci. 1966, 22, 78.
27. Hunter, K.A.; Liss, P.S. Limnol. Oceanogr. 1982, 27, No.2, 322.
28. Lyklema, J. Proceedings of the Engineering Foundation Conference; Sea Island, Georgia; 1985.
29. Thurman, E.M.; Malcolm, R.L. Terrestrial Humic Materials; Christman, R.F.; Gjessing, E.T., Eds.; Ann Arbor Sci.: Michigan; 1983.
30. Fuerstenau, D.W. The Chemistry of Biosurfaces, Vol.1; Hair, M.L., Ed.; Marcel Dekker: New York; 1971.
31. Israelachvili, J.N. Intermolecular and Surface forces; Academic Press: London; 1985.
32. Liang, L.; Morgan, J.J. Aquatic Sciences; 1989; in press.

RECEIVED July 20, 1989

# ADVANCEMENTS IN MODELING: MODELING SENSITIVITIES

# Chapter 24

# Uncertainties in Ground Water Chemistry and Sampling Procedures

## Michael J. Barcelona

**Aquatic Chemistry Section, State Water Survey Division, Illinois Department of Energy and Natural Resources, 2204 Griffith Drive, Champaign, IL 61820**

> The chemistry of ground-water and subsurface systems is a topic of increasing applied and fundamental interest. Assessments of contaminated conditions and evaluations of remedial action activities call for a depth of understanding and substantial confidence in chemical data interpretation. The spatial and temporal resolution of ground-water chemistry investigations is limited due to the physical constraints involved in the location and construction of wells in dynamic hydrogeochemical settings. Direct sources of error which can affect confidence in chemical data include degassing, due to depressurization during pumping or filtration, oxidation on contact with the atmosphere, stagnation within the well bore and interactions with well casing, pump or sealing materials. More subtle sources of uncertainty in chemical interpretations result from sorption and leaching interactions between water samples and various materials. The direct sources of uncertainty can cause gross errors in interpretation though they can be controlled or quantified. The more subtle difficulties demand additional research and care in the development of ground-water sampling protocols.

The ground water environment presents many challenges to detailed chemical investigations. One may expect some degree of spatial and temporal variability in chemical distributions due to coupled chemical and physical processes in aquifers. Added complexities are posed by the transport of colloidal particles, significant vertical and horizontal redox gradients, unknown organic materials, biological catalysis and transformations, as well as the inevitable disturbance involved in the collection of water samples. Criteria for sample representation may be more complicated than for surface water or atmospheric sampling.

The purpose of this paper is twofold. It is intended as an overview of subsurface environmental conditions with an emphasis on controlling various sources of error and identifying causes of variability in ground-water chemistry. One of the major sources of variability or uncertainty that we can either control or identify is the error introduced in ground-water sampling operations. Controllable sources of uncertainty and error in chemical results have been identified which can introduce bias in determinations of chemical constituent concentrations in ground water. These include sampling well location, construction, purging of stagnant water from wells, and the sample collection or handling protocols prior to analysis. Sampling errors may go undetected even in well-planned investigations. This demands that data should be checked for consistency with past experience and actual site conditions. An examination of data reliability should precede any interpretation of the predicted speciation, reactivity or mobility of minor or trace constituents in subsurface systems.

## OVERVIEW OF SUBSURFACE ENVIRONMENTAL CONDITIONS

The ground-water environment is quite variable in its physical conditions of ground-water distribution, occurrence and flow. What was considered a sterile environment, devoid of oxygen, is known to contain dissolved oxygen and considerable microbial biomass ([1]). Distributions of chemical constituents are reflected in the corresponding hydrogeologic conditions and microbial diversity of this environment (2-3). Effective sampling and analysis efforts take into account our developing understanding of subsurface conditions and the influences which contamination and ground-water pumpage have on chemical distributions (4). There are no single "true" values for specific environmental chemical constituents. Rather there are spatial and temporal distributions of species which must be investigated with increasing sophistication and detail in evolving investigations.

Physical Conditions

Temperature, pressure and flow conditions in aquifers vary substantially in space and time (5). Temperature effects on ground-water chemistry include solubility and mixing, as well as thermodynamically preferred reaction paths and rates (2,6). Pressure gradients effect gas solubility and to some extent influence the porosity and permeability of geologic formations (7). Saturated flow and recharge effects influence the transport of ground water and chemical constituents. Approximate ranges of these physical variables for "natural" and "disturbed" (i.e., contaminated) sites from the literature are summarized in Table I. It should be noted that temperature effects can be important at sites downgradient from landfills where microbial processes may elevate temperatures by ten to fifteen degrees celsius as compared to background conditions. Also, there are inevitable depressurization effects on water samples taken from depths which can shift carbonate equilibria, pH and chemical speciation.

Microbial Conditions

Wilson and McNabb (1) have suggested that the population density of microbiota below the root zone in North American probably exceeds the bacterial biomass in the rivers and lakes on the continent. Biochemical measures of the biomass and metabolic activity of subsurface biota indicate that the biomass is largely bacterial (8). Microbial biotransformation potential has been demonstrated in the laboratory for compounds in a number of priority pollutant classes (9-10).
Approximate ranges of subsurface bacterial biomass and metabolic activity from the literature are shown in Table II. There have been few studies of relative measures of biomass or activity at contaminated versus uncontaminated sites. It is clear that certain functional classes of bacteria appear to increase in organic-rich environments (11-12). There are substantial problems associated with determining microbial activity in contaminated environments due to the diversity of types of microorganisms and chemical interferences with assay procedure.

Variability in Concentrations of Chemical Constituents

Evidence of microbial respiration processes and oxygen utilization is apparent in distributions of dissolved oxygen from aquifers contaminated with landfill leachate (13) and organic-rich recharge or effluents (14-15). A review of chemical conditions in a large number of aquifers strongly suggests that major oxidizing and reduced species are not in redox equilibrium (16). Studies of redox-active chemical species have also disclosed the presence of unusual oxidizing chemical species such as hydrogen peroxide and nitric oxide in shallow unconfined aquifer systems (17-18). These chemical oxidants undoubtedly are related to microbially-catalyzed processes and suggest that the assumption of redox equilibrium in the subsurface must be carefully evaluated in modeling efforts. Disequilibrium should be expected in disturbed situations where sedimentary systems, deposited thousands of years ago, may have been disturbed hydrogeochemically by man's activities in the past 50 to 100 years.
The results of recent investigations confirmed that vertical gradients of redox potential and major oxidized and reduced species in the subsurface are quite marked (19). Average vertical profiles of redox potential (Eh) and dissolved oxygen, and Fe(II) concentrations measured in a pristine shallow sand and gravel aquifer are shown in Figure 1. The dissolved oxygen concentrations varied from about 90% saturation near the water table surface at {10 m from the land surface to consistently undetectable levels at 33 m. Corresponding vertical gradients in Fe(II) and Eh are also very strong at the site scale (Table III). These apparent redox gradients clearly show that short (i.e., relative to the saturated thickness) screened well completions at multiple depths will be required to accurately assess chemical conditions (4). The need for vertical definition in sampling network design has been addressed by the development and utilization of

Table I. Ranges of Physical Variables:
Temperature (T), Pressure (P), and Flow Velocity (V)

| Variable | Causes/effects | Natural | Disturbed |
|---|---|---|---|
| T | Mixing, reaction paths and rates, solubility considerations | 3° to 20°C | 3° to 35°C |
| P | Gas solubility, permeability and porosity | 1 to 10 bar | 1 to 1000 bar |
| V ($m \cdot d^{-1}$) | Recharge, pumping mixing | <1 to 1000 | <1 to 1000 |

Table II. Ranges of Biological Variables: Biomass and Metabolic Activity

| | | Natural | Disturbed |
|---|---|---|---|
| Biomass | Catalytic or transformation potential | $10^1$-$10^8$ (cells·$g^{-1}$) | $10^4$-$10^9$ (cells·$g^{-1}$) |
| Activity | Turnover rates | 0.1 $\mu g \cdot L^{-1} \cdot hr^{-1}$ | ? |
| $V_{max}$ Glucose (Specific Gravity) | Metabolic status | (0.03-0.06 x $10^{-9}$ glucose· $h^{-1} \cdot cell^{-1}$) | ? |

Table III. Spatial Variability in Ground-Water Chemistry

| | Site scale (m to Km) | Large scale (Km to $10^3$ Km) |
|---|---|---|
| **Horizontal** | | |
| $O_2$ | -0.01 to +0.5 ($mg \cdot L^{-1} \cdot m^{-1}$) | 0.3 to 1 ($mg \cdot L^{-1} \cdot km^{-1}$) |
| $Fe^{2+}$ | 0.01 to 0.1 ($mg \cdot L^{-1} \cdot m^{-1}$) | 0.02 to 0.2 ($mg \cdot L^{-1} \cdot km^{-1}$) |
| Eh | -3 to 1 ($mV \cdot m^{-1}$) | -0.5 to -180 ($mV \cdot km^{-1}$) |
| **Vertical** | | |
| $O_2$ | -0.2 to +0.77 $mg \cdot L^{-1} \cdot m^{-1}$) | -- |
| $Fe^{2+}$ | -0.01 to 0.05 ($mg \cdot L^{-1} \cdot m^{-1}$ | -- |
| Eh | -2 to -40 ($mV \cdot m^{-1}$) | -- |

multi-level samplers (20). The depth range, suction mechanism and sample volume limitations of this type of sampler limits its use to very shallow depths. Undoubtedly, ongoing research on spatial variability of ground-water chemistry will produce further results which will demand more vertical and horizontal resolution of chemical constituent or contaminant distributions in the subsurface.

Temporal variations in ground-water chemical constituents observed in samples from wells are strongly influenced by the near-field effects of pumping and ground-water recharge caused by irrigation or artificial recharge operations. Short-term chemical variability, extending over time periods of minutes to days, and longer term variability due to seasonal effects or regional recharge processes have been reviewed by Montgomery et al. (21). They point out that the temporal sensitivity of sampling network results is a function of both sampling frequency and the duration of the sampling period. Typical magnitudes of temporal variation of water quality constituents in both the short-and long-term are shown in Table IV. The data in the table indicate that 3- to 100-fold concentration variations, from an arbitrary mean or baseline level, can occur for both major ionic and redox-active chemical constituents in response to the pumping, hydrologic and seasonal effects mentioned above. The most extreme variations have been observed due to short-term effects, particularly during the purging of stagnant water from wells prior to sampling (Barcelona, M. J.; Lettenmaier, D. P.; Schock, M. R. Environmental Monitoring and Assessment, in press July, 1989). Variations have often been observed for pH, conductance, temperature and volatile organic compounds during purging operations. Purging-related chemical variations represent significant uncertainties which must be controlled or accounted for in effective ground-water sampling efforts. They will be dealt with in the second part of this paper.

Longer-term chemical variations in ground water have been observed which may effect the estimates of long-term mean concentrations and complicate comparisons between samples taken from positions upgradient and downgradient from contaminant sources. Figure 2 depicts chloride and methane concentrations in biweekly samples from monitoring wells upgradient (i.e., #5 and #6) and downgradient (#13-8,9,10,11,12) from an anaerobic waste treatment impoundment. The wells are ordered in the figure along the ground-water flow path which traverses the impoundment. In Figure 2a, the chloride time series for the upgradient wells clearly shows concentrations which increase to well above the downgradient levels due to the effects of regional recharge and hydrogeochemical processes. In this case the long-term variability of chloride concentrations would severely bias estimates of flow and dilution of constituents in transport if observations of upgradient chloride determinations were only made early in the sampling period. The methane data in Figure 2b depict the behavior of a waste constituent which increases downgradient from the anaerobic impoundment. This is due to the fact that the rates of microbial generation processes exceed the rate at which the dilution of methane in the leachate occurs in transport. Placing additional wells along ground-water flow paths can provide valuable chemical and hydrologic information on spatial variability. Intensive arrays of wells sampled over time permit the estimation of temporal and spatial variations in both "conservative" and "reactive" chemical constituents.

The preceding discussion has consisted of a brief review of physical, chemical and biological factors which lead to significant uncertainty in determinations of chemical constituents in ground water. Uncertainties in ionic constituents alone could cause major difficulties in the application of equilibrium speciation models even if the subsurface was a homogeneous, well-mixed aquatic environment. The discussion has not treated solid-solution interactions or the influence of dissolved or solid-associated organic matter. In the following section the effects of ground-water sampling design or operational details are discussed with an emphasis on minimizing systematic error.

GROUND-WATER SAMPLING: RECOGNIZING AND CONTROLLING ERROR AND UNCERTAINTY

Sampling errors have received increasing attention in recent years due to the scientific and regulatory activity addressing widespread ground-water contamination problems. There have been a number of useful references published in this regard which supplement this discussion (22-24). Once the initial purpose of the investigation and hydrogeologic conditions at a particular site have been established, a preliminary network design and sampling protocol should be prepared. Network designs should be based on a detailed hydrogeologic study so that sampling well locations lie in the most likely flow path for the constituents of interest. The sampling protocol then consists of a detailed set of written steps to be taken during actual sample collection (25-26). It is imperative that the location, drilling, and design (i.e., casing diameter, screen size, slot interval, gravel pack, etc.) of monitoring wells must be done with the most complete hydrogeologic data

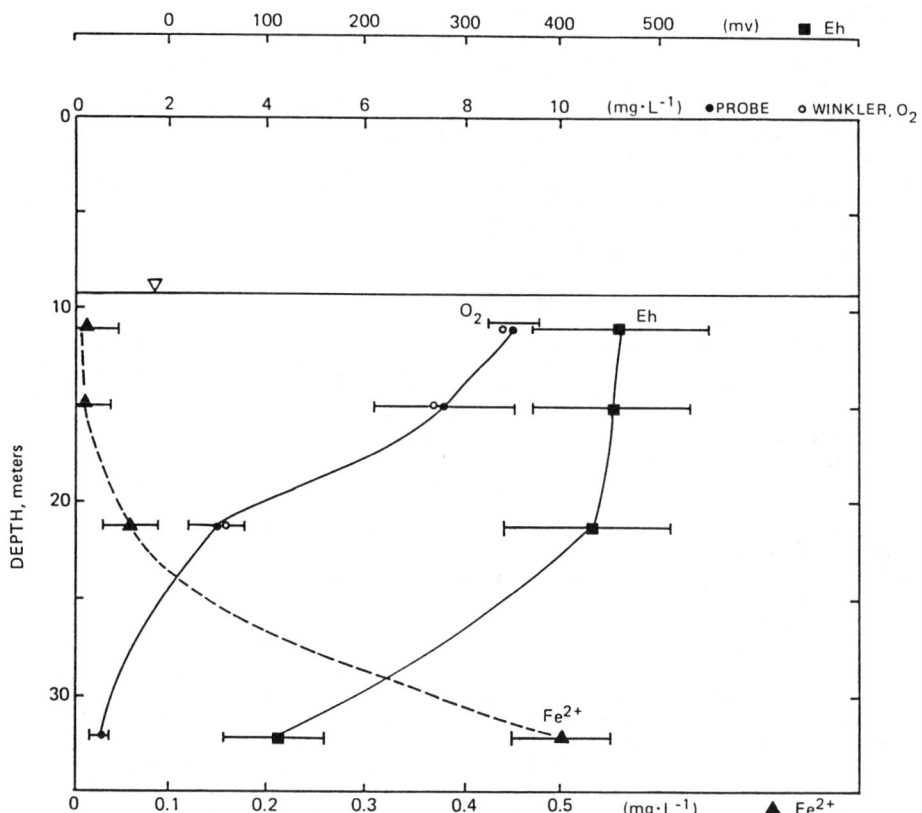

Figure 1. Average vertical distributions of Pt electrode potentials (Eh), dissolved oxygen and Fe(II) concentrations with depth in a shallow unconfined aquifer. (The error bars represent the ± one standard deviations range of thirty-nine biweekly determination.)

Table IV. Chemical Variability: Temporal Concentration Variations
Variations x times above or below an arbitrary baseline level

| Short-term (minutes to days) | | | |
|---|---|---|---|
| $NO_3^-$ | 13x | Fe(II) | 110x |
| $SO_4^=$ | 7x | $HS^-$ | 15x |
| $NH_3$ | 3x | | |
| Long-term (weeks to years) | | | |
| $NO_3^-$ | 6x | Fe(II) | 3x |
| $SO_4^=$ | 7x | | |
| $Cl^-$ | 3x | | |

# 24. BARCELONA  *Uncertainties in Ground Water Chemistry*  315

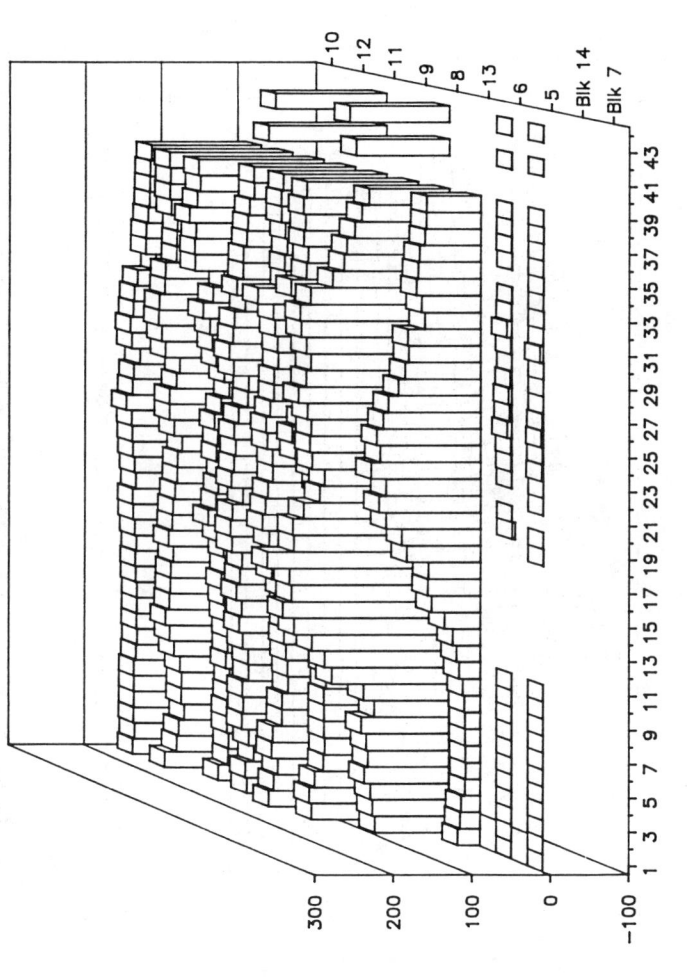

Figure 2. Time series of a, chloride (Cl⁻) concentrations at upgradient (i.e., wells #5 and #6) and downgradient wells along the flow path from an anaerobic treatment impoundment for 39 determinations. *Continued on next page.*

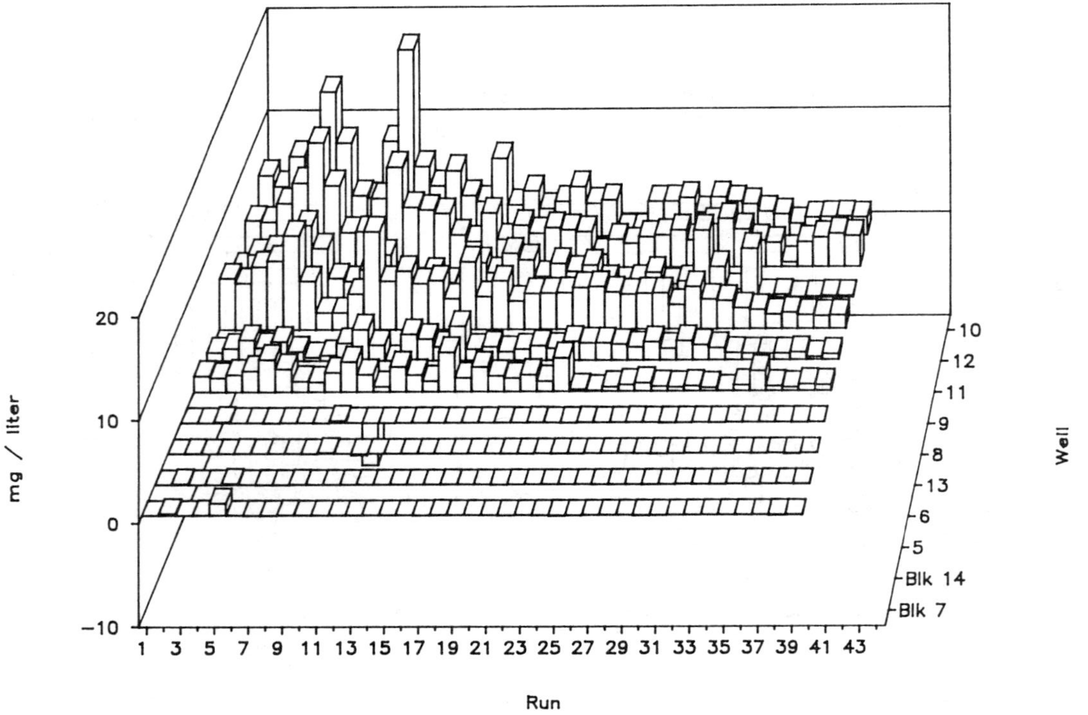

Figure 2. *Continued.* Time series of b, methane ($CH_4$) concentrations at upgradient (i.e., wells #5 and #6) and downgradient wells along the flow path from an anaerobic treatment impoundment for 39 determinations.

possible (27). An improperly located or constructed well will yield neither representative samples nor data of known certainty or confidence level. Given the rapidly expanding state of ground-water science it would be difficult to devise a perfect sampling protocol.

Among the most important steps in the ground-water sampling protocol are well design, purging of stagnant water, selection of sampling device or sampling tubing, sample filtration and preservation. All of these steps have the potential to introduce gross errors into the final result. If a biased result is input to chemical speciation models, the overall prediction of chemical speciation, reactivity, solubility, mobility, and persistence may be false, regardless of the validity of equilibrium assumptions.

Well Design

Presuming that the well is properly located, the design should include a length of screen which is comparable in length to the thickness of the hydrogeologic unit of interest (27-28). Screens which are placed adjacent to several geologic strata of varying aquifer properties (e.g., porosity, hydraulic conductivity, etc.) will yield mixtures of ground water from these strata on pumping. For example, dissolved metal concentrations in samples taken from a long-screened (i.e., 12 m) well have been shown to vary by more than a factor of ten depending on the placement of the pump intake within the screen (29). The importance of monitoring well screen length cannot be underestimated. Also, the design of the screen slot size and sand pack should be carefully considered to minimize water sample turbidity and the introduction of foreign particulates.

The selection of materials for the screen, casing, sand pack, and sealing operations are also important in effective well design but these subjects are beyond the scope of this discussion (30-32).

Well Development and Purging

Presuming that well location, drilling, design and completion steps have been properly conducted it is necessary to develop the monitoring well to remove fine sediments created during the drilling operation. Pumping, swabbing or surging the water in a newly constructed well must be done at sufficient flow rates to remove the particulates and create reliable sand pack/screen contact. The rates of water removal during development represent an upper limit which should not be exceeded during purging stagnant water or sample collection, otherwise further well development or well damage can occur (27-28). The result will be a well which will yield turbid water which may be totally unrepresentative of formation water chemistry regardless of filtration procedures.

Sampling wells must be purged of the water which remains stagnant in the screened interval between sampling intervals in order to collect representative samples (4,33). This is because the water in the well can undergo changes due to exposure to the atmosphere, reduced pressure and reactions with casing materials. Improperly purged wells have been documented to yield ground-water chemical results which may be orders of magnitude different from the composition of the formation water accessed after pumping. This is true of both inorganic and organic constituents over a range of concentrations (34-38).

While methods have been published to estimate the purging requirement to obtain greater than 90 percent formation water (24,39-40) it is common practice to purge three to five stagnant water volumes prior to sampling. This practice ignores the unique hydraulic characteristics of individual wells in saturated geologic formations (41). The recent work of Robbins (42) demonstrates in a convincing manner that the hydraulics of individual wells cannot be ignored in determining the purging requirement for removal of stagnant water. He argues against long screen lengths relative to the degree of vertical contaminant plume spreading. The result of poor well design in this case would be underestimates of the extent of contamination, migration rates, false indications of retardation or degradation and vertical spreading as well as unreliable concentration distributions and transport predictions. These consequences negate the purpose for detective contaminant monitoring efforts.

The preferred purging practice is to estimate the purging requirements based on the hydrogeology of the formation and the hydraulics of the individual well. This volumetric requirement at a given flow rate should then be verified in the field during each sampling operation by measurements of important chemical constituents. The measurements of pH, temperature, conductance, Eh or dissolved oxygen lend themselves to reliable in-line flow cell measurements (43) and they are important master variables for major ion chemistry of natural waters. Inhomogeneous spatial distributions of trace organic contaminants may give rise to erratic measured concentrations in purged waters (37), although the major ionic composition may be representative of that in the formation water. Documented procedures for estimating and verifying the efficiency of purging operations must be a central part of the sampling protocol.

All of the preceding steps in providing a reliable sampling point must be carefully considered, documented and quality-assured by using good professional judgement. Experience, the use of preliminary hydrogeochemical information and a critical reading of the pertinent literature provide minimum assurance that a reliable sampling point has been constructed. Conventional laboratory quality assurance/quality control procedures (i.e., blind control samples, field and procedural blanks, field and calibration standards, replicate samples and standards) only apply to those steps in the sampling protocol which follow well development and purging.

Sampling Devices

The selection of appropriate sampling devices for the chemical constituents of interest and the hydrogeochemical situation has been reviewed recently (44). Preferred pumping devices provide a consistent, steady flow rate of minimally disturbed ground water for in-line chemical measurements and sampling operations under variable hydrologic conditions. Positive displacement bladder pumps insure reproducible sampling performance regardless of operator idiosyncracies as long as they are used under specific lift and hydraulic head conditions. Dedication of the sampling device to a monitoring well is far preferable to involved field cleaning or decontamination procedures. The pitfalls involved in the selection and operation of ground-water sampling devices have been reviewed in the literature (24,45-46).

Sampling Tubing

The selection of appropriate sampling tubing may seem to be a fine detail within the protocol, but tubing connected to the sampling pump has intimate contact with the ground-water sample. For trace organic and inorganic determinations polytetrafluoroethylene (PTFE) tubing is preferable to other flexible polymeric materials (47). In deep installations where the tubing may be exposed to air for long periods of time, oxygen permeation through PTFE tubing may give rise to false indications of Eh, dissolved oxygen and ferrous iron levels (48).

Sample Filtration

The filtration of ground-water samples is a topic of substantial current controversy. While it is clear that water samples for determinations of volatile organic constituents or dissolved gases should never be filtered, publications exist which suggest this to uncritical readers (37,49). For geochemical modeling purposes it is important that water samples for "dissolved" inorganic analyses are filtered with minimal disturbance of the in situ condition. One must recognize, however, that colloidal species can be important in subsurface chemical mass balances and contaminant transport in ground-water systems (50). Presuming that the introduction of foreign particulates resultant from drilling operations, grouts/seals or casing corrosion can be minimized, the filtration of water samples remains necessary for geochemical modeling efforts. These measures may under- or overestimate the actual amount of chemical species transported in a dissolved or suspended state.

Sample Preservation and Handling

This final step of the field sampling protocol should be designed to insure that the integrity of the water sample is maintained before analysis and that handling operations are minimized (24). Recent results have shown that significant problems exist with the common practice of using artificial ice packs to maintain the recommended 4°C storage temperature (51). When these types of containerized frozen solutions are used to avoid the potential loss of labeling information under melting water ice conditions, it is likely that the samples never reach 4°C. Water samples, which cannot be frozen for subsequent analysis, should be chilled with water ice at ~4°C and then transferred to coolers with artificial ice packs for transport. Mechanical refrigeration, of course, would be preferable but this is not always possible given the field conditions at sites of many ground-water investigations.

## CONCLUSIONS

Subsurface physical, chemical and biological conditions are quite variable in space and time. These factors can be the source of considerable "natural" variability in ground-water chemistry at both clean and contaminated sites. Sampling and analytical errors or variability can be controlled if these activities are planned and documented as protocols which take into account the unique characteristics of individual sampling points and conditions. It is critical that a sound hydrogeologic basis exists for the steps in the sampling protocol which precede actual sample collection.

Since ground-water chemistry investigations are under way at a great number of sites for a large number of purposes it is important that one consult the scientific literature critically. In this way, one can gain confidence in various techniques and decisions which reduce uncertainty and improve the reliability of the inputs to chemical modeling.

## ACKNOWLEDGMENTS

The author is grateful to many colleagues for their contributions to the literature and critical discussions with them which made this review possible. He appreciates the efforts of Pam Beavers, John Brother and Linda Riggin in the preparation of the manuscript.

## LITERATURE CITED

1. Wilson, J. T.; McNabb, J. F. EOS 1983, 64, 33, 505-506.
2. Matthess, G. The Properties of Ground Water; J. Wiley & Sons, 1982.
3. Freeze, A. R.; Cherry, J. A. Groundwater; Prentice-Hall, Inc.: Englewood Cliffs, New Jersey, 1979.
4. Barcelona, M. J. In Principles of Environmental Sampling; Keith, L. H., Ed.; American Chemical Society: Washington, DC, 1988; Chapter 1.
5. Davis, S. N.; DeWiest, R. J. M. Hydrology, Second Edition; Wiley: New York, NY, 1967; 463 pp.
6. Stumm, W.; Morgan, J. J. Aquatic Chemistry, Second Edition; Wiley Interscience: New York, NY, 1981; 780 pp.
7. Peyton, G. R.; Gibb, J. P.; LeFaivre, M. H.; Ritchey, J. D. Proc. 2nd Canadian/American Conference on Hydrogeology: "Hazardous Wastes in Ground Water: A Soluble Dilemma, Banff, Alberta, Canada, June 25-29, 1985, pp 101-107.
8. White, D. C.; Smith, G. A.; Gehran, M. J.; Parker, J. H.; Findlay, R. H.; Martz, R. F.; Frederickson, H. L. Dev. Ind. Microbiol. 1983, 24, 201-211.
9. Lee, M. D.; Thomas, J. M.; Borden, R. C.; Bedient, P.B.; Wilson, J. T.; Ward, C. H. CRC Critical Reviews in Environmental Control 1988, 18, 1, 29-89.
10. Kobyashi, H.; Rittmann, B. E. Environmental Science and Technology 1982, 16, 3, 170A-183A.
11. Towler, P. A.; Blakey, N. C.; Irving, T. E.; Clark, L.; Maris, P. J.; Baxter, K. M.; Macdonald, R. M. Hydrology in the Service of Man. Memories of the 18th Congress, International Association of Hydrogeologists: Cambridge, U.K., 1985; pp 84-97.
12. Wilson, J. T.; Leach, L. E.; Henson, M.; Jones, J. N. Ground Water Monitoring Review 1986, 6, 4, 56-64.
13. Baedecker, M. J.; Back, W. Ground Water 17, 5, 429-437.
14. Van Beek, C. G. E. M.; Van Puffelen, J. Water Resources Research 1987, 23, 1, 69-76.
15. Champ, D. R.; Gulens, J.; Jackson, R. E. Can. J. Earth Sciences 1979, 16, 1, 12-23.
16. Lindberg, R. D.; Runnells, D. D. Science 1984, 225, 925-927.
17. Holm, T. R.; George, G. K.; Barcelona, M. J. Analytical Chemistry 1987, 59, 4, 582-586.
18. Barcelona, M. J.; Garske, E. E. Analytical Chemistry 1983, 55, 6, 965-967.
19. Barcelona, M. J.; Holm, T. R.; Schock, M. R.; George, G. K. Water Resources Research, 1989, 25, 5, 991-1003.
20. Cherry, J. A.; Gillham, R. W.; Anderson, F. G.; Johnson, P. E. J. Hydrol. 1983, 63, 31-49.
21. Montgomery, R. H.; Loftis, J. C.; Harris, J. Ground Water 1987, 25, 2, 176-184.
22. Monitoring Ground-Water Quality: Monitoring Methodology, EPA-600/4-76-026, U. S. Environmental Protection Agency, Las Vegas, NV, 1976.
23. Ground Water Monitoring and Sample Bias, Report No. 4367, American Petroleum Institute, Environmental Affairs Department, Washington, DC, June 1983.

24. Practical Guide for Ground Water Sampling, State Water Survey Contract Report No. 374 (EPA 600/S2-85/104), USEPA-RSKERL, Ada, OK, 1985.
25. American Chemical Society Committee on Environmental Improvement. Analytical Chemistry 1980, 52, 2242-2249.
26. Barcelona, M. J.; Gibb, J. P. American Society for Testing and Materials, Philadelphia, PA, ASTM STP-963, 1988, pp 17-26.
27. Driscoll, F. Ground Water and Wells, Second Edition; Johnson Division: St. Paul, MN, 1986; 1089 pp.
28. Monitoring Well Design and Construction, EPA 625/6-87/016, U.S. Environmental Protection Agency, CERI, Cincinnati, OH, 1987.
29. Cowgill, U. In Principles of Environmental Sampling; Keith, L. H., Ed.; ACS Professional Reference Books: Washington, DC, 1988; Chapter 11.
30. A Guide to the Selection of Materials for Monitoring Well Construction and Ground Water Sampling, EPA 600/S2-84-024, U.S. Environmental Protection Agency, RSKERL: Ada, OK, 1983.
31. Barcelona, M. J.; Helfrich, J. A. Proc. of the Ground Water Geochemistry Conference, National Water Well Association, Denver, Colorado, February 16-18, 1988, pp 363-375.
32. Evans, L. G.; Ellingson, S. B. Proc. of the Ground Water Geochemistry Conference, National Water Well Association, Denver, Colorado, February 16-18, 1988, pp 377-389.
33. Barcelona, M. J.; Helfrich, J. A. Environmental Science and Technology 1986, 20, 11, 1179-1184.
34. Marsh, J. M.; Lloyd, J. W. Ground Water 1980, 18, 4, 366-373.
35. Palmer, C. D.; Keely, J. F.; Fish, W. Ground Water Monitoring Review 1987, 7, 4, 40-47.
36. Robin, M. J. L.; Gillham, R. W. Ground Water Monitoring Review 1987, 7, 4, 85-93.
37. Smith, J. S.; Steele, D. P.; Malley, M. J.; Bryant, M. A. In Principles of Environmental Sampling; Keith, L. H., Ed.; ACS Professional Reference Books, American Chemical Society: Washington, DC, 1988; Chapter 17.
38. Panko, A. W.; Barth, P. In Ground Water Contamination Field Methods; Collins, A. G.; Johnson, A. I., Eds.; STP 963; American Society for Testing and Materials: Philadelphia, PA, 1988; pp 232-239.
39. Procedures for the Collection of Representative Water Quality Data from Monitoring Wells; Cooperative Ground Water Report 7; Illinois State Water Survey and State Geological Survey: Champaign, IL, 1981.
40. Barber, C.; Davis, G. B. Ground Water 1987, 25, 5, 581-587.
41. Cohen, R. M.; Rabold, R. R. Ground Water Monitoring Review 1988, 8, 1, 51-59.
42. Robbins, G. A. Ground Water, 1989, 27, 2, 155-162.
43. Garske, E. E.; Schock, M. R. Ground Water Monitoring Review 1986, 6, 3, 79-84.
44. Bibliography of Ground Water Sampling - Internal Report, EPA/600/X-87/235, U. S. Environmental Protection Agency, EMSL: Las Vegas, NV, 1987.
45. Barcelona, M. J.; Helfrich, J. A.; Garske, E. E. In Verification of Sampling Methods and Selection of Materials for Ground-Water Contamination Studies; Collins, A. G.; Johnson, A. I., Eds.; STP 963; American Society for Testing and Materials: Philadelphia, PA, 1988c; pp 221-231.
46. Rehm, B. W.; Stolzenburg, T. R.; Nichols, D. G. Field Measurement Methods for Hydrogeologic Investigations: A Critical Review of the Literature, EPRI-EA-4301, Electric Power Research Institute, Palo Alto, CA, 1985, 328 pp.
47. Barcelona, M. J.; Helfrich, J. A.; Garske, E. E. Analytical Chemistry 1985b, 57, 2, 460-464 (errata: 57, 13, 2752).
48. Holm, T. R.; George, G. K.; Barcelona, M. J. Ground Water Monitoring Review 1988, 8, 3, 83-89.
49. Holden, P. W. Primer on Well Water Sampling for Volatile Organic Compounds; University of Arizona, Water Resources Center Research, 1984, 44 pp.
50. Gschwend, P. M.; Reynolds, M. D. J. of Contaminant Hydrology 1987, 1, 1, 309-327.
51. Kent, R. T.; Payne, K. E. In Principles of Environmental Sampling; Keith, L. H., Ed.; ACS Professional Reference Book, American Chemical Society: Washington, DC, 1988; Chapter 15.

RECEIVED July 20, 1989

# Chapter 25

# Using Chemical Analyses and Assessing Quality in Aqueous Environmental Monitoring Programs

**Thomas R. Wildeman[1], Leslie S. Laudon[1,3], Roger L. Olsen[2], and Richard W. Chappell[2]**

[1]Department of Chemistry and Geochemistry, Colorado School of Mines, Golden, CO 80401
[2]Camp, Dresser, and McKee, Inc., 2300 15th Street, Suite 400, Denver, CO 80202

The U.S. Environmental Protection Agency (EPA) has established an analytical program involving contract laboratories to analyze the thousands of environmental samples collected at Superfund sites each year. Due to the possibility of litigation on many of these sites, the data generated must be of known quality and legally defensible. To accomplish this, a rigorous quality assurance and quality control program has been incorporated into the Contract Laboratory Program (CLP). However, the data user must independently evaluate the quality and usability of the data generated from the CLP. In some cases and for selected parameters, independent analyses outside of the CLP are appropriate.

All thermodynamic modeling, solute transport modeling, and evaluation of geochemical processes require the use of chemical data obtained through laboratory analyses. When calculation of risk to the public/environment and the selection of remedial alternatives is also involved, the resultant information from modeling or other evaluations is critical in determining the action to be taken at hazardous waste sites. These actions may range from additional studies to million dollar remediations. Therefore, credibility of laboratory data must be known. At many hazardous waste sites, the data will also be presented and used in the legal arena. Because of the critical nature of the laboratory results at hazardous waste sites, the Environmental Protection Agency (EPA) established the Contract Laboratory Program (CLP).

The laboratories who do CLP analyses for inorganic constituents have to maintain an extensive Quality Assurance program. This includes analysis of duplicates, laboratory blanks, interference checks, and laboratory control samples. However, the prescribed Quality Assurance does not always transfer to the laboratory bench. Also at times, routine analytical procedures will not yield the chemical information that is necessary for aqueous modeling. The following points out when these problems occur and what steps the researcher needs to take to insure data adequate for modeling.

## THE CONTRACT LABORATORY PROGRAM

The EPA's CLP was established to produce data of known and documented quality which is legally defensible. The program analyzes thousands of samples (in excess of 40,000 in 1985)

[3]Current address: Kennedy, Jenks, and Chilton, 3336 Bradshaw Road, Suite 140, Sacramento, CA 95827

from hazardous waste sites every year for use in characterizing sites, performing risk assessments, and determining remedial alternatives (1-6). Although modeling contaminant transport is often included in these activities, legal defensibility and consistency between sites are the primary objectives of the program. To meet these objectives, a high level of quality assurance and documentation has been incorporated into the laboratory and field programs.

Standardized and specialized analytical services to support a variety of Superfund sampling activities from preliminary site investigation to complex remedial monitoring and enforcement actions are provided by the CLP. To do this, four separate analytical services programs are operated: Organic Routine Analytical Services (RAS), Inorganic RAS, Dioxin RAS, and Special Analytical Services (SAS). The RAS program analyses are performed by a network of laboratories operating under fixed price contracts with the EPA. The SAS program provides analytical service to meet specific analytical requirements which do not fall under the RAS program. Analyses may include use of EPA standard methods that are not RAS, for example metal analyses by AA instead of ICP; or use of non-standardized analyses, for example, forms of sulfur. The request for SAS must include detailed methods, required accuracy and precision, and documentation requirements. The EPA solicits CLP laboratories to perform the SAS request.

All RAS and SAS analyses are administered by the Sample Management Office (SMO) which is staffed by an EPA contractor. All communications with the laboratory are through SMO personnel and direct contact with the CLP laboratory is not permitted. The SMO also administers the award of RAS and SAS contracts which are for three months duration. The time of contracts is short so that other laboratories can gain entrance into the program and so analyses are not biased by one laboratory.

The next section describes the quality of the analytical results obtained through the CLP at the Clear Creek/Central City (CC/CC) Superfund Site in Colorado. Conclusions concerning the data and recommendations for obtaining better quality are discussed.

PROJECT DESCRIPTION

The CC/CC Superfund Site is located in the Central City-Idaho Springs mining district approximately 30 miles west of Denver, Colorado. Several mine drainages in the district (7) were included on the National Priorities List under the Comprehensive Environmental Response, Compensation, and Liability Act of 1980 (CERCLA or Superfund). The authors have been involved in the Remedial Investigation leading to a Feasibility Study report (8). One of the recommendations of this report led to the construction and testing of a pilot wetland at the Big Five Tunnel (9). The majority of water and soil samples were analyzed using the CLP. This paper is based on the results obtained from the inorganic portions of the CLP while conducting these studies. Recently, a study was published examining the organic analysis portion of the CLP (10).

DISCUSSION OF RESULTS

Overall Quality

Once data are received from the CLP, they are carefully reviewed through a formal process called data validation. This includes checking all raw data and adding qualifiers to each result. The qualifiers can range from rejected to estimated. Values are usually rejected based upon poor matrix spike recovery. Values can be qualified as estimates based on calibration problems, duplicates that exceeded control limits, holding time violations, or interferences. During the Remedial Investigation portion of the study, the data validation results in Table I were tabulated.

Table I. Data Validation from the CC/CC Remedial Investigation Program

| Quarter | Number of Analyses Reviewed | Percent Rejected | Percent Qualified |
|---------|------------------------------|------------------|-------------------|
| 1 | 2,700 | 0.0 | 11 |
| 2 | 8,100 | 1.4 | 9 |
| 3 | 5,800 | 0.6 | 13 |
| 4 | 5,100 | 5.0 | 18 |

The amount of rejected and qualified data varied greatly with a large increase in the fourth quarter. Samples from the first three quarters were analyzed by one laboratory, and the samples from the fourth quarter were analyzed by a different laboratory. Based upon the above results and experience at other sites, not all laboratories perform equally. Even though both laboratories were using identical methods, quality control procedures, and statements of work designed to obtain consistent results, the resultant quality of work varied greatly. Apparently, the procedures and associated quality control were not adequately transferred to the laboratory bench.

Problems with Specific Parameters

Although it appears in Table I that the percent of analyses rejected and qualified is low, the problems are concentrated within a few species. Both laboratories appeared to have difficulty analyzing these selected parameters. The most difficult parameters appeared to be selenium in water samples (70 to 100 percent qualified) and antimony in solid samples (46 to 75 percent qualified). Other difficult parameters include silver, tin, and thallium. These parameters were not important in terms of risk at the CC/CC site. If these elements are important at other sites, special, instead of routine, analytical procedures should be requested to obtain better results. For example, for tin, thallium, and silver, furnace atomic absorption maybe more accurate; and for antimony and selenium, cold vapor atomic absorption may be the method of choice.

During selected quarters, some important parameters were qualified. For example, 55 and 40 percent of the sulfate data were qualified in the third and fourth quarters due to calibration and spike recovery problems. Ninety percent of the arsenic in the soil data was qualified in the fourth quarter. Ten dissolved metals in water (Al, Sb, As, Be, Cd, Fe, Pb, Mn, Se, and Th) were rejected or qualified over 90 percent of the time in the fourth quarter. Even within the same laboratory, results for difficult parameters varied significantly from quarter to quarter.

Difficult Matrices

Samples of waters and soils from hazardous waste sites can be quite variable in chemical composition. However, due to the sample load handled by the CLP and the need to use standardized, legally defensible analytical techniques, the chosen methods are most applicable to typical analytes and matrices. Little flexibility in achieving optimum conditions is allowed when analyzing individual parameters in unusual samples (3). As an independent check on laboratory quality, several calculations were performed and compared to the criteria given in Table II. Results of the evaluation are provided in Table III.

Results of the stream samples are typically acceptable. However, samples of discharges had larger negative charge balances. This result, coupled with the high ratios of TDS to SC, indicates that some cationic species was present that was not measured or included in the charge balance. In Table III, the calculated TDS was done using a spreadsheet program. To improve the TDS calculation, the cation and anion species were calculated using MINTEQ and then the TDS calculation was made. For those waters where there was a large charge inbalance, the calculation of TDS using the MINTEQ species considerably improved the balance.

Table II. Acceptability Criteria for CC/CC Waters

| Test | Criteria |
|---|---|
| Charge Balance | < 10% |
| Measured Total Dissolved Solids (TDS)/ | |
|   Specific Conductance (SC) | 0.6 - 0.9 |
| Calculated TDS/SC | 0.6 - 0.9 |
| Field SC/Laboratory SC | 0.9 - 1.1 |
| Field pH/Laboratory pH | 0.9 - 1.1 |
| Calculated TDS/Measured TDS | 0.9 - 1.1 |

Table III. Results of Analytical Acceptability Checks in Selected Mine Drainages (first nine samples) and Stream Water (SW) Samples

| Samples | Chg. Bal.[a] | Cal. TDS | Cal.TDS Fld SC | Ms.TDS Lab SC | Cal.TDS Lab SC | Fld SC Lab SC | Fld pH Lab pH | Cal.TDS Ms.TDS |
|---|---|---|---|---|---|---|---|---|
| AR3-1[c] | +0.8 | 742.5 | 0.79 | 0.70 | 0.77 | 0.97 | 0.99 | 1.09 |
| BF1-1 | -12.1 | 2418.7 | 0.88 | 1.06 | 0.86 | 0.99 | 0.99 | 0.82[d] |
| NT2-1 | +4.9 | 618.0 | 0.73 | 0.79 | 0.73 | 1.00 | 0.94 | 0.92 |
| QH1-1 | -35.0[b] | 4376.8 | 1.12 | 1.46 | 1.12[b] | 1.00 | 0.91 | 0.77[b] |
| GG2-1 | -20.2 | 3152.8 | --- | 1.21 | 1.03 | --- | 0.94 | 0.85 |
| GG2-2 | -8.8 | 2061.0 | --- | 1.01 | 0.77 | --- | --- | 0.76 |
| GG13-1 | -15.6 | 905.2 | 0.78 | 0.87 | 0.73 | 0.94 | 1.12 | 0.84 |
| GG13-2 | -7.0 | 793.2 | 0.81 | 0.99 | 0.74 | 0.92 | 0.93 | 0.75 |
| GG13-3 | -21.5 | 1065.4 | 0.45 | 0.95 | 0.72 | 1.59 | 1.77 | 0.76 |
| SW1-1 | -1.2 | 78.1 | 0.65 | 0.73 | 0.76 | 1.16 | 0.98 | 1.04 |
| SW2-1 | -3.8 | 76.0 | 0.61 | 0.75 | 0.67 | 1.10 | 0.88 | 0.89 |
| SW3-1 | -7.7 | 140.7 | 0.95 | 0.73 | 0.66 | 0.70 | 1.05 | 0.91 |
| SW5-1 | -14.4 | 137.6 | 0.76 | 0.69 | 0.76 | 1.00 | 1.03 | 1.10 |
| SW11-1 | +3.0 | 61.9 | 0.79 | 0.78 | 0.80 | 1.00 | 0.93 | 1.03 |
| SW1-2 | +4.7 | 163.4 | 0.76 | 0.22[b] | 0.46 | 0.60 | 1.03 | 2.04[b] |
| SW2-2 | +3.2 | 175.7 | 0.83 | 0.45[b] | 0.78 | 0.95 | 0.99 | 1.76[b] |
| SW3-3 | +0.4 | 220.6 | 0.70 | 0.89 | 0.98 | 1.40 | 1.04 | 1.10[b] |
| SW5-2 | -6.55 | 224.8 | 0.64 | 0.69 | 0.64 | 1.01 | --- | 0.94[b] |
| SW11-2 | -1.9 | 75.1 | --- | 0.54 | 0.68 | --- | --- | 1.25[b] |
| SW29-2 | +1.6 | 139.9 | 0.77 | 0.51 | 0.71 | 0.92 | --- | 1.40[b] |

[a] Note: "+" sign denotes excess cations. "-" denotes excess anions.
[b] Lab TDS value and sulfate value are qualified as estimates.
[c] Number after hyphen gives quarter of sample.
[d] Lab TDS value is qualified as an estimate.
"---" = Not calculated (data not available).

In addition to the difficult matrix in acid mine drainage (low pH and high dissolved solids), effluent from the wetlands treatment cells contained high levels of dissolved organic compounds. These samples were impossible to filter through a 0.45 micrometer filter. Use of RAS analyses for these samples is not appropriate. In some cases, even a SAS cannot be specified without independent evaluation of appropriate methods.

Results Interpretations

Use of RAS procedures results in specific values using specific procedures. The user of the data must know the procedures used to be able to give the correct interpretation. An example of this is the digestion procedure for water used by the CLP (4). Five possible digestion procedures for waters are available or proposed (4, 11, 12). These are listed in Table IV as Dissolved, Acid Soluble, Total Recoverable, EPA Total, and CLP Total.

Table IV. Concentrations of Metals in a Whole Water Sample Taken from Chase Gulch near Central City, CO. Sample Number MHG-128 from reference (11)

| Digestion Method | Cd | Metal Conc. in microgram/l | | | |
|---|---|---|---|---|---|
| | | Fe | Pb | Mn | Na |
| Dissolved | 5.0 | 150 | 10 | 820 | 5200 |
| Acid Soluble | 9.5 | 2500 | 270 | 1300 | 4600 |
| Tot. Recov. | 9.0 | 43000 | 710 | 1300 | 6400 |
| EPA Total | 1.3 | 45000 | 100 | 1400 | 3600 |
| CLP Total | 11 | 51000 | 2000 | 1500 | 4200 |

The digestions in Table IV are roughly listed from the most mild (Dissolved) to the most vigorous (CLP Total). All digestions call for field preservation to a pH less than 2 with nitric acid. The Dissolved digestion calls for nothing else, the Acid Soluble digestion calls for filtering after 16 hours. Since this is a whole water, addition of a strong acid to the sample could dissolve some of the suspended solids. The more vigorous procedures call for digestions with various amounts of hydrochloric/nitric acids and then filtering. Comparison of results using the various procedures is extremely important because effluent standards and aquatic criteria are usually based on values obtained by Acid Soluble or Total Recoverable digestions.

Results for selected parameters for a stream water with moderate total suspended solids are presented in Table IV. Typically, the concentrations increase as the vigor of the digestion increases. This result is seen for iron and manganese. However, the results for cadmium and lead are unusually low for the EPA Total digestion, and the results for sodium are roughly the same for all digestions. A more extensive study of the results of different digestions (11) show that although the concentration results generally increase with the vigor of the digestion, surprising inconsistencies develop depending upon the element, the total suspended solids, and the location. Use of RAS analyses alone may lead to incorrect conclusions concerning comparison to stream or effluent standards.

### Field Quality Assurance/Quality Control (QA/QC) Checks

In addition to the QA/QC performed by the laboratory, an independent QA/QC program administered by the data users is essential in evalutating data quality. The "Field" QA/QC checks usually consist of travel blanks, field blanks, decontamination blanks, co-located samples, split samples, and blind standards. Split samples are most important for evaluating laboratory precision and blind standards are most important in assessing laboratory accuracy. The relative percent difference between split samples exceeded the 20 percent criteria at least once for Al, Sb, Be, Cd, Co, Cu, Fe, Mo, and Ag. Overall, few values exceeded the 20% criteria.

Results for the blind standards are presented in Table V. As shown, poor results were consistently reported for Ag, Cd, Fe, Hg, K, and Na. In addition, the results show large differences between quarters. The first and third quarter data are generally acceptable while the fourth quarter data is not acceptable. As previously stated, the first three quarter analyses were performed by one laboratory and another laboratory performed the fourth quarter analyses. As the results in the second quarter show, even within one laboratory, analyses can vary.

Table V. Results of Blind Water Standards Analyzed by
the CLP During the Clear Creek - Central City RI/FS

| Analyte | Target Range | Quarter I | II | III | IV | IV |
|---|---|---|---|---|---|---|
| Ca (mg/l) | 24-34 | * | * | - | 0.28 | 66.5 |
| K (mg/l) | 13-19 | <0.25 | <2.17 | - | 0.708 | 1.77 |
| Na (mg/l) | 50-58 | 1.73 | 0.782 | - | 0.859 | 2.6 |
| As (ug/l) | 36-46 | * | 35 | * | * | 35 |
| Ba (ug/l) | 292-364 | * | * | * | 130 | 770 |
| Cd (ug/l) | 27-41 | * | 48 | * | 13 | 92.5 |
| Cr (ug/l) | 44-56 | * | * | 19 | 130 | |
| Fe (ug/l) | 121-159 | * | 162 | * | 183 | 662 |
| Pb (ug/l) | 85-121 | * | * | * | 39 | 300 |
| Mn (ug/l) | 96-124 | * | * | * | 45 | 290 |
| Hg (ug/l) | 1.1-1.9 | * | 0.5 | * | 0.23 | 2.25 |
| Se (ug/l) | 26-46 | * | * | * | * | * |
| Ag (ug/l) | 84-110 | * | 36 | * | 41 | 253 |

*analytical result within target range. "-" = not analyzed

All results from both laboratories for the blind standards were not as accurate as the laboratories report for known performance standards and internal QA/QC samples. That is, if the analyst knows that performance standards are being analyzed, apparently the person is more careful; or perhaps the supervisor will perform the critical analyses. The use of blind standards is essential in evaluating laboratory quality. Based on such analyses, much of the fourth quarter data on the CC/CC project was used with caution or not used at all for modeling and risk assessment.

### Modeling

Inorganic RAS includes 23 metals plus cyanide. A SAS request is required for the analysis of major anions such as sulfate, chloride, and bicarbonate. These parameters will be required in most thermodynamic-based aqueous models. As previously described, anion analyses such as sulfate were qualified a large percent of the time. Therefore, it is critical to verify the analytical results through methods described in the previous sections. In some cases, independent analyses should be performed for critical parameters such as sulfate and selected metals. Analyses for important parameters such as Fe(II) and Fe(III) will definitely have to be analyzed by the data user to insure accuracy. Due to the holding time, the CLP program would not be produce accurate results for such time sensitive analyses.

Examples of Data Use

In many cases, results of the RAS and SAS analyses serve to indicate the direction for more definitive research that is outside the CLP program. An example of this follows.

At the Big Five Pilot wetland in Idaho Springs, the ultimate objective is removal of the heavy metals from the mine drainage. The fate of sulfur in the system is important to this removal. The pilot system consists of three 18.6 m$^2$ cells and a typical metal mine drainage flows into each cell at the rate of 200 gal/min (13). One of the processes that can be quite important for metal removal is the reduction of sulfate to bisulfide catalyzed by sulfate-reducing bacteria. To monitor whether this process was working, analyses of total sulfur and the forms of sulfur in the substrates was requested by a SAS designating ASTM method D-2492 (14). The results of the requested analyses are given in Table VI.

Table VI. Forms of Sulfur in the Initial Substrate Materials Used in the Big Five Tunnel Pilot Wetland and in the Substrates After Two Months of Operation

| Sulfur Forms | Initial[a] Materials | Top Front | Top Back | Bottom Front | Bottom Back |
|---|---|---|---|---|---|
| | | CELL A | | CELL A | |
| Sulfate | 0.03 | 0.42 | 0.56 | 0.28 | 0.25 |
| Pyritic | 0.02 | 0.10 | 0.21 | 0.09 | 0.09 |
| Organic | 0.27 | 0.07 | 0.03 | 0.05 | 0.05 |
| Total | 0.34 | 0.59 | 0.79 | 0.41 | 0.40 |
| | | CELL B | | CELL B | |
| Sulfate | 0.27 | 0.34 | 0.24 | 0.33 | 0.26 |
| Pyritic | 0.13 | 0.23 | 0.33 | 0.25 | 0.30 |
| Organic | 0.16 | 0.02 | 0.02 | 0.05 | 0.05 |
| Total | 0.55 | 0.60 | 0.59 | 0.63 | 0.60 |
| | | CELL C | | CELL C | |
| Sulfate | 0.27 | 0.26 | 0.30 | 0.22 | 0.24 |
| Pyritic | 0.13 | 0.26 | 0.20 | 0.27 | 0.22 |
| Organic | 0.16 | 0.04 | 0.03 | 0.05 | 0.04 |
| Total | 0.55 | 0.56 | 0.54 | 0.54 | 0.50 |

[a]The substrate in Cell A is mushroom compost; the substrate in Cells B and C is a mixture of equal parts of peat/ aged steer manure/ and aged wood chips. The top is the first six inches, the bottom is at three feet, the front is near the inlet, and the back is near the outlet.

When the data for initial materials are compared with that for the substrates after the system had been operating from October 23, 1987 to January 16, 1988, interesting changes have apparently occurred. In all sampling locations in all three cells, there is a change from organic to inorganic sulfur forms. In Cell A, there is a major increase in the amount of total sulfur.

This shows in the results from the ASTM method as sulfate. In cells B and C, the total sulfur remains constant, however there is a significant increase in the pyritic sulfur as measured by the ASTM method.

The results indicate a significant mineralization of sulfur over a short time period. However, the ASTM method for forms of sulfur results in elemental sulfur and sulfur in the form of iron monosulfides being designated as organic fraction. In these substrates, these may be important forms. To obtain detailed analyses on the forms of sulfur, research is now being done using an analytical scheme that has been specially adapted form the work of others who have tried these studies (15, 16).

## SUMMARY

Even though all CLP laboratories follow the same procedures, large differences in the quality of work exist. Depending upon the objectives of the analyses, the use of one laboratory may be necessary. However, even the use of one laboratory does not necessarily assure consistent results. Differences apparently exist between individual analysts. In many cases, the QA/QC plans are not adequately transferred to the laboratory bench. Under a routine analytical schedule, selected parameters yield poor QA/QC results independent of the laboratory. Special analytical program requests should be considered if such parameters are critical.

The CLP is best suited for routine matrices. If complex matrices are anticipated, a SAS request or independent analyses outside of the CLP should be considered.

The data user must know the CLP procedures before accurate interpretations can be made. Differences exist between "CLP Total" values and regulatory criteria values such as "Total Recoverable".

Use of an independent QA/QC program is mandatory to evaluate laboratory quality. Use of blind standards and splits are essential in assessing accuracy and precision. Independent analyses outside the CLP may be necessary to obtain appropriate data for modeling and to more definitively evaluate selected areas.

## ACKNOWLEDGMENTS AND DISCLAIMER

The authors thank all the EPA personnel with whom they have worked over the years; their help has been invaluable. However, this paper is based on the personal observations of the authors and does not necessarily reflect the policies of their employers or the U.S. Environmental Protection Agency.

## LITERATURE CITED

1. Almich, B.P.; Budde, W. L.; Shobe, W. R. Environ. Sci. Technol. 1986, 20, 16-21.
2. Laidlaw, R.H. In Quality Control in Remedial Site Investigations: Hazardous and Industrial Solid Waste Testing, Fifth Volume; ASTM STP 925; Perket, C. L., Ed.; American Society for Testing and Materials: Philadelphia, 1986; pp 67-73.
3. Homer, D.H.; Dirgo, J. A.; Ellis III, H. V.; Morton, E. S. Proc. of 7th National Conference on Management of Uncontrolled Hazardous Waste Sites, 1986, pp 213-216.
4. White, D.K.; In Quality Control in Remedial Site Investigations: Hazardous and Industrial Solid Waste Testing, Fifth Volume; ASTM STP 925; Perket, C. L., Ed.; American Society for Testing and Materials: Philadelphia, 1986; pp 103-111.
5. User's Guide to the Contract Laboratory Program, U. S. Environmental Protection Agency, U.S. Government Printing Office: Washington, DC, 1987.
6. Kovell, S.P. In Quality Control in Remedial Site Investigations: Hazardous and Industrial Solid Waste Testing, Fifth Volume; ASTM STP 925; Perket, C. L., Ed.; American Society for Testing and Materials: Philadelphia, 1986; pp 74-84.
7. Wildeman, T.R.; Cain, D.; Ramirez, A. J. In Water Resources Problems Related to Mining; Proc. No. 18, American Water Resources Association, 1974, pp 219-229.

8. Camp, Dresser and McKee, Inc.; "Final Draft Remedial Investigation Report"; U.S. EPA Contract No. 68-01-6939, 1987.
9. Camp, Dresser and McKee, Inc.; "Draft Feasibility Study Report", U.S. EPA Contract No. 68-01-6939, 1987.
10. Swallow, K.C.; Shifrin, N. S.; Doherty, P. J. Environ. Sci. Technol. 1988, 22, 136-142.
11. Camp, Dresser and McKee, Inc.; "Draft Remedial Investigation Addendum", U.S. EPA Contract No. 68-01-6939, 1987.
12. Methods for Chemical Analysis of Water and Wastewater, EPA-600/4-79-020, U. S. Environmental Protection Agency, U. S. Government Printing Office: Washington, DC, 1979.
13. Howard, E.A.; Emerick, J. C.; Wildeman, T. R. Proc. of the International Conference on Constructed Wetlands for Wastewater Treatment; Ann Arbor Press: Ann Arbor, MI, 1989; pp 761-764.
14. "Standard Test Method for Forms of Sulfur in Coal," Designation: D2492-84; In Annual Book of ASTM Standards, Vol. 5, American Society for Testing and Materials, Philadelphia, 1988.
15. Tuttle, M.L.; Goldhaber, M. B.; Williamson, D. L. Talanta 1985, 33, 953-961.
16. Wieder, R.K.; Lang, G. E.; Granus, V. A. Limnol. Oceanogr. 1985, 30, 1109-1115.

RECEIVED September 8, 1989

# Chapter 26

# Expert Systems To Support Geochemical Modeling

**F. J. Pearson, Jr.[1], B. Skytte Jensen[2], and Andreas Haug[3]**

[1]Ground Water Geochemistry, 10700 Richmond Avenue, Suite 263, Houston, TX 77042
[2]Risø National Laboratory, DK–4000 Roskilde, Denmark
[3]Intera Technologies Inc., 6850 Austin Center Boulevard, Austin, TX 78731

>   Expert judgments on the reliability of water analyses, appropriateness of minerals, suitability of modeling program, and adequacy of thermochemical data are required to successfully model the geochemistry of ground water. Expert systems are computer programs based on the concepts of artificial intelligence which are used to extract, structure and make accessible the knowledge of human experts in a particular field. We are constructing a prototype system to provide expert knowledge to those performing routine geochemical modeling of ground water. The goal of the system is to assess how well the analysis of a water sample represents the chemistry of the ground water from which it was taken. The system includes information required to make that assessment organized in a structure suitable for use with a commercial expert system shell program. It comprises three modules: A main module for data input, routine tests of analytical data, and control of program flow; A carbonate module to evaluate the amount and distribution of dissolved carbonate species; and a saturation index module to test the concentrations of other dissolved constituents against the modeled saturation indices of minerals containing them. Two enhancements to the prototype system are envisioned: coupling to a geochemical model so the system can itself test the effects of adjusting analytical data, and including a facility for the expert system to select minerals likely to influence the chemistry of water in aquifers of interest.

Many computer programs are available for the geochemical modeling of ground water. To use them successfully requires the application of considerable judgment guided by experience. Judgments are needed about the reliability of the water analysis used, the appropriateness of the minerals considered, the suitability of the program chosen for the modeling, and the adequacy of its supporting thermochemical data set.

EXPERT JUDGMENT IN GEOCHEMICAL MODELING

   A ground-water analysis must pass two broad tests before it can be considered reliable. First, is the water analysis itself correct. Analytical data are commonly evaluated using ionic charge balances, and comparisons between calculated sums of analyzed dissolved solids, and measured residues on evaporation. The history of the analytical laboratory's performance with standard and replicate samples is also an important consideration. Second, does the water analyzed adequately represent water as it exists in the water-bearing unit of interest.

The second question can often be answered if the sample can be placed in a spatial and (or) temporal context. For example, differences among samples taken from the same point at different times could indicate either changing ground-water conditions or shortcomings in the sampling process. Likewise, if a sample is inconsistent with regional chemical trends, it may be a poor sample.

The expert system discussed here is intended to assist in judging the quality of single samples which cannot be evaluated as members of sample groups in space or time. Evaluating such samples may require considerable judgment. Some cases are straightforward. For example, if a sample from a limestone appears to be grossly oversaturated with respect to calcite, the water analyzed probably does not accurately represent formation water. On the other hand, oversaturation with respect to quartz in a low temperature water would not necessarily indicate a poor sample or an error in the silica analysis. If such a water were oversaturated with respect to amorphous $SiO_2$, however, the silica analysis would certainly be questionable. This type of consideration leads to a second major type of judgment.

The databases accompanying geochemical codes commonly include a very large number of minerals and it is left to the user to decide whether the saturation indices calculated for each of them are relevant to the problem at hand. Judgment is required to select minerals which are appropriate for modeling the water of interest. Before choosing a given mineral it is necessary to decide whether it is present or likely to be present in the water-bearing formation, whether it is likely to be reactive at the formation temperature during the residence time of the water of interest, and whether the thermodynamic data being used for the mineral are suitable for the problem at hand.

Analytical reliability and the choice of reactive minerals are factors explicitly evaluated by almost anyone doing geochemical modeling, although the depth of the evaluation is likely to vary with the experience of the modeler. There is another group of questions which are also important but which are rarely explicitly considered even by experienced modelers. These concern the appropriateness of the program chosen to perform the modeling and of the database of thermodynamic constants and other values required to support the program.

One concern about the modeling program itself is whether its method for calculating activity coefficients of aqueous species is suitable to the water of interest. For example, one should not try to make calculations on a brine using a model which employs only Davies' equation for activity coefficients. Another concern is whether the program is capable of making reliable calculations at the temperature and pressure of the sample. Few of the codes generally used for modeling ground waters include the effects of pressure, but the amount of error introduced by neglecting pressure in geochemical modeling is rarely addressed. A related question is the reliability at the sample temperature of the method used by the program to adjust thermochemical data for temperatures different from the reference temperature. For example, log K values adjusted using temperature-invariant reaction enthalpies and the van't Hoff equation are decreasingly reliable as temperatures deviate more than 20 or 30 degrees from the reference temperature.

The second question normally not explicitly addressed is whether the thermodynamic data employed are correct and appropriate for the problem at hand. Are the data for mineral and aqueous species internally consistent, or are they simply lists of values from many sources with no consideration given to whether it is correct to combine them. It is also important to question whether the identity of the aqueous species included are consistent with the model used in the program to calculate activity coefficients. The inclusion of a very large number of aqueous ion pairs may be inconsistent with the calculation of activity coefficients using Pitzer's expression, for example. It also important to evaluate whether the data are appropriate at the temperature of the sample of interest. Even though a program may be capable of evaluating a long polynomial expression for the temperature dependence of stability constants, if the value for only one temperature coefficient is available for a given mineral or aqueous species, that entity cannot be considered in waters with temperatures far removed from the reference temperature.

## EXPERT SYSTEMS

There are a number of disciplines in addition to geochemistry where judgment guided by experience is required to reach the desired objective. These include medicine, where a diagnosis of the cause of an illness may be based on a physician's subjective evaluation of symptoms as well as on the objective results of laboratory tests. Likewise, in economic geology, a decision of where to drill for mineral deposits depends on the skill of the exploration geologist as well as on data from cores and seismic lines, and on structural, sedimentologic or tectonic conceptual models.

Computer programs have been developed which embody knowledge distilled from experts in a given area and which, when given suitable data, reach the same conclusions as would human experts. Such programs are called expert systems and have been constructed for a number of purposes including medical diagnosis and mineral exploration. Hushon (1) has assembled information about expert systems for environmental problems, and Waterman (2) discusses the principles and construction of expert systems and describes systems used in a number of fields.

Expert systems fall within the broad computer discipline known as artificial intelligence. They have two essential components - a collection of information about their field of expertise, and a structure for applying that information to the problem at hand. The first component is often referred to as the knowledge or rule base of the system and the second as the inference engine.

A number of computer languages have been written for artificial intelligence applications (2). Expert systems can be constructed in these languages or in any other flexible computer language such as Pascal or C. This system is intended for use on personal computers operating under PC- or MS-DOS. Therefore, the Prolog artificial intelligence language, an inexpensive version of which is available for such systems, was first considered. To use this or any other language for an expert system application requires the programming of an inference engine as well as the development of a knowledge base. Pre-programmed inference engines and shell programs are also available for expert system construction. We chose to use one of those for our initial system so we could focus on the knowledge base component and save the time needed to program an inference engine. The resulting prototype system runs slowly, but is adequate for testing knowledge base structures.

## GEOCHEMICAL EXPERT SYSTEM

Expert systems tend to be very large programs. The present interest is to explore the application of such techniques to the problems of geochemical modeling rather than to develop a full-fledged expert system. Thus, the scope of the present prototype system has been restricted by assuming that both the program chosen for the modeling and its supporting database are correct. An additional simplifying assumption is that all waters to be examined come from carbonate aquifers.

The goal of the present geochemical expert system is to assess whether the chemical analysis of a ground water reliably represents the chemistry of the water in its host formation, and if it does not, to suggest adjustments to the analysis to make it more reliable. An analysis examined by the expert system and modified following its suggestions should be a good description of the ground water in its source formation.

The system associates a quality variable $q\_x$ with each parameter, x, of the analysis. Each quality variable is assigned an initial value of zero. When there is evidence to support or contradict the reliability of the parameter value, the quality variable is incremented or decremented. Consider the calcium concentration in an analysis. Initially its quality variable, **q_calcium**, is set to zero. Assume that the carbonate species concentrations have been shown to be reliable but that the geochemical model shows calcite to be oversaturated by several tenths of a unit. The sample is of a limestone ground water, and so should be calcite saturated. Assuming that the program and thermochemical data used are correct, the calcite oversaturation can be interpreted as indicating an erroneous calcium analysis. **q_calcium** is therefore decremented because of this evidence that the calcium value is incorrect. The overall quality of the analysis is based on the values of the quality variables of its individual constituents.

Data on samples to be examined are given in a table which is read by the expert system program. The table includes sample identification, analytical data, and the results of simple calculations such as for charge balance and the sum of analyzed dissolved constituents. It also includes mineral saturation indices and data on the carbonate system calculated by the geochemical modeling program.

The expert system consists of three modules. The first module, the main module, allows the user to choose the data file of interest and to modify any of the values in it. The module then checks the charge balance and for consistency between the calculated sum of dissolved constituents and the analyzed residue on evaporation. It also checks for such gross input errors as negative concentration values. Should a charge imbalance or difference between sums and residues suggest that an important constituent is missing from the analysis, the program points out the identity of any dissolved constituent which has a missing or zero concentration in the table. The main module then directs program flow to the succeeding two modules.

The second module, the carbonate module, examines data on the concentration and distribution of dissolved carbonate species including the results of geochemical modeling. A variety of carbonate data are available for many analyses. These provide redundancies useful to evaluate the carbonate system but which lead to a large number of permutations and combinations of data by which the evaluations can be made. The complexity of the carbonate system makes it particularly suitable for analysis using the inferential techniques typical of expert systems. Thus, the carbonate module is discussed in detail in the following section.

The chemistry of dissolved carbonate is often disturbed during the sampling of a ground water, by $CO_2$ loss or uptake, for example (3). If the carbonate module detects that such a change may have occurred, it suggests how the user might proceed to correct it. This usually involves an adjustment of the pH followed by rerunning the geochemical model. This step is a logical place for an automatic interface between the expert system and a geochemical modeling program. As discussed below, such coupling is an enhancement which would be included in a more complete geochemical expert system.

The third module, the saturation index module, examines the saturation indices calculated by the geochemical modeling program for consistency with the mineralogy of the water-bearing formation and with the constraints of the phase rule. Because the present system is restricted to water-bearing formations of simple mineralogy, it examines only a few minerals, all of which react rapidly enough that they are unlikely to be oversaturated in a ground water. Thus, if a water is saturated with gypsum and oversaturated with celestite, the module concludes, first, that the sulfate concentration is probably correct and, second, that the strontium content is probably too high. Thus, it increments q_sulfate and decrements q_strontium.

The saturation index module would be the location for additional enhancements to the system to make it applicable to waters from formations of more complex mineralogy, to include uncertainty in the thermochemical data used, and perhaps to consider mineral reaction rates and water residence times. The direction of such enhancements are sketched below.

## CARBONATE MODULE: ILLUSTRATION OF EXPERT SYSTEM OPERATION

The dissolved carbonate system can be described by a number of parameters. Most geochemical models require the pH, which indicates the distribution among the carbonate species, and either the total dissolved carbonate content, the total alkalinity, or the concentrations of bicarbonate and carbonate to indicate the amount of carbonate present. From these data, models calculate the concentrations and thermodynamic activities of the various dissolved carbonate species. The total dissolved carbonate content is the sum of these concentrations. The activity of the carbonate ion is used in calculating the saturation indices of carbonate minerals.

If only the pH and one concentration value are available, no internal checks of the dissolved carbonate system are possible. The saturation indices of carbonate minerals are useful to check carbonate concentrations however. For example, if water from a limestone appears to be oversaturated with calcite, it is more likely that the activity of carbonate is in error than the activity of calcium, because accurate measurements of the parameters from which the carbonate activities are calculated are more difficult than the analysis of dissolved calcium.

In many ground-water investigations, more data are collected than the minimum required to describe the carbonate system. Measurements of dissolved $CO_2$ concentrations are often made as part of the suite of analyses for all dissolved gases or by base titrations in the field. Measurements of total dissolved carbonate concentrations made by precipitation of solid carbonate or evolution of $CO_2$ following acidification may also be available. With such redundant analytical data, the internal consistency of all the carbonate data can be tested.

The carbonate module in the present system compares values of the total dissolved carbonate concentrations from several sources to evaluate the internal consistency of data on the carbonate system. Three total carbonate concentrations may be available: **tdc1**, the sum of the analyzed concentrations of dissolved $CO_2$, bicarbonate and carbonate, **tdc2**, the analyzed total dissolved carbonate concentration, and **tdc3**, the value calculated from the analyzed pH and alkalinity by the geochemical modeling program. Differences among these three concentrations are used to decide the quality of the carbonate analytical data.

Figure 1 shows how the differences among the three total carbonate concentrations lead to conclusions about the quality of the analytical data on which the carbonate concentrations are based. The three columns at the left represent the arithmetic differences between the three total carbonate concentrations. These differences can be positive ("plus" in the table), negative ("minus" in the table), or they may be smaller than some uncertainty (10% is the value used in the prototype system), in which case they are considered "good".

The six columns on the right represent the conclusions which can be drawn about the analytical data from which the total carbonate concentrations were calculated. The entries in each column indicate whether the quality variable corresponding to the data item is to be incremented ("+"), decremented ("-"), or left unchanged ("- -") based on the differences found in the three left columns.

Two examples illustrate the meaning of Figure 1.

First, consider a sample for which an analyzed total dissolved carbonate value is not available. The first and third differences at the left of the table are not defined. The sum of the individual analyzed carbonate species is within 10% of the total carbonate content calculated by the geochemical model so the second difference is designated "good". This example corresponds to the second case in Figure 1. Agreement between the two total dissolved carbonate values suggests that both are correct so their quality variables, **q_tdc1** and **q_tdc3**, are incremented, as indicated by the "+" in the respective columns on the right side of the table. The total carbonate sum includes concentrations of bicarbonate and carbonate, which are usually derived from the measured alkalinity, as well as the analyzed dissolved $CO_2$ content, $CO_{2(aq)}$. Because the sum is taken to be correct, its components should also be correct and their quality variables are incremented. The modeled total carbonate is also taken to be correct which implies that the analytical data for pH and alkalinity from which it was calculated are also valid. Thus, the quality variables for these values are also incremented. When this case occurs, the expert system program displays the following message to the user: "**The calculated and modeled total dissolved carbonate concentrations agree within x%. This suggests that both the values are correct and further indicates that the pH and alkalinity and the dissolved carbon dioxide, bicarbonate, and carbonate concentrations are also correct**". In this display, x is the actual difference between the sum and the modeled dissolved carbonate concentrations.

As a second example, consider the ninth case in Figure 1. Here all three values for total dissolved carbonate are available. The sum of the individual concentrations agrees with the analyzed total dissolved carbonate value so their difference is "good". The modeled concentration is less than the other two however, so their differences with it are "plus". The two values which agree are again taken to be correct and their quality variables, **q_tdc1** and **q_tdc2** are incremented as shown on the right side of the table. The modeled total carbonate content, which disagrees with the other two values, is

| Differences | | | Conclusions | | | | | |
|---|---|---|---|---|---|---|---|---|
| Sum − Anal | Sum − Model | Anal − Model | Sum (tdc1) | Anal (tdc2) | Model (tdc3) | pH | Alkalinity | $CO_2$ (aq) |
| − | − | − | − | − | − | − | − | − |
| − | good | − | + | − | + | + | + | + |
| − | plus/minus | − | − | − | − | − | − | − |
| − | − | good | − | + | + | + | + | + |
| − | − | plus/minus | − | − | − | − | − | − |
| good | good | good | + | + | + | + | + | + |
| plus | good | plus/minus | + | − (low) | + | + | + | + |
| minus | good | plus/minus | + | − (high) | + | + | + | + |
| good | plus | plus | + | + | − | − (high) | + | + |
| good | minus | minus | + | + | − | − (low) | + | + |
| plus | plus | good | − | + | + | + | + | − (high) |
| minus | minus | good | − | + | + | + | + | − (low) |
| plus/minus | plus/minus | plus/minus | − | − | − | − | − | − |

Figure 1. Decision table to evaluate quality of carbonate analytical data from differences between total dissolved carbonate concentrations equal to the sum of analyzed carbonate species, as analyzed directly, and as modeled from pH and alkalinity.

considered to be incorrect and its quality variable, q_tdc3, is decremented as indicated by the "-" on the right side of the table. Because the sum of the analyzed constituents is correct, so probably are the concentrations of the constituents themselves. Thus, the quality variables for the alkalinity and for $CO_{2(aq)}$ are incremented. If the modeled total carbonate is low but the alkalinity is correct, the pH used in the modeling must be too high. The quality variable for the pH is therefore decremented and the message displayed for the user is: "The calculated and measured values for the total dissolved carbonate agree within x% but the modeled value is smaller. This suggests that the calculated and measured total dissolved carbonate concentrations are correct and further indicates that the concentrations used for dissolved carbon dioxide, bicarbonate, and carbonate are also correct. The low modeled value indicates that the pH used is too high". As before, x represents the actual difference between the two agreeing carbonate concentrations.

The programming of the decisions shown in Figure 1, along with the messages returning information to the user, is part of the knowledge base of the prototype expert system. To indicate the size of knowledge bases, note that the code for this rather simple decision table is about 300 lines long.

The second example illustrates the potential for coupling between expert system and geochemical modeling programs. In that example, it was shown that the analyzed pH value was probably high. This being the case, the concentration and thermodynamic activity of carbonate calculated by the modeling program would also be high and would lead to erroneously positive saturation indices for carbonate minerals. Water in carbonate aquifers is generally saturated with calcite. Given a fixed alkalinity value, a change in carbonate activity and hence a change in calcite saturation index, is directly proportional to a change in pH. Thus, an estimate of the correct pH can be made by subtracting the calcite saturation index from the analyzed pH.

Having produced a better pH value, a coupled expert system could call a geochemical modeling program and rerun it with the new pH. The newly modeled total dissolved carbonate values and carbonate mineral saturation indices could then be substituted for the old values in the water data set and the expert system rerun to test for improved consistency among the new water data.

Such a coupled expert system would be able not only to evaluate analytical data, but to adjust them to improve both their internal consistency and the accuracy with which they represent the chemistry of water in the formation. This evaluation and adjustment, if done at all, now requires manual calculation and geochemical modeling by a human expert. Manual calculations are likely to be slower and their results less consistent from expert to expert, or from time to time from a single expert, than results guided by an expert system program.

## SATURATION INDEX MODULE: EXTENDING THE EXPERT KNOWLEDGE BASE

The prototype saturation index module is oriented toward the overall system goal of evaluating water analyses and the extent to which they represent the chemistry of formation waters. The present module examines the saturation indices of a few minerals which are assumed to react rapidly enough that, if present in the formation, they should be in equilibrium with ground water. Positive saturation indices, indicating oversaturation, are interpreted as indicators of erroneously high concentrations of appropriate dissolved constituents. Saturation indices close to zero are interpreted as indicating correct reported concentrations. When the saturation index of a mineral is negative, the user is asked whether that mineral is present. If it is not, no conclusions can be drawn. If it is present, the concentrations of the appropriate dissolved constituents are flagged as possibly being low.

In practice, answers to questions about the extent to which given minerals influence the chemistry of specified ground waters require expert judgment. In addition to the simple test of mineral presence or absence used in the prototype saturation index module, a realistic evaluation would include such criteria as reaction rates, stability relative to other minerals of the same or closely similar composition, uncertainty in mineral stability constants due to compositional variation, temperature and its effect on the previous criteria, and the time likely to have been available for chemical interaction between the water and the rock from which it was sampled.

With the exception of the last, these criteria could be considered as properties of minerals themselves. If so, they could be included in compilations of mineral data such as those that contain the thermodynamic properties used by geochemical models to calculate mineral saturation indices.

There should be little difficulty in developing a program to manage such an enhanced mineral data set which would also be capable of transferring information to both the geochemical modeling programs and an expert system. A program which maintains internal consistency among a set of thermochemical data for aqueous species and minerals was written in Pascal as part of a predecessor effort to the expert system work described here, for example. This program already has the capability to write files of thermochemical data in the formats required by geochemical modeling programs and could readily be extended to include fields for categories describing properties of mineral reactivities. Such data could be written to a file accessible by the expert system program or the database could be read directly by an expert system written in Prolog for example.

The challenge of this enhancement to the knowledge base would come in selecting the categories to be used to describe the reaction behavior of minerals and the information to be contained in each category for individual minerals. To be generally useful, the categories and, particularly, the mineral information should represent a consensus of the opinions of experienced ground-water geochemists and modelers. Developing such a consensus would be difficult and time consuming. Thus, at least the first steps towards this type of enhancement of the knowledge base and the capabilities of geochemical expert systems will probably cover highly specific situations which can be explored with relatively well understood minerals.

The present saturation index module considers only minerals common in carbonate aquifers and assumes that water residence times are sufficient to assure mineral-water equilibria. The next step might consider geothermal systems of simple mineralogy. For this step the expert system might include various chemical geothermometers as indicators of analytical reliability. While high temperatures would promote the attainment of mineral-water equilibria, they would also add to the uncertainty in the thermochemical data used for modeling those equilibria.

A database including categories describing mineral reactivity would also support the development of an expert system with a goal slightly different from the present one. Geochemical modeling is often used to develop the composition of hypothetical waters in equilibrium with certain rock or minerals under specified temperature and pressure conditions. The information contained in such an enhanced mineral database would enable an expert system to evaluate a user's choice of minerals as controls on the chemistry of hypothetical ground waters.

## SUMMARY AND CONCLUSIONS

Expert judgment is needed for successful geochemical modeling. Expert system programs are used to structure such judgments in other scientific disciplines. The prototype of a geochemical expert system is being constructed. Its goal is to evaluate the quality of analyses of samples of ground water and to assess how well the samples represent conditions in their source aquifers. The present system includes a main module which controls program flow and performs tests on the internal consistency of the analyses themselves, a carbonate module which tests for internal inconsistencies among measured and modeled parameters describing the dissolved carbonate system, and a saturation index module which uses criteria of mineral equilibria to test analyses of other dissolved constituents.

The present expert system program discovers possible errors and inconsistencies in a ground-water analysis, but it is restricted to water from carbonate aquifers and it cannot directly interact with geochemical modeling programs. The next step in its development will be to couple the carbonate module to a geochemical model to enable the expert system to adjust an analysis for consistency among the parameters of the carbonate system. The expert system will then be applicable in practice to adjust analyses of carbonate ground waters which have been disturbed during sampling so that they better represent the chemistry of water in the source formation.

In principle, such an expert system could also consider the extent to which a wide variety of minerals would influence ground-water chemistry at various temperatures and reaction times, as well as the natural variability and experimental uncertainties associated with the thermochemical data used to model geochemical equilibrium. To develop a consensus in the geochemical community about the categories of information required to evaluate mineral reactivity and the properties appropriate for each of many minerals, would be a time consuming task. Thus, in the near future at least, such enhancements will probably be limited to consideration of tightly defined problems of simple mineralogy.

## ACKNOWLEDGMENTS

This work is supported in part by the Research Programme for the Management and Storage of Radioactive Waste of the Commission of the European Communities.

## LITERATURE CITED

1. Hushon, J. M. Environmental Science and Technology 1987, 21, 838-841.
2. Waterman, D. A. A Guide to Expert Systems; Addison-Wesley: Reading, Mass., 1986.
3. Pearson, F. J., Jr; Fisher, D. W.; Plummer, L. N. Geochim. Cosmochim. Acta 1978, 42, 1799-1808.

RECEIVED July 20, 1989

# Chapter 27

# Numerical Modeling of Platinum Eh Measurements by Using Heterogeneous Electron-Transfer Kinetics

**J. Houston Kempton[1], Ralph D. Lindberg[2], and Donald D. Runnells[2]**

[1]PTI Environmental Services, 1260 Baseline Road, Suite 102, Boulder, CO 80302
[2]Department of Geological Sciences, University of Colorado, Boulder, CO 80309-0250

> This research evaluates the measurement of the "master" Eh of solutions in terms of heterogeneous electron-transfer kinetics between aqueous species and the surface of a polished platinum electrode. A preliminary model is proposed in which the electrode/solution interface is assumed to behave as a fixed-value capacitor, and the rate of equilibration depends on the net current at the interface. Heterogeneous kinetics at bright platinum in 0.1 m KCl were measured for the redox couples Fe(III)/Fe(II), Fe(CN)$_6^{3-}$/Fe(CN)$_6^{4-}$, Se(VI)/Se(IV), and As(V)/As(III). Of the couples considered, only Fe(III)/Fe(II) at pH 3 and Fe(CN)$_6^{3-}$/Fe(CN)$_6^{4-}$ at pH 6.0 were capable of imposing a Nernstian potential on the platinum electrode.

Potentiometry is an electrochemical technique in which the electrical potential of an "inert" electrode is measured against that of a reference electrode while both are immersed in an aqueous solution. A problem in potentiometry is that the measured potential may be slow to achieve a steady value. This is especially common in attempts to measure the Eh of solutions that are poorly poised, as is the case with most natural waters, and it is not uncommon for measured redox potentials to drift for many hours (1, 2). The long equilibration times, together with published reports of large discrepancies between platinum Eh values and actual solution compositions (3), have led to a great deal of uncertainty and skepticism about the use of Eh measurements.

This paper attempts to model and define the conditions under which platinum Eh measurements are likely to reflect the true electrical potential of aqueous solutions. The double layer at the surface of the electrode is modeled as a fixed capacitor ($C_{dl}$), and the rate at which an electrode equilibrates with a solution (i.e. the rate at which $C_{dl}$ is charged) is assumed to be proportional to the electrical current at this interface. The current across the electrode/solution interface can be calculated from classical electrochemical theory, in which the current is linearly proportional to the concentration and electron-transfer rate constant of the aqueous species, and is exponentially proportional to the potential across the interface.

The phenomena involved in potentiometric measurement are shown graphically in Figure 1. This is a composite of three linear sweep voltammograms in three different solutions: 10$^{-4}$ m ferrocyanide, 10$^{-4}$ m ferricyanide, and a mixture 10$^{-4}$ m in each. In Figure 1, a potential has been applied across the electrode/solution interface and the resulting current measured. In potentiometry, a passive electrode is allowed to drift under the resulting current until its potential equilibrates with the solution, at which time the net current is zero. This potential is called the "rest potential" and is equal to the system Eh for a solution which is at homogeneous equilibrium. Figure 1 shows that the point of zero net current is in fact composed of positive and negative components (so called "partial currents"). At the rest potential, the value of the cathodic current, which equals the absolute value of the anodic current, is defined as the exchange current. Higher exchange currents result in more rapid equilibration and more stable potential.

For a solution not in homogeneous equilibrium, the rest potential will be a "mixed potential" (Figure 7.21 in Reference 4), and will lie between the electrochemical potentials of the most oxidized and the most reduced couples present. The actual value of a mixed potential will depend on the heterogeneous rate constants for the transfer of electrons between the various aqueous couples and the surface of the electrode.

In this study, heterogeneous electron-transfer kinetics were measured for the following: Se(VI)/Se(IV), As(V)/As(III), $Fe(CN)_6^{3-}/Fe(CN)_6^{4-}$, and Fe(III)/Fe(II). All experiments were done at pH 6.0 with the exception of the iron couple, which was done at pH 3.0. Using electron-transfer kinetic constants, aqueous diffusion coefficients, aqueous concentrations, starting potentials, and a constant double-layer capacitance model, values for the change of EMF as a function of time for a platinum electrode were calculated numerically. The result of this simulation was then compared to the observed potentiometric response for a solution of the same concentration.

THEORY

The following brief discussion reviews electrochemical topics which bear upon the numerical model of electrode drift. For additional details the reader is referred to the comprehensive text by Bard and Faulkner (5).

An electrochemical cell is a type of electrical circuit. As such, it may be modeled with an electrical analog circuit. The potentiometric cell can be considered to be an electrical potential applied to a capacitor and a resistor in series. The capacitor represents the interface between the electrode and the solution, the applied potential is the solution Eh, and the resistor represents the heterogeneous kinetics of the aqueous redox species. The term "heterogeneous kinetics" denotes electron transfer between different phases, in this case aqueous species and the noble-metal electrode. The time required for the capacitor to equilibrate with the applied potential depends on the size of the capacitor and the electrical current.

A model predicting electrode response with time must therefore consider the following: (1) the double-layer capacitance, (2) the concentration of electroactive species at the electrode surface (which in turn is affected by the diffusion coefficients), (3) the values of the formal potentials ($E^{0'}$), (4) the heterogeneous rate constants of the redox species (with respect to the electrode material and electrolyte composition), and (5) the electrical potential of the electrode itself.

The capacitance of the double layer, $C_{dl}$, may be described as follows:

$$C_{dl} = Q/E \tag{1}$$

where: $C_{dl}$ = capacitance (coul volt$^{-1}$, or farads)
Q = charge in coulombs
E = electrical potential (volts).

Equation 1 is a simplification because the double-layer capacitance has been shown to be a function of potential, solution composition, electrode material, electrode pretreatment, concentration of adsorbed species, and duration of contact between solution and electrode (6, 7).

The potential of the electrode at time t is:

$$E(t) = E_{t=0} + \int_0^t \frac{i(t) \, dt}{C_{dl}} \tag{2}$$

where: E(t) = potential of the electrode at time t
$E_{t=0}$ = potential of electrode at t=0
i(t) = electrical current at time t (amps)
$C_{dl}$ = capacitance (farads)
t = time (sec).

The current due to electrochemical reactions can be described by the current overpotential equation (Equation. 2.5 in Reference 5). For a single redox couple (such as Fe(II)/Fe(III)) this is:

$$i = nFAk^0 \{Co_{(o,t)}\text{Exp}((-\alpha nF/RT)(E-E^{0'})) - Cr_{(o,t)}\text{Exp}(((1-\alpha)nF/RT)(E-E^{0'}))\} \quad (3)$$

where: 
- $A$ = area of electrode ($cm^2$)
- $i$ = current (coul $sec^{-1}$)
- $n$ = electrons in reaction (coul $mol^{-1}$)
- $F$ = Faraday's constant (96,492 coul $equiv^{-1}$)
- $k^0$ = heterogeneous rate constant (cm $sec^{-1}$)
- $R$ = gas constant (8.314 joule $mol^{-1}$ $K^{-1}$)
- $E$ = potential of electrode (volts)
- $E^{0'}$ = formal potential of the couple (volts)
- $\alpha$ = electron-transfer coefficient (unitless)
- $Co_{(o,t)}$ = concentration of oxidized species at electrode surface (mol $cm^{-3}$)
- $Cr_{(o,t)}$ = concentration of reduced species at electrode surface (mol $cm^{-3}$)

The area of the electrode, A, in Equation 3 was measured in this study by means of chronocoulometry (8) in an aqueous ferricyanide solutions.

In the case of multiple redox reactions, Equation 3 can be extended to include the total current by summing up the partial currents for each redox couple. Equation 3 requires the concentration of each electroactive species at the electrode surface.

In a vigorously stirred solution, the concentration at the electrode surface can be assumed to be equal to the concentration in the bulk solution and the potential at any time can be calculated by numerically integrating Equation 3 over time and substituting it into Equation 2.

A more rigorous interpretation of Equation 3 for a quiescent solution can be done by considering the net diffusion of electroactive species due to electrochemical reactions at the electrode. A general solution to the diffusion equation with a planar source is given by Nicholson and Shain (9) and has been expanded on by Imbeaux and Saveant (10):

$$Co_{(o,t)} = Co^* - Ic(t)/nFAD_o^{1/2} - Ia(t)/nFAD_o^{1/2} \quad (4)$$

$$Cr_{(o,t)} = Cr^* + Ia(t)/nFAD_r^{1/2} + Ic(t)/nFAD_r^{1/2} \quad (5)$$

where: 
- $Co^*$ = bulk concentration of oxidized form (mol $cm^{-3}$)
- $Cr^*$ = bulk concentration of reduced form (mol $cm^{-3}$)
- $D_o$ = oxidized-form diffusion coefficient ($cm^2$ $sec^{-1}$)
- $D_r$ = reduced-form diffusion coefficient ($cm^2$ $sec^{-1}$)
- $Ic(t)$ and $Ia(t)$ (in couls $sec^{-1/2}$), defined below.

$Ic(t)$ and $Ia(t)$ above are defined by Imbeaux and Saveant (10)

$$Ic(t) = \frac{1}{(\pi)^{1/2}} \int_0^t [i_c(v)/(t-v)^{1/2}] \, dv \quad (6)$$

$$Ia(t) = \frac{1}{(\pi)^{1/2}} \int_0^t [i_a(v)/(t-v)^{1/2}] \, dv \quad (7)$$

where: 
- $v$ = variable of integration (sec)
- other symbols as defined before.

Equations 4 and 5 represent the change in concentrations due to consumption and production of reactants and products at the electrode. The difference in signs in Equations 4 and 5 is due the sign convention for currents: cathodic reactions have positive current, so the convolution integral for the cathodic current, Ic(t), is positive, and visa versa.

The relationship between the time and potential is then calculated by numerically integrating the current (11), Equation 3, until diffusional control of the dissolved species causes the potential to approach a constant value. Details of the procedure can be found in Kempton (12).

EXPERIMENTAL METHODS

Electrochemical experiments in this study were performed with a BAS-100 Electrochemical Analyzer (Bioanalytical Systems Inc., Lafayette, Indiana). For potentiometric experiments in calm solutions, platinum disk electrodes were used in the BAS-100 cell stand. For hydrodynamic experiments a Pine Instruments Inc. rotating-disk platinum electrode was interfaced with the BAS-100.

Electrode pretreatment is critical for consistent results, so all electrodes were prepared in an identical manner before each experiment. The platinum surface was polished on a felt pad with 0.05 micron alumina and water. The surface was then rinsed with deionized water, wiped dry with a tissue, and the experiment begun as quickly as possible.

All experiments were run in 0.1 m reagent grade KCl without further purification. The pH was adjusted in the experimental cell with small additions of reagent-grade 1.0 m HCl or 1.0 m NaOH. Except where specifically noted, oxygen was removed by bubbling the cell for 15 to 60 minutes, depending on the volume of the cell, with high-purity nitrogen. Polarographic experiments in this laboratory have shown that solutions purged using this approach contain less than 20 ppb dissolved oxygen. All experiments were thermostatically controlled at $25.0 \pm 0.5$ °C. Reference electrode were calibrated daily so that potentials could be related to the standard hydrogen electrode (SHE). In all experiments, rates were determined in terms of formal potentials and molal concentrations.

KINETIC MEASUREMENTS

A rotating platinum disk electrode was used for gathering data for the calculation of the kinetic constants. Tafel plots (5) were developed (Figure 2), from which the heterogeneous rate constants, $k^o$, and the electron-transfer coefficients, $\alpha$, were determined. In all determinations of kinetic parameters, a blank voltammogram, run under identical conditions, was subtracted from the data to remove current due to background reactions or charging of the double layer.

The rate of reaction was too slow to produce measurable current ($i_k$) for arsenic and selenium species within the stability range of water at platinum. For those two elements an upper limit was estimated for the value of the rate constant by solving the current overpotential equation (Equation 3) for $k^o$ with the assumption of $\alpha = 0.5$ and $i_k = 4 \times 10^{-6}$ amps cm$^{-2}$. This value of $i_k$ was chosen empirically from examination of the data; it represents the lowest current that could clearly be distinguished from background currents with the instruments used. The true value of $k^o$ must therefore be equal to or less than the value that is calculated from the minimum limiting current.

ELECTRODE AREA AND DIFFUSION COEFFICIENTS

Diffusion coefficients were taken from the literature or measured using chronocoulometry (5). The area of the electrode, A, was measured using $Fe(CN)_6^{3-}$ in a solution of 0.004 m $K_3Fe(CN)_6$ in 0.1 m KCl, for which the diffusion coefficient is given as $0.762 \pm 0.01 \times 10^{-5}$ cm$^2$ sec$^{-1}$ (8). It was found from six measurements to be 0.242 cm$^2 \pm 0.005$; this value was then used to determine diffusion coefficients for species of interest (from the integrated Cottrell equation, Equation 5.9.1 in Reference 5).

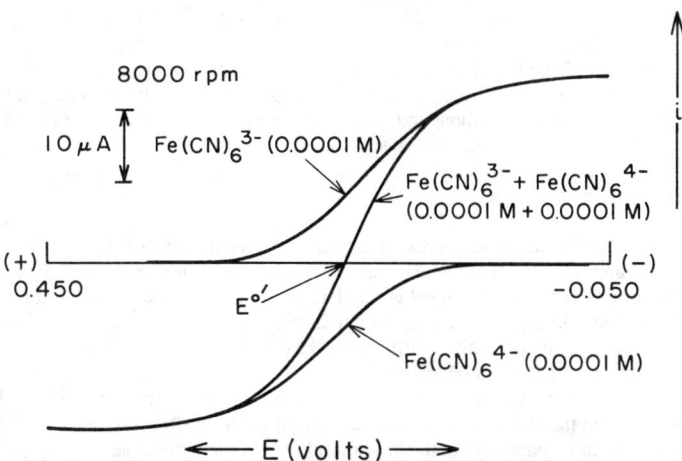

Figure 1. Composite of three independent, experimentally determined, linear sweep voltammograms. One for 0.0001 m $K_3Fe(CN)_6$, one for 0.0001 m $K_3Fe(CN)_6$, and one for a 1:1 mixture of 0.001 m of each. Temperature = 298 ± 1 K.

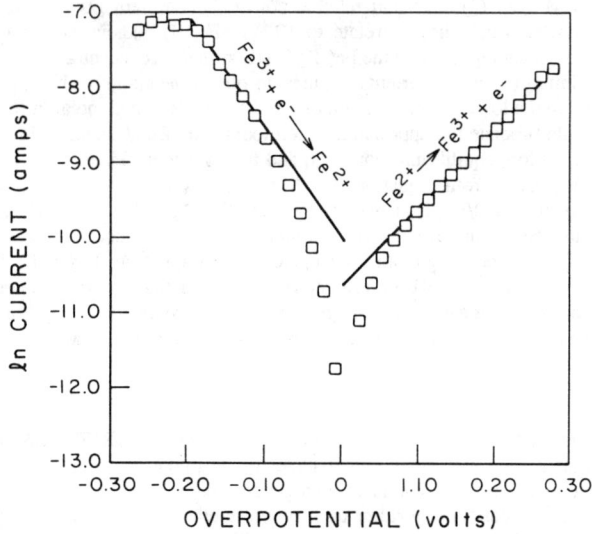

Figure 2. Tafel plots of oxidation and reduction of iron at pH 3 in 0.1 m KCl determined at a rotating platinum disk electrode. Currents are extrapolated values for an infinite rate of rotation. Temperature = 298 ± 1 K.

## CAPACITANCE OF THE DOUBLE LAYER

Cyclic voltammetry was used to measure the capacitance of the interface as a function of potential. The current is the product of the scan rate (volt sec$^{-1}$) times capacitance (coul volt$^{-1}$) (Equation 1.2.15 in Reference 5). The instantaneous capacitance (in farads, or coul volt$^{-1}$) at any potential is the current divided by the scan rate. Figure 3 is the current versus potential signal obtained from a cyclic voltammogram in 0.1 m KCl. A slope is superimposed on the charging current, probably the result of IR drop across the solution, and was corrected by using the average of the positive and the negative scan currents for calculating the capacitance.

## EXPERIMENTAL RESULTS

Figure 4 shows the calculated apparent capacitances versus potential (as Eh) from experiments using cyclic voltammetry. Measurements at high scan rates are more useful for determining capacitance because the background signal is smaller. Based on the determinations at scan rates of 50, 100, and 200 mV sec$^{-1}$ in Figure 4, a constant capacitance of $0.85 \times 10^{-4}$ farad cm$^{-2}$ was chosen for the modeling, which compares satisfactorily with $0.34 \times 10^{-4}$ farad cm$^{-2}$ given by Formaro and Trasatti (7) for platinum in 1.0 m perchloric acid.

The values and sources of diffusion data used in this study are given in Table I. The values of heterogeneous electron transfer rate constants are given in Table II. Rate constants were not measured above pH 3.0 for the iron couple due to the low solubility of ferric hydroxide.

A review of published exchange current densities for iron(14) produced an average $k^o$ of $0.03 \pm 0.05$ cm sec$^{-1}$ (n=5) for Fe(III)/Fe(II) in aqueous HCl, and $0.09 \pm 0.02$ cm sec$^{-1}$ (n=6) for Fe(CN)$_6^{3-}$/Fe(CN)$_6^{4-}$ in aqueous KCl. These are in reasonable agreement with the measured values in Table II. As explained earlier, for selenium and arsenic, no electrochemical reaction was detected between the upper two valence states, so the limiting maximum rate constants were approximated based on the minimum detectable currents.

Stumm and Morgan (4) state that reliable potential measurements require exchange currents above $10^{-7}$ amp cm$^{-2}$, which they correlate to $10^{-6}$ m Fe(III) and Fe(II), but they note that it is difficult to measure a meaningful Eh in the Fe(III)/Fe(II) system at concentrations below about $10^{-5}$ m. In this study, minimum exchange currents required to obtain meaningful Eh values were determined empirically by noting the lowest concentration of a redox couple, equal molal in reduced and oxidized species, which would produce the approximate thermodynamic Eh at platinum in 10 minutes. In this experiment, concentration, equilibrium potential, and heterogeneous kinetics are known, so exchange current can be calculated from Equation 3. Minimum exchange currents were determined for equimolal solutions of Fe(III)/Fe(II) at pH 3.0 and Fe(CN)$_6^{3-}$/Fe(CN)$_6^{4-}$ at pH 6.0. For the Fe(III)/Fe(II) couple a stable Nernstian response at a platinum electrode first occurred at about $3 \times 10^{-5}$ m each in Fe(II) and Fe(III), corresponding to an exchange current of approximately $1 \times 10^{-5}$ amp cm$^{-2}$. In the case of Fe(CN)$_6^{3-}$/Fe(CN)$_6^{4-}$ at pH 6 a reliable response was first observed at concentrations above about $7 \times 10^{-6}$ m, corresponding to an exchange current of approximately $8 \times 10^{-6}$ amp cm$^{-2}$. Both of these are above the minimum value of $10^{-7}$ amp given by with Stumm and Morgan (14).

## NUMERICAL MODEL

The computer code for the electrode equilibration model (EHDRIFT) was written in PASCAL for use on a microcomputer. The program calculates the rest potential which is the EMF value where the currents sum to zero. If the system is in homogeneous equilibrium the rest potential will represent the system Eh. The numerical algorithm uses Eulers method (11) to integrate Equation 2, which involves recalculation of the aqueous concentrations (Equations 4 and 5) at each time step. A full listing of the source code can be found in Kempton (12).

The numerical simulations of the drift of the electrode in a stirred solution, the integration of Equation 3, where concentrations at the electrode surface are assumed to be constant, were numerically sound and matched analytical solutions. The inclusion of diffusion terms (Equations 4 and 5),

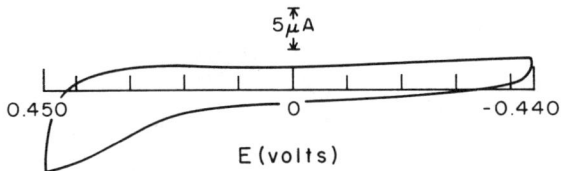

Figure 3. Cyclic voltammogram of 0.1 m KCl at pH 6.0. Platinum working electrode with saturated calomel reference electrode. Approximately constant capacitance between -0.40 and +0.20 volts. Temperature = 298 ± 1 K.

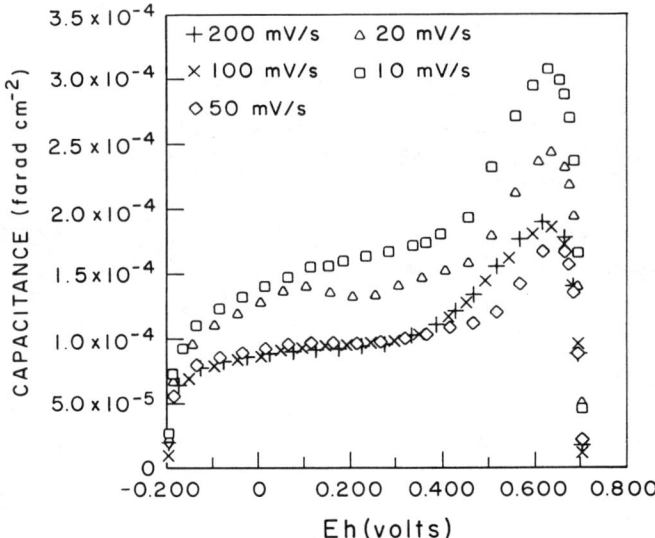

Figure 4. Apparent double-layer capacitance as a function of Eh, determined from cyclic voltammetry. Increase above about +0.45 volts is due to oxidation of platinum. Zero values at ends of scans are artifacts of the method and do not represent true capacitance values. Temperature = 298 ± 1 K.

**Table I.** Diffusion coefficients at 298 K in 0.1m KCl used in numerical modeling of electrochemical response of aqueous species of interest. Values of diffusion have been multiplied by $10^{+5}$ for presentation in this table

| Species | Matrix | Diffusion coefficient, D ($cm^2$/sec) | Source |
|---|---|---|---|
| $Fe^{2+}$ | pH=3 | 0.51 | This Study |
| $Fe^{3+}$ | pH=3 | 0.34 | This Study |
| $Fe(CN)_6^{4-}$ | pH=6 | 0.65 | Sawyer and Roberts, 1974 |
| $Fe(CN)_6^{3-}$ | pH=6 | 0.76 | Sawyer and Roberts, 1974 |

**Table II.** Heterogeneous electron-transfer kinetics at the platinum disk electrode at 298 K. All values determined from linear portions of Tafel plots at rates of rotation extrapolated to infinity. Reaction of Se and As too slow for observation

| Redox Couple | pH | **$k^o$ (cm $sec^{-1}$) | $\alpha$* |
|---|---|---|---|
| Fe(III)-->Fe(II) <br> Fe(II)-->Fe(III) | 3 <br> 3 | 1.0E-2 <br> 5.1E-3 | 0.37 <br> 0.73 |
| $Fe(CN)_6^{3-}$-->$Fe(CN)_6^{4-}$ <br> $Fe(CN)_6^{4-}$-->$Fe(CN)_6^{3-}$ | 6 <br> 6 | 1.5E-2 <br> 1.3E-2 | 0.24 <br> 0.59 |
| Se(VI)/Se(IV) | 6 | 2.3E-29 | 0.5* |
| As(V)/As(III) | 6 | 1.3E-23 | 0.5* |

\* $\alpha$ could not be measured by this method; assumed to be 0.5
\*\* Replicate determinations and published studies suggest that the precision of these $k^o$ determinations is probably not better than 75 to 100%.

however, introduced the following numerical limitations: 1) initial time steps on the order of $1 \times 10^{-6}$ seconds were required to maintain convergence, 2) the use of convolution integrals (Equations 6 and 7) meant that each new step required calculations from all previous steps, resulting in excessive computational time, and 3) increasing the time steps resulted in numerical instability. The net result was that the simulations of potential drift in which diffusion was considered would not converge if the electrode started more than about 200 mV from equilibrium.

Figure 5 shows the modeled and observed changes in potential at platinum over time for $1.3 \times 10^{-5}$ m Fe(III) along with $1.3 \times 10^{-5}$ m Fe(II) chloride solution at pH 3.0. Aqueous diffusion was included in this simulations. The data in Figure 5 indicate that the Eh-drift model predicts equilibration at a rate which is much faster than is observed. Comparable results were obtained for $1 \times 10^{-5}$ m potassium ferrocyanide and ferricyanide at pH 6 in a 0.1 m KCl solution. The discrepancy between the predicted and observed electrode response suggests that the proposed model omits some key, but unidentified, reactions. It is likely that the same non-modeled reactions which prevent Nernstian response below an exchange current of about $1 \times 10^{-5}$ amp cm$^{-2}$ are responsible for the poor fit in Figure 5. The most likely reactions involve the platinum electrode, adsorbed oxygen, and water (15, 16). The present model does not include corrosion reactions involving the platinum electrode. Nevertheless, the model is useful because it illustrates the relationship between heterogeneous kinetics and the measurement of Eh, and it confirms the fact that other surface phenomena occur on a platinum electrode which significantly limit the applicability of potentiometry.

RESULTS FOR INDIVIDUAL REDOX COUPLES

From the kinetic constants reported in Table II, it is possible to determine which redox couples have the ability to yield meaningful Eh values, as follows:

Fe(III)/Fe(II). A pH 3.0 chloride solution equimolar in the two iron species was found to yield a Nernstian response down to $6 \times 10^{-5}$ m total iron, corresponding to an exchange current of about $1 \times 10^{-5}$ amp cm$^{-2}$. It appears that At pH 6.0 Fe(III) is too insoluble to influence a platinum electrode. Potentiometric and kinetic experiments with platinum in Fe(II) solutions at pH 6.0 indicated the formation of a passivating layer, in agreement with Doyle (17).

Fe(CN)$_6^{3-}$/Fe(CN)$_6^{4-}$. Equimolal solutions of ferrocyanide and ferricyanide at pH 6.0 were found to impart Nernstian responses on platinum down to about $7 \times 10^{-6}$ m in each valence. This corresponds to an exchange current of $8 \times 10^{-6}$ amp cm$^{-2}$.

As(V)/As(III). Heterogeneous kinetics of the arsenic couple were immeasurably slow at platinum, and it should not be possible to achieve the exchange current required for a Nernstian response on platinum. Conversely, a measured Eh should not reveal anything about the aqueous speciation of As(V) and As(III).

Se(VI)/Se(IV). As with arsenic, the heterogeneous kinetic constants of oxidized selenium species at platinum are too small to yield a meaningful potentiometric response. Potentiometric measurements in a $1 \times 10^{-3}$ m solution of Se(VI)/Se(IV) at pH 6, as expected, showed no statistical deviation from blank 0.1 m KCl samples.

CONCLUSIONS

Aqueous species with fast electron-transfer kinetics at the surface of a platinum electrode are able to produce Nernstian potentials. However, a theoretical effort to model the rate of equilibration between platinum and a redox solution was only partially successful, probably due to reactions not included in the model. The shape of the theoretical Eh-time drift curve matched the observed shape for iron-bearing solutions, but the modeled rate of approach to equilibrium was much faster than the observed rate.

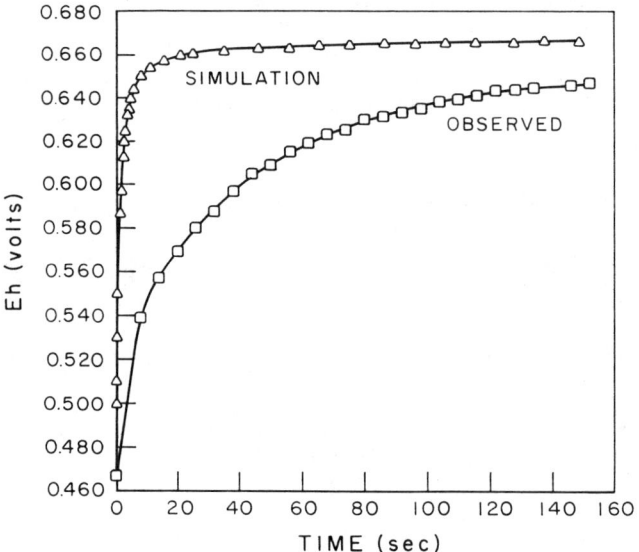

Figure 5. Numerical simulation and observed Eh drift curve for platinum electrode in $1.3 \times 10^{-5}$ m $FeCl_2$ and $1.3 \times 10^{-5}$ m $FeCl_3$ in 0.1 m KCl adjusted to pH 3 with HCl. Diffusion included in numerical simulation. Temperature = $298 \pm 1$ K.

In terms of natural environments, quantitative application of platinum Eh measurements to equilibrium chemical modeling is extremely limited. Of the four systems studied here, only the Fe(III)/Fe(II) couple in acidic solutions is common in natural waters and is capable of yielding a Nernstian response.

Considering the limitations of Eh measurement and the common occurrence of homogeneous disequilibrium (3), it would seem prudent to limit the use of Eh to either specific reactions or to qualitative descriptions when discussing natural systems. It has been suggested (18, 19, 20) that the measurement of dissolved gases such as $O_2$, $H_2S$, $NH_4$, $CH_4$, and $H_2$ may be a preferable method for characterizing the redox status of natural waters.

ACKNOWLEDGMENTS

Research grants from the Basalt Waste Isolation Project of Rockwell International Corporation (Hanford, Washington) and from the Electric Power Research Institute (Contract 8000-16, Dr. Ishwar Murarka) are gratefully acknowledged. Dr. Karl Koval, Department of Chemistry, University of Colorado, gave generously of his time and patience.

LITERATURE CITED

1. Back, W.; Barnes, I. In Equipment for field measurement of electrochemical potentials: U.S. Geol. Survey Research 1961, 1961, p. C366-C368.
2. Runnells, D.D.; Lindberg, R.D.; Kempton, J.K. Scientific Basis for Nuclear Waste Management X, Material Res. Soc. Sympos. Proc. 84, J.K Bates; W.B. Seefeldt, Eds., 1987, p. 723-733.
3. Lindberg, R.D.; Runnells, D.D. Science, 1984, 225, 925-927.
4. Stumm, W.; Morgan, J.J. Aquatic Chemistry, 2nd ed. Wiley-Interscience: New York, 1981; pp. 490-495.
5. Bard, A.J.; Faulkner, L.R. Electrochemical Methods: Fundamentals and Applications, Wiley and Sons: New York, 1980; Chapter 1, 3, and 4.
6. Bockris, J. O.; Reddy, A.K.N. Modern Elecotrochemistry, Vol. II, Plenum Press; New York, 1970; pp. 718-790.
7. Formaro, L.; Trasatti, S. Electrochimica Acta, 1967, 12, 1457-1469.
8. Sawyer, D.T.; Roberts, J.L. Experimental Electrochemistry for Chemists, John Wiley and Sons; New York, 1974, p 77.
9. Nicholson, R.S.; Shain, I. Anal. Chem. 1964, 36, 706-723.
10. Imbeaux, J.C.; Saveant, J.M. Jour. Electroanal. Chem. and Interfac. Electrochem. 1973, 44, 169-187.
11. Maron, J.M. Numerical Analysis: A Practical Approach; Macmillan Inc.: New York, 1982; pp. 377-386.
12. Kempton, J.H. Ms. Thesis, Univ. of Colorado, Boulder, 1987.
13. Heusler, K. E. In Encyclopedia of Electrochemistry of the Elements, Bard, A. J., Ed.; Marcel Dekker, Inc.: New York, 1982; pp. 228-381.
14. Morris, J.C.; Stumm, W. In Equilibrium Concepts in Natural Water Systems, Stumm, W., Ed.; Adv. in Chem. Series No. 67, American Chemical Society: Washington, DC, 1967; pp. 270-285.
15. Hoare, J.P. In The Electrochemistry of Oxygen; Interscience: New York, 1968; pp. 13-46.
16. Whitfield, M. Limnol. and Oceanogr; 1974, 19, 857-865.
17. Doyle, R.W.S. Amer. Jour. Sci., 1968, 266, 840-859.
18. Berner, R.A. Jour. Sed. Petrol., 1981, 51, 359-365.
19. Lindberg, R.D. Ph.D. Dissertation, Univ. of Colo., Boulder, 1983.
20. Stumm, W. Schweiz. Z. Hydrol., 1984, 46, 291-295.

RECEIVED July 20, 1989

# Chapter 28

# Use of Model-Generated $Fe^{3+}$ Ion Activities To Compute Eh and Ferric Oxyhydroxide Solubilities in Anaerobic Systems

Donald L. Macalady[1], Donald Langmuir[1], Timothy Grundl[1,3], and Alan Elzerman[2]

[1]Department of Chemistry and Geochemistry, Colorado School of Mines, Golden, CO 80401
[2]Department of Environmental Systems Engineering, Clemson University, Clemson, SC 29634

> Redox conditions, pH and ferric oxyhydroxide solubilities limit dissolved iron in most natural waters. Laboratory studies were performed in anaerobic systems in which Fe(III) activities were calculated using the computer model PHREEQE (46) with revised formation constants for iron hydroxy and chloride complexes. Lab and field Eh measurements using Pt or wax-impregnated-graphite (WIG) electrodes can provide nernstian potentials in the presence of measurable Fe(II) at pH's as high as 6.6. This allows calculation of $Fe^{3+}$ activities under a wide variety of conditions without difficult measurements of trace (< $10^{-6}$ M) Fe(III) concentrations. Also calculated are ferric oxyhydroxide solubilities, expressed as pQ = $-\log[Fe^{3+}][OH^-]^3$, which range in general from about 37 to 44. Laboratory studies in ferric/ferrous chloride solutions between pH 2 and 7 indicate an overall stoichiometry of 1/(3.06) +/- 0.15, and pQ values from about 38 to 41 for ferric oxyhydroxides after 12-200 hours. The above conclusions were supported by analyses and modeling of groundwater geochemical field data collected at Otis Air Force Base, MA and near Leadville, CO.

The measurement of dissolved Fe(III) and calculation of the activities of aqueous ferric species have challenged geochemists and others in their efforts to understand the chemistry of iron in natural waters (1-5). A new method for determining the activities of dissolved Fe(III) species in anaerobic systems containing Fe(II) is described in this paper (see also 6), along with conclusions about the behavior of redox electrodes in such systems and solubilities of associated ferric-oxyhydroxide solids. Redox conditions in dramatically different anaerobic aquifers, at Otis Air Force Base, MA and near Leadville, CO are also discussed.

The ultimate goal of this research is to develop a methodology for predicting the redox behavior of anaerobic aquifer systems, particularly toward introduced chemicals. The approach described herein is an effort to understand in detail the behavior of one couple,

[3]Current address: Department of Geosciences, University of Wisconsin—Milwaukee, Milwaukee, WI 53207

0097–6156/90/0416–0350$06.00/0
© 1990 American Chemical Society

[Fe(III)/Fe(II)]. Future investigations of additional couples will hopefully provide a more comprehensive understanding of the important parameters affecting redox transformations of anthropogenic chemicals in groundwaters.

## REDOX POTENTIAL MEASUREMENTS

The redox potential (ORP) for a single redox couple is related to the activities of species in solution through the Nernst equation (7) which has the limitations inherent in any thermodynamic relationship. If a measured ORP (corrected to the standard hydrogen electrode, SHE) corresponds to one computed from the experimental activities of a particular redox couple, electrode behavior is said to be nernstian with respect to that couple.

The complex and generally non-nernstian behavior of redox electrodes in natural systems has been discussed by many authors (8-11). Problems include mixed potentials (12-15), poisoning of platinum redox electrodes (16), lack of internal redox equilibrium (8,15,16), and lack of electrochemical equilibrium (17). Several reviews of the use of redox electrodes in geochemical studies have been published (18-20).

Redox electrode behavior in natural systems has been reviewed by Morris and Stumm (8), who conclude that few of the redox couples important in natural waters have sufficient electrode exchange currents to provide nernstian electrode response. Exceptions include the ferrous/ferric iron couple, perhaps $Mn^{2+}/MnO_2$ (21), and certain redox couples involving native sulfur, hydrogen sulfide and polysulfides (22-24). For the iron couple (8), Fe(III) and Fe(II) ion activities of $10^{-5}$ M or greater are said to be necessary for nernstian response, as is the absence of trace dissolved oxygen. Nernstian behavior of redox electrodes in natural systems dominated by iron has been reported or assumed by many authors (25-27, for example). Doyle (26) concluded that the apparently nernstian response of Pt electrodes above a pH of about 4.2 in systems containing ferrous iron is due to the presence of ferric oxyhydroxide solid phases, such as lepidocrocite and maghemite, coating the Pt electrode. This coating, not one or more of the hydrolysis species of $Fe^{3+}$, provides the exchange current necessary for stable and reproducible electrode potentials. The above conclusions have particular relevance to the findings reported herein, and are discussed in some detail below.

## FORMATION CONSTANTS OF IRON HYDROXIDE AND CHLORIDE COMPLEXES

Equilibrium and rate constants for the hydrolysis and chloride complexation of Fe(III) and Fe(II) ions are necessary in a detailed study of iron redox chemistry. Table I lists an internally consistent set of values for the relevant equilibrium constants. Their accuracy is discussed later in the context of a brief sensitivity analysis of the data. The rates of iron hydrolysis and chloride complexation reactions are also mentioned.

## SOLUBILITIES OF THE FERRIC OXYHYDROXIDES

The solubility products of stoichiometric ferric oxyhydroxides can be expressed in terms of ion activities as:

$$pK_{sp} = -\log[Fe^{3+}][OH^-]^3 \qquad (1)$$

TABLE I
Cumulative Formation Reactions and Constants at 25°C
and Zero Ionic Strength for Complexes Important
in This Study

| REACTION | $p^*\beta_n$ | Sources |
|---|---|---|
| 1) $Fe^{3+} + H_2O = Fe(OH)^{2+} + H^+$ | 2.19 | (28,29) |
| 2) $Fe^{3+} + 2H_2O = Fe(OH)_2^+ + 2H^+$ | 5.67 | (1) |
| 3) $Fe^{3+} + 3H_2O = Fe(OH)_3^0 + 3H^+$ | 12.56 | # |
| 4) $Fe^{3+} + 4H_2O = Fe(OH)_4^- + 4H^+$ | 21.6 | (30) |
| 5) $Fe^{2+} + Cl^- = FeCl^+$ | 0.51 | (29,31,32) |
| 6) $Fe^{3+} + Cl^- = FeCl^{2+}$ | 1.48 | (29,30) |
| 7) $Fe^{3+} + 2Cl^- = FeCl_2^+$ | 2.13 | (29,30) |

In $p^*\beta_n$, the n denotes the number of ligands in the complex formed.
#This value was computed from $^*\beta_3$(molal) = 2.4 x $10^{-14}$ measured in 0.68 m NaClO$_4$ solutions by Byrne and Kester (1). Baes and Mesmer (28) suggest $^*\beta_3$ (I=0) < $10^{-12}$, which is consistent with the value tabulated above.

Published values for $pK_{sp}$ based on field and laboratory solubility measurements range from about 37 for freshly precipitated, amorphous, colloidal-sized material, to about 43.5 for more crystalline phases (33-34). In low temperature aqueous systems, goethite ($\alpha$-FeOOH) is probably the most stable well-crystallized oxyhydroxide, with hematite ($\alpha$-Fe$_2$O$_3$) slightly less stable (20,35). Free energies for $Fe^{3+}$ and hematite given by Robie et al.(36) and by Wagman et al. (29) correspond to $pK_{sp}$ values for hematite of 43.9 and 43.8 respectively. More recent hematite data from Robinson et al.(37), combined with Wagman et al.'s(29) free energy for $Fe^{3+}$, yields a $pK_{sp}$ = 44.2. Assuming a free energy difference of +0.53 kcal/mole for 2 $\alpha$-FeOOH (goethite) = $\alpha$-Fe$_2$O$_3$ (hematite) + H$_2$O in water at 25°C (35), and free energies for hematite from Robie et al.(36), Wagman et al.(29), or Robinson et al. (37), respective $pK_{sp}$ values of 44.1, 44.0, and 44.3 result for goethite. These values are consistent with the highest $pK_{sp}$'s for ferric oxyhydroxides measured in aqueous systems (33,34).

Langmuir and Whittemore (33) point out that oxyhydroxide precipitates formed in the laboratory at low temperatures are usually mixtures of colloidal-sized amorphous material and crystalline phases, with individual crystal dimensions ranging from 50 to 2000 A. In such mixtures, the most soluble phase, usually amorphous, determines the pQ, the value calculated for $pK_{sp}$ for non-equilibrium phases. Although the crystalline phases grow at the expense of the amorphous material and each other, colloidal-sized crystalline phases can persist indefinitely in dilute waters low in dissolved Fe(II) or Fe(III).

Particle size effects are probably the chief cause of solubility variations of the crystalline oxyhydroxides (33). The surface free energy of amorphous material is about 100 ergs/cm$^2$, and that of crystalline goethite considerably higher (D. Langmuir, unpublished data). Thus, log K variation due to particle size effects, $\delta$log $K_s$, equals (9.30 x $10^{-11}$). $\epsilon$.S at 25°C, where $\epsilon$ is the average surface free energy of the solid oxyhydroxide in ergs/cm$^2$, and S is its surface area in cm$^2$/mol. Langmuir and Whittemore report that goethite crystal rods about 50 A thick and 300 A long show a pQ of 40.0 (33). S for a mole of such rods is 1.7 x $10^8$ cm$^2$. Assuming a $pK_{sp}$ of 44 for crystalline macroscopic goethite gives $\delta$log $K_s$ = 4, and suggests an average surface energy of 250 ergs/cm$^2$ for goethite, which supports the idea that particle size effects can account for most of the observed range of about 37 to 44 for pQ values in aqueous systems.

## STOICHIOMETRY OF FERRIC OXYHYDROXIDES

The discussion above has assumed a $Fe^{3+}/OH^-$ stoichiometry of 1/3 for the oxyhydroxides. There is convincing evidence, however, that the freshly precipitated oxyhydroxides formed in acid salt solutions, especially above 0.05M, are initially deficient in $OH^-$ (38,39). According to Murphy et al. (40), the initial precipitate upon neutralization is comprised of spherical polycations, 15 to 35 A in diameter. These are presumably the small ferric-hydroxy polymers described by Dousma and deBruyn (41) which form reversibly and rapidly from precursor monomers and dimers. The next step is the formation of large polymers (38,39). The hydroxide deficiency (+ charge) of the small and large polymers is compensated for by adsorbed and/or coprecipitated anions such as $Cl^-$, $NO_3^-$, and $ClO_4^-$. Several studies have shown that such anion-enriched polymers can persist at ambient temperatures, particularly at high anion concentrations in acid solution. Thus, Biedermann and Chow (42) report a two-month old precipitate with the composition $Fe(OH)_{2.7}Cl_{0.3}$ formed from 0.5 M NaCl solutions, and Fox (43) suggests the formula $Fe(OH)_{2.35}(NO_3)_{0.65}$ for a three-month old colloid formed from 0.05 M $Fe(NO_3)_3$ solutions. (See also (44)).

It seems certain that during the crystallization process, the Fe(III)/OH ratio decreases continuously, starting with the ferric-hydroxy monomers and dimers and continuing through small to large polymers, to crystalline oxyhydroxide solids. The crystal growth rate increases with increasing temperature, and is roughly proportional to Fe(II), and less so to Fe(III) present during crystallization (33,34,41). An x-ray or SEM identifiable crystalline solid may form in two hours at ambient temperature (26), or develop in four months (33), or take 15 years (45).

Murphy et al.(40) suggest that $Fe^{3+}$ hydrolysis first gives spherical polycations in all salt solutions, but the dominant anion governs which ferric oxyhydroxide will ultimately form. Sulfate favors goethite and minor lepidocrocite ($\gamma$-FeOOH), which slowly converts to goethite in solution over about one year (see Table II).

$FeCl_3$ solutions first produce akaganeite ($\beta$-FeOOH), which gradually converts to goethite over 15 years (45). Hematite forms in a few days from amorphous material precipitated at 90°C in $Fe(NO_3)_3$ solutions (39). At ambient temperatures, hydrolysis of $Fe(NO_3)_3$ or $Fe(ClO_4)_3$ solutions produces goethite and minor, unstable lepidocrocite (39,40). In natural fresh waters, which are dominantly sulfate-rich when acid and bicarbonate-rich at higher pH's, freshly precipitated oxyhydroxides are chiefly mixtures of amorphous material and goethite (33).

Fox (43) proposes that the colloidal ferric oxyhydroxide first precipitated in the laboratory and present in river waters has a Fe(III)/OH ratio of about 1/2.35. His suggested stoichiometry is based on the slope of a $-\log[Fe^{3+}]$ versus pH plot between pH 1.7 and 6.6. The published measurements which he argues support his stoichiometry involve a narrow pH range from 1.7 to 3.6. Taken separately, the published data are as well fit by a line of 1/3 slope as by one of 1/2.35 slope. Only when Fox's own data, measured in 0.05 M nitrate solutions, are included is the 1/2.35 stoichiometry suggested. Fox's conclusions are discussed below in comparison to experimental results presented here for both added solid oxyhydroxide phases and those precipitated from chloride (or mixed chloride-nitrate) solutions and aged less than 200 hours.

## EXPERIMENTAL APPROACH

The experimental protocol for this research involves several essential features. First, no redox transformations are allowed. Ferrous and ferric iron chlorides are introduced into strictly anaerobic aqueous systems, with careful exclusion of oxygen (6). Second, pH and the potentials at platinum and wax-impregnated graphite (WIG) electrodes are recorded as a function of time, beginning shortly after the addition of ferric chloride.

Table II
Negative Log (pQ) of the Apparent Ion-activity Product, $[Fe^{+3}][OH^-]^3$ for Some Ferric Oxyhydroxide Solids Present in Laboratory Systems and Natural Waters[1]

| pQ | Solid(s) present | Solution composition | Age | pH range | Remarks |
|---|---|---|---|---|---|
| 38.7 +/-0.6 | lepido-crocite | 0.2 M $H_3BO_4$ 0.05 M Cl, minor $SO_4$ | 2 hrs. | 4.5-7.2 | Ppt. formed on Pt electrode |
| 37.3-43.3 | amorph., goethite some lepido-crocite | 0.01 M $SO_4$ added $HCO_3$ at pH's near 7 | 3-302 days | 2.1-7.0 | Smallest pQ's for freshest ppt., highest for highest Fe(II) |
| 36.6-42.7 | unknown probably amorph.+ goethite | Ca-Mg$(HCO_3)_2$ groundwaters | fresh- > $10^3$ years | 4.7-7.8 | Oldest waters with least Fe(II) had lowest pQ's |
| 37.5 +/-0.1 | amorph. | seawater, 0.56 M NaCl | fresh | 3.3-9.4 | Fe(III) (aq) determined with $^{59}Fe$ |
| 39.0 +/-0.7 | amorph. + ? | $NO_3$,Cl,$SO_4$, and $ClO_4$; 0.01-3.0 M | fresh- 3 mo. | 1.7-6.6 | See text |

[1]References are 26,33,34,1,and 43, respectively.

A Ross reference cell is used for ORP measurements, with its potential calibrated after each reading using Light's solution (46). Platinum electrodes were carefully pretreated (6) and kept in an oxygen-free solution for at least ten hours before each run, since initially adsorbed oxygen at the Pt electrode surface can give erroneous potential measurements for up to two hours, even in anaerobic systems (6).

Third, the Fe(III) and Fe(II) activities in each system at each time are calculated by the procedure outlined below, and an ORP calculated using the Nernst equation ($Eh_c$). $Eh_c$ is then compared to $Eh_m$, the Eh recorded by the redox electrodes (corrected to SHE). The solubility product for the precipitated solid ferric hydroxide phase is also calculated for each data point using the pH and computed values of $Fe^{3+}$ or one of its hydrolysis species.

Variations of this basic experimental approach include the addition of various solid phases and dissolved species common in natural waters. The effect of these additional components on the calculated and observed ORP's is then noted. Sodium or potassium nitrate or chloride was also added in some experiments to adjust the ionic strength.

## CALCULATION OF IRON SPECIES IN THE SYSTEM

In the simple systems described above, added $FeCl_3$ first dissolves then rapidly hydrolyses. Ferric hydroxide solids become apparent within a minute of this dissolution. As the solid precipitates, pH drops. Total Fe(II), however, remains constant in solution, as the solubility of Fe(II) solids was never exceeded. This conclusion was verified by the measurement of Fe(II) concentrations(6). At each data point, the remaining Fe(III), in its free and complexed forms, was calculated using the computer program PHREEQE (47).

The calculation was essentially a charge balance (6). Input data included the pH, total Fe(II), total chloride and where relevant, added $Na^+$, $K^+$, and/or $NO_3^-$. Assuming the complexation constants given in Table I (and auxiliary constants in the PHREEQE data base) and using the extended Debye-Huckel equation for activity coefficients, the program added Fe(III) in a quantity necessary to achieve a charge balance. The output included activities and concentrations of each $Cl^-$ and $OH^-$ complex of Fe(III) and Fe(II).

This procedure proved viable for Fe(II) concentrations from 0.02 to 25 mM, initial Fe(III) between 0.1 and 25 mM, and ionic strengths (I's) from 0.005 to 0.150 M. Practical limitations in the technique resulted when Fe(III) levels were sufficiently low that precipitation produced a pH change so small as to magnify experimental uncertainties in concentration data to the point that they were of the same order as the charge imbalance. Ionic strengths below about 0.005 M posed no additional calculation problems, but measurements of pH, redox, and reference potentials became erratic (48). Fe(II) concentrations below 0.02 mM were not investigated.

Variations in the above protocol included the addition of other ions and/or solids. Included were some experiments in which no ferric chloride was added, and Fe(III) was added only as a solid phase such as hematite, goethite or goethite-coated quartz sand. In these experiments, discussed below, PHREEQE could not be used to calculate $[Fe^{3+}]$.

LABORATORY RESULTS AND DISCUSSION

The experiments involving only the ferric/ferrous redox couple and PHREEQE-generated activities spanned pH's from 1.7 to 3.5, ferrous ion activities from $10^{-2}$ to $10^{-5}$ M, ferric ion activities from $10^{-2.7}$ to $10^{-7.1}$ M, and I's between 0.001 and 0.175 M. With the exception of the experiments with I's less than about 0.005 M, all experiments yielded $Eh_c$'s which differed from $Eh_m$'s by less than 5%. Agreement was generally much better than this, and, regardless of initial concentrations, calculated and observed potentials usually agreed within 1% (which was typically +/- 0.005 V). Experiments at low ionic strength gave erratic results, which could generally be brought into agreement by the addition of an inert salt (usually KCl or $NaNO_3$) to raise the ionic strength above about 5 mM (6).

In all, over 100 experiments were conducted with varying Fe(III)/Fe(II) ratios for times up to 10 days. In a typical experiment, the pH and $Eh_m$ dropped steadily from the first data point (ca. 5 min. after addition of Fe(III)) to 3-100 hrs. The time required for pH's and redox potentials to stabilize varied depending upon the initial ferric iron. The highest added Fe(III) showed very little of the early, rapid changes in measured parameters, but exhibited slow changes for the longest times. Figures 1 and 2 illustrate results for two typical experiments. For experiments similar to those summarized in Figure 2, calculation of activity coefficients using the Davies Equation (49) instead of the extended Debye-Huckel Equation increased the differences between measured and calculated values by about 8 mV.

Several features of these results will be highlighted. First, the consistent agreement between observed and calculated values indicates that both the experimental method and the thermodynamic data base are valid. The $Eh_c$'s are not remarkably sensitive, however, to the value for any single equilibrium constant. Typically, a change of +/- 0.2 log units in any constant will change $Eh_c$ by no more than +/- 10 mV, which is approximately the limit of the accuracy of our data. This is somewhat surprising in that most of the Fe(III) is complexed in all of our runs (see Table III).

Second, $Eh_c$'s and $Eh_m$'s generally agree even for the first data point, typically taken about 5 min. after addition of $FeCl_3$. This means that the electrode and solution are in electrochemical equilibrium within this time frame. This, in turn, indicates that all hydrolysis and chloride complexation reactions are essentially complete within 5 min., even at very low iron activities, consistent with the limited literature on this subject (50,51).

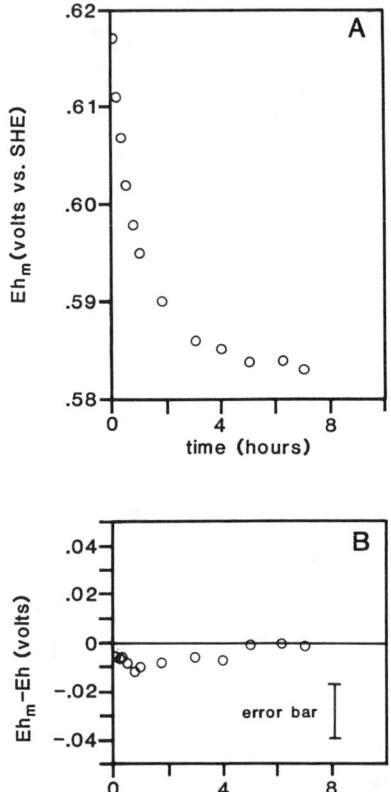

Figure 1.
A) Measured Eh ($Eh_m$) as a function of time for a solution initially containing 0.2 mM $FeCl_3$ and 2.0 mM $FeCl_2$. Final ionic strength is 0.007 M, final pH = 3.2.
B) $Eh_m$ minus $Eh_c$ as a function of time for the same solution. Error bar represents reproducibility. T = 25.0°C.

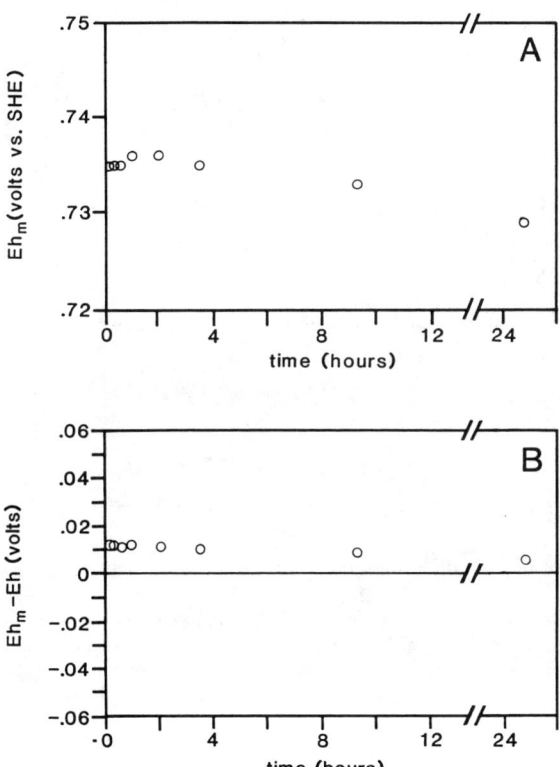

Figure 2.

A) $Eh_m$ as a function of time for a solution initially containing 5.0 mM each of $FeCl_3$ and $FeCl_2$. Final ionic strength is 0.036 M, final pH = 2.2.

B) $Eh_m$ minus $Eh_c$ vs. time for the same solution. Data reproducibility is within the size of the data symbols. T = 25.0°C.

Third, the $Eh_m$'s and related experimental data can be used to calculate the apparent solubility product for the ferric hydroxide solid which is present in the experimental systems. Results of such calculations, expressed in terms of the conditional (non-equilibrium) solubility product, pQ (= $-\log[Fe^{3+}][OH^-]^3$), are illustrated in Figures 3 and 4. Figure 4 shows that, at higher added ferric iron, the initial precipitation reaction is more rapid, and the observable change in pQ seen at lower concentrations (Figure 3) is essentially completed before the first data point (5 min.).

Table III
General Composition and Percent Activities of Iron Species for Some Typical Experimental Solutions

A. General Composition

| Exp. # | pH | Total Cl (molar) | Total Fe (molar) | Total Fe(III) (molar) |
|---|---|---|---|---|
| 43C | 1.80 | 0.123 | $1.44 \times 10^{-2}$ | $4.64 \times 10^{-3}$ |
| 40B | 2.27 | $2.44 \times 10^{-2}$ | $3.81 \times 10^{-3}$ | $1.14 \times 10^{-3}$ |
| 55C | 3.27 | $9.15 \times 10^{-3}$ | $1.47 \times 10^{-4}$ | $1.10 \times 10^{-5}$ |

B. Percent of Total Fe(II) and Fe(III) Activities

| Exp. # | Percent [Fe(II)] | | Percent [Fe(III)] | | | | |
|---|---|---|---|---|---|---|---|
| | $Fe^{2+}$ | $FeCl^+$ | $Fe^{3+}$ | $FeOH^{2+}$ | $Fe(OH)_2^+$ | $FeCl^{2+}$ | $Fe(Cl)_2^-$ |
| 43C | 79.4 | 20.6 | 21.8 | 9.5 | --- | 50.5 | 18.3 |
| 40B | 94.0 | 6.0 | 33.4 | 43.5 | --- | 21.1 | 2.0 |
| 55C | 97.4 | 2.6 | 6.8 | 75.7 | 16.2 | 1.4 | --- |

Thus, the rates of precipitation and "aging" of ferric oxyhydroxide solids are evident from each of our experiments. In all cases, solid is indicated at the first data point, and the pQ of the solid increases and levels off over a period of time depending upon the conditions. In 31 separate experiments, spanning the entire range of iron concentrations listed above, the average "final" value for pQ is 38.3 +/- 0.4. (All uncertainties expressed here and subsequently are +/- one standard deviation). As discussed above, these pQ values probably correspond to a mixture of amorphous ferric hydroxide and akaganeite present 12-200 hours after initial precipitation.

Fourth, the fact that $Eh_m$'s closely correspond to $Eh_c$'s over such a wide range of $[Fe^{3+}]$'s suggests that, in the presence of $10^{-5}$ M or greater Fe(II), $Eh_m$'s are valid indicators of $[Fe^{3+}]$ at lower levels than the limit of $10^{-5}$ M suggested by Morris and Stumm (8). The experiments described indicate nernstian behavior down to at least $10^{-7}$ M $Fe^{3+}$. The total Fe(III) activity in solution, including all hydrolysis and complexation species, is of course much higher than $[Fe^{3+}]$. (When $[Fe^{3+}]$ was $10^{-7}$ M, total dissolved Fe(III), was ca. $10^{-6}$ M). It is probable that all Fe(III) species are electroactive and contribute to the exchange current necessary for nernstian response. Thus, the experiments described indicate stable, nernstian response from Pt and WIG electrodes at activities about 10 times lower than previously indicated. Additions of cations such as $K^+$, $Na^+$, $Ca^{2+}$, and $Mn^{2+}$, or anions such as $NO_3^-$, $Cl^-$, and $SO_4^{2-}$, in concentrations up to 0.01 M did not alter the agreement between Eh's calculated for the iron couple and $Eh_m$'s, nor did such additions measurably affect the relationships between pQ's and time.

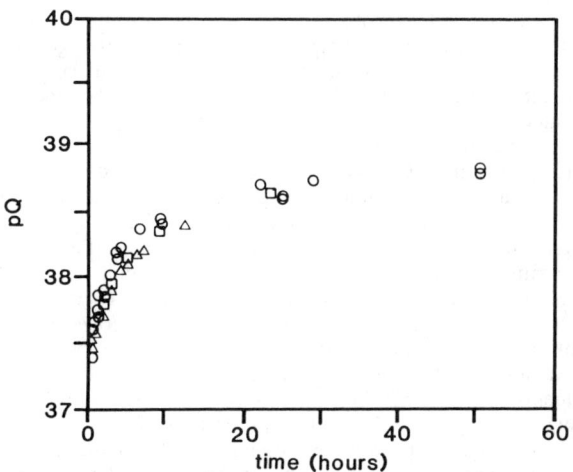

Figure 3.

The apparent solubility product (pQ) versus time for five separate ferric/ferrous chloride solutions. Each symbol represents a different solution. All have initial $FeCl_3$ = 5.0 mM. $FeCl_2$ concentrations vary from 0.02 to 5.0 mM. T = 25.0°C.

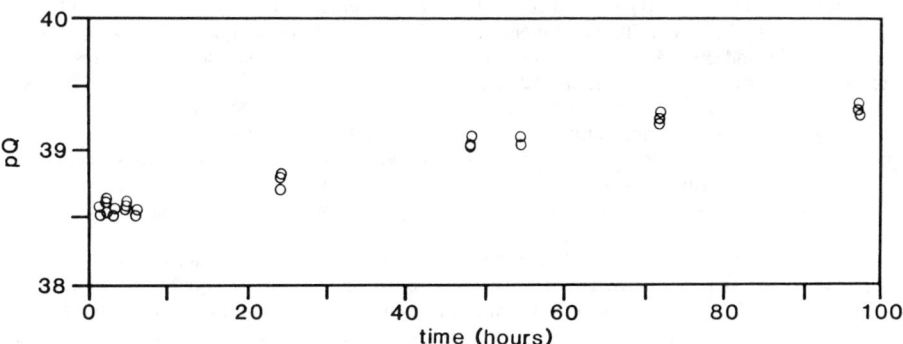

Figure 4.

The apparent solubility product (pQ) at 25.0°C versus time for three separate solutions initially containing 25.0 mM each of $FeCl_3$ and $FeCl_2$.

The final group of laboratory experiments demonstrated apparently nernstian response at even lower Fe(III) activities. These involved addition of solid phases to the experimental systems, and showed $Eh_m$'s consistent with the Nernst equation in anaerobic systems containing ferrous iron and solid ferric oxyhydroxides at pH's up to 6.4.

When quartz sand or goethite-coated sand (6) was added (0.1-1.0 g solid/100ml) to systems in which ferrous and ferric chlorides were the only solutes initially present, $Eh_m$'s drifted over several hours to values slightly higher than the $Eh_c$'s. Alternatively, the observed pH, added Fe(II), and the pQ value of 38.3 (from the previous 31 experiments), give $Eh_c$'s which agree within +/- 10 mV of $Eh_m$ values. Thus, the presence of these solid phases alters the validity of the assumptions inherent in the PHREEQE calculations, but the electrodes still respond in a nernstian manner. The pH, $[Fe^{2+}]$, and $[Fe^{3+}]$ values in these trials were ca. 2.9, 0.6 mM, and 0.055 mM, respectively.

In two similar experiments, illite clay was added instead of sand. Here, pH drifted upward to steady values of 6.4 and 6.1, presumably due to exchange of protons with cations on the clay surfaces. The $Eh_m$'s in the two trials leveled off at 18 and 93 mV for Pt and 38 and 120 mV for WIG electrodes. Calculated Eh's using PHREEQE-generated $[Fe^{3+}]$'s were off by over 400mV. However, again assuming a pQ value of 38.3 for the solid ferric phases present, the $Eh_c$'s calculated from the implied $[Fe^{3+}]$, the pH and the known ferrous activity, were 45 and 94 mV, respectively. The agreement to < 30 mV for both electrode materials is within the range of uncertainty produced by the experimental variation in the pQ values (+/- 0.4). The calculated $pFe^{3+}$ values were 15.5 and 14.6 respectively.

Other experiments in this series involved addition of 1-6 mM $FeCl_2$, but with Fe(III) added only through solid phases. Thus the approach to Fe(III) equilibrium was from undersaturation, in contrast to all previous experiments. Of the 21 trials, six involved goethite-coated sand (6) as the solid phase, six hematite (reagent grade $Fe_2O_3$), and nine (natural) crystalline goethite (ground to pass a 100 mesh screen). In some trials, the pH was adjusted by addition of NaOH or HCl. In each run, the final $Eh_m$ (as expressed by Pt or WIG electrodes) and pH were used to calculate an implied pQ for the solid phase. The results are summarized in Table IV.

The experiments involving goethite-coated sand (at pH's 3.39-6.16) yielded average pQ values (after 95-170 hrs) of 38.6 +/- 0.5 for both Pt and WIG electrodes. Those involving hematite (at pH's 3.06-6.01) showed final pQ's which averaged 40.1 +/- 1.1 for both Pt and WIG. The crystalline goethite experiments (at pH's 3.03-5.73) yielded average pQ's of 39.6 +/- 1.0 for Pt and 39.8 +/- 0.9 for WIG electrodes. Ferrous iron analyses in all cases agreed with ferrous chloride initially added.

These pQ values are consistent with the results of Langmuir and Whittemore (33) for small crystals of goethite and lepidocrocite. The pQ for synthetic- goethite-coated sand is probably indicative of some quite amorphous material on the sand. By itself, agreement of pQ values would hardly be worth mention. The notable result is the complete agreement of electrode-based calculations over a wide pH range for two very different electrode surfaces, independent of solid phase present.

The conclusions of Doyle (26) regarding the reason for apparently nernstian behavior of Pt electrodes at pH's above about 4.2 cannot be expected to apply equally to hydrophobic surfaces of WIG electrodes and relatively hydrophilic surfaces of Pt electrodes.

There are at least three alternate explanations. Perhaps the exchange current provided by aquo-ferric complexes produces a stable electrode response in the presence of millimolar Fe(II) at Fe(III) activities far below those previously believed. The lower limit of total aqueous Fe(III) in these experiments is about $10^{-9}$ M.

Another possibility is that colloidal ferric oxyhydroxide species can provide the necessary exchange current directly from the aqueous phase. The resultant potential would not be nernstian with respect to the $[Fe^{3+}]/[Fe^{2+}]$ couple, but rather to a couple involving Fe(II) ions and the solid phase. This potential should be different for each solid phase, although such differences may not be resolvable in these experiments.

TABLE IV

Results of Experiments in which Equilibrium with Respect to Ferric Oxyhydroxide Solids Was Approached from Undersaturation
(Ferrous Ion Activities Varied from 0.6 to 6.0 mM)

| Trial | Solid[1] | pH | Eh(Pt) (mV) | Eh(WIG) (mV) | pQ(Pt) | pQ(WIG) |
|---|---|---|---|---|---|---|
| 1 | GS | 3.83 | 408 | 398 | 38.9 | 39.1 |
| 2 | GS | 6.16 | -37 | -42 | 39.5 | 39.6 |
| 3 | GS | 6.01 | 67 | 68 | 38.2 | 38.2 |
| 4 | GS | 4.77 | 273 | 250 | 38.3 | 38.7 |
| 5 | GS | 4.21 | 445 | 447 | 38.1 | 38.1 |
| 6 | GS | 3.39 | 573 | 574 | 38.3 | 38.3 |
| 7 | H | 4.74 | 208 | 207 | 39.3 | 39.6 |
| 8 | H | 4.56 | 253 | 215 | 39.5 | 40.1 |
| 9 | H | 6.01 | -27 | -33 | 39.8 | 39.9 |
| 10 | H | 3.06 | 412 | 421 | 42.1 | 42.0 |
| 11 | H | 4.32 | 268 | 305 | 40.7 | 40.2 |
| 12 | H | 4.40 | 361 | 382 | 39.0 | 38.7 |
| 13 | G | 4.54 | 190 | 195 | 40.9 | 40.8 |
| 14 | G | 4.13 | 311 | 309 | 39.9 | 39.9 |
| 15 | G | 3.03 | 478 | 479 | 40.2 | 40.2 |
| 16 | G | 3.32 | 442 | 440 | 40.0 | 40.0 |
| 17 | G | 5.73 | 34 | -3 | 39.6 | 40.1 |
| 18 | G | 4.50 | 194 | 198 | 40.8 | 40.7 |
| 19 | G | 3.91 | 456 | 469 | 38.9 | 38.7 |
| 20 | G | 4.12 | 450 | - | 38.3 | - |
| 21 | G | 4.70 | 361 | 356 | 38.1 | 38.2 |

[1] GS = goethite-coated sand (6), H = reagent grade $Fe_2O_3$ (hematite), G = crystalline goethite. See text.

Alternately, Pt and WIG electrodes may respond in identical ways by completely different electron exchange mechanisms. This seems unlikely. Further work is necessary to determine which, if any, of these explanations is correct.

In any event, Doyle's (26) explanation for Pt electrode behavior at high pH's in the iron system is now in doubt. The results also support the use of Pt electrodes to calculate pQ's for ferric oxyhydroxide phases in aqueous systems dominated by iron, as proposed and utilized by Langmuir and Whittemore (33).

The results of all trials for which solid ferric phases were present, including 56 separate experiments, are illustrated in a graph of $\log[Fe^{3+}]$ vs pH in Figure 5. The slope of the best fit line is 3.06 +/- 0.15. Thus, the complete data set does not show the departure from 1/3 stoichiometry for oxyhydroxide phases observed by Fox (43). Rather, it reaffirms the stoichiometry $Fe(OH)_3$ for such phases.

However, if we analyze only data for freshly precipitated solid phases, which includes all data with pH's < 3.6, excluding six of the experiments for non-freshly precipitated solids shown in Table IV, a slope of 2.3 is indicated. This is in agreement with the conclusions of Fox and others (as outlined above), that freshly precipitated ferric oxyhydroxides may have an $OH^-/Fe$ ratio of less than three, presumably because of anionic substitution for hydroxide in the solid.

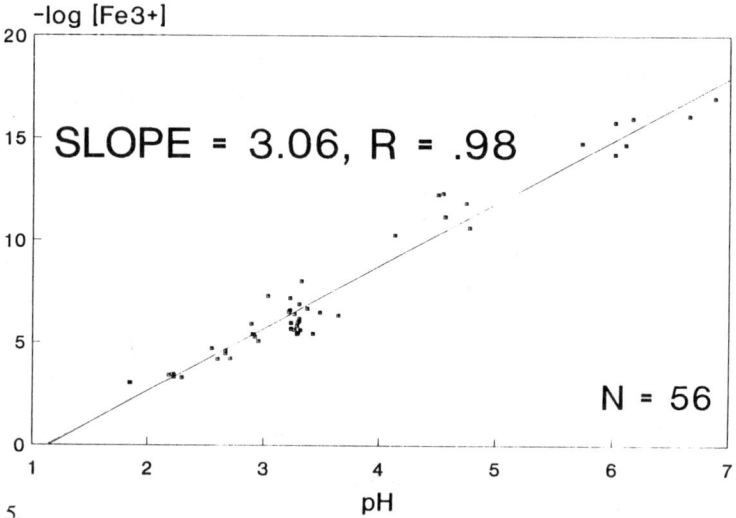

Figure 5. The negative log of ferric ion activity versus pH for the 56 experiments from this study, all at 25.0°C. The slope of the best-fit line is 3.06, with a standard deviation of 0.15.

FIELD DATA, RESULTS AND DISCUSSION

Field data were collected during the summer of 1988 from two different groundwater systems. The first is a groundwater research site on Cape Cod, MA operated by the U.S. Geological Survey and described in detail elsewhere (52). It is located downgradient from the sewage-effluent infiltration beds on Otis Air Force Base. The water table in the sand and gravel aquifer was 5-7 m below land surface in the wells investigated. Groundwater has been contaminated by effluent infiltration, but contains < 2 mg/L organic carbon. Iron is present in only a few wells and depths, so only a limited portion of the data is usable for this review.

At each well and depth, concentrations of dissolved oxygen, Fe(II), $NO_3^-$ and $NO_2^-$ were measured. ORP's at Pt and WIG electrodes were measured against a Ross reference electrode. A Ross combination electrode was used to determine pH. Temperature and conductivity were also recorded. The detection limits for Fe(II) and N were 0.05 mg Fe or N per liter, and accuracies for each were +/- 10%. Sulfide was not measured, but no $H_2S$ odor was discernable in any sample. More detailed descriptions of the sampling and analytical methods will be published elsewhere.

Chemical analyses were all performed using sealed, evacuated ampules manufactured by Chemetrics, Inc. (Calverton, VA). The oxygen technique deserves a brief mention. The most sensitive ampules gave $O_2$ concentrations from 0-10 micrograms per liter, with a precision of +/- 0.5 μg $O_2$/L. The accuracy of this colorimetric method has been established (White, A.F.; Peterson, M.L.; Solbau, R.D. Groundwater, in press). The lack of interferences by 100mg/L of Fe(II), $NO_3^-$, $NO_2^-$, $Cl^-$, and $SO_4^{2-}$ was also confirmed in separate experiments.

Of the 60 individual wells/depths sampled at the Cape Cod site, only 4 contained measurable Fe(II). All others showed at least traces (10-20 μg/L) of dissolved $O_2$ and no measurable Fe(II). For these 56 sampling points, the $Eh_m$ was universally higher at Pt than at WIG electrodes. This is consistent with the observation that Pt shows a potential indicative of a layer of adsorbed $O_2$ when oxygen is present (23).

For sites with measurable Fe(II), its concentration was used with the field T, pH and $Eh_m$ to estimate pQ for the ferric hydroxide in the aquifer. The results are shown in Table V.

Table V
Water Compositions and Calculated Ferric Oxyhydroxide Solubilities at Different Depths in Two Cape Cod Wells

| Well | $O_2$ (μg/L) | Depth (m) | pH | Fe(II) (mg/L) | Field Eh (mV) | | pQ (electrode) | |
|------|------|------|------|------|------|------|------|------|
|      |      |      |      |      | (Pt) | (WIG) | (Pt) | (WIG) |
| 16-17 | 40 | 10.4 | 6.26 | 0.10 | 260 | 176 | 37.9 | 39.4 |
| 16-17 | 50 | 10.6 | 6.24 | 0.10 | 279 | 178 | 37.6 | 39.4 |
| 343 | 0 | 17.4 | 6.46 | 16.0 | 95 | 108 | 38.0 | 37.8 |
| 343 | 0 | 24.1 | 6.60 | 21.0 | 80 | 91 | 37.7 | 37.5 |

Well 16-17 has Fe(II) values near the detection limit of 0.05 mg/L and contains $O_2$ at low levels, which may be the cause of the higher Pt electrode reading. The site is in a transition zone, out of redox equilibrium, and contains nitrite at ca. 0.5 mg N/L. However, the laboratory results discussed above indicate that $Eh_m$ at the WIG electrode may still provide an accurate representation of the distribution of ferric/ferrous species in the system. Thus, the pQ value (39.4) calculated from the $Eh_m$ at the WIG electrode may be reliable. This pQ indicates that a ferric oxyhydroxide with a solubility similar to that of colloid-sized goethite is present.

For well 343, agreement in the $Eh_m$'s for both electrodes at the two depths is consistent with the laboratory results for anaerobic systems containing ferrous iron. Further, the very low pQ values observed were not unexpected for this well. Gschwend and Reynolds (53) reported that, at the 17 m depth, this well water contains colloidal particles, which they showed to be a ferrous phosphate solid. The above results suggest the presence also of amorphous ferric hydroxides, probably colloidal in size. The limited data at this site illustrate the insights which can be provided for field situations where detailed groundwater chemistry is coupled with dual readings of ORP's using two different electrode materials.

The second site is in Tennessee Park, a sub-alpine wetland near Leadville, Colorado. Thirteen wells were drilled here as part of a project to determine the effects of the wetland on the input from an alpine stream heavily contaminated with acid mine drainage (54). The site and the wells have been described in detail (55). The water table is within one meter of the surface in all wells, which have screen depths less than four meters below the land surface. The composition of the well waters changes seasonally and in response to precipitation events.

The wells are sampled monthly, but only the August, 1988 data are presented here. Analytical methods used for $O_2$, Fe(II), ORP, pH, and $NO_3^-$ were the same as at the Cape Cod sites. Sulfide was analyzed potentiometrically (56), and dissolved organic carbon by coulometric detection of $CO_2$ generated by combustion of the sample. Temperature and conductivity were also recorded. Detection limits were 0.01 mg S/L for sulfide and 1.0 mg C/L for dissolved organic carbon. The results for the August, 1988 sampling, summarized in Table VI, indicate a much more complex system than at the Cape Cod site. The presence of measurable $O_2$ and Fe(II) is indicated in several wells (1,7,9,and 14), and three (7,9, and 12) have both $O_2$ and sulfide. The lack of redox equilibrium is striking. Microbial activity is likely to be controlling much of the redox activity in wells at this site, but a detailed discussion of these interactions is beyond the scope of this review.

Intrepretation of the observed ORP's is problematic. The potentials at the Pt electrode are influenced not only by the presence of oxygen in some wells, but also by sulfide in others. WIG electrode response has been shown (Macalady, D.L., unpublished results) to be influenced strongly by the presence of substantial dissolved organic carbon (> ca. 2 mg C/L), and insensitive to the presence of sulfide. Agreement to within 30 mV in the potentials recorded at the two electrodes is limited to three wells (7, 12, and 14).

Detailed analysis of these data, and data collected at the same wells on different dates, will be published elsewhere. Data are presented here only to illustrate the kind of electrode response which can be expected in more complicated redox environments. The use of the two types of redox electrodes again illustrates the utility of a dual electrode approach to ORP measurements. Substantial differences between Pt and WIG electrodes indicate electrode interferences which may obviate calculations based on the Nernst equation.

Where the potentials agree, there is of course the chance that the agreement is due to a fortuitous correspondence of interferences. This is probably the case for well 7, where traces of $O_2$, DOC, and sulfide have unknown effects on electrode potentials, and ORP's recorded at this well on other dates range from 17 to 321 mV. We believe, however, that in most groundwater systems, agreement between WIG and Pt electrodes indicates that $Eh_m$ corresponds to a nernstian Fe(III)/Fe(II) potential.

Calculations based upon this assumption are thus possible only for well 14 (12 has no measurable Fe(II)). The resulting pQ values for a ferric oxyhydroxide phase are 38.3 (Pt) and 38.7 (WIG), which, as expected for this highly dynamic groundwater system, indicates a poorly-crystallized material.

As suggested in a recent paper by Barcelona et al. (57), redox interactions in complex groundwater systems, if they are to be useful in predictions of redox processes, must be represented by detailed analytical information about each redox couple. For example, if a series of pQ's is calculated for all of the observed potentials, a range of values between 36 and 42 is obtained. Although this range is reasonable in the light of previous observations in natural systems, it is suspect when one considers the dynamic nature of this groundwater regime. Also, calculations for a single well (#8) vary from 37.5 to 41.6 depending upon which electrode potential is used. Thus, caution must be attached to interpretations of ferric oxyhydroxide solubilities based solely upon $Eh_m$, Fe(II), and pH measurements.

TABLE VI
Analytical Results for Tennessee Park Wetland Wells, August 1988

| Well # | Depth to water | Depth to screen | $O_2$ ($\mu$g/L) | Fe(II) (mg/L) | DOC (mg/L) | pH | Eh(mV) Pt | Eh(mV) WIG | Sulfide (mg/L) |
|---|---|---|---|---|---|---|---|---|---|
| 1  | 0.12  | 1.50 | 103 | 4.5   | 0    | 5.36 | 171 | 233 | 0 |
| 3  | 0.10  | 2.06 | 0   | 7.3   | 2.1  | 6.04 | 149 | 231 | 0 |
| 4  | -0.30 | 2.06 | 750 | 0     | 2.2  | 6.10 | 148 | 208 | 0 |
| 5  | 0.74  | 2.08 | 0   | 15.   | 26.4 | 6.00 | 130 | 71  | 0.20 |
| 6  | 0.38  | 2.08 | 0   | 17.   | 46.9 | 5.86 | 14  | 101 | 4.0 |
| 7  | 0.45  | 2.08 | 6   | 0.35  | 4.3  | 5.98 | 210 | 201 | 0(+) |
| 8  | 0.22  | 2.08 | 0   | 9.0   | 7.4  | 5.99 | -2  | 231 | 0.86 |
| 9  | 0.74  | 3.56 | 650 | 0.30  | 1.5  | 4.82 | 375 | 284 | 0.06 |
| 10 | 0.75  | 3.56 | 0   | 16.4  | 13.5 | 6.60 | 38  | 159 | 0.80 |
| 11 | 0.67  | 2.97 | 0   | 2.8   | 7.2  | 6.30 | 43  | 147 | 0.31 |
| 12 | 0.74  | 3.56 | 35  | 0     | 0    | 5.73 | 196 | 175 | 0(+) |
| 13 | 0.25  | 2.97 | 0   | 8.0   | 5.2  | 6.21 | 102 | 168 | 0.2 |
| 14 | 0.55  | 2.97 | 5   | 12.0  | 0    | 5.72 | 221 | 196 | 0 |

Note: All depths are in meters. The (+) in the sulfide column indicates a positive sulfide smell observed in the water as it was sampled, even though none was detected by the analyses.

SUMMARY

Laboratory studies of the Fe(III)/Fe(II) system involving a new method of calculating ferric species activities have been used to:

1. Establish the validity of the stability constants for ferric and ferrous hydroxy- and chloro-complexes shown in Table I.

2. Reaffirm the rapid rates at which ferrous and ferric hydrolysis and chloride complexation reactions achieve equilibrium in experimental systems.

3. Provide new data on the rates of formation, stability constants and stoichiometry of ferric phases precipitated by hydrolysis in chloride and chloride/nitrate aqueous media.

4. Extend the range of ferric species activities over which Pt electrode response is nernstian, and, provide evidence for the comparable behavior of another, physically and chemically distinct electrode material, wax-impregnated graphite. Experiments in laboratory and field situations indicate nernstian response of both Pt and WIG electrodes at pH's as high as 6.4, and thus total Fe(III) activities as low as $10^{-9}$ M.

5. When combined with careful chemical characterization of the water chemistry, Eh measurements at Pt and WIG electrodes can be used to establish the nature of the ferric oxyhydroxide phases in (anaerobic) groundwaters. This and previous studies show that a wide range of solid phases, with pQ's from 36.6 to 43.5 can exist in groundwater.

DISCLAIMER

Although the research described in this article has been supported in part by the United States Environmental Protection Agency through cooperative agreement #CR-813077-01 between the Colorado School of Mines and the R.S. Kerr and Athens Environmental Research Laboratories of the EPA, it has not been subjected to Agency review and, therefore, does not necessarily reflect the views of the Agency. No official endorsement should be inferred.

ACKNOWLEDGEMENTS

Laboratory and field analyses and helpful discussions from Vernon T. Tate, Myron Brooks, and Kathryn Walton-Day, all of the Colorado School of Mines, are gratefully acknowledged.

LITERATURE CITED

1. Byrne, R.H.; Kester, D.R. Mar. Chem. 1976, 4, 255-274.
2. Vlek, P.L.G.; Blom, T.J.M.; Beek, J.; Lindsay, W.L. Soil Sci. Soc. Am. Proc. 1974, 38, 429-432.
3. Iwase, M.; Yotsuyanagi, T.; Nagayama, M. Nippon Kagaku Kaishi 1986, 12, 2271-2276.
4. Gayer, K.H.; Wootner, L. J. Phys. Chem. 1956, 60, 1569-1571.
5. Lengweiler, H.; Buser, W.; Feitknecht, W. Helv. Chim. Acta 1961, 44, 796-805.
6. Grundl, T.J. PhD Thesis, Colorado School of Mines, Golden, CO, 1987.
7. Nernst, W. Ber. 1897, 30, 1547-1563.
8. Morris, J.G.; Stumm, W. in Equilibrium Concepts in Natural Water Systems, Stumm, W., Ed.; ACS Advances in Chemistry Series no. 67; American Chemical Society, Washington, DC, 1967; p. 270.
9. Frevert, T. Schweiz Z. Hydrol. 1984, 46, 269-290.
10. Stumm, W.in Advances in Water Pollution Research, Jaag, O.; Leibmann, H., Eds. Water Pollution Control Federation, Washington, DC; 1967, Vol. 1, p. 283.
11. Lindberg, R.D.; Runnells, D.D. Science 1984, 225, 925-927.
12. Bohn, H.L. Soil Sci. Soc. Am. Proc. 1968, 32, 211-215.
13. Bohn, H.L. Soil Sci. Soc. Am. Proc. 1969, 33, 639-640.
14. Bohn, H.L. Soil Sci. 1971, 112, 39-45.
15. Power, G.P.; Ritchie, I.M. J. Chem. Ed. 1983, 60, 1022-1026.
16. Jackson, R.E.; Patterson, R.J. Water Resour. Res. 1982, 18, 1255-1268.
17. Hostettler, J. Am. J. Sci. 1984, 284, 734-759.
18. Back, W.; Barnes, I. U.S. Geol. Surv. Prof. Paper, 1965, 498-C, 16 pp.
19. Back, W.; Barnes, I. U.S. Geol. Surv. Prof. Paper, 1961, 286-C, 366-368.
20. Langmuir, D. in Proceedings in Sedimentary Petrology, Carver, R.E., Ed.; Wiley-Interscience, New York, NY, 1971, pp. 597-634.
21. Bricker, O.P. Am. Mineralogist 1965, 50, 1296-1299.
22. Berner, R.A. Geochim. Cosmochim. Acta 1963, 27, 563-575.
23. Whitfield, M. Limnol. Oceanogr. 1969, 14, 547-558.
24. Boulegue, J.; Michard, G. in Chemical Modeling in Aqueous Systems, Jenne, E.A., Ed.; ACS Symposium Series No. 93; American Chemical Society, Washington, DC, 1979, pp. 25-50.
25. Nordstrom, D.K.; Jenne, E.A.; Ball, J.W. in Chemical Modeling in Aqueous Systems, Jenne, E.A., Ed.; ACS Symposium Series No. 93; American Chemical Society, Washington, DC, 1979, pp. 51-80.
26. Doyle, R.W. Am. J. Sci. 1968, 266, 840-859.
27. Ponnamperuma, F.N.; Tianco, E.M.; Loy, T. Soil Sci. 1967, 103, 374-381.
28. Baes, C.; Mesmer, R. The Hydrolysis of Cations, Wiley-Interscience, New York, NY, 1976, 496 pp.

29. Wagman, D.; Evans, W.; Parker, V.; Halow, I.; Bailey, S.; Schumm, R. U.S. Nat. Bur. Stand. Tech. Note, 1969, 270-4, 141 pp.
30. Smith, R.M.; Martell, A.E. Critical Stability Constants, Vol. 4, Inorganic Complexes; Plenum Press, New York, NY, 1976; 257 pp.
31. Barnes, H. L., Ed. Geochemistry of Hydrothermal Ore Deposits; Wiley-Interscience, New York, NY, 1979; 798 pp.
32. Cobble, J. J. Phys. Chem. 1953, 21, 1446-1450.
33. Langmuir, D.; Whittemore, D.O. in Nonequilibrium Systems in Natural Water Chemistry; Hem, J.D., Ed.; ACS Advances in Chemistry Series No. 106; American Chemical Society, Washington, DC, 1971, pp. 209-234.
34. Whittemore, D.O.; Langmuir, D. Groundwater 1975, 13, 360-365.
35. Langmuir, D. Am. J. Sci. 1971, 271, 147-156 and 1972, 272, 972.
36. Robie, R.A.; Hemingway, B.S. Fisher, J.R. U.S. Geol Surv. Bull. 1978, 1452, 456 pp.
37. Robinson, G.R. Jr.; Haas, J.L.; Schafer, C.M.; Haselton, H.T. Jr. U.S. Geol. Surv. Open File Report 1983, 83-79, 79 pp.
38. Dousma, J.; deBruyn, P.L. J. Colloid Interface Sci. 1978, 64, 154-170.
39. Combes, J.M.; Manceau, A.; Calas, G.; Bottero, J.Y. Geochim. Cosmochim. Acta 1989, 53, 583-594.
40. Murphy, P.J.; Posner, A.M.; Quirk, J.P. J. Colloid Interface Sci. 1976, 56, 312-319.
41. Dousma, J.; deBruyn, P.L. J. Colloid Interface Sci. 1976, 56, 527-539.
42. Beidermann, G.; Chow, J.T. Acta Chem. Scand. 1966, 20, 1376-1388,
43. Fox, L.E. Geochim. Cosmochim. Acta. 1988, 52, 771-777.
44. Spiro, T.G.; Allerton, S.E.; Renner, J.; Terzis, A.; Bills, R.; Saltman, P. J. Am. Chem. Soc. 1966, 88, 2721-2726.
45. Feitknecht, V.W.; Giovanoli, R.; Michaelis, W.; Muller, M. Z. anorg. allg. Chem. 1975, 417, 114-124.
46. Light, T.S. Anal. Chem. 1972, 44, 1038-1039.
47. Parkhurst, D.L.; Thorstenson, D.C.; Plummer, L.N. U.S. Geol. Surv. Water Res. Invest. 1980, 80-96, 209 pp.
48. Kopelove, A.; Franklin, S.; McGaha-Miller, G. American Laboratory June, 1989, 40-47.
49. Davies, C.W. Ion Association; Butterworth and Co.,Ltd., London, 1962.
50. Heymann, E. Z. Anorg. Chem. 1928, 171, 18-41.
51. Lamb, A.B.; Jacques, A.G. J. Am. Chem. Soc. 1938, 60, 1215-1225.
52. LeBlanc, D.L., Ed. U.S. Geol. Surv.Open File Report 1984, 84-475, 180 pp.
53. Gschwend, P.M.; Reynolds, M.D. J. Contam. Hydrol. 1987, 1, 309-328.
54. Mcknight, D.M.; Kimball, B.A.; Bencala, K.E. Science 1988, 240, 637-640.
55. Walton-Day, K.; Briggs, P.H. Proc. 4th U.S.G.S. Toxic Substances Mtg., 1988, p. 76.
56. Baumann, E.W. Anal. Chem. 1974, 46, 1345-1349.
57. Barcelona, M.J.; Holm, T.R.; Schock, M.R.; George, G.K. Water Resour. Res., 1989, 25, 991-1003.

RECEIVED August 24, 1989

## Chapter 29

# Energetics and Conservative Properties of Redox Systems

### Michael J. Scott and James J. Morgan

**Environmental Engineering Science, W. M. Keck Laboratories, California Institute of Technology, Pasadena, CA 91125**

> The redox status of an aqueous system is described by the concentrations of the oxidized and reduced species of all system components. Redox systems, generally not at equilibrium as the result of kinetically slow redox reactions, are poorly characterized by intensity factors ($E_H$ or pE) alone. Capacity factors, which reflect the total concentration of relevant species, are conservative parameters that can be meaningful guides to the redox status of aqueous systems. Oxidative capacity (OXC) is defined as a conservative quantity that incorporates a comprehensive chemical analysis of the redox couples of an aqueous system into a single descriptive parameter. OXC classifies aqueous systems in terms of well-defined geochemical and microbial parameters (e.g., oxic, sulfidic). Examples of model and actual groundwater systems are discussed to illustrate the concept. A redox titration model is another tool that is useful in describing a redox system as it approaches an equilibrium state.

This paper discusses concepts useful in understanding the redox status of aqueous systems. The redox status of an aqueous system is described by the concentrations of the oxidized and reduced species of all components of the system. The oxidation state of a component affects the solubility, adsorption behavior, and inherent aqueous toxicity of the component. For example, the highest oxidized form of selenium, Se(VI), is more soluble and less adsorbing than the more reduced Se(IV) and elemental forms. It is because of its solubility and adsorption behavior that Se(VI) has greater environmental exposure concentrations, and thus, is the most potentially dangerous form of selenium as far as environmental contamination is concerned. The total concentration of each component in the system is also an important controlling factor of the redox status. A simple example is the susceptibility to change of the redox status of two oxic systems. The first, a natural system open to the large oxygen reservoir of the atmosphere (i.e., unlimited or infinite source), is well buffered against oxygen depletion, and therefore its redox status is very stable. The other, a system with a limited supply of oxygen (e.g., a groundwater aquifer or deep lake bottom), has a redox status that will change as the oxygen is depleted by natural processes.

The redox status and the microbial ecology of an aquatic system are interrelated. Microbial populations are distinctly different in oxic and anoxic aqueous environments; depending on the redox status, they may consist of aerobes, denitrifiers, fermenters, sulfate reducers, or methanogens ([1]). In addition, the fate of aqueous contaminants and the rate of transformation is highly dependent on the redox status. For example, naphthalene is microbially degraded only in oxic environments while DDT and other chlorinated hydrocarbons have been efficiently degraded only in anoxic environments ([2,3]). As a result of its influence on aqueous systems it is important that investigators understand more fully the redox status of their systems. Knowledge of the redox status is particularly vital for such systems as aquifers, estuaries, deep lakes, and oceans. These systems may have hydraulic residence times long enough for slow redox reactions to greatly change the concentrations of the oxidants and reductants.

An aqueous natural system needs to be characterized by both intensity and capacity factors. Intensity factors (e.g., pH) reflect activities of <u>free</u> species; capacity factors, on the other hand, reflect the <u>total</u> concentration of relevant species (e.g., total acidity). A capacity factor can be described as a conservative property that is indirectly temperature- and pressure-dependent while an intensity factor is directly dependent on temperature and pressure. Thus for acid-base systems, pH is an intensity factor while alkalinity, or the acid neutralizing capacity, is a capacity factor. An important characteristic of any aqueous system is the total concentrations of the components of interest.

Redox parameters analogous to those for acid-base chemistry can be defined for all aqueous systems. The redox intensity factor pE is an energy parameter in non-dimensional form that describes the ratio of electron acceptors (oxidants) and donors (reductants) in a redox couple. The redox potential ($E_H$) of the system is an alternative and equivalent intensity factor. Table I summarizes the complete thermodynamic analogy between pH and pE. An analogy between acid-base and redox systems can also be made for capacity factors.

Table I: Thermodynamic Analogy between pH and pE

$$HA = H^+ + A^- \quad K_2$$
$$H_2O + H = H_3O^+ \quad K_3 = 1$$
$$HA + H_2O = H_3O^+ + A^- \quad K_1$$

$$\frac{[A^-][H_3O^+]}{[HA][H_2O]} = K_1 = K_2 K_3$$

$$\frac{[H^+][A^-]}{[HA]} * \frac{[H_3O^+]}{[H^+][H_2O]} = K_1 = K_2$$
$$\Downarrow \qquad \Downarrow$$
$$K_2 \qquad K_3 = 1$$

$$\therefore \frac{[H^+][A^-]}{[HA]} = K_1$$

$$pH = pK_1 + \log([A^-]/[HA])$$

$$Fe^{3+} + e^- = Fe^{2+} \quad K_2$$
$$1/2\, H_2(g) + H_2O = H_3O^+ + e^- \quad K_3 = 1$$
$$1/2\, H_2(g) + Fe^{3+} + H_2O = H_3O^+ + Fe^{2+} \quad K_1$$

$$\frac{[Fe^{2+}][H_3O^+]}{[Fe^{3+}][H_2O](P_{H_2})^{1/2}} = K_1 = K_2 K_3$$

$$\frac{[Fe^{2+}]}{[Fe^{3+}][e^-]} * \frac{[H_3O^+][e^-]}{(P_{H_2})^{1/2}[H_2O]} = K_1 = K_2$$
$$\Downarrow \qquad \Downarrow$$
$$K_2 \qquad K_3 = 1$$

$$\therefore \frac{[Fe^{2+}]}{[Fe^{3+}][e^-]} = K_1$$

$$pE = \log K_1 + \log([Fe^{3+}]/[Fe^{2+}])$$
$$\Downarrow$$
$$pE^o{}_{Fe}$$

However, acid-base and redox systems are far from analogous with respect to making analytical measurements of the system's intensity and capacity factors. Aqueous systems of acid-base reactions are typically easily defined. Since acid-base kinetics are rapid (i.e., equilibrium is easily attained) and the reaction species are present in measurable concentrations, determination of pH and other characteristics of a system can be straightforward (1,4). The redox status is usually not as easily defined. Electron transfer reactions are usually kinetically slow and electroactive species are often low in concentration and difficult to measure (4,5). Also, several of the pertinent redox species are not electroactive (e.g., $SO_4^{2-}$, $NO_3^-$, $CH_4$).

A natural aqueous redox system may or may not be at total (internal) equilibrium (6). How the redox status is characterized depends upon the state of the system relative to equilibrium. Regardless of whether the system is at equilibrium or not there is an individual, instantaneous pE for each redox couple in the system. Each pE corresponds to the relative concentrations of the oxidized and reduced species and is defined by Nernstian relationships (1). When a system of oxidants and reductants is at internal chemical equilibrium the pE of every couple is identical and this distinct value is the system pE. Only in this special case can the redox status be characterized by determining two of the following three factors: pE, the total concentration of all components, and the relative concentrations of oxidants and reductants. When two of the factors are known, the third can be determined from Nernstian and mass balance relationships. Thus, only when the chemical system is at internal equilibrium and one of the redox couples of the system is electrochemically active and present in measurable concentrations can the system pE be calculated from an electrode potential.

The concept of a distinct system pE is meaningless if one or more redox couples are not at equilibrium. The pE of each redox couple will only be representative of that couple. Each couple pE will be different. However, the range of the individual values of pE does give an indication of the relative degree of disequilibrium. Systems near equilibrium should have a small range of pE values while systems far from equilibrium should have a broad range. Lindberg and Runnells (6) demonstrated that none of 30 representative groundwaters exhibited redox equilibrium and that there existed a wide range of pE values for each system, up to as much as 17 pE units. On the basis of this and other works noting the lack of a unique pE in most groundwaters, recent investigations have concluded that additional qualitative guides are necessary for a better understanding of the redox status of natural waters (6-8). The qualitative guides require (i) total analysis of all redox couples in the system, (ii) kinetic studies of non-equilibrium reactions, and (iii) in-situ consideration of minerals, gases, water, and bacteria that may be missing from the sample solution.

A major factor in the uncertainties of the redox status of an aqueous system is the time dependency of redox reactions. The fact that the rates of electron transfer are relatively slow and that the electrochemical response of an electrode is also time dependent makes the determination of pE or $E_H$ undependable and of little general value as a single master variable. We wish to define a conservative quantity that will incorporate a comprehensive chemical analysis of the redox couples of an aqueous system into a single descriptive parameter for that redox system. This capacity factor is called the oxidative capacity (OXC) of a redox system and represents the total number of transferable electrons. This concept allows us to classify aqueous redox systems by a conservative quantity as is done with alkalinity and acidity measurements. This parameter will also allow investigators to better characterize the redox status of an aqueous system than is possible with a knowledge of the redox potential alone.

DEFINITIONS

Operationally oxidative capacity can be defined as the equivalent sum of all oxidants that can be reduced with a strong reductant (e.g., $H_2$, H atoms, electrons) to an equivalence point. Similarly, the reductive capacity (RDC) can be determined from the oxidation of reductants by a strong oxidant (e.g., $O_2$, O atoms) to a preselected equivalence point. At every equivalence point a particular electron condition defines a reference level of electrons. OXC and RDC can be defined as

$$OXC = \sum n_i [Ox]_i - \sum n_i [Red]_i = - RDC$$

where $[Ox]_i$ and $[Red]_i$ represent the concentration (molal) of the individual oxidants and reductants of the system and $n_i$ is the number of equivalent electrons that are transferred. For example, in $O_2$

reduction, the reduction can be written as

$$O_2 + 4 H^+ + 4 e^- = 2H_2O$$

and $n_i = 4$; in $MnO_2$ reduction, the reaction can be written

$$MnO_2 + 2 H^+ + 2 e^- = Mn^{2+} + 2H_2O$$

and $n_i = 2$. For solid phases, concentrations are expressed as the mean number of moles of solid per volume of contacting water, i.e., the surface concentration (molal). This is the product of the total solid concentration (g/kg), the mean surface area of the solid (area/g), and the mean mole number of surface sites per unit area (#moles/area).

Some of the dominant redox couples of most aqueous systems are listed in Table II. In a system of initially several oxidants (and potential reductants and catalysts), the most energetically favored species is reduced first and then followed by less energetic species. For example, dissolved oxygen is the preferred (thermodynamically, but not necessarily kinetically) oxidant in the oxidation of organic matter to $CO_2$ and water. Following the consumption of dissolved oxygen, the next strongest oxidant is used. This succession allows us to create a relative electron free energy diagram, or a "redox ladder", (Figure 1) with the oxidant of each redox couple on the left side of the ladder and the reductant on the right. The strongest oxidants are at the top and the strongest reductants are at the bottom. As shown in Figure 1, $O_2(aq)$ is the strongest oxidant and dissolved organic matter is the strongest reductant. The scale of the ladder is the pE of the redox couple at a constant pH. The ratio of oxidant to reductant of each couple is unity. An oxidant of a couple can oxidize a reductant of another couple which occupies a lower rung on the ladder.

Table II: Selected Important Half Reactions in a Groundwater System

| Half Reaction | $pE^o$ | $E_H^o$ (v) |
| --- | --- | --- |
| $1/4\ O_2(g) + H^+ + e^- = 1/2\ H_2O$ | 20.75 | 1.23 |
| $1/5\ NO_3^- + 6/5\ H^+ + e^- = 1/10\ N_2 + 3/5\ H_2O$ | 21.05 | 1.25 |
| $1/2\ MnO_2(s) + 2 H^+ + e^- = 1/2\ Mn^{2+} + H_2O$ | 20.42 | 1.21 |
| $Fe(OH)_3(s) + 3 H^+ + e^- = Fe^{2+} + 3 H_2O$ | 16.00 | 0.95 |
| $1/8\ SO_4^{2-} + 9/8\ H^+ + e^- = 1/8\ HS^- + 1/2\ H_2O$ | 4.25 | 0.25 |
| $1/4\ CO_2(g) + H^+ + e^- = 1/4\ CH_2O + 1/4\ H_2O$ | -1.20 | -0.07 |

The redox ladder is useful in selecting an electron reference level (ERL). For example, if we choose bisulfide ($HS^-$) as the ERL, then the species on the left side above and including sulfate ($SO_4^{2-}$) are the system oxidants and the species on the right side beneath $HS^-$ on the redox ladder are the system reductants. The reference level species is neither a system oxidant nor a system reductant. It is important to note that the ERL is a relative electron level and thus can be chosen at any level; usually the level is chosen as a matter of convenience. If the oxidation of organic matter is the subject of interest, then the choice of $HS^-$ as the ERL is convenient; this choice defines dissolved organic matter as the only system reductant and all other species are either system oxidants or not important for determining OXC. On the other hand, if the presence of oxygen is of interest, $H_2O$ would be the convenient choice for the ERL, thus making dissolved oxygen the only system oxidant.

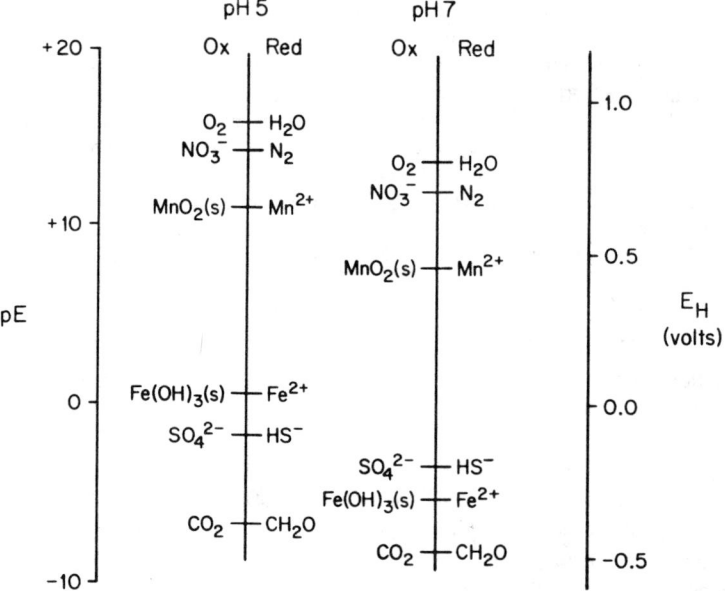

Figure 1. Redox ladder at pH 5 and pH 7 for major redox couples ([Ox]/[Red] = 1).

## EXAMPLES

To illustrate the concept of OXC, a few examples are discussed below. The first example is a simple homogeneous closed system consisting of 1.0 mm $O_2(aq)$ and 0.5 mm $CH_2O(aq)$. By setting $HS^-$ as the ERL, $O_2(aq)$ is the system oxidant and $CH_2O(aq)$ is the system reductant. The oxidative capacity of this system is

$$OXC = 4\,[O_2] - 4\,[CH_2O] = 2.0 \text{ meq/l}$$

It is possible to define several other parameters:

$$OX_T = \Sigma\, n_i\, [Ox]_i$$
$$RD_T = \Sigma\, n_i\, [Red]_i$$

For the example,

$$OX_T = 4\,[O_2] = 4.0 \text{ meq/l}$$
$$RD_T = 4\,[CH_2O] = 2.0 \text{ meq/l}$$

It is readily seen that $OXC = OX_T - RD_T$.

The next example is a hypothetical groundwater system where some of the system oxidants are present in the solid phase. The system is defined as follows:

| | | | |
|---|---|---|---|
| $[O_2]$ | = 0.25 mm | $[NO_3^-]$ | = 0.02 mm |
| $[MnO_2(s)]$ | = 0.01 mm | $[Fe(OH)_3(s)]$ | = 0.01 mm |
| $[SO_4^{2-}]$ | = 0.20 mm | $[CH_2O]$ | = 0.30 mm |
| pH | = 7.0 | | |

By setting $HS^-$ as the ERL we define the system oxidants to be $O_2$, $NO_3^-$, $MnO_2$, $Fe(OH)_3$, and $SO_4^{2-}$, and the system reductant to be $CH_2O$. Thus the oxidative capacity of this system is

$$OXC = 4\,[O_2] + 5\,[NO_3^-] + 2\,[MnO_2] + [Fe(OH)_3] + 8\,[SO_4^{2-}] - 4\,[CH_2O] = 1.53 \text{ meq/l}.$$

The above system can be classified within the geochemical classification system of Berner (9). According to the Berner classification, a system is oxic if $[O_2] > 1$ μm and anoxic if $[O_2] < 1$ μm. Initially the example system is oxic ($[O_2] = 250$ μm). It is of interest to know its condition when the system has reached equilibrium. To determine this it is necessary to define additional parameters based on the concentrations of the redox species of the system. OXC can be defined in terms of the Berner classifications.

$$OXC = o\text{-}OXC + p\text{-}OXC + s\text{-}OXC - m\text{-}RDC$$

where
$$o\text{-}OXC = 4\,[O_2]$$
$$p\text{-}OXC = 5\,[NO_3^-] + 2\,[MnO_2] + [Fe(OH)_3]$$
$$s\text{-}OXC = 8\,[SO_4^{2-}]$$
$$m\text{-}RDC = 4\,[CH_2O] = RDC.$$

The letters o, p, s, and m represent the system classification of the type of environment where the respective oxidants are reduced: oxic, post-oxic, sulfidic, and methanic.

Also, for simplicity, the following combinations can be created:

op-OXC = o-OXC + p-OXC
ops-OXC = op-OXC + s-OXC = $OX_T$.

Therefore an aqueous system will be
 a) oxic
  if (o-OXC) - (m-RDC) > 0.0
  or (o-OXC) > (m-RDC)
 b) anoxic
  if (m-RDC) ≥ (o-OXC).

When the system is anoxic, the Berner classification is further divided into three subgroups:
 i) post-oxic
  if (op-OXC) > (m-RDC)
 ii) sulfidic
  if (m-RDC) ≥ (op-OXC)
  and $OX_T > RD_T$
 iii) methanic
  if $RD_T > OX_T$
  or OXC < 0.0.
In the above example:
  o-OXC = 1.0 meq/l
  p-OXC = 0.13 meq/l
  s-OXC = 1.6 meq/l
  m-RDC = 1.2 meq/l
 Condition A:  m-RDC > o-OXC
The system will be anoxic when the system has fully reacted.
 Condition B:
  i) op-OXC = 1.13 meq/l < m-RDC
  ii) $OX_T > RD_T$
The system will be sulfidic.

A third example of OXC is taken from field observations. In response to the recommendation by Hostettler (7) that the most reliable characterization of the redox status of a natural water is a complete chemical analysis of all redox-active species, the Illinois State Water Survey (10) collected and analyzed groundwater samples from a pristine aquifer on a monthly basis for one year. Table III lists calculated values of $OX_T$, $RD_T$, and OXC for samples taken from depths of 35, 50, and 65 feet.

OXC is greatest in the most shallow samples and it decreases with depth. In addition, the lowest values at a constant depth occur in the winter samples and OXC varies little from April to September. This seasonal effect is the result of a decrease in total reductants; $RD_T$ values for December through February are more than three times those for the spring and summer months. $OX_T$ is relatively constant throughout the entire sampling period. It is uncertain whether or not the seasonal effect is just an artifact; sampling for a longer duration is necessary before a true seasonal effect can be noticed. It should be noted that these values do not represent the true OXC of the system. In the study the quantity of Fe and Mn oxides is not determined; only their presence is acknowledged and no reduced aqueous Fe and Mn ions are detected. Furthermore organic carbon concentrations are reported as purgeable and non-purgeable organic carbon and no oxidation state is given. Organic carbon can have an oxidation state ranging from C(-IV) (methane) to C(III) (formic acid). An "average" oxidation state of the organic carbon present can be obtained if both chemical oxygen demand (COD) and total organic carbon (TOC) are determined (1). Since only TOC is reported (the sum of purgeable and non-purgeable organic carbon), an average oxidation state for organic carbon can only be assumed. For simplicity in calculating $RD_T$, organic carbon is assigned an average oxidation state of zero and a four electron transfer is assumed for the oxidation of the organic carbon in the system.

Incomplete analysis is the main limitation of OXC. Groundwater systems consist of highly non-uniform and non-homogeneous mixtures of solutes, colloids, and varying sized solid particles. At the present, reactive amounts of solid oxidants and reductants in such a system can only be roughly estimated. More information about the type of particles and particle size distributions in groundwater is needed before OXC can become more than a good estimate for such systems.

Table III: Calculated values of $OX_T$, $RD_T$, and OXC from field data (10)

| Sample | $OX_T$ (µeq/l) | | | $RD_T$ (µeq/l) | | | OXC (µeq/l) | | |
|---|---|---|---|---|---|---|---|---|---|
| | 35 ft | 55 ft | 65 ft | 35 ft | 55 ft | 65 ft | 35 ft | 55 ft | 65 ft |
| Dec-84 | 4498 | 3145 | 2723 | 900 | 768 | 968 | 3598 | 2377 | 1755 |
| Jan-85 | 4295 | -ND- | 2647 | 968 | -ND- | 768 | 3327 | -ND- | 1879 |
| Feb-85 | 4817 | 3174 | 2972 | 2568 | 800 | 668 | 2249 | 2374 | 2304 |
| Mar-85 | 5012 | 2990 | 2890 | 600 | 132 | 200 | 4412 | 2858 | 2690 |
| Apr-85 | 5010 | 3120 | 2548 | 68 | 68 | 200 | 4942 | 3052 | 2348 |
| May-85 | 5146 | 3090 | 2743 | 368 | 332 | 200 | 4778 | 2758 | 2543 |
| Jun-85 | 5523 | 3231 | 3028 | 568 | 100 | 32 | 4955 | 3131 | 2996 |
| Jul-85 | 5236 | 3497 | 3396 | 268 | 68 | 132 | 4968 | 3429 | 3264 |
| Aug-85 | 5268 | 3032 | 3063 | 468 | 200 | 332 | 4800 | 2832 | 2731 |
| Sep-85 | 5092 | 2853 | 3004 | 232 | 68 | 32 | 4860 | 2785 | 2970 |
| Oct-85 | 5349 | 3083 | 3002 | -ND- | -ND- | -ND- | -ND- | -ND- | -ND- |
| AVG | 5022 | 3122 | 2911 | 701 | 282 | 353 | 4289 | 2844 | 2548 |
| SD | 365 | 169 | 236 | 715 | 297 | 329 | 927 | 339 | 485 |
| LOW | 4295 | 2853 | 2548 | 68 | 68 | 32 | 2249 | 2374 | 1755 |
| HIGH | 5523 | 3497 | 3396 | 2568 | 800 | 968 | 4968 | 3429 | 3264 |

MODEL REDOX TITRATION

Although we know equilibrium will be a poor approximation in many instances, equilibrium models of redox systems are helpful in making a preliminary assessment of which redox species will interact and what the redox status will be during and after such interactions. A beneficial exercise is to model the titration of a hypothetical system of oxidants with a strong reductant. Figure 2a shows the response of pE to the titration by $CH_2O(aq)$ of a hypothetical system consisting of $O_2(aq)$, $NO_3^-$, $MnO_2(s)$, $Fe(OH)_3(s)$, and $SO_4^{2-}$. Initially pE increases as the dissolved oxygen of the system oxidizes the dissolved organic carbon. The direction of the pE change may seem intuitively backwards at first. However, pE in this case is a function of pH as well as the ratio of oxidants to reductants. Production of $CO_2$ decreases pH. This decrease affects the pE to a greater extent than the ratio of oxidants to reductants, and thus causes the increase in pE. When the dissolved oxygen is depleted the pE steps down and $NO_3^-$ is the next strongest oxidant which reacts with the reductant. Similar steps occur as one oxidant is depleted and another begins to react. The titration of a redox system may be viewed as the process of progressing down the redox ladder to a lower energy state. Figure 2b shows the response of pH to such a titration. In a buffered system ($C_T = 1.0$ mm), the pH decreases slightly during the reduction of dissolved oxygen as $CO_2$ is produced. The pH remains nearly constant during the reduction of nitrate as a result of a balance between the production of $CO_2$ and consumption of protons. During the reduction of the manganese and iron oxides, the pH increases greatly as protons are consumed in ratios of 4 and 8, respectively, to $CO_2$ production. In

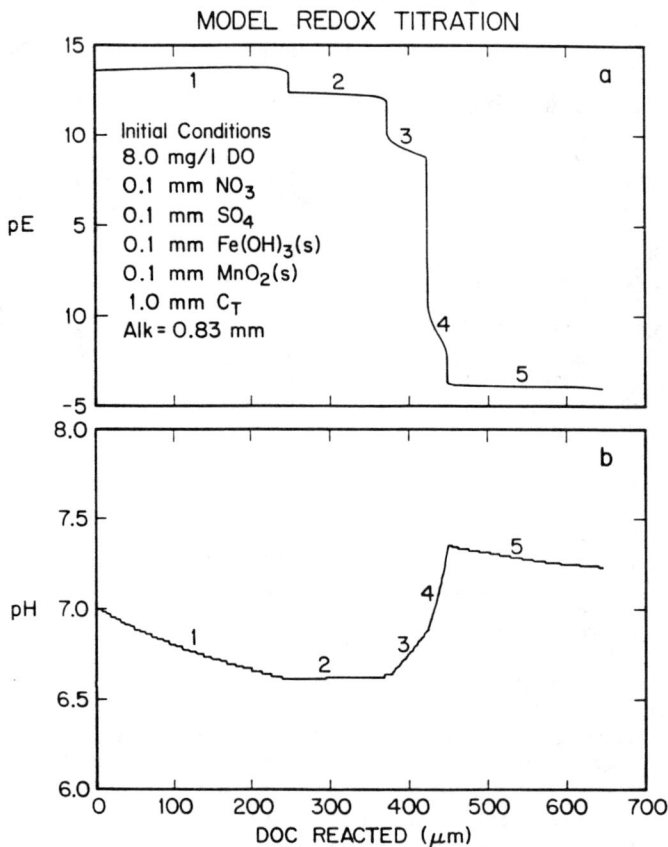

Figure 2. Redox titration curve of a model groundwater system: a) pE response and b) pH response. Numbered segments correspond to sequential reduction of 1) $O_2$(aq), 2) $NO_3^-$ (aq), 3) $MnO_2$ (s), 4) $Fe(OH)_3$ (s), and 5) $SO_4^{2-}$ (aq).

the reduction of sulfate, the ratio of protons consumed to $CO_2$ produced is nearly unity and the change in pH is small.

The progression of this redox system can also be followed on a pH-pE diagram. Figure 3 displays the titration path in pH-pE space. If the progression is well understood, then the behavior of any trace redox species (Se, As, Cr, etc.) may be predicted by the combination of the pH-pE diagram of the trace species and the titration path. The effect of such a titration on selenium speciation is also shown in Figure 3. Initially selenium is in the form selenate [Se(VI)] but as soon as the $MnO_2(s)$ reduction is complete, it is reduced to biselenite [Se(IV)], and then to elemental Se.

Knowledge about the response of trace contaminants to a changing redox status is useful in the assessment and treatment of natural systems. This is especially true in the case of bioreclamation where the introduction of microbes, as catalysts, or dissolved oxygen, as a nutrient, will have a great effect on the redox status. In addition, information on the response of the system to variations in redox status is valuable in designing kinetic experiments on redox species. Determining whether a redox reaction is thermodynamically favorable under the system's redox status is an important initial step.

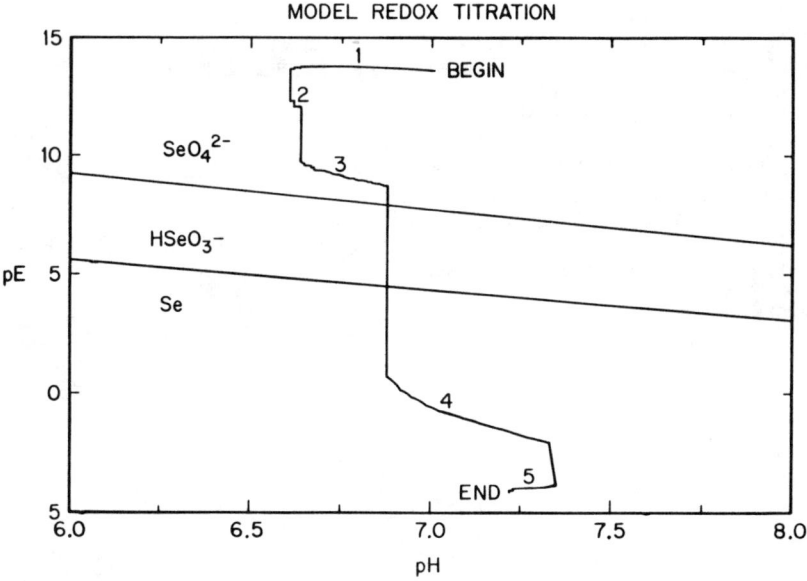

Figure 3. Path of redox titration of a model groundwater system in pH-pE space and the effect on the speciation of a trace element, e.g., selenium ($Se_T = 0.1$ μm).

## CONCLUSIONS

For a better comprehension of the chemistry of a groundwater system the redox status needs to be well-defined. Until recently, most efforts have relied solely upon $E_H$ or pE, intensity factors, as the master variable. However, it is apparent that these intensity factors do not truly represent the redox status of a system because some pertinent redox couples are not electroactive and redox reactions are generally slow and are not at equilibrium. In this paper, the oxidative capacity, a capacity factor, is operationally defined and shown to be a better descriptive parameter of the redox status. Determination of the OXC of an aqueous system allows investigators to classify the system in terms of well-defined geochemical and microbial parameters. This classification combined with other predictive tools, such as a redox titration, allows one to predict the identity and assess the role of chemical reactions and microbial populations within a specific groundwater system. As such, the capacity factor OXC should be determined in water quality assessment.

## ACKNOWLEDGMENTS

This work was supported by a grant from the Andrew W. Mellon Foundation.

## LITERATURE CITED

1. Stumm, W.; Morgan, J. J. Aquatic Chemistry, Wiley: New York, 1981, 538p.
2. Hutchins, S. R.; Tomson, M. B.; Bedient, P. B.; Ward, C. H. CRC Crit. Rev. in Env. Control, 1985, 15, 355-413.
3. Macalady, D. T.; Tratnyek, P. G.; Grundl, T. J. J. Contam. Hydrol., 1986, 1, 1-28.
4. Morgan, J. J.; Stone, A. T. In Chemical Processes in Lakes; Stumm, W. Ed., Wiley-Interscience: New York, 1985; pp. 389-426.
5. Hoffmann, M. R. Environ. Sci. Technol., 1981, 15, 345-53.
6. Lindberg, R. D.; Runnells, D. D. Science, 1984, 225, 925-27.
7. Hostettler, J. D. Am. J. Sci., 1984, 284, 734-59.
8. Stumm, W. Schweiz. Z. Hydrol., 1984, 46, 291-96.
9. Berner, R. A. J. Sed. Petrology, 1981, 51, 359-65.
10. Holm, T. R.; George, G. K.; Barcelona, M. J. CR-811477, U. S. E. P. A.-R. S. Kerr Environmental Research Laboratory, Ada, OK, 1986.

RECEIVED August 18, 1989

# Chapter 30

# Rates of Inorganic Oxidation Reactions Involving Dissolved Oxygen

## L. Edmond Eary and Janet A. Schramke

### Battelle, Pacific Northwest Laboratories, P.O. Box 999, Richland, WA 99352

Measured concentrations of redox couples in low temperature, aerated groundwaters often yield conflicting potentials, making it difficult to define system-wide redox conditions for use in equilibrium geochemical models. Redox disequilibrium is caused, in part, by the slow rates at which the reduced species are oxidized by dissolved $O_2$. To describe these rates, literature data were used to calculate half-lives for the $\Sigma Fe^{2+}$, $\Sigma S^{2-}$, $\Sigma Cu^+$, and $\Sigma Mn^{2+}$ oxygenation reactions as a function of pH at 0.2 atm $O_2$. These data and our experimentally measured rates for $\Sigma As^{3+}$ oxygenation yield half-lives that range from seconds to years, depending on solution conditions. The large differences in half-lives, which are strongly dependent on pH, imply that some reduced species are rapidly oxidized, whereas others may metastably persist in oxidized environments, resulting in redox disequilibrium. The half-life versus pH diagrams provide a convenient means for comparing oxygenation rates and assessing likely conditions for disequilibrium between reduced species and the $O_2(aq)/H_2O$ couple.

One of the major obstacles in applying geochemical models to natural systems is representing redox conditions accurately. Most geochemical models use thermodynamic laws to describe stability relationships for aqueous species and solids under assumed conditions of chemical equilibrium. Inherent in this assumption is that the speciation of redox species can be defined with a single, system-wide potential. Methods for defining this potential have included using electrodes to measure potentials (i.e., the Eh) or analytically determining the activities of both members of specific redox couples. At chemical equilibrium, either method would yield a single value that is representative of the system-wide redox potential. In some systems, such as acid-mine waters, high activities of $\Sigma Fe^{3+}$ and $\Sigma Fe^{2+}$ provide a dominant and reversible couple that poises the redox conditions, allowing accurate representation of the system-wide potential (1). However, many groundwaters are not well poised, and speciation calculations for numerous groundwaters have shown that various redox couples typically exist in a state of disequilibrium (2). Systems in disequilibrium yield mixed potentials that may or may not represent the redox distribution of any specific couple. When disequilibrium exists, geochemical model calculations that are based on system-wide potentials will give incorrect speciation results for the redox species. Alternatively, the dominant redox states for individual couples may be specified during a geochemical modeling exercise, but this practice requires either analytical determinations of redox species or knowledge of the rates of individual redox reactions.

Dissolved $O_2$ is the principal oxidant that controls the oxidation states of many redox-sensitive elements in groundwaters that are exposed to atmospheric $O_2$, and the redox potentials of these systems can usually be expected to evolve towards equilibrium with the $O_2(aq)/H_2O$ couple. This paper summarizes experimental data on the oxidation rates of $\Sigma Fe^{2+}$, $\Sigma S^{2-}$, $\Sigma Cu^+$, $\Sigma Mn^{2+}$, and $\Sigma As^{3+}$ ions by dissolved $O_2$, where the summation signs refer to total concentrations of the species present in the specified oxidation state. The objective is to compile the rate data and present them in plots of the half-life, $t_{1/2}$, of the reduced species versus pH for $PO_2 = 0.2$ atm, where $t_{1/2}$ is the time required for a reaction to reach 50% completion (5 half-lives is equivalent to 97% completion). The half-life diagrams provide a convenient reference frame for comparing reaction rates and can be used to identify the conditions and time frames under which disequilibrium between reduced species and the $O_2(aq)/H_2O$ couple is most likely to occur. It is these disequilibrium conditions for which direct analytical determinations of redox species are crucial to geochemical modeling studies.

OXIDIZING STRENGTH OF DISSOLVED OXYGEN

Theoretically, the oxidizing strength of dissolved $O_2$ is defined by the potential of the $O_2(aq)/H_2O$ couple (shown in Figure 1 as the long dashed line for $PO_2 = 0.2$ atm). If equilibrium with this couple always existed in aerated waters, then the other redox species shown in Figure 1 should exist entirely in their higher oxidation states. The oxygenation rates of $\Sigma Fe^{2+}$, $\Sigma Cu^+$, $\Sigma S^{2-}$, and $\Sigma Mn^{2+}$ are directly proportional to the dissolved $O_2$ concentration (discussed below), suggesting that dissolved $O_2$ exerts a redox potential that is consistent with the $O_2(aq)/H_2O$ couple. However, because of slow reaction rates, equilibrium between $O_2(aq)/H_2O$ and other redox couples may not always be achieved (discussed below).

Oxidation reactions involving dissolved $O_2$ are slow because the diatomic $O_2(aq)$ molecule must be split during the electron transfer process. The initial step in this process involves the molecular collision of the reduced species with $O_2(aq)$ to produce $O_2^-$. Subsequent redox and speciation reactions involving $O_2^-$ and the reduced species, such as transition metals, can produce $H_2O_2$ and $\cdot OH$ radicals (5). However, the peroxides generated by such reactions are metastable relative to both oxidation and reduction reactions because redox species with potentials less than the $O_2(aq)/H_2O_2$ couple are oxidized by $H_2O_2$, whereas species with higher potentials catalyze the decomposition of $H_2O_2$ to $H_2O$ and $O_2(aq)$ and are themselves reduced (Figure 1). For example, both $Fe^{2+}$ and $Cu^+$ are rapidly oxidized by peroxide but catalytic decomposition reactions between peroxide and $Fe^{3+}$ and $Cu^{2+}$ can produce steady-state concentrations of $Fe^{2+}$ and $Cu^+$ in aerated solutions, where they would not be expected according to available oxygenation rate data or thermodynamic data (6-9). Consequently, redox disequilibrium can be caused both by slow reactions between reduced species and $O_2(aq)$ and reactions between redox species and unstable oxygen reaction products such as peroxides and $\cdot OH$ radicals.

KINETICS OF OXYGENATION REACTIONS

The following sections summarize literature data on the kinetics of $\Sigma Fe^{2+}$, $\Sigma S^{2-}$, $\Sigma As^{3+}$, $\Sigma Cu^+$, and $\Sigma Mn^{2+}$ oxidation by dissolved $O_2$.

Ferrous Ion Oxygenation

The kinetics of $\Sigma Fe^{2+}$ oxygenation have been the focus of numerous studies, which are summarized in Table I. Most studies have found or assumed that the $\Sigma Fe^{2+}$ oxygenation rate is linearly dependent on both the partial pressure of $O_2$ and the $\Sigma Fe^{2+}$ concentration. In solutions with pH < 3.5 (Table I), the rate is independent of pH (10), yielding the rate expression

$$-d[\Sigma Fe^{2+}]/dt = k[\Sigma Fe^{2+}] PO_2 \qquad (1)$$

where k is a rate constant, $PO_2$ is the $O_2$ partial pressure in atm, and the brackets denote concentrations. In solutions with pH > 3.5, the rate is pH-dependent, changing transitionally from zero-order to a second-order dependence on $(OH^-)$. This dependence results in the following rate expression for $\Sigma Fe^{2+}$ oxygenation for pH > 4.5 (Table I)

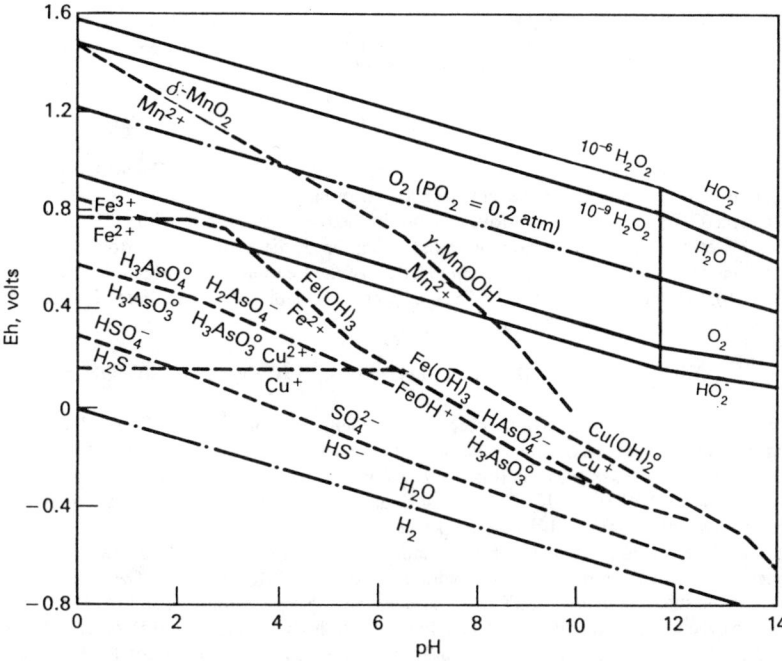

Figure 1. Eh-pH diagram for $O_2$(aq)/$H_2O$ (long-short dashed lines), peroxide species (solid lines), and other redox couples (short dashed lines) at 25°C. Thermodynamic data are from Wagman et al. (3), except for Mn species which are from Bricker (4).

Table I. Rate expressions for $\Sigma Fe^{2+}$ oxygenation

| Rate Expression<br>$-d[\Sigma Fe^{2+}]/dt =$<br>(mol/liter/sec) | log k[a] | Solution (M) | pH | Ref. |
| --- | --- | --- | --- | --- |
| $k[\Sigma Fe^{2+}]PO_2$ | -8.78 | Dilute $HClO_4$ | 1.0-3.5 | (10) |
| $k[\Sigma Fe^{2+}]PO_2(OH^-)^2$ | 12.12 | Dilute $HClO_4$ | 4.5-6.0 | (10) |
| $k[\Sigma Fe^{2+}]PO_2(OH^-)^2$ | 11.3-11.82 | Dilute $HClO_4$ | 6.84 | (11) |
| $k[\Sigma Fe^{2+}]PO_2(OH^-)^2$ | 11.4-11.7 | Fresh Waters | 6.5-7.5 | (12) |
| $k[\Sigma Fe^{2+}]PO_2(OH^-)$ | 5.35-5.43[b] | Seawater | 5.9-6.9 | (13) |
| $k[\Sigma Fe^{2+}]PO_2(OH^-)^2$ | 12.35-12.49[b] | Seawater | 7.1-8.4 | (13) |
| $k[\Sigma Fe^{2+}]PO_2(OH^-)^2$ | 11.22-11.47 | $10^{-1}$ Na salts | 6.50 | (14) |
| $k[\Sigma Fe^{2+}]PO_2(OH^-)^2$ | 11.65-11.77 | Alk < $10^{-2.2}$ eq/l | 5.5-7.2 | (15) |
| $k[\Sigma Fe^{2+}]PO_2[F^-]^2(OH^-)$ | 7.22 | $10^{-1.7}$-$10^{-1}$ F- | 5.6-7.0 | (14) |
| $k[\Sigma Fe^{2+}]PO_2[H_2PO_4^-](OH^-)$ | 7.24-7.07 | $10^{-1.7}$-$10^{-1}$ $H_2PO_4^-$ | 4.5-5.5 | (14) |

[a] Units for k (at 25°C) depend on the rate expression form.
[b] Extrapolated to 25°C from 15°C data, assuming a tenfold increase per 15°C (11).

$$-d[\Sigma Fe^{2+}]/dt = k[\Sigma Fe^{2+}] PO_2 (OH^-)^2 \qquad (2)$$

where the parentheses denote activities.

The pH-dependence for $\Sigma Fe^{2+}$ oxygenation rate is shown by a plot of the $\Sigma Fe^{2+}$ half-life versus pH in Figure 2. At a constant $PO_2 = 0.2$ atm, the half-life for $\Sigma Fe^{2+}$ is given by

$$t_{1/2} = 0.693/k' \qquad (3)$$

where $k' = k(OH^-)^n(0.2)$ with $n = 0$, 1, or 2, depending on pH (Table I). Calculated half-lives for $\Sigma Fe^{2+}$ range from years at pH < 3.5 but decrease to minutes at pH > 6 because the second-order dependence on $(OH^-)$ results in a 100-fold increase in the oxygenation rate with each unit increase in pH. This pH effect is caused by the more rapid oxidation of the hydrolyzed ferrous species, whose concentrations are increased with increase in pH, than of the free ion, $Fe^{2+}$, which dominates in acidic solutions (16). Millero (16) expresses $\Sigma Fe^{2+}$ oxygenation as the sum of the rates for the individual ferrous species, i.e.,

$$-d[\Sigma Fe^{2+}]/dt = k_0[Fe^{2+}] + k_1[FeOH^+] + k_2[Fe(OH)_2(aq)] + k_3[Fe(OH)_3^-] \qquad (4)$$

where $k_0$ through $k_3$ are pseudo-first-order rate constants. A value of $10^{-8.08}$ sec$^{-1}$ for $k_0$ can be obtained from the rate studies of Singer and Stumm (10), which were conducted with $PO_2 = 0.2$ atm in acidic solutions where $Fe^{2+}$ was the dominant species (Table I). In comparison, Millero (16) has determined values of $10^{-1.55}$ and $10^{-3.86}$ sec$^{-1}$ for $k_1$ and $k_2$, respectively, indicating that $Fe(OH)_2(aq)$ is oxidized more rapidly than the other ferrous species.

Millero (16) has also predicted that oxygenation rates in seawater should be 28% slower than in freshwater because of the formation of ion pairs (i.e., $FeCl^+$, $FeCl_2(aq)$, and $FeSO_4(aq)$), which are less rapidly oxidized than the hydrolyzed ferrous species. In contrast, Tamura et al. (14) have shown that the $\Sigma Fe^{2+}$ oxygenation rate is increased in solutions with high phosphate and fluoride concentrations (Figure 2). The increased rate may be caused by more rapid oxygenation of $Fe^{2+}$-phosphate and $Fe^{2+}$-fluoride ion pairs than of the hydrolyzed species, although the formation of these species as active intermediates has not been confirmed (14). However, concentrations of at least $10^{-1.5}$ M $\Sigma F^-$ and $10^{-3.5}$ M $\Sigma PO_4^{3-}$ are needed to produce substantial increases in the oxidation rate (Figure 2). In most groundwaters, the $\Sigma Fe^{2+}$ oxygenation rate is well represented by the expressions for dilute solutions given in Table I. These expressions and the resulting half-lives show that disequilibrium between $\Sigma Fe^{2+}$ and the $O_2(aq)/H_2O$ couple is most likely to occur in acidic solutions where the oxygenation rate is slowest (Figure 2). In neutral to alkaline aerated solutions, $\Sigma Fe^{2+}$ is rapidly oxidized and should not normally persist for extended periods.

Sulfide Ion Oxygenation

The complete oxidation of $\Sigma S^{2-}$ involves the transfer of eight electrons from the sulfide ion to an oxidant

$$H_nS + 4 H_2O = SO_4^{2-} + (n+8) H^+ + 8 e^- \qquad (n = 0 \text{ to } 2) \qquad (5)$$

but it is unlikely that so many electrons would be transferred in a single molecular collision. Even the splitting of the diatomic $O_2$ molecule can consume four electrons at most. Therefore, the complete oxidation of $\Sigma S^{2-}$ to $SO_4^{2-}$ involves the formation of intermediate oxidation products. These products have been detected in numerous experimental studies of $\Sigma S^{2-}$ oxygenation and include $S_x^{2-}$ ($x = 2$-5), $S(s)$, $S_2O_3^{2-}$, and $SO_3^{2-}$. Other polythionates, such as $S_3O_6^{2-}$ and $S_4O_6^{2-}$, may also form initially, but studies by Rolla and Chakrabarti (17) have shown that these species decompose to $S_2O_3^{2-}$ in a few hours in solutions equilibrated with atmospheric $O_2$.

The formation of a variety of intermediate products results in a complex reaction pathway for complete $\Sigma S^{2-}$ oxidation, and at present, the formation mechanisms for each of the intermediate products are not well defined. Most studies of the kinetics of $\Sigma S^{2-}$ oxidation by dissolved $O_2$ have simplified their results (Table II) by reporting the overall rate of $\Sigma S^{2-}$ disappearance caused by a general oxygenation reaction

$$\Sigma S^{2-} + O_2(aq) \rightarrow \text{oxidation products} \qquad (6)$$

where the various intermediate products are assumed to be eventually converted to $SO_4^{2-}$. Most workers (Table II) report a first-order dependence on both [$\Sigma S^{2-}$] and $PO_2$ for the rate of [$\Sigma S^{2-}$] disappearance, i.e.,

$$-d[\Sigma S^{2-}]/dt = k[\Sigma S^{2-}] \, PO_2 \qquad (7)$$

where k depends on pH. However, Chen and Morris (18), report fractional orders of 1.36 and 0.56 for [$\Sigma S^{2-}$] and $PO_2$, respectively (Table II).

Equation 7 implies that the initial rate-determining step involves the collision of a sulfide ion with $O_2(aq)$, followed by oxidation and disproportionation of the products (21,24), e.g.,

$$HS^- + 3/2 \, O_2(aq) \rightarrow SO_3^{2-} + H^+ \qquad \text{(slow)} \qquad (8)$$

$$SO_3^{2-} + 1/2 \, O_2(aq) \rightarrow SO_4^{2-} \qquad \text{(fast)} \qquad (9)$$

$$SO_3^{2-} + HS^- + 1/2 \, O_2(aq) + H^+ \rightarrow S_2O_3^{2-} + H_2O \qquad \text{(fast)} \qquad (10)$$

$$S_2O_3^{2-} + 1/2 \, O_2(aq) \rightarrow SO_4^{2-} + S(s) \qquad \text{(slow)} \qquad (11)$$

In this reaction sequence, $SO_3^{2-}$ oxidation to $SO_4^{2-}$ and the initial formation of $S_2O_3^{2-}$ proceed rapidly, while product $S_2O_3^{2-}$ and colloidal sulfur are slowly oxidized to $SO_4^{2-}$. Avrahami and Golding (24) report a half-life for $S_2O_3^{2-}$ oxygenation of eight days in aerated, alkaline solutions at 25°C. O'Brien and Birkner (19) also found that $S_2O_3^{2-}$ was a metastable reaction product, although their experiments were generally only 72 hours long. They proposed a parallel set of reactions for solutions with high ratios of [$O_2(aq)$] to [$\Sigma S^{2-}$] in which the oxidation of $\Sigma S^{2-}$ produces equimolar amounts of $SO_3^{2-}$, $S_2O_3^{2-}$, and $SO_4^{2-}$, i.e.,

$$4 \, HS^- + 51/2 \, O_2(aq) \rightarrow SO_3^{2-} + S_2O_3^{2-} + SO_4^{2-} + 2 \, H^+ + H_2O \qquad (12)$$

Subsequently, the intermediate products are oxidized to $SO_4^{2-}$, but O'Brien and Birkner (19) did not report rates for these reactions.

Rate expressions from various studies given in Table II were used to calculate half-lives for $\Sigma S^{2-}$ oxygenation as a function of pH for 0.2 atm $PO_2$ and 25°C. (Figure 3). The shortest half-lives were derived from the experimental studies conducted in seawater solutions and range from five hours to 20 minutes (21-23), although more recent studies by Millero et al. (20) indicate that half-lives of 25 hours are more typical of seawater solutions. More rapid oxygenation of $\Sigma S^{2-}$ in seawater compared to freshwaters has been attributed to the primary salt effect (19,20), and to catalysis by various organic compounds and metal ions present in seawater (21,25). In buffered, distilled water solutions, half-lives for $\Sigma S^{2-}$ range from one to 35 days, depending on the pH and the ratio of [$O_2(aq)$] to [$\Sigma S^{2-}$] (Figure 3). The most complex pH dependence was found by Chen and Morris (18), who report two pH regions of increased oxygenation rate in alkaline solutions (Figure 3). The increase in $\Sigma S^{2-}$ oxygenation rate at pH 8.0 was attributed to the formation of polysulfide species produced by reaction between colloidal sulfur and $HS^-$ (18). Chen and Morris (18) observed the formation of polysulfide species spectrophotometrically in solutions with low ratios of [$O_2(aq)$] to [$\Sigma S^{2-}$], and polysulfide formation increased at the same pH at which oxygenation rates were most rapid, implying that these species are oxidized more rapidly than $HS^-$ or $H_2S(aq)$. The decrease in rate at pH > 8.3 (Figure 3) may be caused

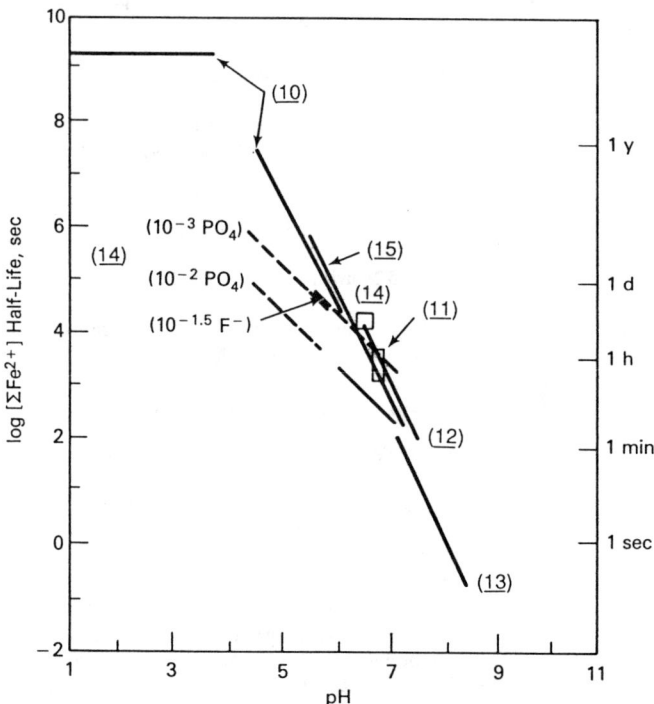

Figure 2. Half-lives for $\Sigma Fe^{2+}$ versus pH at $PO_2 = 0.2$ atm at 25°C. Underlined numbers refer to sources of data used in rate calculations (see Table I).

Table II. Rate expressions for $\Sigma S^{2-}$ oxidation

| Rate Expression $d[\Sigma S^{2-}]/dt =$ (mol/liter/sec) | -log k[a] | pH | $[\Sigma S^{2-}]$ -log(M) | $[O_2(aq)]$ -log(M) | Ref. |
|---|---|---|---|---|---|
| $k[\Sigma S^{2-}]^{1.36}PO_2^{0.56}$ | 4.5-3.9[b] | 6.9-12.5 | 4.3-3.7 | 3.8-3.1 | (18) |
| $k[\Sigma S^{2-}]PO_2$ | 4.9-4.3[b] | 4.0,7.5,10.0 | 4.6-2.9 | 3.7-3.0 | (19) |
| $k[\Sigma S^{2-}]PO_2$ | 6.2-5.3[b] | 1.0-12.0 | 4.6 | Air Sat. | (20) |
| $k[\Sigma S^{2-}]PO_2$ | 2.5 | 7.5-7.8 | 5.7 | 5.0-4.6 | (21)[c] |
| $k[\Sigma S^{2-}]PO_2$ | 2.7 | 8.0,8.5 | 6.0-3.7 | Air Sat. | (22)[c] |
| $k[\Sigma S^{2-}]PO_2$ | 1.7 | 8.2 | 5.9 | Air Sat. | (23)[c] |

[a] The units for k (at 25°C) depend on the rate expression form.
[b] Rate constant is dependent on the pH.
[c] Seawater solutions.

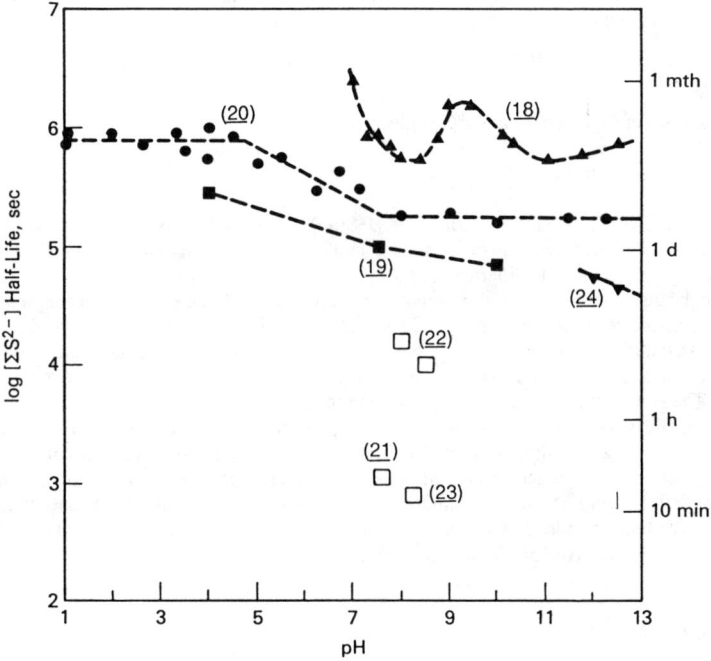

Figure 3. Half-lives for $\Sigma S^{2-}$ versus pH at $PO_2 = 0.2$ atm at 25°C. Underlined numbers refer to sources of data used in rate calculations (see Table II).

by a slowing of the rate of colloidal sulfur formation as the pH is increased, thereby decreasing the rate of polysulfide formation (18). However, Millero (25) has shown that the rate increase between pH 6.0 and 8.3 reported by Chen and Morris (18) can be accounted for by the dissociation of $H_2S(aq)$ to produce $HS^-$ (i.e., $H_2S(aq) = HS^- + H^+$, pK = 6.98 at 25°C (3)). Using the rate data of Chen and Morris (18), Millero (25) expresses the overall rate of $\Sigma S^{2-}$ oxygenation as the sum of the rates of the individual species,

$$-d[\Sigma S^{2-}]/dt = k_1 [H_2S(aq)][O_2(aq)] + k_2 [HS^-][O_2(aq)] \quad (13)$$

and has determined $k_2$ to be substantially greater than $k_1$. In subsequent experimental studies, Millero et al. (20), observed that the rate increased as the pH was raised from 6.0 to 8.0 (Figure 3), confirming that the more reactive species is $HS^-$, and that at higher pH, the rate is independent of pH, in contrast to the complex behavior reported by Chen and Morris (18) (Figure 3). Equation 13 can be rewritten in terms of $PO_2$ using an activation energy of 57 kJ/mol (20) to give

$$-d[\Sigma S^{2-}]/dt = k[\Sigma S^{2-}] PO_2 \quad (14)$$

The rate constant in Equation 14 at 25°C is given by

$$\log k = (10^{-4.46} + 10^{-3.83} K_1/[H^+]) / (1 + K_1/[H^+]) - 0.91 \quad (15)$$

where $K_1$ is the first dissociation constant for $H_2S(aq)$. The results reported by O'Brien and Birkner (19) also indicate no significant pH dependence in alkaline solutions, and yield half-lives that are in good agreement with those of Millero et al. (20) (Figure 3).

Although, the kinetics of $\Sigma S^{2-}$ oxygenation are complex and rate expressions are most developed only for the initial reaction step between $\Sigma S^{2-}$ and $O_2(aq)$ three conclusions can be drawn from the available rate data: 1) Under conditions where $O_2(aq)$ is not a limiting reactant, the initial oxidation step leads to the formation of sulfoxy species such as $SO_3^{2-}$ and $S_2O_3^{2-}$, which are eventually oxidized to $SO_4^{2-}$. The slow steps under such conditions are the initial reaction between $O_2(aq)$ and $H_2S(aq)$ or $HS^-$, with half-lives of one to 10 days, and the subsequent oxidation of $S_2O_3^{2-}$, with a half-life of about eight days. 2) In $O_2(aq)$-limited solutions, the initial step leads to the formation of colloidal sulfur and polysulfide species (18). Under these conditions, the slow steps include the initial one, with a half-life of 10 to 30 days (Figure 3), and the subsequent slower oxidation of polysulfide and colloidal sulfur (18). Rates for these two latter reactions are not well known. 3) Sulfide oxygenation rates may be faster in seawater (half-lives of 20 minutes to five hours) than in dilute solutions because of catalysis by unspecified agents present in seawater.

Arsenite Ion Oxygenation

The $\Sigma As^{5+}/\Sigma As^{3+}$ couple has been suggested as a redox indicator because its potential lies in the intermediate Eh-pH domain of most natural waters (26). Changes in As redox state with change in season have been observed in some freshwater lakes (27), indicating that changes in As oxidation state do occur at observable rates in natural environments. However, the rates of such reactions are not well known.

Consequently, a series of experiments were conducted to measure the rate of $\Sigma As^{3+}$ oxidation by dissolved $O_2$ as a function of pH at 25° and 90°C. Experiments were conducted in 1.0-liter glass kettles, and temperatures were controlled to $\pm 1$°C with heating mantles. Carbon dioxide-free air was continuously bubbled through the solutions to maintain $PO_2 = 0.2$ atm. The initial $\Sigma As^{3+}$ concentration was $10^{-4.0}$ M, and the background electrolyte was $10^{-2.0}$ M NaCl in all experiments. To measure rates, samples were removed periodically from the kettles and total As and $\Sigma As^{5+}$ concentrations were determined by the molybdate-blue method of Johnson and Pilson (28). The concentrations of $\Sigma As^{3+}$ were determined by difference.

Oxygenation rates were observed to be most rapid in acidic solutions, and were decreased as the pH was raised from 2.0 to 5.5 (Figure 4). However, even in the experiments at pH 2.0, only about 9% of the initial $\Sigma As^{3+}$ concentration was oxidized to $\Sigma As^{5+}$ after about 100 days (Figure 4). In alkaline solutions, oxygenations rates increased with pH from the values measured at pH 5.5, peaking between pH 9.0 and 11.0. At pH 12.0 and 12.5, slower oxygenation rates were again apparent (Figure 4). Oxygenation rates at 90°C were similar to those observed at 25°C and showed the same dependence on pH (Figure 4).

These rate data indicate that $\Sigma As^{3+}$ oxidation by dissolved $O_2$ is slow and has a complex dependence on pH. To show this dependence, the rate data from Figure 4 were fit by least-squares regression to second-order polynomials with zero intercepts. The derivatives of the polynomials at time equal to zero were used to determine the initial rates. A plot of these initial rates (solid symbols in Figure 5) as a function of pH shows that oxygenation rates are increased when pH < 4.0 and when pH is from 9.0 to 11.0. Comparisons of our data with a rate measurement reported by Cherry et al. (26), which was calculated from their $\Sigma As^{3+}$ loss with time (initial conditions of: $10^{-6.18}$ M $\Sigma As^{3+}$ and 1.0 atm $PO_2$), indicates about an order of magnitude difference in rate between our two studies (Figure 5). Rates determined by Johnson and Pilson (29) in artificial seawater at 25°C overlap with these measurements, but indicate a greater pH dependence (Figure 5). Johnson and Pilson (29) also noted a dependence on salinity and that rates were increased by exposure to sunlight. These effects may account for the higher rates measured in their experiments at pH 7.5 to 8.2 compared to the rates observed in our experiments. Some bacteria present in seawater have also been reported to oxidize and reduce As (30,31) and do so at rates that are significantly faster than oxidation solely by dissolved $O_2$ (Figure 5).

Arsenite oxygenation rates have previously been reported to be increased in either acidic or alkaline solutions compared to near-neutral solutions (32). The rate increase at pH < 4.0 is not easily explained. The dominant $\Sigma As^{3+}$ species at all pH values less than 9.23 is $H_3AsO_3(aq)$, indicating that a change in speciation with pH is not the probable cause for the increased rate in acidic solutions that was observed in the experiments. The oxidation of $\Sigma As^{3+}$ to $\Sigma As^{5+}$ requires the transfer of two electrons. If oxidation by $O_2(aq)$ proceeds by a series of one electron steps, then it is possible that intermediate $As^{4+}$ species are more stable under acidic conditions, thereby resulting in an increase in oxygenation rate. However, no evidence for this mechanism is available.

The rate dependence in alkaline solutions (Figure 5) suggests that oxidation is affected by $\Sigma As^{3+}$ speciation. The rate increase shown in Figure 5 is centered at the pH where $H_2AsO_3^-$ is the major $As^{3+}$ species ($H_3AsO_3(aq) = H_2AsO_3^- + H^+$; pK = 9.23 at 25°C (3)). A logarithmic plot of oxidation rate versus the $H_2AsO_3^-$ concentration in alkaline solutions gave a linear relationship with a slope of 0.46+0.09. Using this dependence, $\Sigma As^{3+}$ oxygenation in solutions with pH between 8 and 12.5 is given by

$$-d[\Sigma As^{3+}]/dt = k[H_2AsO_3^-]^{0.46\pm0.09} \qquad (16)$$

where k = $10^{-9.86}$ mol$^{0.5}$/liter$^{0.5}$/sec at 25°C and 0.2 atm $PO_2$.

The slow rate of $\Sigma As^{3+}$ oxygenation by dissolved $O_2$ is shown in the half-life versus pH diagram of Figure 6. The oxygenation rate data for seawater from Johnson and Pilson (29) give half-lives that range from several months to a year (Figure 6). These long half-lives suggest that the redox cycling of As in seawater may be controlled as much by microbial processes as by reactions with dissolved $O_2$ (30,31). In freshwaters, our data indicate that $\Sigma As^{3+}$ oxygenation is also very slow, with a half-life in the range of one to three years, although it is possible that rates may be catalyzed by unknown aqueous species. Also, $\Sigma As^{3+}$ oxidation by Mn oxides has been to shown to be rapid (33) and may be more important for affecting As redox chemistry than dissolved $O_2$ in oxidized environments.

<u>Cuprous Ion Oxygenation</u>

Oxidation of $Cu^+$ by dissolved $O_2$ is rapid in most solutions. To slow the reaction rate so that accurate measurements could be made, Gray (34) used the acetonitrile complex to limit the

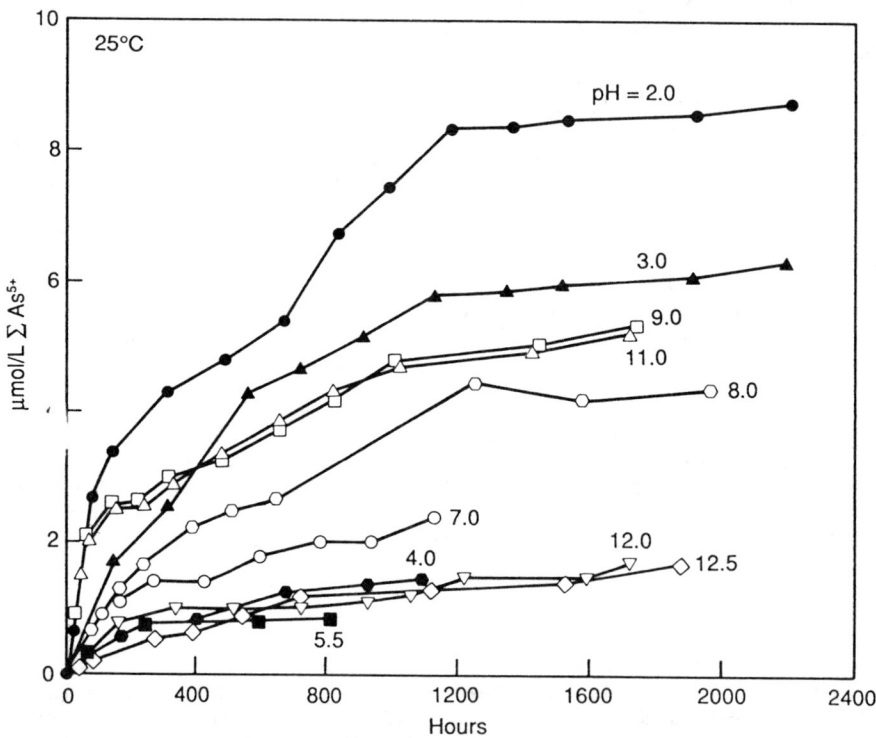

Figure 4. Rates of $\Sigma As^{5+}$ increase with time caused by $\Sigma As^{3+}$ oxidation at $PO_2 = 0.2$ atm at 25°C.

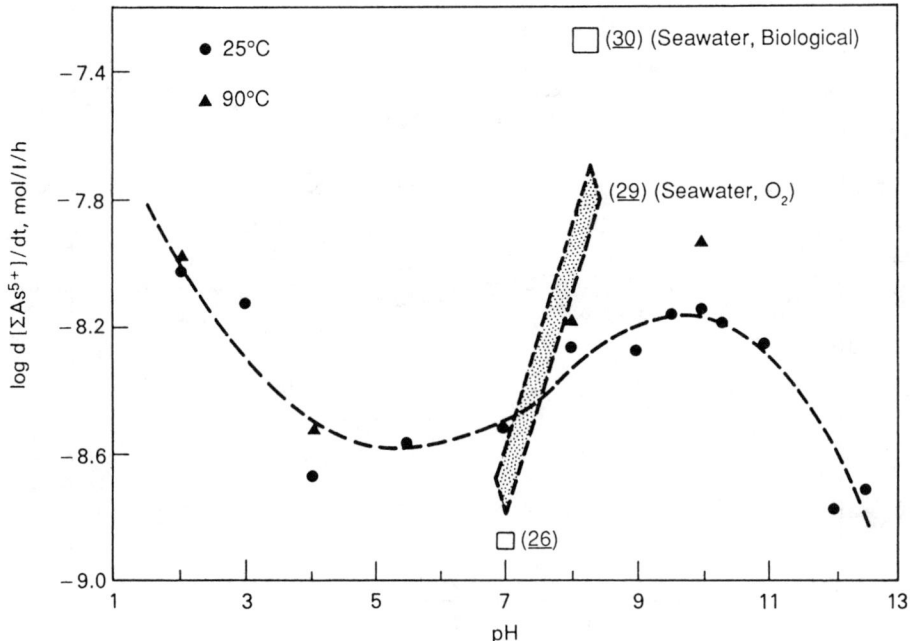

Figure 5. Logarithms of the initial rates of $\Sigma As^{3+}$ oxidation as a function of pH from this study and the literature (referred to by the underlined reference numbers).

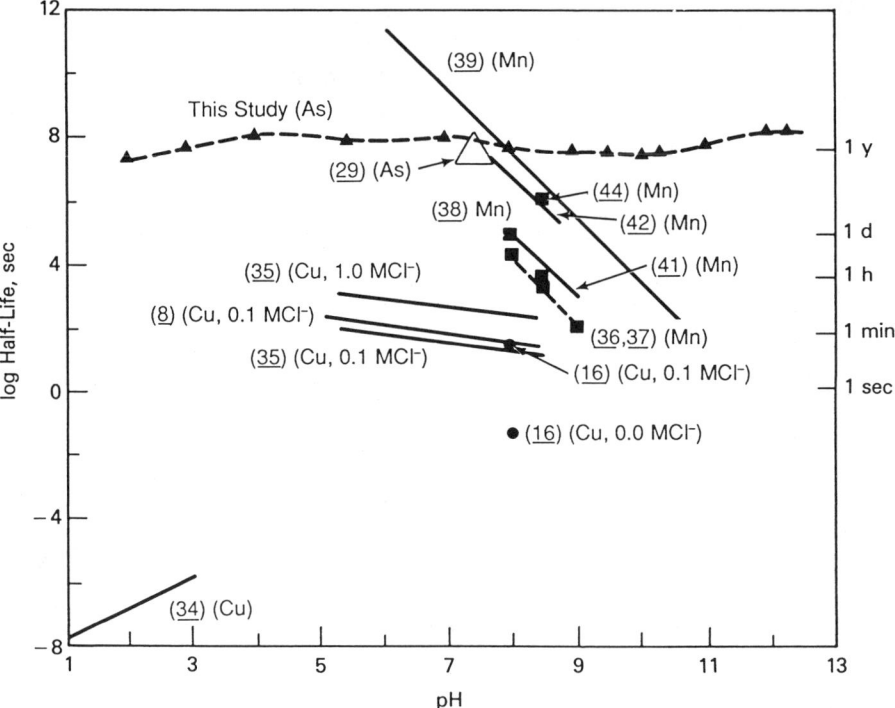

Figure 6. Half-lives for $\Sigma As^{3+}$, $\Sigma Cu^+$, and $\Sigma Mn^{2+}$ versus pH for $PO_2 = 0.2$ atm at 25°C. Solution compositions are given in parentheses. Underlined numbers refer to sources of data used in rate calculations (see Tables III and IV).

concentration of the free $Cu^+$ ion in solution. Gray (34) found that $Cu^+$ oxygenation rate is directly dependent on $[Cu^+]$ and $[O_2(aq)]$ in acidic solutions (Table III) and proposed the mechanism

$$Cu(CH_3CN)_2^+ = Cu^+ + 2\ CH_3CN(aq) \qquad (17)$$

$$Cu^+ + O_2(aq) = CuO_2^+ \qquad (18)$$

$$CuO_2^+ + H^+ \rightarrow Cu^{2+} + HO_2(aq) \qquad (19)$$

$$Cu^+ + HO_2(aq) \rightarrow Cu^{2+} + HO_2^- \qquad (20)$$

$$H^+ + HO_2^- = H_2O_2(aq) \qquad (21)$$

Reactions 17 and 18 were assumed to be rapid so that the involved species maintained equilibrium distributions (34). The third step shown in Reaction 19 was assumed to be rate-limiting, yielding the rate expression:

$$-d[Cu^+]/dt = k[Cu(CH_3CN)_2^+][O_2(aq)][H^+] / [CH_3CN(aq)]^2 \qquad (22)$$

where $k = k_{19}K_{17}K_{18} = 10^{7.84}\ sec^{-1}$ at 30°C, $K_{17}$ and $K_{18}$ refer to equilibrium constants, respectively, and $k_{19}$ is the rate constant for Reaction 19. Substituting the equilibrium constant for the acetonitrile complexation reaction (log $K_{17} = 4.35$) gives the rate equation for oxygenation of uncomplexed $Cu^+$

$$-d[Cu^+]/dt = k'[Cu^+][O_2(aq)][H^+] \qquad (23)$$

where $k' = 10^{5.19}\ liter^2/mol^2/sec$. For $PO_2 = 0.2$ atm, Equation 23 yields half-lives of $10^{-7.8}$ to $10^{-5.8}$ seconds over pH 1.0 to 3.0, showing that $\Sigma Cu^+$ is oxidized very rapidly in acidic solutions (Figure 6).

The overall stoichiometry of the mechanism given in Reactions 17-21 indicates that one mole of $H_2O_2$ should be produced for every two moles of $Cu^+$ oxidized and for every one mole of $O_2(aq)$ reduced. Gray (34) detected $H_2O_2$ in the experimental solutions and determined ratios close to those expected. In other experiments on $Cu^+$ oxidation in air-saturated seawater solutions, Moffett and Zika (8) determined that one mole of $H_2O_2$ was produced for every two moles of $Cu^+$ oxidized, in agreement with Gray's mechanism. Moffett and Zika (8) also determined that the oxygenation rate was first order with respect to $[Cu^+]$ and that it was proportional to $(H^+)^{0.22}$ between pH 5.1 to 8.6 (Table III). Moffett and Zika (8) also observed that the oxidation rate slowed considerably over time because of a reverse reaction in which $Cu^{2+}$ was reduced back to $Cu^+$ by the $H_2O_2$ that was generated by the sequence of redox and speciation reactions involved in splitting of the $O_2(aq)$ molecule (Reactions 17-21). This reverse reaction was more rapid in alkaline solutions and produced steady-state concentrations of $Cu^+$ in aerated seawater solutions where only $Cu^{2+}$ might be expected from both the oxygenation rate information and speciation calculations based on equilibrium with the $O_2(aq)/H_2O$ couple.

Moffett and Zika (8) also found that the rate of $Cu^+$ oxidation is inversely proportional to chloride concentration (Table III), indicating that the free $Cu^+$ ion is oxidized more rapidly than $Cu^+$-chloride ion pairs (Figure 6). However, Millero's (16) analysis of Moffett and Zika's data revealed that the CuCl(aq) is also oxidized by dissolved $O_2$ at an appreciable rate. Millero (16) derived the following rate expression for $Cu^+$ oxidation in seawater for $PO_2 = 0.2$ atm

$$-d\ln[\Sigma Cu^+]/dt = k0\cdot\alpha Cu^+ + k_1\cdot\beta_1\cdot\alpha Cu^+ [Cl^-] \qquad (24)$$

where $k_0 = 14.1\ sec^{-1}$ and $k_1 = 3.9\ sec^{-1}$ are pseudo first-order rate constants for $Cu^+$ and CuCl(aq) oxidation, respectively, $\beta_1$ is the formation constant for CuCl(aq), and $\alpha Cu^+$ is the ratio of free $Cu^+$ to total dissolved Cu. The half-lives for $\Sigma Cu^+$ that are calculated from Equation 24, which is applicable at pH 8.0, range from 0.1 seconds in chloride-free solutions to more than an hour in 0.1 M chloride solutions (Figure 6).

More recently, Sharma and Millero (35) measured the $\Sigma Cu^+$ oxygenation rate over a wide range of solution conditions (Table III) and developed the following rate expression

$$-d[\Sigma Cu^+]/dt = k[\Sigma Cu^+][O_2(aq)] \qquad (25)$$

The rate constant k in kg/mol/min is given by

$$\log k = 10.73 + 0.23 pH - 2373/T - 3.33 I^{0.5} + 1.45 I \qquad (26)$$

where T is in Kelvins and I is ionic strength. The 0.23-dependence on $(H^+)$ is nearly equivalent to the 0.22 dependence determined by Moffett and Zika (8) and the temperature dependence corresponds to an activation energy of 45.6 kJ/mol.

Half-lives for $\Sigma Cu^+$ oxygenation that are calculated from Equations 25 and 26 range from two to 30 minutes, depending on I and pH (Figure 6). In view of the half-lives for $\Sigma Cu^+$ shown in Figure 6 and the increased rate of $\Sigma Cu^{2+}$ reduction by $H_2O_2$ that occurs in alkaline solutions, it is more likely that $Cu^+$ will persist in solutions that contain high concentrations of chloride and that are also alkaline, such as seawater. In most dilute neutral to acidic groundwaters that contain even traces of dissolved $O_2$, $\Sigma Cu^+$ will be rapidly oxidized to $\Sigma Cu^{2+}$.

Manganous Ion Oxygenation

The oxygenation of $\Sigma Mn^{2+}$ results in the formation of an Mn oxide solid, $MnO_x(s)$, in which x ranges from 1.3 to 2.0, depending on solution conditions. The oxygenation rate is most rapid in alkaline solutions and tends to increase over time, indicating that the reaction is catalyzed by the product Mn oxide (36-39). Autocatalysis of the oxygenation reaction leads to the rate expression (Table IV)

$$-d[\Sigma Mn^{2+}]/dt = k_0[\Sigma Mn^{2+}] + k[\Sigma Mn^{2+}][MnO_x(s)] \qquad (27)$$

where $k_0$ and k are rate constants for the aqueous and autocatalytic components of the oxygenation reaction, respectively. After an initial period of $MnO_x(s)$ formation, the rate is dominated by the second term in Equation 27 for which k has a second-order dependence on hydroxyl ion, i.e.,

$$k = k'[OH^-]^2 P_{O_2} \qquad (28)$$

The second-order dependence suggests that the rate-determining step involves a single electron transfer between $Mn(OH)_2(aq)$ and $O_2(aq)$. The $Mn(OH)_2(aq)$ species is present in very low concentrations under most pH conditions. However, the broken oxygen and hydroxyl bonds on the surface of $MnO_x(s)$ may produce species approximating $Mn(OH)_2(aq)$ in the adsorbed layer, thereby catalyzing the oxidation reaction by dissolved $O_2$ (40). The rate is also reported to be increased by lepidocrocite (41,42), montmorillonite, kaolinite, and goethite (39), indicating that the surfaces of other oxide and hydroxide solids may catalyze the oxygenation reaction (Table IV). In fact, some type of surface or possibly bacteria appears to be required to initiate $\Sigma Mn^{2+}$ oxidation, given that oxygenation rates in homogeneous, sterile solutions are 10 to 100 times slower than those measured in heterogeneous solutions and many natural waters (42-44). In related rate studies, Davies (42) found that cations, such as $Mg^{2+}$, decrease the $\Sigma Mn^{2+}$ oxygenation rate by displacing $Mn^{2+}$ from binding sites on the oxide surface, and that rates were also decreased by chloride ions because of the formation of $MnCl^+$ and $MnCl_2(aq)$, which are less rapidly oxidized than the uncomplexed $Mn^{2+}$ species.

Kessick and Morgan (40) found that the stoichiometries of the Mn oxides produced by $\Sigma Mn^{2+}$ oxygenation at pH 8.8 and 9.0 were consistent with $MnOOH(s)$. Hem and Lind (45) found that the initial Mn oxide produced by oxidation at pH 8.5 to 9.5 and 25°C was $Mn_3O_4(s)$ (hausmanite), but at 0°C, the initial oxide was ß-$MnOOH(s)$ (feitknechite), with variable amounts of γ-$MnOOH(s)$

Table III. Rate expressions for $\Sigma As^{3+}$ and $\Sigma Cu^+$ oxidation

| Rate Expression $-d[\Sigma As^{3+},\Sigma Cu^+]/dt =$ (mol/liter/sec) | log k[a] | Solution (M) | $[\Sigma As^{3+},\Sigma Cu^+]$ -log(M) | pH | Ref. |
|---|---|---|---|---|---|
| $k[H_2AsO_3^-]^{0.5}$ | -9.86 | 0.01 NaCl | 4.0 | 8.0-12.5 | b |
| $k[\Sigma As^{3+}]^{0.68}[S]^{0.67}[H^+]^{-0.76}$ | -15.9 | Seawater[c] | 6.9-5.4 | 7-8 | (29) |
| $k[\Sigma Cu^+][H^+]PO_2$ | 9.3[d] | Dilute $HClO_4$, 0.06-0.13 $CH_3CN$ | 6.7-6.0 | 1-3 | (34) |
| $k_0 a Cu^+ + k_1\beta_1\alpha Cu^+[Cl^-]\}[\Sigma Cu^+]$ | $k_0$=14.1 $k_1$=3.9 | 0.1-0.7 NaCl | 6.5 | 8.0 | (16) |
| $k[\Sigma Cu^+][O_2(aq)]$ | e | Seawater[c] | 5.0-7.0 | 5.3-8.6 | (35) |

[a] Units for k (at 25°C) depend the rate expression form.
[b] This study.
[c] Salinity (S) = 5-45 (g/kg)(1000).
[d] log k at 30°C.
[e] $k = 10.73 + 0.23pH - 2373/T - 3.33I^{0.5} + 1.45I$, where T is Kelvins, and $[O_2(aq)]$ is in mol/kg.

Table IV. Rate expressions for $\Sigma Mn^{2+}$ oxidation

| Rate Expression $-d[\Sigma Mn^{2+}]/dt =$ (mol/liter/sec) | k[a] | Solution (M) | $[\Sigma Mn^{2+}]$ -log M | pH | Ref. |
|---|---|---|---|---|---|
| $k_0[Mn^{2+}] + k[Mn^{2+}][MnO_x(s)]$ | $k_0=10^{9.76}[OH^-]^2PO_2$ $k=10^{6.97}[OH^-]^2PO_2$ | 0.01 $H_3BO_3$ | 3.3-3.7 | 9.0-9.8 | (39) |
| $k[\Sigma Fe^{3+}(s)][\Sigma Mn^{2+}]PO_2$[b] | $k=10^{10.56}[OH^-]^2$ | 0.7 NaCl, $10^{-3}$ - $10^{-4}$ $[\Sigma Fe^{3+}(s)]$ | 4.3 | 8.0-8.7 | (41) |
| $k[=FeOH][Mn^{+2}][\alpha]PO_2$[c] | $k=10^{-12.2}[OH^-]^2$ | 0.7 $NaClO_4$ | 6.0 | 7.5-8.5 | (42) |

[a] Units for k (at 25°C) depend on the rate expression form.
[b] $[\Sigma Fe^{3+}(s)]$ refers to molar concentration of ferric iron in solid phases such as lepidocrocite, γ-FeOOH(s).
[c] [=FeOH] is the molar concentration of surface hydroxyl groups and [α] is concentration of Fe oxide in g/l.

(manganite), and α-MnOOH(s) (groutite) depending on whether $MnSO_4(aq)$ was the dominant aqueous Mn species. The initial oxides, $Mn_3O_4(s)$ and ß-MnOOH(s), converted to γ-MnOOH(s) over time (45), and in aerated environments with continued aging they may alter to ∂-$MnO_2$(s) (birnessite).

Figure 6 shows half-lives versus pH for $\Sigma Mn^{2+}$ at 25°C and 0.2 atm $PO_2$ calculated from the rate expressions given in Table IV (solid lines) and experimental data (symbols). Extrapolation of Wilson's (39) rate expression to near-neutral pH yields half-lives on the order of years (Figure 6). However, as the pH is increased to the experimental conditions of 9.0 to 9.8, the half-life is decreased to minutes because of the second-order dependence on ($OH^-$). The half-lives calculated from the rate expression given by Wilson (39) are generally longer than those determined from the other studies, although a comparable half-life is calculated from the rate measurements made by Delfino and Lee (44) in a lake water with pH 8.5. The expression reported by Davies (42) for $\Sigma Mn^{2+}$ oxidation in the presence of lepidocrocite (Table IV) yields shorter half-lives than either Delfino and Lee (44) or Wilson (39). Sung and Morgan (41) also report increased oxidation rates in the presence of lepidocrocite and a first-order dependence on [$\Sigma Fe^{3+}$] (Table IV). Half-lives determined from the experimental data of Hem (36,46) and Coughlin and Matsui (38) are in good agreement with those of Sung and Morgan (41), and indicate half-lives that range from about a day at pH 8.0 to a few minutes at pH 9.0 (Figure 6). Overall, the half-lives shown in Figure 6 indicate that $\Sigma Mn^{2+}$ may be oxidized rapidly in seawaters with pH near 8.0, but that $\Sigma Mn^{2+}$ may persist for long periods in aerated fresh waters with pH from 5.5 to 7.5.

SUMMARY

The above discussion of oxygenation rates clearly indicates that various reduced species are oxidized by dissolved $O_2$ at considerably different rates and that the oxidation rate for an individual reduced species can vary significantly as a function of solution conditions. For example, oxygenation rates are slow for $\Sigma Fe^{2+}$ at pH < 4.0, for $\Sigma S^{2-}$ at pH < 5.0, for $\Sigma Mn^{2+}$ at pH < 8.0, and for $\Sigma As^{3+}$ at all pH values, as evidenced by half-lives that range from months to years (Figures 2, 3, and 6). The long half-lives that are calculated for these pH values indicate that the reduced species can exist in aerated solutions for long periods. In contrast, oxygenation rates are more rapid for $\Sigma Fe^{2+}$ at pH > 4.0, for $\Sigma S^{2-}$ at pH > 5.0, for $\Sigma Mn^{2+}$ at pH > 8.0, and for $\Sigma Cu^+$ at all pH values (Figures 2, 3, and 6). Under these conditions, the calculated half-lives range from minutes to weeks, indicating that the oxidized species will normally be dominant in aerated solutions. It should be noted that these half-life calculations were carried out for conditions of 0.2 atm $PO_2$. Because the rates of each of the oxygenation reactions discussed are all directly dependent on $PO_2$ (Tables I-IV), half-lives at lower $PO_2$ would be proportionately longer and the potential for redox disequilibrium greater. Additionally, processes such as peroxide formation by photochemical reactions or by reactions between $O_2$(aq) and transition metals (8,9,34), can also produce steady-state redox distributions that are in apparent disequilibrium with the $O_2$(aq)/$H_2O$ couple. Even with these caveats, the half-life versus pH diagrams are useful for describing the solution conditions and time frames under which redox equilibrium between the reduced species discussed here (i.e., $\Sigma Fe^{2+}$, $\Sigma S^{2-}$, $\Sigma Cu^+$, $\Sigma Mn^{2+}$, and $\Sigma As^{3+}$) and the $O_2$(aq)/$H_2O$ couple might or might not be expected, and also provide a convenient means for comparing reaction rates. This information can provide insight on the probable oxidation states of redox-sensitive elements as a function of $PO_2$ and pH conditions for input to geochemical modeling studies of systems for which redox conditions are not well characterized by direct analytical determinations or might be expected to change over time.

ACKNOWLEDGMENTS

We thank Ken Krupka and Randy Arthur for their helpful suggestions for improving the technical content of this paper and Laurel Grove for her editorial comments. A portion of this work was funded by the Electric Power Research Institute under Contract RP2485-03, titled "Chemical Attenuation Studies".

## LITERATURE CITED

1. Nordstrom, D.K.; Jenne, E.A.; Ball, J.W. In *Chemical Modeling of Aqueous Systems*; Jenne, E.A., Ed.; Am. Chem. Soc. Symp. Series 93, Am. Chem. Soc.: Washington, D. C., 1979; pp 50-79.
2. Lindberg, R.D.; Runnels, D.D. *Science* 1984, 225, 925-927.
3. Wagman, D.P.; Evans, W.H.; Parker, V.B.; Schumm, R.H.; Halow, I.; Bailey, S.M.; Churney, K.L.; Nuttall, R.L. *J. Phys. Chem. Ref. Data* vol. 11, Supplement 2, Am. Inst. of Physics: Washington, D. C., 1982.
4. Bricker, O.P. *Am. Mineral.* 1965, 50, 1296-1354.
5. McAuley, A. *Inorg. Reaction Mechanisms* 1977, 5, 107-118.
6. Barb, W.G.; Baxendale, J.H.; George, P.; Hargrave, K. R. *Trans. Faraday Soc.* 1951, 47, 591-617.
7. Eary, L.E. *Metall. Trans.* 1985, 16B, 181-186.
8. Moffett, J.W.; Zika, R.G. *Marine Chem.* 1983, 13, 239-251.
9. Moffett, J.W.; Zika, R.G. *Environ. Sci. Tech.* 1987, 21, 804-810.
10. Singer, P.C.; Stumm, W. *Science* 1970, 167, 1121-1123.
11. Sung, W.; Morgan, J.J. *Environ. Sci. Tech.* 1980, 14, 561-568.
12. Davison, W.; Seed, G. *Geochim. Cosmochim. Acta* 1983, 47, 67-79.
13. Roekens, E.J.; van Grieken, R.E. *Marine Chem.* 1983, 13, 195-202.
14. Tamura, H.; Goto, K.; Nagayama, M. *J. Inorg. Nucl. Chem.* 1976, 38, 113-117.
15. Ghosh, M.M. In *Aqueous-Environmental Chemistry of Metals*; Rubin, A.J., Ed.; Ann Arbor Science Publishers: Ann Arbor, Michigan, 1974; pp 193-217.
16. Millero, F.J. *Geochim. Cosmochim. Acta* 1985, 49, 547-553.
17. Rolla, E.; Chakrabarti, C.L. *Environ. Sci. Tech.* 1982, 16, 852-857.
18. Chen, K.Y.; Morris, J.C. *Environ. Sci. Tech.* 1972, 6, 529-537.
19. O'Brien, D.J.; Birkner, F.B. *Environ. Sci. Tech.* 1977, 11, 1114-1120.
20. Millero, F.J.; Hubinger, S.; Fernandez, M.; Garnett, S. *Environ. Sci. Tech.* 1987, 21, 439-443.
21. Cline, J.D.; Richards, F.A. *Environ. Sci. Tech.* 1969, 3, 838-843.
22. Almgren, T.; Hagström, I. *Water Res.* 1974, 8, 395-400.
23. Östlund, H.G.; Alexander, J. *J. Geophys. Res.* 1963, 68, 3995-3997.
24. Avrahami, M.; Golding, R.M. *J. Chem. Soc.* 1968, A, 647-651.
25. Millero, F.J. *Marine Chem.* 1986, 18, 121-147.
26. Cherry, J.A.; Shaikh, A.U.; Tallman, D.E.; Nicholson, R.V. *J. Hydrol.* 1979, 43, 373-392.
27. Aggett, J.; O'Brien, G.A. *Environ. Sci. Tech.* 1985, 19, 231-238.
28. Johnson, D.L.; Pilson, M.E. *An. Chim. Acta* 1972, 58, 289-299.
29. Johnson, D.L.; Pilson, M.E. *Environ. Letters* 1975, 8, 157-171.
30. Scudlark, J.R.; Johnson, D.L. *Estuarine, Coastal and Shelf Sciences* 1982, 14, 693-706.
31. Johnson, D.L. *Nature* 1972, 240, 44-45.
32. Ferguson, J.F.; Gavis, J. *Water Res.* 1972, 6, 1259-1274.
33. Oscarson, D.W.; Huang, P.M.; Liaw, W.K. *J. Environ. Qual.* 9, 1980, 700-703.
34. Gray, R.D. *J. Am. Chem. Soc.* 1969, 91, 56-62.
35. Sharma, V.K.; Millero, F.J. *Environ. Sci. Tech.* 1988, 22, 768-771.
36. Hem, J.D. *Chemical Equilibria and Rates of Manganese Oxidation*; U.S.G.S. Water-Supply Paper 1667-A, U. S. Gov. Printing Office: Washington, D. C., 1963.
37. Morgan, J.J.; Stumm, W. In *Advances in Water Pollution Research*; Jaag, O., Ed.; Pergamon: New York, 1964; Vol. 3, pp. 103-131.
38. Coughlin, R.W.; Matsui, I. *J. Catalysis* 1976, 41, 108-123.
39. Wilson, D.E. *Geochim. Cosmochim. Acta* 1980, 44, 1311-1317.
40. Kessick, M.A.; Morgan, J.J. *Environ. Sci. Tech.* 1975, 9, 157-159.
41. Sung, W.; Morgan, J.J. *Geochim. Cosmochim. Acta* 1981, 45, 2377-2383.
42. Davies, S.H.R. In *Geochemical Processes at Mineral Surfaces*; Davis, J.A.; Hayes, K.F., Eds.; Chem. Soc. Symp. Ser. 323, Am. Chem. Soc.: Washington, D. C., 1985; pp 487-502.
43. Diem, D.; Stumm, W. *Geochim. Cosmochim. Acta* 1984, 48, 1571-1573.

44. Delfino, J.J.; Lee, G.F. Environ. Sci. Tech. 1968, 2, 1094-1100.
45. Hem, J.D.; Lind, C.J. Geochim. Cosmochim. Acta 1983, 47, 2037-2046.
46. Hem, J.D. Geochim. Cosmochim. Acta 1981, 45, 1369-1374.

RECEIVED July 20, 1989

# ADVANCEMENTS IN MODELING: THERMODYNAMIC AND KINETIC ADVANCES

## Chapter 31

# Revised Chemical Equilibrium Data for Major Water–Mineral Reactions and Their Limitations

**Darrell Kirk Nordstrom[1], L. Niel Plummer[2], Donald Langmuir[3], Eurybiades Busenberg[2], Howard M. May[4], Blair F. Jones[2], and David L. Parkhurst[4]**

[1]U.S. Geological Survey, Water Resources Division, 345 Middlefield Road, Mail Stop 420, Menlo Park, CA 94025
[2]U.S. Geological Survey, 432 National Center, Reston, VA 22092
[3]Department of Chemistry and Geochemistry, Colorado School of Mines, Golden, CO 80401
[4]U.S. Geological Survey, Denver Federal Center, Mail Stop 418, Lakewood, CO 80225

>   A revised, updated summary of equilibrium constants and reaction enthalpies for aqueous ion association reactions and mineral solubilities has been compiled from the literature for common equilibria occurring in natural waters at 0-100°C and 1 bar pressure. The species have been limited to those containing the elements Na, K, Li, Ca, Mg, Ba, Sr, Ra, Fe(II/III), Al, Mn(II,III,IV), Si, C, Cl, S(VI) and F. The necessary criteria for obtaining reliable and consistent thermodynamic data for water chemistry modeling is outlined and limitations on the application of equilibrium computations is described. An important limitation is that minerals that do not show reversible solubility behavior should not be assumed to attain chemical equilibrium in natural aquatic systems.

Chemical modeling results for aqueous systems is dependent on the primary thermodynamic and kinetic data needed to perform the calculations. For aqueous equilibrium computations, a large number of thermodynamic properties of solute-solute, solute-gas and solute-solid reactions are available for application to natural waters and other aqueous systems. Unfortunately, an internally consistent thermodynamic data base that is accurate for all modeling objectives, has not been achieved. Nor is it likely to be achieved in the near future. The best that can be hoped for is a tolerable level of inconsistency, with continuing progress toward the utopian goal through national and international consensus.

An essential attribute of accurate thermodynamic data is its internal consistency (see next section). Another characteristic of such data is that it has been reproduced by different investigators using different techniques and/or methods of evaluation. The tremendous need for such evaluations has been stressed by Stockmayer (1), and Lide (2), because the use of erroneous numerical values can have severe consequences for a highly technological society. Aqueous chemical models, for example, are finding increased use by water quality specialists, geochemists, hydrologists and engineers as an important tool for the interpretation of natural water chemistry.

Research investigators within the Water Resources Division of the U.S. Geological Survey have developed a series of computer programs that can perform various types of equilibrium computations for chemical reactions in natural waters. Over fifteen years ago the original programs WATCHEM (3), WATEQ (4), and SOLMNEQ (5) were introduced to perform speciation calculations for selected major components in natural waters. Since then, a series of U.S.G.S. reports have provided modifications of the original WATEQ program (6 - 13). The original thermodynamic database for the WATEQ series was published in the first paper (4) as a table of equilibrium constants and enthalpies of reaction. Later reports revised or added selected values to the database but did not reproduce unadjusted numbers. Consequently, no single document contains all values of this extensively revised database. One objective for this paper is to provide such documentation, for both major constituents (Ca, Mg, Na, K, $HCO_3$, $CO_3$, Cl, $SO_4$, F, $Si(OH)_4$, H and OH) and selected minor and trace elements (Ba, Sr, Ra, Li, Al, Fe(II), Fe(III) and Mn). Further revisions are being planned to include more trace elements. The experience gained from numerous applications of the WATEQ series of programs has affected earlier decisions regarding which components and reactions should be in the program and how the reactions should be portrayed. Furthermore, progress in thermodynamic data evaluation, and in understanding the behavior of mineral dissolution/precipitation reactions and of redox species, has affected not only the values in the database but also how they are used. The main objectives of this paper are: (1) to document the revised equilibrium constants and their temperature dependencies found most reliable for applications to natural waters; (2) to explain why some mineral reactions are best left out of routine applications of chemical modeling; (3) to explain some of the difficulties inherent in the interpretation of aquatic geochemical processes; and (4) to describe examples of the difficulty of achieving thermodynamic consistency in equilibrium data (e.g., calorimetric vs. solubility data).

REQUIREMENTS FOR THERMODYNAMIC CONSISTENCY

Thermodynamic consistency is achieved when the following criteria are met (14):

- The fundamental thermodynamic relationships and their consequences are obeyed. This criterion permits the comparison of calorimetric and solubility data.

- Common scales are used for temperature, energy, atomic mass and the fundamental physical constants.

- Conflicts and inconsistencies among measurements are resolved.

- An appropriate mathematical model is chosen to fit all the temperature and pressure dependent data.

- An appropriate aqueous chemical model is chosen to fit all aqueous solution data.

- An appropriate choice of standard states is made and applied to all similar substances.

Numerous discrepancies can be found in the literature when comparing measurements of the same system reported by different investigators or when comparing solubility data with calorimetric, electrochemical or vapor pressure data, etc. There is no universally-acccepted aqueous chemical model. There is no universally-accepted model for temperature or pressure dependence of thermodynamic functions. Often the only available measurement of some property does not adequately characterize the solid or the aqueous phase. Inconsistencies are common among aqueous chemical models and they can be very difficult to resolve. One inconsistency is that non-ideality can be interpreted using different electrolyte theories such as the ion-association theory (15), the specific-ion interaction theory (16, 17), or the ion hydration theory (18). Further inconsistency can arise from neglecting to fit simultaneously all types of solution data (heat measurements, vapor pressure measurements, density measurements, electrochemical measurements, freezing and boiling point measurements) with a single, reliable model utilizing the best available data for the density and dielectric constant of the solvent. An excellent example of a comprehensive approach to resolving such inconsistencies is given in papers by Ananthaswamy and Atkinson (19, 20), in their evaluation of the properties of aqueous calcium chloride. Similar evaluations need to be done for other solutes relevant to natural water chemistry and then correlated where common ions occur. Evaluated aqueous solute data must also be fitted together with solubility and solid phase data in a thermodynamic network (21, 22). In fact, this is the approach used by the Committee on Data for Science and Technology (CODATA). After this evaluation has been done, the entire thermodynamic network must be refitted if a new value of an important property is reported. As a result progress is slow and only a tolerable level of inconsistency can be hoped for. Fortunately, many equilibrium constants reported for the same reaction can be in good agreement in spite of these inconsistencies. All of these inconsistencies need not be totally resolved for the objectives of chemical modeling, although no one has really defined how well thermodynamic properties must be known. The ion-pair data presented in this paper is restricted to the ion-association model which is limited to an upper concentration of about 1 molal because it uses a modified, extended Debye-Hückel expression for the activity coefficients (4). The advantage of this approach is that much more data are available for use in multicomponent, multiphase systems of interest to geochemists and the precision of the data is often better for low ionic strength solutions -- the majority of natural waters.

## THE USE OF FREE ENERGIES VS EQUILIBRIUM CONSTANTS

The thermodynamic database for an aqueous chemical model is generally presented as a tabulation of free energies or equilibrium constants. The use of free energies both for the database and to calculate equilibrium constants has been avoided as much as possible in the present compilation because such an approach can introduce much larger errors than the use of equilibrium constants. As examples, free energy-based solubility product constants, for the common minerals quartz, calcite and gypsum, will be compared to values for the same constants based on highly reliable solubility data. Free energy data for these minerals and their associated solutes in the dissolution reactions:

$$SiO_2 + 2H_2O = Si(OH)_4^0$$

$$CaCO_3 = Ca^{2+} + CO_3^{2-}$$

$$CaSO_4 \cdot 2H_2O = Ca^{2+} + SO_4^{2-} + 2H_2O$$

are shown in Table 1, obtained from six important sources: The National Bureau of Standards (NBS, 23); three U.S. Geological Survey (USGS) sources in which the data on quartz are from the recent evaluation by Hemingway (24), the data on calcite are from Robinson et al. (25) and the remaining USGS data are from Robie et

Table 1. Gibbs free energies of formation from the elements for species in the dissolutions of quartz, calcite and gypsum, and derived solubility product constants at 298.15 K

| Species | $G_f^0$ (kJ/mol) | | |
|---|---|---|---|
| | CODATA | NBS | USGS |
| Quartz | --- | -856.64 | -856.288 |
| Calcite | -1129.074 | -1128.79 | -1130.610 |
| Gypsum | -1797.359 | -1797.28 | -1797.197 |
| $Si(OH)_{4(aq)}$ | --- | -1316.6 | -1308.0 |
| $Ca^{2+}_{(aq)}$ | -552.803 | -553.58 | -553.54 |
| $CO_3^{2-}{}_{(aq)}$ | -527.898 | -527.81 | -527.90 |
| $SO_4^{2-}{}_{(aq)}$ | -744.002 | -744.53 | -744.630 |
| $H_2O_{(l)}$ | -237.141 | -237.129 | -237.141 |
| $logK_{sp}$ Quartz | --- | -2.51 | -3.95 |
| $logK_{sp}$ Calcite | -8.47 | -8.30 | -8.61 |
| $logK_{sp}$ Gypsum | -4.60 | -4.36 | -4.34 |

al. (26); and two CODATA Recommended Key Values sources in which the data on calcite and gypsum are from Garvin et al. (27) and the remaining CODATA values are from Cox et al. (28).

Comparison of the NBS and the USGS log Ksp values shows that quartz solubility is discrepant by more than an order of magnitude. The generally accepted solubility at 298.15 K is about 6 mg/L, equivalent to log Ksp = -3.98 (29), if we make the safe assumptions that no activity coefficient corrections are needed and that molarity equals molality. Although several percent error may be attached to this solubility value, it cannot possibly be as high as the NBS data implies. The main source of error is the free energy value for silicic acid. The value, discussed in Hemingway (30), must be considered the more reliable because the resulting log $\overline{K}$ is closer to the measured value.

Calcite solubility product constants range over 0.3 log units. The major tabulated differences are in the free energies of calcite and of the calcium ion. The most reliable measurement and evaluation of calcite solubility is that of Plummer and Busenberg (31). They found log Ksp = -8.48(± 0.02) at 298.15 K which agrees excellently with the CODATA value of -8.47. The main source of error can be traced to a 2 kJ/mol difference between the CODATA and USGS values for the enthalpy of formation of calcite from the elements. The recent CODATA revisions of the calcium ion and calcite values take into account many different properties including the Plummer and Busenberg solubility value (31). Hence, they are the most reliable values for this system.

The gypsum solubility product constant, log Ksp = -4.58(± 0.015), is known with high precision and accuracy at 298.15 K due to the careful measurements of Lilley and Briggs (32), as well as good agreement with many other measurements (cf. 33). The CODATA free energies are the only ones compatible with the solubility determinations since they were based on several high-quality solubility experiments and on calorimetric data. The USGS free energies are based on NBS data that pre-dates the Wagman et al. (23) and CODATA revisions.

The main point of these examples is that the most reliable thermodynamic property is the one obtained by the most direct path, i.e. the one closest to the actual measurement. Free energies of individual phases or species are always derived values, never directly measured ones. Only certain properties, such as heat capacities, heat contents, entropies and volumes are directly measured for a single species or phase. Free energy measurements are measurements of processes and reactions. Reported free energies of individual species are nearly always derived from free energy measurements of processes such as an EMF measurement or a solubility measurement. Reversing the calculation to obtain a solubility product constant from free energies can introduce additional errors. Hence, the best data for use in chemical modeling will be those values based on reaction equilibria and not those based on free energies of individual reactants and products. When individual free energies have to be used (i.e. when no reaction equilibria data exist) then it becomes very important to tie the values to reaction equilibria that are well-established, as Cox et al. (28) have done. This is not to say that errors and inconsistencies don't appear when interpreting solubility data, especially reducing data from high ionic concentrations, but just that the more direct the measurement, the more reliable the thermodynamic properties are likely to be.

DEBYE-HÜCKEL SOLVENT PARAMETERS

The calculation of activity coefficients for aqueous species requires Debye-Hückel theory to represent long-range electrostatic interactions among ions. These interactions are a function of the density, $\rho$, the dielectric constant, $\varepsilon$, and the temperature of the solvent. All other parameters are either fundamental physical constants or empirical fitting parameters. For example, the Debye-Hückel solvent parameters, A and B, appear in the extended Debye-Hückel equation (14). A and B are both a function of the $\rho$ and $\varepsilon$ of water. New data and recent evaluations for water and revisions in the fundamental physical constants postdate the original values used in the WATEQ program. Gildseth et al. (34) have evaluated the density of water, from both their measurements (5-80°C) and those of others, to an accuracy of 3 ppm. The function that gives the best fit is:

$$\rho = 1 - \frac{(t - 3.9863)^2(t + 288.9414)}{508929.2(t + 68.12963)} + 0.011445 e^{-374.3/t}$$

where t is in degrees Celsius. Uncertainties in this function, over the range 0-100°C, are overshadowed by uncertainties in the value of $\varepsilon$.

There have been four recent evaluations of the dielectric constant for water. The earliest is that of Helgeson and Kirkham (35), who fit a single equation to measurements for the pressure and temperature ranges of 1-5000 bars and 0-600°C. Bradley and Pitzer (36) developed an equation for the dielectric constant up to 350°C and 500 bars. The most comprehensive evaluation appears to be that of Uematsu and Franck (37), in which errors were weighted according to temperature range for the total range of 0-350°C and up to 5 kbar. Finally, Khodakovsky and Dorofeyeva (38) evaluated the dielectric constant from 0-300°C and up to 5 kbar. Ananthaswamy and Atkinson (19) point out that the Bradley and Pitzer (36) equation agrees excellently with the IUPAC recommended values (39), it does not depend on the density or saturation pressure of water as do other equations, and seems a reasonable compromise compared to other values. On the other hand, comparing results from the four procedures over the range of 0-100°C, deviations are not greater than 0.1%. All of these equations are quite lengthy because of the large range of temperature and pressure to which they have been fitted. The temperature range is limited to 0-100°C in this paper and we have chosen the Uematsu and Franck (37) equation, modified as follows:

$$\varepsilon = 2727.586 + 0.6224107T - 466.9151 \ln T - 52000.87/T$$

This fits to within 0.01 units of the empirical dielectric constant (about 0.013%) up to 100°C and agrees quite well with the results of other published evaluations.

REVISED EQUILIBRIUM DATA

The thermodynamic data cited in Table 2 (at the end of this discussion) are restricted to 0-100°C and 1 bar (100 kPa) pressure, standard state conditions for solids and infinite dilution reference state for aqueous species. The mineral and aqueous species have been limited to those applicable to natural waters that contain the following elements: Na, K, Li, Ca, Mg, Ba, Sr, Ra, Fe, Al, Mn, Si, C, Cl, S and F. Only sulfate is considered for sulfur species, but both Fe(II) and Fe(III) species are tabulated. No solid solution models are considered. A range of solubility product constants is given for minerals whose solubilities depend significantly on the "degree of crystallinity," i.e. particle size effects, order-disorder phenomena and defect structures. These minerals are dolomite, siderite, rhodocrosite, gibbsite, ferrihydrite/goethite and quartz/chalcedony. [It now appears that the reported range of solubilities for quartz and chalcedony (microcrystalline quartz) reflects crystal ordering and particle size effects for the same basic structure (S.R. Gislason and R.O. Fournier, pers. comm.)]. Kaolinite, sepiolite and kerolite are also expected to be affected by the degree of crystallinity, but inadequate data exist to describe these effects at this time. Enthalpies of reaction are given in kcal/mol because the programs were originally set up with these units. Equilibrium constants are generally given to one more figure than is significant for purposes of avoiding round-off errors.

In the past, speciation computations applied to water analyses often included ion activity product (IAP) values and saturation indices for minerals that have never displayed reversible solubility behavior either in laboratory studies or in natural waters. Some of these minerals are unstable at 298 K and 1 bar, others dissolve incongruently, and still others are not thermodynamically identifiable phases in the traditional phase rule sense. If reversibility has never been shown and there is good reason to believe that they do not attain equilibrium solubility, they should be deleted from equilibrium-based modeling computations and from the interpretation of low-temperature equilibrium mineral assemblages (41). For example, mineral groups such as smectites, illites and micas have never been shown to control water composition as reflected by a constant IAP for a known composition of that mineral in an aquifer where water composition has significantly varied. Such demonstrations, however, are plentiful for minerals such as gypsum and calcite. Consequently, the following minerals or mineral groups are being deleted from the present compilation: smectites, illites, chlorites, micas, feldspars, amphiboles, pyroxenes and pyrophyllite. Talc also is deleted because it is only known to form in brines at low temperatures and such high ionic strength solutions are outside of the range of applicability of the chosen chemical model. It is important to remember that although natural water systems may not achieve equilibrium saturation with respect to this list of silicates, these minerals may still affect the overall water-rock mass balance relationship along a flow path, as might be described by the models developed by Garrels and Mackenzie (42) and Parkhurst et al. (43). Important chemical components can always be added or removed from a water body without achieving reversible solubility control. This partial equilibrium condition exists when the chemical potentials of some components in a system reach equilibrium while others do not (e.g. calcite and barite may reach equilibrium solubility but co-existing biotite or plagioclase may never reach this state). The advantages of both approaches should

Table 2. Summary of Revised Thermodynamic Data. I. Fluoride and Chloride Species

| Reaction | $\Delta H_r^0$ (kcal/mol) | log K | Ref. | Reaction | $\Delta H_r^0$ (kcal/mol) | log K | Ref. |
|---|---|---|---|---|---|---|---|
| $H^+ + F^- = HF^0$ | 3.18 | 3.18 | (a) | $Al^{3+} + F^- = AlF^{2+}$ | 1.06 | 7.0 | (54) |
| $H^+ + 2F^- = HF_2^-$ | 4.55 | 3.76 | (7) | $Al^{3+} + 2F^- = AlF_2^+$ | 1.98 | 12.7 | (54) |
| $Na^+ + F^- = NaF^0$ | ---- | -0.24 | (49) | $Al^{3+} + 3F^- = AlF_3^0$ | 2.16 | 16.8 | (54) |
| $Ca^{2+} + F^- = CaF^+$ | 4.12 | 0.94 | (50) | $Al^{3+} + 4F^- = AlF_4^-$ | 2.20 | 19.4 | (54) |
| $Mg^{2+} + F^- = MgF^+$ | 3.2 | 1.82 | (b) | $Al^{3+} + 5F^- = AlF_5^{2-}$ | 1.84 | 20.6 | (54) |
| $Mn^{2+} + F^- = MnF^+$ | ---- | 0.84 | (23) | $Al^{3+} + 6F^- = AlF_6^{3-}$ | -1.67 | 20.6 | (54) |
| $Fe^{2+} + F^- = FeF^+$ | --- | 1.0 | (c) | | | | |
| $Fe^{3+} + F^- = FeF^{2+}$ | 2.7 | 6.2 | (50) | $Si(OH)_4 + 4H^+ + 6F^-$ | -16.26 | 30.18 | (55) |
| $Fe^{3+} + 2F^- = FeF_2^+$ | 4.8 | 10.8 | (50) | $= SiF_6^{2-} + 4H_2O$ | | | |
| $Fe^{3+} + 3F^- = FeF_3^0$ | 5.4 | 14.0 | (50) | $Fe^{2+} + Cl^- = FeCl^+$ | ---- | 0.14 | (d) |
| $Mn^{2+} + Cl^- = MnCl^+$ | ---- | 0.61 | (23) | $Fe^{3+} + Cl^- = FeCl^{2+}$ | 5.6 | 1.48 | (52) |
| $Mn^{2+} + 2Cl^- = MnCl_2^0$ | ---- | 0.25 | (23) | $Fe^{3+} + 2Cl^- = FeCl_2^+$ | ---- | 2.13 | (52) |
| $Mn^{2+} + 3Cl^- = MnCl_3^-$ | ---- | -0.31 | (23) | $Fe^{3+} + 3Cl^- = FeCl_3^0$ | ---- | 1.13 | (56) |

| Mineral | Reaction | $\Delta H_r^0$ (kcal/mol) | log K | Ref. |
|---|---|---|---|---|
| Cryolite | $Na_3AlF_6 = 3Na^+ + Al^{3+} + 6F^-$ | 9.09 | -33.84 | (e) |
| Fluorite | $CaF_2 = Ca^{2+} + 2F^-$ | 4.69 | -10.6 | (f) |

| Redox Potentials | $\Delta H_r^0$ (kcal/mol) | $E^0$ (volts) | log K | Ref. |
|---|---|---|---|---|
| $Fe^{2+} = Fe^{3+} + e^-$ | 9.68 | -0.770 | -13.02 | (g) |
| $Mn^{2+} = Mn^{3+} + e^-$ | 25.8 | -1.51 | -25.51 | (h) |

| Reaction | Analytical Expressions for Temperature Dependence | Ref. |
|---|---|---|
| $H^+ + F^- = HF^0$ | $\log K_{HF} = -2.033 + 0.012645T + 429.01/T$ | (a) |
| $CaF_2 = Ca^{2+} + 2F^-$ | $\log K_{FLUORITE} = 66.348 - 4298.2/T - 25.271 \log T$ | (f) |

References for fluoride and chloride species

a.) log K, $\Delta H_r^0$ and temperature dependence from Naumov et al. (47) in agreement with the critical evaluations by Bond and Hefter (48) and Garvin et al. (27); b.) log K from Sillen and Martell (51), $\Delta H_r^0$ from Smith and Martell (52); c.) estimated from a measurement of 0.83 at I = 1 M and the tendency for divalent fluorides to have log K ≈ 1 (53); d.) based on Davison (108) which agrees well with Turner et al. (109); e.) log K from Roberson and Hem (57) and $\Delta H_r^0$ from (58) for cryolite and from (28) for ions; f.) based on reference (50) but forced to go through logK = -10.6 at 298.15 K to be in agreement with the solubility data of Macaskill and Bates (59) and Brown and Roberson (60); g.) $E^0$ and log K from Whittemore and Langmuir (61), $\Delta H_r^0$ from V. Parker, personal communication; h.) based on (23) in agreement with Bard et al. (62).

*Continued on next page*

Table 2. Summary of Revised Thermodynamic Data. II. Oxide and Hydroxide Species

| Reaction | $\Delta H_r^0$ (kcal/mol) | log K | Ref. | Reaction | $\Delta H_r^0$ (kcal/mol) | log K | Ref |
|---|---|---|---|---|---|---|---|
| $H_2O = H^+ + OH^-$ | 13.362 | -14.000 | (a) | $Fe^{3+} + H_2O = FeOH^{2+} + H^+$ | 10.4 | -2.19 | (65) |
| $Li^+ + H_2O = LiOH^0 + H^+$ | 0.0 | -13.64 | (65) | $Fe^{3+} + 2H_2O = Fe(OH)_2^+ + 2H^+$ | 17.1 | -5.67 | (c) |
| $Na^+ + H_2O = NaOH^0 + H^+$ | 0.0 | -14.18 | (65) | $Fe^{3+} + 3H_2O = Fe(OH)_3^0 + 3H^+$ | 24.8 | -12.56 | (c) |
| $K^+ + H_2O = KOH^0 + H^+$ | --- | -14.46 | (65) | $Fe^{3+} + 4H_2O = Fe(OH)_4^- + 4H^+$ | 31.9 | -21.6 | (c) |
| $Ca^{2+} + H_2O = CaOH^+ + H^+$ | --- | -12.78 | (b) | $2Fe^{3+} + 2H_2O = Fe_2(OH)_2^{4+} + 2H^+$ | 13.5 | -2.95 | (65) |
| $Mg^{2+} + H_2O = MgOH^+ + H^+$ | --- | -11.44 | (65) | $3Fe^{3+} + 4H_2O = Fe_3(OH)_4^{5+} + 4H^+$ | 14.3 | -6.3 | (65) |
| $Sr^{2+} + H_2O = SrOH^+ + H^+$ | --- | -13.29 | (65) | $Al^{3+} + H_2O = AlOH^{2+} + H^+$ | 11.49 | -5.00 | (54) |
| $Ba^{2+} + H_2O = BaOH^+ + H^+$ | --- | -13.47 | (65) | $Al^{3+} + 2H_2O = Al(OH)_2^+ + 2H^+$ | 26.90 | -10.1 | (54) |
| $Ra^{2+} + H_2O = RaOH^+ + H^+$ | --- | -13.49 | (66) | $Al^{3+} + 3H_2O = Al(OH)_3^0 + 3H^+$ | 39.89 | -16.9 | (54) |
| $Fe^{2+} + H_2O = FeOH^+ + H^+$ | 13.2 | -9.5 | (65) | $Al^{3+} + 4H_2O = Al(OH)_4^- + 4H^+$ | 42.30 | -22.7 | (54) |
| $Mn^{2+} + H_2O = MnOH^+ + H^+$ | 14.4 | -10.59 | (65) | | | | |

| Mineral | Reaction | $\Delta H_r^0$ (kcal/mol) | log K | Ref. |
|---|---|---|---|---|
| Portlandite | $Ca(OH)_2 + 2H^+ = Ca^{2+} + 2H_2O$ | -31.0 | 22.8 | (65) |
| Brucite | $Mg(OH)_2 + 2H^+ = Mg^{2+} + 2H_2O$ | -27.1 | 16.84 | (65) |
| Pyrolusite | $MnO_2 + 4H^+ + 2e^- = Mn^{2+} + 2H_2O$ | -65.11 | 41.38 | (d) |
| Hausmanite | $Mn_3O_4 + 8H^+ + 2e^- = 3Mn^{2+} + 4H_2O$ | -100.64 | 61.03 | (d) |
| Manganite | $MnOOH + 3H^+ + e^- = Mn^{2+} + 2H_2O$ | ---- | 25.34 | (23) |
| Pyrochroite | $Mn(OH)_2 + 2H^+ = Mn^{2+} + 2H_2O$ | ---- | 15.2 | (65) |
| Gibbsite (crystalline) | $Al(OH)_3 + 3H^+ = Al^{3+} + 3H_2O$ | -22.8 | 8.11 | (69) |
| Gibbsite (microcrystalline) | $Al(OH)_3 + 3H^+ = Al^{3+} + 3H_2O$ | (-24.5) | 9.35 | (e) |
| $Al(OH)_3$ (amorphous) | $Al(OH)_3 + 3H^+ = Al^{3+} + 3H_2O$ | (-26.5) | 10.8 | (f) |
| Goethite | $FeOOH + 3H^+ = Fe^{3+} + 2H_2O$ | ---- | -1.0 | (72) |
| Ferrihydrite (amorphous to microcrystalline) | $Fe(OH)_3 + 3H^+ = Fe^{3+} + 3H_2O$ | ---- | 3.0 to 5.0 | (g) |

| Reaction | Analytical Expressions for Temperature Dependence | Ref. |
|---|---|---|
| $H_2O = H^+ + OH^-$ | $\log K_w = -283.9710 + 13323.00/T - 0.05069842T + 102.24447 \log T - 1119669/T^2$ | (a) |
| $Al^{3+} + H_2O = AlOH^{2+} + H^+$ | $\log K_1 = -38.253 - 656.27/T + 14.327 \log T$ | (54) |
| $Al^{3+} + 2H_2O = Al(OH)_2^+ + 2H^+$ | $\log \beta_2 = 88.500 - 9391.6/T - 27.121 \log T$ | (54) |
| $Al^{3+} + 3H_2O = Al(OH)_3^0 + 3H^+$ | $\log \beta_3 = 226.374 - 18247.8/T - 73.597 \log T$ | (54) |
| $Al^{3+} + 4H_2O = Al(OH)_4^- + 4H^+$ | $\log \beta_4 = 51.578 - 11168.9/T - 14.865 \log T$ | (54) |

References for oxide and hydroxide species

a.) refitted from Olafsson and Olafsson (63), in good agreement with Marshall and Franck (64); b.) CODATA compatible (27), in good agreement with (65); c.) log K from (65) except log$\beta_3$ is corrected to I = 0 from Kester et al. (67) and enthalpies are estimated from free energies of reaction and entropies estimated from a correlation plot; d.) Robie and Hemingway (68) using ion values from (23); e.) Hem and Roberson (70) for log K and enthalpy estimated by assuming that it changes by the same amount as the free energy; f.) Feitknecht and Schindler (71) for log K and enthalpy derived as above and considered highly uncertain; g.) data based on the range of reported values from (72), Schwertmann and Taylor (73) and Norvell and Lindsay (74).

*Continued on next page*

Table 2. Summary of Revised Thermodynamic Data. III. Carbonate Species

| Reaction | $\Delta H_r^0$ (kcal/mol) | log K | Ref. | Reaction | $\Delta H_r^0$ (kcal/mol) | log K | Ref. |
|---|---|---|---|---|---|---|---|
| $CO_2(g) = CO_2(aq)$ | -4.776 | -1.468 | (31) | $Ca^{2+} + CO_3^{2-} = CaCO_3^0$ | 3.545 | 3.224 | (31) |
| $CO_2(aq) + H_2O = H^+ + HCO_3^-$ | 2.177 | -6.352 | (31) | $Mg^{2+} + CO_3^{2-} = MgCO_3^0$ | 2.713 | 2.98 | (80) |
| $HCO_3^- = H^+ + CO_3^{2-}$ | 3.561 | -10.329 | (31) | $Sr^{2+} + CO_3^{2-} = SrCO_3^0$ | 5.22 | 2.81 | (76) |
| $Ca^{2+} + HCO_3^- = CaHCO_3^+$ | 2.69 | 1.106 | (31) | $Ba^{2+} + CO_3^{2-} = BaCO_3^0$ | 3.55 | 2.71 | (77) |
| $Mg^{2+} + HCO_3^- = MgHCO_3^+$ | 0.79 | 1.07 | (75) | $Mn^{2+} + CO_3^{2-} = MnCO_3^0$ | --- | 4.90 | (81) |
| $Sr^{2+} + HCO_3^- = SrHCO_3^+$ | 6.05 | 1.18 | (76) | $Fe^{2+} + CO_3^{2-} = FeCO_3^0$ | --- | 4.38 | (53) |
| $Ba^{2+} + HCO_3^- = BaHCO_3^+$ | 5.56 | 0.982 | (77) | $Na^+ + CO_3^{2-} = NaCO_3^-$ | 8.91 | 1.27 | (82) |
| $Mn^{2+} + HCO_3^- = MnHCO_3^+$ | --- | 1.95 | (78) | $Na^+ + HCO_3^- = NaHCO_3^-$ | --- | -0.25 | (15) |
| $Fe^{2+} + HCO_3^- = FeHCO_3^+$ | --- | 2.0 | (79) | $Ra^{2+} + CO_3^{2-} = RaCO_3^0$ | 1.07 | 2.5 | (66) |

| Mineral | Reaction | $\Delta H_r^0$ (kcal/mol) | log K | Ref. |
|---|---|---|---|---|
| Calcite | $CaCO_3 = Ca^{2+} + CO_3^{2-}$ | -2.297 | -8.480 | (31) |
| Aragonite | $CaCO_3 = Ca^{2+} + CO_3^{2-}$ | -2.589 | -8.336 | (31) |
| Dolomite(Ordered) | $CaMg(CO_3)_2 = Ca^{2+} + Mg^{2+} + 2CO_3^{2-}$ | -9.436 | -17.09 | (a) |
| Dolomite(Disordered) | $CaMg(CO_3)_2 = Ca^{2+} + Mg^{2+} + 2CO_3^{2-}$ | -11.09 | -16.54 | (b) |
| Strontianite | $SrCO_3 = Sr^{2+} + CO_3^{2-}$ | -0.40 | -9.271 | (76) |
| Siderite(crystalline) | $FeCO_3 = Fe^{2+} + CO_3^{2-}$ | -2.48 | -10.89 | (c) |
| Siderite(precipitated) | $FeCO_3 = Fe^{2+} + CO_3^{2-}$ | --- | -10.45 | (d) |
| Witherite | $BaCO_3 = Ba^{2+} + CO_3^{2-}$ | 0.703 | -8.562 | (77) |
| Rhodocrosite(crystalline) | $MnCO_3 = Mn^{2+} + CO_3^{2-}$ | -1.43 | -11.13 | (e) |
| Rhodocrosite(synthetic) | $MnCO_3 = Mn^{2+} + CO_3^{2-}$ | --- | -10.39 | (e) |

| Reaction | Analytical Expressions for Temperature Dependence | Ref. |
|---|---|---|
| $CO_2(g) = CO_2(aq)$ | $\log K_H = 108.3865 + 0.01985076T - 6919.53/T - 40.45154 \log T + 669365/T^2$ | (31) |
| $CO_2(aq) + H_2O = H^+ + HCO_3^-$ | $\log K_1 = -356.3094 - 0.06091964T + 21834.37/T + 126.8339 \log T - 1684915/T^2$ | (31) |
| $HCO_3^- = H^+ + CO_3^{2-}$ | $\log K_2 = -107.8871 - 0.03252849T + 5151.79/T + 38.92561 \log T - 563713.9/T^2$ | (31) |
| $Ca^{2+} + HCO_3^- = CaHCO_3^+$ | $\log K_{CaHCO_3^+} = 1209.120 + 0.31294T - 34765.05/T - 478.782 \log T$ | (31) |
| $Mg^{2+} + HCO_3^- = MgHCO_3^+$ | $\log K_{MgHCO_3^+} = -59.215 + 2537.455/T + 20.92298 \log T$ | (75) |
| $Sr^{2+} + HCO_3^- = SrHCO_3^+$ | $\log K_{SrHCO_3^+} = -3.248 + 0.014867T$ | (76) |
| $Ba^{2+} + HCO_3^- = BaHCO_3^+$ | $\log K_{BaHCO_3^+} = -3.0938 + 0.013669T$ | (77) |
| $Ca^{2+} + CO_3^{2-} = CaCO_3^0$ | $\log K_{CaCO_3^0} = -1228.732 - 0.299444T + 35512.75/T + 485.818 \log T$ | (31) |
| $Mg^{2+} + CO_3^{2-} = MgCO_3^0$ | $\log K_{MgCO_3^0} = 0.9910 + 0.00667T$ | (80) |
| $Sr^{2+} + CO_3^{2-} = SrCO_3^0$ | $\log K_{SrCO_3^0} = -1.019 + 0.012826T$ | (76) |
| $Ba^{2+} + CO_3^{2-} = BaCO_3^0$ | $\log K_{BaCO_3^0} = 0.113 + 0.008721T$ | (77) |
| $CaCO_3 = Ca^{2+} + CO_3^{2-}$ | $\log K_{CALCITE} = -171.9065 - 0.077993T + 2839.319/T + 71.595 \log T$ | (31) |
| $CaCO_3 = Ca^{2+} + CO_3^{2-}$ | $\log K_{ARAGONITE} = -171.9773 - 0.077993T + 2903.293/T + 71.595 \log T$ | (31) |
| $SrCO_3 = Sr^{2+} + CO_3^{2-}$ | $\log K_{STRONTIANITE} = 155.0305 - 7239.594/T - 56.58638 \log T$ | (76) |
| $BaCO_3 = Ba^{2+} + CO_3^{2-}$ | $\log K_{WITHERITE} = 607.642 + 0.121098T - 20011.25/T - 236.4948 \log T$ | (77) |

*Continued on next page*

Table 2. Summary of Revised Thermodynamic Data. IV. Silicate Species

| Reaction | $\Delta H_r^0$ (kcal/mol) | log K | Ref. |
|---|---|---|---|
| $Si(OH)_4^0 = SiO(OH)_3^- + H^+$ | 6.12 | -9.83 | (87) |
| $Si(OH)_4^0 = SiO_2(OH)_2^{2-} + 2H^+$ | 17.6 | -23.0 | (88) |

| Mineral | Reaction | $\Delta H_r^0$ (kcal/mol) | log K | Ref. |
|---|---|---|---|---|
| Kaolinite | $Al_2Si_2O_5(OH)_4 + 6H^+ = 2Al^{3+} + 2Si(OH)_4^0 + H_2O$ | -35.3 | 7.435 | (a) |
| Chrysotile | $Mg_3Si_2O_5(OH)_4 + 6H^+ = 3Mg^{2+} + 2Si(OH)_4^0 + H_2O$ | -46.8 | 32.20 | (b) |
| Sepiolite | $Mg_2Si_3O_{7.5}(OH)\cdot 3H_2O + 4H^+ + 0.5\ H_2O = 2Mg^{2+} + 3Si(OH)_4^0$ | -10.7 | 15.76 | (c) |
| Kerolite | $Mg_3Si_4O_{10}(OH)_2\cdot H_2O + 6H^+ + 3H_2O = 3Mg^{2+} + 4Si(OH)_4^0$ | --- | 25.79 | (90) |
| Quartz | $SiO_2 + 2H_2O = Si(OH)_4^0$ | 5.99 | -3.98 | (29) |
| Chalcedony | $SiO_2 + 2H_2O = Si(OH)_4^0$ | 4.72 | -3.55 | (29) |
| Amorphous Silica | $SiO_2 + 2H_2O = Si(OH)_4^0$ | 3.34 | -2.71 | (29) |

| Reaction | Analytical Expressions for Temperature Dependence | Ref. |
|---|---|---|
| $Si(OH)_4^0 = SiO(OH)_3^- + H^+$ | $\log K_1 = -302.3724 - 0.050698T + 15669.69/T + 108.18466 \log T - 1119669/T^2$ | (87) |
| $Si(OH)_4^0 = SiO_2(OH)_2^{2-} + 2H^+$ | $\log \beta_2 = -294.0184 - 0.072650T + 11204.49/T + 108.18466 \log T - 1119669/T^2$ | (88) |
| $Mg_3Si_2O_5(OH)_4 + 6H^+ = 3Mg^{2+} + 2Si(OH)_4^0 + H_2O$ | $\log K_{CHRYSOTILE} = 13.248 + 10217.1/T - 6.1894 \log T$ | (b) |
| $SiO_2 + 2H_2O = Si(OH)_4^0$ | $\log K_{QUARTZ} = 0.41 - 1309/T$ | (29) |
| $SiO_2 + 2H_2O = Si(OH)_4^0$ | $\log K_{CHALCEDONY} = -0.09 - 1032/T$ | (29) |
| $SiO_2 + 2H_2O = Si(OH)_4^0$ | $\log K_{AMORPHOUS\ SILICA} = -0.26 - 731/T$ | (29) |

References for silicate species

a.) (41) for log K, (26) for enthalpy; b.) log K, obtained from (89) data after conversion using our $K_w$ equation and least squares fitting, is consistent with (26) data; c.) log K from (90); $\Delta H_r^0$ obtained from 273-373 K fit of (90) data.

References for carbonate species

a.) (26), using ion values from (23); b.) from Helgeson et al. (83) using ion values of (23); c.) log K of Smith (84) recalculated using the present aqueous model at 303 K, adjusted to 298 K using $\Delta H_r^0$ calculated using ion values from (23) and Robie et al. (85) for solid; d.) Singer and Stumm (86) recalculated to be consistent with the present aqueous model; e.) log K from Garrels et al. (82) and $\Delta H_r^0$ from (23) and Robie et al. (85) for the solid.

*Continued on next page*

Table 2. Summary of Revised Thermodynamic Data. V. Sulfate Species

| Reaction | $\Delta H_r^0$ (kcal/mol) | log K | Ref. | Reaction | $\Delta H_r^0$ (kcal/mol) | log K | Ref. |
|---|---|---|---|---|---|---|---|
| $H^+ + SO_4^{2-} = HSO_4^-$ | 3.85 | 1.988 | (91) | $Mn^{2+} + SO_4^{2-} = MnSO_4^0$ | 3.37 | 2.25 | (99) |
| $Li^+ + SO_4^{2-} = LiSO_4^-$ | ---- | 0.64 | (52) | $Fe^{2+} + SO_4^{2-} = FeSO_4^0$ | 3.23 | 2.25 | (c) |
| $Na^+ + SO_4^{2-} = NaSO_4^-$ | 1.12 | 0.70 | (a) | $Fe^{2+} + HSO_4^- = FeHSO_4^+$ | --- | 1.08 | (102) |
| $K^+ + SO_4^{2-} = KSO_4^-$ | 2.25 | 0.85 | (b) | $Fe^{3+} + SO_4^{2-} = FeSO_4^+$ | 3.91 | 4.04 | (c) |
| $Ca^{2+} + SO_4^{2-} = CaSO_4^0$ | 1.65 | 2.30 | (96) | $Fe^{3+} + 2SO_4^{2-} = Fe(SO_4)_2^-$ | 4.60 | 5.38 | (c) |
| $Mg^{2+} + SO_4^{2-} = MgSO_4^0$ | 4.55 | 2.37 | (97) | $Fe^{3+} + HSO_4^- = FeHSO_4^{2+}$ | --- | 2.48 | (102) |
| $Sr^{2+} + SO_4^{2-} = SrSO_4^0$ | 2.08 | 2.29 | (98) | $Al^{3+} + SO_4^{2-} = AlSO_4^+$ | 2.15 | 3.02 | (54) |
| $Ba^{2+} + SO_4^{2-} = BaSO_4^0$ | --- | 2.7 | (52) | $Al^{3+} + 2SO_4^{2-} = Al(SO_4)_2^-$ | 2.84 | 4.92 | (54) |
| $Ra^{2+} + SO_4^{2-} = RaSO_4^0$ | 1.3 | 2.75 | (66) | $Al^{3+} + HSO_4^- = AlHSO_4^{2+}$ | --- | 0.46 | (104) |

| Mineral | Reaction | $\Delta H_r^0$ (kcal/mol) | log K | Ref. |
|---|---|---|---|---|
| Gypsum | $CaSO_4 \cdot 2H_2O = Ca^{2+} + SO_4^{2-} + 2H_2O$ | -0.109 | -4.58 | (d) |
| Anhydrite | $CaSO_4 = Ca^{2+} + SO_4^{2-}$ | -1.71 | -4.36 | (d) |
| Celestite | $SrSO_4 = Sr^{2+} + SO_4^{2-}$ | -1.037 | -6.63 | (106) |
| Barite | $BaSO_4 = Ba^{2+} + SO_4^{2-}$ | 6.35 | -9.97 | (d) |
| Radium sulfate | $RaSO_4 = Ra^{2+} + SO_4^{2-}$ | 9.40 | -10.26 | (66) |
| Melanterite | $FeSO_4 \cdot 7H_2O = Fe^{2+} + SO_4^{2-} + 7H_2O$ | 4.91 | -2.209 | (107) |
| Alunite | $KAl_3(SO_4)_2(OH)_6 + 6H^+ = K^+ + 3Al^{3+} + 2SO_4^{2-} + 6H_2O$ | -50.25 | -1.4 | (e) |

| Reaction | Analytical Expressions for Temperature Dependence | Ref. |
|---|---|---|
| $H^+ + SO_4^{2-} = HSO_4^-$ | $\log K_2 = -56.889 + 0.006473T + 2307.9/T + 19.8858 \log T$ | (91) |
| $CaSO_4 \cdot 2H_2O = Ca^{2+} + SO_4^{2-} + 2H_2O$ | $\log K_{GYPSUM} = 68.2401 - 3221.51/T - 25.0627 \log T$ | (d) |
| $CaSO_4 = Ca^{2+} + SO_4^{2-}$ | $\log K_{ANHYDRITE} = 197.52 - 8669.8/T - 69.835 \log T$ | (d) |
| $SrSO_4 = Sr^{2+} + SO_4^{2-}$ | $\log K_{CELESTITE} = -14805.9622 - 2.4660924T + 756968.533/T - 40553604/T^2 + 5436.3588 \log T$ | (106) |
| $BaSO_4 = Ba^{2+} + SO_4^{2-}$ | $\log K_{BARITE} = 136.035 - 7680.41/T - 48.595 \log T$ | (d) |
| $RaSO_4 = Ra^{2+} + SO_4^{2-}$ | $\log K_{RaSO4} = 137.98 - 8346.87/T - 48.595 \log T$ | (66) |
| $FeSO_4 \cdot 7H_2O = Fe^{2+} + SO_4^{2-} + 7H_2O$ | $\log K_{MELANTERITE} = 1.447 - 0.004153T - 214949/T^2$ | (107) |

References for sulfate species

a.) log K from Rhigaletto and Davies (92), $\Delta H_r^0$ from Austin and Mair (93); b.) log K from Truesdell and Hostetler (94), $\Delta H_r^0$ from Siebert and Christ (95) refitting of (94); c.) log K values are in good agreement between (52), (95) and Stipp (103); enthalpies are derived from the Fuoss fitting method of Siebert and Christ (95) except for the iron(III) sulfate ion triplet which assumes a value equivalent to that for the aluminum sulfate ion triplet; d.) Langmuir and Melchior (33), where the gypsum data is refitted from Blount and Dickson (105) and is in excellent agreement with the highly precise data of Lilley and Briggs (32) at 298 K; e.) log K from Adams and Rawajfih (100) and $\Delta H_r^0$ calculated from enthalpies of formation found in Kelley et al. (101) and Robie et al. (26).

be clear -- one is needed to determine the tendency for mineral solubility constraints on water composition and the other is needed to describe mass transfer sources and sinks in a geochemically reacting flow system (44).

It may also be argued that the simulation of water-rock interactions should allow for solubility equilibria involving feldspars, micas, etc. For such studies the choice of solublity product constants and free energies must and should be made by the investigators. We cannot propose such values here when an enormous range of values and properties (solid-solutions, interlayering, defects, surface areas, etc.) is known to exist for these minerals and reversible solubility behavior has not been demonstrated.

A brief summary of the status of the thermodynamic properties for water-mineral reactions using the ion-association theory and revised data is:

1. These computations are reliable for the range 0-100°C and up to 1 molal ionic strength for major univalent and divalent ions, a limited set of minor and trace elements, and iron and manganese redox species.

2. Major carbonate mineral solubilities and their associated ion pairs are reliable except for dolomite, siderite and rhodocrosite, for which ranges of Ksp values are estimated.

3. Oxide and hydroxide solubilities are generally reliable for calcium, magnesium, aluminum and iron, but the Ksp values can range over several orders of magnitude depending on degree of crystallinity, especially particle size affects. There are continuing controversies regarding the actual reactive phases being measured in solubility studies, and further refinements have been proposed (45), (46).

4. Quartz, kaolinite, chrysotile, sepiolite and kerolite solubilities are reliable for estimates in chemical modeling at low temperatures. Ksp values for these minerals are also strongly influenced by degree of crystallinity. Other silicate mineral solubilities are either unreliable, or do not describe the behavior of these minerals in natural waters.

5. Common sulfate mineral solubilities and their associated ion pairs are reliable within other restrictions of the model.

## LITERATURE CITED

1. Stockmayer, W. H. Science 1978, 201, 577.
2. Lide, D. R., Jr. Science 1981, 212, 1343-1349.
3. Barnes, I.; Clarke, F. E. U. S. Geol. Survey Professional Paper 498-D, 1969.
4. Truesdell, A. H.; Jones, B. F. J. Res. U. S. Geol. Survey 1974, 2, 233-248.
5. Kharaka, Y. K.; Barnes, I. National Tech. Infor. Serv. Tech. Report PV214-899, 1973.
6. Plummer, L. N.; Jones, B. F.; Truesdell, A. H. U. S. Geol. Survey Water-Resources Invest. Report 76-13, 1976.
7. Ball, J. W.; Jenne, E. A.; Nordstrom, D. K. In Chemical Modeling in Aqueous Systems; Jenne, E. A., Ed.; ACS Symposium Series No. 93; American Chemical Society: Washington, DC, 1979; pp 815-835.
8. Ball, J. W.; Nordstrom, D. K.; Jenne, E. A. U. S. Geol. Survey Water-Resources Invest. Report 78-116, 1980.
9. Ball, J. W.; Jenne, E. A.; Cantrel, M. W. U. S. Geol. Survey Open-File Report 81-1183, 1981.

10. Parkhurst, D. L.; Thorstenson, D. C.; Plummer, L. N. U. S. Geol. Survey Water-Resources Invest. Report 80-96, 1980.
11. Ball, J. W.; Nordstrom, D. K.; Zachmann, D. W. U. S. Geol. Survey Open-File Report 87-50, 1987.
12. Kharaka, Y. K.; Gunter, W. D.; Aggarwal, P. K.; Perkins, E. H.; DeBraal, J. D. U. S. Geol. Survey Water-Resources Invest. Report 88-4227, 1988.
13. Plummer, L. N.; Parkhurst, D. L.; Fleming, G. W.; Dunkle, S. A. U. S. Geol. Survey Water-Resources Invest. Report 88-4153, 1988.
14. Nordstrom, D. K.; Munoz, J. L. Geochemical Thermodynamics; Blackwell Scientific: Palo Alto, 1986.
15. Garrels, R. M.; Thompson, M. E. Am. J. Sci. 1962, 260, 57-66.
16. Pitzer, K. S. In Thermodynamic Modeling of Geological Materials: Minerals, Fluids and Melts; Carmicael, I. S. E.; Eugster, H. P., Eds.; Reviews in Mineralogy; Mineralogical Society of America: Washington, DC; Vol. 17, pp 97-142.
17. Weare, J. H. In Thermodynamic Modeling of Geological Materials: Minerals, Fluids and Melts; Carmichael, I. S. E.; Eugster, H. P., Eds.; Reviews in Mineralogy; Mineralogical Society of America: Washington, DC; Vol. 17, pp 143-176.
18. Wolery, T. J.; Jackson, K. D. 1989; this volume.
19. Ananthaswamy, J.; Atkinson, G. J. Chem. Eng. Data 1984, 29, 81-87.
20. Ananthaswamy, J.; Atkinson, G. J. Chem. Eng. Data 1985, 30, 120-128.
21. Haas, J. L., Jr.; Fisher, J. R. Am. J. Sci. 1976, 276, 525-545.
22. Pearson, J. 1989; this volume.
23. Wagman, D. D.; Evans, W. H.; Parker, Y. B.; Schumm, R. H.; Halow, I.; Bailey, S. M.; Churney, K. L.; Nuttall, R. L. J. Phys. Chem. Ref. Data 1982, 11, Supp. 2.
24. Hemingway, B. Am. Mineral. 1987, 72, 273-279.
25. Robinson, G. R., Jr.; Haas, J. L., Jr.; Schafer, C. M.; Hazelton, H. T., Jr. U. S. Geol. Survey Open-File Report 83-79, 1982.
26. Robie, R. A.; Hemingway, B. S.; Fisher, J. R. U. S. Geol. Survey Bull. 1452, 1979 (reprinted with corrections).
27. Garvin. D.; Parker, Y. B.; White, H. J., Jr. CODATA Thermodynamic Tables -- Selections for Some Compounds of Calcium and Related Mixtures: A Prototype Set of Tables; Hemisphere Publishing Corporation: Washington, DC, 1987.
28. Cox, J. D.; Wagman, D. D.; Medvedev, V. A. CODATA Key Values for Thermodynamics; Hemisphere Publishing Corporation: Washington, D. C., 1989, in press.
29. Fournier, R. In Geology and Geochemistry of Epithermal Systems; Berger, B. R.; Bethke, P. M., Eds.; Reviews in Economic Geology Vol. 2; Society of Economic Geologists, 1985, pp 45-61.
30. Hemingway, B. S.; Robie, R. A.; Kittrick, J. A. Geochim. Cosmochim. Acta 1978, 42, 1533-1534.
31. Plummer, L. N.; Busenberg, E. Geochim. Cosmochim. Acta 1982, 46, 1011-1040.
32. Lilley, T. H.; Briggs, C. C. Proc. Royal Society London, Ser. A 1976, 349, 355-368.
33. Langmuir, D.; Melchior, D. Geochim. Cosmochim. Acta 1985, 49, 2423-2432.
34. Gildseth, W.; Habenschuss, A.; Spedding, F. H. J. Chem. Eng. Data 1972, 17, 402-409.
35. Helgeson, H. C.; Kirkham, D. H. Am. J. Sci. 1974, 274, 1089-1198.
36. Bradley, D. J.; Pitzer, K. S. J. Phys. Chem. 1979, 83, 1599-1603.

37. Uematsu, M.; Franck, E. U. J. Phys. Chem. Ref. Data 1980, 9, 1291-1306.
38. Khodakovsky, I. L.; Dorofeyeva, V. A. Geokhimiya 1981, 8, 1174.
39. Kienitz, H.; Marsh, K. N. Pure Appl. Chem. 1981, 53, 1847.
40. Cohen, E. R.; Taylor, B. N.; CODATA Bull. No. 63 1986.
41. May, H. M.; Kinniburgh, D. G.; Helmke, P. A.; Jackson, M. L. Geochim. Cosmochim. Acta 1986, 50, 1667-1677.
42. Garrels, R. M.; Mackenzie, F. T. In Equilibrium Concepts in Natural Water Systems; Stumm, W., Ed.; Advances in Chemistry Series No. 67; American Chemical Society: Washington, DC, 1967; pp 222-242.
43. Parkhurst, D. L.; Plummer, L. N.; Thorstensen, D. C. U. S. Geol. Survey Water-Resources Invest. Report 82-14, 1982.
44. Plummer, L. N.; Parkhurst, D. L.; Thorstenson, D. C. Geochim. Cosmochim. Acta 1983, 47, 665-685.
45. Apps, J. A.; Niel, J. M. 1989; this volume.
46. Hemingway, B. S. In Advances in Physical Geochemistry; Saxena, S. K., Ed.; Springer-Verlag: New York, 1982; Vol. 2, pp 285-316.
47. Naumov, G. B.; Ryzhenko, B. N.; Khodakovsky, I. L. Handbook of Thermodynamic Data; Atomizdat: Moscow, 1971; p 327.
48. Bond, A. M.; Hefter, G. T. IUPAC Chemical Data Series No. 27; Pergamon Press: Oxford, 1980; p 71.
49. Duer, W. C.; Robinson, R. A.; Bates, R. G. J. Chem. Soc. Faraday Trans. I 1972, 68, 716-722.
50. Nordstrom, D. K.; Jenne, E. A. Geochim. Cosmochim. Acta 1977, 41, 175-188.
51. Sillen, L. G.; Martell, A. E. Stability Constants of Metal-Ion Complexes; Spec. Publ. No. 17; The Chemical Society: London, 1964; p 754.
52. Smith, R. M.; Martell, A. E. Critical Stability Constants; Plenum Press: New York, 1976.
53. Langmuir, D. In Chemical Modeling in Aqueous Systems; Jenne, E. A., Ed.; ACS Symposium Series No. 93; American Chemical Society: Washington, DC,. 1979; pp 353-387.
54. Nordstrom, D. K.; May, H. M. In The Environmental Chemistry of Aluminum; Sposito, G., Ed.; CRC: Baca Raton, 1989; PP 29-53.
55. Roberson, C. E.; Barnes, R. B. Chem. Geol. 1978, 21, 239-256.
56. Yatsimirskii, K. B.; Vasil'ev, V. P. Instability Constants of Complex Compounds; Pergamon Press: Oxford, 1960; p 220.
57. Roberson, C. E.; Hem, J. Geochim. Cosmochim. Acta 1968, 32, 1343-1351.
58. Stull, D. R.; Prophet, H. 1971 JANAF Thermochemical Tables, second edition; U. S. Gov't Print. Office, Washington, DC.
59. Macaskill, J. B.; Bates, R. G. J. Phys. Chem. 1977, 81, 496-498.
60. Brown, D. W.; Roberson, C. E. J. Res. U. S. Geol. Survey 1977, 5, 509-517.
61. Whittemore, D. O.; Langmuir, D. J. Chem. Eng. Data 1972, 17, 288-290.
62. Bard, A. J.; Parsons, R.; Jordan, J. Standard Potentials in Aqueous Solution; Marcel Dekker: New York, 1985; p 834.
63. Olafsson, G.; Olafsson, I. J. Chem. Thermodyn. 1981, 13, 437-440.
64. Marshall, W. L.; Franck, E. U. J. Phys. Chem. Ref. Data 1981, 10, 295-304.
65. Baes, C. F., Jr.; Mesmer, R. E. The Hydrolysis of Cations; Wiley-Interscience: New York, 1976; p 489.
66. Langmuir, D.; Riese, A. C. Geochim. Cosmochim. Acta 1985, 49, 1593-1601.

67. Kester, D. R.; Byrne, R. H., Jr.; Liang, Y.-J. In *Marine Chemistry in the Coastal Environment*; Church, T. M., Ed.; ACS Symposium Series No. 18; American Chemical Society: Washington, DC, 1975; pp 56-79.
68. Robie, R. A.; Hemingway, B. S. *J. Chem. Thermodyn.* 1985, 17, 165-181.
69. May, H. M.; Helmke, P. A.; Jackson, M. L. *Geochim. Cosmochim. Acta* 1979, 43, 861-868.
70. Hem, J. D.; Roberson, C. E. *U. S. Geol. Survey Water-Supply Paper* 1827-D, 1972.
71. Feitknecht, W.; Schindler, P. W. *Pure Appl. Chem.* 1963, 6, 130-199.
72. Langmuir, D.; Whittemore, D. O. In *Nonequilibrium Systems in Natural Water Chemistry*; Hem, J. D., Ed.; ACS Advances in Chemistry Series No. 106; American Chemical Society: Washington, DC, 1971: pp 209-234.
73. Schwertmann, U.; Taylor, R. M. In *Minerals in Soil Environments*; Dixon, J. B.; Weed, S. B., Eds.; Soil Science Society of America: Madison, Wisconsin, 1977; pp 145-180.
74. Norvell, W. A.; Lindsay, W. L. *Soil Sci. Soc. Am. J.* 1982, 46, 710-715.
75. Siebert, R. M.; Hostetler, P. B. *Am. J. Sci.* 1977, 277, 697-715.
76. Busenberg, E.; Plummer, L. N.; Parker, V. B. *Geochim. Cosmochim. Acta* 1984, 48, 2021-2035.
77. Busenberg, E.; Plummer, L. N. *Geochim. Cosmochim. Acta* 1986, 50, 2225-2233.
78. Morgan, J. J. In *Principles and Applications of Water Chemistry*; Faust, S. D.; Hunter, J. V., Eds.; John Wiley and Sons: New York, 1967; pp 561-624.
79. Fouillac, C.; Criaud, A. *Geochemical J.* 1984, 18, 297-303.
80. Siebert, R. M.; Hostetler, P. B. *Am. J. Sci.* 1972, 277, 716-734.
81. Palmer, D. A.; van Elkik, R. *Chem. Rev.* 1983, 83, 651-731.
82. Garrels, R. M.; Thompson, M. E.; Siever, R. *Am. J. Sci.* 1960, 258, 402-418.
83. Helgeson, H. C.; Delaney, J. M.; Nesbitt, H. W.; Bird, D. K. *Am. J. Sci.* 1978, 278-A, p 229.
84. Smith, H. J. *J. Am. Chem. Soc.* 1918, 40, 879-883.
85. Robie, R. A.; Hazelton, H. T., Jr.; Hemingway, B. S. *Am. Mineral.* 1984, 69, 349-357.
86. Singer, P. C.; Stumm, W. *J. Am. Water Works Assoc.* 1970, 62, 198-202.
87. Busey, R. H.; Mesmer, R. E. *Inorg. Chem.* 1977, 16, 2444-2450.
88. Volosov, A. G.; Khodakovsky, I. L.; Ryzhenko, B. N. *Geochem. Internat.* 1972, 9, 362-377.
89. Hostetler, P. B.; Christ, C. L. *Geochim. Cosmochim. Acta* 1968, 32, 485-497.
90. Stoessell, R. K. *Geochim. Cosmochim. Acta* 1988, 52, 365-374.
91. Marshall, W. L.; Jones, E. V. *J. Phys. Chem.* 1966, 70, 4028-4040.
92. Rhigaletto, E. C.; Davies, C. W. *Trans. Faraday Soc.* 1930, 26, 592-600.
93. Austin, J. M.; Mair, A. D. *J. Phys. Chem.* 1962, 66, 519-521.
94. Truesdell, A. H.; Hostetler, P. B. *Geochim. Cosmochim. Acta* 1968, 32, 1019-1022.
95. Siebert, R. M.; Christ, C. L. 1976; unpublished data on Fuoss fitting of stability constant data.
96. Bell, R. P.; George, J. H. B. *Trans. Faraday Soc.* 1953, 49, 619-627.
97. Nair, V. S. K.; Nancollas, G. H. *J. Chem. Soc.* 1958, 3706-3710.
98. Reardon, E. J. *Geochim. Cosmochim. Acta* 1983, 47, 1917-1922.

99. Nair, V. S. K.; Nancolas, G. H. J. Chem. Soc. 1959, 3934-3939.
100. Adams, F.; Rawajfih, Z. Soil Sci. Soc. Am. J. 1977, 41, 686-692
101. Kelley, K. K.; Shomate, C. H.; Young, F. E.; Naylor, B. F.; Salo, A. E.; Huffman, E. H. U. S. Bur. Mines Tech. Report 688, 1946.
102. Mattigod, S. V.; Sposito, G. Soil Sci. Soc. Am. J. 1977, 41, 1092-1097.
103. Stipp, S. M.S. Thesis, University of Waterloo, Waterloo, Ontario, 1983.
104. Akitt, J. W.; Greenwood, N. N.; Lester, G. D. J. Am. Chem. Soc. 1969, 5, 803-807.
105. Blount, C. W.; Dickson, F. W. Am. Mineral. 1973, 58, 323-331.
106. Reardon, E. J.; Armstrong, D. K. Geochim. Cosmochim. Acta 1987, 51, 63-72.
107. Reardon, E. J.; Beckie, R. D. Geochim. Cosmochim. Acta 1987, 51, 2355-2368.
108. Davison, W. Geochim. Cosmochim. Acta 1979, 43, 1693-1696.
109. Turner, D. R.; Whitfield, M.; Dickson, A. G. Geochim. Cosmochim. Acta 1981, 45, 855-881.

RECEIVED October 6, 1989

# Chapter 32

# Solubilities of Aluminum Hydroxides and Oxyhydroxides in Alkaline Solutions

## Correlation with Thermodynamic Properties of $Al(OH)_4^-$

### John A. Apps and John M. Neil[1]

**Lawrence Berkeley Laboratory, 1 Cyclotron Road, Berkeley, CA 94720**

>The solubilities of gibbsite, boehmite and diaspore in alkaline solutions between 20 and 350°C are evaluated and their thermodynamic properties reconciled. The thermodynamic properties of the aluminate ion, $Al(OH)_4^-$, are derived over the same temperature range and compared with predictions based on the revised Helgeson-Kirkham-Flowers equation of state. Preliminary thermodynamic properties of bayerite and $\Delta G^\circ_{f,298}$ for nordstrandite are also derived from solubility data in alkaline solutions. Log $K_{s4}$ values for gibbsite, bayerite, boehmite and diaspore between 0 and 350°C, and thermodynamic data for $Al(OH)_4^-$ or $AlO_2^-$, are tabulated for use in distribution-of-species computer codes.

During the last decade, there has been an increasing interest in mathematically simulating the evolution of groundwaters. The purpose of such simulations is not only to reconcile observed with predicted groundwater compositions, but also to correlate the physical, mineralogical and isotopic changes of participating mineral and aqueous species with time, (1,2). Such modeling is required in a variety of programs for the deep burial of radioactive waste, toxic waste disposal, geothermal energy, and water resource management and conservation.

A complete description of any groundwater system necessitates consideration of reactions between rock forming minerals and the aqueous phase. This cannot be achieved without accurate thermodynamic properties of both the participating aluminosilicate minerals and aqueous aluminum species. Most computer codes used to calculate the distribution of species in the aqueous phase utilize the "reaction constant" approach as opposed to the "Gibbs free energy minimization" approach (3). In the former, aluminosilicate dissolution constants are usually written in terms of the aqueous aluminum species, $Al^{+++}$, which is related to other aqueous aluminum species by appropriate dissociation reactions.

[1]Current address: U.S. Geological Survey, Federal Building, 2800 Cottage Way, Sacramento, CA 95825

The choice of $Al^{+++}$ is unfortunate. The most recent CODATA compilation (4) gives $\Delta H^°_{f,298} = -128.68 \pm 0.36$ kcal.mol$^{-1}$ and $S^°_{298} = -77.7 \pm 2.4$ cal.mol$^{-1}$.K$^{-1}$ for $Al^{+++}$. These values are based in part on an earlier Russian analysis of calorimetric data and gibbsite solubility measurements (5). Some groundwater modelers (6,7) prefer $Al^{+++}$ thermodynamic values determined by Hemingway et al. (8) where $\Delta H^°_{f,298} = -126.91 \pm 0.96$ kcal.mol$^{-1}$ and $S^°_{298} = 73.6 \pm 3.6$ cal.mol$^{-1}$.K$^{-1}$. These are consistent with several gibbsite solubility measurements in dilute acid aqueous solutions (9–12) yielding similar log *$K_{s0,298}$ values near 8.11 (7,12) for the reaction

$$Al(OH)_3(\text{gibbsite}) + 3H^+ = Al^{+++} + 3H_2O$$

The CODATA thermodynamic properties for $Al^{+++}$ are based in part on log *$K_{s0,298}$ = 7.95 ± 0.44 for gibbsite. Both are higher than those of two recent solubility studies using synthetic gibbsite that was treated to remove adhering foreign material, i.e., log *$K_{s0,298}$ = 7.55 ± 0.055 (13) and 7.74 ± 0.14 (14), respectively.

Couturier et al. (15) found that log *$\beta_{4,298}$ for the reaction

$$Al^{+++} + 4H_2O = Al(OH)_4^- + 4H^+$$

derived from the earlier gibbsite solubility measurements (9,11,16) in dilute acid and basic aqueous solutions, was ≈−23.2 ± 0.2, which differs from the value of −22.20 derived from their own homogeneous aqueous solution measurements between 20 and 70°C. They concluded that either the gibbsite solubility measurements in acid media or those in basic media were incorrect. Using other evidence from natural systems, they determined that those in basic media were more reliable and accepted $\Delta H^°_{f,298} = -126.6$ kcal.mol$^{-1}$ and $S^°_{298} = -77.0$ cal.mol$^{-1}$.K$^{-1}$ for $Al^{+++}$.

It is clear that there is no consensus regarding the values assigned to the thermodynamic properties of $Al^{+++}$, and that present discrepancies in the recommended values are too large for any confidence to be placed in their use in groundwater modeling. Additional studies will be needed to close this issue. In contrast, there is a substantial and reconcilable literature for refinement of the thermodynamic properties of the aluminate ion, $Al(OH)_4^-$ or $AlO_2^-$. Many studies have been conducted on the solubilities of aluminum hydroxides in alkaline solutions, including gibbsite, whose thermodynamic properties have been well characterized by calorimetry. Much of this literature is in response to development of the Bayer process for refining bauxitic aluminum ores. The evaluation of aluminum hydroxide solubilities in alkaline solution is facilitated by relatively simple speciation. Although some questions have been raised as to the presence of polynuclear species in extremely concentrated aluminate solutions (17–20), sodium aluminate solutions generally contain only one mononuclear aluminum species, the aluminate ion. This contrasts with acid solutions where extensive hydrolysis of aluminum species occurs as a function of pH and temperature, and many real and imagined polynuclear hydroxy aluminum species have been identified (21–25).

There is therefore a strong argument in favor of defining aluminosilicate dissolution reactions in reaction constant codes in terms of the aluminate ion, because of the precision by which its thermodynamic properties might be determined in relation to well defined calorimetric standards, such as gibbsite and corundum. It is also preferable to describe aluminosilicate dissolution reactions in terms of the aluminate ion, because the need for recalculating the dissolution reaction constants for all aluminosilicate species can be avoided when refined properties for $Al^{+++}$ become available. Finally, it should be noted that most groundwaters are neutral to slightly alkaline, where $Al(OH)_4^-$, rather than $Al^{+++}$ dominates in the absence of fluoride complexation.

In this paper, the procedures are summarized by which the solubilities of gibbsite, boehmite, and diapore in alkaline solution are evaluated and used to compute thermodynamic properties of boehmite and the aluminate ion. Published data are then used in conjunction with the derived properties of the aluminate ion given in this paper to calculate preliminary thermodynamic properties for bayerite, and the Gibbs free energy of formation of nordstrandite at 25°C. Aluminum hydroxide and oxyhydroxide, and corundum solubility reaction constants, and other data suitable for modifying the thermodynamic data bases in reaction constant computer codes are also tabulated. With this information, an investigator will be able to model more precisely neutral to alkaline groundwaters.

Readers who desire a complete tabulation of the data, a description of experimental methods used in acquiring previously unpublished data and a critical discussion regarding the accuracy of the data sets employed in the evaluation, should request a copy of (26) from the senior author.

SOLUBILITIES OF ALUMINUM HYDROXIDES

The equilibration rates of aluminum hydroxide and oxyhydroxides with the aqueous phase are greatly accelerated in strongly alkaline solutions and with increasing temperature (26). With care, and using a variety of techniques (26), it is possible to establish whether or not equilibration has been attained with respect to a given solid phase.

Available solubility measurements on gibbsite, boehmite and diaspore span a temperature range from 20-350°C thereby permitting, in principle, the calculation of the Gibbs free energy of formation, $\Delta G_f^o$ of $Al(OH)_4^-$, over that temperature range. To do this, however, also requires the heat capacity, $C_P^o(T)$, the entropy, $S_{298}^o$, and the enthalpy of formation, $\Delta H_{f,298}^o$, of the participating solid phases. These properties have been fully determined by calorimetric means only for gibbsite and corundum, although even the value of $\Delta H_{f,298}^o$ of the latter has been questioned (27). It is possible to compute $\Delta H_{f,298}^o$ for diaspore from corundum with available phase equilibria data (28-31), which, together with $C_P^o(T)$ and $S_{298}^o$ from Perkins et al. (32), completely characterizes this phase. If the solubility of boehmite is reconciled with those of gibbsite and diaspore, the thermodynamic properties of boehmite can be refined and the consistency of the published properties of gibbsite and corundum can be tested. The thermodynamic data for the solid phases used in, and derived through this study, are summarized in Table I.
Aluminum hydroxides are assumed to dissolve according to the following reactions:

$$Al(OH)_3 + OH^- = Al(OH)_4^- \qquad K_{s4} = \frac{[Al(OH)_4^-]}{[Al(OH)_3][OH^-]} \qquad (1)$$

and

$$AlOOH + OH^- + H_2O = Al(OH)_4^- \qquad K_{s4} = \frac{[Al(OH)_4^-]}{[AlOOH][OH^-][H_2O]} \qquad (2)$$

where [ ] denote activities. Solubility data from the investigations listed on Figures. 1-4 (11,12,38-66) were evaluated using the EQ3 distribution of species code (67) and log $K_{s4}$ computed. The results were plotted graphically against computed ionic strength, I, and extrapolated empirically to I = 0. The empirical extrapolation procedure was necessary because the electrolyte model was used outside its range of validity, and because no better models were or are available for evaluation of high ionic strength alkaline aluminate solutions. The resulting log $K_{s4}(I = 0)$ values for the respective hydroxides are plotted against the reciprocal of absolute temperature in Figures 1–4.

Discrepancies between the data sets were reconciled as discussed in (26). The increasing scatter in $K_{s4}$ for gibbsite with falling temperature is attributed primarily to contamination of Bayer process gibbsite with bayerite or other surface reactive precipitates, and is discussed further below. Those measurements which are believed to approach most closely the true solubility of gibbsite are reported by by Kittrick (11) and Russell et al. (45).

Table I. Summary of thermodynamic properties of gibbsite, boehmite, diaspore, and corundum

| Mineral | Formula | $C_{P,298}^{\circ}$ | Maier-Kelley Function[a] parameters | | | $S_{298}^{\circ}$ | $\Delta H_{f,298}^{\circ}$ | $\Delta G_{f,298}^{\circ}$ | $\log K_{f,298}^{\circ}$ |
|---|---|---|---|---|---|---|---|---|---|
| | | | a | $b \times 10^3$ | $c \times 10^{-5}$ | | | | |
| | | cal.mol$^{-1}$·K$^{-1}$ | | | | cal.mol$^{-1}$·K$^{-1}$ | kcal.mol$^{-1}$ | kcal.mol$^{-1}$ | |
| Bayerite[c] | Al(OH)$_3$ | 22.246 | 8.65 | 45.6 | 0.0 | 18.97±1.43[d] | −307.83±0.31[d] | −275.57±0.32[d] | 202.00±0.24[d] |
| Gibbsite | Al(OH)$_3$ | 22.246 | 8.65[b] | 45.6[b] | 0.0[b] | 16.36±0.03[f] | −309.06±0.28[f] | −276.02±0.29[f] | 202.33±0.21[f] |
| Nordstrandite | Al(OH)$_3$ | – | – | – | – | – | – | −275.83±0.31[f] | 202.18±0.23[d] |
| Boehmite | AlOOH | 15.696 | 12.905[e] | 20.700[e] | −3.005[e] | 8.99±1.03[d] | −237.89±0.31[d] | −219.29±0.30[d] | 160.74±0.22[d] |
| Diaspore | AlOOH | 12.771 | 12.149[g] | 13.273[g] | −2.965[g] | 8.45±0.02[g] | −238.83±0.16[h] | −220.08±0.16[h] | 161.32±0.12[h] |
| Corundum | Al$_2$O$_3$ | 19.006 | 21.742[f] | 11.065[f] | −5.365[f] | 12.18±0.03[f] | −400.50±0.31[f] | −378.16±0.32[f] | 277.20±0.23[f] |

Uncertainties are = 2σ when based upon a statistical evaluation. Note that the uncertainties in $\Delta H_{298}^{\circ}$, $\Delta G_{298}^{\circ}$ and $\log K_{298}^{\circ}$ of the aluminum hydroxides and oxyhydroxides with respect to corundum and liquid water are between 2 and 5 times smaller than those of the corresponding properties with respect to the elements.

[a] $C_P^{\circ} = a + bT + cT^{-2}$.
[b] Kelley (33).
[c] Heat capacity assumed to be the same as that of gibbsite.
[d] This paper and Apps et al. (26).
[e] Mukaibo et al. (34).
[f] Calculated from Hemingway and Robie (35), Hemingway et al. (36), Robie et al. (37)
[g] Calculated from Perkins et al. (32)
[h] Calculated using phase equilibria data by Haas (30) as reported by Helgeson et al. (27).

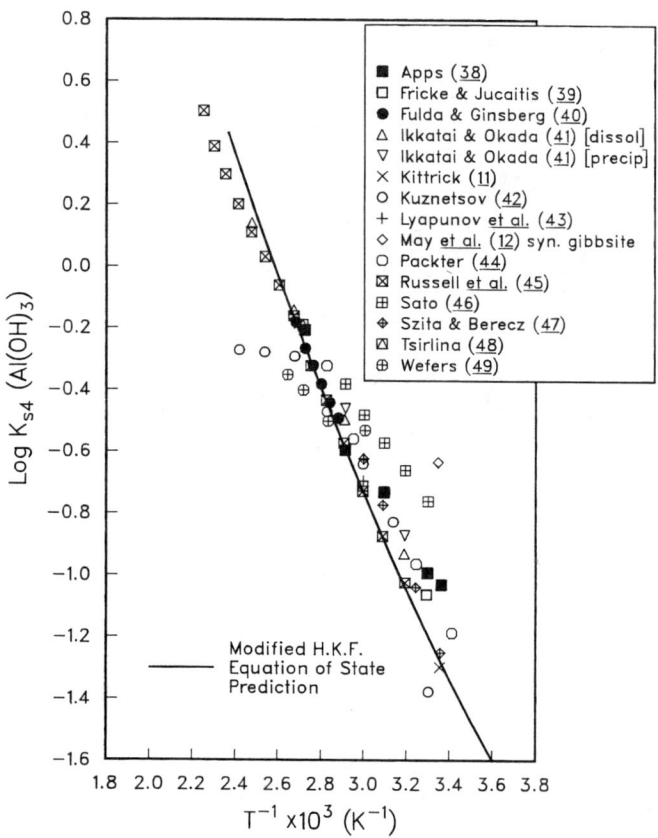

Figure 1. Plot of log $K_{s4}(Al(OH)_3)$ versus the reciprocal of absolute temperature, in which solubility measurements attributed to gibbsite are compared. The experimental precision of the solubility measurements produces a maximum $2\sigma$ uncertainty of $\leq \pm$ 0.1 in log $K_{S4}$ and $\leq \pm 4K$ in temperature.

Calculation of the Thermodynamic Properties

Boehmite, AlOOH. Log $K_T^o$ for the univariant reactions: gibbsite = boehmite + $H_2O(l)$, and boehmite = diaspore, were calculated from the differences in log $K_{s4}$ for the respective aluminum hydroxides at discrete temperatures, and where the solubility determinations overlapped. Log $K_f^o$ for boehmite was then calculated using calculated values of log $K_f^o$ for gibbsite, diaspore and water. The log $K_f^o$ data for boehmite were regressed in conjunction with the Maier-Kelley heat capacity function determined by Mukaibo et al. (34), and the respective values of $\Delta G_{f,298}^o$, $\Delta H_{f,298}^o$ and $S_{298}^o$ derived for boehmite, as given in Table I. The most notable feature of these properties, is that $S_{298}^o$ = 8.99 ± 1.03 cal.mol$^{-1}$.K$^{-1}$. This differs substantially from 11.58 ± 0.05 cal.mol$^{-1}$.K$^{-1}$, determined for an aluminum monohydrate by Shomate and Cook (68), and often erroneously assigned to boehmite.

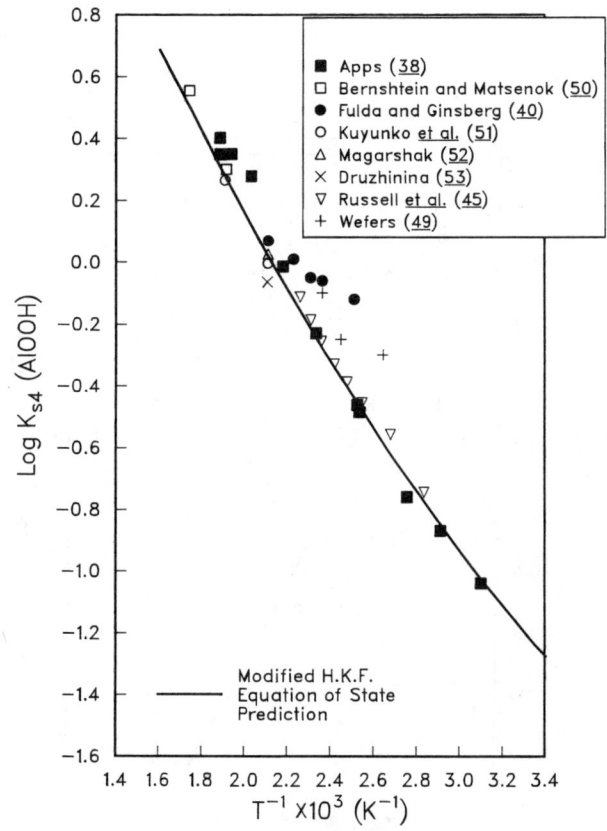

Figure 2. Plot of log $K_{s4}(AlOOH)$ versus the reciprocal of absolute temperature in which solubility measurements attributed to boehmite are compared. The experimental precision of the solubility measurements produces a maximum $2\sigma$ uncertainty of $\leq \pm$ 0.1 in log $K_{S4}$ and $\leq \pm$ 4K in temperature.

**The Aluminate Ion, $Al(OH)_4^-$.** Calculation of $\Delta \bar{G}_f^o$ or log $\bar{K}_f^o$ of the aluminate ion from gibbsite, boehmite and diaspore in alkaline solutions using equations (1) or (2) requires the corresponding standard state partial molal properties of the hydroxyl ion, $OH^-$. These may be determined from the dissociation constant for water, $K_w$ if it is accurately known over the range of temperatures and pressures investigated. The experimental measurements of diaspore solubility, reported here, extend to 350°C on the aqueous phase saturation surface, but experimental determinations of $K_w$ by Sweeton et al. (69) attain only 300°C. An electrolyte model is therefore needed to extrapolate $K_w$ another 50°C. The recently revised version of the Helgeson, Kirkham, Flowers (H.K.F.) equation of state by Tanger and Helgeson (70) was used to perform this extrapolation.

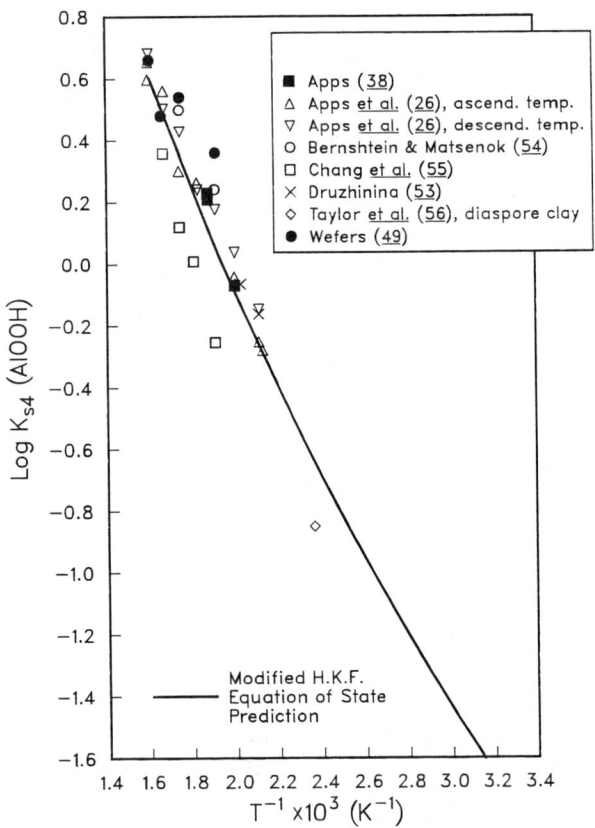

Figure 3. Plot of log $K_{s4}(AlOOH)$ versus the reciprocal of absolute temperature, in which solubility measurements attributed to diaspore are compared. The experimental precision of the solubility measurements produces a maximum $2\sigma$ uncertainty of $\leq \pm 0.2$ in log $K_{S4}$ and $\leq \pm 4K$ in temperature.

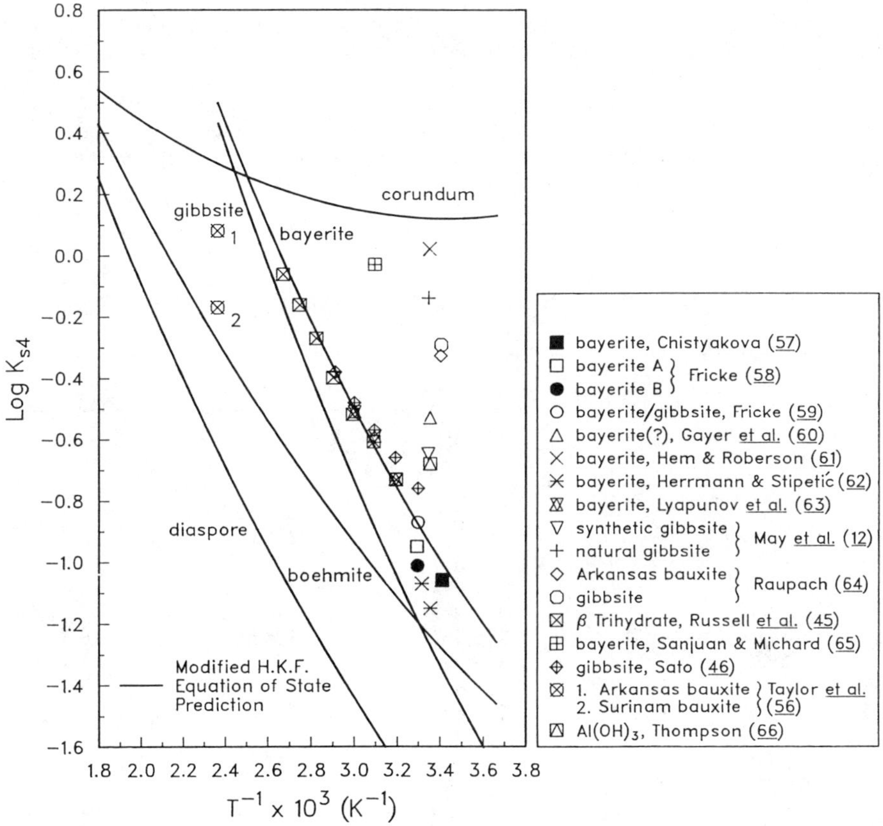

Figure 4. Plots of log $K_{s4}$ versus the reciprocal of absolute temperature, comparing the corresponding dissolution constants for bayerite, gibbsite, boehmite, diaspore and corundum. The experimental precision of the plotted solubility measurements produces a maximum $2\sigma$ uncertainty of $\leq \pm 0.2$ in log $K_{S4}$ and $\leq \pm 4K$ in temperature.

Calculation of $\log \overline{K}_f^o$, and $\overline{\Delta G}_{f,T}^o$ for $Al(OH)_4^-$ was conducted in two steps; an initial regression of selected experimental values of $\log K_{s4}$ for gibbsite, boehmite and diaspore between 20°C and 350°C, followed by calculation of $\log \overline{K}_{f,T}^o$, and $\overline{\Delta G}_{f,T}^o$ for $Al(OH)_4^-$ at 25°C intervals between 0°C and 350°C. These results are presented in Table II. $\overline{C}_{P,298}^o$, $\overline{S}_{298}^o$ and $\overline{\Delta H}_{f,298}^o$ for $Al(OH)_4^-$ may be calculated from $\overline{\Delta G}_f^o$ variation with temperature, but are very sensitive to experimental error. This calculation was performed between 20° and 70°C assuming that $\overline{C}_P^o$ remains constant over that temperature interval, and the resulting values incorporated in Table III. $\overline{C}_P^o = +21.08 \pm 1.65$ cal.mol$^{-1}$.K$^{-1}$, which is in good agreement with the 23.06 cal.mol$^{-1}$.K$^{-1}$ value obtained by Hovey et al. (71) using calorimetry. The procedures recommended by Shock and Helgeson (72), were used to calculate the revised H.K.F. equation of state parameters for $AlO_2^-$, which are summarized in Table III. The predicted values of $\log \overline{K}_{f,T}^o$ for $Al(OH)_4^-$ using this equation of state, are in excellent agreement with those obtained from solubility measurements, the maximum deviation being 0.07 at 250°C, (Table II). The H.K.F. equation of state prediction may be slightly more precise; however, differences are essentially within the limits of experimental error.

Table II. $\log \overline{K}_f^o(Al(OH)_4^-)$ and $\Delta \overline{G}_f^o(Al(OH)_4^-)$ between 0 and 350°C Along the Saturation Surface of Water

| T,°C | P, bars | $\Delta \overline{G}_f^o(Al(OH)_4^-)$, kcal.mol$^{-1}$ | $\log \overline{K}_f^o(Al(OH)_4^-)$ | | |
|---|---|---|---|---|---|
| | | | This work | Predicted by the H.K.F. Equation of State | Difference |
| 0 | 0.006 | −315.667 | 252.556 | 252.580 | −0.02 |
| 25 | 0.032 | −311.862 | 228.590 | 228.609 | −0.02 |
| 50 | 0.123 | −308.020 | 208.307 | 208.322 | −0.01 |
| 60 | 0.199 | −306.476 | 201.042 | 201.055 | −0.01 |
| 75 | 0.386 | −304.158 | 190.925 | 190.932 | 0.00 |
| 100 | 1.013 | −300.281 | 175.863 | 175.858 | 0.00 |
| 125 | 2.320 | −296.380 | 162.679 | 162.662 | +0.02 |
| 150 | 4.957 | −292.446 | 151.036 | 151.013 | +0.03 |
| 175 | 8.918 | −288.471 | 140.672 | 140.637 | +0.04 |
| 200 | 15.536 | −284.460 | 131.387 | 131.333 | +0.06 |
| 225 | 25.478 | −280.375 | 123.001 | 122.940 | +0.06 |
| 250 | 39.735 | −276.202 | 115.380 | 115.314 | +0.07 |
| 275 | 59.425 | −271.901 | 108.403 | 108.339 | +0.06 |
| 300 | 85.832 | −267.428 | 101.969 | 101.910 | +0.06 |
| 325 | 120.447 | −262.676 | 95.971 | 95.926 | −[a] |
| 350 | 165.212 | −257.436 | 90.283 | 90.251 | −[a] |

[a]This work is not independent of the H.K.F. Equation of State above 300°C.

Bayerite, $Al(OH)_3$. Bayerite occurs very rarely in nature (26), but is easily synthesized. Little is known regarding its thermodynamic properties. Some heats of decomposition have been measured in relation to gibbsite (73–76). One HF calorimetry study (77) and

Table III. Summary of thermodynamic properties of the aluminate ion

| Formula | $\overline{C}^\circ_{P,298}$ cal.mol$^{-1}$.K$^{-1}$ | $\overline{S}^\circ_{298}$ cal.mol$^{-1}$.K$^{-1}$ | $\Delta\overline{H}^\circ_{f,298}$ kcal.mol$^{-1}$ | $\Delta\overline{G}^\circ_{f,298}$ kcal.mol$^{-1}$ | $\log \overline{K}^\circ_{f,298}$ | $\overline{V}^\circ$ cm$^3$.mol$^{-1}$ |
|---|---|---|---|---|---|---|
| Al(OH)$_4^-$ | +21.08±1.65 | +29.59±1.03 | −357.56±0.31 | −311.88±0.33 | +228.61±0.24 | +46.3±0.3[a] |
| AlO$_2^-$ | −14.89 | −3.81 | −220.93 | −198.51 | +145.51 | +10.2 |

Uncertainties are 2σ when based upon a statistical evaluation.
[a] Hovey et al. (71).

H.K.F. Equation of State Parameters for AlO$_2^-$

| $a_1 \times 10^1$ cal.mol$^{-1}$.bar$^{-1}$ | $a_2 \times 10^{-2}$ cal.mol$^{-1}$ | $a_3$ cal.K.mol$^{-1}$.bar$^{-1}$ | $a_4 \times 10^{-4}$ cal.K.mol$^{-1}$ | $c_1$ cal.mol$^{-1}$.K$^{-1}$ | $c_2 \times 10^{-4}$ cal.K.mol$^{-1}$ | $\omega \times 10^5$ cal.mol$^{-1}$ |
|---|---|---|---|---|---|---|
| 3.1586 | 3.0566 | −2.1559 | −2.9054 | 13.331 | −6.075 | 1.6866 |

several alkaline solubility measurements (45,57–63,65) have also been made. Apps et al. (26) evaluated these measurements as a function of temperature between 20 and 100°C, making the approximation that $C_{P,T}^o$ of bayerite is the same as that of gibbsite. A plot of log $K_{s4}$ for bayerite versus reciprocal temperature, using the H.K.F. equation of state is given in Figure 4. Comparison of the difference in $\Delta H_{f,298}^o$ between gibbsite and bayerite, derived from calorimetric studies (77) with that determined from solubility measurements is excellent, i.e., 1.24 ± 0.11 kcal.mol$^{-1}$, versus +1.20 ± 0.26 kcal.mol$^{-1}$, lending credence to the validity of the solubility measurements.

Bayerite is a frequent but unsuspected contaminant in Bayer process gibbsite (78–80). This material is often used in experimental determinations of gibbsite solubility in dilute acid and alkaline solutions (11–14,38,65,81) leading with rare exceptions (13,14) to misleading interpretations of gibbsite solubility. Bayerite is only slightly more soluble than gibbsite between 20 and 100°C. A comparison of Figure 1 with Figure 4, suggests that many of the plotted gibbsite measurements in Figure 2 may be similarly affected by bayerite contamination. This is supported by many observations in the literature (79,82–86). Several reported solubility measurements on bayerite and gibbsite included in Figure 4, fall well above those values accepted in this paper as correct. While it is not possible to infer in all cases, the causes of the discrepancies, it is probable that equilibrium was measured with respect either to finely crystalline particles with a significant surface free energy contribution to the solubility, or to surface contamination. When combined with the short duration of the experiments, a temperature near 25°C, and the dilute nature of some of the alkaline solutions, it would not be surprising to find that equilibrium was not attained with respect to coarsely crystalline bayerite or gibbsite.

Other Gibbsite Polymorphs. Apart from bayerite, two other gibbsite polymorphs have also been observed in nature. The most common is nordstrandite, first reported to occur naturally in 1962 (87,88), but subsequently found in four distinctive low temperature, i.e. less than 100°C, environments (26,89,90). A summary of reported occurrences is given by Apps et al. (26). Both laboratory (91) and field evidence (92) suggest that nordstrandite solubility at earth surface temperatures falls between those of bayerite and gibbsite. Apps et al. (26) attempted to quantify the nordstrandite solubility at 25°C by taking into account small contributions to Gibbs free energy due to the variable surface areas of nordstrandite coexisting with gibbsite and bayerite. This lead to a calculated $\Delta G_{f,298}^o$ = -275.83 ± 0.31 kcal.mol$^{-1}$ for nordstrandite.

Another recently discovered polymorph is doyleite (93). Chao et al. (93) believe that other polymorphs may also occur depending on the stacking order of the $Al(OH)_3$ sheets, e.g., see (94). A notable feature is the small differences in Gibbs free energy separating the gibbsite, bayerite and nordstrandite polymorphs at 25°C. Their persistence in nature may consequently be governed by their particle size, by slight differences in environmental conditions, or by slow reaction kinetics.

APPLICATION OF THERMODYNAMIC DATA TO MODELING

Table IV summarizes log $K_{s4}$ values for aluminum hydroxides and oxyhydroxides, and corundum between 0 and 350°C. They were calculated using the modified H.K.F. equation of state (70) together with the data given in Tables I and II. These values are suitable for incorporation into distribution of species codes such as EQ3 (67), provided that $Al(OH)_4^-$ (or $AlO_2^-$) is made a basis species. Calculation of dissolution constants for other aluminosilicates can be made using the Gibbs free energy data for $Al(OH)_4^-$ or $AlO_2^-$ provided in Table II.

Table IV. Log $K_{s4}$ for Aluminum Hydroxides and Oxyhydroxides, and Corundum taken along the Water Saturation Surface

| T,°C | Log $K_{s4}$ | | | | |
|---|---|---|---|---|---|
| | Gibbsite | Bayerite | Boehmite | Diaspore | 0.5 Corundum |
| 0   | −1.679 | −1.263 | −1.462 | −2.108 | +0.127 |
| 25  | −1.276 | −0.943 | −1.240 | −1.821 | +0.119 |
| 50  | −0.889 | −0.626 | −1.022 | −1.548 | +0.138 |
| 60  | −0.740 | −0.502 | −0.937 | −1.445 | +0.149 |
| 100 | −0.184 | −0.032 | −0.623 | −1.059 | +0.210 |
| 150 | +0.433 | +0.499 | −0.279 | −0.638 | +0.298 |
| 200 | | | +0.018 | −0.269 | +0.391 |
| 250 | | | +0.276 | +0.057 | +0.481 |
| 300 | | | +0.500 | +0.348 | +0.568 |
| 350 | | | +0.689 | +0.604 | +0.642 |

ACKNOWLEDGMENTS

This work was supported by the U.S. Nuclear Regulatory Commission, through NRC FIN No. B 3040-6 under Interagency Agreement DOE-50-80-97, through U.S. Department of Energy Contract No. DE-AC03-76SF00098.

We are indebted to many of our colleagues and associates for their help through critical discussions, making available data prior to publication, and reviewing unpublished versions of a Lawrence Berkeley Laboratory report from which much of this information is drawn. In particular, we would like to express our appreciation to Dr. Chi-Hyuck Jun, who assisted the senior author with many of the calculations. Finally, the senior author would like to acknowledge the debt he owes to his former teacher at Harvard University, Professor R. M. Garrels, under whom some of the research reported in this paper was originally conducted.

LITERATURE CITED

1. Plummer, L. N.; Parkhurst, D. L.; Thorstenson, D. C. Geochim. Cosmochim. Acta 1983, 47, 665-686.
2. Plummer, L. N. In this volume.
3. Zeleznik, F. J.; Gordon, S. Ind. Eng. Chem. 1968, 60, 27-57.
4. Cox, J. D.; Wagman, D. D.; Medvedev, V. A. Codata Key Values for Thermodynamics. Hemisphere Publishing Corporation, New York, 1989.

5. S. Khodakovskiy, I. L.; Katorcha, L. V.; Kuyunko, S. S. Geokhim. 1980, 1606 – 1624.
6. Nordstrom, D. K.; May, H. M. In The Environmental Chemistry of Aluminum; Sposito, G., Ed.; CRC: Baca Raton, 1989 (in press).
7. Nordstrom, D. K.; Plummer, L. N.; Langmuir, D.; Busenberg, E.; May, H. M.; Jones, B. F.; Parkhurst, D. L. In this volume.
8. Hemingway, B. S.; Robie, R. A. Geochim. Cosmochim. Acta 1977, 41, 1402-1404.
9. Singh, S. S. Soil Sci. Soc. Amer. Proc. 1974, 38, 415 – 417.
10. Frink, C. R.; Peech, M. Soil Sci. Soc. Amer. Proc. 1962, 26, 346-347.
11. Kittrick, J. A. Soil Sci. Soc. Amer. Proc. 1966, 30, 595-598.
12. May, H. M.; Helmke, P. A.; Jackson, M. L. Geochim. Cosmochim. Acta 1979, 43, 861-868.
13. Bloom, P. R.; Weaver, R. M. Clays Clay Miner. 1982, 50, 281-286.
14. Peryea, F. J.; Kittrick, J. A. Clays Clay Miner. 1988, 36, 391-396.
15. Couturier, Y.; Michard, G.; Sarazin, G. Geochim. Cosmochim. Acta 1984, 48, 649-659.
16. Hemingway, B. S.; Robie, R. A.; Kittrick, J. A. Geochem. Cosmochim. Acta 1978, 42, 1533-1543.
17. Dibrov, I. A.; Mal'tsev, G. A.; Mashovets, V. P. Zh. Prikl. Khim. 1964, 37, 1920-1929.
18. Szita, L.; Berecz, E. Mag. Kém. Foly. 1975, 81, 386-392.
19. Eremim, N. I.; Volokihov, A; Mironov, V. E. Russ. Chem. Rev. Engl. Transl. 1974, 43, 92-106.
20. Akitt, J. W.; Gessner, W. J. Chem. Soc. Dalton Trans. 1984, 147-148.
21. Baes, C. F.; Mesmer, R. E. The Hydrolysis of Cations, John Wiley and Sons: New York, 1976.
22. Mesmer, R. E.; Baes, C. F. Jr. Inorg. Chem. 1971, 10, 2290-2296.
23. MacDonald, D. D.; Butler, P.; Owen, D. J. Phys. Chem. 1973, 77, 2474-2479.
24. Akitt, J. W.; Greenwood, N. N.; Khandelwal, B. L.; Lester G. D. J. Chem. Soc. Dalton Trans. 1972, 604-610.
25. Brown, P. L.; Sylva, R. N.; Batley, G. E.; Ellis, J. J. Chem. Soc. Dalton Trans. 1985, 1967-1970.
26. Apps, J. A.; Neil, J. M.; Jun, C.-H. Lawrence Berkeley Laboratory Rep. 21482, 1988.
27. Helgeson, H. C.; Delany, J. M.; Nesbitt, H. W.; Bird, D. K. Amer. J. Sci., 1978, 278-A.
28. Fyfe, W. S.; Hollander, M. A. Amer. J. Sci., 1964, 262, 709-712.
29. Matsushima, S.; Kennedy, G. C.; Akella, J.; Haygarth, J. Amer. J. Sci., 1967, 265, 28-44.
30. Haas, H. H. Amer. Mineral. 1972, 57, 1375-1385.
31. Wefers, K. Z. Erzbergbau Metallhuettenwes. 1967, 20, 13-19; 71-75.
32. Perkins, D.; Essene, E. Jr.; Westrum, E. F.; Wall, V. Amer. Mineral. 1979, 64, 1080-1090.
33. Kelley, K. K. Bull. U.S. Bur. Mines 584, 1960.
34. Mukaibo, T.; Takahashi, Y.; Yamada, K. Int. Conf. Calorim. Thermodyn. First, 1969, 375-380.
35. Hemingway, B. S.; Robie, R. A. J. Res. U.S. Geol. Surv. 1977, 5, 413-429.
36. Hemingway, B. S.; Robie, R. A.; Fisher, J. R.; Wilson, W. H. J. Res. U.S. Geol. Surv. 1977, 5, 797-806.
37. Robie, R. A.; Hemingway, B. S.; Fisher, J. R. U.S. Geol. Surv. Bull. 1452, 1979.
38. Apps, J. A. Ph.D Thesis, Harvard University, Cambridge, Massachusetts, 1970 .
39. Fricke, R.; Jucaitis, P. Z. Anorg. Allg. Chem. 1930, 191, 129-149.

40. Fulda, W.; Ginsberg, H. In Tonerde and Aluminum Walter de Gruyter and Co., Berlin, 1951, Vol. 1 p 31.
41. Ikkatai, T.; Okada, N. In Extractive Metallurgy of Aluminum; Gerhard, G.; Stroup, P. T., Eds.; Interscience: New York 1963; p 159.
42. Kuznetsov, S. J. Zh. Prikl. Khim. 1952, 25, 748-751.
43. Lyapunov, A. N.; Khodakova, A. G.; Galkina, Z. G. Tsvetn. Met. (N.Y.) 1964, 37, 48-51.
44. Packter, A. Colloid and Polym. Sci. 1979, 257, 977-980.
45. Russell, A. S.; Edwards, J. D.; Taylor, C. S. J. Met. 1955, 7, 1123-1128.
46. Sato, T. Kogyo Kagaku Zasshi 1954, 57, 805-808.
47. Szita, L.; Berecz, E. Báyászati és Koházati Lapok, Koházat 1970, 103, 37-44.
48. Tsirlina, S. M. Legk. Met. 1936, 5, No. 3, 28-37.
49. Wefers, K. Metall (Berlin) 1967, 25, 422-431.
50. Bernshtein, V. A.; Matsenok, Ye. A. Zh. Prikl. Khim. 1961, 38, 1935-1938.
51. Kuyunko, N. S.; Malinin, S. D.; Khodakovskiy, I. L. Geochem. Int. 1983, 20, 76-86.
52. Magarshak, G. K. Legk. Met. 1938, 7, No. 2, 12-16.
53. Druzhinina, N. K. Tsvetn. Met. (Moscow.) 1955, 28, 54-56.
54. Bernshtein, V. A.; Matsenok, Ye. A. Zh. Prikl. Khim. 1965, 38, 1935-1938.
55. Chang, B.-T.; Pak, L.-H.; Li, Y.-S. Bull. Chem. Soc. Japan 1979, 52, 1321-1326.
56. Taylor, C. S.; Tosterud, M.; Edwards, J. D. Aluminum Company Amer. Rep. R-95, 1927, 75 p.
57. Chistyakova, A. A. Tsvetn. Met. (N.Y.) 1964, 37, 58-63.
58. Fricke, R. Z. Anorg. Allg. Chem. 1928, 175, 249-256.
59. Fricke, R. Kolloid Z. 1929, 49, 229-243.
60. Gayer, K. H.; Thompson, L. C.; Zajicek, O. T. Can. J. Chem. 1958, 36, 1268-1271.
61. Hem, J. D.; Roberson, C. E. U.S. Geol. Surv. Water-Supply Paper 1827-A, 1967.
62. Herrmann, E.; Stipetic, J. Z. Anorg. Chem., 1950, 262, 258-287.
63. Lyapunov, A. N.; Grigor'eva, G. N.; Varzina, A. G. Tr. Vses. n.–i. iproekt. in–ta alymin., magn. i electrod. prom-sti 1973, 85, 24-32.
64. Raupach, M. Aus. J. Soil Res. 1963, 1, 28-35.
65. Sanjuan, B.; Michard, G. Geochim. Cosmochim. Acta 1987, 51, 1825-1831.
66. Thompson, L. C. A. Dissertation, Wayne State University, Detroit, Michigan, 1955.
67. Wolery, T. J. Lawrence Livermore Laboratory Rep. UCRL-S3414, 1983.
68. Shomate, C. H.; Cook, O.A. J. Amer. Chem. Soc. 1946, 68, 2140-2142.
69. Sweeton, F. H.; Mesmer, R. E.; Baes, C. F., Jr. J. Solution Chem. 1974, 3, 191-214.
70. Tanger, J. C.; Helgeson, H. C. Amer. J. Sci. 1988, 288, 19-98.
71. Hovey, J. K.; Hepler, L G.; Tremaine, P. R. J. Phys. Chem. 1988, 92, 1323-1332.
72. Shock, E.; Helgeson, H. C. Geochim. Cosmochim. Acta 1988, 52, 2009-2036.
73. Fricke, R.; Severin, H. Z. Anorg. Allg. Chem. 1932, 205, 287-308.
74. Eyraud, C.; Goton, R.; Trambouze, Y.; Tran Huu The; Prettre, M. C. R. Hebd. Seances Acad. Sci. 1955, 240, 862-864.
75. Michel, M. C. R. Hebd. Seances Acad. Sci. 1957, 244, 73-74.
76. Strobel, U.; Henning, O. Wiss. Z. Hochsch. Archit. Bauwes. Weimar 1972, 19, 383-385; Chem. Abstr. 1973, 79, 10527.
77. Fricke, R.; Wullhorst, B. Z. Anorg. Allg. Chem. 1932, 205, 127-144.
78. Calvet, E.; Boivinet, P.; Thibon, H.; Maillard, A. Bull. Soc. Chim. France 1951, 402-416.
79. Van Straten, H. A.; Holtkamp, B. T. W.; de Bruyn, P. L. J. Colloid Interface Sci. 1984, 98, 342-362.

80. Van Straten, H. A.; de Bruyn, P. L. J. Colloid Interface Sci. 1984, 102, 260-277.
81. Hitch, B. F.; Mesmer, R. E.; Baes, C. F., Jr.; Sweeton, F. H. Oak Ridge National Laboratory Rep. 5623, 1980.
82. Geiling, S.; Glocker, R. Z. Elektrochem. 1943, 49, 269-273.
83. Ginsberg, H.; Hüttig, W.; Stiehl, H. Z. Anorg. Allg. Chem. 1962, 318, 238-256.
84. Brosset, C. Acta Chem. Scand. 1952, 6, 910-940.
85. Bye, G. C.; Robinson, J. G. Kolloid Z. für Polym. 1964, 198, 53-60.
86. Barnhisel, R. I.; Rich, C. I. Soil Sci. Soc. Amer. Proc. 1965, 29, 531-534.
87. Wall, J. R. D.; Wolfenden, E. B.; Beard, E. H.; Deans, T. Nature 1962, 196, 261-265.
88. Hathaway, J. C.; Schlanger, S. O. Nature 1962, 196, 265-266.
89. Milton, C.; Dwornik, E. J.; Finkelman, R. B. Amer. Mineral. 1975, 60, 285-291.
90. Chao, G. Y.; Baker, J. Can. Mineral. 1982, 20, 77-85.
91. Schoen, R.; Roberson, C. E. Amer. Mineral. 1970, 55, 43-77.
92. Davis, C. E.; DeFour, J. Myrie; Adams, J. A; Lyew-Ayee, P. A. Trav. Com. Int. Etude Bauxites Alumine Alum. 1976, No. 13, 171-181.
93. Chao, G. Y.; Baker, J.; Sabina, A. P.; Roberts, A. C. Can. Mineral. 1985, 23, 21-28.
94. Saalfeld, H. Neues Jahrb. Mineral. Abh. 1960, 95, 1-87.

RECEIVED August 31, 1989

# Chapter 33

# Aluminum Hydrolysis Reactions and Products in Mildly Acidic Aqueous Systems

### John D. Hem and Charles E. Roberson

### U.S. Geological Survey, Water Resources Division, Mail Stop 427, 345 Middlefield Road, Menlo Park, CA 94025

Early stages of the hydrolysis and hydroxide polymerization reactions of aluminum have been studied using a constant slow aluminum perchlorate flux into a solution held at constant pH by addition of dilute NaOH from a pH stat, observing rates of base addition and periodically determining concentrations of the various species of dissolved aluminum. Total final Al concentrations generally were between $2 \times 10^{-4}$ and $5 \times 10^{-4}$ molal and holding pH's ranged from 4.75 to 5.20. Polymeric ionic species began to form when supersaturation with respect to microcrystalline gibbsite reached an affinity of reaction ($\underline{A}$) value near 1.0 kcal at pH 4.75, and the $\underline{A}$ value remained nearly constant during the remainder of the titration. Similar behavior, but with somewhat higher $\underline{A}$ values, was observed at higher pH's with a maximum $\underline{A}$ near 1.6 kcal at pH 5.20. Conditional first order rate constants were determined for the formation of polymeric ionic units with OH/Al molar ratios from 2.0 to 2.4. Average log k" values ($sec^{-1}$) ranged from -4.59 at pH 4.75 to -3.25 at pH 5.20 for experiments done at 25°C. Rate constants about 0.6 log units less negative were measured at 35°C, and about 0.9 log units more negative at 10°C. Species whose OH/Al ratio was greater than 2.4 were formed at the higher pH's, and developed into microcrystalline gibbsite. Polymeric aluminum hydroxide ions and colloidal microcrystalline gibbsite may occur in river and lake waters, especially those affected by low-pH precipitation, but positive identification of such material is difficult.

An understanding of the mechanisms of aluminum hydrolysis and the formation of crystalline species of aluminum hydroxide has been viewed as important in various fields of pure and applied chemistry, biochemistry, and geochemistry. In part, this interest results from the unique properties of certain hydrolysis species of aluminum that appear to be present as polymeric or polynuclear macro-ions. These ions have a strong positive charge and may interact with specific charge sites on surfaces they encounter. The polymeric species also may grow by accretion, and they may persist metastably for months or years under some conditions ([1]).

This chapter not subject to U.S. copyright
Published 1990 American Chemical Society

Aluminum hydrolysis reactions are particularly sensitive to pH and temperature, and to the concentrations of other ligands that might compete with $OH^-$ in complex formation. The most important inorganic competitor is $F^-$ (2). Certain organic ligands also can form strong complexes with aluminum (3). Hydrolysis reactions of aluminum also are influenced by sulfate, and hydroxysulfate solids may control aluminum solubility in acidic solutions (4).

This paper presents results of some open-system laboratory experiments in which aluminum hydrolysis behavior was studied in detail between pH 4.75 and 5.20, at 10°, 25°, and 35°C. Results are compared with those of our earlier work and that of others on the hydrolysis of aluminum in dilute solutions below the pH of minimum aluminum solubility.

## PREVIOUS WORK

In earlier studies of aluminum hydrolysis reactions (1, 5) test solutions in which $OH_T/Al_T < 3.0$ were prepared by batch-wise mixing of stock solutions of $Al(ClO_4)_3$, $HClO_4$, $NaClO_4$, and $NaOH$. The analysis procedure separated the aluminum into three forms based on the kinetics of color development with ferron (8-hydroxy, 7-iodo, 5-quinoline sulfonic acid). A related technique using 8-quinolinol was used earlier by others (6-7). More recently, the ferron procedure was used by Parthasarathy and Buffle (8) and by Bersillon et al. (9) to study aluminum hydrolysis at higher $Al_T$ concentrations. Jardine and Zelazny (10) described in detail the nature of interactions between the ferron reagent and aluminum species. The three forms of aluminum are designated (1) as: $Al_a$, very rapidly reacting, presumably consisting of uncomplexed $Al^{3+}$ and monomeric $Al(OH)_x^{(3-x)+}$ complexes; $Al_b$, a form reacting with pseudo - first order kinetics having a half-time of 12 to 20 minutes, and presumed to consist of polynuclear $Al_n(OH)_x^{(3n-x)+}$ ionic species in which Al ions are bound together by double OH bridges; and the third form, $Al_c$, that did not react with ferron at a significant rate over a period of several hours. $Al_c$ was shown by electron microscopy to consist of small crystals of gibbsite.

Smith and Hem (1) concluded that the amount of the $Al_b$ polymer formed in their experimental solutions was, in part, a function of the rate of addition of the NaOH solution during preparation. Fast addition tended to produce larger proportions of $Al_c$. Slower addition produced less $Al_c$ and more $Al_b$. This behavior is probably related to the higher initial degree of local supersaturation that occurs during rapid addition of base.

## EQUILIBRIUM SOLUBILITY OF GIBBSITE

In the absence of complexing agents other than $OH^-$ equilibrium solubility can be represented as a summation of the concentrations of 4$Al_a$ species, $Al^{3+}$, $AlOH^{2+}$, $Al(OH)_2^+$, and $Al(OH)_4^-$. Concentrations of individual solute species can be derived from thermodynamic activities used in equilibrium computations by means of the Debye-Hückel equation and appropriate values for ionic strength. It should be emphasized that the polymeric material, $Al_b$, is a metastable reaction intermediate, and does not directly participate in the reversible equilibria controlling gibbsite equilibrium solubility.

Table I contains standard free energies of formation for monomeric aluminum, and aluminum hydroxide complexes and two forms of crystalline gibbsite. Chemical equilibria and stability data in Tables II and III can be used with the $C_{Al_a}$ summation to compute equilibrium aluminum solubility as a function of pH at several temperatures and ionic strengths. The results of such calculations for microcrystalline gibbsite for the pH range of 4.0 to 6.0 are given in Figure 1. Over the 30°C range from 5.0°C to 35.0°C the solubility at pH 4.00 changes by more than 2 log units, with the highest solubility at the lowest temperature. The temperature effect is still nearly as great at pH 5.00 but is much less at pH 6.0, where the solubility at 5.0°C is only about a factor of 2 greater than the value for 25.0°C.

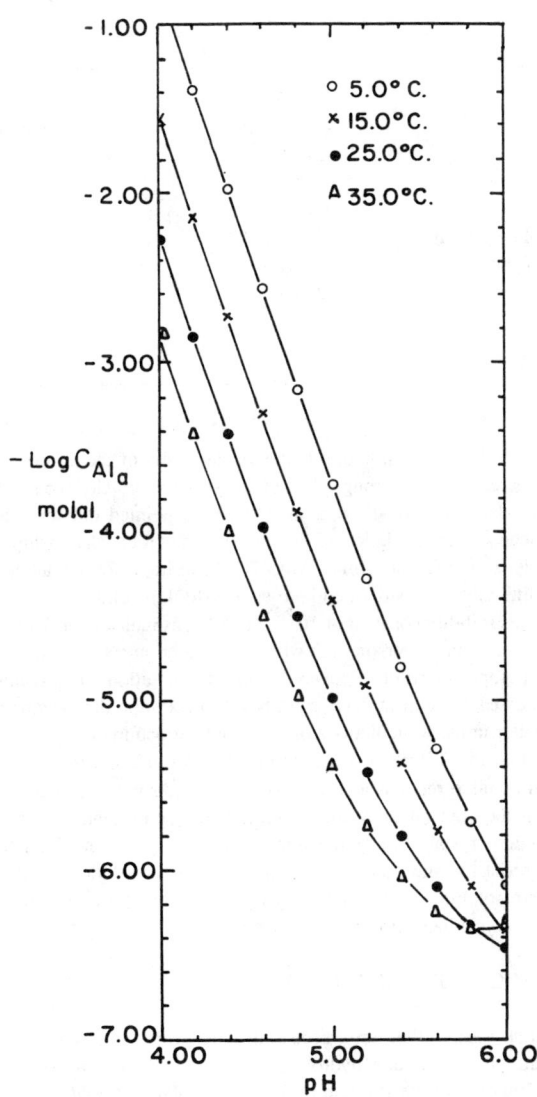

Figure 1. Equilibrium solubility of microcrystalline gibbsite at four different temperatures and 1 atm. pressure. Ionic strength 0.010.

Table I. Chemical Thermodynamic Data for Aluminum Species

| Species | $\Delta G_f^\circ$ kcal | Source of data |
|---|---|---|
| $Al^{3+}$ | $-117.0 \pm 0.3$ | (40) |
| $AlOH^{2+}$ | $-166.9 \pm 0.3$ | Calc. from data in (41) |
| $Al(OH)_2^+$ | $-216.6$ | Calc. from data in (48) |
| $Al(OH)_4^-$ | $-312.8$ | Calc. from data in (48) |
| $Al(OH)_{3c}$ | | |
| microcrystalline gibbsite | $-274.3 \pm 0.3$ | (12) |
| synthetic gibbsite | $-276.1$ | (42) |
| bayerite | $-275.6$ | (43) |
| $H_2O$ l | $-56.69$ | (44) |
| $OH^-$ | $-37.594$ | (44) |

Another feature indicated by this plot is the displacement of the minimum solubility to lower pH's as temperature increases. The strong effect of temperature on $Al(OH)_3$ solubility and aluminum solute speciation was also noted by Seip et al. (11), who pointed out that the effect could have environmental significance. Extrapolation of the gibbsite solubility relationship to pH's above 6.0 is probably not justifiable, as a different form of $Al(OH)_{3c}$, bayerite, is formed in alkaline solutions (12).

"Microcrystalline gibbsite" is defined here as an $Al(OH)_3$ solid having the crystalline form of gibbsite and a hydrolysis stability constant of $10^{-9.35}$ at 25°C, as indicated in Table II. This is based on experimental data of Hem and Roberson (12) who assigned an uncertainty of ± 0.3 log unit to their value. Electron micrographs of solids recovered from their solutions by passing them through 0.10 μm porosity filters showed some hexagonal crystals with maximum dimensions near 0.10 μm, along with many much smaller units. Calculations of edge and surface interfacial energies for the smaller particles (1) showed that the difference in solubility between microcrystalline and macrocrystalline synthetic gibbsite can be interpreted as a particle-size effect. As noted by Parks (13) the solubility of such material tends to be controlled by the smallest particles present. Under some conditions of solution chemistry, microcrystalline gibbsite alters very slowly and material having the solubility indicated here may persist for more than 3 years of aging (1) in solution at 25°C. Data in Table II demonstrate that where solubility is controlled by hydrolysis equilibria this microcrystalline form is more soluble than well-crystallized gibbsite by a factor of 20.

EXPERIMENTAL PROCEDURES AND RESULTS

The new experimental results that are discussed here were obtained by modified laboratory procedures. A controlled steady-state hydrolysis at constant ionic strength was accomplished by adding, with an autotitrator, a dilute solution ($4.53 \times 10^{-4}$ molal of $Al(ClO_4)_3$ at a very slow constant rate to a reaction vessel containing at the start 150 ml of background electrolyte solution (0.010 molal $NaClO_4$). The solution in the reactor was continuously maintained at a pre-selected pH by a second autotitrator, operating in a pH-stat mode that added 0.010 molal NaOH through capillary teflon tubing on demand. Both titrating solutions contained enough of the background electrolyte to bring their ionic strength to 0.01, and the $Al(ClO_4)_3$ solution contained enough $HClO_4$ to maintain its pH near 3.5. The reaction vessel (a covered teflon beaker) was immersed in a water bath that held the temperature constant (±0.1°C), and the solution was rapidly stirred with a mechanical stirrer. Amounts of reagents added and rates were recorded continuously on strip charts. A record of amounts of solution removed was also maintained as samples for Al species analysis were withdrawn using a calibrated syringe. A glass-calomel combination electrode (Radiometer GK 2401C) was used as pH sensor. Equipment used included Radiometer ABU80 autoburettes, PHM84 pH meters, TTT 80 titrators, and REC 80 servographs (use of brand names is for identification only and does not imply endorsement by the U.S. Geological Survey).

Table II. Equilibrium Constants for Aluminum Hydroxide Precipitation Reaction, $Al^{3+} + 3H_2O \rightleftarrows Al(OH)_{3c} + 3H^+$, at Temperatures Indicated

| Solid species | log $^*K_{so}$ | T,°C | Source of data |
|---|---|---|---|
| Microcryst. gibbsite | -10.40 | 10 | a/ |
| Microcryst. gibbsite | -9.35 | 25 | (12) |
| Microcryst. gibbsite | -8.72 | 35 | a/ |
| Synthetic gibbsite | -8.11 | 25 | (45) |
| Synthetic gibbsite | -8.04 | 25 | Calc. from data in (42) |
| Synthetic gibbsite | -8.04 | 25 | Calc. from data in (46) |
| Synthetic gibbsite | -8.60 | 15 | Calc. from data in (46) |
| Synthetic gibbsite | -7.52 | 35 | Calc. from data in (46) |
| Synthetic gibbsite | -7.97 | 25 | Calc. from data in (47) |

a/ Calculated from Van't Hoff equation using values of $\Delta H°_{Al(OH)_3\ amorph} = -305$ kcal/mol, $\Delta H°_{Al^{3+}} = -127$ kcal/mol, and $\Delta H°_{H_2O\ell} = -70.41$ kcal/mole given by Wagman et al. (44).

From the titration and analysis data, the concentrations of $Al_a$, $Al_b$, and $Al_c$ can be calculated for any specific time. The concentrations of $OH^-$ over and above the amount used in reacting with free $H^+$ in the aluminum perchlorate titrant solution can be apportioned among these aluminum species by assuming that: 1) $Al_a$ consists of $Al^{3+}$, $AlOH^{2+}$, $Al(OH)_2^+$, and $Al(OH)_4^-$, all of which are at equilibrium with one another at the pH of the solution in the reactor; 2) $Al_c$ is $Al(OH)_{3c}$, hence $OH_c$ is three times the determined value of $Al_c$; and 3) the remaining OH is combined with $Al_b$.

Titrations were performed at pH 4.75, 4.90, 5.00, and 5.20 at temperatures of 25°C, and for some of these pH's titrations were made at 35° and 10°C. Generally, the titration phase of each experiment was run for a total of 20 to 30 hours, but titration was not continuous. The equipment was operated for 5 or 6 hours each day for about a week and was shut down overnight except for the temperature control bath. Samples were taken at the beginning and end of each day's run, and sometimes more frequently. The final total aluminum concentrations reached in the experimental runs ranged from about $2.0 \times 10^{-4}$ to about $5.0 \times 10^{-4}$ molal, but in most of the titration period in each experiment the $Al_a$ fraction concentration was near or below $10^{-4.0}$ molal. Characteristically there was a short period of adjustment at the start of each day's run, after which the rate of base addition stabilized and remained nearly constant.

Table III. Aluminum Hydrolysis Equilibrium Constants at Various Temperatures

| Reaction | T,°C | Log K | Source of data |
|---|---|---|---|
| $Al^{3+} + H_2O \rightleftarrows AlOH^{2+} + H^+$ | 10 | -5.44 | (48) |
|  | 25 | -5.00 | (48) |
|  | 35 | -4.73 | (48) |
| $Al^{3+} + 2H_2O \rightleftarrows Al(OH)_2^+ + 2H^+$ | 10 | -11.2 | (48) |
|  | 25 | -10.1 | (48) |
|  | 35 | -9.5 | (48) |
| $Al^{3+} + 4H_2O \rightleftarrows Al(OH)_4^- + 4H^+$ | 10 | -24.3 | (48) |
|  | 25 | -22.7 | (48) |
|  | 35 | -21.7 | (48) |
| $H_2O \rightleftarrows H^+ + OH^-$ | 10 | -14.53 | (49) |
|  | 25 | -14.00 | (49) |
|  | 35 | -13.68 | (49) |

Figure 2 is a graph showing the concentrations of $Al_a$ and $Al_b$ at the end of each day's run during a total titration time of 1,240 minutes in experiment Al 12, done at pH 4.90 and 25°C. The tendency for $Al_a$ concentration to be maintained at a nearly constant level was observed in all the titrations. However, the steady-state concentration was somewhat lower at higher pH's. The concentration of $Al_b$ increased at a nearly steady rate during this experiment. No $Al_c$ could be detected in this solution, but some $Al_c$ was produced in titrations made at pH 5.00 and 5.20 at 25°C and larger amounts of $Al_c$ were produced at 35°C. Detection limits for the procedure are near $10^{-6.0}$ molal.

In the open system maintained during these titrations the hydrolysis reactions are driven by the continuous addition of $Al^{3+}$ ions. The rate of the hydrolysis reactions can be inferred from the rate of $OH^-$ addition, with appropriate adjustments, and the reaction kinetics can be evaluated. Early in the titration period the formation of monomeric aluminum hydroxy-ions ($Al_a$) is the only reaction thermodynamically feasible as the solution is below saturation with respect to microcrystalline gibbsite. The solid point in Figure 2 represents the equilibrium concentration of $Al_a$ with respect to that solid at 25°C, I = 0.01 and pH 4.90, which was reached in about 50 minutes.

The second stage, polymerization of monomers to form $Al_b$, did not begin in any of these experiments until after substantial super-saturation with respect to microcrystalline gibbsite was reached. Supersaturation is expressed here in terms of reaction affinity, $\underline{A}$, in kilocalories (kcal) from the relationship: $\underline{A} = -2.303RT (\log Q - \log K)$ where R is the gas constant, T is temperature in the kelvin scale, Q is the activity quotient, and K is the equilibrium constant for the reaction being considered. Positive values of $\underline{A}$ are required for the chemical reaction to proceed to the right as written. The threshold for polymerization to begin is near 1.0 kcal per mole of Al at pH 4.75 and 4.90. Threshold $\underline{A}$ values were somewhat higher at higher pH.

The third stage in the reaction represents conversion of $Al_b$ to $Al_c$ by growth of individual units to a size displaying solid-state behavior. At 25°C the critical level for $\underline{A}$ for $Al_c$ formation was near 1.30 kcal. Table IV summarizes the affinity data for all the experiments considered here. The reaction affinities are affected by pH, temperature and rate of $Al^{3+}$ addition, and these factors need to be considered in any interpretation of the data.

Table IV. Reaction Affinities Observed in Titrations at Various pH's and Temperatures

| Experiment | pH | T,°C | Reaction affinities, precipitation of microcrystalline gibbsite, kcal | | |
|---|---|---|---|---|---|
| | | | 1/ | 2/ | 3/ |
| Al 14[4/] | 4.75 | 25.0 | 0.89 | 1.01 | 0.95 |
| Al 17[4/] | 4.75 | 35.0 | 1.13 | 1.14 | 1.18 |
| Al 12[4/] | 4.90 | 25.0 | .97 | 1.20 | 1.13 |
| Al 15[4/] | 4.90 | 35.0 | 1.26 | 1.10 | 1.25 |
| Al 19[4/] | 5.00 | 25.0 | 1.20 | 1.34 | 1.28 |
| Al 20[5/] | 5.00 | 25.0 | 1.47 | 1.50 | 1.49 |
| Al 13[4/] | 5.20 | 25.0 | 1.38 | 1.28 | 1.34 |
| Al 21[5/] | 5.20 | 25.0 | 1.62 | 1.55 | 1.60 |
| Al 16[4/] | 5.20 | 10.0 | 1.11 | 1.18 | 1.15 |

1/ $\underline{A}$ at first appearance of Al$_b$.
2/ $\underline{A}$ at end of titration.
3/ Mean $\underline{A}$ for titration period.
4/ Al addition rate 0.055 µmole/min.
5/ Al addition rate 0.122 µmole/min.

The three stages of the hydrolysis reaction can be represented schematically by one equilibrium and two irreversible reactions as follows:

$$Al^{3+} + H_2O \rightleftarrows AlOH^{2+} + H^+ \tag{1}$$

$$AlOH^{2+} + H_2O \rightarrow 1/6\, Al_6(OH)_{12}^{6+} + H^+ \tag{2}$$

$$1/6\, Al_6(OH)_{12}^{6+} + H_2O \rightarrow Al(OH)_{3c} + H^+. \tag{3}$$

The sum of reactions, 1, 2, and 3, is: $Al^{3+} + 3H_2O \rightleftarrows Al(OH)_{3c} + 3H^+$. Values of K for the summarizing reaction are given in Table II, and K$_1$ values are given in Table III. A continuous state of equilibrum for 1 implies that $\underline{A}_1 = 0$. Any departure from zero for the observed values for $\underline{A}$ for the summarizing reaction then can be assigned to the sum of reactions 2 and 3 and is designated $\underline{A}_{23}$. The formula Al$_6$(OH)$_{12}^{6+}$ represents Al$_b$ in equations 2 and 3 above and is equivalent to a single hexagonal ring of 6Al$^{3+}$ ions bound together by six (OH)$_2$ bridges. The fundamental distinction between Al$_a$ and Al$_b$ species, as earlier studies pointed out, is that the OH$^-$ in the Al$_b$ species is present in structural bridging positions.

KINETICS OF Al$_b$ FORMATION

Data in Table IV for experiments Al 19 and 20 and for Al 13 and 21 indicate the effect of rate of addition of Al on the reaction affinity. In both pairs of experiments, when Al addition rate was increased the value of $\underline{A}$ increased. This difference in $\underline{A}$ results from the higher steady-state concentration of Al$_a$ that is maintained by the faster rate at which reactants are supplied. The rate of Al$_b$ formation also must increase to maintain the steady-state condition.

In effect this is a self-organized system that maintains a steady state, although not at chemical equilibrium, by a balance of reaction rates against reactant fluxes, a condition commonly seen in real-world systems. The titration experiments permit a rather direct measurement of the rate of nucleation and polymerization reactions that transform Al$_a$ monomers to Al$_b$ polymers. The titrations at pH 4.75 and 4.90 yielded only Al$_a$ and Al$_b$. Although some Al$_c$ was formed in titrations at pH 5.00, the polymeric product was about 90% Al$_b$. Hence, the data obtained in these experiments should be useful for determining the effect of pH on the rate of the polymerization process leading from Al$_a$ to Al$_b$. At least a part of the information from titrations made at pH 5.20 should also be useful because during much of those experiments the principal product also was Al$_b$.

The concentration of reaction product, Al$_b$, is assumed not to affect the forward rate of reaction 2 by direct reversal, but it could influence the overall observed rate if Al$_b$ rapidly converts to Al$_c$. This process could become coupled to the first step and alter the mechanism. Preliminary titrations made at pH 5.40 produced mostly Al$_c$ and the data are not included here.

As the addition of unreacted Al continues, the concentration of Al$_a$ should stabilize at a value governed by the relationship between the rate of formation of Al$_b$ and the entering Al flux. The formation of Al$_b$ under these conditions should be a psuedo first-order process whose rate will be indicated by the rate at which OH$^-$ must be added to react with H$^+$ produced in the conversion of Al$_a$ to Al$_b$. The determined concentration of AlOH$^{2+}$ and the properly adjusted rate of OH$^-$ addition over a suitable time interval permits calculating a conditional rate constant.

The rate of OH$^-$ addition taken from the recorder trace of the pH-stat must first be adjusted to account for the free H$^+$ that is added with the aluminum perchlorate solution. A further adjustment is required for the OH$^-$ taken up by monomeric Al$_a$ species, assuming all these species are at equilibrium, controlled by the titration pH.

The three equilibrium expressions for the formation of monomers are:

$$[AlOH^{2+}] = [Al^{3+}] * K_1 [H^+]^{-1} \qquad (4)$$

$$[Al(OH)_2^+] = [Al^{3+}] * K_{12} [H^+]^{-2} \qquad (5)$$

$$[Al(OH)_4^-] = [Al^{3+}] * K_{14} [H^+]^{-4} \qquad (6)$$

The concentration of Al$_a$, C$_{Ala}$ is represented by the species summation:

$$C_{Ala} = [Al^{3+}] \gamma^{-1}_{Al^{3+}} + [AlOH^{2+}] \gamma^{-1}_{AlOH^{2+}} + \qquad (7)$$
$$[Al(OH)_2^+] \gamma^{-1}_{Al(OH)_2^+} + [Al(OH)_4^-] \gamma^{-1}_{Al(OH)_4^-}$$

where γ terms are activity coefficients and bracketed quantities are thermodynamic activities. Concentrations of OH$_a$ are computed from

$$C_{OHa} = C_{AlOH^{2+}} + 2C_{Al(OH)_2^+} + 4C_{Al(OH)_4^-}. \qquad (8)$$

From these equations and data in Table III the value of the molar ratio C$_{OHa}$/C$_{Ala}$ for any specified pH can be computed. Because all the added Al must pass through the monomeric hydrolysis stage, it follows that the flux, $\phi$Al$_T$, being added to the reactor by the autoburette, can be used in place of C$_{Ala}$ to calculate an equivalent OH requirement analogous to C$_{OHa}$.

## 33. HEM & ROBERSON  *Aluminum Hydrolysis Reactions and Products*

The remaining OH flux, $\phi OHb$, can be assigned to production of Alb. The concentration of Alb is known from the analyses of the solutions. If the ratio $C_{OHb}/C_{Alb}$ has a value of 2.0, the stoichiometry for conversion of Ala to Alb can be represented directly by equation 2, and one mole of added OHb produces one mole of Al in the form of Alb. If the OHb/Alb ratio is greater or less than 2.0 the rate of OH addition requires a further correction to give a rate constant in terms of aluminum that is reacting to form Alb. This correction is designated n in the final rate equation, which is written:

$$n\phi OHb = dC_{Alb}/dt = k_2" C_{AlOH^{2+}} \qquad (9)$$

and n is computed from the relationship:

$$n = (C_{OHb}/C_{Alb} - 1.0)^{-1}. \qquad (10)$$

$AlOH^{2+}$ is the predominant form of monomeric aluminum over most of the pH range where the kinetic experiments were made. The OHb flux is converted to moles $L^{-1}$ $sec^{-1}$ by dividing the calculated flux in moles per second by the volume of solution in the reactor at the midpoint of the time interval being considered. Rate constants were calculated using time intervals of 1 to 2 hours.

Results of the titration experiments are summarized in Table V. Figure 3 is a plot of the conditional rate constant data for these experiments vs the controlled pH. The observed range in $k_2"$ for a given pH is ±0.2 log units. There is a clear indication in Figure 3 of a well defined pH dependence of the nucleation and polymerization rate. The dashed line on the graph represents a third-order pH dependency - that is, a change of 0.5 unit in pH causes a change in $k_2"$ of 1.5 log units.

Table V. Typical Titration Data Used for Computing Kinetics of Al(OX)x Polymerization

| Expt. | T, °C | pH | Reactant flux $\mu mol$ $min^{-1}$ $Al_T$ | $\mu mol L^{-1}$ $min^{-1}$ $OH_T$ | $OH_b$ | $A_{23}$ | $\frac{OH_b}{Al_b}$ | n | $mol L^{-1}$ x $10^{-4}$ $C_{AlOH^{2+}}$ | $\mu mol L^{-1}$ $min^{-1}$ $\Delta Al_b/\Delta t$ | log $k_2"$ sec$^{-1}$ | log $k_2"$ mean for ex pt. |
|---|---|---|---|---|---|---|---|---|---|---|---|---|
| | | | | | (kcal) | | | | | | | |
| A114 | 25.0 | 4.75 | 0.055 | .088 | .107 | .96 | 2.0 | 1.00 | .570 | .107 | -4.50 | } -4.59 |
| | | | | .093 | .123 | 1.01 | 2.2 | .83 | .619 | .102 | -4.56 | |
| A112 | 25.0 | 4.90 | .055 | .115 | .216 | 1.20 | 2.3 | .77 | .431 | .166 | -4.19 | -4.12 |
| A119 | 25.0 | 5.00 | .055 | .136 | .281 | 1.30 | 2.0 | 1.00 | .317 | .281 | -3.83 | } -3.81 |
| | | | | .138 | .285 | 1.34 | 2.1 | .91 | .340 | .259 | -3.90 | |
| A120 | 25.0 | 5.00 | .122 | .315 | .822 | 1.50 | 2.4 | .71 | .453 | .587 | -3.67 | } -3.60 |
| | | | | .312 | .806 | 1.50 | 2.3 | .77 | .453 | .621 | -3.64 | |
| A121 | 25.0 | 5.20 | .122 | .366 | .871 | 1.58 | 2.4 | .71 | .209 | .618 | -3.31 | -3.25 |
| A113 | 25.0 | 5.20 | .055 | .177 | .412 | 1.35 | 2.4 | .71 | .140 | .293 | -3.46 | } -3.40 |
| | | | | .176 | .409 | 1.28 | 2.4 | .71 | .125 | .290 | -3.41 | |
| A117 | 35.0 | 4.75 | .055 | .084 | .188 | 1.23 | 2.0 | 1.00 | .368 | .188 | -4.07 | -3.98 |
| A115 | 35.0 | 4.90 | .055 | .165 | .408 | 1.24 | 2.2 | .83 | .190 | .339 | -3.53 | -3.54 |
| A116 | 10.0 | 5.20 | .055 | .111 | .221 | 1.17 | 2.2 | .83 | .431 | .183 | -4.15 | -4.17 |

438  CHEMICAL MODELING OF AQUEOUS SYSTEMS II

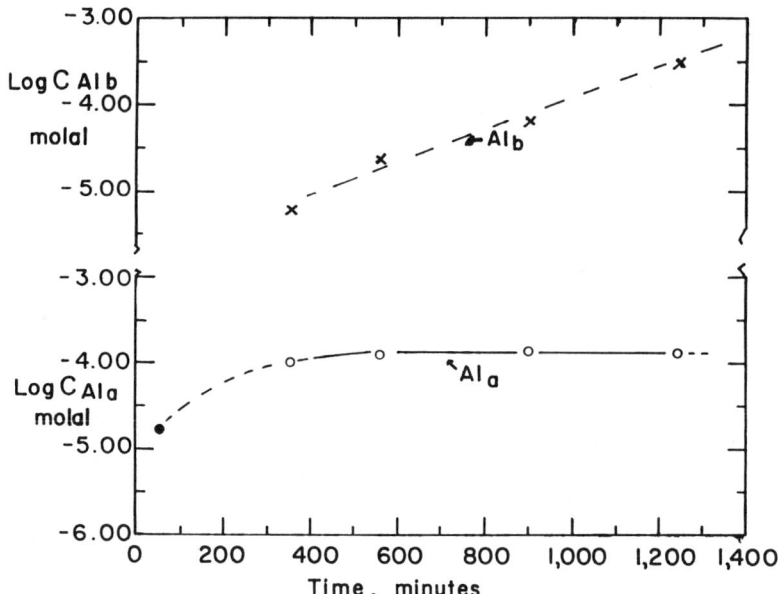

Figure 2. Total concentrations of $Al_a$ and $Al_b$ species measured during titration, experiment Al 12, pH 4.90, T = 25.0°C.

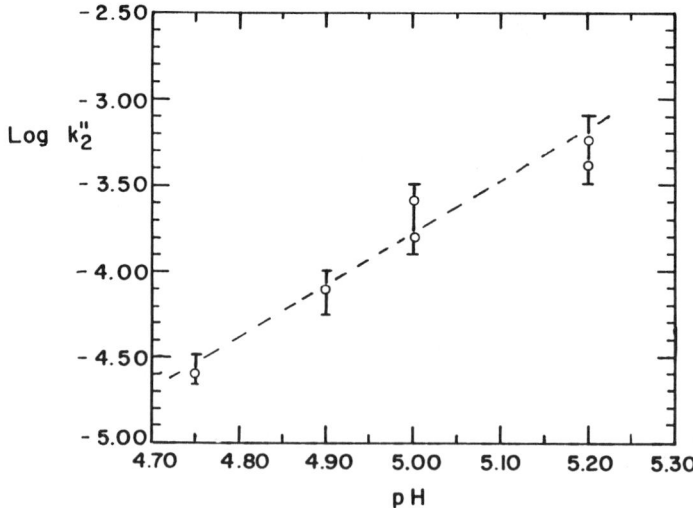

Figure 3. Kinetics of formation of $Al_a$ at 25.0°C from pH 4.75 to 5.20. Open circles are mean values of log $k_2''$ measured in six experimental runs; vertical lines represent range of log $k_2''$ values computed for runs at pH's indicated. Units for $k_2''$ are sec.$^{-1}$.

There is also a strong temperature effect on the rate constant. Applying the relevant data in Table V to the Arrhenius relationship:

$$E_a = \frac{(\log k_2 - \log k_1) \times 2.303 \, R \, T_1 T_2}{T_2 - T_1} \tag{11}$$

where $E_a$ is the energy of activation, k and T represent rate constants and kelvin temperatures, respectively, and R is the gas constant, gives $E_a$ values of 25.6, 27.3 and 21.6 kcal for pH's of 4.75, 4.90 and 5.20, respectively. The thermodynamic significance of these numbers is obscured by the strong effect of temperature on $H^+$ activity, and the empirical nature of the rate constant. However, the low $E_a$ obtained at pH 5.20 may be related to the formation of some $Al_c$ at that pH during titration.

## DISCUSSION OF KINETIC RESULTS

The effect of using different rates of Al addition can be observed in the data given in Tables IV and V. It was assumed that the calculated rate constant would not be significantly affected by using different Al addition rates. This required that the rate chosen be adequate to achieve the necessary reaction affinity but not so fast that it would prevent a steady-state from being reached. Supersaturation that may occur in the solution in close proximity to the reagent delivery points proably does not have an important effect and the measured pH is assumed to represent conditions throughout the solution during titration. The slow addition rates, the dilute nature of the added solutions and continuous mechanical stirring all help minimize local supersaturation. In addition, the formation of $Al_a$ species is fast and reversible, while the irreversible steps to form $Al_b$ and $Al_c$ are probably not fast enough to progress significantly during the short time any given parcel of the solution is passing through the area affected by incomplete mixing of reagents.

Approximately doubling the rate of Al addition in titration increased the average value of $k_2''$ by 0.21 log unit at pH 5.00 and by 0.15 log unit at pH 5.20 (Table V). This suggests that the system may not have been totally self-compensating for this factor. However, the effect on the rate constants is probably within uncertainty limits that are imposed by various other factors that affect the experimental results.

In titrations at pH 5.20 and 5.00 the size of $Al_b$ units, as indicated by the $OH_b/Al_b$ ratio seems to reach a maximum at 2.4, which could be interpreted as a boundary in the transition of $Al_b$ to $Al_c$, microcrystalline gibbsite. Some additional support for this hypothesis can be obtained from experiments of Parthasarathy and Buffle ([8]), who used ultrafiltration to separate polymeric $Al_b$ from $Al_c$. They observed that the polymeric material passed through a filter having a nominal porosity of 20 Å. Preliminary experiments we have made with this technique indicated that our $Al_b$ fraction also can pass through this membrane, but $Al_c$ does not.

Figure 4 includes a structure comprising four rings each having 6 $Al^{3+}$ ions connected by $(OH)_2$ bridges that would have the formula $Al_{16}(OH)_{38}^{10+}$. This unit has a diameter somewhat less than 20 Å ([1]), and its OH/Al ratio is 2.38. It is also of interest to note that such a unit has 10 $Al^{3+}$ ions in external sites where an unsatisfied unit of positive valence is available. An additional shared $(OH)_2$ bridge could be established at such sites as the unit grows laterally. There are geometric constraints on the ways in which the six-membered ring units can combine. The $Al_{16}(OH)_{38}$ unit can be formed from 2 rings and 2 dimeric $Al_2(OH)_2$ units as shown in the drawing by adding 4 more $(OH)_2$ bridges, and other combinations could be imagined but it cannot form by the interaction of 4 $Al_6(OH)_{12}$ units. The net charge per Al in the $Al_{16}(OH)_{38}$ unit is + 0.63. Parthasarathy and Buffle ([8]) determined that their $Al_b$ polymer had a net charge of 0.53 or 0.56, depending on the method of determination. They favored the composition $Al_{13}O_4(OH)_{24}^{7+}$ (net charge + 0.54 per Al). As will be noted later, we do not believe this form of the polymer is compatible with a reaction path leading to crystalline gibbsite.

## CHARACTERIZATION OF Al$_b$

The differing chemical behavior of hydroxide bound in monomeric species from that bound in Al$_b$, the polymeric material, is evidenced in various ways. From the viewpoint of coordination chemistry, an explanation for this difference is the development of a bridging structure in which hydroxide ions in pairs link adjacent Al$^{3+}$ ions. The structure of gibbsite can be represented in three dimensions by a double layer of close-packed spheres, each representing an OH$^-$ ion. Smaller spheres representing Al$^{3+}$ ions occupy octahedral coordination positions in the interstices between layers. Each Al is coordinated with 6 OH ions, but to maintain a charge balance each Al$^{3+}$ ion shares the total 6 negative charge units on these OH ions with 3 other Al$^{3+}$ ions. The resulting structural pattern can extend indefinitely in the a and b directions. The individual Al$^{3+}$ ions occupy only 2/3 of the octahedral sites and the structural units can be schematically represented as an hexagonal ring with the sides representing pairs of bridging OH ions and points marking the positions of Al$^{3+}$ (Fig. 4a). The single six-membered ring would have the formula Al$_6$(OH)$_{12}^{6+}$ with each Al still having a single unsatisfied positive charge.

This structural unit for polymeric material was suggested by Hsu and Bates ([14]), and was adopted in our earlier work ([12]). It also was used by Stol et al. ([15]) as a basis for explaining precipitate growth in hydrolyzing Al solutions. It is the basis for the representation of Al$_b$ as Al$_6$(OH)$_{12}^{6+}$ in the equation for conversion of Al$_a$ to Al$_b$ (equation 2). In effect, the rate constants determined here represent the rate of converting free or monomeric OH into bridging OH.

Growth mechanisms represented by Figure 4b show how Al$_6$(OH)$_{12}^{6+}$ units might be joined by establishing new (OH)$_2$ bridges (represented by dashed lines) between the rings. The gibbsite structure of Al$_c$ entails stacking of the layers in the c direction, following an hexagonal close packing arrangement, where hydrogen bonding between adjacent double layers aligns them in an ab, ba, sequence (Fig. 4c, after ([16])). Estimates by Smith and Hem ([1]) of the thickness of Al$_c$ particles suggest they would contain three or more double layers.

The structural model for the principal polynuclear aluminum hydroxide ion given by Baes and Mesmer ([17]) is for a species having the composition Al$_{13}$O$_4$(OH)$_{24}^{7+}$. It involves 3 coalesced Al$_6$(OH)$_{12}$ rings with an additional Al$^{3+}$ ion in the center of the structure that is in tetrahedral coordination with 4 oxygen ions that are included in the Al$_6$ rings. Presumably these originally were OH$^-$ ions but for some reason have been deprotonated. This structure was originally proposed by Johansson ([18]), who identified units of this type by X-ray diffraction in precipitates of basic aluminum sulfate and selenate prepared by adding Na$_2$SO$_4$ or Na$_2$SeO$_4$ solutions to a partly hydrolized AlCl$_3$ solution (OH$_T$/Al$_T$ = 2.5). No gibbsite was reported in these precipitates.

The Al$_{13}$O$_4$(OH)$_{24}^{7+}$ unit would appear unlikely to assimilate readily into a gibbsite structure, as the latter does not include O$^{2-}$ ions or 4 coordinated Al. Furthermore, OH ions of the complex are arranged in cubic close packing rather than the hexagonal close-packing of gibbsite. Aluminum oxyhydroxides might form under some experimental conditions and at higher Al concentrations than we have used. Parthasarathy and Buffle ([8]) postulated that the polymeric Al(OH)$_x$ species they synthesized was predominantly Al$_{13}$O$_4$(OH)$_{24}^{7+}$. The $^{27}$Al$^X$ NMR spectra of solutions containing relatively high concentrations of Al$_b$ (0.1 molal in total Al) from which Al$_c$ had been removed by ultrafiltration gave a peak they attributed to tetrahedrally coordinated AlO$_4$ groups.

Work by Bertsch, Anderson and Layton ([19]) that included NMR studies indicates that the Al$_{13}$ polymer is formed in solutions having Al concentrations greater than $10^{-3.0}$ molal or that have high OH$_T$/Al$_T$ ratios. These investigators suggested that Al(OH)$_4^-$ is a required precursor for the formation of the Al$_{13}$O$_4$(OH)$_{24}^{7+}$ unit.

It seems unlikely that significant amounts of the Al$_{13}$ polymeric species cited above would have been produced under our experimental conditions. Perhaps the reluctance of Al$_b$ to convert to Al$_c$ during aging of some of the solutions could be attributed to the extensive structural rearrangements that conversion of the Al$_{13}$ polymer to gibbsite would entail.

$Al_6(OH)_{12}^{6+}$

Figure 4(a). Schematic representation of $Al_6(OH)_{12}^{6+}$ complex. Al ions at corners; connecting lines represent pairs of bridging $OH^-$ ions (not to scale).

$Al_{16}(OH)_{38}^{10+}$

(b). Schematic representation of $Al_{16}(OH)_{38}^{10+}$ polymeric ion. Solid lines outline original polymeric units and dimers; dashed lines are new double $OH^-$ bridges formed between original units.

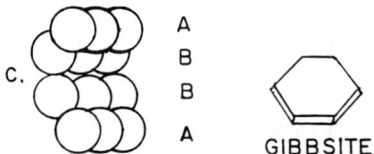

(c). Gibbsite double-layer stacking arrangement.

A study of polymerization kinetics and species structure was made by Stol et al. (15) using a batch titration technique and characterization of polymers by light-scattering. Although their experimental conditions differed substantially from those used in this study, many of their conclusions are in general accord with our own. They suggested that above a critical value for the ratio $C_{OHb}/C_{Alb}$ of ≈2.5 the positive charge on the edges of hexagonal $Al_b$ is no longer a strong deterrent to aggregation of $Al_b$ species to form crystalline solids with the gibbsite-type layered structure. By this interpretation, the development of $Al_c$, or microcrystalline gibbsite does not become a major factor in the hydrolysis process in titration done between pH 4.75 and 5.20. Values for the $OH_b/Al_b$ ratio given in Table V reached a maximum of 2.4.

A later study by DeHek et al. (20) explored the effect of $SO_4^{2-}$ ions on the precipitation mechanism. These investigators suggested that $SO_4^{2-}$ could form surface groups on the $Al_b$ structures that catalyze the formation of precipitates. The role of $SO_4^{2-}$ in aluminum hydrolysis in natural systems also was thought by Nordstrom (4) to be an important one.

## ENVIORNMENTAL AND GEOCHEMICAL SIGNIFICANCE OF AlOH POLYMERIZATION

### Effect of Mineral Surfaces on Al Behavior

It seems likely that there are some natural aqueous systems, especially those impacted by acidic precipitation or discharges of acid mine drainage, in which polymeric Al species like those studied in this work will be formed. Most of these systems will contain solid surfaces, and a tendency for aluminum polymeric species to nucleate and precipitate on certain types of mineral surfaces was documented by Brown and Hem (21). Conclusions relevant to the probable course of the aluminum hydroxide polymerization process in natural water may be summarized as follows:

1. The behavior of monomeric ($Al_a$) species is explainable, in general, as a gradual approach to gibbsite equilibrium except for solutions that contained excessive concentrations of dissolved silica that had been leached from the mineral substrates. The reaction affinities for gibbsite precipitation are similar to those observed at similar aging times in the absence of surfaces. In the silica-rich solutions, precipitation of clay-mineral precursors could have occurred.

2. Rate of the $Al_b$ polymer growth to become $Al_c$ appears not to be affected significantly by the presence of mineral surfaces. The same general rate and first-order disappearance of $Al_b$ occurred in both the presence and absence of surfaces, after an initial induction period. However, in the absence of surfaces the growth process did not attain a steady well-defined rate until after about 3 weeks of aging. In the presence of surfaces there was an almost immediate decrease in $Al_b$ that continued until the end of the 100 day aging period.

### Observations in Natural and Anthropogenicaly-Influenced Systems

Removal of color and turbidity caused by particulate matter is commonly accomplished, during the treatment of water for municipal and industrial purposes, by adding a solution of $Al^{3+}$, to the raw water in a holding tank or mixing facility while maintaining the pH at a level favorable for aluminum hydrolysis. The particulate material in the raw water is coagulated by the aluminum hydrolysis species and can be removed by filtration. Much of the literature on aluminum hydrolysis is directly or indirectly related to the chemistry involved in this process.

Depending on the various changing properties of the input water, there can be an aluminum residual probably consisting of polymeric aluminum hydroxide in the treated water that is not removed by filtration. Miller et al. (22) compared total aluminum concentrations (determined by an A-A procedure) in treated and untreated water from 184 United States public water supply systems and found substantially more Al in the treated than in the untreated water in many instances. Similar conditions can be seen in data collected earlier on the chemical composition of water supplies of the

100 larger U. S. cities (22). The additional aluminum in these treated waters probably was a form of Al$_b$ or Al$_c$, but identification of specific forms was not attempted in these studies. The failure of ion-exchange column demineralizer units to remove aluminum from tap water furnished by public utilities (24) does suggest that it is polymeric, and possibly in the colloidal size range. A later survey (25) provides additional information on possible causes of residual Al in treated water.

A commonly observed effect of acidic precipitation is an increased concentration of aluminum in water of streams and lakes. Water of lakes that received acidic runoff may increase in transparency (26) owing to coagulation and settling of suspended organic and inorganic particulates. This effect is probably related to aluminum hydrolysis, as it is the same as the coagulation step in water-supply treatment cited earlier.

A study of the chemistry of a small California stream that receives acid mine water (27) showed that above pH 4.9 the dissolved aluminum concentration appeared to be controlled by aluminum hydroxide solubility equilibria, but below pH 4.6 aluminum hydroxy-sulfates apparently predominated. This study also cited data for acid mine drainage in the Appalachian region and lakes affected by acid precipitation that showed a similar pattern of aluminum behavior.

A procedure developed by Barnes (28), and later modified by Driscoll (29), has been used to separate monomeric reactive species of Al from polymeric and colloidal forms. In tests of the procedure, (28) extractable Al concentrations in a series of natural-water samples that had been filtered through 0.10 µm pore-size membranes, acidified to pH 2.0 and stored for several weeks, were compared with extractable Al in aliquots of the same filtered samples that were not acidified and on which the extraction was done immediately after filtration. The results indicated that acidification and storage increased the extractable Al concentration by nearly a factor of 2 in these samples.

All the samples included in the Barnes (28) study were from groundwater sources and most had pH's near neutrality. The acid soluble fraction passing through the 0.1 µm pores of the filter probably was microcrystalline gibbsite and polymeric aluminum hydroxide macro ions. The kinetics of dissolution of polymeric Al(OH)$_x$ species in acid were studied by Hem (30). Material passing through 0.10 µm porosity filters showed pseudo first-order behavior when dissolution pH was held constant, and the half-time for dissolution was near 10 hours at pH 3.0 and about 12 minutes of pH 2.0. The polymeric species studied in the kinetic experiments were prepared by a batch mixing technique similar to that used by Smith and Hem (1). The solutions were aged before filtration for periods of a few days up to two weeks. Probably a larger fraction of the aluminum present in surface water could have characteristics of Al$_b$ than is to be expected in ground water. Currently, available data are not sufficient to make any quantitative statement. Campbell et al. (31), developed procedures for aluminum speciation that utilize filtration and chelating resin treatment of surface water samples and suggested that aluminum hydroxide polynuclear cations less than 60 Å in diameter may be taken up by the resin along with monomeric complex species.

Kennedy et al. (32), reported the effect of using different filtration and preservation techniques on the aluminum concentration determined in the filtrate of a sample collected at high flow from the Mattole River of northern coastal California. The aluminum concentration determined in the first 2 liters of water filtered through a 0.45 µm porosity filter and subsequently acidified with ultrapure nitric acid (final pH 1.5 to 1.7) and stored for 90 days before analysis was more than 2.2 µmolal. An aliquot of the same original water filtered through a 0.10 µm porosity filter and given the same subsequent treatment yielded an aluminum concentration somewhat less than 0.6 µmolal. This concentration also was measured within an analytical uncertainty of about ±0.2 µmolal in a portion of the 0.10 µm filtrate that was not acidified but from which Al was extracted by an organic reagent (8-quinolinol dissolved in methyl isobutyl ketone) immediately after filtration. In a companion study, Jones et al. (33) calculated that their Mattole River water sample was more than an order of magnitude supersaturated with respect to kaolinite, which should have been the most stable aluminum bearing solid. An electron micrograph of particulate matter from a Mattole River sample that had passed through a 0.45 μm filter was published by Hem et al. (34). Hexagonal units resembling gibbsite are visible in this micrograph, but their composition was not determined.

Whether the Mattole River sample of Kennedy et al. (32) contained a significant amount of $Al_b$ cannot be determined from the data they presented. Colloidal particulates that perhaps contained $Al_c$ were probably present in material passing through the 0.45 μm filter. The extraction procedure used in their experiments involved a 15 minute contact time of the 8 quinolinol reagent with the sample making it likely that some, at least, of the $Al_b$ fraction would be extracted. The method of Barnes (28) uses the same extractant but only a 10 sec. contact time, in order to minimize extraction of polymeric Al species.

Effects of aluminum in low pH solutions on aquatic biota, especially fish, and on land plants growing in acidic soil have been studied by various investigators (35-38). Although much of the work has assigned the observed toxicity to monomeric dissolved forms of Al, the behavior of polymeric species in biochemical systems is probably in need of more careful study than it has so far received. Schindler (39) in a review of the topic of environmental impacts of acid rain noted that aluminum is highly toxic to fish and quoted literature references to the effect that polymeric and colloidal aluminum hydroxide species may physically obstruct gill membranes and cause asphyxiation. The maximal pH range for this effect was reported as 5.2 to 5.4.

## Implications of Polymerization Mechanisms and Kinetics for Occurrence of $Al_b$ in Streams

The experimental data given and reviewed here show that there are critically important changes in the hydrolysis behavior of aluminum between pH 4.5 and 5.5. Early stages of polymerization and precipitation of aluminum hydroxide and related secondary minerals occur near pH 5.0 and can be greatly accelerated or retarded by small pH changes.

It would appear likely that small amounts of $Al_c$ could be present as colloidal solid particles in many river waters. Conditions favorable for stabilizing polymeric $Al_b$ ionic species are readily attainable in the laboratory, as our study has demonstrated. The maximum $Al_b$ yields were in solutions titrated at pH 5.00, and total monomeric aluminum concentrations were held near $10^{-4.0}$ molal. This high an $Al_a$ concentration in natural water could be reached through dissolution of soil minerals at a pH near or below 4.50.

The requirement that the solution be supersaturated with respect to gibbsite to form $Al_b$ ($A_{23} >$ 1.0 kcal) would be difficult to meet at pH's much below 4.75, owing to the rapid increase in equilibrium solubility of Al and decrease in proportion of the $AlOH^{2+}$ species as pH decreases. For significant amounts of unreacted polymeric $Al_b$ to be present in surface water requires that there be a prior history of low-pH inflow and a supply of dissolved Al, along with low levels of aluminum-complexing species other than OH. Once formed, the smaller polymeric units might well persist for a considerable time. The kinetics of $Al_b$ formation in the open-system laboratory experiments are directly relevant toward improving our understanding of the environmental behavior of this element.

## LITERATURE CITED

1. Smith, R. W.; Hem, J. D. U.S. Geol. Surv. Water-Supply Pap. 1827-D, 1972, D1-D51.
2. Hem, J. D. U.S. Geol. Surv. Water-Supply Pap. 1827B, 1968, B1-33. p.
3. Lind, C. J.; Hem, J. D. U.S. Geol. Surv. Water-Supply Pap., 1827G, 1975, G1-G83.
4. Nordstrom, D. K. Geochim et Cosmochim Acta, 1982, 46, 681-692.
5. Smith, R. W. In Nonequilibrium, Systems in Natural Water Chemistry; Hem, J. D., Ed.; Amer. Chem. Soc., Washington, D.C, 1971, 250-279
6. Okura, T.; Goto, K.; Yotuyanagi, T. Analytical Chem., 1962, 34, 581-582.
7. Turner, R. C. Canadian J. of Chem., 1969, 47, 2521-2527.
8. Parthasarathy, N.; Buffle, J. Water Research, 1985, 19, No. 1, 25-36.

9. Bersillon, J. L.; Hsu, P. H.; Fiessinger, F. Soil Sc. Soc. Amer. J., 1980, 44, 630-634.
10. Jardine, P. M.; Zelazny, L. W. Soil Sci. Soc. Amer. J., 1986, 50, 895-900.
11. Seip, H. M.; Muller, L.; Naas, A. Water, Air and Soil Pollution, 1984, 23, 81-95.
12. Hem, J. D.; Roberson, C. E. U.S. Geol. Surv. Water-Supply Pap., 1827A, 1967, A1-A55.
13. Parks, G. A. Amer. Mineralogist, 1972, 57, 1163-1189.
14. Hsu, P. H.; Bates, T. F. Mineralogical Magazine, 1964, 33, 749-768.
15. Stol, R. J.; Van Helden, A. K.; De Bruyn P. L. J. of Colloid and Interface Sci., 1976, 57, 115-131.
16. Wefers, K.; Misra, C. Alcoa Tech. Pap. No. 19, revised, 1987, 1-92.
17. Baes, C. F.; Mesmer, R. E. Wiley-Interscience: New York, 1976, 489.
18. Johansson, G. Svensk Kemisk Tidskrift, 1963, 2, 6-7.
19. Bertsch, P. M.; Anderson, M. A.; Layton, W. J. Preprint, Env. Chem. Div. 194 Amer. Chem. Soc. Mtg., 1987, 142-143.
20. DeHek, H.; Stol, R. J.; Debruyn, P. L. J. Colloid Interface Sci., 1978, 64, 72-89.
21. Brown, D. W.; Hem, J. D. U.S. Geol. Surv. Water-Supply Pap. 1827-F, 1975, F1-F48.
22. Miller, R. G.; Kopfler, F. C.; Kelty, K. C.; Stober, J. A.; Ulmer, N. S. Amer. Water Works Asso. J., 1984, 76, 84-91.
23. Durfor, C. N.; Becker, E. U.S. Geol. Surv. Water-Supply Pap. 1812, 1964, 364 p.
24. Kerr, D. N. S.; Ward, M. K.; Arze, R. S.; Ramos, J. M.; Grekas, D.; Parkinson, I. S.; Ellis, H. A.; Owen, J. P.; Simpson, W.; Dewar, J.; Martin, A.; McHugh, M. F. Kidney International, 1986, 29, Supp. 18, S-58-S-64.
25. Letterman, R. D.; Driscoll, C. T. Amer. Water Works Asso. J., 1988, 80, No. 4, 154-158.
26. Effler, S. W.; Schofran, G. C.; Driscoll, C. T. Canadian J. of Fisheries and Aquatic Sci., 1985, 42, 1707-1711.
27. Nordstrom, D. K.; Ball, J. W. Science, 1986, 232, 54-56.
28. Barnes, R. B. Chem. Geol 1975, 15, 177-191.
29. Driscoll, C. T. Inter. Nat. J. of Enviro. Analytical Chem. 1984, 16, 267-283.
30. Hem, J. D. In Leaching and Diffusion in Rocks and Their Weather-Products; SS. Augustithis, Ed.; Theophrastus Publications SA, Athens, Greece, 1983, 51-62.
31. Campbell, P. G. C.; Bisson, M.; Bougie, R.; Tessier, A.; Villenuve, J. P. Analytical Chem., 1983, 55, 2246-2252.
32. Kennedy, V. C.; Zeleger, G. W.; Jones, B. F. Water Resources Research, 1974, 10, 785-790.
33. Jones, B. F.; Kennedy, V. C.; Zellweger, G. W. Water Resources Research, 1974, 10, 791-793.
34. Hem, J. D.; Roberson, C. E.; Lind, C. J.; Polzer, W. L. U.S. Geol. Surv. Water-Supply Pap. 18227-E, 1973, E54.
35. Baker, J. K. In Acid Rain/Fisheries; Johnson, E. E., Ed.; Amer. Fisheries Soc.: Bethesda, Maryland, 1982; p. 165.
36. Driscoll, C. T.; Baker, J. P.; Bisogni, J. J.; Schofield, C. L. Nature, 1980, 284, 161-164.
37. Foy, C. D. In The Plant Root and its Environment; Carson, E. W., Ed.; University of Virginia Press: Charlottesville, Virginia, 1974, p 601.
38. Parker, D. R.; Zelazny, L. W.; Kinraide, T. B. Proc. 194 Amer. Chem. Soc. Mtg., 1987, p. 369-372.
39. Schindler, D. W. Science, 1988, 239, 149-157.
40. Hemingway, B. S.; Robie, R. A. Geochim. Cosmochim. Acta, 1977, 41, 1402-1404.
41. Raupach, M. Australian J. of Soil Res., 1963, 1, 28-35.
42. Hemingway, B. S.; Haas, J. L. Jr.; Robinson, G. R. Jr. U.S. Geol. Surv. Bull. 1544, 1982, 70 pp.
43. Hemingway, B. S.; Robie, R. A.; Kittrick, J. A. Geochim. Cosmochim. Acta, 1978, 42, 1533-1543.
44. Wagman, D. D.; Evans, W. H.; Parker, V. B.; Halow, I.; Bailey, S. M.; Schumm, R. H. Nat. Bureau of Standards Tech. Note 270-3, 13, 207-208.

45. May, H. M.; Helmke, P. A.; Jackson, M. L. Geochim. Cosmochim. Acta, 1979, 43, 861-868.
46. Singh, S. S. Proc. Soil Sci. Soc. of Amer., 1974, 38, 415-417.
47. Kittrick, J. A. Proc. Soil Sci. Soc. of Amer., 1966, 30, 595-598.
48. Nordstrom, D. K.; Plummer, L. N.; Langmuir, D.; Busenberg, E.; May, H. M.; Jones, B. F.; Parkhurst, D. L. This volume, 1989.
49. Ackerman, T. Zeitschrif für Elektrochemie 1958, 62, 411-419

RECEIVED August 31, 1989

# Chapter 34

# Effect of Ionic Interactions on the Oxidation Rates of Metals in Natural Waters

## Frank J. Millero

Rosenstiel School of Marine and Atmospheric Science, University of Miami, 4600 Rickenbacker Causeway, Miami, FL 33149

> The reduced valence of metals in natural waters can be formed by photochemical, biochemical and geochemical processes. The longevity of these reduced metals will be influenced by the rates of oxidation with $O_2$ and $H_2O_2$. This paper will examine how ionic interactions affect the rates of Cu(I) and Fe(II) oxidation in natural waters using measurements made as a function of pH, ionic strength and composition. The oxidation of Cu(I) is affected by the strong interactions of $Cu^+$ with $Cl^-$. The formation of $CuCl_2^-$ and $CuCl_3^{2-}$ ion pairs causes the rates to decrease. At a constant $Cl^-$ concentration, the increase in the concentration of $Mg^{2+}$ causes the rates to decrease and the increase in the concentration of $HCO_3^-$ causes the rates to increase. The oxidation of Fe(II) is affected by the strong interactions of $Fe^{2+}$ with $OH^-$. The formation of $FeOH^+$ and $Fe(OH)_2$ ion pairs causes the rates to increase. Most other anions, with the exception of $HCO_3^-$, that form ion pairs cause the rates to decrease.

Trace metals in natural waters can participate in a wide variety of chemical reactions. Some of thes metals are needed, while others can be toxic to organisms. The bioavailability and toxicity of metals is strongly affected by the redox state and speciation of a given metal. Copper, for example, is toxic to many marine organisms when it is in the free or uncomplexed form. Iron and manganese are needed by phytoplankton. The oxidized forms of these metals are insoluble in seawater and are quickly scavenged from surface waters. Although thermodynamic speciation calculations can yield the most probable redox form of a given metal, the form is normally controlled by the rates of oxidation and reduction of the metals. Recently a number of studies have been made on the rates of oxidation and reduction of copper and iron in natural waters (1-9). These measurements provide reliable rate equations for the oxidation of Cu(I) and Fe(II) as a function of pH, temperature, ionic strength and ionic composition. In the present paper, these results will be used to examine how ionic interactions affect the rates of Cu(I) and Fe(II) oxidation in natural waters.

The effect of ionic interactions on the reaction rates of metals can be affected by the anions in solution ($Cl^-$, $SO_4^{2-}$, $OH^-$, etc.); while the anion reaction rates (e.g., $HS^-$) can be affected by the cations in solution ($H^+$, $Na^+$, $Mg^{2+}$, etc.).

Examples of the effect anions have on the rates of the reactions of cations are the formation of ion pairs

$$Cu^+ + Cl^- \longrightarrow CuCl^0 \tag{1}$$
$$Fe^{2+} + OH^- \longrightarrow FeOH^+ \tag{2}$$

The formation of CuCl causes the rates of oxidation of Cu(I) to decrease (1-2,5-9), while the formation of $FeOH^+$ causes the rates of oxidation of Fe(II) to increase (2-5,10).

Examples of the effect of cations on the rates of reactions of anions are

$$HS^- + H^+ \longrightarrow H_2S \tag{3}$$
$$O_2^- + Mg^{2+} \longrightarrow MgO_2^+ \tag{4}$$

The formation of $H_2S$ causes the rates of sulfide oxidation to decrease (11), while the formation of the ion pair $MgO_2^+$ causes the rates of $O_2^-$ to $H_2O_2$ disproportionation to decrease (12).

Our interest in the effects of ionic interactions on the rates of reactions has largely been for redox processes occurring in the oceans. These redox processes occur at oxic-anoxic interfaces and in the surface waters of the oceans due to photochemical processes (1). The formation of $H_2O_2$ in surface waters (1,13,14) is thought to occur by the following reaction scheme

$$C + h\nu \longrightarrow C^* \tag{5}$$
$$C^* \longrightarrow C^+ + e^-(aq) \tag{6}$$
$$O_2 + e^-(aq) \longrightarrow O_2^- \tag{7}$$
$$2O_2^- + 2H^+ \longrightarrow H_2O_2 + O_2 \tag{8}$$

The $O_2^-$ can react with the oxidized form of metals to produce reduced species

$$Fe^{3+} + O_2^- \longrightarrow Fe^{2+} + O_2 \tag{9}$$
$$Cu^{2+} + O_2^- \longrightarrow Cu^+ + O_2 \tag{10}$$

The longevity of these reduced metals will depend on the rates of oxidation of Fe(II) and Cu(I) with $O_2$ and $H_2O_2$

$$Fe(II) + O_2 \longrightarrow Products \tag{11}$$
$$Fe(II) + H_2O_2 \longrightarrow Products \tag{12}$$
$$Cu(I) + O_2 \longrightarrow Products \tag{13}$$
$$Cu(I) + H_2O_2 \longrightarrow Products \tag{14}$$

This present paper will examine how the oxidation of Fe(II) and Cu(I) in natural waters can be affected by changes in the composition. These results will be explained in terms of ionic interactions. The results are taken from measurements made in this laboratory (2,3,5-9,14,15). These papers should be examined for details of the methods used.

## CU(I) RATES OF OXIDATION

The rates of oxidation of Cu(I) with $O_2$ and $H_2O_2$ have been made in seawater, NaCl and sea salts (6-9). The measurements were made as a function of pH and temperature. A comparison of the results for seawater and NaCl solutions at 25°C and pH = 8.0 are shown in Figures 1 and 2(6-9). The large ionic strength dependence in NaCl and seawater solutions for the oxidation of Cu(I) with $O_2$ and $H_2O_2$ is related to changes in the Cl⁻ concentration. This chloride dependence can be analyzed by assuming that the various Cu(I) chloro complexes have different rates of oxidation (2). This gives

$$Cu^+ + H_2O_2 \xrightarrow{k_0} Products \tag{15}$$
$$CuCl + H_2O_2 \xrightarrow{k_1} Products \tag{16}$$

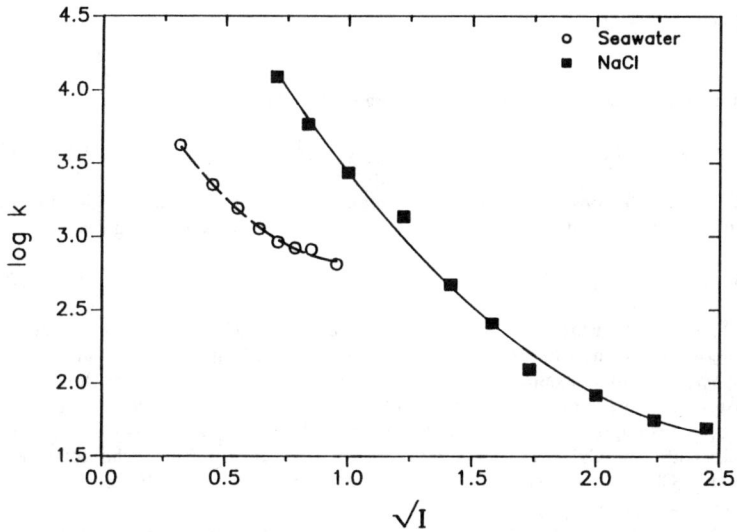

Figure 1. The rate constant, (k, mol$^{-1}$ kg H$_2$O s$^{-1}$), for the oxidation of Cu(I) with O$_2$ in NaCl and seawater at 25°C and pH = 8.0.

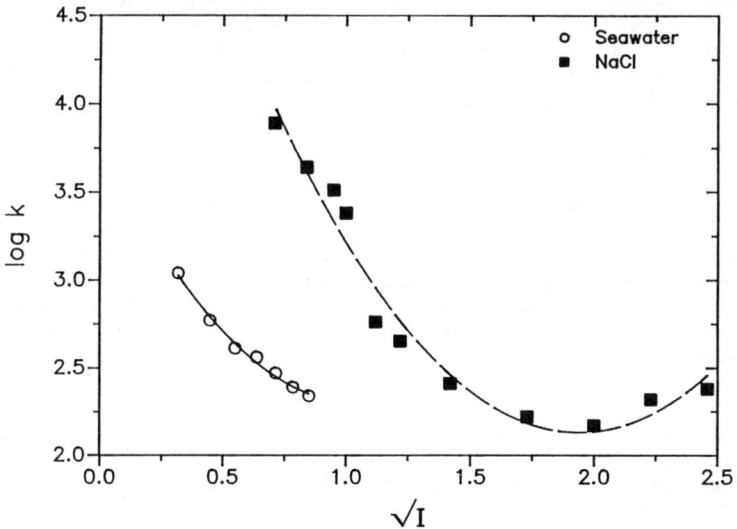

Figure 2. The rate constant, (k, mol$^{-1}$ kg H$_2$O s$^{-1}$), for the oxidation of Cu(I) with H$_2$O$_2$ in NaCl and seawater at 25°C and pH = 8.0.

$$CuCl_2^- + H_2O_2 \xrightarrow{k_2} \text{Products} \qquad (17)$$

$$CuCl_3^{2-} + H_2O_2 \xrightarrow{k_3} \text{Products} \qquad (18)$$

The observed rate constant is given by ($k$, mol$^{-1}$ kgH$_2$O sec$^{-1}$)

$$k = k_0\alpha_{Cu} + k_1\alpha_{CuCl} + k_2\alpha_{CuCl_2} + k_3\alpha_{CuCl_3} \qquad (19)$$

where the various rate constants, $k_i$, are for the molar fractions $\alpha$ of species $\underline{i}$. The substitution of the stepwise stability constants, $\beta_i^*$, for the formation of the various ion pairs gives

$$k/\alpha_{Cu} = k_0 + k_1\beta_1^*[Cl^-] + k_2\beta_2^*[Cl^-]^2 + \ldots \qquad (20)$$

Values of $k/\alpha_{Cu}$ for measurements of Cu(I) and O$_2$ made in NaCl-NaClO$_4$ mixtures at a constant ionic strength are shown in Figure 3. The linear behavior indicates that Cu$^+$ and CuCl are the reactive species. Similar results for the measurements of Cu(I) + H$_2$O$_2$ are shown in Figure 4 (9). Measurements made over a wider range of ionic strength (to 6m) indicate that the oxidation of CuCl$_2^-$ becomes important (7). Measurements in NaBr-NaClO$_4$ and NaI-NaClO$_4$ indicate that the ion pairs CuBr and CuI have similar rates of oxidation with CuCl (8). The differences of the overall rates in Cl$^-$, Br$^-$ and I$^-$ solutions ($k_{Cl} > k_{Br} > k_I$) is related to the differences in the stability constants ($\beta_{CuI} > \beta_{CuBr} > \beta_{CuCl}$).

The results for the oxidation of Cu(I) in seawater at the same Cl$^-$ concentration are lower than the values in NaCl (see Figures 1 and 2). To elucidate these effects, measurements have been made on the rates of oxidation in NaCl, NaMgCl, NaCaCl, NaClHCO$_3$, NaClSO$_4$ and NaMgClSO$_4$ solutions at their concentrations in seawater. The results are shown relative to the measurements in NaCl in Figure 5. The addition of SO$_4^{2-}$ shows no effect, while the addition of Mg$^{2+}$ and Ca$^{2+}$ causes the rates to decrease. The addition of HCO$_3^-$ causes the rates to increase. The solution containing Na$^+$, Mg$^{2+}$, Cl$^-$ and HCO$_3^-$ gives rates in agreement with the seawater results. Measurements of the Mg$^{2+}$ and HCO$_3^-$ effects have been made over a wide range of ionic strengths (7). These results indicate that the effects are diminished at higher ionic strengths.

The decrease in the rate of oxidation caused by the addition of Mg$^{2+}$ or Ca$^{2+}$ can be attributed to the slow exchange of MgL complexes with Cu$^{2+}$

$$MgL + Cu^{2+} \longrightarrow CuL + Mg^{2+} \qquad (21)$$

This slow exchange may cause the overall oxidation rates of Cu(I) to be slower due to the back reactions of Cu(II) with H$_2$O$_2$ (1). NTA, CO$_3^{2-}$, EDTA and B(OH)$_4^-$ give similar rates at concentration levels sufficient to complex Cu$^{2+}$ in Na-Mg-Cl solutions. Thus, this exchange reaction may be common for natural ligands able to complex Mg$^{2+}$ and Cu$^{2+}$, such as humic material.

The increase in the rate of oxidation due to the addition of HCO$_3^-$ and CO$_3^{2-}$ may be related to the formation of CuHCO$_3^\circ$ which slows down the back reaction. A preferable possibility is that the increase is due to the formation of Cu(I) carbonate ion pairs

$$Cu^+ + HCO_3^- \longrightarrow CuHCO_3 \qquad (22)$$
$$Cu^+ + CO_3^{2-} \longrightarrow CuCO_3^- \qquad (23)$$

These ion pairs may be more reactive than CuCl. Differences in the back reactions of CuEDTA and CuCO$_3$

$$CuEDTA + O_2^- \xrightleftharpoons Cu^+ + O_2 + EDTA \qquad (24)$$
$$CuCO_3 + O_2^- \xrightleftharpoons Cu^+ + O_2 + CO_3^{2-} \qquad (25)$$

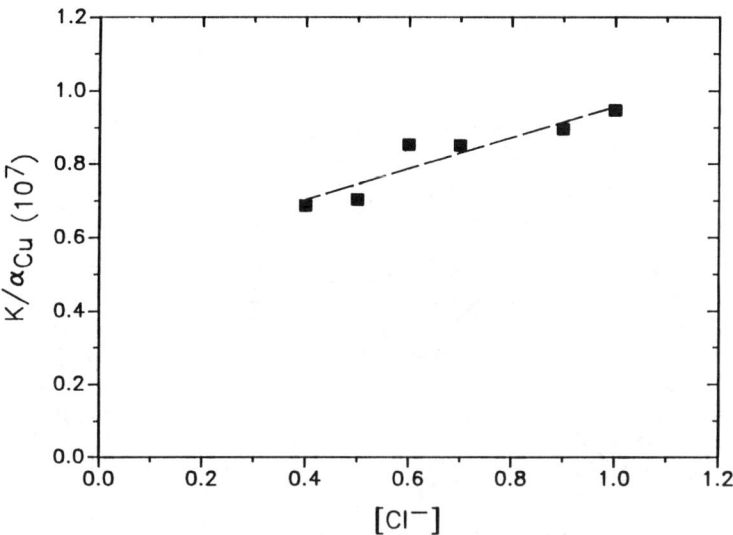

Figure 3. Values of $k/\alpha_{Cu}$, (k, mol$^{-1}$ kg H$_2$O s$^{-1}$), for the oxidation of Cu(I) with O$_2$ in NaCl-NaClO$_4$ mixtures at I = 1.0 at 25°C.

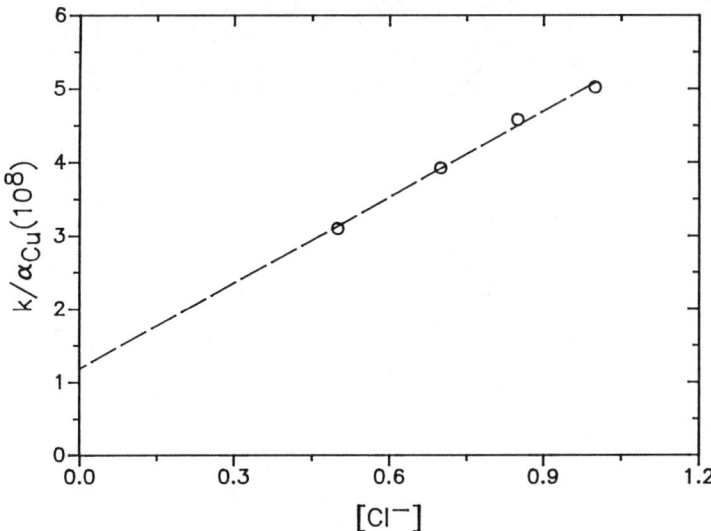

Figure 4. Values of $k/\alpha_{Cu}$, (k, mol$^{-1}$ kg H$_2$O s$^{-1}$), for the oxidation of Cu(I) with H$_2$O$_2$ in NaCl-NaClO$_4$ mixtures at I = 1.0 at 25°C.

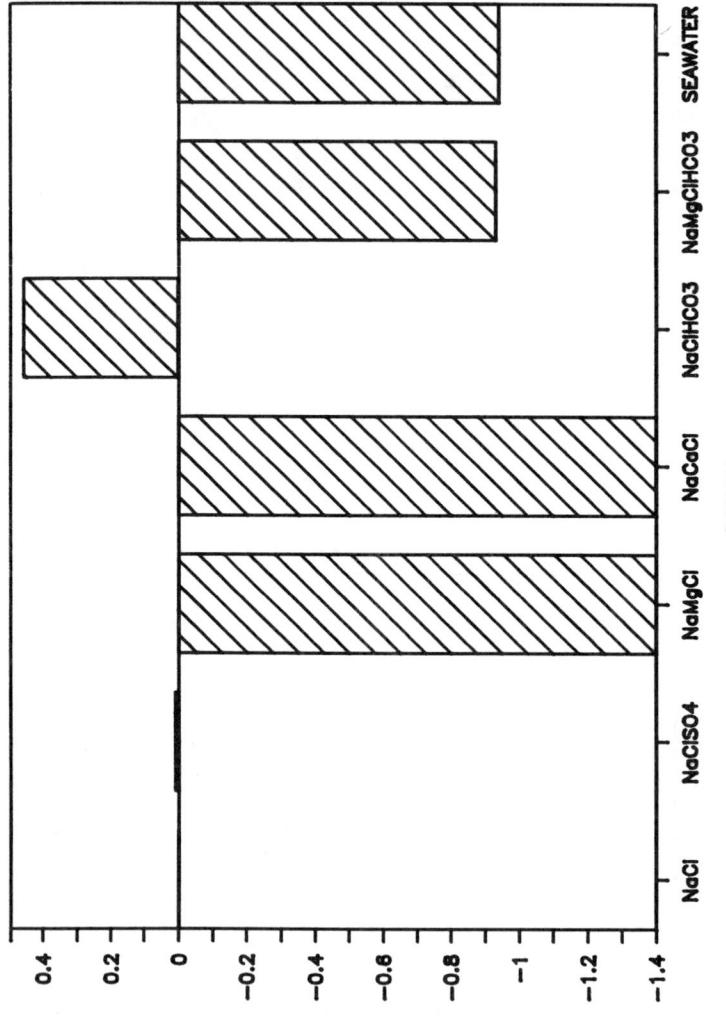

Figure 5. The effect of various ions on the oxidation of Cu(I) with $O_2$ at $I = 0.7$ at 25°C.

with $O_2^-$ or $H_2O_2$ could also have different rates. If reaction (24) is faster than reaction (25), the overall rates of Cu(I) will be faster in $CO_3^{2-}$ solutions. Further kinetic and thermodynamic data are needed to elucidate the $HCO_3^-$ or $CO_3^{2-}$ effects.

Even though the causes of how $Mg^{2+}$, $Ca^{2+}$ and $HCO_3^-$ affect the rates of oxidation of Cu(I) is uncertain, the correction of the seawater results for these effects yields rate constants that agree with the measured values in NaCl (see Figure 6).

## FE(II) RATES OF OXIDATION

The rates of Fe(II) oxidation with $O_2$ in natural waters have been studied by a number of workers (see references in 3). Measurements have also been made in artificial media (16,17,18). Stumm and Lee (10) showed that from pH = 5 to 7, the rate of oxidation of Fe(II) in water was given by (k, mol$^{-3}$ kg $H_2O^3$ min$^{-1}$)

$$d[Fe(II)]/dt = -k[Fe(II)][OH^-]^2[O_2] \tag{26}$$

The overall rate constant is given by

$$\log k = 21.56 - 1545/T - 3.29\ I^{1/2} + 1.52\ I \tag{27}$$

Millero et al. (3) have shown that this equation is valid in water and seawater from 0 to 50°C and salinity, S = 0 to 40. This second order dependence of the oxidation with $OH^-$ or $H^+$ is shown in Figure 7 for the pseudo first order rate constant ($k_1$)

$$d[Fe(II)]/dt = -k_1[Fe(II)] \tag{28}$$

More recently, Millero and Izaguirre (16) have examined the pH dependence of the oxidation of Fe(II) in NaCl and NaClO$_4$ solutions as a function of ionic strength and temperature. The results at I = 6.0 are shown in Figure 8. These results clearly demonstrate that over a wide range of temperature (0 to 50°C) and ionic strength (0 to 6m), the rate of oxidation of Fe(II) is second order with respect to $H^+$ or $OH^-$. As discussed elsewhere (2), this is related to the rate determining step for the oxidation being

$$Fe(OH)_2 + O_2 \longrightarrow Fe(OH)_2^+ + O_2^- \tag{29}$$

The $O_2^-$ produced reacts very quickly with the more predominant $Fe^{2+}$

$$Fe^{2+} + O_2^- + 2H_2O \longrightarrow Fe(OH)_2^+ + H_2O_2 \tag{30}$$
$$Fe^{2+} + H_2O_2 + H_2O \longrightarrow Fe(OH)_2^+ + OH\cdot + H^+ \tag{31}$$
$$Fe^{2+} + OH\cdot + H_2O \longrightarrow Fe(OH)_2^+ + H^+ \tag{32}$$

The concentration of $Fe(OH)_2$ is given by

$$Fe(OH)_2 = [Fe(II)]\beta_2\ [OH^-]^2/\alpha_{Fe} \tag{33}$$

where

$$\alpha_{Fe} = (1 + \beta_1[OH] + \beta_2[OH]^2 + \beta_3[OH]^2)^{-1} \tag{34}$$

and $\beta_i$ is the stepwise association constant. Thus, the overall rate equations can be simplified to

$$d[Fe(II)]/dt = -k'[Fe(OH)_2][O_2] \tag{35}$$

where k' is related to k of Equation 26, $\beta_2$ and $\alpha_{Fe}$.

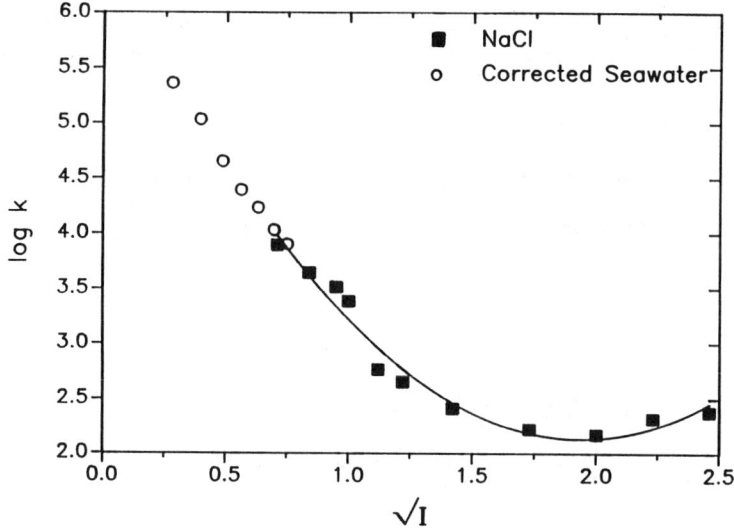

Figure 6. The rate constants for the oxidation of Cu(I) with $H_2O_2$ where the seawater results have been corrected for $Mg^{2+}$, $Ca^{2+}$ and $HCO_3^-$ ion effects at 25°C.

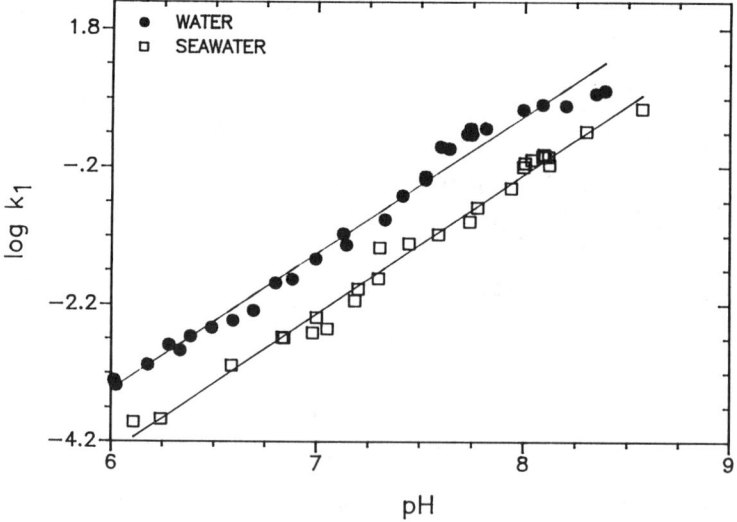

Figure 7. Values for the pseudo first-order rate constants for the oxidation of Fe(II) with $O_2$ in water and seawater as a function of pH at 25°C.

The effect of pH on the oxidation of Fe(II) with $H_2O_2$ is shown in Figure 9. The slope is near zero below a pH = 4.0 and has a value of 1.0 from 6 to 9. These results indicate that both $Fe^{2+}$ and $FeOH^+$ are reactive to oxidation by $H_2O_2$.

From the comparisons shown in Figure 7, the rate of oxidation of Fe(II) in seawater is lower in seawater than in water at a given pH (on the free scale). As discussed elsewhere (2), this decrease has been attributed to the formation of ion pairs

$$Fe^{2+} + Cl^- \longrightarrow FeCl^+ \tag{36}$$
$$Fe^{2+} + SO_4^{2-} \longrightarrow FeSO_4^\circ \tag{37}$$

that are less reactive to oxidation. Measurements of the overall rate constant in seawater and NaCl as a function of ionic strength (Figure 10) indicated that at higher ionic strengths, the seawater results are lower than the NaCl results. To investigate this decrease, we have made some measurements as a function of composition at a constant pH and ionic strength (0.7m). The results are given in Figure 11. The replacement of $Na^+$ with $Mg^{2+}$ (0.0547m) causes a decrease in the rate. This effect was not expected. This decrease in the rate due to $Mg^{2+}$ may be related to the decrease in the concentration of $CuHCO_3^\circ$ or $CuCO_3^-$ species which, as discussed later, cause the rates to increase.

The replacement of $Cl^-$ by $SO_4^{2-}$ (0.0293m) causes the rate of oxidation of Fe(II) to decrease. This can be attributed to the formation of the $FeSO_4$ ion pair. The decrease can be used to estimate the stability constant for the formation of $FeSO_4$

$$k_{NaCl}/k_{NaClSO_4} = 1 + \beta_{FeSO_4}[SO_4^{2-}] \tag{38}$$

Our results yield $\beta_{FeSO_4} = 500 \pm 125$ which is in reasonable agreement with the infinite dilution value ($\beta = 132$), but larger than expected at I = 0.7 $\beta = 10$ (19). More reliable estimates will be made later.

A solution made up of $Na^+$, $Mg^{2+}$, $Cl^-$ and $SO_4^{2-}$ gives the same results as an artificial seawater solution ($Na^+$, $Mg^{2+}$, $Ca^{2+}$, $K^+$, $Cl^-$, $SO_4^{2-}$, $Br^-$, $HCO_3^-$) and real seawater (3).

To determine the effect of ionic strength and temperature on the rates of oxidation of Fe(II), measurements were made in NaCl and $NaClO_4$ solutions (16). The 25°C results are shown in Figure 12. In dilute solutions the results are the same (16.4) within experimental error. At higher ionic strengths the results in NaCl are lower than in $NaClO_4$. The decrease in the rates at low ionic strengths is similar to the seawater results (3) and the results in $NaClO_4$ of Sung and Morgan (18). This decrease is related to the effect of ionic strength on $\beta_2$ as well as the rate constant in pure water. The slope varies from -3.3 in seawater to -1.6 in NaCl or $NaClO_4$. Since one would not expect the rate constant for Equation 29 to show an ionic strength dependence, one is forced to attribute these negative slopes to the reaction of $Fe^{2+}$ and $O_2^-$.

The lower rate constants in NaCl at higher ionic strengths can be attributed to the formation of $FeCl^+$

$$k_{ClO_4}/k_{Cl} = 1 + \beta_{FeCl}[Cl^-] \tag{39}$$

The results from 4 to 6 m give $\beta_{FeCl} = 1.2 \pm 0.5$, which is in reasonable agreement with the literature value (19) of 1.0.

The effect of various anions on the oxidation of Fe(II) (19) has been determined by making measurements in NaCl-NaX at a constant ionic strength. The results (16) are shown for log k as a function of the anion (X) in Figure 13. The addition of $HCO_3^-$ causes the rate to increase, while the addition of $SO_4^{2-}$ and $B(OH)_4^-$ causes the rate to decrease. The anions, $Br^-$, $NO_3^-$ and $ClO_4^-$ cause the rate to increase slightly. The overall order of the rate constants are in the order $HCO_3^- >> Br^- > NO_3^- > ClO_4^- > Cl^- >> SO_4^{2-} >> B(OH)_4^-$.

The large increase due to the addition of $HCO_3^-$ can be attributed to the formation of $FeCO_3$

$$Fe^{2+} + CO_3^{2-} \longrightarrow FeCO_3 \tag{40}$$

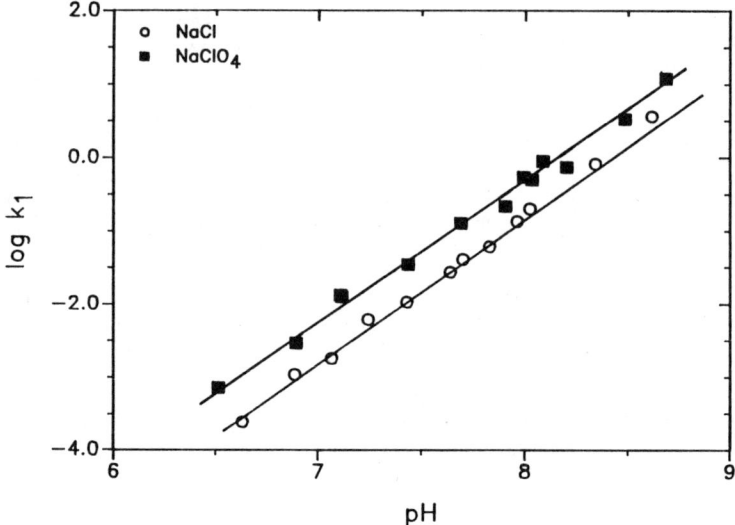

Figure 8. Values for the pseudo first-order rate constants for the oxidation of Fe(II) with $O_2$ in NaCl and $NaClO_4$ (I = 6.0 m) as a function of pH at 25°C.

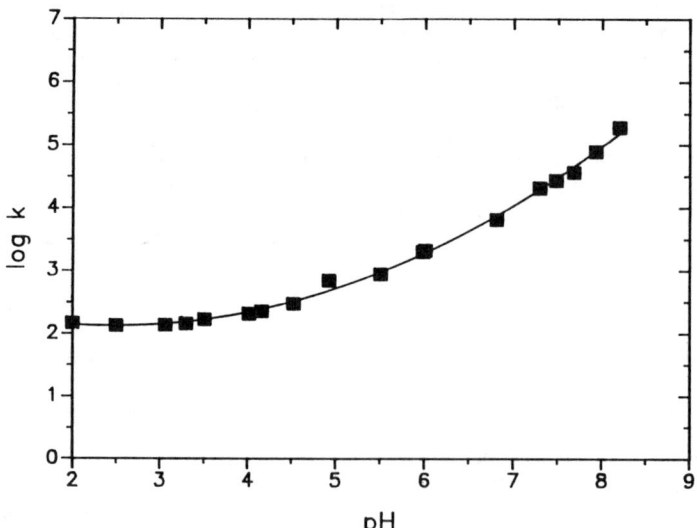

Figure 9. The effect of pH on the rate constant for the oxidation of Fe(II) with $H_2O_2$ in seawater at 25°C.

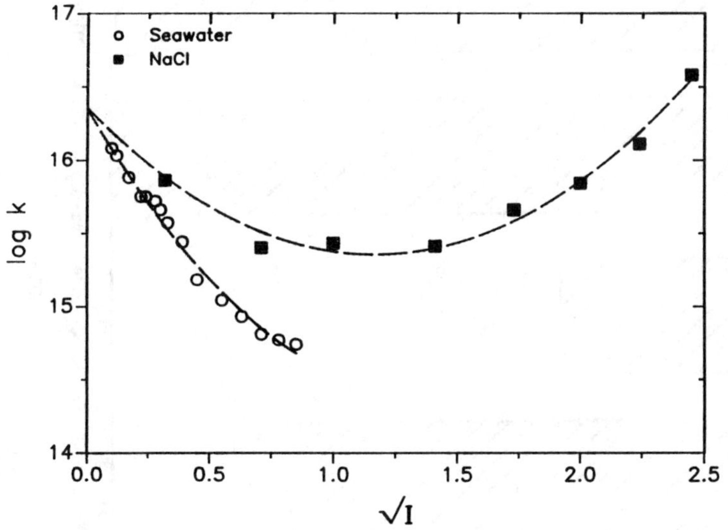

Figure 10. The rate constant for the oxidation of Fe(II) with $O_2$ in NaCl and seawater as a function of ionic strength at 25°C.

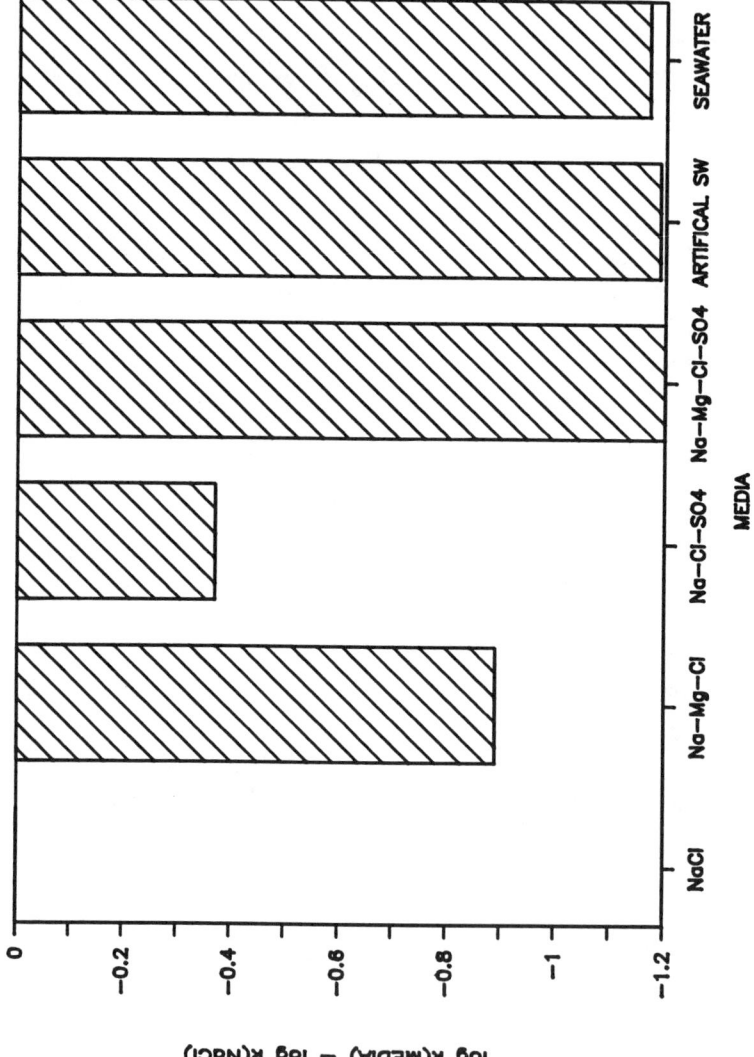

Figure 11. The effect of various ions on the oxidation of Fe(II) with $O_2$ at I = 0.7 at 25°C.

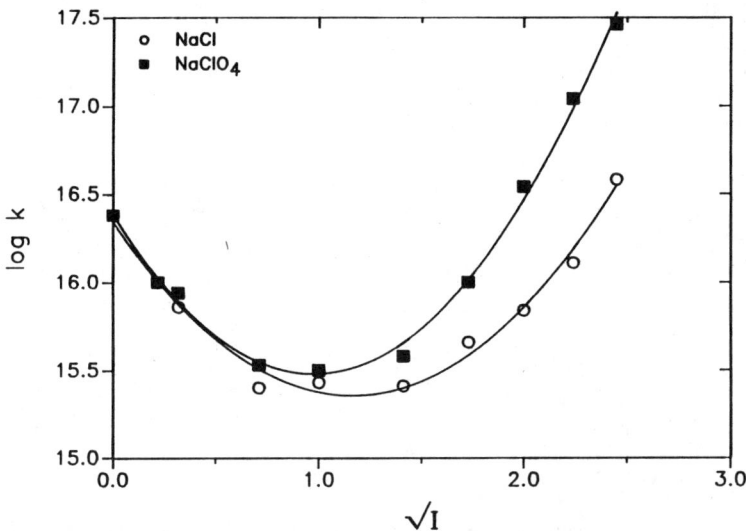

Figure 12. The rate constant for the oxidation of Fe(II) with $O_2$ in NaCl and $NaClO_4$ solutions as a function of ionic strength at 25°C.

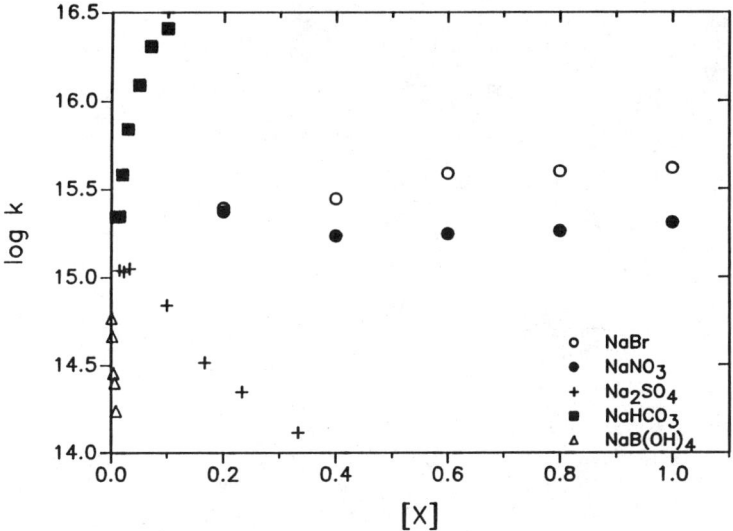

Figure 13. The effect of various anions (X) on the oxidation of Fe(II) with $O_2$ in NaCl-NaX solutions at I = 1.0 and 25°C.

which has a faster rate of oxidation than $Fe(OH)_2$. Since the value of $\beta_{FeCO_3}$ is not available at the present time, it is not possible to make a reliable estimate of the rate of oxidation of $FeCO_3$. The slight increase upon the addition of $Br^-$, $NO_3^-$ and $ClO_4^-$ is related to the formation of $FeCl^+$ ion pairs. The other anions have weaker interactions of these anions with $Fe^{2+}$ compared to $Cl^-$.

The strong decreases in the rates due to the addition of $SO_4^{2-}$ and $B(OH)_4^-$ can be attributed to the formation of $FeSO_4$ and $FeB(OH)_4^+$ ion pairs. If one assumes that the ion pairs cannot be oxidized, the decrease in the rates can be given by

$$K_{CL}/k_X = 1 + \beta_{FeX}[X^-] \qquad (41)$$

The experimental values of $K_{Cl}/k_X$ give $\log \beta_{FeSO_4} = 1.8 \pm 0.1$ and $\log \beta_{FeB(OH)_4} = 3.2 \pm 0.1$ at $I = 1.0$ and 25°C. The value for $FeSO_4$ is in good agreement with the literature infinite dilution value ($\log \beta_{FeSO_4} = 2.1$ ([19])). To the best of our knowledge, values are not available for $\beta_{FeB(OH)_4}$. Our estimates, however, are in reasonable agreement with the $\beta$'s for the formation of $Cu^{2+}$ and $Pb^{2-}$ complexes with $B(OH)_4^-$ ($\log \beta = 3.7$ and 2.3, respectively ([20])).

In conclusion, our recent studies of the rates of oxidation of Cu(I) and Fe(II) clearly demonstrate how the formation of ion complexes can affect the rates. Future rate studies for other metals are needed to examine how changes in the composition of natural waters can affect their rates of oxidation and reduction.

ACKNOWLEDGMENT

The author wishes to acknowledge the support of the Office of Naval Research (N00014-87-G-0116) and the Oceanographic section of the National Science Foundation (OCE86-00284) for this study.

LITERATURE CITED

1. Moffett, J.W.; Zika, R.G. Mar. Chem. 1983, 13 239-51.
2. Millero, F.J. Geochim. Cosmochim. Acta 1985, 49, 547-53.
3. Millero, F.J.; Sotolongo, S.; Izaguirre, M. Geochim. Cosmochim. Acta 1987, 51, 793-801.
4. Moffett, J.W.; Zika, R.G. Environ. Sci. Technol. 1987, 21, 804-10.
5. Millero, F.J.; Izaguirre, M.; Sharma, V.K. Mar. Chem. 1988, 22, 179-91.
6. Sharma, V.K.; F.J. Millero. Environ. Sci. Technol. 1988, 22, 768-71.
7. Sharma, V.K.; Millero, F.J. J. Solution Chem. 1988, 17, 581-99.
8. Sharma, V.K.; Millero, F.J. J. Inorg. Chem. 1988, 27, 3256-59.
9. Sharma, V.K.; Millero, F.J. Geochim. Cosmochim. Acta, in press.
10. Stumm, W.; Lee, F.F. Environ. Sci. Technol. 1961, 53, 143-6.
11. Millero, F.J.; Hubinger, S.; Fernandez, M.; Garnett, S. Environ. Sci. Technol. 1987, 21, 439-43.
12. Millero, F.J. Geochim. Cosmochim. Acta 1987, 51, 351-53.
13. Zika, R.G.; Saltzman, E.S.; Cooper, W.J. Mar. Chem. 1985, 17, 265-75.
14. Zika, R.G.; Moffett, J.W.; Cooper, W.J.; Petasne, R.G.; Saltzman, E.S. Geochim. Cosmochim. Acta 1985, 49, 1173-84.
15. Millero, F.J.; Sotolongo, S. Geochim. Cosmochim. Acta, in press.
16. Millero, F.J.; Izaguirre, M. J. Solution Chem. 1988, 17, 581-99.
17. Tamura, H.; Goto, K.; Nagayama, M. J. Inorg. Nucl. Chem. 1976, 38, 113-117.
18. Sung, W.; Morgan, J.J. Environ. Sci. Technol. 1980, 14, 561-68.
19. Kester, D.R.; Byrne, R.H.; Liang, Y. In Marine Chemistry of the Coastal Environment; Church, T.M., Ed.; American Chemical Society: Washington, DC, 1975; p 56.
20. van den Berg, M.G. Geochim. Cosmochim. Acta 1984, 48, 2613-17.

RECEIVED July 20, 1989

# Chapter 35

# Role of Reactive-Surface-Area Characterization in Geochemical Kinetic Models

### Art F. White and Maria L. Peterson

**U.S. Geological Survey, Water Resources Division, Mail Stop 420, 345 Middlefield Road, Menlo Park, CA 94025**

Modeling of kinetic and sorption reactions requires estimates of the surface area of natural substrates which are often difficult to measure directly. A synthesis of data indicates that BET measurements on fresh surfaces exceed geometric estimates by a mean roughness factor of 7 over a wide range in particle sizes. Surface roughness factors for naturally weathered silicates are shown to approach 200 and are strongly dependent on mineral composition. Fractal analysis indicates a dimension of 2.0, and a self similarity comparable to a smooth spherical geometry. Estimates of reactive surface areas are commonly related through transition state theory to the surface defect density. Data indicate, however, that the actual surface dislocation density is lower than commonly assumed and the number of surface dislocations that actually represent potential reaction sites must be extremely low. A compilation of available kinetic models indicates that reactive surface areas commonly are one to three orders-of-magnitude lower than physical surface areas, with closer fits for geochemical systems having short residence times.

Major advances in incorporating reaction kinetics into geochemical models have occurred since the publication of the last Proceedings of the Symposium on Chemical Modeling in Aqueous Systems, ten years ago. These advances include greater understanding of reaction mechanisms, determination of experimental reaction rates, and development of computational methods linking kinetic expressions with reaction path and coupled transport processes. A review of the recent literature, however, indicates that advances in applying and validating kinetic models for natural systems have proceeded much more slowly.

The difficulty in applying reaction kinetics to natural systems is inherent in the integrated rate equation,

$$M = \frac{A}{m_s} kt, \qquad (1)$$

where the mass transfer, M (moles/kg solution), is equal to the product of the ratio of reactive surface area, A ($m^2$), to solution mass, $m_s$ (kg), the kinetic rate constant, k (moles/$m^2$/s), and reaction time, t (s). For most natural systems, M is known from solution analyses, k is assumed equal to experimentally determined rate constants, and t is estimated from fluid residence times using age dating techniques or hydraulic head and conductivity measurements. The surface area to mass ratio in such models usually is calculated from geometric or surface area measurements (1-3).

This chapter not subject to U.S. copyright
Published 1990 American Chemical Society

Because these parameters are linearly related in Equation 1, the statistical errors introduced by incorrect estimates are comparable for each term. The greatest cumulative error in many kinetic models lies in the estimating of natural systems' reactive surface areas. This paper will review previous work, present additional data, and provide an estimate of the errors involved in using surface-area parameters in geochemical models.

## CHARACTERIZATION OF PHYSICAL SURFACES

The simplest method of estimating mineral reactive surface area is to equate it to the physical surface area, usually reported in $m^2/gm$ of substrate. Specific surface area, $A/V_m$ ($m^2/cm^3$ of substrate), will be used here for directly comparing surface areas of minerals of different densities. Geochemical models commonly define the physical surface area in terms of solution mass ($m_s$), as in Equation 1, rather than solid phase volume ($V_m$). The two ratios are related by the expression

$$\frac{A}{m_s} = \rho_s \times 10^{-3} \frac{n}{f} \frac{A}{V_m}, \tag{2}$$

where $\rho_s$ is the solution density ($cm^3/kg$), n is the percent porosity, and f is the volume percent of a reacting phase in the host rock.

The definition and magnitude of the physical surface area is closely associated with the scale and type of the measurement technique. Geometric measurement includes size fractionation by sieving or settling, optical methods including microscopy, particle counters, photocorrelation spectroscopy (4), and small angle neutron scattering (SANS) (5). These techniques provide minimum, maximum or average dimensions which then employ a geometric model, i.e. sphere, cube, plane, and so forth, to calculate the surface area. Such geometric estimates are often the only available approach for some complex systems, such as fracture surface areas. The validity of such estimates depends on the correctness of the assumed macroscopic shape, accuracy of the particle size measurements, and microscopic smoothness of the surfaces.

Without detailed descriptions of individual particle geometry, mineral grains are normally characterized by length or diameter, $\partial$, which represents a range of lengths. The specific surface area is then defined as

$$\frac{A}{V_m} = \int_0^\infty \frac{K}{\partial} P(\partial) d\partial, \tag{3}$$

where $P(\partial)$ is the probability of a mineral grain having a diameter between $\partial$ and $(\partial + d\partial)$. Two commonly used statistical functions for describing the probable distribution frequency of $\partial$ are the Gaussian and log-normal distributions. The commonly observed skewed distribution with tailing toward finer particles is fitted by the latter log-normal law (6). K is an empirical factor used to correct for deviations from the spherical or cubic form. For natural rounded sand particles, K was found to equal 6.1 (7), close to the theoretical value of 6.0. For crushed quartz, K ranges from 14 to 18 (8).

If the particles are of the same size or are assumed to be all of a mean size diameter, $\partial_o$, the specific surface area is

$$\frac{A}{V_m} = \frac{K}{\partial_o}. \tag{4}$$

Methods for micro-measurement of surface areas include the Brunauer, Emmett, and Teller (BET) method (9), which relies on the adsorption of monolayers of gas, commonly nitrogen or argon, the adsorption of organic molecules such as ethylene glycol and ethylene glycol monoethyl ether (EGME) (10), and the use of infrared internal reflectance spectroscopy (11) which characterizes bonding of sorbed water. These last two techniques have been confined principally to surface areas of clay minerals.

In BET measurements, the physical surface area is commonly assumed to be covered by nitrogen atoms, of radius 1.62 nm (12). When using surface areas in geochemical modeling, this area is assumed to be comparable to the physical interface existing between aqueous solution and the mineral surface. For the literature data plotted in Figure 1 (13-16), water BET surface areas are comparable to or exceed those calculated for nitrogen. A number of additional studies involving water sorption found that adsorption isotherms did not fit the BET equation because of surface chemical interactions. Excessively high BET water surface areas, such as exhibited by the smectite data (14), result from the fact that water, with its larger dipole moment (17) and higher bonding strength, can penetrate into expandable clay lamellae and smaller pore spaces than can nitrogen (16). In such expandable clays, van Olphen (18) estimated that BET water vapor surface areas may be in error between 50 and 100 percent. Except for such clays, the assumption that the commonly employed BET surface area is representative of the substrate physically available to water molecules appears valid.

The measured BET surface area reflects the sum of both the internal and external surfaces. Internal surfaces are composed of pores which may vary greatly both in size and shape. The International Union of Pure and Applied Chemistry (6) classified pores according to their average width: micropores <2 nm, mesopores 2 to 50 nm, and macropores >50 nm. These ranges reflect, in part, the methods by which internal surfaces areas are measured. Internal surfaces in the micropore and mesopore range are calculated from deviations from the BET isotherm in the form of differing hysteresis loops between the adsorption and desorption isotherms (19). The distribution of macropores are usually determined by mercury porosimetry (20). Because of specific pore morphology, the BET surface area may not reflect the actual surface area involved in chemical reactions (21, 22). Internal pore spaces may exhibit reaction mechanisms controlled by local equilibrium and diffusional transport at the same time as the external surface is controlled by surface reactions. A lack of correlation between reactivity and BET surface area, for example, is shown for biogenic marine carbonates in which the dissolution rates are more accurately estimated from the grain morphology and pore microstructure (23).

## QUANTITATIVE RELATIONSHIP BETWEEN GEOMETRIC AND BET SURFACES

A number of individual studies have concluded that BET surface areas are greater than geometric estimates of surface areas (24-27). However no systematic attempt has been made to correlate geometric and BET surface areas over a wide range of mineral types and particle sizes to determine if measurement methods are interchangeable in kinetic models. The logarithmic plot in Figure 2 shows such a synthesis of about 40 references (28-67) in which both grain size ranges and BET values are reported. The ranges in particle size, shown along the horizontal axis, vary from sand to clay size. Data are divided into six general compositions and include both natural and synthetic minerals. Biogenic material was excluded on the premise that surface characteristics were controlled by processes different than those for inorganic substrates. As expected, the regression fit lies above the geometric line indicating that measured BET surface areas are almost always greater than the geometric estimates. However a striking and somewhat unexpected feature is the parallel nature of the two lines indicating that the differences in BET and geometric surface areas are consistent over a wide range in grain size.

The classical treatment of particle surface area and surface irregularity has been to regard it as a deviation from the ideal reference, i.e. Equations 3 and 4, or to model it as a superposition of a regular periodic function using Fourier transforms (68). In geochemical models, the roughness factor, R, is most often employed (69). R is defined as the ratio of the actual surface, assumed in this case to be the BET surface, and the estimated geometric surface. Reported roughness factors frequently range from close to unity (1.08 for mica (70)) to relatively large values such as 15.0 for biogenic carbonate sediments (23). Based on the data in Figure 2, the best estimate of the mean roughness of reported minerals is the ratio of the y-intercepts of the two lines, or 7.

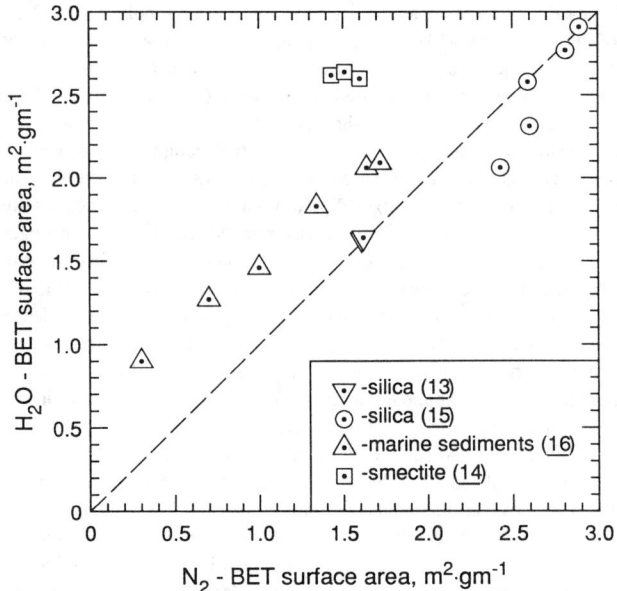

Figure 1. Comparison of surface areas calculated from BET sorption isotherms for nitrogen and water.

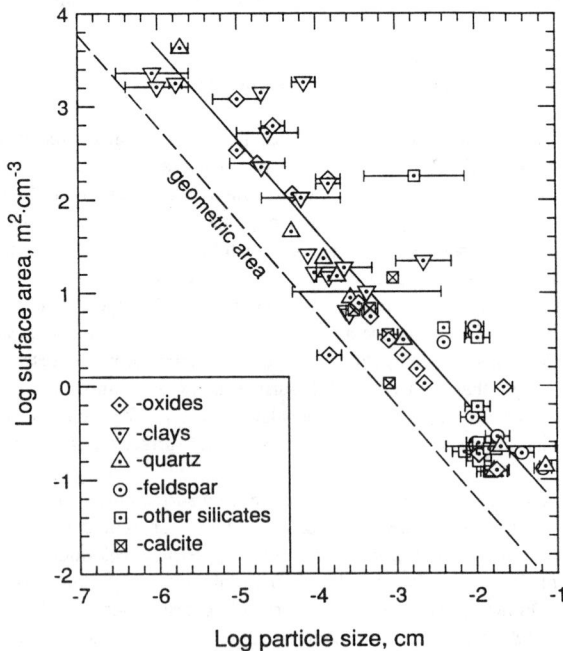

Figure 2. BET surface areas as a function of particle diameter. The dashed line is solution to Equation 4 (K=6) assuming spherical geometry, with slope and intercept of -1.00 and -3.22 respectively. The solid line is the linear regression fit to reported BET data (28-67), with R-squared coefficient of 0.86 and respective slope and intercept of -0.99 and -2.38.

Mandelbrot (24) has suggested a different approach to surface irregularity by using fractal dimensions. A recent proliferation of studies has substantiated the hypothesis of self similarity for a number of natural systems (71-73). By this approach, surface irregularity scaling is given by the fractal dimension D, whose range is defined as $2 < D < 3$ and which is related to the surface area by the proportionality

$$A \propto \partial^{D-3}. \tag{5}$$

D equals 2 when the scaling of A is defined by a simple geometric relationship (Equation 4). As D approaches 3, the volume dimension, the surface area is no longer dependent on the diameter of the particle. This latter limit implies that the surface irregularity is so great that it completely fills any volume defined by D. The fractal dimension is then an intensive parameter which describes the change in surface topography or roughness as a function of grain size. Extensive roughness and pronounced pore structure, as in zeolites for example, can still yield D values close to 2.0 (74). The fractal dimension manifests itself in the size distribution of roughness rather than the presence of roughness. This approach has been applied to the surface analysis of a small number of geologic materials over a limited particle size range including synthetic periclase (3-22µm, D=2.0), crushed quartz (0.5-12.0µm, D=2.0 to 2.2), and crushed limestone (50-2600µm, D=2.1 to 2.8) (71).

As demonstrated by Pfeifer (74), the range of self similarity defining the fractal dimension is

$$f_o < f < f_{max}, \tag{6}$$

where $f_o$ is the diameter of the yardstick measuring the self similarity, i.e. the nitrogen atom (1.62 nm), and $f_{max}$ is defined as

$$f_{max} = f_o (\partial_{max}/\partial_{min}), \tag{7}$$

where $\partial_{max}$ and $\partial_{min}$ are the maximum and minimum particle diameters.

The ratio of particle diameters ($\partial_{max} \backslash \partial_{min}$) for the data presented in Figure 2 is very large ($10^5$) compared to the previous studies, making the particle size - surface area relationship much more sensitive to fractal analysis. The regression slope for all of the reviewed data in Figure 2 is very close to -1.0, implying a dimension of 2.0 (Equation 5). Based on this analysis, the mineral classes treated together in Figure 2 do not exhibit a fractal dimension but possess self similarity comparable to two dimensional geometric surfaces over a size range from clay to sand size particles.

Most of the literature data cited in Figure 2, particularly for the larger diameters which constitute the range of many primary minerals, were determined for fresh surfaces prepared by crushing. Few data were available on natural surfaces which may exhibit increased surface areas caused by pitting and etching during weathering, or by surface coatings or surface mineral precipitates. In the limited number of kinetic studies in which BET surface areas were measured before and after experimental reaction, surface area changes ranged from only minor increases for quartz and feldspars (<25%) (75) to significant increases for olivine (>400%) (76).

The effect of weathering on mineral surface areas is addressed in this paper through BET surface measurements made on a number of naturally weathered silicates over a 40 to 500µm size range. Material was obtained from a grus-type granitic weathering profile from the Sierra Nevada Mountain Range, California, a deeply weathered granitic profile from Montara Mountain, California, a beach dune deposit from Half Moon Bay, California, and glacial outwash sand from Cape Cod, Massachusetts. The mineralogy of all four samples was dominated by quartz and feldspars with lesser amounts of mafic minerals, mostly hornblende and biotite.

The materials were ultrasonically cleaned and wet sieved to maximize the removal of fines and then treated with hydrogen peroxide and sodium dithionite to remove organic compounds and any iron oxyhydroxide coating which could contribute to increased surface areas. This clean size fraction was further split by isodynamic magnetic separation into predominantly quartz-feldspar and mafic fractions. Characterization of the cleaned material by scanning electron microscopy and energy dispersive x-ray analysis revealed mineral grains with a homogeneous grain size distribution and a lack of clay minerals and other adhering fine material. Individual grains varied markedly in the extent of pitting and etching for each material. In general, quartz and K-feldspar exhibited the least pitting and hornblende and mica the most. However, even quartz grains were extensively etched in the highly weathered Montara sample.

Single point BET surface areas were determined on uncleaned, cleaned, and magnetically separated sample splits using a Micrometrics Flowsorb system and a 30% nitrogen - 70% helium gas mixture. (The use of brand names in this paper is for identification purposes only and does not constitute endorsement by the U.S. Geological Survey.) Figure 3 shows a cumulative plot of the surface areas for splits of the weathered Tahoe and Montara granites. Several conclusions can be reached from these data: (a) The total surface areas of both materials are from one to two orders of magnitude greater than their geometric estimates. (b) The more highly weathered and surficially pitted Montara material exhibits larger surface areas, particularly for the mafic minerals. (c) Even though quartz and feldspars are the dominant mineral phases, they constitute only a small portion of the total surface area for each size fraction. (d) The total surface area is dominated by surfaces of the mafic minerals and their iron oxyhydroxide weathering products.

The BET surface areas of the quartz-feldspar fractions of each of the four materials are plotted in Figure 4. These surface areas, unlike the mafic fractions, give reasonably similar results, with the Montara sample exhibiting somewhat larger values. The weathered surface areas are compared to reported surface areas of fresh feldspar (75) and quartz (25) and are generally one half to one and a half orders of magnitude higher. Comparisons with geometric estimates result in roughness factors (R) of approximately 20 to 200. These roughness factors, based on BET measurements, are substantially higher than roughness factors of ten calculated by Helgeson, Murphy, and Aagaard (69) based on SEM analysis of surface pitting of weathered feldspars. This discrepancy arises due to differences in the resolution of the two techniques.

CHARACTERIZATION OF REACTIVE SURFACES

An equation relating the kinetic reaction rate to the physical surface area is

$$\frac{dM}{dt} \propto \left(\frac{A}{m}\right)_s^b, \qquad (8)$$

where b is the exponential power function. In most kinetic studies, Wenzel's law is invoked which states that the rate of reactions between solids and liquids is directly proportional to the area of contact (71). Actual experimental studies, in which the physical surface area was varied either by increasing the mineral/water mass using particles of the same size or by using the same mass composed of particles of differing sizes, have demonstrated that b exhibits considerable variation. For example, Walter and Morse (77) found a direct relationship between reaction rate and surface area (b = 1, Equation 8) for Iceland spar dissolution at 25°C. In contrast, White (78) and Lagache (79) found that the respective reaction rates for rhyolitic glass at 25°C and alkali feldspar at 200°C varied approximately with the square root of the surface area (b = 0.5, Equation 8). Holdren and Speyer (75) found even lower correlations between reaction rates and surface areas for alkali feldspar dissolution at 25°C in which b varied between -0.02 and 0.18.

The lack of a direct proportionality between surface area and reaction rate in the latter cases requires a distinction between the physical surface area and that surface area which participates directly in chemical reactions. This paper uses the term 'reactive surface area' to denote the surface containing chemically reactive sites, i.e. the effective surface area as defined by Helgeson et al. (69),

35. WHITE & PETERSON  *Reactive-Surface-Area Characterization*  467

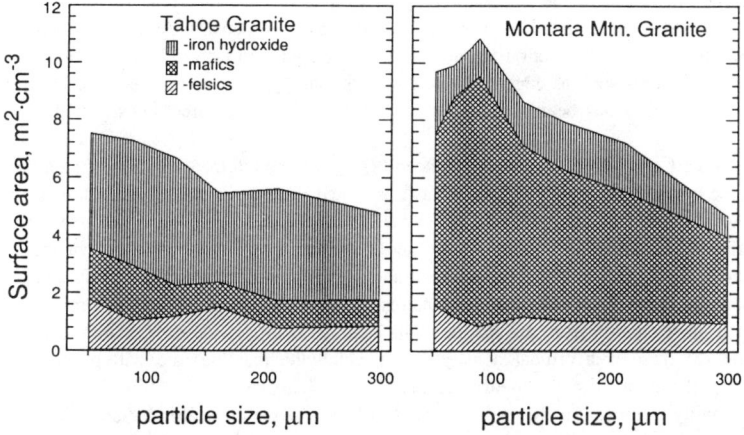

Figure 3. Cumulative surface areas for sample splits of weathered granite including mafic and felsic minerals and iron hydroxide coatings.

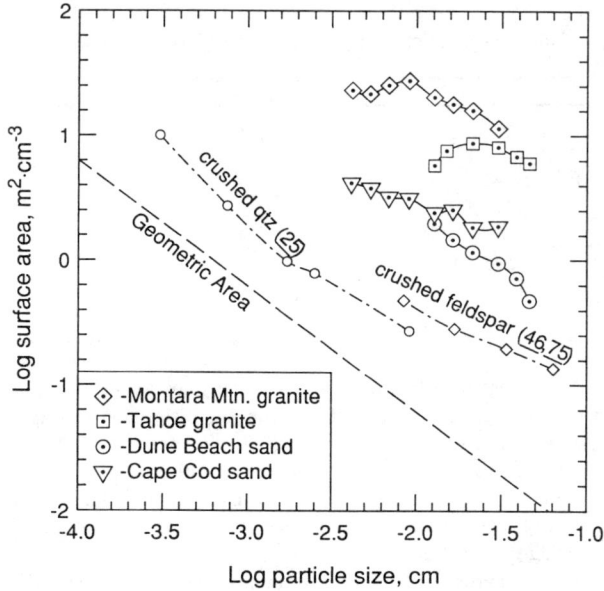

Figure 4. BET surface areas of weathered quartz + feldspar compared to fresh crushed quartz (25) and feldspar (46, 75). Geometric line is solution to Equation 4 (K=6).

which is also actively coupled physically and hydrologically to the geochemical system being modeled.

Recent advances in applying transition state theory to geochemical kinetics (80, 81) have emphasized the interaction of the activated complex with specific surface reaction sites. The rate of reaction is assumed to be a function of the surface reaction site density. A correspondence is also observed between surface dissolution features such as etch pits, and crystallographically controlled extended defect features such as edge and screw dislocations (82). Based on these lines of evidence, the reactive surface area has been proposed to be proportional to the defect density within minerals (75).

A Monte Carlo model has been advanced (83,84) to explain the evolution of etch pits based on the balance between decreases in free energy associated with the destruction of defect structures versus the increase in free energy associated with the corresponding increase in surface area. The model requires input parameters including surface free energies, Poisson ratios, shear moduli, and Burges vectors which can only be reliably estimated for a small class of compositionally simple minerals, such as quartz. Nevertheless, two interesting conclusions result from the model: (a) The extent of etch pit formation from dislocations will increase as a function the reaction affinity of the solution, and (b) there is a critical density above which the reaction rate will be controlled by dissolution at dislocation sites. This density is estimated to be approximately $10^9$ dislocations per cm$^2$ (83) and is roughly equivalent to $10^4$ kilometers of dislocations per cm$^3$ (85). A dislocation density of $10^{10}$/cm$^2$ is required to significantly affect the enthalpy and solubility of quartz (86). Helgeson, Murphy, and Aagaard (69) assumed a dislocation density of $10^8$/cm$^2$ in their calculations on the free energy of etch pit formation on surface activated complexes.

Actual dislocation density data for natural minerals are quite limited and are skewed toward physically deformed phases with high numbers of dislocations (Table I).

Table I. Dislocation Densities of Natural Minerals (cm$^{-2}$)

| Mineral | Dislocation density | Reference |
|---|---|---|
| Plagioclase/deformed | $10^7 - 10^{12}$ | (87) |
| Quartz/mylonites | $10^9$ | (88) |
| K-feldspar/greenschist | $10^8 - 10^9$ | (89) |
| Olivine/peridotite | $10^7 - 10^8$ | (90) |
| Plagioclase/granulite | $10^7 - 10^8$ | (91) |
| Quartz/undeformed | $10^6$ | (82) |
| Calcite/undeformed | $10^6$ | (92) |
| Quartz/undeformed | $10^3$ | (93) |

High resolution TEM imaging of dislocations, which is the principal measurement technique, is difficult below a density of approximately $10^6$/cm$^2$. The low dislocation density for undeformed quartz, $10^3$/cm$^2$, reported by Wegner and Christie (93), was determined by chemical etching techniques. In reality, the defect density of many undeformed minerals may be less than $10^6$/cm$^2$. Comparison of the limited data (Table I) with the density required for dislocation-controlled dissolution in the model of Blum and Lasaga (83), implies that dislocation densities in all but the most deformed rocks may not directly control surface reactions.

The lack of a strong correlation between dislocation densities and reaction rates has also been demonstrated experimentally, as shown in Table II. In these studies the dislocations were mechanically induced, creating dislocation densities which varied by several orders of magnitude. The upper limit for defect densities exceeded the $10^9$/cm$^2$ threshold density proposed by Blum and Lasaga (83). However, the large range in dislocation densities produced only factor of two increases in the reaction rates.

Table II. Effect of Dislocation Densities on Reaction Rates

| Density Increase | Rate Increase | Material | Reference |
|---|---|---|---|
| $10^7$ to $10^{11}$/cm | x2 | Chalcopyrite | (94) |
| $10^6$ to $10^{10}$/cm | x2 | Calcite | (92) |
| $10^6$ to $10^{11}$/cm | x2 | Rutile | (94) |

Holdren and Speyers (75) also invoked defect densities to explain the lack of significant positive correlation between reaction rate and BET surface areas of feldspars. They proposed that for larger grain sizes where the spacing between surface defect sites is much less than the typical individual grain dimensions, the reaction rate should be proportional to the surface area (b = 1, Equation 8). However, as the grain size of the experimental material is reduced, the grain dimensions must eventually approach the size of spacing between adjacent defects, and dissolution rates would become increasingly independent of surface area. By using the reported surface area and grain-size distributions of their data, and assuming a moderate dislocation density of $10^6/cm^2$, a total surface density of between $1.5 \times 10^3$ and $1.8 \times 10^5$ dislocations per particle results. Such a defect distribution does not meet their criteria that the reaction site spacings must become comparable to or greater than the dimensions of the individual mineral grains. This implies that the number of surface dislocation sites that actually represent potential reaction sites must be extremely low.

## COMPARISONS OF PHYSICAL AND REACTIVE SURFACE AREAS IN NATURAL SYSTEMS

Reaction kinetics represented by the general form of Equation 1 have been employed in a number of quantitative chemical models of natural systems. Under ideal conditions, the four parameters, total mass transfer, kinetic rate constants, time, and the reactive surface area can be determined independently, permitting the unique definition of the model. In most cases, at least one of the variables, most often surface area, is treated as a dependent term. This nonuniqueness arises when the reactive surface area of a natural system cannot be estimated, or because such estimates made either from geometric or BET measurements do not produce reasonable fits to the other parameters. Most often the calculated total mass transfer significantly exceeds the observed transfer based on measured aqueous concentrations.

Estimated physical surface areas based on geometric or BET measurements are compared in Table III and Figure 5 with calculated reactive surface areas using specific kinetic models (1, 96-107). With a single exception (107), they all model silicate dissolution in ground water systems. The accuracy of the surface area ranges reported in both Table III and Figure 5 varies depending on the manner in which that model's data were presented and, in some cases, due to possible subjective interpretation by the present authors. Several papers present specific comparisons between physical and reactive surface areas (98-100, 103). Other tabulations were compiled after additional written or oral communication with the authors (104-107). Two models (102, 105) lacked explicit physical surface areas, and these were calculated for the present tabulation based on stated particle sizes and geometry.

The method of estimation of physical surface area for each model is included as footnote 1 in Table III. Some models (1, 96, 97, 102) used generic geometric considerations. The surface area of a smooth parallel fracture, $A/V_s$, is assumed proportional to $2/w$, the fracture width, and for close packed spheres of a given diameter, $A/V_s$ approximates $17.10/\partial$. An increase in geometric specificity is apparent in models which characterize particle size distributions and system porosity (100, 104, 105). More accurate estimates of physical surface areas are contained in models where the reactive phase was isolated and BET surface measurements were conducted (98, 106). Some models (99, 101) used BET surface areas which were determined on in situ core material where the original particle textures and porosity were maintained.

TABLE III. Comparisons of Physical and Reactive Surface Areas ($m^2$ /kg $H_2O$)

| Physical Surface Area | | Reactive Surface Area | | Geochemical System | $(n)^3$ % | $(f)^4$ % | Ref. |
|---|---|---|---|---|---|---|---|
| $1.9 \times 10^0$ | $(1)^1$ | -- | | Nepheline weathering | -- | 25 | (96) |
| $10^2 - 10^4$ | (1) | -- | | Quartz precipitation; fine grained sediments | -- | -- | (1) |
| $4 \times 10^0$ | (1) | -- | | Microcline dissolution | 30 | -- | (97) |
| $4 \times 10^{-1} - 1 \times 10^1$ | (3) | $2 \times 10^{-3} - 3 \times 10^{-2}$ | $(a)^2$ | Opal dissolution; marine sediments | 84 | 10 | (98) |
| $2-6 \times 10^3$ | (3) | $6 \times 10^1 - 1.8 \times 10^2$ | (a) | Glass dissolution; tuff | 40 | 100 | (99) |
| $9 \times 10^1 - 2 \times 10^2$ | (2) | $4-9 \times 10^0$ | (a) | Plagioclase dissolution in gneiss | 23 | 15 | (100) |
| $4 \times 10^0 - 9 \times 10^1$ | (3) | $8 \times 10^{-1} - 4 \times 10^0$ | (b) | Tuff dissolution; experimental | 98 | 100 | (101) |
| $7 \times 10^{-1} - 3 \times 10^0$ | (1) | $8 \times 10^{-2} - 5 \times 10^{-1}$ | (a) | Plagioclase dissolution in saprolite | 25 | 8.3 | (102) |
| $10^0 - 10^1$ | (2) | $1-6 \times 10^{-2}$ | (b) | Silicate dissolution; hydrothermal | 3 | 100 | (103) |
| $5 \times 10^0 - 5 \times 10^1$ | (2) | $2 \times 10^{-4} - 2 \times 10^{-2}$ | (b) | Silicate dissolution | 30 | 100 | (104) |
| $4-7 \times 10^{-1}$ | (2) | $6 \times 10^{-1} - 5 \times 10^0$ | (a) | Glass dissolution; basalt | 20 | 100 | (105) |
| $3-7 \times 10^{-1}$ | (3) | $7 \times 10^{-3} - 2 \times 10^{-2}$ | (c) | Magnetite dissolution in sand | 30 | 0.5 | (106) |
| $3-5 \times 10^{-3}$ | (2) | $2 \times 10^{-4} - 3 \times 10^{-2}$ | (a) | Stream bed calcite precipitation | 100 | 100 | (107) |

$^1$ Method of surface area estimation: (1) generic geometry, (2) specific geometry, (3) BET.
$^2$ Model: (a) analytical, (b) EQ3NR/EQ6, (c) Dynamix.
$^3$ Porosity.  $^4$ Reactive mineral fraction.

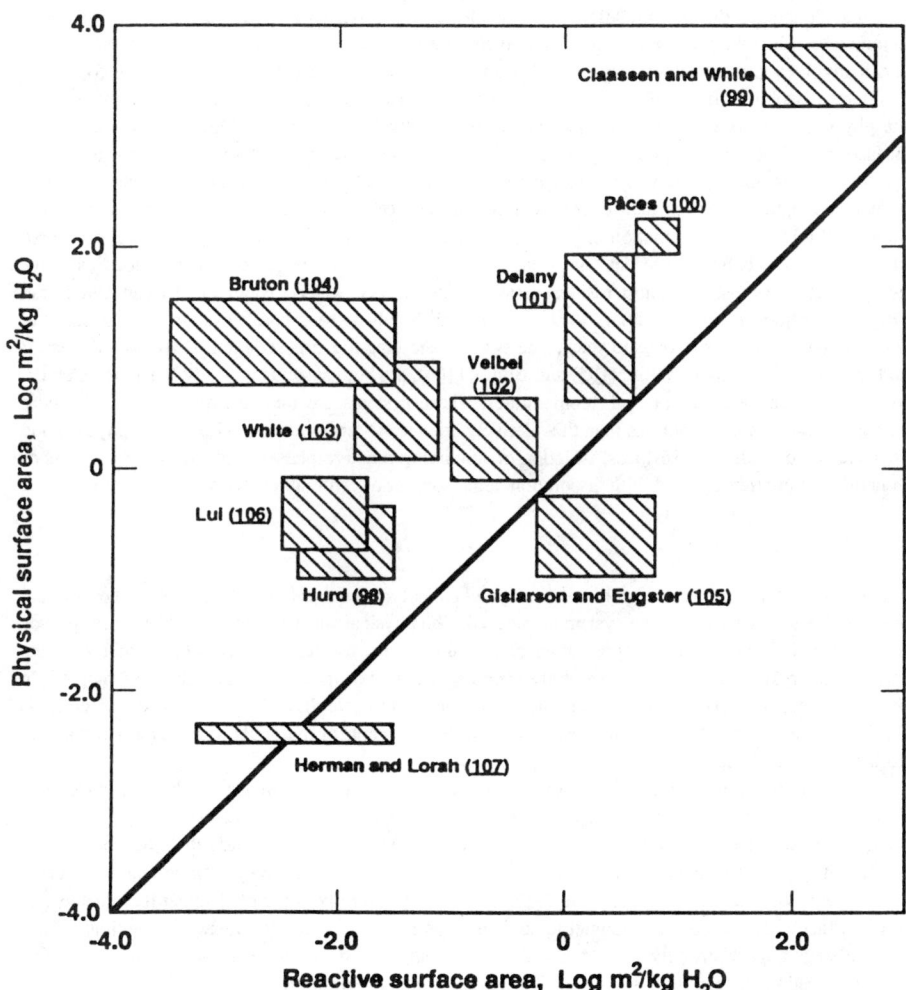

Figure 5. Comparison of estimated physical surface areas to calculated reactive surface areas for models tabulated in Table III.

The range in reactive surface areas for each model given in Table III was calculated assuming that the surface area depends on the experimentally determined rate constant. Mathematically, both parameters are linked and either can be calculated as a dependent variable in Equation 1. The methods of calculation are contained as footnote 2 in Table III. Estimated values for n and f (Equation 2) are also reported.

As indicated by Figure 5, only the work of Herman and Lorah (107) overlies the diagonal line that denotes a direct correlation between physical and reactive surface areas. Their model is unique in that it is the only carbonate system, it simulates mineral precipitation rather than dissolution, and it involves surface rather than ground water. The model of Gislason and Eugster (105), which simulates dissolution of basaltic glass, produces reactive surface areas which are slightly greater than the physical surface area. The model is characterized both by low reaction rates and low estimated surface areas. The reported geometry is comparable to pebble sized material with high porosity.

The remaining models produced reactive surface areas that range between one and three orders of magnitude lower than estimated physical surface areas. There is no obvious correlation between the method of estimation of physical surface area, i.e. BET versus geometry, and the degree of the deviation between physical and reactive surface areas. A closer correlation does exist for geochemical systems with short residence times. The surface water system of Herman and Lorah (107) has residence times of hours to days while the highly porous basalt system of Gislason and Eugster (105) has times of days to years. The systems showing the largest deviations between physical and reactive surface areas, White (103) and Bruton (104), are both systems with residence times of $10^4$ to $10^6$ years and are also at high temperatures. These systems are undoubtedly subjected to more extensive and complex reactions than the relatively open short-term systems, with greater physical and chemical modification of surfaces, isolation of remaining reactive phases, and extensive amounts of mineral precipitation coupled with dissolution being sources of errors in the model calculations.

CONCLUSIONS

To quantitatively model reaction kinetics of geochemical systems, reliable estimates of the physical and reactive surface areas of the system are needed. The physical surface areas have been measured on the basis of either the macroscopic nature of the surface, i.e. estimates of its bulk geometry, or the microscopic nature, i.e. the areal extent of coverage by atoms or molecules, as in the BET method. In the latter case, comparisons with water sorption isotherms indicate that BET-determined surface areas produce reliable estimates of the mineral/water interface, except for materials with high microporosity such as expandable clays.

A correlation between unweathered surfaces and particle diameter results in a roughness factor of approximately 7 determined in this paper, which is applicable over a large range of grain sizes. Analysis of self similarity using the methods of Mandelbrot indicates that the compilation covered in Figure 2 does not process a fractal dimension. Its self similarity is comparable to that of a two dimensional surface. Surface area determinations of naturally weathered minerals indicate that BET surface areas can exceed geometric surface areas by more than two orders of magnitude. Such surface areas appear strongly dependent on mineral composition. In kinetic models of such systems, geometric estimates from bulk grain size distributions are poor assumptions.

In a number of kinetic studies, the reaction rate is not directly proportional to the physical surface area. This discrepancy leads to the concept of the reactive surface area which has been closely linked to the reaction site density and the dislocation density. Theoretical analysis indicates that reaction rates will become directly proportional to defect densities only in highly stressed and deformed minerals. However, experiments using high dislocation-density minerals still produce a low correspondence between reaction rates and total defect density. Energies associated with surface dislocations must be highly heterogeneous and the number of dislocations that actually represent potential reaction sites must be extremely low.

A compilation of available kinetic models shows that, in most cases, the calculated reactive surface areas are one to three orders of magnitude less than the estimated physical surface areas. Commonly, geometric and BET surface areas are used interchangeably in kinetic studies to measure physical surface areas. The models that did produce closer fits were for open systems with short residence times. Comparisons assumed experimentally correct reaction rates and dependent reactive surface areas. In reality, the reaction rate and the reactive surface area are explicitly linked on the basis of surface controlled reactions. The product of these two terms determines the mass transfer for a specific system.

LITERATURE CITED

1. Rimstidt, J. D.; Barnes, H. Geochim. Cosmochim. Acta 1980, 44, 1683.
2. Lasaga, A. C. J. Geophys. Res. 1984, 89, 4009.
3. Helgeson, H. C.; Murphy, W.; Aaggaard, P. Geochim. Cosmochim. Acta 1984, 48, 2405.
4. Cummins, P. G.; Staples, E. Langmuir 1987, 3, 1109.
5. Hall, P.L.; Churchman, G.; Theng, B. Clay and Clay Minerals 1985, 33, 345.
6. Gregg, S. J.; Sing, K. Adsorption, Surface Area and Porosity; Academic Press: London, 1982.
7. Fair, G. M.; Hatch, L. J. Amer. Waterworks Assoc. 1933, 25, 1551.
8. Cartwright, J. Ann. Occup. Hygiene 1962, 5, 163.
9. Brunauer, S.; Emmett, P.; Teller, E. J. Amer. Chem. Soc. 1938, 60, 309.
10. Heilman, M. D.; Carter, D.; Gonzalez, C. Soil Sci. 1965, 100, 409.
11. Mulla, D. J.; Low, P.; Roth, C. Clay and Clay Minerals 1985, 33, 391.
12. McClellan, A. L.; Harnsberger, H. J. Coloid Interf. Sc. 1967, 23, 577.
13. Basset, D. R.; Boucher, E.; Zettlemoyer, A. J. Colloid Interf. Sc. 1968, 27, 649.
14. Mooney, R. W.; Keenan, A.; Wood, L. J. Amer. Chem Soc. 1952, 74, 1367.
15. Hagymassy, J.; Brunauer, S. J. Colloid Interf. Sc. 1970, 33, 317.
16. Slabaugh, W. H.; Stump, A. J. Geophs. Res. 1964, 69, 4773.
17. Klier, K.; Shen, J.; Zettlemoyer, A. J. Phys. Chem., 1973, 77, 1458.
18. van Olphen, H. In Surface Area Determination; Everett, D. H.; Ottewill, R. H., Eds.; Butterworths: London, 1970; p 255.
19. Bruauer, S.; Mikhail, R.; Bodor, E. J. Colloid Interf. Sci. 1967, 24, 451.
20. Conner, W. C.; Lane, A.; Hoffman, A. J. Colloid Interf. Sci. 1984, 100, 185.
21. Stakebake, J. L.; Fritz, J. J. Coloid. Interf. Sci. 1984, 100, 34.
22. Berner, R. A.; Holdren, G. Geology 1977, 5, 369.
23. Walter, L. M.; Morse, J. J. Sed. Petrol. 1984, 54, 1081.
24. Mandelbrot, B. B. The Fractal Geometry of Nature; Freeman: San Francisco, 1982.
25. Leamnson, R. N.; Thomas, J.; Ehrlinger, H. Ill. State Geol. Survey, Cir. 444, 1969.
26. Knauss, K. G.; Wolery, T. Geochim. Cosmochim. Acta 1986, 50, 2481.
27. Siegel, D. I.; Pfannkuch, H. Geochim. Cosmochim. Acta 1984, 48, 197.
28. Ambe, S. Langmuir 1987, 3, 489-493.
29. Bloom, P. R.; Weaver, R. Clays and Clay Minerals 1982, 30, 281-286.
30. Brown, D. W.; Hem, J. USGS Water Sup. Paper 1984, 2187, 35 p.
31. Carlson, L.; Schuetmann Clays and Clay Minerals 1980, 28, 272-279.
32. Cartwright, J.; Whealtey, K.; Sing, K. J. Appl. Chem. 1958, 8, 259.
33. Choi, I.; Malghen, S.; Smith, R. Proc. Int. Symp. on Rock/Water Interaction 1974.
34. Chou, L.; Wollist, R. Geochim. Cosmochim. Acta 1984, 48, 2205-2217.
35. Clark, C. S. J. Sanitary Eng. 1966, 92, 127-145.
36. Cosey, W. H.; Carr, M.; Graham, R. Geochim. Cosmochim. Acta 1988, 52, 1545-1556.
37. Dandurend, J. L.; Gaut, R.; Schott, J. Tectonophys. 1982, 83, 365-386.

38. Dayal, R. Geochim. Cosmochim. Acta 1977, 41, 135-141.
39. Fenet, J.; Gaut, R.; Kihn, V.; Sevely, J. Phys. Chem. Minerals 1987, 15, 163-170.
40. Fleming, B. A. J. Colloid and Interface Sci. 1986, 110, 40-65.
41. Gengis, B. S.; Gengis, L. Powder Technology 1973, 7, 85.
42. Grandstaff, D. E. Geochim. Cosmochim. Acta 1977, 41, 1097-1103.
43. Grandstaff, D. E. In Rates of Chemical Weathering of Rocks and Minerals; Colman, S.; Nethier, D., Eds.; Academic Press: New York, 1986; pp 41-59.
44. Hall, P. L; Churchman, G.; Theng, B. Clays and Clay Minerals 1985, 33, 345-349.
45. Hempson, J. W.; Cornell, O.; Meeid, T., Soil Sci. Am. J. 1986, 50, 1150-1154.
46. Holdren, G. R.; Speyer, P. Amer. J. Sci. 1985, 285, 994-1026.
47. Holdren, G. R.; Speyer, P. In Rates of Chemical Weathering; Coleman, S.; Nethier, D., Eds.; Academic Press: New York, 1986, pp 61-81.
48. Krauss, K. G.; Wolery, T. Geol. Soc. Am. Bull. 1988, 52, 43-53.
49. Lin, F. C.; Clemeney, C. Clay and Clay Minerals 1981, 29, 101-106.
50. Luce, R. W.; Bartlett, R.; Parks, G. Geochim. Cosmochim. Acta 1972, 36, 35-50.
51. May, H. M.; Kinniburgh, D.; Helmke, P.; Jackon, M. Geochim. Cosmochim. Acta 1986, 50, 1617-1677.
52. McKibben, M. A.; Barnes, H. Geochim. Cosmochim. Acta 1986, 50, 1509-1520.
53. Meyer, D. E.; Haderman, N. J. Phys. Chem. 1970, 70, 2077-2080.
54. Olphen, H. V.; Fripait, J. Data Handbook for Clay Materials and Other Non-metalic Minerals; Pergamon Press: Oxford, 1970, 346 p.
55. Plummer, L. N.; Wigley, T. Geochim. Cosmochim. Acta 1976, 40, 191-202.
56. Renden, J. L.; Carnejo, J.; Arembani, P.; Serra, C. J. Colloid Interface Sci. 1983, 92, 508-516.
57. Schivertmann, V.; Caqmbia, P.; Murod, E. Clay and Clay Minerals 1985, 33, 369-378.
58. Schott, J.; Berner, R.; Sjoberg, E. Geochim. Cosmochim. Acta 1981, 45, 2123-2135.
59. Schott, J.; Berner, R. Geochim. Cosmochim. Acta 1983, 47, 2233-2240.
60. Senger, A.; Stoffers, P.; Heller, P.; Kallai; Szcfrenels Clay and Clay Minerals 1984, 32, 375-383.
61. Shayer, A. Clay and Clay Minerals 1984, 32, 272-278.
62. Sjoberg, E.L. Geochim. Cosmochim. Acta 1976, 40, 441-447.
63. Talibudeen, O.; Goulding, K. Clay and Clay Minerals 1983, 31, 137.
64. Tamura, S.; Shubasaki, Y.; Miguta, H. Clay and Clay Minerals 1983, 31, 413-421.
65. Walter, L. M.; Messe, J. Geochim. Cosmochim. Acta 1985, 49, 1503-1513.
66. Werth, G. S.; Gieshes, J. J. Colloid Interface Sci. 1979, 68, 492-506.
67. Wiersm, C. L.; Rimstidt, J. Geochim. Cosmochim. Acta 1984, 43, 85-92.
68. Greenwood, J. A.; Williamson, J., Proc. R. Soc. London Ser. A, 1984, 393, 133.
69. Helgeson, H. C.; Murphy, W.; Aagaard, P. Geochim. Cosmochim. Acta 1984, 48, 2405.
70. Nonaka, A. J. Colloid. Interf. Sci. 1984, 99, 335.
71. Avnir, D.; Farin, D.; Pfeifer, P. J. Colloid. Interf. Sci. 1985, 103, 1121.
72. Orford, J. D.; Whalley, W. Sedimentology 1983, 30, 655.
73. Burrough, P. A. Nature 1981, 294, 240.
74. Pfeifer, P.; Avnir, D. J. Chem. Phys. 1983, 79, 3558.
75. Holdren, G. R.; Speyer, P. Geochim. Cosmochim. Acta 1985, 49, 675.
76. Grandstaff, D. E. Geochim. Cosmochim. Acta 1978, 42, 1899.
77. Walter, L. M.; Moorse, J. Geochim. Cosmochim. Acta 1985, 49, 1503.
78. White, A. F.; Claassen, H. Chem. Geology 1980, 28, 91.
79. Lagache, M. Geochim. Cosmochim. Acta 1976, 40, 157.
80. Aagaard, P.; Helgeson, H. Amer. J. Sci. 1982, 282, 237.

81. Lasaga, A. C. In Kinetics of Geochemical Processes; Lasaga, A. C.; Kirkpatrick, R. J., Eds.; Reviews in Mineralogy, 1981, 8, 135-168.
82. Brantley, S. L.; Crane, S.; Crerar, D.; Hellmann, R.; Stallard, R. Geochim. Cosmochim. Acta 1986, 50, 2349.
83. Blum, A. E.; Lasaga, A. In Aquatic Surface Chemistry; Stumm, W., Ed.; Wiley and Sons: New York, 1987.
84. Lasaga, A. C.; Blum, A. Geochim. Cosmochim. Acta. 1986, 50, 2363.
85. Buseck, P. R.; Veblen, D. Geochim. Cosmochim. Acta 1978, 42, 669.
86. Wintsch, R. P.; Dunning, J. J. Geophys. Res. 1985, 90, 3649.
87. White, S. Contrib. Mineral. Petrol. 1975, 50, 287.
88. Knipe, R. J.; Wintsch, R. Adv. Phys. Geochem. 1985, 4, 325.
89. Sacerdoti, M.; Labernardiere, H.; Gandais, M. Bull. Mineral. 1980, 103, 148.
90. Buiskool Toxopeus, J. M.; Boland, J. Tectonophys. 1976, 32, 209.
91. Olsen, D. S.; Kohlstedt, D. Phys. Chem. Mineral. 1984, 11, 153.
92. Brantley, S. L.; Schott, J.; Guy, C.; Crerar, D. EOS 1987, 68, 1490.
93. Wegner, M. W.; Christie, J. Contrib. Mineral. Petrol. 1976, 59, 1976.
94. Casey, W. H.; Carr, M.; Graham, R. Geochim. Cosmochim. Acta 1988, 52, 1545.
95. Murr, L. E.; Hiskey, J. Metall. Trans. 1981, 12B, 255.
96. Lasaga, A. C. J. Geophys. Res. 1984, 89, 4009.
97. Helgeson, H. C.; Murphy, W. Math. Geol. 1984, 15, 109.
98. Hurd, D. C. Geochim. Cosmochim. Acta 1973, 37, 2257.
99. Claassen, H. C.; White, A. In ACS Symposium Series No. 93, Jenne, E. A., Ed.; American Chemical Society: Washington, DC, 1979, p 771.
100. Paces, T. Geochim. Cosmochim. Acta. 1983, 47, 1855.
101. Delany, J. M. Livermore Natl. Lab. Report UCRL-53631, 1985.
102. Velbel, M. A. Amer J. Sci. 1985, 285, 904.
103. White, A. F. Proc. 5th Internat. Symp. Water-Rock Reaction, 1986, p 626.
104. Bruton, C. J. Proc. Lawerence Livermore Nat. Lab. Workshop on Geochemical Models, 1986.
105. Gislason, S. R.; Eugster, H. Geochim. Cosmochim. Acta 1987, 51, 2827.
106. Liu, C. W. Ph.D. Thesis, University of California, Berkeley, California, 1987.
107. Herman, J. S.; Lorah, M. Chem. Geol. 1987, 62, 251.

RECEIVED August 24, 1989

# ADVANCEMENTS IN MODELING: ORGANIC COMPOUNDS

# Chapter 36

# Quantitative Structure—Activity Relationship Models for Predicting Aqueous Solubility

## Comparison of Three Major Approaches

**Nagamany N. Nirmalakhandan and Richard E. Speece**

**Environmental and Water Resources Engineering, Vanderbilt University, Nashville, TN 37235**

> Three major approaches to the prediction of aqueous solubility of organic chemicals using Quantitative Structure Activity Relationship (QSAR) techniques are reviewed. The rationale behind six QSAR models derived from these three approaches, and the quality of their fit to the experimental data are summarized. Their utility and predictive ability are examined and compared on a common basis. Three of the models employed octanol-water partition coefficient as the primary descriptor, while two others used the solvatochromic parameters. The sixth model utilized a combination of connectivity indexes and a modified polarizability parameter. Considering the ease of usage, predictive ability, and the range of applicability, the model derived from the connectivity- polarizability approach appears to have greater utility value.

Several excellent QSAR models for predicting aqueous solubility have been proposed during the past few years. Many of them covered small sets of selected classes of congeneric compounds, while few covered a wide variety of compounds. The first major attempt in developing QSAR model for aqueous solubility was by Hansch et al ([1](#)), who used the octanol-water partition coefficient, p, to derive a simple linear equation for the solubility of 156 organic liquid solutes. Since then, several semi-theoretical and empirical models using log p have been reported, though for smaller numbers and particular classes of compounds ([2-5](#)). The second approach, developed by Kamlet and co-workers ([6,7](#)), was of a more fundamental nature. Known as the Linear Solvation Energy Relationships (LSER), this approach uses solvatochromic parameters to model the solute-solvent interactions in the solution process. The third approach, developed by the current authors ([8-10](#)), uses molecular connectivity indexes, $\chi$, and a modified polarizability parameter, $\underline{\Phi}$, to model the solute-solvent interactions.

In terms of statistical considerations, these models have been reported to be very strong as shown by their high regression coefficient, r. However, except in a very few cases, the utility value and the predictive ability of many of these models have not been demonstrated by their respective authors. Such features of QSAR models should be made available so that an appropriate model could be selected by end-users depending on ease of usage, reliability, degree of accuracy required, and their own expertise. This paper is an attempt in providing a comparative analysis of six important models for solubility developed from the three major approaches as reported by four groups of workers. These six models were selected for evaluation because all of them employed 100 or more compounds in their training sets and reported $r > 0.95$ with standard error $< 0.3$, which make all of them strong candidates for many practical applications.

## THE THREE MAJOR APPROACHES

In this section, the reasoning behind each of the three approaches is outlined, and the models derived from the respective approaches are presented.

### The log p Approach.

The rationale for this approach was based on the similarity between the dissolution of an organic solute in water, and its partitioning between two solvents. Thus, the equilibrium between an organic solute and its saturated aqueous solution was thought to be similar to that of the partitioning of the solute between itself and water, and a linear relationship between log S (S = solubility in moles/lit) and log p was sought. This study by Hansch et al (1) reported the following QSAR model:

[Model 1]:
for Aliphatic and Aromatic liquid solutes
$$\log (1/S) = 1.339 \log p - 0.978 \tag{1}$$
n = 156; r = 0.935.

where n is the number of solutes evaluated. A thermodynamic justification for this model has been presented by Hansch et al (1). By considering the chemical potential of the solute and ignoring non-ideality, they derived a theoretical equation of the form:

$$\log (1/S) = \log p + [(\mu^o{}_{(oct)} - \mu_{(l)})/2.303 \cdot RT] \tag{2}$$

where $\mu^o{}_{(oct)}$ is the chemical potential of the solute in one mole of ideal octanol solution, and $\mu_{(l)}$ is that of the pure liquid solute. Reasonable agreement between equations (1) and (2) can be seen in their form and the coefficient and sign of the log p term.

Following the approach introduced by Hansch et al, Yalkowski and Valvani (11) and Yalkowski et al (12) developed a semi-theoretical equation to cover both liquid and solid solutes, using an entropy of fusion term, $\Delta S_f$, and a melting point term, MP:

[Model 2]:
for Aliphatic and Aromatic liquid and solid solutes
$$\log S = - \log p - 1.11 \Delta S_f \cdot (MP-25)/1364 + 0.54 \tag{3}$$
n = 167; r = 0.994.

[Model 3]:
for Aromatic liquid and solid solutes
$$\log S = -0.944 \log p - 0.01 MP + 0.323 \tag{4}$$
n = 164; r = 0.977.

The coefficients of the log p term in the above two models, are in very good agreement with that of the theoretically derived equation (equation 2).

### The LSER Approach.

Kamlet et al (6,7), who introduced and developed this approach, use a linear combination of three energy terms to model solubility related properties, SP, in solute-solvent systems. The parameters used to quantify these energy terms are called the solvatochromic parameters. The first energy term, called the 'cavity term', is a measure of the free energy necessary to separate the solvent molecules (by overcoming the solvent-solvent interactions) to provide a suitably sized cavity for the solute. The second term, called the 'dipolar term', is a measure of the exoergic energy associated with the solute-solvent interactions (e.g. dipole-dipole). The third term, 'hydrogen bonding term' accounts for the exoergic effects of complexation between systems capable of taking part in hydrogen bonding.

Using appropriate solvatochromic parameters, these three energy terms are modeled to relate to solubility related property, SP of various solutes in a given solvent (e.g. aqueous solubility):

$$SP = SP_0 + m\ V_I/100 + s\ \pi^* + a\ \alpha_m + b\ \beta_m \qquad (5)$$

where, $SP_0$, m, s, a and b are the constant coefficients, and $V_I$, $\pi^*$, $\alpha_m$, and $\beta_m$ are the solvatochromic parameters relating to the solute.

The LSER approach has been very successfully applied by Kamlet et al to model many physicochemical properties and biological activities. Their model for aqueous solubility was (6,7):

[Model 4]:
for Aliphatic liquid solutes
$$\log S = 0.05 - 5.85\ V_I/100 + 1.09\ \pi^* + 5.23\ \beta_m \qquad (6)$$
n = 115; r = 0.994.

[Model 5]:
for Aromatic liquid and solid solutes
$$\log S = 0.57 - 5.58\ V_I/100 + 3.85\ \beta_m - 0.011(MP-25) \qquad (7)$$
n = 70; r = 0.991.

The quality of fit of these LSER models is excellent and the error is comparable to the uncertainties in experimental methods of determining solubility, and has been claimed to have "reached the level of exhaustive fit" (6).

The Connectivity-Polarizability Approach.

In this approach, the solute-solvent interactions are modeled using polarizability and the molar volume of the solute. Polarizability, $\Phi$, is in turn modeled by Ketelaar's method (13), where an atomic contribution scheme is employed. Molar volume is in turn modeled by molecular connectivity indices, $\chi$, which are calculated using slightly modified algorithms (9), originally proposed by Kier and Hall (14,15). These indices encode information on the molecular topology and its heteroatom content. They have been shown to correlate well with the solutes' molar volume, and polarizability (14,15). Since polarizability information is duplicated by $\chi$ and $\Phi$, in this approach, a combination of these two parameters is used to model aqueous solubility, by optimizing the atomic contributions to $\Phi$, and deriving a modified polarizability parameter, $\underline{\Phi}$. The basic QSAR model derived on this basis was (8):

[Model 6]:
for Aliphatic and Aromatic liquid and solid solutes
$$\log S = 1.465 + 1.758\ ^0\chi - 1.465\ ^0\chi^v + 1.01\ \underline{\Phi} \qquad (8)$$
n = 145; r = 0.975.

where $\underline{\Phi}$ = - 0.963 (N° of Cl) - 0.361 (N° of H) - 0.767 (N° of Double Bonds). The above model has been verified on many testing sets of miscellaneous compounds, and the model fitted a total of 470 compounds including ethers, esters, PCBs, PNAs, PCDDs, etc. with an r of 0.99 and standard error of 0.33 (8-10).

## COMPARISON BETWEEN THE SIX MODELS

An overall summary of the above six models is shown in Table I. In this Table, the descriptors are classified into four "types"- experimental, assigned, estimated and calculated. In some models, descriptors are assigned numerical values depending on the solute, which if used incorrectly, could lead to erroneous results. The difference between "estimated" and "calculated" is that an error may be associated with the estimated value while rigid algorithms are used in determining the "calculated" values" yielding firm values. In multiple regression analyses, if the basic assumption of error-free independent variables is violated, invalid models may result. Therefore, models using "calculated" descriptors would be more preferable to those using experimental, assigned or estimated ones.

Statistical Qualities Of The Models.

Considering the unexplained variance, UV, in the experimental data, the goodness of fit is seen to be superior (UV < 5%) in Models 2, 3, 4, 5 and 6. The variance in Model 1 (UV > 10%), though weakest, is quite remarkable in that only one descriptor was used to cover 156 compounds, while the other models employ 2 or more descriptors. Adjusted $r^2$, is a statistical indicator which can be used to compare data sets of different numbers of chemicals and descriptors on an equitable basis. Models 2 and 5, which rank high on this basis, need an experimental input, the errors of which might affect not only the accuracy of the result, but also, the statistical validity of the model itself. The last two columns of Table I give a direct indication of the reliability of the different models. They show the number and percent of compounds for which the error in fitting is greater than 0.3 log units (i.e. factor of 2). On this basis of comparison, the LSER approach (Models 4 and 5) ranks very high, while the log p approach for the aromatic compounds (Model 3) appears to be the least reliable.

Utility Value Of The Models.

The utility value of a QSAR model depends primarily on the nature of the descriptors used in developing the model. The availability of the descriptors, ease of calculation, accuracy or consistency of their values, applicability to new compounds, and ability to represent the structural and atomic variations in the chemicals are some of the relevant factors to be considered in comparing different descriptors. Based on the above factors, and considering the fact that the connectivity-polarizability approach does not require any experimental data, Model 6 appears to be the most desirable, along with the log p approach to a lesser degree. In the connectivity-polarizability approach, simple and rigid algorithms are used which can be applied to all classes of chemicals. The major limitation of the parameters in this approach is that they can not differentiate between isomeric members of certain congeneric serieses.

Methods for estimating log p have been firmly established and fragment constants and substituent factors are available for most atomic combinations. However, these estimation methods ignore effects of substituent interactions, and corrections for such effects are not currently available. Thus, estimated log p can be used confidently for compounds containing mono-functional substitutions, while for others with mixed substitution, the results may be questionable. Because of this, the developers of these estimation methods have recommended that experimental log p values, rather than estimated ones, be used wherever possible (16). However, due to the nonavailability of experimental values, many QSAR workers continue to use the estimated ones. In fact, in deriving Model 1, Hansch et al (1) used estimated log p values for 133 compounds and experimental log p values for only 23 compounds. Yalkowski and Valvani (11,12) used estimated log p values for all compounds in deriving their two Models.

In the LSER approach, values have been assigned for some compounds; for others, rules are becoming available for estimating the parameters. In many cases, established values are not yet available, severely limiting the utility of this otherwise attractive model. The solvatochromic parameters are in an evolving stage, and currently, established values for about 600 chemicals are believed to be available. The rules for their estimation are far from being rigid or complete, and many exceptions and special cases have to be taken into account, which demands considerable expertise and insights. The practical implementation of this approach would need a relatively large database of

TABLE I. Summary of Solubility Models Derived From Three Major Approaches

| MODEL Study By (Year) | Chemicals in Data Set | | Type and Number of Descriptors Used in the Model * | | | | | Statistics of the Model | | | Range ** of log S | Predictions where error > 0.3 | |
|---|---|---|---|---|---|---|---|---|---|---|---|---|---|
| | Nº | Type | Ex | As | Es | Ca | TOTAL | Corr. Coeff. r | Adjusted r squared | Standard Error | | Nº | % |
| MODEL 1 Hansch et al, (1968) | 156 | Mixed Liquids | - | - | 1 (log p) | - | 1 | 0.935 | 0.873 § | n/r | 6 | 47 | 30 |
| MODEL 2 Yalkowski and Valvani, (1980) | 167 | Mixed Solids & Liquids | 1 (MP) | 1 (ΔSf) | 1 (log p) | - | 3 | 0.994 | 0.987 § | 0.242 | 8 | 43 | 26 |
| MODEL 3 Yalkowski et al, (1983) | 164 | Aromatic Solids & Liquids | 1 (MP) | - | 1 (log p) | - | 2 | 0.977 | 0.953 § | n/r | 11 | 94 | 57 |
| MODEL 4 Kamlet et al, (1986) | 115 | Aliphatic Liquids | - | 3 (SCP) | - | - | 3 | 0.994 | 0.987 § | 0.153 | 8 | 8 | 7 |
| MODEL 5 Kamlet et al, (1986) | 70 | Aromatic Solids & Liquids | 1 (MP) | 2 (SCP) | - | - | 3 | 0.992 | 0.979 § | 0.216 | 8 | 12 | 17 |
| MODEL 6 Nirmalakhandan and Speece, (1988) | 470 | Mixed Solids & Liquids | - | - | - | 3 oX,oXv,ø | 3 | 0.990 | 0.976 | 0.327 | 12 | 122 | 25 |

\* Types of Descriptors: Ex- Experimental; As- Assigned; Es- Estimated; Ca- Calculated
MP- Melting Point; p- Octanol/water partition coefficient; SCP- Solvato chromic parameters;
ΔSf- Entropy of fusion; oX,oXv- molecular connectivity indexes; Ø - polarizability parameter.
\*\* Range shown in orders of magnitude;   n/r - Not reported in original study;   § Not reported in original study, but calculated in this study.

parameters as well as rigid parameter estimation rules. However, if established values or consistent rules for their estimation become available, log S could be calculated using hand-calculators, whereas, all the other approaches would need a computer program, with some form of graphic input capability to describe the structure.

Range Of Applicability

Considering the heterogeneity of the training set, Model 6, (connectivity-polarizability approach) covering the largest number of aliphatic and aromatic solid and liquid solutes with just one equation, appears to be very broad-based. In terms of the range of numerical values of S covered, Model 6 again ranks high, covering over 12 orders of magnitude. The LSER approach requires two markedly different equations for the aliphatic and aromatic compounds. However, within each class, the LSER models are the only ones reported containing amines, nitro compounds, etc. in the training set. Even though log p can be used to predict solubility for aliphatic and smaller aromatic molecules, as shown by Models 1 and 2, a separate equation, as shown by Model 3, fits the larger aromatic molecules with substitutions better.

Physical Significance Of The Descriptors.

When comparing different QSAR models, one of the important points to be considered is the physical significance of the descriptors. Among those used in QSAR models for solubility, the solvatochromic parameters seem to be the most fundamental, and physically significant. They are useful in understanding the solution process (at a molecular level) by identifying and resolving it into three steps. Such knowledge could be beneficial in understanding other solute-solvent related phenomena (e.g. solubility in blood). A significant feature of this approach is that, for the first time, solubility has been resolved quantitatively into more fundamental properties of the solute. Further, since the solvatochromic parameters ( $V_I/100$, $\pi^*$, $\beta_m$) are scaled to be roughly the same order of magnitude, the relative importance of the appropriate parameter in governing solubility is clearly shown by their respective coefficients in these equations. The fact that the same solvatochromic parameters are used to model a variety of solubility related properties shows that this approach is sound and fundamentally very strong.

The connectivity indices and the polarizability parameters, however, relate a solutes solubility directly to its molecular structure, and thus could be more useful in the design and evaluation of new chemicals. A particular drawback of the polarizability parameter used here is that, unlike the LSER descriptors, it is not universally applicable to all solute-solvent interactions. It has to be defined and optimized for each property being studied. The log p descriptor is purely empirical, and does not portray any direct mechanistic significance in relation to the solutes molecular structure. Further, since Model 1 is significantly improved by including melting point data, it can be noted that log p alone does not encode sufficient information relating aqueous solubility.

Predictive Ability Of The Models.

The applicability of the models to estimate aqueous solubility of new compounds is one of the important points to check when comparing different models. In reality, the models can be verified by testing on existing compounds which were not included in deriving the original model. In this regard, the predictive ability of the connectivity-polarizability model and the flexibility of the approach in accommodating new compounds have been amply demonstrated (8-10). One of the main features of this approach is that compounds with mixed substitutions and structures could also be satisfactorily modeled, provided the substituents and structures are adequately represented individually in the training set.

To evaluate the predictive ability of the models discussed above, we have assembled a testing set of ten compounds not included in the original training sets, but similar to them in structure and heteroatom content. The performance of each of the three approaches could thus be compared on a common basis. The rationale for picking these particular ten compounds is as follows: the first three are alcohols, which have been adequately represented in the training sets of all three approaches. The

TABLE II. Results of Predictive Tests on Six Models

| Compound | Exp. log S * | Log p Approach ||||||| LSER Approach ||||| Con-Pol Approach ||
| | | Model 1 || Model 2 || Model 3 || Model 4 || Model 5 || Model 6 ||
| | | Pred. log S * | Error | Pred. log S * | Error | Pred. log S * | Error | Pred. log S * | Error | Pred. log S * | Error | Pred. log S * | Error |
|---|---|---|---|---|---|---|---|---|---|---|---|---|---|
| 1,1 Diethyl pentanol | -2.42 | -2.82 | 0.40 | -2.30 | -0.12 | n/a | n/a | -1.94 | -0.48 | n/a | n/a | -2.47 | 0.05 |
| 3,5,5 Trimethyl hexanol | -2.50 | -2.43 | -0.07 | -2.01 | -0.49 | n/a | n/a | -2.14 | -0.36 | n/a | n/a | -2.37 | -0.14 |
| 2,6 Dimethyl 3 heptanol | -2.55 | -2.61 | 0.06 | -2.14 | -0.41 | n/a | n/a | -2.23 | -0.32 | n/a | n/a | -2.38 | -0.17 |
| 2 Bromo ethyl acetate | -0.63 | -0.41 | -0.22 | -0.42 | -0.21 | n/a | n/a | -1.18 | 0.55 | n/a | n/a | -0.85 | 0.22 |
| 2 Chloro ethyl acetate | -0.61 | -0.37 | -0.24 | -0.51 | -0.10 | n/a | n/a | -1.28 | 0.67 | n/a | n/a | -0.68 | 0.07 |
| 2, 6 Cl. biphenyl | -5.63 | n/a | n/a | -4.73 | -0.90 | -4.67 | -0.96 | | | -5.47 | -0.16 | -5.98 | 0.34 |
| 2, 2', 4, 5' Cl. biphenyl | -7.25 | n/a | n/a | -7.17 | -0.08 | -6.84 | -0.41 | | | -6.77 | -0.48 | -7.05 | -0.21 |
| 2, 2', 3, 3', 6, 6' Cl biphenyl | -7.78 | n/a | n/a | -7.44 | -0.34 | -7.05 | -0.73 | | | -8.56 | 0.78 | -8.12 | 0.34 |
| 2, 2', 3, 3', 4, 4', 6 Cl. biphenyl | -8.26 | n/a | n/a | -7.61 | -0.65 | -7.20 | -1.06 | | | -9.15 | 0.89 | -8.65 | 0.39 |
| 2, 2', 3, 3', 4, 5, 5', 6, 6' Cl. biphenyl | -9.42 | n/a | n/a | -9.89 | 0.47 | -9.20 | -0.22 | | | -10.82 | 1.40 | -9.71 | 0.29 |
| Average error = | | | -0.01 | | -0.28 | | -0.68 | | 0.01 | | 0.49 | | 0.12 |
| Absolute average error = | | | 0.20 | | 0.38 | | 0.68 | | 0.48 | | 0.74 | | 0.22 |

* S in Moles/L; n/a- Not applicable

next two compounds are halogenated esters which carry combination of structures, atoms and functional groups represented in the training sets of all three approaches. Finally, the five PCBs picked represent typical compounds of environmental concern, a class for which there is a severe lack of data. The results of this predictive test are shown in Table II.

While this test is not expected and cannot be considered to be the ultimate test for a universal solubility model, it is hoped that this would help in evaluating the different approaches and in identifying future research areas in this and related topics. Models 2 and 6 are applicable to all the ten compounds selected while the other four models are applicable only to sub-groups. Of the former two, Model 6 can be seen to predict reasonably well with the error averaging 0.12 log units. Models 3 and 5 performed poorly for all the compounds tested. Model 1 predicted very well for the five compounds tested. Although the average error of Model 4 was 0.01, its predictions were inconsistent.

CONCLUSIONS

The six models discussed here have their own merits and demerits. From the analysis reported above, the following general conclusions can be drawn. Model 1, based on the log p approach appears to be the simplest to use if either experimental or estimated log p values are readily available. In the absence of experimental log p data, one has to accept the result with a high degree of uncertainty, because, estimated log p values can introduce additional errors. If log p and melting point data are available, then Models 2 and 3 could yield more accurate results, particularly if the solute is a solid at room temperature. The LSER approach (Models 4 and 5) has great potential in predicting solubility, but at this point of time, its practical applicability is limited by the non-availability of appropriate parameters or a firm set of rules for their estimation. If further research is focused in rectifying these shortcomings, and if larger data sets could be tested, the LSER approach would probably emerge as the most suitable tool for predicting aqueous solubility. Model 6, based on the connectivity-polarizability approach, has the advantage of requiring neither experimental data or in-depth knowledge of solute-solvent interaction parameters. Further, the wide coverage and the good quality of fit of this model implies greater confidence in its predictive ability.

Literature Cited

1. Hansch et al, J. Org. Chem., 1968, 33, 347-350.
2. Miller et al, J. Chem. Eng. Data, 1984, 29, 184-190.
3. Mackay et al, Chemosphere, 1980, 9, 701-711.
4. Baker et al, Phys. Chem. Liq., 1987, 16, 279-292.
5. Baker et al, Quant. Struct. Act. Relat., 1984, 3, 10-16.
6. Kamlet et al, Jour. Pharm Sci., 1986, 75, 338-349.
7. Kamlet et al, Jour. Phy Chem., 1987, 91, 1996-2004.
8. Nirmalakhandan, N.; Speece, R. E., Env. Sci.& Technol.,1988, 22,328-338.
9. Nirmalakhandan, N. Ph.D.Thesis, Drexel University, Philadelphia, 1988.
10. Nirmalakhandan, N.; Speece, R. E., Env. Sci.& Technol., 1989, 23, 708-713.
11. Yalkowsky, S. H.; Valvani, S. C., Jour. Pharm. Sci.,1980,69, 912-922.
12. Yalkowsky et al, Residue Review, 1983, 85, 43-55.
13. In Horvath A. L. Halogenated Hydrocarbons, Marcel Dekker, Inc., NY 1982.
14. Kier, M. J.; Hall, L. H., Molecular Connectivity in Chemistry and Drug Design, Academic Press, NY 1976.
15. Kier, M. J.; Hall, L. H., Molecular Connectivity in Structure Activity Analysis, Research Studies Press Ltd., England, 1986.
16. Leo et al, J. Med. Chem. 1975, 18, 865-868.

RECEIVED August 24, 1989

# Chapter 37

# Equilibrium Model for Organic Materials in Water

## Frank R. Groves, Jr., and Majd El-Zoobi

### Chemical Engineering Department, Louisiana State University, Baton Rouge, LA 70803

> This chapter reviews thermodynamic principles governing aqueous solubility and Henry's law constants for mixtures of sparingly organic compounds. Special consideration is given to the effect of completely miscible cosolvents such as methanol and ethanol. A new approximate method of predicting cosolvent effects is presented. The results should be useful in supplying necessary phase equilibrium data to complex computer programs for modeling transport and fate of sparingly soluble organics in the environment.

Release of organic material to the environment presents a growing problem for environmental protection. Leaks of gasoline or solvents from underground storage tanks is one example of this situation. Contamination of ground water by organics from chemical manufacturing sites is another instance of current concern.

Phase equilibrium considerations, specifically solubility in water and partial pressure of organic compounds in air over a water solution, are key factors in assessing transport and fate of these materials. This chapter reviews thermodynamic principles governing phase equilibrium with special emphasis on environmental problems. Emphasis is placed on aqueous solubility and Henry's law constants for mixtures of sparingly soluble organic liquids. Special consideration is given to the effect of completely miscible cosolvents and a new method for prediction of cosolvent partition coefficients is presented. A brief treatment of solubility of organic solids is included. The discussion concludes with a summary of effects of emulsification and humic acids on organic solubility.

## SOLUBILITY AND HENRY'S LAW CONSTANTS FOR ORGANIC LIQUIDS IN WATER

### Basis Thermodynamic Equations

Consider an organic liquid phase (H) in equilibrium with an aqueous phase (W). The basic thermodynamic law governing the phase equilibrium is the equality of fugacities of all components in the two phases.

$$\gamma_i^H = \gamma_i^W, \qquad i = 1, ..., N \qquad (1)$$

In terms of activity coefficients, $\gamma_i$ and mole fractions, $x_i$, this can be written:

$$\gamma_i^H x_i^H = \gamma_i^W x_i^W, \qquad i = 1, ..., N \qquad (2)$$

Thus the partition coefficient, $K_i$, defined as $x_i^W/x_i^H$, is related to the activity coefficients.

$$K_i = \gamma_i^H/\gamma_i^W \qquad (3)$$

0097–6156/90/0416–0486$06.00/0
© 1990 American Chemical Society

At a fixed temperature the activity coefficients are functions of solution composition. Their composition dependence can be expressed via various empirical equations such as the familiar Van Laar equations or the theoretically well grounded UNIQUAC equations (1).

Using one of these activity coefficient equations it is possible to calculate liquid-liquid equilibrium (LLE) behavior of multicomponent liquid systems. Consider, for example, the ternary system of Figure 1. A system of overall composition A splits into two liquid phases B and C. The calculation of compositions of B and C is analogous to the flash calculation of vapor-liquid equilibrium problems. By using the UNIQUAC equations to obtain the partition coefficients, $K_i^L$, this problem can be solved for any composition A of the overall system. The calculations are lengthy but computer programs for this purpose (2) have been published. In this paper simpler approximate methods for phase equilibrium problems of environmental interest is sought. For the moment it is sufficient to note that the activity coefficients provide the means of complete liquid-liquid equilibrium computations.

It is also of interest to compute the partial pressure of an organic material in air (A) in equilibrium with a liquid (L) solution. At ambient pressure, equation (1) reduces to:

$$p_i = \gamma_i^L x_i^L P_i^* \tag{4}$$

where $p_i$ is the partial pressure of substance i in the air and $p_i^*$ = vapor pressure of i at the solution temperature. In terms of a Henry's law constant, $H_i$ :

$$p_i = H_i x_i^L \quad \text{where} \quad H_i = \gamma_i^L P_i^* \tag{5}$$

If the liquid composition and pure component vapor pressure are known, the activity coefficient can be obtained from a correlating equation and the Henry's law constant can be calculated. Since Henry's law applies to dilute solutions, it is sufficient to evaluate the activity coefficient in infinite dilution, $x_i^L \to 0$.

All of the activity correlating equations contain empirical constants. The UNIQUAC equations have two constants for each pair of components in a multicomponent system. For liquid-liquid equilibrium purposes these constants can be determined from binary mutual solubility and ternary tie line data. Anderson and Prausnitz (3) show that constants determined in this way can be used to predict approximately the LLE behavior of multicomponent systems.

Simplified Equations for Sparingly Soluble Liquids

Many organic materials of environmental interest exhibit low mutual solubilities with water. This permits simplifications in the governing thermodynamic equations. For pure hydrocarbon component i in contact with water, equation (2) can be solved for the solubility, $x_{is}^W$:

$$x_{is}^W = (\gamma_i^H x_i^H)/\gamma_i^W \cong 1/\gamma_i^W \tag{6}$$

where we have recognized that the hydrocarbon phase is essentially pure hydrocarbon so that $x_i^H \cong 1.0$ and $\gamma_i^H \cong 1.0$. Since pure hydrocarbon solubilities are often known or easily determined, equation (6) is more often used to calculate the activity coefficient of the hydrocarbon in the saturated aqueous solution.

The activity coefficient is also useful in predicting the Henry's law constant, $H_i$. Applying equation (5) to hydrocarbon component i:

$$H_i = \gamma_i^W P_i^* = (1/x_{is}^W)/P_i^* \tag{7}$$

Mackay and Shiu, (4), use this method to calculate Henry's Law constants for a large number of organic compounds. They show that the results are in satisfactory agreement with experimental data.

Another use of the saturated solution coefficients is the prediction of solubilities of sparingly soluble hydrocarbon mixtures in water. Applying equation (6) to hydrocarbon component i of the mixture:

$$x_{is}^W = (\gamma_i^H x_i^H)/\gamma_i^W \tag{8}$$

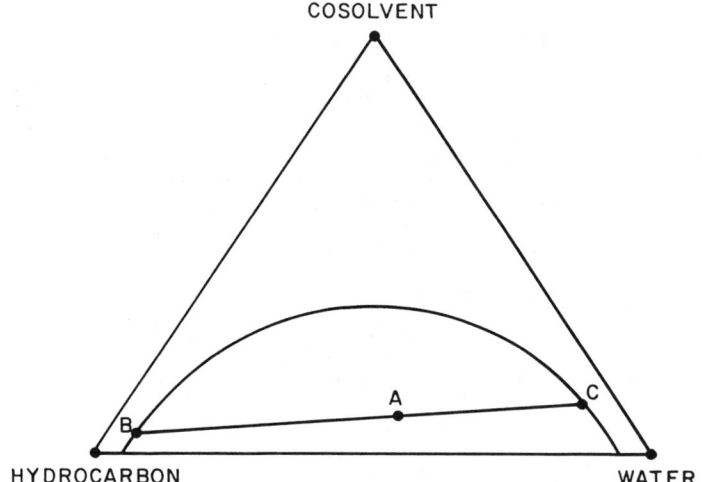

Figure 1. Equilibrium Phase Separation

Since the solubility of the hydrocarbons is small, interactions between hydrocarbon components in the aqueous phase will be small. Thus the hydrocarbon activity coefficients, $\gamma_i^W$, in the aqueous phase should be approximately the same as in an aqueous solution saturated with the pure hydrocarbon and equal to the reciprocal of the pure component solubility. The hydrocarbon phase composition is assumed to be known so the activity coefficient $\gamma_i^H$ can be obtained from one of the empirical correlation equations, neglecting the small water content of the phase.

Leinonen and Mackay (5) used this method with the Redlich-Kister equation for the hydrocarbon phase activity coefficients to predict mixture solubilities. Agreement with experiment was good for mixtures of benzene with n-hexane, 1-hexane, and 2-methylpentane and for cyclohexane mixtures with the same substances.

Bannerjee (6) determined the solubilities of mixtures of several chlorobenzenes with one another and with toluene and benzyl alcohol. The results agreed satisfactorily with predictions from equation (8). The water phase activity coefficients were obtained from the pure component solubilities via equation (6). The organic phase activity coefficients for mixtures of chlorobenzenes with hydrocarbons were predicted by the UNIFAC group contribution method. Mixtures containing only chlorobenzenes were essentially ideal, $\gamma_i^H = 1.0$. When benzyl alcohol was used as a cosolute, the agreement with equation (8) was improved by using UNIFAC to predict the activity coefficient in the aqueous phase.

Burris and MacIntyre (7) measured solubilities of several liquid hydrocarbon mixtures: n-octane/methylcyclohexane, n-octane/1,4-dimethylnaphthalene, n-octane/naphthalene, 1-methylnaphthalene/n-octane, 1-methlnaphthalene/methylcyclohexane, 1-methylnaphthalene/1,4-dimethylnaphthalene, and 1-methylnaphthalene/naphthalene. For structurally similar materials the mixtures behaved ideally and hydrocarbon phase activity coefficients could be taken as unity yielding good agreement with equation (8). For alkane-aromatic mixtures, the assumption of unit activity coefficient failed. When one of the pure components was a solid an estimated supercooled liquid solubility was used to compute the water phase activity coefficient.

Effects of Cosolvents

An important question is the effect of a cosolvent completely miscible with water on the solubility of a sparingly soluble material. An example is the effect on hydrocarbon solubility of methanol or ethanol added to gasoline as an octane enhancer. Equation (8) still applies but the key question is the effect of the cosolvent on the activity coefficient, $\gamma_i^W$, of the hydrocarbon in the water phase. We now outline a simplified approach (8) to solving this problem.

Consider a hydrocarbon mixture of known composition containing a known fraction of a cosolvent (CS). The problem is to determine the hydrocarbon solubility in water in the presence of the cosolvent. Using an experimental or predicted partition coefficient, $K_{cs}$, for the cosolvent the equilibrium mole fraction of cosolvent in the aqueous phase can be obtained.

$$x_{cs}^W = K_{cs} x_{cs}^H \qquad (9)$$

Then a correlating equation; e.g., UNIQUAC, can be used to compute the activity coefficient, $\gamma_i^W$, for the hydrocarbon in the water phase. In this calculation the hydrocarbon is assumed to be infinitely dilute in the water phase, a reasonable approximation because of its limited solubility. The solubility then follows from equation (8). Table I shows the results of this approximate calculation for the benzene-water-ethanol system. The UNIQUAC equation with the constants of Brandani et al. (9) was used to predict the hydrocarbon activity coefficients.

Prediction of the Cosolvent Partition Coefficient

An important parameter needed in the above calculation is the partition coefficient, $K_{cs}$, of the cosolvent. We now propose an approximate procedure for predicting this parameter. The method is based on the basic equilibrium equation (3) applied to cosolvent:

$$K_{cs} = \gamma_{cs}^H / \gamma_{cs}^W \qquad (10)$$

The two activity coefficients are computed from an activity coefficient correlation. In this computation the small amount of water in the hydrocarbon phase is neglected as is the trace of hydrocarbon in the water phase. Thus the aqueous phase is treated as a mixture of water and completely miscible cosolvent while the organic phase is considered a mixture of hydrocarbon with miscible cosolvent.

The one constant version of the Wilson equation is used to correlate the activity coefficients for the cosolvent. The Wilson equation for a multicomponent mixture can be written as follows: (10).

$$\ln \gamma_k = - \ln \left( \sum_{j=1}^{N} x_j \Lambda_{kj} \right) + 1 - \sum_{i=1}^{N} \left[ x_i \Lambda_{ik} / \sum_{j=1}^{N} x_j \Lambda_{ij} \right] \qquad (11)$$

where

$$\Lambda_{ji} = (v_i / v_j) \exp[-(\lambda_{ij} - \lambda_{jj})/RT] \text{ and } v_i, v_j = \text{molar volume of liquid i,j.} \qquad (12)$$

Tassios (11) proposed to use $\Delta U_j^v$, the energy of vaporization of pure component j, as the energy of interaction $\lambda_{jj}$ for the like pairs of molecules in equation (12). This leaves a single constant, $\lambda_{ij}$, the energy of interaction between the unlike molecules, i and j. Tassios found that this procedure gave satisfactory correlations for vapor-liquid equilibria for a number of systems.

The use of the Wilson equation for liquid-liquid equilibrium is unusual because the equation cannot predict phase separation. In the present application the equation is not being used to predict immiscibility--the organic and water phase are assumed to be immiscible. The Wilson equation is used instead to correlate activity coefficients in the miscible range for the hydrocarbon-cosolvent and water-cosolvent systems.

The single constant Wilson equation was fitted to experimental partition coefficient data for ethanol for the ternary systems benzene-water-ethanol (9) and n-hexane-water-ethanol (12). The partition coefficient was calculated from equation (12) using the Wilson equation to compute the activity coefficients. The constants, $\lambda_{ij}$, were chosen to give a best fit in the least squares sense to the experimental partition coefficients. Table II shows the resulting constants.

The constants were then used to predict partition coefficients for the four component benzene-hexane-water-ethanol system. This prediction was accomplished in the following way. The hydrocarbon phase composition was assumed to be known. The activity coefficient $\gamma_{cs}^H$ was calculated by the Wilson equation. Equation (10) was then solved simultaneously with the partition coefficient equation:

$$x_{cs}^W = K_{cs} x_{cs}^H \qquad (13)$$

to obtain the partition coefficient, $K_{cs}$, and the cosolvent mole fraction in the water phase, $x_{cs}^W$. The equations were solved iteratively starting with an estimate of $x_{cs}^W$. From the estimate a trial value of $K_{cs}$ was obtained from equation (10). A calculated value of $x_{cs}^W$ was then computed from equation (13). The calculations were repeated until the calculated $x_{cs}^W$ matched the estimate. An interval halving convergence scheme provided a reliable and stable iterative approach to the final solution. The solution procedure was started from various initial $x_{cs}^W$ both above and below the correct value. In the examples studied so far the iterations always converged to the same value. There was no evidence of multiple solutions.

Table II compares predictions of $K_{cs}$ for the benzene-hexane-water-ethanol system with the experimental data of El-Zoobi (13). The Wilson constant for the benzene-hexane pair for these calculations was taken from the compilation of Holmes and Van Winkle (14) derived from vapor liquid equilibrium data. The agreement of the prediction with experiment is satisfactory.

Similar predictions could be made using other activity coefficient correlations, e.g., UNIQUAC. The Wilson treatment of the four component system uses only three adjustable constants--for the benzene-ethanol, hexane-ethanol, and ethanol-water systems. The constants were determined by fitting ternary liquid-liquid equilibrium data. The UNIQUAC correlation would require six constants.

SOLUBILITY OF SOLIDS

Pure Solids

The solubility of organic solids is discussed by Prausnitz et al. (10). The basic equilibrium condition is still the equality of fugacities in the two phases, solids (s) and water (w), equation (1). For a pure sparingly soluble organic solid this can be written:

$$f_2^{os} = \gamma_2^W x_2^W f_2^o \qquad (14)$$

where the solid has been taken as component 2, solvent as component 1. The fugacity of the pure solid is denoted by $f_2^{os}$ and $f_2^o$ is the standard state fugacity for component 2. Pure liquid component 2 at the temperature and pressure of the system is chosen for the standard state so:

Table I. Approximate Calculation of Cosolvent Effect

Benzene-Water-Ethanol System

| Mole Fraction Ethanol in Benzene Phase | Water Phase Hydrocarbon Mole Fraction | |
|---|---|---|
| | Predicted | Experimental |
| 0.0056 | 0.00066 | 0.0005 ± 0.00004 a |
| 0.015 | 0.0011 | 0.0010 ± 0.00001 b |
| 0.053 | 0.0035 | 0.0030 ± 0.00003 b |
| 0.123 | 0.0105 | 0.0150 ± 0.00002 b |

(a) Groves (8)  (b) Brandani, et al. (9)


Table II. Prediction of Cosolvent Partition Coefficient Using One Constant Wilson Equation

Benzene(1) - Hexane(2) - Water(3) - Ethanol(4) System

| Hydrocarbon Phase Composition | | | $K_E = x_E^W/x_E^H$ | |
|---|---|---|---|---|
| $x_1^H$ | $x_2^H$ | $x_4^H$ | Predicted | Experimental |
| 0.935 | 0.0 | 0.053 | 2.56 | 2.5 ± 0.2 |
| 0.0 | 0.988 | 0.0112 | 17.3 | 17.5 ± 1.4 |
| 0.540 | 0.457 | 0.0022 | 8.51 | 9.1 ± 0.7 |
| 0.606 | 0.379 | 0.0128 | 5.63 | 6.0 ± 0.5 |
| 0.555 | 0.376 | 0.0588 | 3.16 | 4.0 ± 0.3 |

Wilson Constants: (calories/mole)

$\lambda_{11} = -8098$ a         $\lambda_{14} = -8810$ b
$\lambda_{22} = -7664$ a         $\lambda_{24} = -8815$ b
$\lambda_{33} = -10,552$ a       $\lambda_{34} = -10,435$ b
$\lambda_{44} = -10,162$ a       $\lambda_{12} = -7700$ c

(a) energy of vaporization
(b) fitted to ternary data
(c) Holmes and Van Winkle (14)

$$f_2^\circ = f_2^{ol} \tag{15}$$

where $f_2^{ol}$ is the fugacity of the pure liquid. This is a hypothetical but well defined standard state since the pure material is a solid at the temperature and pressure of the system. Equation (14) can be solved for the solid solubility:

$$x_{2s}^W = (1/\gamma_2^W)(f_2^{os}/f_2^{ol}) \tag{16}$$

Prausnitz, et al. (10) show how the pure component fugacity ratio ($f_2^{os}/f_2^{ol}$) can be estimated from readily available data. Choi and McLaughlin (15) use this approach to correlate experimental data on aromatic hydrocarbon solubilities in various solvents, using the Scatchard-Hildebrand solubility parameter correlation to compute the liquid phase activity coefficient. For aqueous solutions of sparingly soluble hydrocarbons equation (16) is useful in solving for the activity coefficient, $\gamma_2^W$ from a known pure component solubility.

Mixtures of Solids

The solubility of mixtures of solids is a complex question which is discussed by Bannerjee (6). If two solids do not form a solid solution equation (20) applies separately to each. Assuming low solubilities, the unlike molecules should exhibit negligible interaction in the aqueous phase and the water phase activity coefficients, $\gamma_i^W$, should be the same as for pure solid. Under these conditions the solubility of the mixture should be the sum of the (small) solubilities of the pure solids. If the solids form a solid solution, the equilibrium equations take the form:

$$\gamma_i^H x_i^H f_i^{os} = \gamma_i^W x_i^W f_i^{ol} \tag{17}$$

If the solid solution is ideal, $\gamma_i^H = 1.0$, and the solubility can be computed using the pure component solubility to approximate the aqueous phase activity coefficient $\gamma_i^W$, via equation (16).

Practical Considerations for Field Applications

The methods described above yield results expected in the laboratory when a pure organic phase is equilibrated under controlled conditions with pure water. In the field, the solubility may be greater for several reasons.

Formation of Microemulsions

When an organic phase is equilibrated with water in the laboratory care is taken to avoid vigorous shaking which causes formation of emulsions. If shaking is used, various filtration techniques can be employed to eliminate emulsions and obtain reproducible values of the solubility. If these precautions are not observed, the measured solubility will be higher than the true equilibrium value.

In the field, emulsion formation may often be the rule rather than the exception. The presence of small quantities of natural surfactants and the possibility of vigorous mixing as in surface waters will promote emulsification. The resulting organic content of the water phase is difficult to predict since it depends on so many factors. The solubility measured in the laboratory or predicted by the methods discussed above can serve only as a lower limit on the actual organic content of natural waters.

Effect of High Molecular Weight Dissolved Organic Matter

Natural waters frequently contain high molecular weight dissolved organic matter (DOM) - typically humic and fulvic acids derived from decaying plant material. Lower molecular weight organic compounds; e.g. hydrocarbons, can complex with this dissolved organic matter. The resulting hydrocarbon content of the water is then the sum of the solubility as measured under laboratory conditions and the complexed hydrocarbon.

This complexation phenomenon has been studied experimentally and progress has been made on predicting its effect. Chiou et al. (16) describe one approach which correlates the complexed low molecular weight organic material, $S_H^C$, with its solubility in pure water, $S_W$, and the DOM content, X, of the water.

$$S_H^C = K_{dom} \times S_w \tag{18}$$

The apparent solubility of the organic solute, $S_w^*$ then equals the sum of the dissolved solute and the complexed solute.

$$S_w^* = S_w + K_{dom} \times S_w^H \tag{19}$$

This approach yielded an acceptable correlation of solubility for DDT, certain PCB's, trichlorobenzene, and lindane.

The use of this method for field predictions is subject to difficulty since $K_{dom}$ is expected to vary depending on the sources of the dissolved humic and fulvic acids. However, it represents an interesting fundamental approach to the complexation problem. A similar enhancement of apparent solubility occurs via adsorption if finely divided undissolved particulate matter is present in the water.

CONCLUSION

The methods discussed here should be useful is large scale modeling of transport and fate of organic compounds in the environment. For example, if a liquid mixture of light organic materials floats at the top of an aquifer, there will be mass transfer between the organic layer and the aqueous phase. This transport process is a complex phenomenon involving bulk flow of both phases, diffusion, and interaction with the solid rock as well as simple solubility considerations. Lengthy computer programs for modeling transport and fate are being developed. The programs require solubility or partition coefficient data along with other physical constants such as rock permeability, liquid viscosity, and mass transfer coefficients. Because the programs involve integration of nonlinear partial differential equations it is important to minimize the computations required for the physical property data. Thus the simplified approximate methods discussed here offer special advantages. They can be used as short subroutines in the complex transport and fate programs, supplying the necessary partition coefficient data.

ACKNOWLEGEMENTS

This work was supported by the Louisiana State University Hazardous Waste Research Center via a grant from the U.S. Environmental Protection Agency. However, it does not necessarily reflect the views of the agency and no official endorsement should be defined.

LITERATURE CITED

1. Abrams, D.S.; Prausnitz, J.M. AIChE Journal 1975, 21, 116-128.
2. Prausnitz, J.; Anderson, T; Grens, E; Eckert, C.; Hsieh, R; O'Connell, J. Computer Calculations for Multicomponent Vapor-Liquid and Liquid-Liquid Equilibria; Prentice-Hall: Englewoods Cliffs, New Jersey, 1980.
3. Anderson, T.F.; Prausnitz, J.M. Ind. Eng. Chem. Process Des. Dev. 1978, 17, 561-567.
4. Mackay, D; Shiu, W.Y. J. Phys. Chem. Ref. Data 1981, 10, 1175-1199.
5. Leinonen, P.J.; Mackay, D. Canadian Journal of Chemical Engineering 1973, 51, 230-233.
6. Bannerjee, S. Environ. Sci. Technol, 1984, 18, 587-91.
7. Burris, D.R.; MacIntyre, W.G. Environmental Toxicology and Chemistry 1985, 4, 371-377.
8. Groves, F.R. Environ. Sci. Technol, 1988, 22, 282-286.
9. Brandani, V; Chianese, A; Rossi, M.J. Chem. Eng. Data 1985, 30, 27-29.
10. Prausnitz, J.M.; Lichtenthaler, R.N.; Gomez de Azevedo, E. Molecular Thermodynamics of Fluid Phase Equilibria; Second Edition, Prentice Hall; Englewood Cliffs, New Jersey, 1986, p. 259.
11. Tassios, D. AIChE Journal 1971, 17, 1367-1371.
12. Ross, S; Patterson, R.E. J. Chem. Eng. Chem. 1979, 24, 111.
13. El-Zoobi, M. M.S. Thesis, Louisana State University, Baton Rouge (1988).
14. Holmes, M.J.; Van Winkle, M. Ind. Eng. Chem. 1970, 62, 21-31.
15. Choi, P.B.; McLaughlin, E. AIChE Journal 1985, 29, 150.
16. Chiou, C.T.; Malcolm, R.L.; Brinton, T.I.; Kile D.E. Environ. Sci. Technol. 1986, 20, 502-508.

RECEIVED July 20, 1989

## Chapter 38

# Importance of Organic–Inorganic Reactions to Modeling Water–Rock Interactions During Progressive Clastic Diagenesis

### Donald B. MacGowan[1] and Ronald C. Surdam[2]

[1]Enhanced Oil Recovery Institute, Box 3006, University of Wyoming, Laramie, WY 82071
[2]Department of Geology and Geophysics, University of Wyoming, Laramie, WY 82071

> Diagenesis of clastic rocks is a function of the chemical interaction between formation water and rock; thus, the evolution of diagenetically active species in formation waters is important to modeling clastic diagenesis. Several organic-inorganic interactions play key roles in establishing dynamic diagenetic pathways or diagenetic facies of reservoir sands. Carboxylic acid anions (CAA) evolve as a product of microbial action, thermal maturation of sedimentary organic material, abiotic sulfate reduction, and mineral oxidation of dissolved organics. CAA's are important to buffering both formation water Eh and pH and have the capacity to transport Al and Si from aluminosilicate dissolution. At temperatures from about 100 to 140°C, abiotic sulfate reduction of hydrocarbons to hydrogen sulfide and bicarbonate begins to become diagenetically important in deep sedimentary sequences with a sulfate source. $H_2S$ is corrosive to feldspars and carbonates and also provides a sink for Fe via pyrite precipitation. The relative timing of these species' generation allows construction of a generalized organic-inorganic diagenetic model, which predicts generalized diagenetic mineralogy and organic-inorganic reaction pathways.

Recently, the modeling of water rock interactions has become quantitatively possible, using either equilibrium "pathcalc" modeling (cf. 1,2) or a basinal time-temperature history, kinetic approach (3,4). Introduction of the reaction pathway concept has allowed these two approaches to be linked (3). Therefore, to underscore the importance of organic-inorganic interactions during clastic diagenesis and in chemical modeling of water-rock interactions, a review of the nature, origin, distribution, destruction and reactivity of some important aqueous organic species is presented. New data and interpretations on some aspects of organic-inorganic water-rock interactions are included where appropriate. Presented below is a diagenetic model, which is based upon these observations, and which explains the relative timing of diagenetic events observed in reservoir sands the world over (3).

Dissolved organic species have been known to exist in sedimentary basin formation waters since before the turn of the century (5,6,7). A host of aqueous organic species have been identified in sedimentary formation waters including: hydrocarbons, mono-, di- and tri-carboxylic acid anions, keto and hydroxy-acids, amino acids, phenols, cresols, and hydroxybenzoic acids (8,9,10,11,12).

0097–6156/90/0416–0494$06.00/0
© 1990 American Chemical Society

Initially, carboxylic acid anions (CAA) in formation waters were suggested as the precursors to, and possibly the agents of the primary migration of petroleum (13,14) and as proximity indicators of petroleum reservoirs (6,8,14,15). They also have been identified as an important component of formation water alkalinity (16) and have been proposed as the aqueous precursors of natural gas (17). CAA have been shown to be the product of the thermal maturation of kerogen from both field and laboratory studies (11,18,19,20,21,22,23), and possibly the product of mineral oxidation of sedimentary organic material (24). Additionally, CAA have been shown to be the product of abiotic sulfate reduction (25). Several authors have proposed CAA as the ligands for metal transport in forming ore deposits (26,27,28,29,30,31). Short-chain CAA (particularly the difunctional species) have been shown to be capable of transporting Al and Si due to silicate mineral dissolution (11,32,33,34,35). This process has been invoked as a mechanism for developing porosity enhancement due to aluminosilicate mineral dissolution. Their ability to buffer solution pH is orders of magnitude more effective than the carbonate species, thus they exert a major control on the stability of carbonate minerals over the temperature range in which they dominate solution alkalinity (32,36,37,38). 50 to 100% of porosity in some clastic sequences has been attributed to carbonate cement and/or aluminosilicate mineral dissolution (39). In natural waters at pH>5, carboxylic acids exist primarily as their anions, since their $pK_a$'s range from 4.2 to 5.4 (40). These acid anions are present, at times in high concentration, in sedimentary formation waters from petroleum reservoirs, natural gas reservoirs, deep non-petroliferous basins, creek, lake and rain waters, and hot and cold springs (7). Therefore, models of clastic diagenesis must recognize the importance of CAA in buffering formation water pH and in diagenetic mass transfer.

ORGANIC COMPOSITION OF FORMATION WATERS

Concentrations of CAA in formation waters can reach as high as 10,000 ppm acetate (32); however, values of less than 5000 ppm total CAA are more common (35). The monofunctional CAA generally in greatest concentration in formation waters are $C_1$ - $C_5$. The difunctional CAA in greatest concentration are the $C_2$ - $C_4$ anions. o-hydroxybenzoic acid is the hydroxybenzoic acid in highest concentration (11).

Table I shows the maximum reported concentrations of CAA from recent literature. Figure 1 shows the thermal distribution of CAA from recently published analyses. The maximum concentrations are from water samples from basins presently undergoing intense CAA generation and expulsion (1,7,10,16,17,32,35,41,42,43,44,45).

In sedimentary basins currently undergoing intermediate burial diagenesis, concomittant with generation and expulsion of CAA, produced formation water samples will have CAA concentrations similar to those shown in Table I. Diagenetic modeling, and geologic assessment of the reservoir conditions of produced formation water samples demonstrates that this condition is very restricted in time-temperature space (3). Since most produced formation water samples are not taken from this zone, they show CAA concentrations much lower than those in Table I. Diagenetic models which are based upon relict diagenetic fluids, or those which are immature with respect to CAA, will underestimate the importance these species may play during clastic diagenesis.

Sedimentary formation waters typically show a dominance of monofunctional species over difunctional species; also, ethanoate is generally dominant over other monofunctional species in produced formation fluids unaffected by formation fluid mixing or bacterial degradation (35).

The difunctional acid anions have only been recognized to exist in very high concentrations in a few instances. Propanedioic (malonic) and Z-butenedioic (maleic) acid anions and the $C_4$ to $C_{10}$ difunctional CAA have been reported in formation waters (1,32); however, reported concentrations were low. Barth (44) reported the concentration of ethanedioic (oxalic) and propanedioic (malonic) acid anions in formation waters in concentrations up to 38 and 102 ppm respectively. MacGowan and Surdam (35) report concentrations of difunctional CAA in excess of 2500 ppm in California formation waters, although most concentrations of total difunctional CAA are well below 200

Table I.  MAXIMUM REPORTED CONCENTRATIONS OF CARBOXYLIC ACID ANIONS IN PRODUCED FORMATION WATERS

| Acid Anion | PPM | References |
|---|---|---|
| Methanoate (Formate) | 174 | 35 |
| Ethanoate (Acetate) | 10000 | 35 |
| Propanoate (Propionate) | 4400 | 35 |
| Butanoate | 682 | * |
| Pentanoate | 371 | * |
| Hexanoate | 107 | * |
| Heptanoate | 98.7 | * |
| Octanoate | 41.7 | * |
| Ethanedioate (Oxalate) | 494 | 35 |
| Propandioate (Malonate) | 2540 | 35 |
| Butandioate | 63 | 1 |
| Pentandioate | 36 | 1 |
| Hexandioate | 0.5 | 1 |
| Heptandioate | 0.6 | 1 |
| Octandioate | 5.0 | 1 |
| Nonandioate | 6.0 | 1 |
| Decandioate | 1.3 | 1 |
| Cis-Butenedioate (Maleic) | 26 | 35 |
| O-Hydroxybenzoate (Salicylate) | 64.7 | * |

* MacGowan and Surdam, unpublished data, manuscript in prep.

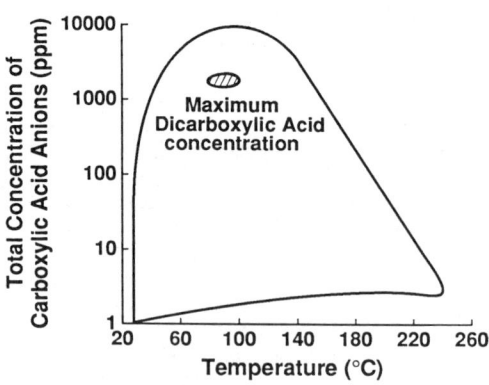

Figure 1. An envelope diagram illustrating the distribution of total concentration of organic acid anions with temperature from recently published data.

ppm. The presence of difunctional CAA in formation waters has been attributed to contamination by drilling fluids (44); however, aqueous organic geochemistry of drilling fluids show little or no difunctional CAA in drilling muds or scale soap solutions. The maximum concentration of difunctional CAA are seen to coincide with the maximum total CAA concentration, generally in formation waters between 80-100°C (Figure 1).

Controls on Concentration and Distribution of Carboxylic Acid Anions in Formation Waters

The thermal distribution of CAA in formation waters has been broken into three temperature zones (1,17,43); Zone 1 (20 to 90°C) is typified by increasing CAA concentrations with increasing temperature. In Zone 2 (90 to 260°C), total CAA concentrations decrease with increasing temperature. Zone 3 extends from 260°C on and is characterized by a lack of CAA. The decrease in concentration in Zone 2 and absence from Zone 3 was attributed to the thermal decarboxylation of the CAA. However, hydrous pyrolysis and kinetic studies indicate that the concentration and distribution of CAA cannot simply be temperature, indeed some researchers feel that it is totally independent of temperature (46).

Origin of Carboxylic Acid Anions in the Diagenetic Environment. Aqueous CAA species can either be produced biogenically by the incomplete metabolism of organic material by sulfate reducing or methanogenic bacteria (7), thermogenically by the thermal maturation of sedimentary organic material (32,48), by the mineral oxidation of dissolved organic material (24), and by the abiotic oxidation of hydrocarbons during sulfate reduction (25).

Anaerobic bacteria can metabolize hydrocarbon material to CAA and hydrogen gas (47). Additionally, methanogenic and sulfate-reducing bacteria produce CAA from carbohydrates, amino acids and higher CAA (47). CAA are common metabolic products in the biosphere (47), but the balance between their production and destruction by microbial processes is such that the concentrations observed in formation waters are not likely to occur by these processes (7).

Biogenic degradation of petroleum and its precursors may lead to the production of CAA; this can occur at temperatures of up to 82°C, providing oxygen or sulfate and water are present (7,53), and may lead to the production of significant quantities of CAA (7). Water-washing of crude petroleum by relatively-fresh meteoric water may displace CAA from the oil phase to the aqueous phase (8,35). This was tested in a series of experiments (54), however, the compositions of the water washes were fairly dissimilar from those of the formation waters from the same basin (54). The CAA analyzed in this study comprised much less than 1% of the crude petroleum sample and do not represent the total CAA present. By comparison, CAA have been reported up to 3% in some crude oils (55,56).

The production of CAA from the thermal evolution of disseminated sedimentary organic material depends upon the chemical make-up of that material and its rate of heating. Kerogen models show both mono- and difunctional CAA functional groups as side chains on the kerogen molecule (32,57,58). During diagenesis, these side chains are cleaved from the molecule and CAA are released. Studies of the evolution of kerogen molecules during diagenesis and catagenesis (58) show that carboxyl groups are continually being released to the diagenetic environment until the end of catagenesis. Figure 2 shows this trend for Types I, II and III kerogen (58). Although poorer in total organic O, a Type II kerogen has nearly as many carboxyl groups as does a Type III kerogen. Van Krevelen diagrams (atomic O/C vs atomic H/C) have shown that most kerogens exposed to elevated temperatures first show a decrease in atomic O/C as O-bearing functional groups are released. At higher temperature, and concomitant with hydrocarbon generation, they show a decrease in the H/C ratio. This indicates that during thermal maturation of source rocks, significant quantities of O-bearing species are being released from the kerogen prior to the expulsion of hydrocarbons.

13C nuclear magnetic resonance and infrared spectroscopy studies also suggest that with increasing temperature the kerogen first releases O-bearing functional groups (i.e., carboxyl-groups and phenol compounds) then releases liquid hydrocarbons (37,58).

Other reactions which may generate CAA in the subsurface are the mineral oxidation of dissolved organic material to CAA by such inorganic species as ferric iron (released during clay diagenesis) (24), as well as abiotic sulfate reduction/hydrocarbon oxidation (25) which may take place at temperatures as low as 100°C (59).

Destruction of Carboxylic Acid Anions. Destruction of CAA has been postulated to occur by two processes; bacterial degradation at low temperature and thermal decarboxylation at higher temperature (17). Little is known about the distribution of CAA substrate bacteria in formation waters, so their affect upon the distribution and concentration of CAA in formation waters is difficult to effectively model. However, bacterial degradation is important in many formation waters (7,43,45). Although they don't coexist, both anaerobic and aerobic bacteria have been reported in different petroliferous environments at temperatures in excess of 90°C (60,61); it should be noted that at temperatures in excess of about 45°C they are metabolically inactive (7). At surface conditions, CAA are so readily metabolized by aerobic bacteria, produced formation water samples require immediate preservation at the time of sampling (35,45).

Anaerobic microbial degradation of CAA to bicarbonate and methane can proceed by two reactions, each mediated by different bacteria. Long-chain fatty-acids are broken down to ethanoate and bicarbonate by hydrocarbon substrate bacteria (7). Ethanoate can be further metabolized by methanogenic bacteria to bicarbonate and methane. These reactions will cause ethanoate to dominate formation water organic alkalinity in the presence of hydrocarbon-substrate bacteria, causing low propanoate/ethanoate ratios. In formation waters where methanogenic bacteria are present, the longer-chain CAA will be relatively enriched with respect to ethanoate. In formation waters experiencing bacterial degradation due to both groups of bacteria, concentrations of CAA will be low (7).

When exposed to high temperature in the presence of clay minerals, thermal decarboxylation of the CAA to $CO_2$ and hydrocarbons can occur (62,63,65). If the reaction occurs in the presence of calcium carbonate, however, ß-cleavage of the carboxyl group will occur (65) producing CAA and hydrocarbons. Decarboxylation of CAA may be a thermal, kinetically controlled process (3,66,67), however, differing mineral matricies have an effect on the rate of decarboxylation (30,69). It has been suggested that the rates of CAA destruction are completely controlled by mineralogical thermodynamics, and not biologically or thermally (46). Additionally, pH has been demonstrated to have a pronounced effect upon CAA decarboxylation reaction rates (66,68).

Formation Water Mixing. The mixing of deep, ethanoate-rich formation waters with shallower, ethanoate-depleted waters has been called upon to explain the ethanoate-dominance of some shallow, relatively cool waters (10,42,43,45). It is difficult, however, to separate the effects of mixing hot, deep ethanoate-rich formation waters with cooler formation waters on the basis of ethanoate dominance or propanoate/ethanoate ratios alone. Degradation of CAA in formation waters by hydrocarbon-substrate bacteria can also explain the relative ethanoate enrichment with respect to the longer-chain CAA of some formation waters.

Kerogen Composition. A further control upon the concentration and distribution of CAA is the proclivity of the sedimentary organic material in a basin to generate CAA, and the kinetics of that generation. This is currently being studied using hydrous pyrolysis of source rock material (3,11,21,22,23). Between 5 and 30 wt.% of immature kerogen is comprised of oxygen-rich functional groups (7). Type III (terrigenous) is generally the most oxygen-rich kerogen, while Type II (marine) is intermediate in oxygen content and Type I (liptinitic) is a relatively oxygen-poor kerogen. Figure 2 demonstrates that nearly as many CAA will be available to the diagenetic environment from the thermal maturation of a Type II kerogen as from a Type III kerogen.

Organic Acid Anions and Feldspar Dissolution
Since perhaps 90% of formation waters from Jurassic-Pleistocene reservoirs presently in the 80-120°C thermal window from first-cycle basins have significant concentrations of CAA, it is

important to gain an understanding of their affect upon reservoir material. Certainly, high concentrations of CAA recently reported in formation waters (35), and the CAA concentrations reported in hydrous pyrolysis experiment products (3,11,21,22,23) coupled with evidence of their effect on aluminosilicate mineral dissolution suggests that they are of great importance to the diagenesis of clastic sediments.

Significant quantities of Al and Si must somehow be mobilized to explain the widespread aluminosilicate dissolution seen in clastic reservoir sequences. Dissolved carbon dioxide cannot complex and transport Al or Si from the site of aluminosilicate dissolution (32,71,72). However, CAA have been demonstrated to have a pronounced affect upon the stability of aluminosilicate minerals at surface temperatures due to their ability to complex and transport Al and Si (34,70). Laboratory experiments at 100°C have shown that monofunctional CAA can, by organo-metallic complexation, raise the concentration of total Al in water (from plagioclase dissolution) by at least one order of magnitude above the inorganic solubility of gibbsite (assuming gibbsite is the stable dissolution product) (32), and that difunctional CAA at concentrations similar to recently reported values (35) can raise total Al concentration three to four orders of magnitude above the apparent inorganic solubility of gibbsite. Experiments which simulated reservoir conditions in the San Joaquin basin of California, and the Louisiana Gulf Coast (35) using whole-core material, mineral grains, formation waters from these areas and artificial brines demonstrated that feldspar dissolution textures noted in thin-section could be artificially created in the lab. The results of these experiments indicate that a significant amount of feldspar mineral grains or whole core material had dissolved and that the Al and Si probably were present as organo-metallic complexes with the CAA in both the synthetic solutions and in the natural formation waters. The Al and Si must have been carried as a complex of some weak base, in as much as addition of dilute nitric acid (as a preservative) to the experimental solutions caused precipitation of an Al-Si-rich gel (35,54). The only weak base present in the artificial solutions were CAA. By analogy, it is thought that the CAA in formation waters in petroleum reservoirs are complexing Al and Si, as well. Etching of quartz grains observed in a sand aquifer through which a petroleum spill had flowed were reproduced experimentally using CAA and quartz grains (34). It was concluded that oxidation of the petroleum had produced CAA in the subsurface, and these had produced the quartz grain dissolution. Indeed, the existence of Si-citrate, Si-ethanedioate and Si-2, al-propanoate electron-donor/acceptor organometalic complexes resulting from quartz dissolution in CAA-rich solutions was demonstrated (34). Etched garnet textures observed in the Morrison Formation were reproduced in experiments utilizing CAA as organometalic chelates. As the etch-textures could not be reproduced with inorganic acids, it was concluded that CAA in formation waters were responsible for those textures (33).

Some aluminosilicate dissolution experiments have been run with crude petroleum in addition to the formation water phase (54). These experiments showed a much smaller increase in Al and Si concentration in the water phase over that in a blank NaCl solution than did those with no crude oil phase. However, scanning electron microscope analyses of the solid phase from these experiments show that a great deal of feldspar dissolution had occurred over the course of these experiments (54). The solid phases in the experiments without a crude oil phase showed precipitation of reaction products on the grain surface due to quenching (100°-25°C) the pressure vessel. The solid phase from the experiments with a crude petroleum phase showed few of these precipitates upon quenching (which produces chemical and mineralogical results similar to those occurring in reservoir fluids during production when temperature and pressure drop markedly). This is evidence that the metal complexes are more hydrophobic than the CAA and may be partitioned into the crude oil phase when temperature and pressure are dropped. The petroleum phase was thinned with n-hexane, filtered to 0.20 μ, and analyzed for Al by x-ray fluorescence both prior to and after the dissolution experiments. No Al was detected in the filtered petroleum prior to the dissolution experiments, but up to 109 ppm Al was detected in the filtered oil after the dissolution experiments. Schematic x-ray diffraction patterns of the > 0.20 μ fraction filtrates are shown in Figure 3; there exists a broad "hump" in the post-experiment filtrate (Figure 3b) indicating the

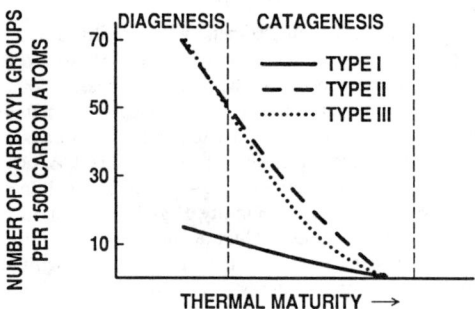

Figure 2. Thermal evolution of kerogen carboxyl groups with increasing thermal maturity (from the data of 58).

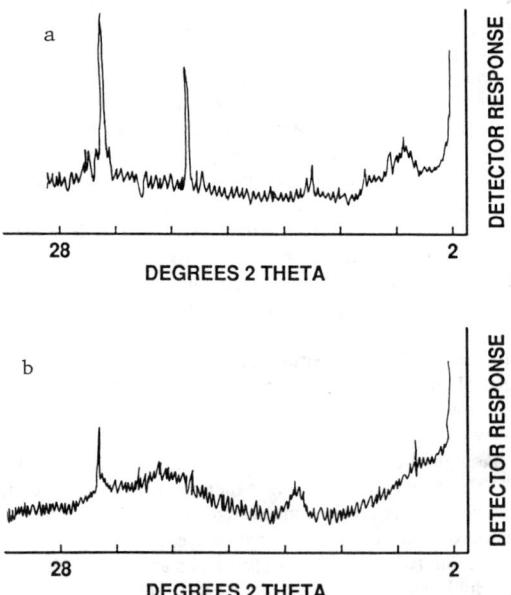

Figure 3. Schematic x-ray diffraction patterns for the > 0.2 μ fraction of petroleum filtrates; (a) prior to the dissolution experiments and (b) after the dissoluton experiments. Note the broad "hump" in the post-experiment filtrate indicating the presence of amorphous Al-Si gel in the oil.

presence in the oil of amorphous, early-formed, pre-precipitate alumino-silicate gels as well. This is further evidence that the metal complexes may be more hydrophobic than the CAA.

Observations indicate that Ca-rich feldspars are more susceptible to organo-ligand promoted dissolution than are the Na-or K-feldspars in experiments, soils and lithified sediment (32,70,73), although the proton-promoted dissolution rates of albite, anorthite and K-feldspar are approximately equal (77). The rate of feldspar dissolution promoted by organo-ligands is proposed to increase when inner-sphere adsorption of organo-ligands on aluminosilicates weakens critical crystal lattice bonds at the site of adsorption (77).

For organo-ligand promoted dissolution, the activated site of ligand formation is the Al, rather than the Si site. This is supported by experimental data in which dissolution rates of aluminosilicates by organo-ligand complexation increased with the increasing Al content of the feldspar (70,76,77). Furthermore, no increase has been measured in dissolution rates for chrysotile (which lacks Al) in solutions containing complex-forming organic compounds over the proton-promoted dissolution rate (77). Although there appears to be no correlation between grain-size and bulk reaction rate (74,75,76,77), the distribution of crystal defects in the experimental solid phase has a strong influence on those reaction rates (77).

Because of the overall low Al concentrations reported in most formation waters to date, it has been suggested that Al as well as Si may be lost by the waters during production or sampling (35). Changes in pH, exsolution of dissolved $CO_2$ and $H_2S$, temperature and pressure changes during production and sampling of well fluids may cause the destruction of the organo-metallic complexes, which in turn may cause Al and Si to precipitate out of solution as hydrous aluminosilicate gels, or colloidal kaolinite and clay mineral precursors (32,78), either within the casing, or near the casing, but within the formation. Even if an alumino-silicate gel or colloidal kaolinite were swept along with the reservoir fluids, they would be removed by filtration during sampling, or prior to analysis and thus not be present in laboratory analyses (35). Additionally, if the complexes, when formed, are more hydrophobic than the organic acid anions, their aqueous solubility will decrease with decreasing temperature. Because of this, as the temperature drops and the organic phase (petroleum) exsolves during production of the formation fluid, the majority of the organo-metallic complexes could be partitioned into the organic (petroleum) phase, and thus are not observed in the analyses.

Use of low and perhaps unrepresentative analytical values for total Al and difunctional CAA concentrations may be one reason why mass balance calculations and computer model simulations of formation waters produce results that de-emphasize the role of organo-metallic complexes in the subsurface transportation of Al and Si (1,2,72,79). In addition, the thermodynamic constants for organo-metallic complexation by these species are unknown at elevated temperatures, and van't Hoff-type approximations used in many models may not be valid as the temperature extrapolation is too large. Complexation and transportation of Al and Si by polar organic compounds remain the only viable mechanism to explain the widespread Al and Si mobility noted in many clastic hydrocarbon reservoirs (3,35,39).

Carboxylic Acid Anions and Carbonate Mineral Stability

CAA can control carbonate mineral stability over the range in which they dominate formation water alkalinity (generally ≈80° -120°C) (32,36,37,80). Initially, at low $PCO_2$, carbonate minerals are not stable in the presence of CAA. However, as pH remains constant when CAA dominate alkalinity, carbonate minerals may actually be stabilized as $PCO_2$ increases during kerogen maturation. Recently, this process has been modeled by "path-calc" type equilibrium simulators (80). Although applicable to carbonate reservoirs into which acetic acid has been pumped (as a well stimulation), flaws exist in this simulation which render it of negligible utility in studying organic-inorganic interactions in clastic reservoirs. A complete discussion of this appears elsewhere (3).

## Thermal Hydrocarbon Oxidation by Sulfate Reduction

Reduction of sulfate by hydrocarbons (either biotic or thermal) occurs because their association in the aqueous diagenetic environment is thermodynamically unstable (25). Depending upon the exact reaction (Table II) and the presence or absence of geocatalysts, abiotic or thermal sulfate reduction/hydrocarbon oxidation can occur at temperatures as low as 100-140°C (25).

Thermocatalytic sulfate reduction probably is the main source of $H_2S$ in the deep subsurface (25,81). The reactions in Table II, based on the evolution of $H_2S$ and $CO_2$ by sulfate reduction of hydrocarbons in clastic sequences where there is available $SO_4^{2-}$ (i.e. sequences with evaporite beds or cements), illustrate the importance of this process to diagenesis. Not only does it provide a diagenetic sink for Fe via pyrite, but depending on the relative availability of reactants, can either cause precipitation or dissolution of calcite and alter feldspars to clay minerals.

In the presence of alkali metals, thermal sulfate reduction will result in precipitation of carbonate cements (mainly calcite and dolomite) or carbonate replacement of dissolving sulfates (gypsum/anhydrite) (25). The reaction of polysulfides with bicarbonate has been suggested as a cause of calcite precipitation. Further, transition and base metals present in formation waters during thermal sulfate reduction could lead to the deposition of disseminated or stratiform base

Table II. Clastic chemical reactions based upon hydrocarbon reduction of sulfate (25,81)

| Reaction | |
|---|---|
| $CaSO_4 + CH_4 \rightarrow CaCO_3 + H_2S + H_2O$ | (1a) |
| $3SO_4^{2-} + 4CH_4 + 6H^+ \rightarrow 3H_2S + 4H_2O + 4CH_2OOH$ | (1b) |
| $CaSO_4 + 3H_2S + CO_2 \rightarrow CaCO_3 + 4S^o + 3H_2O$ | (2) |
| $Fe^{2+} + S^o + H_2S \rightarrow FeS_2 + 2H^+$ | (3) |
| $Fe^{3+} + 1.5 S^o + 1.5 H_2S \rightarrow FeS_2 + 3H^+$ | (4) |
| $CaCO_3 + 2H^+ \rightarrow Ca^{2+} + H_2O + CO_2$ | (5) |
| $CaAl_2Si_2O_8 + CO_2 + 2H_2O \rightarrow Al_2Si_2O_5(OH)_4 + CaCO_3$ | (6) |
| $2NaAlSi_2O_8 + CO_2 + Ca^{2+} \rightarrow Al_2Si_2O_5(OH)_4 + 4SiO_2$ $+ CaCO_3 + 2Na^+$ | (7) |

metal or Mississippi Valley-type deposits (25). $H_2S$ also is highly corrosive to feldspars, although it lacks the ability to complex and transport Al and Si from the site of aluminosilicate dissolution. This, like bicarbonate-promoted feldspar dissolution, leads to insitu alteration of feldspar to clay minerals. As seen in reaction (1b) in Table II, the oxidation of some hydrocarbons by sulfate reduction could lead to the production of carboxylic acid anions in formation waters. If indeed this reaction takes place at temperatures as low as 100°C (25,59), this would be within the stability range of the CAA (Figure 1, and 35); there exists abundant geologic evidence of simultaneously raised $PCO_2$ and carbonate precipitation associated with sulfate reduction of hydrocarbons (25). Figure 3 represents a chemical divide model based on interpretation of published sulfate reduction reaction series (3,25,81). Following the schematic pathway of the relative abundances of major reactants is useful in predicting the type of diagenesis in a clastic sequence of interest and the gas composition of reservoir fluids.

Generalized Organic-Inorganic Diagenetic Model

The evolution of the organic alkalinity of formation waters reflects the interaction of a number of factors (excluding mixing of formation waters), which can be typified by the following generalizations.

1) From 20 to about 80°C, the total concentration of CAA generally increases with increasing temperature; ethanoate is generally the dominant CAA; bicarbonate dominates water alkalinity and early calcite cements may form depending upon critical diagenetic conditions (3).

2) The CAA maximum concentration typically occurs over the 80 to 120°C interval; in this thermal interval, CAA can dominate the formation water alkalinity.

3) The highest concentrations of difunctional CAA are generally associated with the CAA maximum concentration (35); during this time aluminosilicate and carbonate dissolution and mass-transfer of Al and Si occur.

4) Because of their lower thermal stability, the difunctional CAA are the first to be decarboxylated, typically at temperatures of 100 to 110°C; the resultant increase in $PCO_2$ causes stabilization of the carbonate mineral phase (35). Destabilization of Al-difunctional CAA complexes may result in the precipitation of kaolinite. The destabilization of Fe-CAA complexes would provide a source of Fe for ferroan carbonate precipitation.

5) With rising temperature (110 to 130°C), the monofunctional CAA also begin to show a decline in concentration (Figure 1). This results in a fluid alkalinity that is dominated by bicarbonate, and late-stage carbonate dissolution may occur. Additionally, Fe-monofunctional CAA complexes will be destabilized at this time, leading to a pulse of Fe which may result in chlorite formation.

6) As temperatures increase (onset between 100 and 140°C), abiotic sulfate reduction by oxidation of hydrocarbons becomes kinetically important. This process leads to the release of $H_2S$ and bicarbonate to formation waters (25,81). At this time, aluminosilicate alteration to clay minerals may occur; carbonate mineral stability will be a function of $CO_2$, Fe and S availability, as will pyrite precipitation (25,81 and Figure 4).

This qualitative model allows an understanding of the relative timing of diagenetic events related to organic-inorganic interactions during progressive burial diagenesis of sandstone-shale sequences. Because the model is process-oriented, it can be incorporated into "path-calc" type equilibrium simulations (1,2) and time-temperature, kinetic models (3). The understanding of the relative importance of organic-inorganic interactions throughout the burial history of clastic units highlighted by this model is of great importance to the overall modeling of water-rock reactions, and porosity evolution, and to understanding the dynamic nature of clastic diagenesis in sedimentary basins.

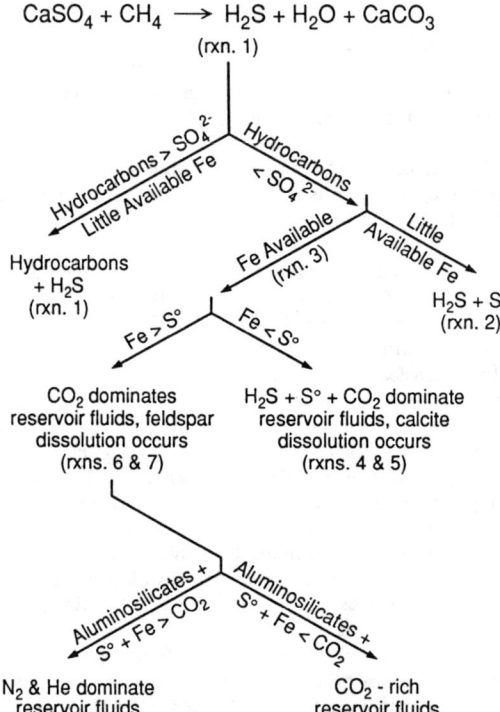

Figure 4. Chemical divide model for sulfate reduction of hydrocarbons, and their effect on diagenetic reactions. Numbers refer to reactions in Table II (based on *3, 25, 81*).

## LITERATURE CITED

1. Kharaka, Y.K.; Hull, R.W.; Carothers, W.W. In Short Course Notes #17, Relationship of Organic Matter and Mineral Diagenesis; S.E.P.M., 1985; pp. 79-176.
2. Harrison, W.J. In Quantitative Dynamic Stratigraphy; Cross, T.A., Ed.; Springer-Verlag; 1989.
3. Surdam, R.C.; Dunn, T.L.; Heasler, H.P.; MacGowan, D.B. In Short Course in Burial Diagenesis; Mineralogical Association of Canada, Hutcheon, I.E., Ed., Min. Assoc. Canada, 1989; pp. 61-134.
4. Berner, R.A.; In Kinetics of Geochemical Processes; Rev. Min. vol. 8, LASAGA, A.C.; Kirkpatrick, R.J., Eds., Min. Soc. Am., 1981; pp. 111-130.
5. Rogers, G.S.; U.S.G.S. Bull., v. 653, 1917; 119p.
6. Zinger, A.S.; Kravchik, T.E. Akad. Nauk SSSR Doklady (C.C.C.P.), 1973; v. 202, pp. 218-221
7. Hatton, R.S.; Hanor, J.S. In, Technical report for geopressured-geothermal activities in Louisiana; final geological report for the period 1 November to 31 October, 1982. Hanor, J.S., Ed.; DOE report no. DOE/NV/10174-3, 1984; pp. 348-454.
8. Collins, A.G., Geochemistry of Oil-field Waters; In Dev. Petrol. Geol. #1; Elsevier, New York, 1975; 496p.
9. Keeley, D.F.; Meriweather, J.R. In Geopressured-Geothermal Energy, proc.; 6th U.S. Gulf Coast Geopressured-Geothermal Energy Conf.; Dorfman, H.; Morton, A., Eds.; Pergammon Press, New York, 1985; pp. 105-113.
10. Kharaka, Y.K.; Law, L.M.; Carothers, W.W.; Goerlitz, D.F. In Roles of Organic Matter in Sediment Diagenesis; Gautier, D.L., Ed.; S.E.P.M. Spec. Pub. no. 38, 1986; pp. 111-122.
11. Surdam, R.C.; MacGowan, D.B. Appl. Geochem., 1987; v. 2, pp. 613-619.
12. Branthaver, J.F.; Thomas, K.P.; Logan, E.R.; Barden, R.E. Fuel Sci. Tech. Int'l., 1988; v. 6, pp. 525-539.
13. Cordell, R.J. A.A.P.G. Bull., 1972; v. 56, pp. 2029-2067.
14. Kartsev, A.A. Nat. Tech. Info. Ser. Report TT73-580022, 1974; 323p.
15. Matusevich, V.M.; Prokopeva, R.G. ISV. Vyssh. Ucheb. Zavendenii NEFT. GAZ.,(C.C.C.P), 1977; v. 2, pp. 7-10.
16. Wiley, L.M.; Kharaka, Y.K.; Presser, T.S.; Rapp, J.P.; Barnes, I Geochim. Cosmochim. Acta 39, 1975; pp. 1707-1710.
17. Carothers, W.W.; Kharaka, Y.K. A.A.P.G. Bull., 1978; v. 62, pp. 2241-2253.
18. Vandenbroucke, M. In Kerogen; Durand, B., Ed.; Editions Technip, Paris, 1980; pp. 415-444.
19. Rouxhet, P.G.; Robin, P.L; Nicaise, G. In Kerogen; Durand, B., Ed.; Editions Technip, Paris, 1980; pp. 163-190.
20. Vandergrift, G.F.; Winans, R.E.; Scott, R.G.; Horwitz, E.P. Fuel, 1980; v. 59, pp. 627-633.
21. Kawamura, K.E.; Tannebaum, E.; Huizinga, B.J.; Kaplan, I.R. Geochem. Jour., 1986; v. 20, pp. 51-59.
22. Kawamura, K.E.; Kaplan, I.S. Geochim. Cosmochim. Acta 81, 1987; pp. 3201-3207.
23. Lundegard, P.D.; Senftle, J.T. Appl. Geochem. 1987; v. 2, pp. 605-612.
24. Crossey, L.J.; Surdam, R.C.; Lahann, R. In Roles of Organic Matter in Sediment Diagenesis; Gautier, D.L., Ed.; S.E.P.M. Special Pub. no. 38; Tulsa, Ok., 1986; pp. 147-155.
25. Machel, H.G.; In Diagenesis of Sedimentary Sequences; Marshall, J.D., Ed.; Geol. Soc. Spec. Pub. No. 36, 1987; pp. 15-28.
26. Barton; Econ. Geol. Mon. 3, 1967; pp. 371-378.
27. Nissenbaum, A; Swaine, D.J. Geochim. Cosmochim. Acta 40, 1976; pp. 809-816.

28. Saxby, J.D.; In Handbook of Strata-bound and Strata-form Ore Deposits; 1976; v. 2, Geochemical Studies, Chap. 5, Wolfe, K.H., Ed.; Elsevier, pp. 111-113.
29. Giordano, T.H.; Barnes, H.L. Econ. Geol. 1981; v. 76, pp. 2200-2211.
30. Drummond, S.E.; Palmer, D.A. Geochim. Cosmochim. Acta 50, 1986; pp. 825-833.
31. Hennet, R.J.; Crerar, D.A.; Schwartz, J. Econ. Geol., 1988; v. 83, pp. 742-764.
32. Surdam, R.C.; Boese, S.; Crossey, L.J. In Clastic Diagenesis; MacDonald, D.A.; Surdam, R.C., Eds.; A.A.P.G. Mem. 37, 1984; pp. 127-150.
33. Hansley, P.L. Jour., Sed. Pet., 1987; v. 57, pp. 666-681.
34. Bennett, P.; Melcer, M.E.; Siegel, D.I.; Hassett, J.P. Geochim. Cosmochim Acta 52, 1988; pp. 1521-1530.
35. MacGowan, D.B.; Surdam, R.C. Organ. Geochem., 1988; v. 12, pp. 245-259.
36. Meshri, I.D. In Roles of Organic Matter and Sediment Diagenesis; Gautier, D.L., Ed., S.E.P.M. Special Pub. no. 38, Tulsa, Ok., 1988; pp. 123-128.
37. Surdam, R.C.; Crossey, L.J. Philos. Trans. R. Soc. Lond. Ser. A., 1985; v. 315, pp. 172-232.
38. Surdam, R.C.; Crossey, L.J. In Relationship of Organic Matter and Mineral Diagenesis; S.E.P.M. Short Course Notes #17, 1985; pp. 177-232.
39. Surdam, R.C.; Crossey, L.J. Ann. Rev. Earth Plan. Sci., 1987; v. 15, pp. 141-170.
40. Martell, A.E.; Smith, R.M. Critical Stability Constants, vol. III: Other Organic Ligands; Plenum Press, New York, 1977; 495p.
41. Workman, A.L.; Hanor, J.S. Trans. Gulf Coast Assoc. of Geol. Soc., 1985; v. 35, 293-300.
42. Hanor, J.S.; Workman, A.L. Appl. Geochem., 1986; v. 1, pp. 37-46.
43. Fisher, J.B. Geochim. Cosmochim. Acta 51, 1987; pp. 2459-2468.
44. Barth, T. Chemomet. Intel. Lab. Sys., 1987; v. 2, pp. 155-160.
45. Means, J.L.; Hubbard, N.J. Org. Geochem. 1987; v. 11, pp. 374-386.
46. Schock, E.L. Geology, 1988; v. 16, pp. 886-890.
47. Kaspar, H.F.; Wuhrmann, K. Appl. Envir. Microbiol., 1978; v. 36, pp. 1-7.
48. Tissot, B.P.; Welte, D.H. Petroleum Formation and Occurence; 2nd Ed., Springer-Verlag, New York, 1984; 699p.
49. Jerris, J.S.; McCarty, P.L. Jour. Water Poll. Cont. Fed., 1965; v. 37, pp. 178-192.
50. Metzler, D.E.; Biochemistry: Chemical Reactions of Living Cells; Academic Press, New York, 1977; 435 p.
51. Fenchel, T.M.; Jorgensen, B.B. Adv. Microb. Ecol., 1977; v. 1, pp. 1-58.
52. Zehnder, A.J.B. Water Pollution Microbiology; Mitchell, R., Ed.; Wiley, New York, 1978; pp. 349-376.
53. Hunt, J.M. Petroleum Geochemistry and Geology; Freeman, W.H., San Francisco, 1979; 617p.
54. MacGowan, D.B.; Surdam, R.C.; Ewing, R.E. Proc. 4th Enhanced Oil Recovery Symp., Ewing, R.E., Copeland, D., Eds.; SPE # 7802, 1988; pp. 621-630.
55. Siefert, W.K.; Howells, W.G. Analyt. Chem., 1969; v. 41, pp. 554-556.
56. Siefert, W.K. In Progress in the Chemistry of Organic Natural Products; Herz, Grisbach and Kirby, Eds.; 1975; pp. 1-50.
57. Murphy, R.C.; Bieman, K.; Djuricic, M.V.; Victorovio, D. Bull. Chem. Soc. Belgrade, 1971; v. 36, p. 281.
58. Behar, F.; Vandenbroucke, M. Org. Geochem., 1987; v. 11, pp. 15-24.
59. Rye, D.M.; Williams, N. Econ. Geol., 1981; v. 76, 1-26.
60. Levorson, A.I. Geology of Petroleum; Freeman, W.H., San Francisco, 1967; 395p.
61. Brock, T.D. Thermophilic Microorganisms and Life at High Temperatures; Plenum, N.Y. 1978; 478p.
62. Eisma, E.; Jurg, J.W. 7th World Petrol. Cong. Proc. 1967; v.2, pp. 61-72.
63. Shimoyama, A.; Johns, W.D. Nature Phys. Sci., 1971; v. 232, pp. 140-144.
64. Johns, W.D.; Shimoyama, A. A.A.P.G. Bull., 1973; v. 56, pp. 2160-2167.

65. Jurg, J.W.; Eisma, E. Science, 1964; v. 144, pp. 1451-1452.
66. Kharaka, Y.K.; Carothers, W.W.; Brown, P.M. Proc. 53rd Am. Conf. Soc. Pet. Eng. of AIME, SPE #7505, 1978; 8 p.
67. Palmer, D.A.; Drummond, S.E. Geochim. Cosmochim. Acta 50, 1986; pp. 813-823.
68. Boles, J.S.; Crerar, D.A.; Grisom, G.; Key, T.C. Geochim. Cosmochim. Acta 52, 1988; pp. 341-344.
69. Kharaka, Y.K.; Carothers, W.W.; Rosenbauer, R.J. Geochim. Cosmochim. Acta 47, 1983; pp. 387-402.
70. Huang, W.H.; Keller, W.D. Amer. Min. 1970; v. 55, pp. 2076-2094.
71. Lundegard, P.; Land, L.S. In Roles of Organic Matter in Sediment Diagenesis; Gautier, D.L., Ed.; S.E.P.M. Spec. Publ. No. 38, Tulsa OK., 1986; pp. 129-146.
72. Giles, M.R.; Marshall, J.O. Marine Petrol. Geol., 1986; v. 3, pp. 243-255.
73. Erich, M.S.; Bloom, P.R. Agron. Abstr. Amer. Assoc. Agron., Madison, WI, 1987.
74. Holdren, G.R., Jr.; Speyer, P.M. Geochim. Cosmochim. Acta 49, 1875; pp. 675-681.
75. Holdren, G.R., Jr.; Speyer, P.M. Geochim. Cosmochim. Acta 51, 1987; pp. 2311-2318.
76. Stumm, W.; Furrer, G.; Wieland, E.; Zinder, B. In The Chemistry of Weathering; Drever, J.I., Ed.; Reidel Pub. Co., 1985; pp. 55-74.
77. Amrhein, C.; Suarez, D.Z. Geochim. Cosmochim. Acta 52, 1877; pp. 2785-2794.
78. Bilinsky, H.; Horvath, L.; Ingri, N.; Sjorberg, S. Geochim. Cosmochim. Acta 50, 1986; pp. 1911-1922.
79. Bjørlykke, E.; In Clastic Diagenesis; MacDonald, D.A.; Surdam, R.C., Eds.; A.A.P.G. Mem. 37, 1984; pp. 277-286.
80. Lundegard, P.D.; Land, L.S. Chem. Geol., 1989; v. 74, pp. 277-287.
81. Siebert, R.M.; In SEPM, Gulf Coast Sect. Prog. and Abstr. 1985; v. 6, pp 30-31.

RECEIVED August 18, 1989

# Chapter 39

# Copper Complexation by Natural Organic Matter in Ground Water

### Thomas R. Holm and Charles D. Curtiss III

**Aquatic Chemistry Section, State Water Survey Division, Illinois Department of Energy and Natural Resources, 2204 Griffith Drive, Champaign, IL 61820**

Complexation of $Cu^{2+}$ by natural organic matter in ground water was studied by complexometric titrations monitored by ion-selective electrode potentiometry (ISE), fluorescence quenching (FQ), and cathodic stripping voltammetry (CSV). The titrations were analyzed using single- and multi-ligand models. At the natural ground-water pH values there was good agreement between model ligand concentrations and conditional stability constants determined by the three methods. At lower pH values the FQ and ISE results diverged. The ISE results are roughly comparable to published ISE results. The FQ results, however, show stronger complexation than published results. For total Cu concentrations from $10^{-8}$ to $10^{-5}$ M, Cu binding could be described by three model ligands at concentrations of $5 \times 10^{-8}$, $2 \times 10^{-7}$, and $1 \times 10^{-5}$ M having log stability constants of 10.1, 8.5, and 5.5, respectively. For this range of total Cu concentrations, complexes formed with natural organic matter are predicted to dominate Cu speciation. The empirical complexation parameters are strictly valid only for the ground waters, conditions (e.g. pH), and total Cu concentrations used in the present work. However, they should also be useful for exploratory modeling.

Metal complexation may affect ground-water quality in many ways, including increasing the solubility and mobility of toxic metal ions and controlling the availability of metal ions that are toxic to or essential nutrients for microorganisms. As a result, the metal-complexing properties of ground waters may be an important factor in determining the transport and fate of contaminants in aquifer systems.

The solubilities of metals in natural waters are controlled by the precipitation and dissolution of slightly soluble compounds such as carbonates, sulfides, and hydrous oxides (1) or, in waters that are undersaturated with respect to these compounds, sorption by hydrous oxides (2) or other minerals. These reactions determine the concentrations of uncomplexed or "free" metal ions with the specific controlling reaction depending on pH, oxidation-reduction conditions, and the concentrations of precipitating anions. Complexation increases the total dissolved concentrations or solubilities of metal ions in equilibrium with precipitated or adsorbed metals. The free metal ion concentration in equilibrium with a precipitating ligand or adsorbing surface is fixed. However, if strong complex-forming ligands are also present, the soluble metal concentration will be the sum of the free metal concentration and the concentrations of complexes.

The mobilities of metal ions in aquifer systems are proportional to their solubilities. Therefore, complexation may enhance the mobility of a metal ion in an aquifer system by increasing its solubility. The solubilities of several metal ions were computed for a range of conditions representative of ground waters (3). A mixture of low-molecular-weight compounds was used to simulate natural organic ligands. Characterization of the metal-complexing properties of the organic matter in ground water, as in the present work, would allow more accurate estimates of metal solubilities.

Metal complexation may influence the transformation of organic contaminants in ground waters by controlling the availability of toxic and essential metal ions to the microorganisms that mediate the transformations. For example, free (uncomplexed) metal ion concentrations, rather than total dissolved metal concentrations, have been shown to control bacterial growth rates (4,5). Metal ions have also been shown to inhibit many microbial processes of ecological importance, including nitrogen fixation, mineralization of carbon, nitrogen, sulfur, and phosphorus, and litter decomposition (6).

While numerous studies (7) have shown that the organic matter in surface waters and soils forms stable complexes with $Cu^{2+}$ and other metal ions, there have been few studies of metal complexation by ground water. The organic matter in ground water was probably carried along with recharge waters and, therefore, probably initially had properties similar to soil or surface-water organic matter. However, the chemical properties of the organic matter may change during the relatively long residence times of ground waters due to adsorption of high-molecular-weight fractions on aquifer materials and microbial respiration (9). Therefore, while studies of the metal-complexing properties of surface waters may provide a first approximation to the corresponding properties of ground waters and a guide to experimental methods for studying ground-water chemistry, ground waters should be studied directly for quantitative understanding. In the present work, the complexation of $Cu^{2+}$ by natural organic matter in ground water was studied.

Because ground waters, like other natural waters, are dilute solutions of many compounds, metal speciation measurements are difficult. Therefore, the metal-complexing properties of natural waters are operationally defined by many factors, including the analytical method used for speciation, the conceptual and mathematical models used to analyze the data, the range of titrant metal concentrations used, and conditions such as pH, ionic strength, and temperature. The analytical methods used to determine metal speciation all have inherent assumptions and limitations. Most published studies of metal complexation in natural waters have used one analytical method. However, confirmation of results (e.g. stability constants, ligand concentrations) by independent methods would add confidence to such results. In the present work, three independent methods were used.

Some published complexation studies have used fulvic acid which was isolated from surface water as the complexing agent (9,10,11). Because fulvic acid is defined by the procedure used to isolate it from water (12), its use in metal-complexation studies may add a degree of uncertainty to the application of the results to natural systems. McKnight et al. (9) did find that the Cu-complexing properties of fulvic acid isolated from a river water were very similar to the river water. However, to avoid potential artifacts in the isolation of an organic fraction, natural ground waters were used without isolation or concentration of organic matter in this work.

Some published $Cu^{2+}$ complexation studies have been performed at a pH of 6 (9,10), probably to avoid hydrolysis and carbonate complexation, and at an ionic strength of 0.1. The pH and ionic strength both affect metal complexation by natural organic matter (11). Therefore, for realistic studies of metal complexation in natural waters, it is desireable to keep both parameters as close to their natural values as possible. In this work, the experiments were performed at the natural pH values of the ground waters (approximately 8.0) as well as at lower values for comparison with published results. The ionic strength used was 0.02 eq $L^{-1}$, which was as close as possible to the natural values of approximately 0.01 eq $L^{-1}$.

MATERIALS AND METHODS

Ground-water samples were collected from three wells (Wells 1, 3, and 4) in the Sand Ridge State Forest in central Illinois. The water samples were filtered (0.4 μm), acidified (0.02% v/v $H_2SO_4$), and stored at 4°C. The oxidation-reduction conditions ranged from oxic near the water table to post oxic (undetectable dissolved oxygen, high iron concentration) in the deepest well. Total organic carbon concentrations were approximately 1 mg $L^{-1}$. The pH values were 8.3, 8.0, and 7.8 for water from wells 1, 3, and 4, respectively. Other details of well construction, well purging, sample collection methods, and water chemistry are described by Holm et al. (13). (See also Barcelona, M. J.; Holm, T.R.; Schock, M. R.; George, G. K. Spatial and Temporal Gradients in Aquifer Oxidation-Reduction Conditions, Wat. Resour. Res., in press.)

The complexometric titrations were performed at the natural pH values of the ground-water samples and also at pH 7.0 and 6.0. The pH was buffered using HEPES (N-2-Hydoxyethylpiperazine-N'-2-hydroxyethanesulfonic acid), PIPES (piperazine-N-N'-bis[2-ethanesulfonic acid], or MES (2-[N-Morpholino]ethanesulfonic acid), which have p$K_a$ values of 7.5, 6.8, and 6.1, respectively and which have negligible affinities for metal ions (14). The buffers were found to be non-fluorescent, so there was no interference in the FQ titrations. The buffer concentrations used were 6.7 mM. The buffers and $NaNO_3$ (used for ionic strength adjustment in calibrating the $Cu^{2+}$ ion-selective electrode) were treated with $MnO_2$ to reduce metal contamination (15). The natural ionic strength of the ground waters was approximately 10 meq $L^{-1}$. Acidification and addition of the buffer raised the ionic strength to approximately 20 meq $L^{-1}$.

Before each titration, buffer was added to the acidified ground water and the pH was adjusted to approximately 5. The solution was sparged. The total inorganic carbon concentration (TIC) was reduced to less than 0.01 mg $L^{-1}$. (TIC was determined by coulometric titration. (16)) The pH was then adjusted to the desired value using $CO_2$-free NaOH. Ion-selective electrode and cathodic-stripping-voltammetric titrations were performed with the surface of the water sample blanketed with nitrogen. Fluorescence-quenching titrations used screw-cap vials to isolate the waters from the atmosphere. Thus, inorganic carbon concentrations were kept low enough that carbonate complexation was negligible.

The response of the Cu ion-selective electrode (ISE)(Orion) was found to be Nernstian to at least pCu 13 in pCu buffers. The electrode was polished and calibrated daily. Before each titration the ISE was conditioned with pCu 10 buffer, and then with $1 \times 10^{-7}$ M Cu in buffered ground water. The lowest total Cu concentration used in the ISE titrations was $1 \times 10^{-7}$ M. The response of the electrode was unreliable below that concentration. The titration was stopped when successive Cu additions resulted in a potential change of less than 28 mV $(\log Cu)^{-1}$. It was assumed that $Cu(OH)_2$ began to precipitate at that point.

The ISE potential was recorded when its rate of change was less than 0.1 mV $min^{-1}$. At the lowest total Cu concentrations, up to two hours was required to reach a steady potential. The ISE equilibration time decreased as the total Cu concentration increased. Using a model of the rate of change of ISE potentials (17) and extrapolating to infinite time, it was estimated that at low total Cu concentrations the potential may have been low by up to 2 mV. If this were the case, then the "true" free $Cu^{2+}$ activity would have been higher than the measured value by a factor of up to 1.18.

Fluorescence was measured with a Turner model 111 filter fluorometer. The excitation filter was a Corning 7-60 (365 nm primary wavelength). The emission filters were Wratten 65-A (495 nm primary wavelength) and 2-A (sharp-cutoff below 415 nm). A digital multimeter was connected to the recorder terminals of the fluorometer to provide digital readout. Fluorescence-quenching (FQ) titrations were performed in batches. Preliminary experiments indicated that quenching was independent of time (at least 26 hours) after 30 minutes. Equilibration times of 60 minutes were used.

The CSV titrations were performed using a BAS 100 electrochemical system and an EG&G Princeton Applied Research 303A static mercury drop electrode. The procedure was essentially that of Van den Berg (18) except that the catechol concentration was $1 \times 10^{-5}$ M.

All titrations were performed at least in duplicate to confirm the results. For the ISE titrations, the lowest total Cu concentration for which reliable potential measurements could be made was $1 \times 10^{-7}$ M. For the FQ titrations, there was no observable quenching below $1 \times 10^{-7}$ M total Cu. The highest possible total Cu concentration for titrations performed at pH 7.0 and natural pH values (approximately 8.0) was approximately $1 \times 10^{-5}$ M. At $10^{-5}$ M total Cu, the rate of change of the ISE potential with added Cu became less than 29 mV $(\log Cu^{2+})^{-1}$ (Nernstian value) and chemical equilibrium calculations indicated that the solutions became oversaturated with respect to $Cu(OH)_2$. The lowest total Cu concentration used in CSV titrations was $1 \times 10^{-8}$ M. The upper limit of Cu concentrations for CSV was approximately $2 \times 10^{-7}$ M due to loss of linearity of the calibration above that value.

For two of the ISE titrations, total Cu concentrations were determined by furnace atomic absorption to check for sorptive losses.

Modeling Titration Curves

Natural aquatic organic matter is comprised of many compounds, only a small fraction of which have been identified (8). Because of the diversity and low concentrations of natural organic compounds, metal complexation is probably the result of competition between many ligands. Many published studies have used the simplifying assumption that all organic ligands in a natural water sample are equivalent to a single ligand (7).

Complexometric titrations of aquatic organic matter have been analyzed by many different models that are more complicated than the single-ligand model. The multi-ligand model assumes a small number of discrete ligands that form 1:1 complexes. Critical comparisons with multidentate (19), electrostatic (20), normal-distribution (21), and affinity-spectrum (22) models have shown that the multi-ligand model provides the best fit to potentiometric titration curves of natural organic matter (23,24,25). Also, the parameters of the multi-ligand model are in a form that can be readily incorporated in chemical equilibrium models. The single- and multi-ligand models were used to analyze experimental data in the present study.

Single- and multi-ligand models contain parameters that are used the same as ligand concentrations and conditional stability constants. However, the model ligand concentrations and conditional stability constants are strictly parameters that describe complexation of $Cu^{2+}$ over the range of total Cu concentrations and other conditions (pH, ionic strength) used in the experiment. That is, the degree of Cu binding is the same as in a solution containing the model ligands.

Natural organic matter fluoresces (26). The fluorescence is quenched by the formation of complexes with paramagnetic metal ions, including $Cu^{2+}$ (27). Fluorescence quenching, therefore, would seem to be complementary to ISE and CSV because fluorescence is proportional to free ligand concentrations. However, it is necessary to assume the quantum efficiencies of all ligands are equal and those of all complexes and fluorophores not affected by complexation are also equal but less than those of the free ligands. This assumption was tested by plotting fluorescence as a function of the concentration of bound Cu (determined by ISE). Only fluorescence values for which there was an approximately linear relationship to bound Cu were used in curve fitting.

A model for fluorescence quenching has been derived assuming one model ligand (28). However, the derivation tacitly assumed that complexation by inorganic ligands is negligible. The one-ligand model was modified (this work) to account for inorganic complexation:

$$F = 1 - \frac{1-B}{2K\alpha L_{tot}} \{K\alpha L_{tot} + K\alpha Cu_{tot} + 1 - [(K\alpha L_{tot} + K\alpha Cu_{tot} + 1)^2 - 4K^2\alpha^2 L_{tot} Cu_{tot}]^{1/2}\} \quad (1)$$

where F is normalized fluorescence, B is background fluorescence (at ligand saturation), and $\alpha$ is the fraction of inorganic Cu in the free $Cu^{2+}$ form.

The background fluorescence has been treated as an adjustable parameter in analyzing FQ data (28). However, complexation parameters were found to be very sensitive to the background fluorescence. To estimate background fluorescence, plots of $1/F$ vs $1/Cu_{tot}$ were extrapolated to $1/Cu_{tot} = 0$ (29). For the titration of a single fluorescent ligand, the fluorescence is (28):

$$F = 1 - (1-B)\frac{[CuL]}{L_{tot}} \quad (2)$$

Generalizing equation 1 gives a model for the fluorescence of multi-ligand systems:

$$F = 1 - (1-B)\frac{\Sigma [CuL_i]}{\Sigma L_{i,tot}} \quad (3)$$

The models for ISE, CSV, and FQ were fit to the titration curves by simplex optimization (30) treating the conditional stability constants and model ligand concentrations as adjustable parameters. Standard deviations of the parameters were estimated using Monte Carlo simulation (30,31). Fitting the multi-ligand FQ model required an iterative solution of the chemical equilibrium problem at each point to compute the concentrations of the complexes.

The stability constants used to calculate Cu complexation by inorganic ligands and catechol are listed in Table I.

## RESULTS AND DISCUSSION

Sunda and Hanson (36) found that sorption removed Cu from solution in some of their ISE titrations of river water at pH values greater than 7. They found it necessary to determine the total Cu concentration by atomic absorption for every titration point. In the present work, no significant Cu losses were observed in ISE titrations performed at pH 8. A possible explanation for this difference may be that the dissolved organic matter in the Sand Ridge ground water is less easily sorbed than surface-water organic matter. The ground water at the Sand Ridge site had been in contact with quartz sand for at least 20 years as estimated by tritium age dating (R. Buddemeier, Lawrence Berkeley Laboratory, personal communication, 1985) before being pumped out of the sampling wells. The easily sorbed component of organic matter may have been removed from the ground water shortly after being recharged.

All experiments were analyzed using one- and two-ligand models. The optimal model for each data set was usually apparent. For example, a one-ligand model sometimes gave a negative ligand concentration, a meaningless result, for FQ titrations. In these cases, a two-ligand model was used. For other FQ titrations and some ISE titrations at low pH, the two-ligand model gave nearly identical values of the two ligand concentrations. This implied that only one ligand was needed to fit the data. The $Cu^{2+}$-complexing parameters of the three ground-water samples at the natural pH values and lower pH values are listed in Table II.

The concentrations of model ligands were related to the ranges of total Cu concentrations used in the titrations. For example, for CSV, which used the lowest total Cu concentrations, the model ligand concentrations were lower than for the ISE and FQ titrations. The conditional stability constants tended to increase as model ligand concentrations decreased. These results are similar to those for Cu complexation by fulvic acid isolated from surface waters (9,11,37). The relative standard deviations of the complexation parameters ranged from approximately three percent to eighty six percent.

The one-ligand model fit the ISE titrations of Well 1 at pH 7.0 and Well 4 at pH 6.0. The two-ligand model had to be invoked to fit the ISE titrations at the natural pH values. There was no detectable binding of Cu in the Well 1 water at pH 6.0.

For the FQ titrations, a two-ligand model was necessary to fit the data for the Well 3 pH 8.0 and the Well 4 pH 6.0 titrations. The one-ligand model yielded negative model ligand concentrations for these titrations. Similar results have been reported for one-ligand models of FQ titrations (38,39,40). The one-ligand model fit the data adequately for all other FQ titrations in the present study.

The FQ complexation parameters were much less sensitive to pH than the ISE parameters. For Well 1, ISE showed no significant binding at pH 6 but strong binding at pH 8.3. The FQ titration curves for Well 1, on the other hand, showed approximately the same amount of quenching at pH 6 as at pH 8.3. This implies that FQ may not be a reliable independent metal speciation method. Cabaniss and Shuman (41) have suggested that FQ may be a useful speciation method if fluorescence is calibrated by another method, such as ISE. The results of the present work suggest that FQ may not be useful without calibration.

The Cu complexation parameters for Wells 1 and 4 showed the most similarity in both model ligand concentrations and conditional stability constants. These waters also showed the greatest differences in oxidation-reduction conditions (oxic and post-oxic), while the ground water with the intermediate dissolved oxygen concentration seemed to be the outlier. However, all complexation parameters agreed within an order of magnitude for all three ground waters.

The complexation parameters determined by ISE, FQ, and CSV for Well 3 at pH 8.0 are compared in Figure 1. The parameters lie in a narrow band on the graph and fall into three groups based on the conditional stability constant. For the "intermediate-strength" ligand, the three analytical methods yielded similar conditional stability constants, while the range of model-ligand concentrations was approximately a factor of five. The concentration of the intermediate ligand is in the range of overlap of total Cu concentrations used by the three analytical methods. For the weakest ligand there was good agreement in conditional stability constants as determined by ISE and FQ. The estimates of ligand concentration differed by approximately an order of magnitude. The strongest ligand was detected only by CSV in the low range of total Cu.

The three analytical methods produce a consistent explanation of Cu complexation by Well 3 water. For total Cu concentrations between $10^{-8}$ M and $10^{-5}$ M, Cu complexation can be described by three model ligands. These results are similar to the findings of Dzombak et al. (25) who found that Cu complexation by fulvic acid could be described by one ligand per order of magnitude in total Cu concentration. As empirical parameters, the complexation parameters determined in the present work are strictly applicable only in the range of total Cu concentrations used in the titrations ($10^{-8}$ to $10^{-5}$ M). The total Cu concentrations in the ground waters were less than $10^{-8}$ M. Therefore, the complexation parameters determined in the present work probably do not provide a reliable estimate of Cu speciation at natural concentrations. Because conditional stability constants increased as model ligand concentrations decreased, the complexation parameters would probably underestimate Cu complexation at natural Cu concentrations. However, the parameters should be useful in predicting Cu speciation in contaminated ground water.

The speciation of Cu in ground water from Well 3 at pH 8 was computed using the complexation parameters determined in the present work. The total Cu concentration was varied

Table I. Stability Constants Used for Computing Cu Speciation

| Reaction | Log K | Reference |
|---|---|---|
| $Cu^{2+} + H_2O = CuOH^+$ | -8.0 | 32 |
| $Cu^{2+} + 2H_2O = Cu(OH)_2^0$ | -16.0 | 33 |
| $Cu^{2+} + SO_4^{2-} = CuSO_4^0$ | 2.31 | 34 |
| $Cu^{2+} + CO_3^{2-} = CuCO_3^0$ | 6.73 | 32 |
| $Cu^{2+} + 2CO_3^{2-} = Cu(CO_3)_2^{2-}$ | 9.83 | 32 |
| $Cu^{2+} + Cat^{2-} = CuCat^0$ | 13.9 | 35 |
| $Cu^{2+} + 2Cat^{2-} = Cu(Cat)_2^{2-}$ | 24.9 | 35 |
| $H^+ + Cat^{2-} = HCat^-$ | 13.0 | 35 |
| $2H^+ + Cat^{2-} = H_2Cat$ | 22.2 | 35 |

Note: $Cat^{2-}$ = catecholate

Table II. Copper Complexation Parameters of Ground Water

| Well | pH | Method[a] | Log $K_1$ | $SD^b$ | Log $L_{1,Tot}$ | $SD^b$ | Log $K_2$ | $SD^b$ | Log $L_{2,Tot}$ | $SD^b$ |
|---|---|---|---|---|---|---|---|---|---|---|
| 1 | 6.0 | FQ | 6.4 | 0.1 | -6.4 | 0.2 | | | | |
| 1 | 6.0 | ISE | No significant binding | | | | | | | |
| 1 | 7.0 | FQ | 6.7 | 0.1 | -5.6 | 0.1 | | | | |
| 1 | 7.0 | ISE | 6.7 | 0.2 | -6.1 | 0.0[c] | | | | |
| 1 | 8.3 | FQ | 6.7 | 0.1 | -6.5 | 1.7 | | | | |
| 1 | 8.3 | ISE | 9.2 | 0.3 | -6.4 | 0.1 | 5.8 | 0.1 | -4.2 | 0.2 |
| 3 | 7.0 | FQ | 6.5 | 0.1 | -5.9 | 0.1 | | | | |
| 3 | 7.0 | ISE | 8.0 | 0.1 | -6.3 | 0.0[c] | 5.5 | 0.5 | -5.6 | 0.2 |
| 3 | 8.0 | FQ | 8.1 | 0.2 | -6.6 | 0.1 | 5.5 | 0.1 | -6.1 | 0.1 |
| 3 | 8.0 | ISE | 8.5 | 0.1 | -6.8 | 0.1 | 5.6 | 0.1 | -5.0 | 0.2 |
| 3 | 8.0 | CSV | 10.1 | 0.3 | -7.3 | 0.2 | 8.7 | 0.1 | -6.1 | 0.1 |
| 4 | 6.0 | FQ | 6.4 | 0.1 | -6.6 | 0.1 | 5.8 | 0.3 | -6.7 | 0.3 |
| 4 | 6.0 | ISE | 6.1 | 0.1 | -5.7 | 0.1 | | | | |
| 4 | 7.8 | FQ | 6.4 | 0.1 | -6.0 | 0.1 | | | | |
| 4 | 7.8 | ISE | 9.1 | 0.1 | -6.3 | 0.0 | 6.6 | 0.1 | -5.3 | 0.0[c] |

Notes: [a] FQ fluorescence quenching, ISE ion-selective electrode, CSV cathodic stripping voltammetry
[b] SD standard deviation
[c] Standard deviation was less than 0.05.

over the range of concentrations used in the CSV, ISE, and FQ titrations ($10^{-8}$ to $10^{-5}$ M). The results are shown in Figure 2. The total organically-bound Cu concentration is shown because the concentrations of the individual species are not meaningful. The concentrations of $CuOH^+$, $Cu(OH)_2^0$, $Cu(CO_3)_2^{2-}$, and $CuSO_4^0$ were less than the concentration of $Cu^{2+}$. Complexes with natural organic matter dominate Cu speciation at low total Cu concentrations and comprise a significant fraction of total Cu over the entire range of Cu concentrations considered.

Comparison with Published Results

All of the metal complexation models described in the materials and methods section, including the multi-ligand model used in the present study, have been criticized for yielding parameters that "have no physicochemical meaning, are extremely operational, and, consequently, cannot even serve as a basis for comparing the properties of different samples even when experimental conditions and interpretation assumptions are identical." (42) It is true that the curve-fitting parameters from the various complexation models are operationally defined. The parameters are functions of the range of total metal concentrations used to determine them (25) and the statistical properties of the analytical methods (10). It is possible, perhaps even tempting, to over-interpret them as different classes of binding sites. However, empirical complexation parameters are not true thermodynamic stability constants or concentrations of actual ligands. The parameters are not independent. The conditional stability constant and the "concentration" of a model ligand are usually highly correlated. Therefore, it is not appropriate to compare complexation parameters determined by different methods or for different water samples. Comparison of metal speciation computed using model parameters, however, is appropriate (43).

Figures 3 and 4 compare Cu speciation computed from complexation parameters determined in the present study with published studies. The parameters used to compute the curves in Figure 3 were determined by ISE, while the parameters used for Figure 4 were determined by FQ. The curves for wells 1, 3, and 4 of the present work were computed using parameters determined at the natural pH values. In Figure 3, the Newport River curve is for a surface water (TOC 15 mg $L^{-1}$, natural pH of 5.95) (36). The Yuma Canal curve is for fulvic acid isolated from a mixture of ground waters (TOC 3.6 mg $L^{-1}$). The Biscayne Aquifer curve is for fulvic acid isolated from ground water collected from a carbonate aquifer (TOC 15.0 mg $L^{-1}$). The Yuma Canal and Biscayne Aquifer titrations were performed at pH 6.25 (9). Cu speciation was computed using other published complexation parameters (9,23,36,37). The curves shown in Figure 3 are representative. The Oyster River Curve in Figure 4 is for a surface water (TOC 8.6 mg $L^{-1}$, pH 6.7) (44). The curves in Figures 3 and 4 show the ranges of total Cu concentrations covered in the experiments.

All curves in Figure 3 show that there was measurable Cu complexation. Free $Cu^{2+}$ was less than total Cu at all points except for the Yuma Canal curve for $10^{-5}$ to $10^{-4}$ M total Cu. For any value of total Cu, the maximum range of computed free $Cu^{2+}$ was approximately a factor of 20. The Well 1 and Newport River curves were practically identical. The Well 4 curve was very close to the Well 1 and Newport River curves for $10^{-7}$ to $10^{-6}$ M total Cu, but diverged at higher concentrations. The Biscayne Aquifer curve was fairly close to the Well 1 curve for $10^{-6}$ M total Cu and approached the Well 3 curve at $10^{-5}$ M. Cu speciation has been shown to depend strongly on pH and TOC (36,41). Despite differences in TOC, pH, and the concentrations of other ions, the free $Cu^{2+}$ concentrations computed for the ground waters of the present study were similar (within a factor of 20 or less) to those computed using parameters determined for diverse waters. One notable difference was for well 1 at pH 6. In the present study, there was no measurable Cu complexation in ground water from well 1 at pH 6. Several published studies have found complexation of Cu complexation by natural waters (36) and fulvic acid isolated from natural waters at pH 6 (9,36,37,41).

For the FQ curves (Figure 4), the maximum range in computed free $Cu^{2+}$ concentrations for any value of total Cu was a factor of approximately 300. The Well 1 and Oyster River curves were very similar with apparently weak Cu binding. In reference 44, four fresh surface water samples and one sediment pore water sample were analyzed by FQ. In all cases, Cu binding was apparently weak. Only one example from reference 44 is shown in Figure 4. Cu binding quenched fluorescence in ground water from wells 3 and 4 of the present work more effectively than in the waters studied in reference 44. This difference in quenching may have been due to stronger Cu binding by the ground water from wells 3 and 4. However, the FQ results from the present work

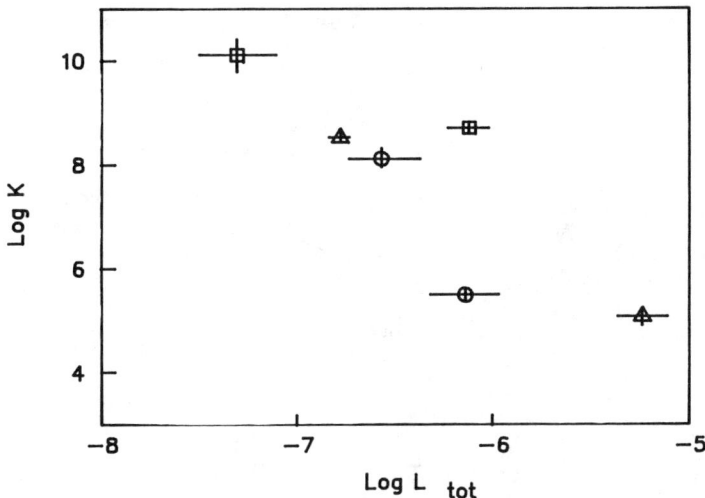

Figure 1. Comparison of Complexation Parameters Determined by Different Analytical Methods for Ground Water from Well 3, pH 8.0.   △ Ion-Selective Electrode, ○ Fluorescence Quenching, ☐ Cathodic Stripping Voltammetry. The lines through the symbols indicate mean values plus or minus one standard deviation.

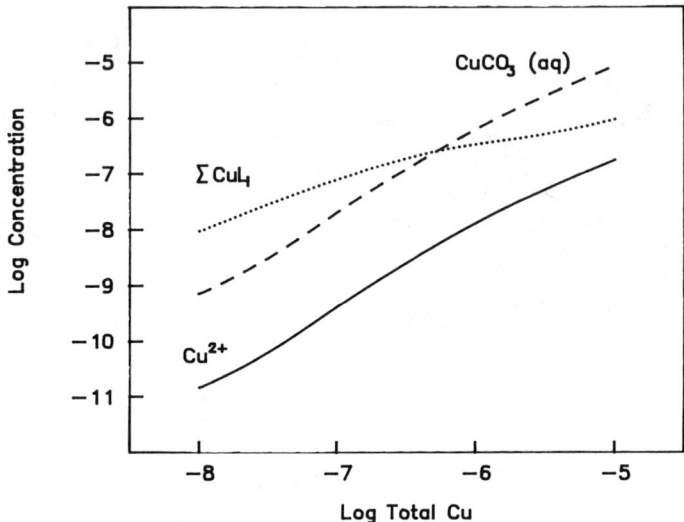

Figure 2.   Copper Speciation in Ground Water from Well 3 Computed Using the Complexation Parameters Determined by ISE and CSV. $\Sigma\ CuL_i$ is Organically Complexed Cu.

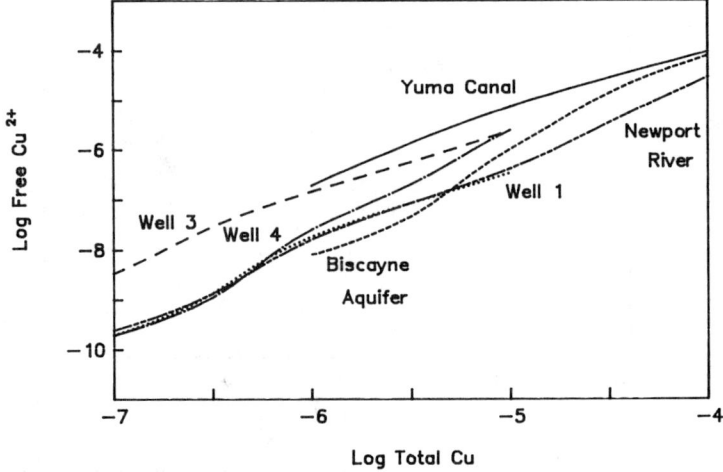

Figure 3. Comparison of Ion-Selective Electrode Titrations Performed at Natural pH Values of Ground Waters with Published Results.

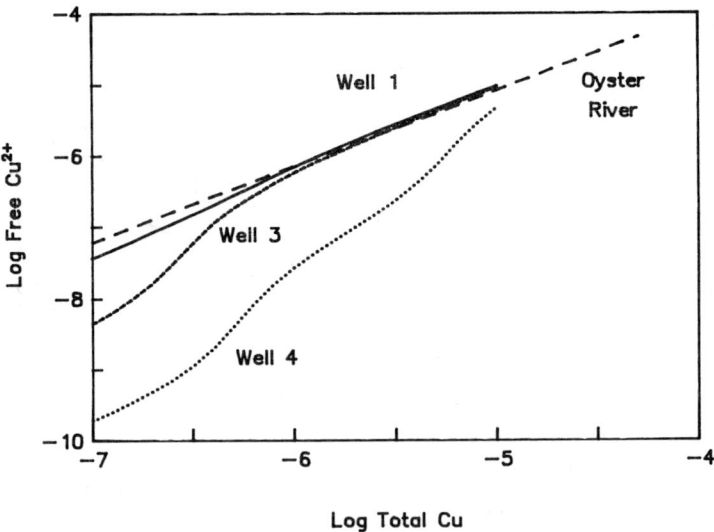

Figure 4. Comparison of Fluorescence-Quenching Titrations Performed at Natural pH Values of Ground Waters with Published Results.

suggest that there may also have been difference in the fluorescence properties of the different waters.

SUMMARY AND CONCLUSIONS

Copper is strongly bound by the natural organic matter in the shallow ground water from the Sand Ridge site. One- or two-ligand models fit the complexometric titrations well. Three independent analytical methods yielded fairly consistent sets of complexation parameters for one ground-water sample, which tends to confirm the interpretation of the data. The complexation parameters can be used as total ligand concentrations and conditional stability constants in chemical-equilibrium computations. The parameters are strictly valid only for the ground waters and under the conditions (pH, temperature) for which they were determined. However, they should be useful in sensitivity analyses of Cu speciation in ground water.

Chemical equilibrium computations show that complexes formed with the natural organic matter are the dominant Cu species for total dissolved Cu concentrations on the order of or less than the model ligand concentrations. The Cu concentrations used in the complexometric titrations were much higher than the natural Cu concentrations. Therefore, the complexation parameters probably do not provide an accurate estimate of Cu speciation at natural concentrations. Cu complexation would probably be underestimated at natural Cu concentrations.

Comparisons of Cu speciation computed for the ground waters of the present work with other natural waters are difficult because of differences in pH, ionic strength, concentrations of major ions, and the use of fulvic acid isolated from the waters. Despite these differences, free Cu concentrations computed for the present study differed from those computed for other waters by a factor of 20 or less. However, the shallowest ground water in the present study differed from many surface waters in that it did not significantly bind Cu at pH 6. Therefore, it is not always valid to assume that ground-water organic matter has the same properties as organic matter in surface water.

The fluorescence of the well 1 ground water was quenched to the same degree at pH 6.0 as at pH 8.3, even though the ISE results showed that there was no significant binding at the lower pH value. This implies that fluorescence quenching probably should not be used as an independent metal speciation method. Cu complexation determined by FQ for well 1 water was similar to published FQ results. However, for wells 3 and 4, Cu complexation was much stronger than for the published FQ results. Apparently natural waters differ significantly with respect to their fluorescence properties.

ACKNOWLEDGMENTS

The authors thank M. E. Caughey, G. K. George, and N. L. Holm for comments on the manuscript and O. E. Smith for inorganic carbon determinations. This research was supported by the R. S. Kerr Environmental Research Laboratory, U. S. Environmental Protection Agency. Although the research described in this article was funded wholly or in part by the United States Environmental Protection Agency, through its cooperative agreement program, it has not been subjected to the Agency's peer and policy review and, therefore, does not necessarily reflect the views of the Agency, and no official endorsement should be inferred.

LITERATURE CITED

1. Stumm, W.; Morgan, J. J. Aquatic Chemistry, 2nd ed.; Wiley-Interscience: New York, 1981; pp. 230-320.
2. Jenne, E. A. In Trace Inorganics in Water, Baker, R. A., Ed.; ACS Symposium Series No. 73; American Chemical Society: Washington, D.C., 1967; pp. 337-387.
3. Brown, D. S.; Carlton, R. E.; Mulkey, L. A. Development of Land Disposal Decisions for Metals Using Minteq Sensitivity Analyses, EPA/600/3-86/030, NTIS PB86-233186.
4. Gillespie, P. A.; Vaccaro, R. F. Limnol. Oceanogr., 1978, 23, 543-548.
5. Sunda, W. G.; Klaveness, D.; Palumbo, A. V. In Complexation of Trace Metals in Natural Waters; Kramer, C. J. M.; Duinker, J. C., Eds.; Nijhoff/Junk: The Hague, 1984; pp. 393-409.
6. Babich, H.; Stotzky, G. Environ. Res. 1985, 36, 111-137.
7. Neubecker, T. A.; Allen, H. E. Wat. Res. 1983, 17, 1-14.

8. Thurman, E. M. Organic Geochemistry of Natural Waters; Nijhoff/Junk: The Hague, 1986; pp. 14-15.
9. McKnight, D. M.; Feder, G. L.; Thurman, E. M.; Wershaw, R. L. Sci. Tot. Environ. 1983, 28, 65-76.
10. Fish, W.; Morel, F. M. M. Can. J. Chem. 1985, 63, 1185-1193.
11. Cabaniss, S. E.; Shuman, M. S. Geochim. Cosmochim. Acta 1988, 52, 185-193.
12. Thurman, E. M.; Malcolm, R. L. Environ. Sci. Technol. 1981, 15, 463-466.
13. Holm, T. R.; George, G. K.; Barcelona, M. J. Dissolved Oxygen and Oxidation-Reduction Potentials in Ground Water; EPA/600/S2-86/042. NTIS No. PB 86-179 678/AS.
14. Good, N. E.; Winget, G. D.; Winter, W.; Connolly, T. N., Izawa, S.; R. M. M. Singh. Biochem., 1966, 5, 467-477.
15. Van den Berg, C. M. G. Anal. Chim. Acta. 1984, 164, 195-207.
16. Huffman, E. W. D., Jr. Microchem. J. 1977, 22, 567-573.
17. Shatkay, A. Anal. Chem. 1976, 48, 1039-1050.
18. Van den Berg, C. M. G. Mar. Chem. 1984, 15, 1-18.
19. Buffle, J.; Greter, F.; Haerdi, W. Anal. Chem. 1977, 49, 216-222.
20. Wilson, D. E.; Kinney, P. Limnol. Oceanogr. 1977, 22, 281-289.
21. Perdue, E. M.; Lytle, C. R. Environ. Sci. Technol. 1983, 17, 654-660.
22. Thakur, A. K.; Munson, P. J.; Hunston, D. L.; Robard, R. Anal. Biochem. 1980, 103, 240-254.
23. Turner, D. R.; Varney, M. S.; Whitfield, M.; Mantoura, R. F. C.; Riley, J. P. Geochim. Cosmochim. Acta 1986, 50, 289-297.
24. Cabaniss, S. E.; Shuman, M. S.; Collins, B. J. In Complexation of Trace Metals in Natural Waters; Nijhof/Junk: The Hague, 1984, pp 165-179.
25. Dzombak, D. A.; Fish, W.; Morel, F. M. M. Environ. Sci. Technol. 1986, 20, 669-675.
26. Black, A. P.; Christman, R. R. J. Am. Wat. Works Assoc. 1963, 55, 753-770.
27. Saar, R. A.; Weber, J. H. Anal. Chem. 1980, 52, 2095-2100.
28. Ryan, E. K.; Weber, J. H. Anal. Chem. 1982, 54, 986-990.
29. Chen, R. F. Anal. Lett. 1986, 19, 963-977.
30. Cacechi, M. S.; Cacheris, W. P. Byte 1984, 9, 340-362.
31. Press, W. H.; Flannery, B. P.; Teukolsky, S. A.; Vetterling, W. T. Numerical Recipes. The Art of Scientific Computing; Cambridge: Cambridge, 1986, pp 529-532.
32. Baes, C. F.; Mesmer, R. E. The Hydrolysis of Cations; Wiley: New York, 1976.
33. Gulens, J.; Leeson, P. K.; Seguin, L. Anal. Chim. Acta 1984, 156, 19-31.
34. Nordstrom, D. K. Ph.D. Thesis, Stanford University, Stanford, CA, 1977.
35. Smith, R. M.; Martell, A. E. Critical Stability Constants. Vol. 3. Other Organic Ligands; Plenum: New York, 1975. P 200.
36. Sunda, W. G.; Hanson, P. J. In Chemical Modeling in Aqueous Systems; Jenne, E. A., Ed.; ACS Symposium Series No. 93; American Chemical Society: Washington, DC, 1979; pp 147-180.
37. Fish, W.; Dzombak, D. A.; Morel, F. M. M. Environ. Sci. Technol. 1986, 20, 676-683.
38. Berger, P.; Ewald, M.; Liu, D.; Weber, J. H. Mar. Chem. 1984, 14, 289-295.
39. Ryan, D. K.; Thompson, C. P.; Weber, J. H. Can. J. Chem. 1983, 61, 1505-1509.
40. Holm, T. R.; Barcelona, M. J. Proc. Nat. Wat. Well Assoc. Ground Wat. Geochem. Conf., 1988, 245-267.
41. Cabaniss, S. E.; Shuman, M. S. Anal. Chem. 1986, 58, 398-401.
42. Altmann, R. S.; Buffle, J. Geochim. Cosmochim. Acta 1988, 52, 1505-1519.
43. Shuman, M. S. In Metal Speciation: Theory, Analysis and Application; Kramer, J. R.; Allen, H. E., Eds.; Lewis: Chelsea, MI, 1988; pp 125-133.
44. Ryan, D. K.; Weber, J. H. Environ. Sci. Technol. 1982, 16, 866-872.

RECEIVED July 20, 1989

# Chapter 40

# Kinetics of Rare Earth Metal Binding to Aquatic Humic Acids

## Sue B. Clark[1] and Gregory R. Choppin

### Department of Chemistry, Florida State University, Tallahassee, FL 32306–3006

Complexation of Eu(III) to humic acid was found to be fast but, initially, only a small fraction is strongly bound. This fraction increases with time over a 2-day period. The dissociation kinetics can be described by an equation with 5 first order terms with rate constants from $1.80 \times 10^{-2}$ to $1.67 \times 10^{-3}$ m$^{-1}$ (pH 4.2). The percent dissociation by the slowest paths is the same as the fraction found to be strongly bound in the complex formation studies after the initial 2 days. These results indicate that the extent of effects due to metal-humate interaction as water moves through a soil may vary with the rate of flow. Particularly for slow waters, humate interactions may be important to include in modeling metal migration.

Humic substances are collections of large organic molecules with a variety of functional groups arranged in no regular, recurring pattern. The diversity of functional groups serve to allow binding to a large number of metal ions. Since these humic substances are ubiquitous, they can have significant roles in the geochemical speciation of metals. Not only can they serve to retard metal ion migration (e.g., when the humics are sorbed to surfaces) or to promote it (e.g., by metal complexation to soluble or colloidal humics), these substances can also influence the oxidation state (e.g., $PuO_2^{2+}$ is rapidly reduced to $Pu^{4+}$ by humics). The concentration of humics depends on many factors such as climate, soil and water pH, substrate material, topography, depositional environment and time. However, in many natural systems, the concentration is sufficient to affect metal ion speciation and, consequently, migration. To model, properly and successfully, the geochemical behavior of metals in many systems requires inclusion of data on the metal-humate interactions.

Humics span a great range of sizes - from less than 1000 daltons to greater than 500,000 daltons. In this paper, we are concerned only with the soluble humics - i.e., the humic and fulvic acids. The particular humic acid studied is ca. 30,000 daltons, but this is only an average value of a wide spectrum of sizes. In molecules of this size, there can be well over 100 units of each functional group. Metal ions may be attracted to the molecule because of the net charge of the molecule (territorial binding) (1,2), or to individual functional groups (site binding) (3). In the former case, the cations retain some degree of translational freedom to move around the outer surface of the humic macromolecule. Both types of binding neutralize some of the charge of the macromolecule, reducing the intramolecular repulsion. As a result, upon metal binding, the humic molecule may undergo changes in its conformation, becoming less extended.

Lanthanide cations, Ln(III), are widely present in nature and based on thermodynamic data, Torres and Choppin (4) have calculated that complexation of Ln(III) by humics could predominate in some organic rich environmental systems and affect the geochemical transport properties of these cations. Ln(III) cations interact predominantly with oxygen donor groups which, in the case of humics, means they react with ionized carboxylate groups. At the pH of natural systems, the phenol groups remain protonated and Ln(III) cations would only interact with them if they (the Ln) are already bound to a neighboring carboxylate.

[1]Current address: Savannah River Laboratory, Interim Waste Technology, Building 773–43A, Room 213, Aiken, SC 29808

The kinetics of the formation and dissociation of these complexes may also be significant in the geochemical behavior of metals. Although studies of complexation kinetics of metal-humate interactions have been reported (e.g., ref. 5,6,7), little insight has been gained into the mechanism of the nature of the binding of the metal ion. The secondary effect of conformational rearrangement subsequent to the initial binding can be expected to affect the kinetics. The migration through a soil could be controlled by either thermodynamic kinetic stability or by a combination. To better evaluate the need to include kinetic parameters in modeling of humic interactions with metals, we have studied the kinetics of trivalent lanthanide (Sm(III) and Eu(III)) interactions with humic acid for which the thermodynamics of binding have been reported previously (1). These cations are representative of Ln(III) behavior as well as of the behavior to be expected of trivalent actinide elements such as Am(III).

## EXPERIMENTAL

Ultrafiltration and ion exchange techniques were used to study characteristics of the formation path of Ln(III) binding to humic acid. A competitive ligand exchange technique was employed to study the complex dissociation. In the latter, arsenazo III (Ars) was used as the competitive ligand. Eu(III) was used in the methods requiring radioactivity while Sm(III) was used in the other studies; however, the chemistry of Sm(III) and Eu(III) can be considered identical within the uncertainties of these methods.

### Reagents

Reagent grade arsenazo III (3,6-bis[2-arsonophenyl]-azo- 4,5-dihydroxy-2,7-naphthalene disulfonic acid, 99.9%) from Aldrich Chemical Co. was used. Radioactive Eu-152 was obtained from Oak Ridge National Laboratory. Stock solutions of Sm(III) and Eu(III) were prepared by dissolving the oxide in concentrated $HClO_4$, followed by filtration and dilution. The stock solutions were standardized by titration with EDTA using xylenol orange as an indicator. Humic acid extracted from sediments collected in Lake Bradford (Tallahassee, Florida) had been characterized as having 3.86 meq/g carboxylate capacity and an average molecular weight of 28,000 daltons (4). A stock solution was prepared by dissolving the humic acid in a minimum volume of 0.1 m NaOH followed by dilution with distilled water. Stock solutions of acetate buffer and MES (4-morpholine-ethane-sulfonic acid) were prepared with Aldrich reagents. All solutions were 0.001 m in buffer, and their ionic strengths were adjusted to 0.10 m with $NaClO_4$.

### Ultrafiltration

Ultrafiltration was used to determine the extent of europium binding to humic acid. While stirring, Eu(III) + Eu- 152 tracer was added to a 10 ml solution of humate at pH 4.2 (acetate buffer) in an Amicon model #8050 ultrafiltration cell. After a 500 $\mu$L aliquot was withdrawn for counting, the cell was immediately pressurized. An Amicon YM2 filtration membrane (1000 molecular weight cut-off) would retain the humic acid plus the bound europium while unbound Eu(III) would elute. One ml of filtrate was collected and tested for the presence of humate by measuring the visible spectrum of the solution with a Milton Roy Spectronic 1201. On cessation of filtration, 500 $\mu$L aliquots were removed from the filtrate and the residual solution above the membrane. The samples were counted for the presence of Eu-152 in a well-type NaI(Tl) crystal connected to a single channel analyzer.

### Ion Exchange

Dowex 50 X 4, 100-200 mesh cation exchange resin was prepared by washing successively with concentrated HCl, distilled water, 0.1 m EDTA, distilled water, and 0.1 m NaOH. The $Na^+$ form resin was loaded into a glass column (1.2 cm ID and 5.7 cm length) and equilibrated with acetate buffer of the desired pH prior to the experiment. Glass wool was placed on top of the resin to minimize disturbance of the resin upon addition of the reactive solutions. These columns had been found to rapidly and quantitatively remove free metal ions from solution (8).

A solution of humic acid was mixed with the Eu(III). At fixed time intervals, aliquots of this solution were withdrawn and passed through the column of resin. Five ml of acetate buffer followed to elute the Eu-Humate complex. This all took $\leq$ 30 s. The free Eu(III) on the resin was then eluted by passage of 0.1 M HCl solution. The two eluant fractions were counted for Eu-152.

Ligand Exchange Technique

The rate of Sm(III) dissociation from humic acid was studied using a ligand exchange technique. A solution of the Sm(III)-humate complex was equilibrated for a fixed time, then an aliquot of arsenazo III was added. The Sm(III)-arsenazo III complex is so stable (9) that it prevents reformation of the Sm(III)-humate. The rate of Sm(III)-Ars complex formation is quite rapid (9) and we found that the observed formation of the Sm(III)-arsenazo complex was determined by the rate of dissociation of the Sm(III)-humate. The formation of the Sm(III)-Ars complex was monitored spectrophotometrically by the increase in absorbance at 665 nm. Reactions were monitored either with a Milton Roy Spectronic 1201 ($t_{1/2} > 0.5$ m) or with a Durrum-Gibson D-109 stopped- flow spectrophotometer interfaced to an IBM-PC XT ($t_{1/2} < 0.5$ m). It had been found (10) that increasing the concentration of arsenazo III did not affect the rate of dissociation of metal ions from humic acid, confirming that arsenazo III does not catalyze dissociation of the metal ions from humic acid.

Data Analysis

All computational work was performed on a Zenith ZF-158-42 personal computer. To determine the number of first order components in the dissociation, analysis by the kinetic spectrum method was used. This technique was been used for Eu-Humate kinetics (11) and utilizes a unitless distribution function, H(k,t), which is defined as:

$$H(k,t) = \frac{\delta^2 (\Delta \text{Absorbance})}{\delta (\ln t)^2} - \frac{(\Delta \text{Absorbance})}{\delta (\ln t)} \quad (1)$$

A plot of H(k,t) vs ln t yields a curve with a maxima for each first-order process. The position of the maximum yields an estimate of the rate constant and the area under the curve, an estimate of the percent of metal ion dissociated by that process.

The kinetic spectra were calculated using a cubic spline smoothing calculation (12). Because the domain of the kinetic spectrum is the natural logarithm of time, small errors in peak position correspond to large errors in rate constants. The kinetic spectrum analysis was used only to obtain preliminary estimates of the rate constants and the percent dissociation by each path. These were then used as the initial values in a simplex non-linear regression (12) fit of the original data. The results of this latter treatment are the reported values.

RESULTS

Ultrafiltration studies on Eu(III)-humic acid systems showed that the Eu(III) cation has associated completely with the humate within the time to complete the filtration (1 m.) In the resin experiments, Eu(III) was added to the humate solution, and, after 15 minutes, the solution was passed rapidly through the resin. Only 4% of the Eu(III) was present in the eluant (i.e., as Eu-Hu). When the contact time was 48 hours before passage through the resin, 38% of the Eu(III) eluted as Eu-Humate. These data must represent a different binding than the 100% binding of the ultrafiltration.

The dissociation of Sm(III) from humic acid followed a first order decay with several terms. Two "fast" processes with half-lives of a few seconds, and three "slow" terms with half-lives of minutes (Fig. 1a and 1b) were found. Thus, the decay data shows 5 distinguishable processes as described by Equation 2:

$$\text{Rate} = \sum_{i=1}^{5} k_i [Sm(Hu)]_i \qquad (2)$$

Three sets of data using the stopped-flow method were collected for solutions of pH 4.2 after the Sm(III) and humic acid had been mixed for at least 48 hours. The averaged results of the three runs are presented in Table I.

The primary emphasis in this paper is on the three "slow" dissociation paths. In Table II, the variation of these terms with pH is given. These values correspond to hydrogen ion dependencies of 0.102, 0.172, and 0.173, respectively, for $k_3$, $k_4$, and $k_5$. The percent dissociation for each path varied when the mixing time prior to dissociation was less than 2 days; for longer times, they were invariant.

Table I. Decay Constants and Percentage Dissociation for the Dissociation of Sm-Hu

$T = 25°C, I = 0.10 \text{ M (NaClO}_4\text{), pH 4.2}$

| Rate Constant | % Dissociation |
|---|---|
| $k_1 = 1.08 \pm 0.17 \text{ s}^{-1}$ | $45.1 \pm 0.70$ |
| $k_2 = 0.132 \pm 0.039 \text{ s}^{-1}$ | $20.2 \pm 0.42$ |
| $k_3 = 1.43 \pm 0.17 \text{ m}^{-1}$ | $6.55 \pm 0.45$ |
| $k_4 = 0.169 \pm 0.014 \text{ m}^{-1}$ | $9.47 \pm 0.34$ |
| $k_5 = 0.00167 \pm 0.00082 \text{ m}^{-1}$ | $15.4 \pm 0.40$ |

Table II. Dependence of Rate Constants on pH

$T = 25°C, I = 0.10 \text{ M (NaClO}_4\text{)}$

| pH | $k_3$ | $k_4$ | $k_5$ |
|---|---|---|---|
| 4.22 | 1.43 | 0.169 | 0.00167 |
| 4.59 | 1.11 | 0.131 | 0.00116 |
| 4.97 | 1.15 | 0.152 | 0.00102 |
| 5.45 | 1.02 | 0.0954 | 0.00100 |

## DISCUSSION

The ultrafiltration experiments confirmed the rapid, complete binding of lanthanides to humate. The ion exchange experiments indicate that initially most of the Eu(III) is very weakly bound as the cation exchange resin removed most of the Eu(III) from the humate. After 2 days of

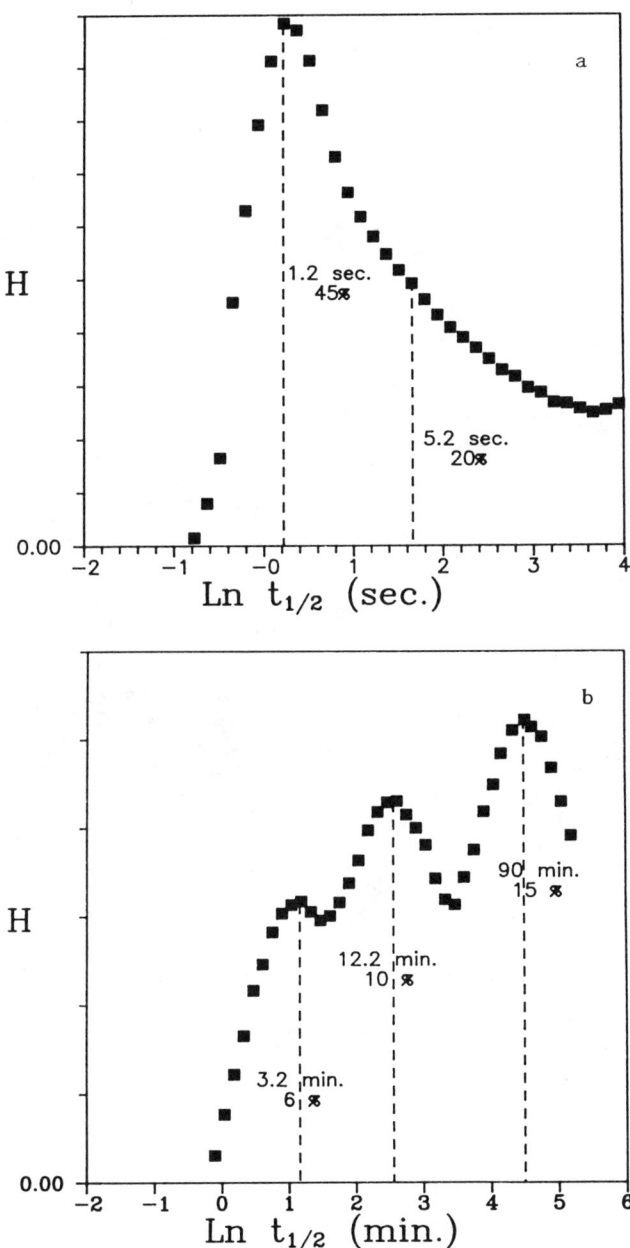

Figure 1. Kinetic spectrum for Sm(III) dissociation from humic acid as measured by a, stopped-flow and b, conventional spectrophotometry. T = 25 °C, I = 1.10 M (NaClO$_4$), pH 4.2, 5% metal loading of the polyelectrolyte.

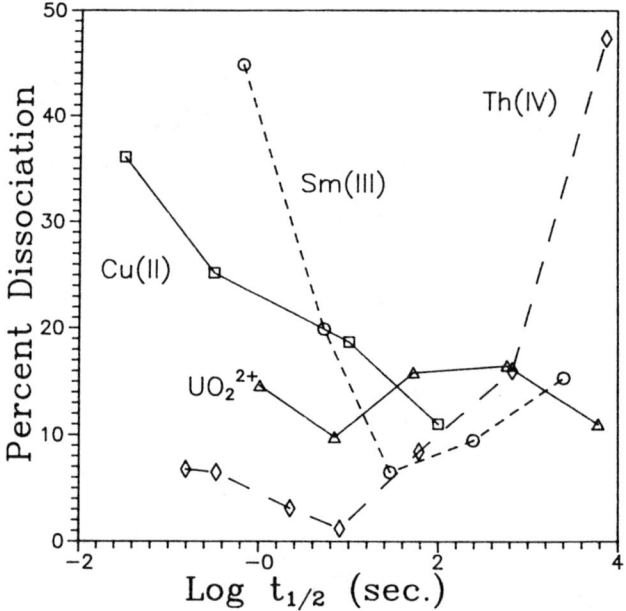

Figure 2. Log (half-life) versus % metal ion dissociation by each decay process. T = 25 °C, I = 0.10 M (NaClO$_4$), pH 4.2. Data for Th(IV) taken from ref. 12, UO$_2^{2+}$ data from reference 13, and Cu(II) data from reference 6.

complexation, about 40% of the Eu(III) is bound so strongly that the resin can no longer compete for the bound metal ions. This 40% of strongly bound lanthanide corresponds well to the amount which dissociates by the "slow" paths after two days of binding time.

Figure 2 shows the relationship between the percent dissociation by a path and the half-life of that path for several metal ions, including Sm(III). The values for half-lives of the three decay processes shown for Cu(II) ([6]) are the median values from the reported ranges of rate constants. Cu(II) interacts with the nitrogen donors as well as the carboxylates while Ln(III), Th(IV) and $UO_2^{2+}$ are complexed by the carboxylates. In general, the higher the cationic charge, the greater the percent dissociation by the slowest pathways.

These observations are consistent with a model in which the cations rapidly associate with the humic acid. Initially, the largest fraction is weakly bound, perhaps by territorial binding and site binding to, surface groups. With time, more of the cations become more strongly bound, probably by a combination of cation migration to interior sites of the humic molecule and of contraction and folding of the humic molecule whereby surface bound cations become trapped internally. Whatever the mechanism, it is obvious that kinetics may play a significant role in metal migration. For fast water movement through humic rich soils, the retardation may be minimal, as there is insufficient contact time for strong binding to occur. However, with slower movement, the metals can become retained by the humics and their migration slowed or even stopped.

ACKNOWLEDGEMENTS
This research was supported by a grant from the USDOE-OBES, Division of Chemical Sciences. We also express our gratitude to Ms. Sussan Sandberg of The Royal Institute of Technology, Stockholm, Sweden for assistance in the ultrafiltration and ion exchange experiments.

LITERATURE CITED

1. Manning, G. S. Radiochim. Acta 1969, 24, 519.
2. Manning, G. S. J. Chem. Phys. 1981, 85, 870.
3. Marinsky, J. A. Coord. Chem. Rev. 1976, 19, 125.
4. Torres, R. A.; Choppin, G. R. Radiochim. Acta 1984, 35, 143.
5. Mak, M. K. S.; Langeford, C. H. Canadian J. Chem. 1975, 60, 2023.
6. Olson, D. L.; Shuman, M. S. Geochim. Cosmochim. Acta 1985, 49, 1371.
7. Plankey, B. J.; Patterson, H. H. Environm. Sci. Technol. 1987, 21, 595.
8. Choppin, G. R.; Williams, M. K. R. J. Inorg. Nucl. Chem. 1973, 35, 4255.
9. Nepomnyashchaya, N. A.; Men'kov, A. A.; Lenskii, A. S. Zh. Neorg. Khim. 1975, 20, 1810.
10. Cacheris, W. P.; Choppin, G. R. Radiochim. Acta 1986, 122, 551.
11. Olson, D. L.; Shuman, M. S. Anal. Chem. 1983, 55, 1103.
12. Caceci, M. S.; Cacheris, W. P. BYTE 1984, 5, 340.
13. Clark, S. B. M.S. Thesis, Florida State University, Tallahassee, Florida, 1987.

RECEIVED October 6, 1989

# Chapter 41

# Microscale Processes in Porous Media

## Transport of Chlorinated Benzenes in Porous Aggregates

### Roger C. Bales and James E. Szecsody[1]

**Department of Hydrology and Water Resources, University of Arizona, Tucson, AZ 85721**

Sorption and desorption of chlorinated benzenes were investigated in a series of column experiments using surface-modified silica of known chemical composition. The resulting breakthrough curves were fit to equilibrium (two-parameter), first-order (three-parameter), and mobile-immobile region (four parameter) one-dimensional advection, dispersion and retardation mathematical models. Slow sorption and desorption, evidenced by tailing of the breakthrough curves, was attributed primarily to slow binding and release rather than diffusion through immobile liquid or diffusion through an organic phase. An equilibrium model could be used to fit breakthrough curves with slow sorption/desorption, but the magnitude of the resulting dispersion coefficient includes effects of both media-specific intra-aggregate transfer and sorbate/sorbent-specific binding/release processes. Using the first-order model with independently determined dispersion coefficients gave a fitted first-order rate coefficient for media- and sorbate/sorbent-specific processes. Comparison of first-order model results from particles with different geometries and experiments at different temperatures suggested that release of sorbates from strong binding sites was the rate-limiting desorption step. The binding involved strong van der Waals interactions, as would be found in many natural waters. The additional parameter provided by the mobile-immobile model could not be physically defined in the experimental system studied, and provides no additional insight for interpreting microscale sorption-desorption processes. The same lack of physical definition for additional parameters should apply to many similiar solute-transport situations.

The rates of microscale processes involved in sorption and desorption of organic solutes in subsurface media, together with time scales for advection and dispersion, determine whether equilibrium or nonequilibrium descriptions can be used for transport modeling. Diffusion through immobile fluid (i.e. intra-aggregate diffusion), transport through a bound organic phase, or binding/release (Figure 1) can contribute to slow sorption and desorption. Resistances to sorption in both subsurface and chromatographic media have also been described in terms of film resistance, intraparticle diffusion, and physical or chemical adsorption (1,2).

The governing equation for one-dimensional solute transport in a porous media with both mobile and immobile regions and reversible sorption in both regions is (3) :

[1]Current address: Battelle, Pacific Northwest Laboratories, P.O. Box 999, Richland, WA 99352

$$\theta_e \frac{\partial C_e}{\partial t} + \theta_i \frac{\partial C_i}{\partial t} + f\rho_b \frac{\partial S_e}{\partial t} + (1-f)\rho_b \frac{\partial S_i}{\partial t} = \theta_e D \frac{\partial^2 C_e}{\partial x^2} - u_e \theta_e \frac{\partial C_e}{\partial x} \quad (1)$$

where $C_e$ and $C_i$ are the solute concentrations ($\mu g\, cm^{-3}$) in the mobile and immobile regions respectively; $S_e$ and $S_i$ are the respective bound concentrations ($\mu g\, g^{-1}$); $\theta_e$ and $\theta_i$ are the volume fractions of the mobile and immobile liquid regions, $\rho_b$ is the dry bulk density ($g\, cm^{-3}$) of the porous media, $f$ is the fraction of sorbent in the mobile region; $D$ is the longitudinal hydrodynamic dispersion coefficient ($cm^2\, s^{-1}$); and $u_e$ is the average interstitial velocity ($cm\, s^{-1}$).

Diffusion through the immobile region is described as slow mass transfer between mobile and immobile water. The solute concentrations in the two regions are related by:

$$\theta_i \frac{\partial C_i}{\partial t} + \rho_b (1-f) \frac{\partial S_i}{\partial t} = \alpha_e (C_e - C_i) \quad (2)$$

where $\alpha_e$, the mass transfer coefficient ($s^{-1}$), depends on both the solute diffusion coefficient and sorbent geometry (4). The relation can also be formulated with a diffusion model for planar, rectangular, spherical or cylindrical geometries (5). At equilibrium, the sorbed and aqueous concentrations in each region are related by:

$$S = K_p C \quad (3)$$

where $K_p$ is the equilibrium partition coefficient ($cm^3\, g^{-1}$). Transport through a bound organic phase can be formulated in a similar manner:

$$\rho_b \frac{\partial S_i}{\partial t} = \alpha_i (K_p C_i - S_i) \quad (4)$$

where $\alpha_i$ is the mass transfer coefficient between water and the bound organic layer. For kinetically limited binding and release the governing equation is:

$$\rho_b \frac{\partial S_i}{\partial t} = k_f \theta_i C_i - k_b S_i \quad (5)$$

where $k_f$ and $k_b$ are the first-order forward and reverse rate coefficients ($s^{-1}$), respectively.

In terms of the dimensionless variables and parameters of Table I, equation (1) becomes:

$$\beta R \frac{\partial C_1}{\partial T} + (1-\beta) R \frac{\partial C_2}{\partial T} = \frac{1}{P} \frac{\partial^2 C_1}{\partial Z^2} - \frac{\partial C_1}{\partial Z} \quad (6)$$

where the dimensionless parameters are retardation factor ($R$), Peclet number ($P$) and mobile-immobile fraction parameter ($\beta$). For the equilibrium model $C_1 = C_2$. Considering a single step of Figure 1 to be rate limiting, equations (2), (4) and (5) can each become, in dimensionless form:

$$(1-\beta) R \frac{\partial C_2}{\partial T} = \omega (C_1 - C_2) \quad (7)$$

where $\omega$ is the dimensionless mass-transfer coefficient. Analytical and numerical solutions for various boundary conditions can be found in the literature.

The purpose of the work reported in this paper is to investigate sorption/desorption rates in well-characterized model systems and distinguish physical versus chemical processes that determine the degree of equilibrium versus non-equilibrium behavior.

## METHODS

Column experiments consisted of feeding a KCl solution containing an organic contaminant into a 1.0 cm by 14 cm column packed with silica or bonded silica until breakthrough occurred, then feeding a lower conductivity, contaminant-free solution until all of the contaminant was removed from the column. The organic solutes used were chlorobenzene (CLB), 1,4-dichlorobenzene (DCB), 1,2,4-trichlorobenzene (TCB), 1,2,4,5-tetrachlorobenzene (TeCB) and pentachlorobenzene (PeCB). All experiments were done under saturated conditions. The resulting breakthrough curves were then evaluated using the above models and sorption parameters ($R$, $P$, $\omega$, $\beta$) estimated. Sorbents were porous aggregates of silica, which in most cases had been partially coated with an aliphatic ($C_1$, $C_8$, or $C_{18}$) or aromatic (phenyl) organic group. Organic modifiers were chosen to simulate portions of natural organic matter that are important in the sorption of the hydrophobic compounds. The experimental procedure has been described previously (6).

Table I. Model parameters and dimensionless variables [a]

| Model | $C_1$ | $C_2$ | $T$ | $Z$ | $P$ | $R$ | $\omega$ | $\beta$ |
|---|---|---|---|---|---|---|---|---|
| Equilibrium | $\dfrac{C}{C_o}$ | — | $\dfrac{tu_e}{L}$ | $\dfrac{x}{L}$ | $\dfrac{u_e L}{D}$ | $1+\dfrac{\rho_b K_p}{\theta}$ | — | — |
| First–order kinetic | $\dfrac{C}{C_o}$ | $\dfrac{S}{K_p C_o}$ | $\dfrac{tu_e \theta_e}{\theta L}$ | $\dfrac{x}{L}$ | $\dfrac{u_e L}{D}$ | $1+\dfrac{\rho_b K_p}{\theta}$ | $\dfrac{k_b \rho_b K_p L}{\theta u_e}$ | $\dfrac{1}{R}$ |
| Bound–organic diffusion | $\dfrac{C}{C_o}$ | $\dfrac{S}{K_p C_o}$ | $\dfrac{tu_e \theta_e}{\theta L}$ | $\dfrac{x}{L}$ | $\dfrac{u_e L}{D}$ | $1+\dfrac{\rho_b K_p}{\theta}$ | $\dfrac{\alpha_i K_p L}{\theta u_e}$ | $\dfrac{1}{R}$ |
| Immobile–fluid diffusion | $\dfrac{C}{C_o}$ | $\dfrac{C_i}{C_o}$ | $\dfrac{tu_e \theta_e}{\theta L}$ | $\dfrac{x}{L}$ | $\dfrac{u_e L}{D}$ | $1+\dfrac{\rho_b K_p}{\theta}$ | $\dfrac{\alpha_e L}{\theta_e u_e}$ | $\dfrac{\theta_e + f \rho_b K_p}{\theta + \rho_b K_p}$ |

[a] cf. equations (1) – (7).

Aggregates with different particle size (20-425 µm) and internal pore diameter (10-27.5 nm average) were used to examine the effect of intra-aggregate diffusion. The importance of diffusion through a bound organic layer was examined by changing the thickness of the bound organic phase. The binding/release step was examined by doing experiments with different surface functional groups (aliphatic vs. phenyl vs. polar). Sorbent properties are summarized in Table II. Experiments at different temperatures (3-48°C) were used to determine if the rate limiting step is binding/release or diffusion.

$R$ was determined from the area under the breakthrough curves using a planimeter. Mass eluted compared well with mass injected, indicating that mass balance was achieved. Dispersion ($D$) for a conservative tracer was determined by fitting the KCl breakthrough curve to the equilibrium model; the fitted parameters were $R$ and $P$. A nonlinear least squares method was used for parameter estimation (7). The sum of the squares of the deviation between model and data ($ssq$) was used as a measure of total error in the model fit.

Solute breakthrough data were also fit to the equilibrium model, yielding $R$ and $P$. The resulting dispersion coefficient was used as an indicator of nonequilibrium. That is, if $D_{solute} \gg D_{KCl}$, then intra-aggregate rather than simply mobile-phase factors were contributing to the dispersion of the sorbing solute. Data were also fitted to the first-order and mobile-immobile models, with $P$ fixed; the parameters estimated were then $R$, $\omega$, and for the mobile-immobile model, also $\beta$. $D_{solute}$ was assumed to be proportional to $D_{KCl}$, and values calculated based on differences in the molecular diffusion coefficients of the two species (1). The calculated ratios for 100 µm aggregates are: CLB: 1.3, DCB: 1.5, TCB: 1.9, TeCB: 3.7, and PeCB: 5.6. Although simultaneously estimating all parameters provided a somewhat better fit to the data (i.e. lower $ssq$), interpretation would not correspond to the conceptual model of Figure 1.

RESULTS

Breakthrough Curves

Breakthrough curves for unbonded silica, and for aliphatic ($C_8$ and $C_{18}$) bonded silica showed little tailing (Figures 2-3), whereas those for phenyl-polymer bonded silica exhibited significant tailing

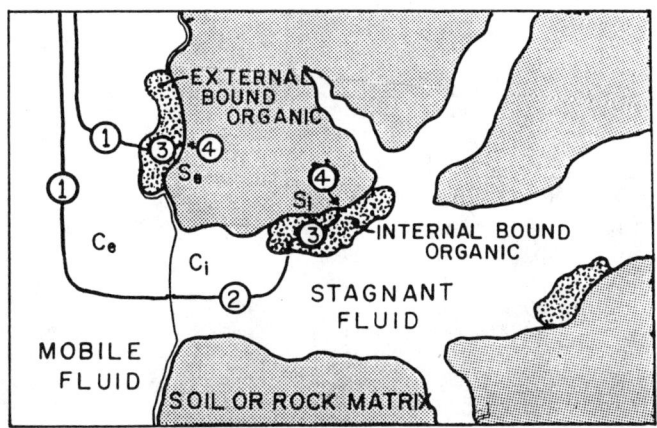

Figure 1: Conceptual model of sorption in porous aggregate: 1) solute transport in mobile fluid and stagnant boundary layer, 2) intraparticle diffusion in stagnant fluid, 3) transport in bound organic phase, and 4) binding and release within the bound organic phase and/or at the mineral surface. $c$ is aqueous and $s$ sorbed concentration; subscripts $e$ and $i$ are for mobile and immobile regions, respectively.

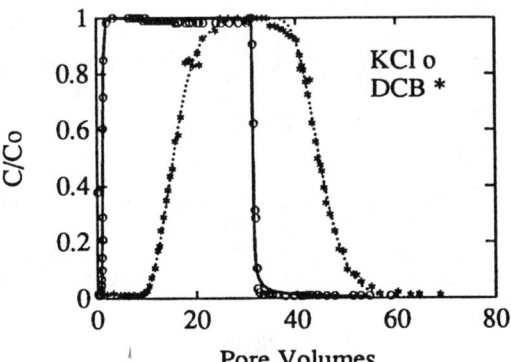

Figure 2. Breakthrough curves for KCl and 1,4-DCB on $C_8$-polymer surface (B-3); $C_o = 110$ ng cm$^{-3}$, diameter = 100 $\mu$m, pore diameter = 27.5 nm; equilibrium model fit.

(Figures 4-5). Sorption was reversible, and was not affected by a lag of up to 60 days between sorption and desorption (e.g. Figure 4). Fitted parameters for 29 column experiments for which good convergence was obtained for the first-order and mobile-immobile models are presented in Table III. Equilibrium-model parameters are presented for an additional 11 experiments. The full data sets for these experiments can be found in Szecsody (8).

Table II. Bonded Silica Properties

| Ident | Dia $\mu m$ | Surface area[a] $m^2g^{-1}$ | Ave pore dia[b] nm | coating | $f_{oc}$ | percent coverage |
|---|---|---|---|---|---|---|
| A | 100 | 180 | 27.5 | $C_1$ | 0.014 | 8.7 |
| B | 100 | 200 | 27.5 | $C_8$ | 0.0068 | 2.1 |
| C | 100 | 190 | 27.5 | $C_{18}$ | 0.011 | 2.1 |
| D | 100 | 220 | 27.5 | PhM | 0.0075 | 0.033 |
| E | 100 | 200 | 27.5 | PhP | 0.014 | 5.1 |
| F | 20 | 350 | 11.0 | PhP | 0.011 | 2.0 |
| G | 20 | 190 | 27.5 | PhP | 0.010 | 3.7 |
| H | 425 | 390 | 10.0 | PhP | 0.011 | 1.9 |
| I | 100 | 200 | 27.5 | — | — | — |
| N | 40 | 150[b] | 12 | $C_{18}$ | 0.18 | 20.9 |
| O | 1.9 | 5.8 | — | $C_1$ | 0.0022 | 0.95 |
| P | 13.5 | 250 | 13.5 | — | — | — |
| Q | 100 | 220 | 27.5 | PhM | 0.0016 | 0.006 |
| R | 100 | 210 | 27.5 | diol | 0.0087 | 0.091 |
| T | 100 | 200 | 27.5 | amine | 0.0081 | 0.084 |
| AE | 8.8 | 1.6 | — | — | — | — |
| AU | 13 | 140[c] | —[c] | $C_{18}$ | 0.019 | — |

[a] by single point $N_2$ adsorption.
[b] Reported by Analytichem Int., Harbor City, CA.
[c] Reported by J. M. Huber Corp., Etowah, TN; pore diameter not reported.

Conservative Tracer Dispersion

The effects of velocity, column size, and packing on dispersion of a conservative tracer were investigated in order to separate the effects of slow sorption and desorption from hydrodynamic dispersion. $D$ values were obtained by fitting an equilibrium model to the breakthrough curves (Table III). There was no correlation between dispersion and column length or diameter, as expected. Dispersion and velocity are related, however.

A plot of $D$ divided by $D_o$ (molecular diffusion coefficient) versus $\frac{u_e d}{D_o}$ (Peclet number, where d is aggregate diameter) shows that for KCl (Figure 6a), dispersion was one to two orders of magnitude above a 1:1 line. The 1:1 line is near the theoretical value for this range of Peclet number (9). No relations between aggregate diameter, pore size, or type of bonded surface and $D$ are apparent for the KCl data. Seven column experiments performed on the same column with KCl and 100-μm unbonded aggregates at velocities of 0.0025-0.2 cm s$^{-1}$ resulted in a power fit of:

$$D = 0.27 u_e^{1.3} \qquad r^2 = 0.995 \qquad (8)$$

which has a slope greater than one. This slope and the relative position of the points above a 1:1 line on Figure 6a are consistent with previous observations of porous particles (10).

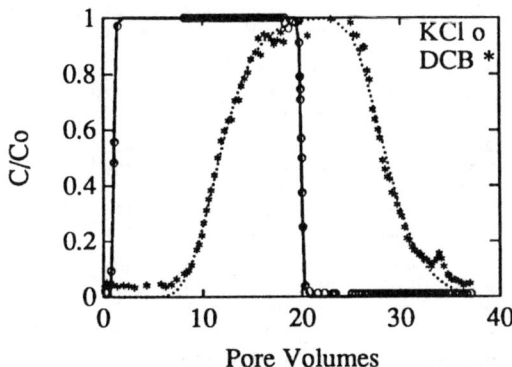

Figure 3. Breakthrough curves for KCl and 1,4-DCB on $C_{18}$-polymer surface (C-1); $C_o$ = 270 ng cm$^{-3}$, diameter = 100 μm, pore diameter = 27.5 nm; equilibrium model fit.

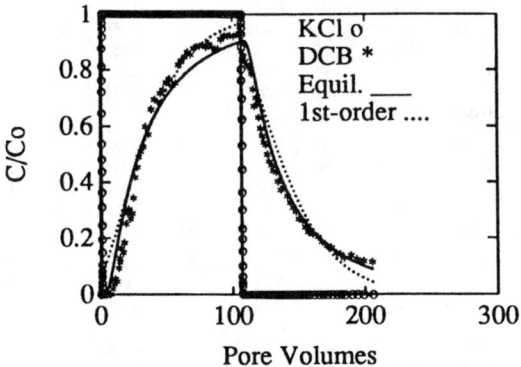

Figure 4. Breakthrough curves for KCl and 1,4-DCB on phenyl–polymer surface (E-4); $C_o$ = 400 ng cm$^{-3}$, diameter = 100 μm, pore diameter = 27.5 nm.

Table III. Results of Column Experiments

| Exp[a] | Surface | vel cm s$^{-1}$ | $D_{KCl}$ cm$^2$s$^{-1}$ | $K_p$ int[b] | Equil model $D$ cm$^2$s$^{-1}$ | $K_p$ | ssq[c] | First-order kinetic model $K_p$ | $N_D$ | $k_b$ | ssq[c] |
|---|---|---|---|---|---|---|---|---|---|---|---|
| Un- | I-1 | 0.0098 | 0.00093 | 4.5 | 0.0025 | 4.0 | 0.138 | 3.9 | 64 | 0.024 | 0.163 |
| bonded | I-7 | 0.031 | 0.0069 | 5.0 | 0.0086 | 3.4 | 0.27 | -- | -- | -- | -- |
|  | I-8 | 0.0049 | -- | 3.9 | 0.0012 | 1.7 | 0.188 | -- | -- | -- | -- |
|  | I-9 | 0.0049 | 0.0011 | 2.8 | 0.0028 | 3.6 | 0.242 | -- | -- | -- | -- |
| $C_1$ | A-1 | 0.047 | 0.011 | 460 | 0.055 | 480 | 0.806 | 410 | 8.3 | 0.00042 | 0.172 |
| bonded | A-2 | 0.052 | 0.029 | 390 | 0.046 | 360 | 0.523 | 340 | 210 | 0.014 | 1.046 |
|  | A-3 | 0.14 | -- | 240 | 0.059 | 170 | 0.364 | -- | -- | -- | -- |
|  | A-4[d] | 0.19 | -- | 130 | 1.2 | 180 | 0.806 | -- | -- | -- | -- |
|  | A-4[e] | 0.066 | -- | 300 | 0.42 | --[f] | --[f] |  |  |  |  |
|  | O-1 | 0.42 | 0.0042 | 53 | 0.038 | 78 | 0.322 | 72 | 7.8 | 0.00085 | 0.395 |
|  | O-2 | 0.092 | 0.35 | 93 | 0.051 | 110 | 1.061 | -- | -- | -- | -- |
| $C_8$ | B-1 | 0.038 | 0.014 | 27 | 0.019 | 28 | 0.080 | 27 | 530 | 0.12 | 0.087 |
| bonded | B-2 | 0.090 | 0.012 | 26 | 0.032 | 24 | 0.108 | 30 | 210 | 0.12 | 0.897 |
|  | B-3 | 0.026 | 0.0069 | 25 | 0.0092 | 36 | 0.022 | 25 | 88 | 0.016 | 0.377 |
|  | B-4 | 0.019 | 0.0061 | 28 | 0.011 | 33 | 0.174 | 30 | 30 | 0.0036 | 0.311 |
|  | B-5 | 0.0090 | 0.00073 | 28 | 0.0047 | 30 | 0.148 | -- | -- | -- | -- |
| $C_{18}$ | C-1 | 0.0070 | 0.00095 | 24 | 0.0028 | 25 | 0.106 | 23 | 110 | 0.0059 | 0.443 |
| bonded | C-2 | 0.075 | 0.013 | 24 | 0.086 | 24 | 0.017 | 23 | 13 | 0.0070 | 0.024 |
|  | C-3[d] | 0.023 | -- | 18 | 0.013 | 22 | 0.051 | -- | -- | -- |  |
|  | C-3[e] | 0.041 | -- | 21 | 0.087 | 26 | 0.045 | -- | -- | -- |  |
| Phenyl | Q-1 | 0.0045 | 0.00020 | 3.2 | 0.00024 | 3.9 | 0.049 | 3.7 | 2200 | 1.5 | 0.054 |
| monomer | Q-2 | 0.0069 | 0.00034 | 4.6 | 0.00045 | 4.0 | 0.138 | 3.9 | 14000 | 10. | 0.118 |
|  | D-1 | 0.0092 | 0.00071 | 3.1 | 0.0029 | 3.5 | 0.092 | 3.3 | 18 | 0.012 | 0.105 |
|  | D-2 | 0.0097 | 0.00075 | 3.4 | 0.0032 | 2.9 | 0.081 | 2.8 | 16 | 0.010 | 0.088 |
|  | D-3[d] | 0.0028 | 0.00020 | 6.1 | 0.00056 | 5.9 | 0.187 | -- | -- | -- | -- |
|  | D-3[e] | 0.0037 | 0.00027 | 5.7 | 0.00074 | 6.0 | 0.187 | -- | -- | -- | -- |
|  | D-4[d] | 0.012 | 0.0023 | 4.5 | 0.0065 | 4.0 | 0.179 | -- | -- | -- | -- |
| Phenyl | E-1 | 0.021 | 0.0063 | 23 | 0.14 | 19 | 0.212 | 15 | 2.7 | 0.00042 | 0.553 |
| polymer | E-2 | 0.015 | 0.00089 | 56 | 0.025 | 54 | 0.059 | 46 | 3.6 | 0.00045 | 0.286 |
|  | E-3 | 0.016 | 0.0012 | 65 | 0.045 | 85 | 0.251 | 68 | 3.2 | 0.00038 | 0.587 |
|  | E-4 | 0.019 | 0.0016 | 95 | 0.064 | 110 | 0.272 | 82 | 2.8 | 0.00027 | 0.464 |
|  | E-5[d] | 0.19 | 1.27 | 8.0 | 1.02 | 10 | 0.129 | -- | -- | -- | -- |
|  | E-5[e] | 0.041 | 0.095 | 9.9 | 0.076 | --[f] | --[f] | -- | -- |  |  |
|  | F-1 | 0.041 | 0.0052 | 25 | 0.59 | 35 | 0.088 | 25 | 2.0 | 0.00054 | 0.172 |
|  | F-2 | 0.037 | 0.012 | 30 | 0.46 | 20 | 0.315 | 14 | 2.3 | 0.00049 | 0.052 |
|  | F-3 | 0.014 | 0.0010 | 110 | 0.018 | 120 | 0.283 | 110 | 6.2 | 0.00039 | 0.398 |
|  | F-4 | 0.041 | 0.0041 | 25 | 0.067 | 22 | 0.182 | 20 | 3.9 | 0.00032 | 0.235 |
|  | F-5 | 0.041 | 0.0089 | 54 | 0.086 | 47 | 0.232 | 39 | 4.1 | 0.0016 | 0.323 |
|  | G-1 | 0.029 | 0.00086 | 30 | 0.068 | 31 | 0.115 | 28 | 4.8 | 0.00099 | 0.238 |
|  | G-2 | 0.035 | 0.0034 | 19 | 0.11 | 20 | 0.068 | 16 | 2.7 | 0.0011 | 0.541 |
|  | G-3 | 0.0333 | 0.0032 | 20 | 0.12 | 17 | 0.169 | 15 | 3.8 | 0.0014 | 0.273 |
|  | H-1 | 0.039 | 0.0042 | 24 | 0.24 | 22 | 0.100 | 21 | 3.0 | 0.00069 | 0.314 |
|  | H-2 | 0.035 | 0.0030 | 17 | 0.11 | 15 | 0.194 | 11 | 6.2 | 0.0018 | 0.445 |
|  | H-3 | 0.040 | 0.0042 | 26 | 0.22 | 26 | 0.066 | 22 | 3.0 | 0.00066 | 0.244 |
| Diol | R-1 | 0.043 | 0.0069 | 5.2 | 0.012 | 4.1 | 0.089 | — | — | — | — |
|  | R-2 | 0.010 | 0.026 | 2.9 | 0.0040 | 2.6 | 0.069 | 2.3 | 450 | 0.36 | 0.081 |
|  | R-3 | 0.015 | 0.0066 | 1.8 | 0.0065 | 1.7 | 0.024 | 1.5 | 10000 | 18 | 0.030 |
| Amine | T-1[g] | 0.024 | 0.049 | 3.6 | 0.030 | 5.2 | 0.608 | 5.0 | 6.9 | 0.0060 | 0.619 |
|  | T-2[g] | 0.015 | 0.032 | 4.6 | 0.071 | 2.8 | 0.163 | 2.6 | 10$^6$ | 1000 | 0.151 |
|  | T-3[g] | 0.015 | 0.0046 | 3.1 | 0.012 | 2.8 | 0.182 | 2.5 | 350 | 0.38 | 0.186 |
|  | T-4[g] | 0.015 | 0.0047 | 3.9 | 0.018 | 2.8 | 0.178 | 2.3 | 61 | 0.070 | 0.199 |

[a] Letters refer to surface (Table 1).
[b] From area under breakthrough curves.
[c] Sum of square errors.
[d] Sorption only.
[e] Desorption only.
[f] Fitted to sorption and desorption data together.
[g] pH values for T-1 through T-4 were 7.0, 3.9, 2.3, and 1.0, respectively.

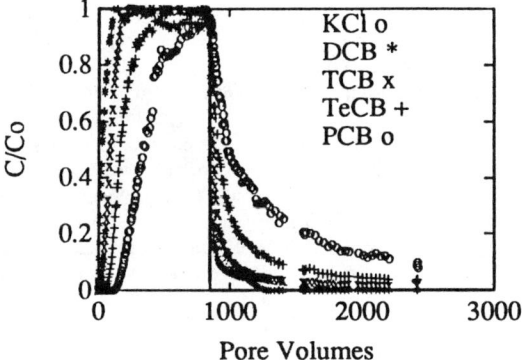

Figure 5. Breakthrough curves for KCl and multiple chlorinated benzenes on phenyl–polymer surface (E-9); $C_o$ = 550, 85, 140, and 28 ng cm$^{-3}$ for DCB, TCB, TeCB, and PCB, respectively; diameter = 100 μm, pore diameter = 27.5 nm.

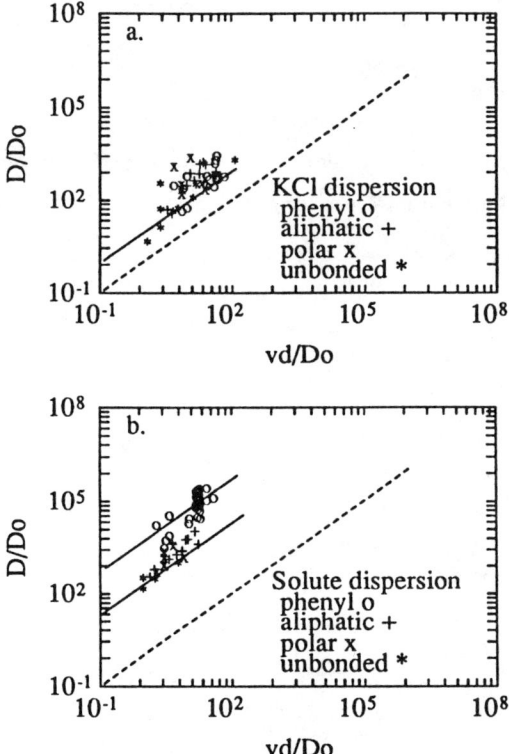

Figure 6. Dispersion versus Peclet number for a, KCl and b, 1,4-DCB; from experiments with silica aggregates with different surface preparations.

## Solute Dispersion

For the chlorinated benzene solutes, $D$ values were 2½ to four orders of magnitude above the 1:1 line (Figure 6b), and thus ½ to three orders of magnitude above the values for KCl on Figure 6a. Further, the fitted $D$'s increase in going from unbonded, to polar-bonded (amine and alcoholic), to aliphatic-bonded, to phenyl-bonded surfaces. The magnitude of the differences between $D_{KCl}$ and $D_{solute}$, especially for the phenyl-polymer surface, suggests that slow sorption and desorption rather than mobile-fluid dispersion or immobile-fluid diffusion is contributing to $D_{solute}$.

To separate slow kinetics due to mobile-immobile transfer, binding-release or intraparticle diffusion from dispersion in the mobile phase, it is necessary to have an independent estimate of $D$ for the organic solutes (as defined by equation 1). Horvath and Lin (11) analyzed hydrodynamic dispersion of organic solutes versus ionic tracers in liquid chromatography using an empirical approach:

$$D = D_o + \frac{\lambda u_e d}{\gamma \left[ +w \left[ \frac{u_e d}{D_o} \right]^{-1/3} \right]} + \frac{wf_\phi^2 \theta_i (u_e d)^{5/3}}{9\gamma D_o^{2/3} (1+k_o)^2} + \frac{k_o u_e^2 d^2}{30\tau D_o (1+k_0)} \qquad (9)$$

where the terms represent molecular diffusion, eddy diffusion (mixing between pores), film resistance and intra-particle diffusion, respectively. In the eddy diffusion term (12), $\lambda$ is a spreading factor (1.0), $d$ is the effective particle diameter, $\gamma$ is an empirical parameter (0.6), $w$ is an obstruction factor (2.75). Film resistance is based on a free surface model (13), where $f_\phi$ is the pore volume accessible to solute molecules (1.0) and $k_o$ is a porosity factor, equal to $\frac{\theta_i(1-\theta_e)}{\theta_e}$. The intraparticle diffusion term is an empirical estimate of spherical diffusion; $\tau$ is the tortuosity (0.66), from Rose (10). $\theta_i$ (0.46), $\theta_e$ (0.40), $d$ and $u_e$ are from experimental data; parameter values for $\lambda$, $\gamma$, $w$ and $\tau$ were taken from Horvath and Lin (11). $D_o$ values of $2.0 \times 10^{-5}$ and $8.1 \times 10^{-6} s^{-1}$ were used for KCl and 1,4-DCB, respectively. As noted above, calculated values of the ratio $\frac{D_{solute}}{D_{KCl}}$ for chlorinated benzenes and 100-μm aggregate particles are 1.6-2.0. Thus slow binding/release rather than these three physical phenomena contributes to the large apparent dispersion for the organic solutes.

The same conclusion follows from the development of *Parker and Valocchi* (14), who gave an analysis considering intraparticle diffusion:

$$D_{total} = \frac{\theta_e}{\theta} D_{mobile} + \frac{(1-\frac{\theta_e}{\theta}) d^2 u_e^2}{15 D_o \tau} \qquad (10)$$

where $D_{mobile}$ is mobile-phase dispersion, $D_o \tau$ is the effective intraparticle diffusion coefficient and $D_{total}$ is the resulting hydrodynamic dispersion coefficient. Assuming that $D_{mobile}$ and $\tau$ are the same for KCl and DCB (absence of anion exclusion), $D_{total}$ will differ only insofar as the $D_o$'s differ. For the 100-μm porous particles used in this research, $D_{total}$ for DCB is calculated to be 1.1 times that for KCl.

In subsequent analyses of non-equilibrium sorption data, $D$ for each chlorinated benzene was fixed at a multiple (see above) of the value for KCl (8). The fitted values of ω then reflect the rate of approach to equilibrium.

The sensitivity of the fitted ω to the fixed $P$ was determined for one solute breakthrough curve (G-2) by fixing the ratio $\frac{D_{DCB}}{D_{KCl}}$ at 1.0, 1.3, 1.7 and 3.0 and determining ω. Increasing the ratio from 1.0 to 3.0 gave only a slight increase in $ssq$, giving good fits for all four cases. However, increasing $D_{DCB}$ by three fold resulted in a small increase in ω (i.e., a faster reaction). Modeling a breakthrough curve exhibiting fast sorption (B-1) also showed the sensitivity to $D$. The best fit was for a $\frac{D_{DCB}}{D_{KCl}}$ of 1.5, with higher values giving greater spreading than experimental data and a large $ssq$.

Fixing $K_p$ at the planimetered value rather than having it fit by the model gave poorer fits (larger *ssq*). Twelve breakthrough curves with varying degrees of slow sorption gave fitted $K_p$'s within 10 percent of planimetered values.

The curve fitting procedure used (*7*), has one other adjustable parameter, pulse input, or number of pore volumes for which the contaminant was fed. As column pore volume was known from independently measurements, pulse input was fixed rather than fit. That is, particle and bulk density of the packing were known from prior measurements, pore volume was determined for each column by weighing dry versus wet, and flow velocity was measured for each experiment. Further, fitted conservative-tracer breakthrough curves gave $R = 1$, suggesting that pulse input and pore volume measurements were consistent.

Although the mobile-immobile model provides another parameter ($\beta$ or $f$), it does not have a clear interpretation in the experimental system studied. Particles used in this work have 99 percent of the surface internal and only 1 percent external, suggesting that $f$ should be small. As $f$ approaches 0, the mobile-immobile model becomes equivalent to the first-order model. Larger values of $f$ could be interpreted as indicating that there is fast exchange of solute with internal pores. In fitting breakthrough curves, consistent values of $f$ between experiments were not often obtained.

Sorption at Different Temperatures

A set of experiments run at different temperatures on the phenyl-bonded surface indicated that $\Delta H^o$ for sorption is on the order of -2.7 to -6.2 kcal mol$^{-1}$, which does not indicate strong binding at the surface (Table IV and Figure 7). $K_p$'s used in the analysis were from the equilibrium model.

Table IV. Sorption onto surface E at different temperatures

| Exp. | T °C | $D_{KCl}$ cm$^2$s$^{-1}$ | $K_p$ int$^a$ | Equilibrium Model | | | | First Order Model | | | |
|---|---|---|---|---|---|---|---|---|---|---|---|
| | | | | $K_p$ | $D$ cm$^2$s$^{-1}$ | $D/D_{KCl}$ | ssq | $K_p$ | $N_D$ | $k_b$ | ssq |
| *1,4-dichlorobenzene* | | | | | | | | | | | |
| E-6 | 3 | 0.043 | 27 | 23 | 0.46 | 10.9 | 0.069 | 17 | 1.8 | 0.0043 | 0.252 |
| E-7 | 11 | 0.033 | 36 | 28 | 0.25 | 7.5 | 0.140 | 21 | 3.6 | 0.0067 | 0.218 |
| E-8 | 17 | 0.47 | 41 | 24 | 0.65 | 1.4 | 0.083 | 15 | 1.4 | 0.0035 | 0.209 |
| E-10 | 35 | 0.0098 | 14 | 11 | 0.65 | 66 | 0.110 | 8.1 | 0.93 | 0.0047 | 0.189 |
| E-11 | 48 | 0.059 | 36 | 16 | 1.3 | 21 | 0.088 | 7.9 | 0.64 | 0.0034 | 0.184 |
| *1,2,4-trichlorobenzene* | | | | | | | | | | | |
| E-6 | 3 | 0.043 | 80 | 97 | 1.1 | 26 | 0.050 | 45 | 1.8 | 0.0016 | 0.493 |
| E-7 | 11 | 0.033 | 120 | 160 | 0.87 | 27 | 0.436 | 64 | 3.0 | 0.0018 | 0.820 |
| E-8 | 17 | 0.47 | 70 | 79 | 1.0 | 2.2 | 0.123 | 41 | 1.4 | 0.0013 | 0.280 |
| E-10 | 35 | 0.0098 | 48 | 45 | 1.2 | 120 | 0.213 | 24 | 1.3 | 0.0023 | 0.350 |
| E-11 | 48 | 0.059 | 55 | 54 | 1.8 | 30 | 0.045 | 23 | 1.1 | 0.0021 | 0.236 |
| *1,2,4,5-tetrachlorobenzene* | | | | | | | | | | | |
| E-6 | 3 | 0.043 | 250 | 300 | 1.5 | 35 | 0.062 | 110 | 1.6 | 0.00056 | 0.570 |
| E-7 | 11 | 0.033 | 270 | 700 | 1.4 | 42 | 0.385 | 200 | 2.4 | 0.00046 | 1.064 |
| E-8 | 17 | 0.47 | 210 | 250 | 1.5 | 3.1 | 0.485 | 100 | 1.3 | 0.00045 | 0.338 |
| E-10 | 35 | 0.0098 | 100 | 100 | 1.3 | 130 | 0.298 | 46 | 1.5 | 0.0014 | 0.331 |
| E-11 | 48 | 0.059 | 88 | 110 | 1.8 | 31 | 0.318 | 50 | 1.2 | 0.00099 | 0.359 |

DISCUSSION

The most likely causes of non-equilibrium behavior in groundwater (as in chromatography) are illustrated on Figure 1. Two approaches to modeling slow sorption-desorption were presented in fitting breakthrough curves: 1) use of an equilibrium model with a dispersion coefficient that includes both

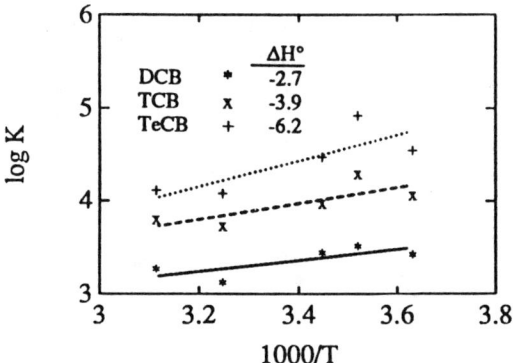

Figure 7. Partition coefficient versus temperature for three chlorinated benzenes on surface E; K ($cm^3_{water}$ $cm^{-3}_{organic\ matter}$) is $K_p$ times the density of the surface phase divided by the mass fraction of organic matter in the solid.

physical and chemical phenomena and 2) separation of slow sorption-desorption from mobile-phase dispersion and use of a first-order kinetic model. In the second case, the time scale for sorption/desorption is related to following model parameters: 1) the two-region-model mass-transfer coefficient ($\alpha_e$) for intraparticle diffusion, 2) the first-order-model mass-transfer rate ($\alpha_i$; equivalent to $\phi k_b$) for diffusion through a bound-organic layer and 3) the first-order-model desorption rate ($k_b$) for a binding step. Note that the term $\dfrac{\alpha_e}{\phi K_p}$ is dimensionally equivalent to $k_b$.

Interpreting the slow intra-aggregate or binding-release process in terms of additional dispersion in an equilibrium model, one can calculate tortuosity values from a model such as equation (10). The result, for 1,4-DCB values given in Table III, gives $\tau$ values (Table V) that are up to three orders of magnitude lower than $\tau$ values based on intra-aggregate diffusion alone (10). Interpreted in terms of slow release gives the first-order rate coefficients of Table III.

Table V. Calculated tortuosities for 100-μm particles with 27-nm pore dia

| Surface | $\tau$ | n |
|---|---|---|
| $C_8$ | 0.16 | 3 |
| $C_{18}$ | 0.013 | 3 |
| Phenyl polymer | 0.00033 | 4 |
| Dialcoholic | 0.020 | 3 |
| Amine | 0.0028 | 4 |

Sorption of DCB onto porous silica and silica modified with $C_8$ and $C_{18}$ aliphatic chains and with phenyl polymer can be adequately described as an equilibrium process since first-order rates are relatively fast (0.1 - 0.01 $s^{-1}$) compared to most residence times in most flow systems. Sorption onto phenyl polymer and $C_1$ surfaces exhibits slow sorption, with $\omega < 5$ and desorption rates less than 0.001 $s^{-1}$ (Table III).

Slow rearrangement of molecules can be due to long diffusional paths, constricted diffusion channels in a solid matrix or diffusion through a dense or highly viscous material. Diffusion of KCl and chlorinated benzenes through immobile fluid for the unmodified versus organic-modified porous silicas should be about the same. The large differences in $\dfrac{D}{D_o}$ for the organic solute on different surfaces (Figure 6b) versus lack of difference for KCl (Figure 6a) suggest that diffusion is not the cause of the slow behavior. The lack of differences between $k_b$ for a 425-μm (surface H) versus 20-μm (surface F) phenyl-modified silica further support this point (Table III).

The thickness of the bound organic surface varies from 0.5 nm for $C_1$ to 2.5 nm for $C_{18}$. If thickness affects the diffusion path length, longer-chain modified surfaces should exhibit slower sorption. $C_8$ and $C_{18}$ surfaces both had fast sorption/desorption, with $C_1$ exhibiting kinetically limited behavior due to polymerization during surface preparation. The chlorinated benzene compounds used in this research are disk-like molecules with a thickness of 0.35 nm and a diameter of 0.8 to 1.1 nm. The thickness of the bound aliphatic surfaces therefore represents less than three solute molecular layers, so describing the rate-limiting step as diffusion through a bound organic layer may not be appropriate. The slow approach to equilibrium, which undoubtedly does involve translation over a small distance, can be described as a binding step.

The slow approach to equilibrium appears to be related to the nature of the surface phase as equilibration times were 5 to 100 times greater on phenyl-modified than on aliphatic-modified surfaces. Sorption or desorption that exhibits a chemical-kinetic limitation is expected in cases of strong chemical binding rather than in cases of non-specific sorption. Strong binding is expected to result in slow breakup of a complex. In the present case, strong van der Waals binding, involving pi-pi electron interactions between solute molecules placed between closely spaced surface phenyl groups could be responsible (15). The relatively small $\Delta H^o$'s are consistent with this weak binding.

## CONCLUSIONS

In the experiments reported here, slow sorption and desorption, evidenced by tailing in breakthrough curves, was attributed primarily to slow binding and release rather than diffusion through immobile liquid, mobile/immobile liquid transfer, or diffusion through an organic phase. An equilibrium model could be used to fit breakthrough curves with slow sorption/desorption, but the magnitude of the resulting dispersion coefficient includes effects of media-specific intra-aggregate transfer and sorbate/sorbent-specific binding/release processes. Using the first-order model with independently determined dispersion coefficients gave a fitted first-order rate coefficient for media- and sorbate/sorbent-specific processes. The first-order model is conceptually appealing for the system studied, as the slow step is apparently released from a sorption site. A conservative tracer (no sorption) is needed to estimate dispersion. Many $D$ values calculated from the literature will be too low for the type of porous aggregates used in this work. Use of the first-order-model $k_b$ rather than an effective $D$ or a $\tau$ parameter provides a kinetic parameter related to the binding-release step. For the compounds and surfaces studied, the time scale for binding/release ($k_b^{-1}$) is independent of media geometry.

The one-dimensional column experiments and parameter-estimation procedure described in this work provides rate constants, or time scales, for sorption/desorption that are independent of the flow and the large-scale media geometry. By using model sorbents the rate constants can be related to the dominant binding interaction, which helps define their applicability.

## ACKNOWLEDGEMENTS

The research was supported in part by grant ECE-8504663 from the National Science Foundation, and in part by the Petroleum Research Fund, administered by the American Chemical Society.

## LITERATURE CITED

1. Horvath, C. and H. Lin, J. Chromatogr. 1978, 149, 43-70.
2. Rubin, J., Water Resour. Res. 1983, 19, 1231-1252.
3. VanGenuchten, M., J. Davidson, and P. Wierenga, Soil Sci. Soc. Amer. Proc. 1974, 38, 29-35.
4. Coates, K. and B. Smith, J. Soc. Petrol. Eng. 1964, 4, 73-84.
5. VanGenuchten, M., in Hydrogeology of Rocks of Low Permeability, IAH Memoires, 1985, vol. 17(2), 513-526.
6. Szecsody, J.E. and R.C. Bales, J. Contam. Hydrol. 1989, 4, 181-203.
7. VanGenuchten, M. Th., Research Report 119, U.S. Salinity Lab, Riverside, CA, 1981.
8. Szecsody, J.E., Ph.D. Dissertation, Department of Hydrology and Water Resources, University of Arizona, Tucson, AZ, 1988.
9. Bear, J., Dynamics of Fluids in Porous Media, American Elsevier, New York, 1972, p. 764.
10. Rose, D., Soil Sci. 1977, 123, 277-283.
11. Horvath, C. and H. Lin, J. Chromatogr. 1976, 126, 401-420.
12. VanDeemter, J., F. Zuiderweg, and A. Klinkenberg, Chem. Eng. Sci. 1956, 5, 271.
13. Pfeffer, R., Ind. Eng. Chem. Fundam. 1964, 3, 380.
14. Parker, J. and A. Valocchi, Water Resour. Res. 1986, 22, 399-407.
15. Jinno, K. and M. Okamoto, Chromatographia 1985, 20, 242-248.

RECEIVED October 6, 1989

# INDEXES

# Author Index

Aggarwal, Pradeep K., 87
Anderson, Paul R., 272
Apps, John A., 414
Bales, Roger C., 526
Barcelona, Michael J., 310
Bassett, R. L., 1
Benjamin, Mark M., 272
Bourcier, William L., 104
Brimblecombe, Peter, 58
Brookins, Douglas G., 154
Bruton, Carol J., 104
Busenberg, Eurybiades, 398
Carnahan, Chalon L., 234
Chappell, Richard W., 321
Choppin, Gregory R., 519
Clark, Sue B., 519
Clegg, Simon L., 58
Curtiss, Charles D., III, 508
DeBraal, Jeffrey D., 117
Delany, Joan M., 104
Eary, L. Edmond, 379
El-Zoobi, Majd, 486
Elzerman, Alan, 350
Glynn, Pierre D., 74
Groves, Frank R., Jr., 486
Grundl, Timothy, 350
Gunter, William D., 87,117
Haug, Andreas, 330
Healy, Richard W., 202
Hem, John D., 429
Holm, Thomas R., 508
Jackson, Kenneth J., 16,104
Janecky, David R., 226
Jensen, B. Skytte, 330
Jones, Blair F., 398
Kempton, J. Houston, 339
Kharaka, Yousif K., 87,117,169
Kipp, Kenneth L., 243
Knauss, Kevin G., 104
Langmuir, Donald, 350,398
Lasaga, Antonio C., 212
Laudon, Leslie S., 321

Liang, Liyuan, 293
Lindberg, Ralph D., 339
Longmire, Patrick, 154
Lundegard, Paul D., 169
Macalady, Donald L., 350
MacGowan, Donald B., 494
Machesky, Michael L., 282
May, Howard M., 398
Melchior, Daniel C., 1
Millero, Frank J., 447
Morgan, James J., 293,368
Neil, John M., 414
Nirmalakhandan, Nagamany N., 478
Nordstrom, Darrell Kirk, 398
Olsen, Roger L., 321
Pabalan, Roberto T., 44
Parkhurst, David L., 30,128,398
Parks, George A., 260
Pavlik, Hannah F., 140
Pearson, F. J., Jr., 330
Perkins, Ernest H., 117
Peterson, Maria L., 461
Pitzer, Kenneth S., 44
Plummer, L. Niel, 128,398
Rea, Rebecca L., 260
Roberson, Charles E., 429
Runnells, Donald D., 140,339
Schramke, Janet A., 379
Scott, Michael J., 368
Speece, Richard E., 478
Steefel, Carl I., 212
Stollenwerk, Kenneth G., 243
Striegl, Robert G., 202
Surdam, Ronald C., 494
Szecsody, James E., 526
Thomson, Bruce M., 154
Toran, Laura, 190
Viani, Brian E., 104
White, Art F., 461
Wildeman, Thomas R., 321
Wolery, Thomas J., 16,104

# Affiliation Index

Alberta Research Council, 87,117
Battelle Memorial Institute, 87
Battelle, Pacific Northwest Laboratories, 379
California Institute of Technology, 293,368
Camp, Dresser, and McKee, Inc., 321
Clemson University, 350
Colorado School of Mines, 321,350,398
EBASCO Services, 1
F. G. Baker Associates, 140
Florida State University, 519
Illinois Department of Energy and Natural Resources, 310,508
Intera Technologies Inc., 330
Lawrence Berkeley Laboratory, 234,414
Lawrence Livermore National Laboratory, 16,104
Los Alamos National Laboratory, 226
Louisiana State University, 486
Oak Ridge National Laboratory, 190
PTI Environmental Services, 339
Penn State University, 282
Plymouth Marine Laboratory, 58
Risø National Laboratory, 330
Roy F. Weston, Inc., 154
Stanford University, 260
U.S. Geological Survey, 30,74,87,117,128,169,202,243,398,429,461
University of Arizona, 1,526
University of California—Berkeley, 44
University of California—Lawrence Berkeley Laboratory, 234
University of California—Los Alamos National Laboratory, 226
University of Colorado, 140,339
University of East Anglia, 58
University of Miami, 447
University of New Mexico, 154
University of Washington, 272
University of Wyoming, 494
Unocal Science and Technology, 169
Vanderbilt University, 478
Yale University, 212

# Subject Index

## A

Acetic acid
  calculation of *meta/para* quotient, 63
  dissociation constants, 63$t$
Acid-leach tailings
  analytical methods, 155
  average concentrations, 159,161$t$
  distribution of total mass of elements, 159,161$f$,162
  equilibrium constants, 158$t$,163,164$t$,165
  geochemical modeling, 163,164–166$t$,167
  major environmental problem, 158
  mobility, 166–167
  results of column leach tests, 162$t$
  sampling methods, 155
  saturation indices, 165$t$
  speciation calculations, 165,166$t$
  thermodynamics, 158–159,160$f$
Acid-leach uranium mill tailing pond raffinates, concentrations of constituents, 156,157$t$
Activity coefficient
  calculation, 60
  scaling conventions in PHRQPITZ, 132,133$t$,134

Activity coefficients in aqueous salt solutions
  hydration theory, 17–28
  model requirements, 16–17
Activity coefficients of aqueous species, description by using EQ3/6 software package, 108
Adsorption of organic solutes, control, 260
Adsorption of surfactants
  cation enhancements, 261
  conceptual models, 261
Adsorption option in SOLMINEQ.88, description, 124–125
Adsorptive additivity, description, 272
$Al^{3+}$, thermodynamic properties, 415
Aluminate ion
  calculation of thermodynamic properties, 419,422–423$t$
  log $K$ and $\Delta G$, 422$t$
Aluminosilicate dissolution reactions, definition of terms, 415
Aluminum hydrolysis
  equilibrium solubility of gibbsite, 430–434
  importance of understanding mechanisms, 429
  influencing factors, 430

Aluminum hydrolysis reactions
  experimental procedures, 432–433
  kinetics of formation, 435–439
  previous work, 430
  reaction affinities, 434,435$t$
  stages, 434–435
  total concentration of Al species measured during titration, 434,438$f$
Aluminum hydroxide(s)
  application of thermodynamic data to modeling, 424,425$t$
  calculation of thermodynamic properties, 418–419,422–424
  dissolution reactions, 416,418
  solubilities, 416
Aluminum hydroxide reactions
  characterization of Alb, 441–442
  effect of mineral surfaces on Al behavior, 441
  implications of polymerization mechanisms and kinetics for occurrence of Alb in streams, 444
  observations in natural and anthropogenically influenced systems, 442–444
Aluminum oxyhydroxides, application of thermodynamic data to modeling, 424,425$t$
Ammonia
  activity coefficient in aqueous salt solutions, 65–65
  interaction parameters, 63,67$t$
  solubility, 63
Anion adsorption, temperature effects, 287,288–290$t$
Aqueous, environmental monitoring programs
  acceptability criteria, 323–324$t$
  data validation, 322$t$,323
  difficult matrices, 323–324$t$,325
  examples of data use, 327$t$,328
  field quality assurance and quality control checks, 326$t$
  interpretation of results, 325$t$
  modeling, 326–327
  problems with specific parameters, 322$t$,323
  project description, 322
Aqueous solubility, prediction via QSAR models, 478
Aqueous species, effect of pressure on activities, 98$t$
Aqueous systems, overview of chemical modeling, 1–10
Aqueous systems of acid–base reactions, definition, 369
Aqueous theory for chemical modeling
  components, 5,6$f$
  ion association approach, 3–4

Aqueous theory for chemical modeling—Continued
  ion hydration approach, 4
  ion interaction approach, 4–5
  model documentation, 5
Arsenite ion oxygenation
  half-life vs. pH, 387,390$f$
  initial rate vs. pH, 387,389$f$
  kinetics, 386–390
  rate increase vs. time, 386–387,388$f$
As(V)/As(III), platinum Eh measurements, 347
Atmosphere, definition, 59

B

Bayerite, calculation of thermodynamic properties, 422,424
Boehmite, calculation of thermodynamic properties, 418

C

$Ca^{2+}$ adsorption on quartz, comparison of simulated and experimental data, 267,269–270$f$
Calcite equilibrium model, $^{14}CO_2$ transport in unsaturated sediments, 206
Capacity factor of aqueous natural system, description, 369
Carbon dioxide, measurement of partial pressure, 203
Carbon dioxide option in SOLMINEQ.88, description, 121
Carbon dioxide retention model, $^{14}CO_2$ transport in unsaturated sediments, 206
Carbon-14 dioxide transport in unsaturated glacial and eolian sediments
  calcite equilibrium model, 206
  $CO_2$ retention model, 206
  concentrations of exchangeable C per volume of unsaturated zone, 206–207
  development of transport equations, 208–209
  gas transport, 207–209
  geochemical modeling, 203,206
  lithology of cross section of site, 202–203,204$f$
  mean partial pressures of $CO_2$, 203,205$t$
  numerical modeling, 209
  physical properties of unsaturated sediments, 203,204$t$
  reasons for study, 202
  simulated and mean partial pressures of $^{14}CO_2$, 203,205$t$
  site study, 202–203,204$f,t$,205$t$

# INDEX

Carbon isotope mass-transfer measurement of ground water contamination
 calculated percent dilution water vs. pH, 199$t$
 effect of siderite precipitation, 197,198$t$
 equilibrium constants, 198
 histogram of cations and anions, 191,193$f$
 hypothetical mixing model, 198,199$t$
 intersection of mixing curves, 199,200$f$
 isotope ratio vs. pH, 196$t$
 modeling methods, 195–196
 modeling results, 196–198$t$
 reaction path modeling, 191,194$f$,195
 sensitivity analysis for isotope ratio calculations, 197$t$
 site description, 191,192$f$
 sulfate vs. alkalinity, 191,194$f$,195
Carbon isotope modeling
 applications, 195
 factors influencing isotopic composition of carbon, 195–196
 methods, 195–196
Carbonate module of geochemical expert system
 function, 333
 illustration of expert system operation, 333–334,335$f$,336
Carbonate species, revised thermodynamic data, 403,406$t$
Carboxylic acid anions in formation waters
 controls on concentration and distribution, 497–498,500$f$
 destruction of carboxylic acid anions, 498
 effect on carbonate mineral stability, 495
 envelope diagram of concentration distribution with temperature, 495,496$f$
 identification, 495
 kerogen composition, 498
 mixing of formation waters, 498
 organic composition of formation waters, 495,496$f$,$t$,497
 origin in diagenetic environment, 497–498,499$f$
 thermal evolution of carboxyl groups vs. thermal maturity, 497,500$f$
Cathodic current, definition, 339–340
Cation adsorption, temperature effects, 287,288–290$t$
Chemical modeling of aqueous systems
 advancements in thermodynamic and kinetic data, 8–9
 applications, 1,5
 categories, 3$f$
 code improvement, 3
 coupled hydrology–chemistry, 1
 Debye–Hückel solvent parameters, 402–403
 development, 1

Chemical modeling of aqueous systems— *Continued*
 documentation of chemical model development, 1,2$f$,3
 examples of computer models and software, 7,13–14
 historical framework, 1,2$f$,3$t$
 influencing factors, 398
 models of aqueous theory, 3–5,6$f$
 new concepts, 7
 organic compounds, 9–10
 requirements for thermodynamic consistency, 399–400
 revised equilibrium data, 403–409
 sensitivities of models, 7–8,13–14
 solubility product constants, 401
 surface chemistry model, 5,7
 thermodynamic data
  carbonate species, 403,406$t$
  fluoride and chloride species, 403,404$t$
  oxide and hydroxide species, 403,405$t$
  silicate species, 403,407$t$
  sulfate species, 403,408$t$
 use of free energies vs. equilibrium constants for thermodynamic data base, 400,401$t$,402
Chloride salts, determination of mean activity coefficient, 37–38,39$f$
Chloride species, revised thermodynamic data, 403,404$t$
Chlorinated benzenes in porous aggregates, transport, 527–537
$Cl^-$ activity, definition, 32
Coadsorption of ionic surfactants with inorganic ions on quartz
 adsorption model, 261–270
 adsorption of surfactants, 261
 $Ca^{2+}$ on quartz, 267,268$f$
 cation enhancement of surfactant adsorption, 261
 dodecyl sulfate on corundum, 263,267,268$f$
 dodecylamine on quartz, 263,265$f$
 $Na^+$- and $Cl^-$-binding constants, 262–263,264–265$f$
 $Na^+$ and $Cl^-$ on corundum, 263,266$f$
 $Na^+$ on quartz, 263,264$f$
 surface ionization, 262$t$
Coagulation, stability ratios, 293
Coagulation of iron oxide particles in the presence of organic materials
 adsorption equilibrium constant, 304
 adsorption measurement, 295
 coagulation experiments, 295
 critical coagulation concentration
  vs. mole of carboxyl functional groups, 305,307$t$
  vs. number of carbon atoms, 305,306$f$
 effect of fatty acids, 302–306

Coagulation of iron oxide particles in the presence of organic materials—*Continued*
  effect of higher molecular weight organic materials, 298,301f,302
  effect of poly(aspartic acid) on coagulation rate, 298,301f,302
  effect of small organic adsorbates, 296,297f,298,299f
  experimental procedure, 294
  hematite adsorption density, mobility, and stability ratio vs. total phthalate concentration, 296,297f,298,299f
  influencing factors, 294
  lauric acid adsorption vs. lauric acid concentration, 304,306f
  measurement of electrophoretic mobility, 295
  mobility vs. concentration and carbon number of molecules, 304–305
  particle characterization, 295t
  particle preparation, 295
  potential, surface species distribution, and experimental stability ratio vs. pH, 298,300f
  potential vs. phthalate concentration, 298,299f
  quantitative treatment of surface speciation, 298
  rates, 293
  stability ratio
    vs. concentration and carbon number of molecules, 304–305
    vs. fatty acid concentration, 302,303f,304
    vs. polyelectrolyte concentration, 298,301f
  study objectives, 294
  use of surface complex formation model, 296
Connectivity–polarizability approach for prediction of aqueous solubility
  description, 480
  model for aliphatic and aromatic liquid and solid solutes, 480
  physical significance of descriptors, 483
  predictive ability, 483,484t,485
  range of applicability, 483
Conservation equations for flow, mass transport, and reaction in porous media
  conservation of momentum, 213–214
  conservation of solute mass, 214–215
  continuity equation, 213
  mathematical formulation, 213–215
Conservation of momentum, equation, 213–214
Conservation of solute mass, equation, 214–215
Constant-capacitance model of silica–iron binary oxide suspension interactions
  adsorbent characteristics and parameters used, 277,278t
  computer programs, 273,277
  fractional removal of Cd, 280f

Constant-capacitance model of silica–iron binary oxide suspension interactions—*Continued*
  fractional removal of $PO_4$, 277,279f
  fractional removal of $SeO_3$, 277,279f,280
  silica sorption reactions, 277
  surface complexation constants, 277
Constant-capacitance surface complexation model, applications, 272
Continuity, equation, 213
Contract laboratory program
  acceptability criteria, 323–324t
  analytical service programs, 322
  data validation, 322t,323
  difficult matrices, 323–324t,325
  examples of data use, 327t,328
  field quality assurance and quality control checks, 326t
  function, 321–322
  interpretation of results, 325t
  modeling, 326–327
  problems with specific parameters, 322t,323
  project description, 322
Copper complexation by natural organic matter in ground water
  comparison of fluorescence-quenching titrations with published results, 514,516f,517
  comparison of ion-selective electrode titrations with published results, 514,516f
  complexation parameters determined by different analytical methods, 512,515f
  copper concentration vs. pH, 511–512
  copper speciation determination by using complexation parameters, 512,514,515f
  $Cu^{2+}$ complexation parameters, 512,513t
  experimental materials, 509
  experimental methods, 509–510
  modeling titration curves, 510–511
  stability constants used for speciation calculations, 511,513t
Coupled hydrology–chemistry, use in modeling geochemical reactions, 10
Coupled sulfur isotopic and chemical mass-transfer modeling
  application to sea floor hydrothermal processes, 229–233
  approach, 227–229
  EQPS computation implementation, 227–229
  hydrothermal fluid mixing with sea water, 229,230f,231
  sulfate reduction model in hydrothermal stockwork, 231,232f,233
Coupling of precipitation and dissolution reactions to mass diffusion via porosity changes
  gypsum accumulation at boundary, 238,240f

# INDEX

Coupling of precipitation and dissolution reactions to mass diffusion via porosity changes—*Continued*
  gypsum precipitation at constant temperature, 238,240$f$
  initial and boundary conditions for simulation
    of calcium and sulfate ion diffusion, 238,240$f$
    of silicic acid diffusion, 237,239$f$
  limitations, 238,241
  porosity at boundary
    for gypsum precipitation, 238,240$f$
    for quartz precipitation, 238,239$f$
  quartz accumulation at boundary, 238,239$f$
  silica transport along temperature gradient, 237–238,239$f$
  THCC computer program, 235–236
  variable porosity in THCC, 235–237
Critical mixing point data, estimation of thermodynamic mixing parameters, 84
Cu(I) oxidation rates
  effect of ions, 450,452$f$,453
  rate constant corrected for ion effects, 453,454$f$
  rate constant for $H_2O_2$ oxidation, 448,449$f$,450
  rate constant for $O_2$ oxidation, 448,449$f$
  rate constant/molar fraction ratio for $H_2O_2$ oxidation, 450,451$f$
  rate constant/molar fraction ratio for $O_2$ oxidation, 450,451$f$
Cuprous ion oxygenation
  kinetics, 387,391–392,393$t$
  rate constant, 392
  rate expressions, 387,391–392,393$t$

## D

Damkohler number, equation, 216
Davies equation, applications, 16
Debye–Hückel charging function, equation, 18–19
Debye–Hückel osmotic function, equation, 18–19
Debye–Hückel parameter, implementation in PHRQPITZ, 129
Debye–Hückel solvent parameters, calculation of activity coefficients for aqueous species, 402–403
Derjaguin–Landau–Verwey–Overbeek theory
  application, 293–294
  coagulation of iron oxide particles in the presence of organic materials, 296–307
Diagenesis, role of organic acids, 181–187
Diffusion of solute through immobile water
  effect of interstitial-water velocity, 243
  rate-controlling mechanisms, 243–244

Dissolution pattern evolution
  channel length vs. Damkohler number, 221,223$f$
  characteristic length scales for channel formation, 221,224
  comparison of numerical simulation results with linear stability analysis, 217,219,220$f$
  contours of permeability vs. Damkohler numbers, 221,222$f$
  initial and boundary conditions, 215
  kinetic rate-controlled regime, 219,221,222–223$f$
  mathematical formulation of conservation equations, 213–215
  nondimensional parameters, 216–217
  numerical methods, 215–216
  steady-state aspect ratio vs. nondimensional parameter, 219,220$f$
  transport-controlled regime, 217,218$f$,219,220$f$
  unstable growth in transport-controlled regime, 217,218$f$
Dissolution–precipitation option in SOLMINEQ.88, description, 121
Dissolved Fe(III) species, activity determination, 350
Dissolved $O_2$
  function in ground water, 379–380
  oxidizing strength, 380,381$f$
  oxygenation kinetics
    arsenite ion, 386–390,393
    cuprous ion, 387,391–392,393$t$
    ferrous ion, 380,381$t$,382,384$f$
    manganous ion, 392,393$t$,394
    sulfide ion, 382–386
Dissolved organic species, identification, 494–495
Distribution coefficients, estimation of thermodynamic mixing parameters, 85
Dodecyl sulfate on corundum, surfactant adsorption, 263,267,268$f$
Dodecylamine on quartz, surfactant adsorption, 263,266$f$
Dynamic hydrothermal processes, coupled sulfur isotopic and chemical mass-transfer modeling, 226–233

## E

EQ3/6 software package for geochemical modeling
  activity coefficients of aqueous species, 108
  advantages, 107–108
  current directions, 112,113–114$f$
  description, 104

EQ3/6 software package for geochemical modeling—*Continued*
  development, 104,107
  EQ3NR codes, 105
  EQ6 codes, 105
  EQLIB library, 104
  extrapolation of data, 106–107
  limitations of submodels, 106
  modeling options, 107
  modeling pitfalls, 105–107
  numerical methods and data structure, 110–112
  pH buffer calculations, 112–113
  principal components, 104
  rate law(s), 110
  rate law integration, 112
  reaction of spent fuel and ground water, 113–114f
  redox disequilibrium in aqueous phase, 110
  solid solutions, 109–110
  thermodynamic data base, 105,109
EQ3NR codes, description, 105
EQ6 codes, description, 105
EQLIB library, description, 104
EQPS
  data files, 227–228
  description, 227
  implementation, 227–229
  operation, 228–229
Equilibrium computations for chemical reactions
  computer programs, 398–399
  requirements of thermodynamic consistency, 399–400
Equilibrium model for organic materials in water
  determination of Henry's law constant, 489
  determination of partial pressure, 487
  effect of cosolvents, 489–490,491t
  effect of high-molecular-weight dissolved organic matter on solubilities, 493–494
  equilibrium phase separation, 487,488f
  microemulsion formation, 492
  practical considerations for field applications, 492
  prediction of cosolvent partition coefficient, 490,491t,492
  simplified equations for sparingly soluble liquids, 487,489
  solubility of mixtures of solids, 492
  solubility of pure solids, 492
  solubility prediction for sparingly soluble hydrocarbon mixtures in water, 487,489
  thermodynamic equations, 486–487,488f
Equilibrium solubility of gibbsite
  chemical thermodynamic data for aluminum species, 430,432t

Equilibrium solubility of gibbsite—*Continued*
  effect of temperature, 430,431f,432
  equilibrium constant vs. temperature, 430,434t
  equilibrium constants for precipitation reaction, 430,433t
Expert judgment in geochemical modeling
  analytical reliability, 330–331
  correctness and appropriateness of thermodynamic data, 331
  effect of water composition on method for calculation of activity coefficients of aqueous species, 331
  mineral selection, 331
Expert systems
  components, 332
  computer languages, 332
  examples of disciplines, 332
  geochemical system, 332–333

F

Fatty acids, effects on coagulation of iron oxide particles in the presence of organic materials, 302–306
Fe(II) oxidation rates
  effect of anions, 455,459f,460
  effect of ionic strength on rate constant, 455,457f,459f
  effect of ions, 455,458f
  effect of pH on rate constant, 455,456f
  effect of temperature on rate constant, 455,459f
  pseudo-first-order rate constants, 453,454f,456
  rate equations, 453
Fe(III)/Fe(II), platinum Eh measurements, 347
$Fe^{3+}$ ion activities for solubility comparison in anaerobic systems
  apparent solubility product vs. time, 358,359f
  calculation of ferric oxide solubilities, 360,361t
  calculation of ion species in system, 354
  Eh values, 364,365t
  Eh vs. time, 355,356–357f,358
  experimental procedure, 353–354
  field data, 363
  general composition of iron species, 355,358t
  limitations, 364–365
  negative log of ferric ion activity vs. pH, 361,362f
  percent activities of iron species, 355,358t
  reasons for Nernstian behavior of Pt electrodes, 360–361

# INDEX

$Fe^{3+}$ ion activities for solubility comparison in anaerobic systems—*Continued*
  water compositions and ferric oxyhydroxide solubilities, 363$t$,364
$Fe^{3+}/OH^-$ stoichiometry, ferric oxyhydroxides, 353
$Fe(CN)_6^{3-}/Fe(CN)_6^{4-}$, platinum Eh measurements, 347
Ferric oxyhydroxides
  effect of particle size on solubility, 352
  $Fe^{3+}/OH^-$ stoichiometry, 353
  negative log of apparent ion activity product 353,354$t$
  solubilities, 351–352
Ferrous ion oxygenation
  half-life vs. pH, 382,384$f$
  kinetics, 380,381$t$,382,384$f$
  rate expressions, 380,381$t$,382
FITEQL, modeling of silica–iron binary oxide suspension interactions, 273,277–280
Fluoride salts, determination of mean activity coefficient, 38
Fluoride species, revised thermodynamic data, 403,404$t$
Formation constants, iron hydroxide and chloride complexes, 351,352$t$
Formic acid, dissociation constants, 63$t$
Froth flotation, description, 267

## G

Gas fractionation option in SOLMINEQ.88
  description, 121
  typical data, 121,124$t$
  typical results, 121,124,125$t$
Gas solubility in electrolyte solutions
  association equilibria, 67,69,70$f$
  atmospheric context, 64
  comparison of salt effects, 67,68$f$
  solubility of weak electrolyte in salt solutions, 64–65,66$f$
  variation of temperature, 69,71$f$
Gas solubility in natural geochemical systems, calculation, 58
Geochemical expert system
  assumptions, 332
  carbonate module, 333
  main module, 333
  modules, 333
  procedure, 332–333
  saturation index modules, 333
Geochemical kinetic models
  role of reactive-surface-area characterization, 461–472
  statistical errors, 462

Geochemical modeling
  acid-leach tailings, 163,164–166$t$,167
  $^{14}CO_2$ transport in unsaturated sediments, 203,206
Geochemical modeling of ground water, expert system, 330–337
Geochemical modeling of water–rock interactions by using SOLMINEQ.88
  carbon dioxide option, 121
  computational details, 118–119
  limitations, 125–126
  lost-gas option, 119,121
  modeling capabilities, 119–125
Gibbsite, equilibrium solubility, 430–434
Gibbsite polymorphs, calculation of thermodynamic properties, 424
Ground water
  determination of mixing, 190
  geochemical modeling via expert system, 330–337
  mathematical simulation of evolution, 414
Ground water environment
  chloride and methane concentrations vs. time, 313,315–316$f$
  physical conditions, 311,312$t$
  problems with investigations, 310
  sampling errors, 313,317–318
  spatial variability, 311,312$t$,313
  temporal variation of water quality constituents, 313,314$t$
  variabilities, 310–311
  vertical profiles of redox potential, dissolved oxygen, and Fe(II) concentrations, 311,314$f$
Ground water sampling
  sample filtration, 318
  sample preservation and handling, 318
  sampling devices, 318
  sampling protocol, 313,317–318
  sampling tubing, 318
  well design, 317
  well development and purging, 317–318
Gypsum, comparison of solubilities calculated with various sulfate salt concentrations, 40

## H

$H_2O_2$ formation in surface waters, reaction scheme, 448
Hazardous waste sites, importance of information for action determination, 321
Henry's law constants
  strong acids, 61,62$t$
  weak electrolytes, 62,63$t$,66$f$,67$t$

High-molecular-weight dissolved organic matter, effect on solubility, 493
Higher molecular weight organic materials, effect on coagulation of iron oxide particles in the presence of organic materials, 298,301f,302
Higher order electrostatic terms, implementation in PHRQPITZ, 129
Humic acids
  kinetics of binding to Ln(III), 520–525
  range of sizes, 519
Humic substances
  description, 519
  roles in geochemical speciation of metals, 519
HYDRAQL
  description, 260
  surfactant adsorption modeling, 261
Hydration theory for approximation of activity coefficient
  Debye–Hückel functions, 18–20
  description, 17–19
  development, 17
  disadvantages of model, 28
  equations for mean activity coefficients, 19,21
  excess Gibbs energy, 20
  fit of model to osmotic coefficient data, 25,26f
  fitting of 1:1 aqueous electrolytes to model, 25,28t
  fitting of 2:1 and 3:1 electrolytes to model, 28
  ion size averaging, 20–21
  mean molal activity coefficient, 25,27f
  preliminary model development, 24–28
  problems with ion size parameter, 19
  proposed model, 22–24
  thermodynamic framework, 19–20
Hydrothermal fluid mixing with sea water
  calculated amounts of mineral precipitates and reaction pathways, 229,230f
  coupled sulfur isotopic and chemical mass-transfer modeling, 229,230f,231
Hydrothermal stockwork, sulfate reduction model, 231,232f,233
Hydrous metal oxides, effect of temperature on ion adsorption, 282–290
Hydroxide salts, determination of mean activity coefficient, 40
Hydroxide species, revised thermodynamic data, 403,405t

I

Individual-ion activity coefficients
  equations, 31–32
  salts used to fit parameters, 32,33t

Inorganic oxidation reactions involving dissolved oxygen, rates, 379–394
Intensity factor of aqueous natural system, description, 369
Ion adsorption by hydrous metal oxides
  influencing factors, 282–283
  isosteric adsorption enthalpies, 288t
  proton adsorption enthalpies, 285,287t
  residual solution concentrations, 289t
  solution concentration ratios, 290t
  temperature effects on cation and anion adsorption, 287,288–290t
  temperature effects on solution equilibria, 283–284
  temperature effects on zero point of charge of oxides, 284–285,286–287t
Ion association approach to chemical modeling of aqueous systems
  applications, 3
  description, 3–4
Ion association aqueous models
  applications, 30
  components, 30
  equations for individual-ion activity coefficients, 31–32,33f
  limitations, 42
  mean activity coefficient(s)
    chloride salt, 37–38,39f
    determination, 30
    fluoride salt, 38
    hydroxide salt, 40
    perchlorate salt, 38
    sulfate salt, 40,41f
  modifications, 30
  technique for fitting ion association parameters, 32,34–36t,37
Ion association parameters
  data base, 32,34–36t
  fitting technique, 32,34–36t,37
  individual-ion activity coefficients, 32
  mean activity coefficients, 37
  stability constants for salts, 32,37
Ion-exchange option in SOLMINEQ.88, description, 124–125
Ion hydration approach to chemical modeling of aqueous systems, description, 4
Ion interaction approach to chemical modeling of aqueous systems
  comparison to ion association approach, 4–5
  description, 4
  development, 4
  development of general equations, 45
  excess Gibbs energy, 45
  prediction of mineral solubilities in electrolyte mixtures, 45
  solubilities at fixed molalities, 45,47f
  solubility vs. temperature, 45,46f
Ion size averaging, determination, 20–21

# INDEX

Ionic charge balance, indicator of sampling error, 7
Ionic interactions, effect on oxidation rates of metals in natural waters, 447–460
Ionization constants, effect of pressure, 88–89,90f
Iron chloride complexes, formation constants, 351,352t
Iron hydroxide complexes, formation constants, 351,352t
Iron oxide particles, coagulation in the presence of organic materials, 293–307

## J

Jarosite–alunite–potassium feldspar–gibbsite–goethite system, activity diagram, 159,160f

## K

Kinetics of Al formation
 concentration of Ala, 436
 effect of concentration of reaction product on forward rate of reaction, 436
 effect of rates of Al addition, 436
 effect of temperature on rate constant, 439
 equilibrium expressions for monomer formation, 436
 formation kinetics for Ala, 437,438f
 schematic representations of complexes, 439,440f
 typical titration data for polymerization reaction, 437t
Kinetics of Alb formation, effect of rate of Al addition on reaction affinity, 435–436
Kinetics of Ln(III)–humic acid binding
 data analysis, 521
 decay constants for Sm(III) dissociation, 522t
 dependence of rate constants on pH, 522t
 distribution function, 521
 half-life vs. percent metal ion dissociation, 524f,525
 ion exchange, 520–521
 ligand-exchange technique, 521
 number of first-order components in dissociation, 521
 percent dissociation for Sm(III), 522t
 reagents, 520

Kinetics of Ln(III)–humic acid binding—*Continued*
 spectrum for Sm(III) dissociation
  measured by conventional spectrometry, 522,523f
  measured by stopped flow, 522,523f
 ultrafiltration, 520

## L

Lanthanide cations
 interaction with humic substances, 519
 kinetics of binding to humic acids, 520–525
Light-stable isotopes, applications, 226
Linear solvation energy relationship approach, description, 478
Linear solvation energy relationship approach for prediction of aqueous solubility
 calculation of solubility-related property, 480
 cavity term, 479
 dipolar term, 479
 hydrogen-bonding term, 479
 model for aliphatic liquid solutes, 480
 model for aromatic liquid and solid solutes, 480
 physical significance of descriptors, 483
 predictive ability, 483,484t,485
 range of applicability, 483
 solvatochromic parameters, 479
 utility value, 481,483
Local equilibrium sorption transport model, description, 246–247
log $p$ approach for aqueous solubility prediction
 model for aliphatic and aromatic liquid and solid solutes, 479
 model for aromatic liquid and solid solutes, 479
 physical significance of descriptors, 483
 predictive ability, 483,484t,485
 range of applicability, 483
 rationale, 479
 utility value, 481
Lost-gas option in SOLMINEQ.88, description, 119,121

## M

Main module of geochemical expert system, function, 333
Manganous ion oxygenation
 half-life vs. pH, 390f,394
 rate expressions, 392,393t

Margules expansion model for aqueous systems
  activity coefficient, 48–49
  advantages, 45,48
  apparent molal volumes, 50,51f
  applications, 49
  Debye–Hückel parameters, 49
  excess Gibbs energy, 48–49
  Gibbs energy, 48
  ionic strength, 48
  parameters evaluated from apparent molal volumes, 52,53t
  parameters evaluated from isothermal–isobaric apparent molal volumes, 50,51t
  parameters evaluated from osmotic coefficients, 52,53t
  recalculation of experimental values to single reference point, 50
  solubility calculations in binary system, 52,54f
  solubility calculations in ternary systems, 52,55f
  solubility in water, 52,54f
  theoretical basis, 48
Mean activity coefficient for salt, calculation by using ion association models, 31
Metal(s), factors influencing bioavailability and toxicity, 447
Metal complexation
  concentrations of cations and anions, 182,184t
  effect on ground water quality, 508–509
  role of organic acid anions, 181–183,184t
  saturation states of minerals with and without organic species, 182,184t
Metal-complexing properties of natural waters, influencing factors, 509
Metastability, effect on chemical model sensitivity, 8
Microbial conditions, ground water environment, 311,312t
Microbial populations, dependence on redox status, 368–369
Microcrystalline gibbsite, definition, 432
MICROQL, modeling of silica–iron binary oxide suspension interactions, 273,277–280
Microscale processes in porous media
  diffusion through immobile region, 527
  equation for one-dimensional solute transport, 526–527
  model parameters and dimensionless variable, 527,528t
  transport of chlorinated benzenes in porous aggregates, 527–537
Mineral solubilities in concentrated brines, prediction, 58

Mineral stability lines, development, 147
MINTEQ, evaluation of reaction pathways of processed oil shale leachate, 140–151
Miscibility gap data, estimation of thermodynamic mixing parameters, 82,83t
Mixed side-pore diffusion transport model, description, 247–248
Mixing option in SOLMINEQ.88, description, 121
Mobilities of metal ions in aquifer systems, relationship to solubilities, 508
Model for activity coefficient approximation
  predictive means for parameter estimation, 16–17
  properties, 16
Modeling of water–rock interactions, See Water–rock interaction modeling
Molal volume change in ionization reactions of aqueous complexes, examples, 91,97t
Molar volume change in ionization reactions, calculation, 88–89
Molybdate sorption, description, 244,245f,246

## N

Natural aquatic organic matter, composition, 510
Nonequilibrium transport of solutes through porous media, conditions for occurrence, 243
Numerical modeling, $CO_2$ transport, 209

## O

One-dimensional reaction transport, equation, 216
Organic acids in subsurface waters
  atomic O/C ratio vs. thermal maturity, 172,173f
  comparison of relative acetate and propionate yields in hydrous pyrolysis experiments, 172,174f,175t
  contributions of anions and bicarbonate to total alkalinity, 183,185f
  effect of acetic acid addition on calcite solubility, 183,184t
  effect of pH buffering, 183,186f
  experimental decarboxylation of acetic acid and acetate, 174,176t
  geochemical interest, 169
  mathematical model of generation, 177,179–180f,181
  mean organic alkalinity vs. reservoir age, 174,177,178f
  metal complexation, 181–183,184t

# INDEX

Organic acids in subsurface waters—*Continued*
  modeling generation, 177,179–180f,181
  occurrence, 170,171f
  origin of major species, 170–175
  relative yields of acetate and propionate, 172,173f
  relative yields of acetate for hydrous pyrolysis experiments, 174,175f
  role in diagenesis, 181–187
  survivability, 174,176t,177,178f
  temperature dependence of acetic crossover concentration, 183,187f
Organic compounds in aqueous systems
  cosolvents, 9
  importance, 9
  macromolecules, 9
  partitioning, 9–10
Organic–inorganic diagenetic model
  generalizations, 502–503
  schematic representation, 503,504f
Organic materials in water, equilibrium model, 486–493
Organic solid(s), solubility, 492
Organic solid mixtures, solubility, 492
Organic solutes, control of adsorption, 260
Osmotic coefficient, calculation, 59
Oxidation rates of metals in natural waters
  Cu(I) rates, 448–454
  effect of anions on cation reactions, 447–448
  Fe(II) oxidation rates, 453
  redox processes in oceans, 448
Oxidative capacity
  definition, 370–371
  examples, 373–374,375t
  limitation, 375
Oxide species, revised thermodynamic data, 403,405t

## P

Partial currents, definition, 339
$pE$, thermodynamic analogy with pH, 369t
Peclet number, equation, 216
Perchlorate salts, determination of mean activity coefficient, 38
pH
  effect of pressure, 98t
  thermodynamic analogy with $pE$, 369t
pH effects, role of organic acids, 183–187
PHRQPITZ
  additions to data base, 130
  brine sample speciation on different activity coefficient scales, 134–135
  data base, 128–129
  Debye–Hückel parameter, 129

PHRQPITZ—*Continued*
  description, 128
  evaporation of sea water, 136
  examples, 134–136
  higher order electrostatic terms, 129
  implementation of Pitzer equations, 128
  interactive input code, 129
  invariant point in evaporation of sea water, 135
  limitations, 134
  maintenance of internal consistency, 130,131f,132t
  precautions, 134
  scale convention of activity coefficients, 132,133t,134
Physical conditions, ground water environment, 311,312t
Physical surfaces
  BET measurement of areas, 462–463,464f
  characterization, 462–463,464f
  specific surface area, 462
PITZINPT, description, 129
Platinum Eh measurements by using heterogeneous electron-transfer kinetics
  apparent double-layer capacitance vs. Eh, 344,345f
  As(V)/As(III), 347
  capacitance of double layer, 344,345f
  diffusion coefficients, 342
  diffusion coefficients by using numerical modeling, 344,346t
  electrode area, 342
  experimental methods, 342
  experimental results, 344,345f,346t
  $Fe(CN)_6^{3-}/Fe(CN)_6^{4-}$, 347
  Fe(III)/Fe(II), 347
  graphical representation, 339–340,343f
  heterogeneous electron-transfer kinetics at Pt disk electrode, 344,346t
  influencing factors, 340–342
  kinetic measurements, 342,343f
  numerical model, 344,347,348f
  relationship between time and potential, 340–342
  Se(VI)/Se(IV), 347
  Tafel plots, 342,343f
  theory, 340–342
Poly(aspartic acid)
  effect on coagulation of iron oxide particles in the presence of organic materials, 298,301f,302
  structure, 298,302
Potentiometry, description, 339
Pressure
  conditions at which sea water becomes saturated, 99,100f
  effect on activities of aqueous species, 98t

Pressure—*Continued*
  effect on calculated pH, 98t
  effect on ionization constants, 88–95
  effect on saturation state, 91,98t,99f
  effect on solubility constants, 88
  effect on stability constants, 91,96f
  estimation of effect on saturation state of minerals, 87
  estimation of effects, 88–89,90f
Pressure dependence of ionization constants, comparison with experimental data, 91t
Pressure effects, applications in geochemistry, 91–101
Primary saturation solubilities, fundamental principles, 81–82
Primary saturation states of solid–solution aqueous–solution systems, determination, 77
Products, definition, 105
Profile side-pore diffusion transport model, description, 248–249

## Q

Quantitative structure–activity relationship (QSAR) prediction
  comparison between models, 481–485
  connectivity–polarizability approach, 480
  development, 478
  linear solvation energy relationship approach, 479–480
  log $p$ approach, 479
  physical significance of descriptors, 483
  predictive ability, 483,484t,485
  predictive value, 478
  range of applicability, 483
  structural qualities, 481
  summary, 481,482t
  utility value, 478,481,483

## R

Rate-controlled sorption transport model, description, 247
Rates of inorganic oxidation reactions involving dissolved oxygen
  oxidizing strength of dissolved oxygen, 380,381f
  oxygenation kinetics
    arsenite ion, 386–390,393
    cuprous ion, 387,391–392,393t
    ferrous ion, 380,381t,382,384f
    manganous ion, 392,393t,394
    sulfide ion, 382–386
Reactant, definition, 105

Reaction constant approach, calculation of species distribution in aqueous phase of ground water, 414–415
Reaction-induced porosity
  examples, 212–213
  regimes, 212
Reaction kinetics applied to natural systems
  advances, 461
  integrated rate equation, 461
Reactive surface(s)
  characterization, 466,468–469t
  dislocation densities of natural minerals, 468t
  effect of dislocation densities on reaction rates, 468,469t
Reactive-surface-area characterization in geochemical kinetic models
  BET surface area vs. particle diameter, 463,464f
  BET surface areas of quartz plus feldspar, 466,467f
  clean-size-fraction characterization, 465–466
  comparisons of physical and reactive surface areas in natural systems, 469,470t,471f,472
  cumulative surface areas for sample splits of granite, 466,467f
  effect of weathering, 465
  physical-surface characterization, 462–463,464f
  quantitative relationship between geometric and BET surfaces, 463–467
  reactive-surface characterization, 466,468–469t
  surface irregularity scaling, 465
Reconstruction of reaction pathways in rock–fluid system by using MINTEQ
  calcium solubility controls and recarbonation of L2 leachate, 144,145f,t,146f
  characterization of L2 oil shale leachate, 141,143t
  composition of Uinta sandstone, 141
  functions of model, 140–141
  magnesium and silica solubility controls, 147,148f,150f
  mineral stability lines
    basis for talc, sepiolite, and diopside, 147
    for solid phases potentially controlling Ca activity, 144,145f
    for solid phases potentially controlling Mg activity, 148f
    for talc, sepiolite, and diopside after five days of reaction, 150f
  MINTEQ simulation of system behavior, 149,150–151f
  partitioning of lithium and fluoride, 147

Reconstruction of reaction pathways in rock–fluid system by using MINTEQ—*Continued*
  preliminary screening of plausible reactions, 141–148,150
  properties of Uinta sandstone, 141,142*t*
  results of reaction path simulation, 149,151*f*
  saturation indices for minerals in L2 leachate, 144,145*t*
  schematic representation showing modified reaction path simulation of system, 149,150*f*
  stages evaluated, 141
Redox electrode behavior in natural systems, Nernstian electrode response, 351
Redox intensity factor, definition, 369
Redox potential
  description, 369
  effect on chemical model sensitivity, 8
  methods for definition, 379
Redox potential for single redox couple, measurement, 351
Redox status of aqueous systems
  definition, 369–370,371*t*,372*f*
  dependence on state of system relative to equilibrium, 370
  examples of oxidative capacity, 373–374,375*t*
  half-reactions of redox couples, 371*t*
  influencing factors, 368
  model redox titration, 375,376–377*f*
  path of redox titration of model ground water system, 377*f*
  range of p$E$ values, 370
  redox ladder, 371,372*f*
  redox titration curve of model ground water system, 375,376*f*,377
  relationship with microbial ecology, 368–369
  time dependence of reactions, 370
Reductive capacity, definition, 370–371
Rest potential, definition, 339

S

Saturation index module of geochemical expert system
  extension of expert knowledge base, 336–337
  function, 333
Saturation states, effect of pressure, 98*t*
Se(VI)/Se(IV), platinum Eh measurements, 347
Sea water, compositions for illustration of pressure effects on aqueous equilibria, 91,97*t*

Secondary porosity, *See* Reaction-induced porosity
Sensitivities of chemical models of aqueous systems
  analytical error, 7
  computational errors, 7–8
  effect of metastability, 8
  effect of redox potential, 8
  sampling error, 7
Silica–iron binary oxide suspensions
  adsorbent characteristics and parameters for constant-capacitance model, 277,278*t*
  bulk and surface characteristics, 273,275*t*
  characteristics, 273,274*f*,275*t*,276*f*
  fractional removal of Cd, 273,274*f*,280*f*
  fractional removal of $PO_4$, 273,276*f*,277,279*f*
  fractional removal of $SeO_3$, 277,279*f*,280
  histograms of particle size distribution, 273,274*f*
  interaction model, 273
Silicate species, revised thermodynamic data, 403,407*t*
Simulation of molybdate transport with rate-controlled mechanisms
  breakthrough curve for Br, 249,250*f*
  comparison of models used for column breakthrough simulation, 246,250*t*
  equilibrium sorption model for sewage-contaminated ground water, 249,251*f*
  Freundlich plot of sorption parameters, 244,245*f*,246
  ground water chemistry, 244,245*f*
  local equilibrium sorption model, 246–247
  mixed side-pore diffusion model, 247–248
  mixed side-pore diffusion model for sewage-contaminated ground water, 252,253*f*
  modeling results, 249–255
  molybdate sorption, 244,245*f*,246
  numerical simulations, 246–249
  physical and chemical parameters for column experiment, 244,245*t*
  procedure for column experiment, 244,245*t*
  profile side-pore diffusion model, 248–249
  profile side-pore diffusion model for sewage-contaminated ground water, 252,254*f*
  rate-controlled sorption model, 247
  rate-controlled sorption model for sewage-contaminated ground water, 249,251*f*
  transport equation, 246
  uncontaminated ground water, 252,255*f*

Small organic adsorbates, effect on coagulation of iron oxide particles in the presence of organic materials, 296,297f,298,299f
Solid–solution(s), EQ3/6 software package, 109–110
Solid–solution aqueous–solution systems
 comparison of solid–solution and pure-phase solubilities, 78–82
 estimation of thermodynamic mixing parameters, 82,83t,84–85
 primary saturation states, 77
 stoichiometric saturation states, 77–78
 thermodynamic equilibrium states, 75,76f
Solid–solution solubilities
 prediction hypotheses, 78–79
 primary saturation and thermodynamic equilibrium solubilities, 81–82
 stoichiometric saturation solubilities, 79,80f,81
SOLMINEQ.88
 adsorption option, 124–125
 computational details, 118–119
 description, 118
 development, 118
 dissolution and precipitation option, 121,123f
 flow chart, 119,120f
 gas fractionation option, 121,124–125t
 geochemical modeling of water–rock interactions, 117–126
 ion-exchange option, 124–125
 limits of modeling, 125–126
 mixing option, 121,122f
Solubility, ferric oxyhydroxides, 351–352,363t
Solubility of aluminum hydroxides
 log $K$ vs. reciprocal of absolute temperature, 418–421f
 thermodynamic properties of solid phases, 416,417t
Solubility of metals in natural waters, controlling factors, 508
Solubility of volatile electrolytes in multicomponent solutions
 calculation of activity coefficient, 59–60
 calculation of gas solubility, 64–71
 calculation of osmotic coefficient, 59
 calculation of water activity, 60
 equilibrium reactions, 58–59
 evaluation of Henry's law constant, 61,62–63t,66f,67t
 Pitzer model parameters, 60,61t
 theory, 58–60,61t
Solubility of weak electrolytes in salt solutions
 activity coefficient of $NH_3$ in aqueous salt solutions, 64–65,66f

Solubility of weak electrolytes in salt solutions—*Continued*
 salt effect of KCl on $NH_3$, 65,66f
Solute dispersion
 curve-fitting procedure, 534
 hydrodynamic dispersion, 534
 intraparticle diffusion, 534
Sorption in porous aggregate, conceptual model, 526,529f
Spinodal-gap data, estimation of thermodynamic mixing parameters, 84
Stability constants, effect of pressure, 88
Stability ratio for coagulation, definition, 293
Stoichiometric saturation, definition, 77
Stoichiometric saturation solubilities
 contour plots, 79,81
 estimation of thermodynamic mixing parameters, 85
 saturation index value vs. stoichiometric saturation, 79,80f
Stoichiometric saturation states of solid–solution aqueous–solution systems
 definition, 77–78
 representation, 78
 saturation curves, 78
Strong acid
 equilibrium between aqueous and gas phases, 58
 Henry's law constants, 61,62t
Subsurface environmental conditions
 microbial conditions, 311,312t
 overview, 310–316
 variability in concentrations of chemical constituents, 311–316
Sulfate salts, determination of mean activity coefficient, 40,41f
Sulfate species, revised thermodynamic data, 403,408t
Sulfide ion oxygenation
 half-life vs. pH, 383,385f,386
 kinetics, 382–386
 rate constant, 386
 rate expression, 382–383,384t
Sulfide oxidation in carbonate environment, reactions, 190–191
Sulfur isotopic processes, applications, 226
Surface chemical model, coagulation of iron oxide particles in the presence of organic materials, 296–307
Surface chemistry in modeling of aqueous systems
 applications, 5
 effect of temperature, 5,7
Surfactant adsorption
 cation enhancement, 261
 conceptual models, 261
 dodecyl sulfate on corundum, 263,267,268f

# INDEX

Surfactant adsorption—*Continued*
  dodecylamine on quartz, 263,266f
Surfactant adsorption model, evaluation of parameters, 261–262
Suspended solids
  rates of coagulation of iron oxide particles, 293
  role in geochemical cycles of metals, 293

## T

Tailing leachate, impact on soil, 163,164t
Tailing pore fluid, chemistry and mineralogy, 155–156,157t,158
Tailing pore water, analyses, 156,157t
Tailing raffinates, definition, 155–156
Tailing salts, determination of chemical and mineralogical properties, 156,158
Temperature, effect on ion adsorption by hydrous metal oxides, 282–290
THCC program
  description, 234–235
  examples, 237–238,239–240f
  numerical solution, 236
  transport equations, 235
  variable porosity, 235–237
  volumetric relations, 236–237
Thermal hydrocarbon oxidation, via sulfate reduction, 502,503t
Thermodynamic activities of ionic species in aqueous solutions, calculation by using ion pair model, 87
Thermodynamic consistency, requirements, 399–400
Thermodynamic data, factors influencing accuracy, 398
Thermodynamic data base, EQ3/6 software package, 109
Thermodynamic data base for chemical modeling of aqueous systems
  construction of reference files, 8–9
  source of error for modeling, 8
Thermodynamic equilibrium solubilities
  estimation of thermodynamic mixing parameters, 85
  fundamental principles, 81–82
Thermodynamic equilibrium states of solid–solution aqueous–solution systems
  activity coefficient fitting, 75
  equations for law of mass action, 75
  Lippmann diagram, 75,76f
  solidus equation, 75
Thermodynamic Henry's law constant, definition, 59

Thermodynamic-mixing-parameter estimation for solid–solution aqueous–solution systems
  data for critical mixing point, 84
  data for miscibility gap, 82,83t
  data for spinodal gap, 84
  distribution coefficient measurements, 85
  stoichiometric saturation solubilities, 85
  thermodynamic equilibrium solubilities, 85
Thermodynamic modeling of aqueous geochemical systems, influencing factors, 16
Trace element distribution among multicomponent solids, modeling, 272
Trace metals in natural waters, chemical reactions, 447
Transport of chlorinated benzenes in porous aggregates
  breakthrough curves, 528,529f,530,531f,533f
  calculated tortuosities, 537t
  causes of nonequilibrium behavior in ground water, 535,537
  conservative tracer dispersion, 530,533f
  dispersion vs. Peclet number, 530,533f
  experimental methods, 527–528,530t
  fitted parameters for column experiments, 530,532t
  influencing factors, 535,537t
  partition coefficient vs. temperature, 535,536f
  properties of bonded silica, 528,530t
  solute dispersion, 534–535
  sorption at different temperatures, 535t,536f
  theory, 526–527

## U

Uranium mill tailings, safe disposal, 154–155

## V

Virial coefficient approach, *See* Ion interaction approach to chemical modeling of aqueous systems
Volatile electrolytes, solubility in multicomponent solutions, 58–71

## W

Water activity, calculation, 60
Water–mineral reactions, status of thermodynamic properties, 398–409

Water–rock interaction(s)
  fluid phases, 118
  geochemical modeling, 117–126
  importance, 117–118
Water–rock interaction modeling
  approaches, 494
  carboxylic acid anions and carbonate
    mineral stability, 501
  controls on concentration and distribution
    of carboxylic acid anions, 497
  destruction of carboxylic acid anions, 498
  generalized organic–inorganic diagenetic
    model, 502–503,505f
  kerogen composition, 498
  mixing of formation waters, 498
  organic acid anions and feldspar
    dissolution, 498–499,500f,501
  organic composition of formation waters,
    495,496f,t,497

Water–rock interaction modeling—*Continued*
  origin of carboxylic acid anions in
    diagenetic environment, 497–498,500f
  thermal hydrocarbon oxidation by sulfate
    reduction, 502,503t
  X-ray diffraction patterns, 499,500f,501
Weak electrolytes
  equilibrium between aqueous and gas
    phases, 58
  Henry's law constants, 62,63t,66f,67t
  solubility in salt solutions, 64–65,66f

Z

Zero point of charge of oxides, temperature
  effects, 284–285,286–287t

*Production: Paula M. Bérard*
*Indexing: Deborah H. Steiner*
*Acquisition: Cheryl Shanks*

*Elements typeset by Hot Type Ltd., Washington, DC*
*Printed and bound by Maple Press, York, PA*

*Paper meets minimum requirements of American National Standard for Information Sciences—Permanence of Paper for Printed Library Materials, ANSI Z39.48–1984* ∞

## Other ACS Books

*Chemical Structure Software for Personal Computers*
Edited by Daniel E. Meyer, Wendy A. Warr, and Richard A. Love
ACS Professional Reference Book; 107 pp;
clothbound, ISBN 0–8412–1538–3; paperback, ISBN 0–8412–1539–1

*Personal Computers for Scientists: A Byte at a Time*
By Glenn I. Ouchi
276 pp; clothbound, ISBN 0–8412–1000–4; paperback, ISBN 0–8412–1001–2

*Biotechnology and Materials Science: Chemistry for the Future*
Edited by Mary L. Good
160 pp; clothbound, ISBN 0–8412–1472–7; paperback, ISBN 0–8412–1473–5

*Polymeric Materials: Chemistry for the Future*
By Joseph Alper and Gordon L. Nelson
110 pp; clothbound, ISBN 0–8412–1622–3; paperback, ISBN 0–8412–1613–4

*The Language of Biotechnology: A Dictionary of Terms*
By John M. Walker and Michael Cox
ACS Professional Reference Book; 256 pp;
clothbound, ISBN 0–8412–1489–1; paperback, ISBN 0–8412–1490–5

*Cancer: The Outlaw Cell,* Second Edition
Edited by Richard E. LaFond
274 pp; clothbound, ISBN 0–8412–1419–0; paperback, ISBN 0–8412–1420–4

*Practical Statistics for the Physical Sciences*
By Larry L. Havlicek
ACS Professional Reference Book; 198 pp; clothbound; ISBN 0–8412–1453–0

*The Basics of Technical Communicating*
By B. Edward Cain
ACS Professional Reference Book; 198 pp;
clothbound, ISBN 0–8412–1451–4; paperback, ISBN 0–8412–1452–2

*The ACS Style Guide: A Manual for Authors and Editors*
Edited by Janet S. Dodd
264 pp; clothbound, ISBN 0–8412–0917–0; paperback, ISBN 0–8412–0943–X

*Chemistry and Crime: From Sherlock Holmes to Today's Courtroom*
Edited by Samuel M. Gerber
135 pp; clothbound, ISBN 0–8412–0784–4; paperback, ISBN 0–8412–0785–2

---

For further information and a free catalog of ACS books, contact:
American Chemical Society
Distribution Office, Department 225
1155 16th Street, NW, Washington, DC 20036
Telephone 800–227–5558